Rare Earths-Transition Metals-Boron Compounds

Emil Burzo

Rare Earths-Transition Metals-Boron Compounds

Basic Properties to Technical Applications

Springer

Emil Burzo
Faculty of Physics
Babeş-Bolyai University
Cluj-Napoca, Romania

ISBN 978-3-030-99247-7 ISBN 978-3-030-99245-3 (eBook)
https://doi.org/10.1007/978-3-030-99245-3

This Springer imprint is published by the registered company Springer Nature Switzerland AG
The registered company address is: Gewerbestrasse 11, 6330 Cham, Switzerland

Contents

Chapter 1
Introduction

The book is devoted to the analysis of physical properties of R-M-B ternary compounds, where R is a rare-earth or yttrium and M = Sc, Ti, V, Cr, Mn, Fe, Co and Ni, as well as of their technical applications. The R-M-B phase diagrams are firstly analysed and the ternary compounds are listed. No ternary R-M'-B phases are found when M' = Sc and Ti. For these systems, the physical properties of M'-M-B ternary compounds are described, some of them having crystal structures similar with those present in R-M-B phase diagrams. The binary R-M' compounds, are also mentioned.

The crystal structure types, evidenced in R-M-B phase diagrams, increases in the sequence: 2 (M = V), 3 (M = Cr, Mn), 12 (M = Fe), 20 (M = Co) and 27 (M = Ni). From those formed with iron, four are metastable phases. For a given structure type, the number of compounds depends on the R partner. Thus, in R-Fe-B system, only La_2FeB_6 compound is formed, their number increasing to 13 for $R_2Fe_{14}B$ series. Four crystal structure types were found with only one rare-earth in R-Co-B phase diagrams (La_2CoB_2, $Sm_4Co_3B_3$, $Nd_2Co_5B_3$, $Nd_5Co_{21}B_4$) and six in R-Ni–B system ($La_5Ni_{19}B_6$, $ErNi_{7.9}B_2$, $LaNi_3B$, YNi_3B_2, $Ce_3Ni_5B_6$, Ho_4NiB_{14}), while the RCo_4B_4 or RNi_4B compounds are formed with all R elements and Y. The presence of other ternary R-M-B compounds, in addition to those listed in tables, has been mentioned in some reports, but their crystal structures either were not determined or latter not found. Some R-Ni–B compounds were not obtained as single phase, their presence being observed only in multiphase alloys. The rare-earths-nickel-based amorphous alloys are also shortly analysed. The presentation of the data follows the lines of R-M-B series in order to ensure their coherence.

The tables with the R-M-B compounds are given at the beginning of each chapter, before analysing the methods used for their elaboration and physical properties, particularly in connection with possible technical uses. The crystal structures and physical properties of (R,R')-M-B and R-(M,M')-B pseudoternary series are analysed in correlation with the composition ranges in which solid solutions are formed and the site locations of substituting R' and M' elements. There are compounds where

E. Burzo, *Rare Earths-Transition Metals-Boron Compounds*,
https://doi.org/10.1007/978-3-030-99245-3_1

only the crystal structures were determined. The opportunity for further investigations of their physical properties is outlined.

The evolution over time of knowledge in the field is presented, as new results have been reported or new models developed in describing the experimental results. In this context, the physical properties of the R-M-B series reported in literature, are evaluated and contradicting data clarified. The basic properties are listed in tables or included in figures. In addition to crystalline structures and elastic properties of parent and hydrogenated compounds, data obtained by neutron diffraction (ND), magnetic measurements, nuclear magnetic resonance (NMR), Mössbauer spectroscopy (MS), specific heat, resistivity, thermopower etc. or those resulting from band structure calculations, are presented. In tables are included the data reported by different groups even on the same systems, in order to evidence the influence of preparation methods and measurement conditions (method, temperature, external field etc.) on their physical properties. Some general trends can be evidenced concerning the influence of R and/or M partners on the physical and technical parameters. The book is based on an extended list of references. The computed lattice parameters of some R-M-B compounds, included in Tables from Materials Research Project [19M1, 20M1], are those available on databases in 2020 year [1].

The review includes also the physical properties of RM_2B_2C borocarbides with M = Co or Ni, having crystal structure related to that of RM_2B_2 series. Depending on the R partner, the RNi_2B_2B series show Pauli type paramagnetism (R = La, Ce), superconductivity (R = Lu, Y), heavy fermion features (R = Yb), magnetic ordering (R = Pr, Nd, Sm, Gd, Tb), as well as both magnetic ordering and superconductivity (R = Dy, Ho, Er, Tm).

The compounds with high boron content are superhard or hard materials. The R-M-B compounds have also a large spectrum of technical applications, as high energy permanent magnets, magnetoelectric, magnetostrictive, magnetocaloric devices, sensors, catalysts, hydrogen storage etc. The high energy permanent magnets are multiphase systems, their useful properties being determined, in a large extent, by those of $R_2Fe_{14}B$ phase. Some new results in development of permanent magnets, particularly connected with grain boundary composition, are also presented.

Through the book, the SI system of units is used, except for magnetic susceptibilities, when in original papers the CGS system was used. In this case, the conversion factor to SI units is given. The superconducting transition temperature is denoted by T_s, different from notation of Curie temperature T_c, to avoid confusion when the same compound is both magnetically ordered and superconductor.

Reference

[1] A. Jain, S.P. Ong, G. Hautier, W. Chen, W.D. Richards, S. Dacek, S. Cholia, D. Gunter, D. Skinner, G. Ceder, K.A. Persson, The materials project: a materials genome approach to accelerating materials innovation, APL Mater. **1**(1) 0,111002 (2013)

Chapter 2
Scandium-Transition Metals (Rare-Earths)-Boron Compounds

The phase diagrams of ternary Sc-M-B systems with M = Cr [04M1], M = Mn [05M1], M = Fe [77S1, 81S1, 88Z1, 90K1, 02R1], M = Co [77S1, 79S1, 90K1], M = Ni [81S1, 90K1, 03S1], as well as of Sc-Ho-B [92G1] and Sc-La-B [90K1] were reported.

The ScB_2 crystallizes in AlB_2-type structure, space group *P6/mmm*. The Young modulus is 680 GPa and Debye temperature $\theta_D = 1020$ K [06L1]. The Sc-R-B-systems, where R is a rare earth, are characterized by the absence of ternary borides and the formation of pseudo-binary solid solutions, when atomic radii of R and Sc are close [98M1]. Thus, in Sc-Ho-B system, the $Sc_xHo_{1-x}B_2$, solid solutions are formed for $x \leq 0.4$ and $x \geq 0.6$. Two phases region was shown in the composition range $0.4 \leq x \leq 0.6$—Fig. 2.1a [92G1]. In the Sc-Mn-B system, solid solutions are found in the composition range $Sc_{0.10–0.30}Mn_{0.90–0.70}B_2$. These crystallize also in AlB_2–type structure, space group *P6/mmm* [05M1]—Table 2.1a. In the structure, there are directly superposed close-packed Sc(R) layers. The B atoms are situated in trigonal prismatic voids and form a planar hexagon-mesh layers. The 3D framework is formed by fused Sc_6B trigonal prisms.

In the Sc-Ho-B system, $Sc_xHo_{1-x}B_4$ solid solutions are also formed up to x = 0.6 [92G1]—Fig. 2.1b. These crystallize in a tetragonal lattice, space group *P4/mbm* [54B1, 81G1]. In this structure, boron covalent sublattice is formed by chains of B_6 octahedra, parallel to the **c**-axis and linked by pairs of boron atoms in the (001) plane. The resulting three-dimensional skeleton contains tunnels parallel to the [001] direction, which are filled by metal atoms. No presence of ScB_4 borides was shown in Sc-B system [70P1]. Recently ScB_4 was identified as stable phase and ScB_6 as a metastable phase [22Z1]. The solubility of ScB_6 in LuB_6 is up to $Lu_{0.9}Sc_{0.1}B_6$ composition [90K1].

The $Sc_xHo_{1-x}B_{12}$ solid solutions are formed in all the composition range, having as HoB_{12}, a cubic type structure, space group $Fm\overline{3}m$ [92G1]—Fig. 2.1c. The ScB_{12} boride crystallizes in a tetragonal structure, having space group *I4/mmm* [65M1]. The cell is pseudo-cubic and dimensionally related to the cubic dodecaboride series. As a result, a cubic type structure was reported for the series, in which the metallic

E. Burzo, *Rare Earths-Transition Metals-Boron Compounds*,
https://doi.org/10.1007/978-3-030-99245-3_2

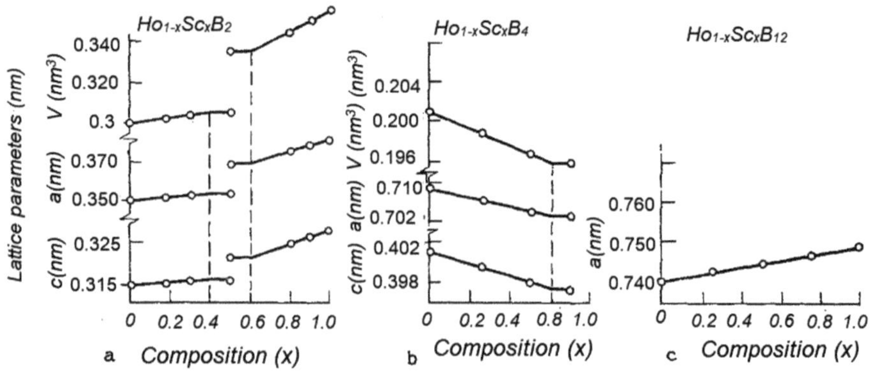

Fig. 2.1 Composition dependences of lattice parameters and volumes for $Ho_{1-x}Sc_xB_2$ (**a**), $Ho_{1-x}Sc_xB_4$ (**b**) and $Ho_{1-x}Sc_xB_{12}$ (**c**) series [92G1]

atoms are located in interstitial openings among the close-packed cubo-octahedral B_{12} clusters.

Ternary compounds were evidenced in the Sc-M-B- phase diagrams, when M is a 3d transition metal. The $ScMB_4$ with M = Fe, Co, Ni borides [88Z1], crystallize in $YCrB_4$ type structure, having space group *Pbam* [70K1]—Table 2.1b. In the RMB_4 series, a and b lattice parameters are mainly governed by the sizes of M and R atoms, while the c parameter by the larger R atoms, separating the B_4 networks. A topological analysis of the electron density revealed the differences between the bonding situation of the Sc atoms, in the heptagonal prismatic environment and the Ni atoms in pentagonal prismatic environment [13E1].

The Sc_2CrB_6 [04M1] and $Sc_{2.28}Mn_{0.72}B_6$ [05M1] borides crystallize in Y_2ReB_6-type lattice, space group *Pbam*—Table 2.1c. The crystal structure is of two layers type, in which one layer is formed by 7-, 6- and 5-membered boron atom rings. The two Sc sites are located between the networks of boron atoms, centering the 7-member and 6-member rings of boron atoms, respectively, while the, Cr(Mn) atoms center the 5-member rings. The corresponding coordination numbers are CN = 23, 20, for the two Sc sites and CN = 17 for Cr (Mn). Since the Sc radius is close to Mn one, (r_{Sc}/r_{Mn} = 1.17), these atoms occupy statistically positions that correspond to coordination numbers CN = 20 and 17, which is the basic feature of $Sc_{2.28}Mn_{0.78}B_6$ boride structure.

The $Sc_4Ni_{29}B_{10}$ boride crystallizes in *I*4$_1$/*amd* type structure [88K1, 10V1]. There are BNi_6Ni monocaped trigonal prisms and pairs of face -linked BNi_6Ni_2 bicapped trigonal prisms, where B_2 pairs, in part, are replaced by a Ni atom. The prisms share atoms to form a 3D-framework. The Sc is situated in large voids. There are present B_2 dumbbells and single B atoms. This structure is a partly disordered derivative of $ErNi_7B_3$ one [88K1, 10V1].

The $Sc_{0.75}Cr_{0.1875}B_x$ (x = 0.0629) metastable phase has superconducting properties [76V1].

Table 2.1 Atomic sites and their coordinates

(a) $Sc_{1-x}Ho_xB_2$ having hexagonal structure, space group *P6/mmm* [06V1]

Atom	Sites	Atomic coordinates			Atomic environment
		x	y	z	
Sc (Ho)	1*a*	0	0	0	pseudo Frank-Kasper $B_{12}Sc_8$
B	2*d*	1/3	2/3	½	coplanar triangle B_3

(b) $ScNi_4B$ having *Pbam* space group [13E1]

Atom	Sites	Atomic coordinates		
		x	y	z
Sc	4*g*	0.12732(8)	0.15036(3)	0
Ni	4*g*	0.13500(6)	0.41114(2)	0
B1	4*h*	0.2897(4)	0.3135(2)	1/2
B2	4*h*	0.3660(3)	0.4679(2)	1/2
B3	4*h*	0.3846(5)	0.0491(2)	1/2
B4	4*h*	0.4781(4)	0.1945(2)	1/2

(c) Sc_2CrB_6 having *Pbam* type structure [04M1]

Atom	Sites	Atomic coordinates		
		x	y	z
Sc1	4*g*	0.82230(8)	0.08623(6)	0
Sc2	4*g*	0.44456(8)	0.12870(6)	0
Cr1	4*g*	0.14313(7)	0.18254(6)	0
B1	4*h*	0.0541(5)	0.0651(4)	1/2
B2	4*h*	0.2555(5)	0.0800(4)	1/2
B3	4*h*	0.2973(5)	0.2378(4)	1/2
B4	4*h*	0.1295(5)	0.3197(4)	1/2
B5	4*h*	0.4805(5)	0.2875(4)	1/2
B6	4*h*	0.0973(5)	0.4737(4)	1/2

(d) $Sc_2Ni_{21}B_6$ having cubic structure, $Fm\overline{3}m$ space group [08G1]

Atom	Sites	Atomic coordinates		
		x	y	z
Sc	8*c*	1/4	1/4	1/4
Ni1	4*a*	0	0	0
Ni2	32*f*	0.38564(3)	0.38564(3)	0.38564(3)
Ni3	48*h*	0	0.16967(3)	0.16967(3)
B	24*e*	0.2691(5)	0	0

(e) $Sc_2FeIr_5B_2$ having tetragonal structure, space group *P4/mbm* [98N1]

Atom	Sites	Atomic coordinates		
		x	y	z

(continued)

Table 2.1 (continued)

Sc	4*g*	0.3234(3)	0.8235(3)	0
Ir1	8*j*	0.21664(6)	0.06927(7)	1/2
Ir2	2*c*	0	1/2	1/2
Fe	2*a*	0	0	0
B	4*g*	0.124(2)	0.624(2)	0

The $Sc_2Ni_{21}B_6$ ternary boride crystallizes in a cubic type structure, $Fm\overline{3}m$ space group—Table 2.1d [08G1, 08G2]. The early proposed compositions [67K1, 81S1], for this compound, were successively corrected. The presence of $Sc_3Ni_{20}B_6$ boride, having $Cr_{23}C_6$-type structure was initially assumed [67K1]. There is some homogeneity range, at T = 800 °C, around the above composition, $Sc_{3\text{-}4}Ni_{20\text{-}19}B_6$, as suggested latter [81S1], the boride being finally identified as $Sc_2Ni_{21}B_6$ [08G2]. The crystal structure is characterized by the presence of tetragonal antiprisms BNi_8 and empty cubes $\square Ni_8$, parallel to the fourfold axes, which are separated by cubo-octahedra Ni_1Ni_{12}. The Sc atoms have the largest coordination number CN = 16, those of Ni, at *f* and *h* sites, being 13 and 14, respectively.

The $ScMIr_5B_2$ compounds with M = Ti, V, Cr, Mn, Fe, Co, Ni and Cu, crystallize in a tetragonal structure of $Ti_3Co_5B_2$-type, having space group P_4/mbm [98N1, 01N1]—Fig. 2.2 and Table 2.1e. In the direction of short **c**-axis, planar nets of iridium atoms, at z = ½, composed of triangles, squares and pentagons, alternate with layers containing B, Sc and M atoms, at z = 0. By the stacking of iridium nets, columns of triangular, tetragonal and pentagonal prisms, running along [001] are formed. The triangular prisms are centered by boron atoms. The pentagonal and the tetragonal prisms in the basic compound, $Sc_3Ir_5B_2$, accommodate the Sc atoms. In the quaternary compounds $Sc_2MIr_5B_2$, the Sc atoms, residing in the tetragonal iridium prims, are completely replaced by M atoms. The columns of scandium-centered pentagonal iridium prisms in $Sc_2MIr_5B_2$ compounds may be also regarded as channels filled by chains of Sc atoms. The M atoms are arranged in rows along [001] direction, with M-M distances of ≅ 0.3 nm in the rows and ≅ 0.66 nm, between the rows.

The $Sc_2FeRu_{5\text{-}x}Rh_xB_2$ borides also crystallize in a substitutional variant of the $Ti_3Co_5B_2$-type structure [07F1]. No superstructure lines were observed, suggesting that Rh and Ru atoms are randomly distributed at their sites. The structure contains trigonal, tetragonal and pentagonal prisms, of M = Ru, Rh atoms, sharing the same site, but with different ratios, stacked on top of each other, thereby building channels along the [001] direction. The trigonal prisms are centered by boron atoms, the pentagonal prisms accommodate the Sc atoms and the tetragonal prisms contain the Fe atoms. As in $ScMIr_5B_2$ borides, the Fe and Sc atoms are arranged in chains with the same intrachain distance (0.302 nm), the interchain separations being ≅ 0.487 nm for Sc atoms and ≅ 0.66 nm for Fe atoms.

The Sc_2Ir_6B structure, as Ti_2Rh_6B ones can be described as a vacancy contained double perovskite-like structure, in which boron ordering is present [06F1]. In this structure, there are stacking of intersecting Kagomé Ir-layers along [111] directions,

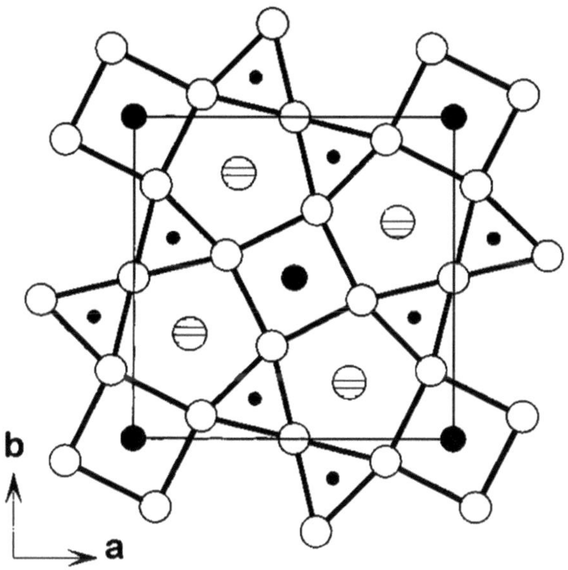

Fig. 2.2 $Sc_2MIr_5B_2$: crystal structure projected along [001] [98N1]; ○-Ir at z = 1/2, all others at z = 0, •-B, ⊜-Sc, ●-M atoms

the Sc occupying the center of each hexagon. The boron is located in interstitial sites between two of these layers. The boron atoms and vacancies alternate, resulting in a different stacking sequence, which leaves every other interstitial layer between the Ir layers unoccupied. In this structure, the Sc is located at 8*c*, Ir at 24*e* and B in 4*a* sites. Superstructure reflections were shown; the space group is *Fm*3*m*.

The $Sc_2Ir_{6-x}M_xB$ series, with M = Ni and Pd, crystallize in a substitutional variant of the above described perovskite structure, replacing Ir by M = Ni, Pd atoms. As a consequence, empty $(IrM)_6$ and B filled $(IrM)_6B$ octahedra, which are corner connected to a 3D network, result. The Ir/M mixture is coordinated by a cubo-octahedron of four Sc and eight Ir/M atoms and one B atom centering one quadratic face. Sc also has a coordination sphere in the form of a cubo-octahedron of Ir/M atoms only [15H1, 20S1]. Superstructure reflections were present for compositions $0 \leq x \leq 5$, having valence electron counts (VEC) from 63 to 68. For x = 6, where VEC = 69, no superstructure reflection was shown. The lattice parameters of Sc-Mi-B compounds are given in Table 2.2.

The band structure calculation on Sc_2M_6B (M = Ir, Pd, Ni) showed a pseudogap near the Fermi level, E_F, for $61 \leq VEC \leq 68$. In the $Sc_2Ir_{6-x}Pd_xB$ system, the hardness decreases monotonically from 9.27(2) GPa (x = 0) to 5.88(6) GPa and then it levels off [20S1]. The calculated Vickers hardness in Sc_2Ir_6B (13.5 GPa), is somewhat higher than the experimental value. The substitution of Ir by Pd in $Sc_2Ir_{6-x}Pd_xB$ would shift E_F toward the upper edge of the pseudo-gap. The shifting of the E_F within pseudo-gap toward its upper edge, as effect of Pd substitution, will add more antibonding states and reduced the bond strength. The transition at x = 5 was shown to be of order–disorder type and accompanied by precipitation of small amounts of boron rich phases [20S1].

Table 2.2 Space groups and lattice parameters

Boride	T (K)	Space group	Lattice parameters (nm)			Reference
			a	b	c	
ScB_2	RT	*P6/mmm*	0.314820(3)		0.351483(5)	[06L1]
$Sc_{0.6}Ho_{0.4}B_2$	RT	*P6/mmm*	0.3225(1)		0.3708(3)	[92G1]
$Sc_{0.4}Ho_{0.6}B_2$	RT	*P6/mmm*	0.3155(1)		0.3540(3)	[92G1]
$Sc_{0.95}Cr_{0.05}B_2$	RT	*P6/mmm*	0.29802(7)		0.3113(1)	[04M1]
$Sc_{0.80}Cr_{0.20}B_2$	RT	*P6/mmm*	0.31205(6)		0.3461(3)	[04M1]
$Sc_{0.13}Mn_{0.87}B_2$	RT	*P6/mmm*	0.3016(1)		0.30350(2)	[05M1]
$Sc_{0.31}Mn_{0.69}B_2$	RT	*P6/mmm*	0.3051(2)		0.30770(2)	[05M1]
$Sc_{0.6}Ho_{0.4}B_4$	RT	*P4/mbm*	0.7033(4)		0.3970(2)	[92G1]
$Sc_{0.25}Ho_{0.75}B_4$	RT	*P4/mbm*	0.7059(4)		0.3977(2)	[92G1]
$ScCo_4B_4$	RT	*P4₂/mmc*	0.4902(5)		0.6961(5)	[79S1]
$ScNi_4B_4$	RT	*I4₁/amd*	0.775		0.1597	[84R1]
$Sc_4Ni_{29}B_{10}$	RT	*I4₁/amd*	0.7686(1)		0.15382(3)	[88K1]
$ScFeB_4$	RT	*Pbam*	0.5884(15)	1.1318(21)	0.3342(8)	[88Z1]
$ScCoB_4$	RT	*Pbam*	0.5762(5)	1.1172(8)	0.3250(3)	[88Z1]
$ScNiB_4$	RT	*Pbam*	0.57853(7)	1.1208(1)	0.32737(3)	[88Z1]
$ScNiB_4$	RT	*Pbam*	0.5787(2)	1.1207(3)	0.3272(1)	[13E1]
Sc_2CrB_6	RT	*Pbam*	0.87909(4)	1.10541(6)	0.32996(1)	[04M1]
$Sc_{2.28}Mn_{0.72}B_6$	RT	*Pbam*	0.87810(5)	1.10641(6)	0.33674(1)	[05M1]
$ScCo_3B_2$	RT	*P6/mmm*	0.4889(3)		0.2977(2)	[69K1]
$ScCo_3B_2$	RT	*P6/mmm*	0.4864(5)		0.2997(1)	[73R1]
$Sc_2Co_{21}B_6$	RT	*Fm3m*	1.0537(5)			[67K1]
$Sc_2Ni_{21}B_6$	RT	*Fm3m*	1.0585(5)			[67K1]
$Sc_2Ni_{21}B_6$	RT	*Fm3m*	1.05940(3)			[08G2]
$Sc_2Ir_{3.9}Ni_{2.1}B$	RT	*Fm3m*	0.78940(5)			[15H1]
$Sc_{2.15}Cr_{0.85}B_x$ ($x \leq 0.01$)	RT	*Fm3m*	1.233			[76V1]
Sc_2Ir_6B	RT	*Fm3m*	0.79653(5)			[15H1]
$Sc_2Ir_{3.9}Ni_{2.1}$	RT	*Fm3m*	0.78940(5)			[15H1]
$ScNi_3B_{0.5}$	RT	*Pm3m*	0.37760(7)			[03S1], [04S1]
$Sc_2TiIr_5B_2$	RT	*P4/mbm*	0.9329(1)		0.3103(1)	[98N1]
$Sc_2VIr_5B_2$	RT	*P4/mbm*	0.9290(1)		0.3116(1)	[98N1]
$Sc_2CrIr_5B_2$	RT	*P4/mbm*	0.9373(1)		0.3064(1)	[98N1]
$Sc_2MnIr_5B_2$	RT	*P4/mbm*	0.9405(1)		0.3043(1)	[98N1]
$Sc_2FeIr_5B_2$	RT	*P4/mbm*	0.9373(1)		0.3028(1)	[98N1]
$Sc_2CoIr_5B_2$	RT	*P4/mbm*	0.9303(1)		0.3031(1)	[98N1]

(continued)

Table 2.2 (continued)

Boride	T (K)	Space group	Lattice parameters (nm)			Reference
			a	b	c	
$Sc_2NiIr_5B_2$	RT	*P4/mbm*	0.9221(1)		0.3095(1)	[98N1]
$ScCuIr_5B_2$	RT	*P4/mbm*	0.9290(1)		0.3067(1)	[98N1]
$Sc_2TiRh_5B_2$	RT	*P4/mbm*	0.9297(1)		0.3089(1)	[01N1]
$Sc_2VRh_5B_2$	RT	*P4/mbm*	0.9278(1)		0.3095(1)	[01N1]
$Sc_2CrRh_5B_2$	RT	*P4/mbm*	0.9352(1)		0.3063(1)	[01N1]
$Sc_2MnRh_5B_2$	RT	*P4/mbm*	0.9342(1)		0.3058(1)	[01N1]
$Sc_2FeRh_5B_2$	RT	*P4/mbm*	0.9284(1)		0.3059(1)	[01N1]
$Sc_2CuRh_5B_2$	RT	*P4/mbm*	0.9267(1)		0.3029(1)	[01N1]
$Sc_2FeRu_{4.05}Ir_{0.95}B_2$	RT	*P4/mbm*	0.9329(3)		0.30103(9)	[14H1]
$Sc_2FeRu_{3.2}Ir_{1.8}B_2$	RT	*P4/mbm*	0.9345(3)		0.30038(9)	[14H1]
$Sc_2FeRu_{2.04}Ir_{2.96}B_2$	RT	*P4/mbm*	0.9380(2)		0.30020(6)	[14H1]
$Sc_2FeRu_{1.2}Ir_{3.8}B_2$	RT	*P4/mbm*	0.9398(3)		0.29962(9)	[14H1]
$Sc_2FeRu_4RhB_2$	RT	*P4/mbm*	0.93225(4)		0.30117(5)	[07F1]
$Sc_2FeRu_2Rh_3B_2$	RT	*P4/mbm*	0.93489(15)		0.30192(6)	[07F1]
$Sc_2FeRuRh_4B_2$	RT	*P4/mbm*	0.93491(17)		0.30187(6)	[07F1]
$ScNi_2B_2C$(metastable)	RT	*I4/mmm*	0.335		1.068	[04K1]
$Sc_{0.9}Y_{0.1}Ni_2B_2C$	RT	*I4/mmm*	0.3463		1.0675	[04K1]
$Sc_{0.8}Y_{0.2}Ni_2B_2C$	RT	*I4/mmm*	0.3477		1.0646	[04K1]
$Sc_{0.7}Y_{0.3}Ni_2B_2C$	RT	*I4/mmm*	0.3484		1.0595	[04K1]

The $ScNi_3B_{0.5}$ compound crystallizes in a cubic structure, space group *Pm*3*m* [03S1, 04S1]. In this structure, the Sc is located at the corners of cube, Ni is at the six face-centered sites and B at the body-centered site. The micro-Vickers hardness of this compound is 4.6 (1) GPa [04S1].

According to [01T1], the as-cast Sc-Ni–B-C samples were found to contain two tetragonal phases with different lattice constants, one of them being superconducting. The superconducting properties of the last phase can be improved by annealing at T = 600 °C. Later on, was shown [04K1] that the as-cast alloys having a nominal composition $ScNi_2B_2C$, is a metastable phase and disappears upon annealing. The structure stability of $Sc_{1-x}Y_xNi_2B_2C$ series with ($0 \leq x \leq 0.3$) has been investigated [04K1]. The superconducting phase was stabilized by appropriately tuning the R-site radius, by substituting Y^{3+} (r = 0.09 nm) at the Sc^{3+} (r = 0.073 nm) site and thereby increasing the average R site radius. A review on crystal structures of R-M-B compounds was also published [84P1].

First principle calculations in the framework of DFT, using the pseudopotential method, showed that Sc_2CrB_6 has a mixed covalent and metallic nature of the bonding [22S1]. The ferromagnetic ground state is predicted for metallic bonded Sc_2CrB_6.The magnetization arises mainly from Cr atom with enhanced contribution from the remaining elements. Band crossing was seen at the Fermi level, with spin polarization indicating the existence of Dirac cone. A transition from ferromagnetic to non-magnetic state, has been also predicted at p = 30.3 GPa.

The $ScNi_3B_{0.5}$ [04S1] and $Sc_2Ni_{21}B_6$ [08G2] are Pauli-type paramagnets. The temperature dependence of the electrical resistivity in $ScNi_3B_{0.5}$ is typical for a metallic system [03S1, 04S1].

The magnetic properties of $Sc_2MIr_5B_2$ compounds with M = Mn, Fe, Co, Ni show interesting features [98N1, 01N1] (Table 2.3). The $Sc_2NiIr_5B_2$ is a Pauli-type paramagnet. The samples with M = Mn, Fe and Co are magnetically ordered. There are strong direct exchange interactions in the rows and weak indirect interactions via the RKKY mechanism, between the rows [98N1]. As a result, there is a ferromagnetic coupling in the rows, al low temperatures. The magnetic coupling between the rows was AFM when M = Fe and FM for borides with M = Co, Mn. Thus, the $Sc_2FeIr_5B_2$ is antiferromagnetically ordered, while the borides with M = Mn, Co are ferromagnetic. Above the magnetic ordering temperatures, the magnetic susceptibilities follow a modified Curie–Weiss type law, as shown as example in case of $Sc_2FeIr_5B_2$—Fig. 2.3. The effective transition metal moments are little smaller than the free ion values [98N1, 01N1].

The temperature dependences of the resistivities in $Sc_2MIr_5B_2$ with M = Fe, Co are different when cooling or warming the samples [98N1]. More investigations are necessary in order to obtain a consistent explanation for this behavior.

In the $Sc_2FeRu_{5\text{-}x}Ir_xB_2$ series, the magnetic ordering changes from antiferromagnetic ($Sc_2FeRu_5B_2$) to a ferromagnetic ($Sc_2FeRuIr_4B_2$) one and finally to a metamagnetic type behavior ($Sc_2FeIr_5B_2$) [14H1]. The borides with x = 2 and 3 have hard magnetic properties. The magnetic behavior of this system was directly related to the evolution of valence electron count (VEC) and in particular to the Ru/Ir site preference. A high Ir amount on site 8*j*, favors ferromagnetic interactions and increases the magnetic moment, whereas a high Ru amount favors antiferromagnetic interactions and lowers the magnetization.

The overall magnetic behavior of $Sc_2MnRh_5B_2$ is typical for a metamagnet [01N1]. In low field, an AFM ordering was shown. By increasing the field, there is a transition to a FM state. The magnetization at T = 1.7 K it is not saturated even in a field of 5 T. The $Sc_2FeRh_5B_2$ is FM ordered. At $T > T_N$, the reciprocal susceptibility for the sample with M = Mn follow a Curie–Weiss dependence. A more complex thermal variation, at $T > T_c$, was shown when M = Fe, suggesting the presence of magnetic impurities.

The magnetic properties, of $Sc_2FeRu_{5\text{-}x}Rh_xB_2$ series, were also correlated with the valence electron count, VEC [07F1]. The samples order ferromagnetically ($3 \leq x \leq 5$), as VEC = 63–65 and antiferromagnetically ($0 \leq x \leq 2$), when VEC = 60–62. It was shown that as the VEC count increases from 60 to 62, the strength of Fe–Fe antiferromagnetic interactions decreases significantly. The ferromagnetic interactions increase when increasing the VEC count from 62 to 65. Band structure calculations were also made on $Sc_2FeRu_{5\text{-}x}Rh_xB_2$ system [02D1, 07S1]. Continuous substitution of Ru by Rh changes the ground state from AFM to FM, as well as increases the effective moment caused by filling the bands with addition electrons

Table 2.3 Magnetic properties

Compound	Magnetic structure and magnetic moments	M_s (μ_B/f.u.)	T_c (T_N) (K)	C (emuK/f.u.)	M_{eff} (μ_B/f.u.)	θ (K)	H_c at T = 5 K (T)	Reference
$ScNi_3B_{0.5}$	Pauli paramagnet, $\chi_0 = 2 \cdot 10^{-2}$emu/f.u.							[04S1]
$Sc_2Ni_{21}B_6$	Pauli paramagnet, $\chi_0 = 1.7 \cdot 10^{-3}$emu/f.u.							[08G2]
$Sc_2MnIr_5B_2$	FM	≅ 1.2	≅ 115 (T_N)	2.59(1)	3.76(1)[c]	184(2)		[98N1]
$Sc_2FeIr_5B_2$	Metamagnetic; T > T_c: $\chi = \chi_0 + C/(T\text{-}\theta)$ $\chi_0 = 4.9(1) \cdot 10^{-2}$emu/f.u.		≅ 190 (T_N)	3.18(4)	4.17(5)[c]	241(1)		[98N1]
$Sc_2CoIr_5B_2$	FM, T > T_c $\chi = \chi_0 + C/(T\text{-}\theta)$ $\chi_0 = 4.0(2) \cdot 10^{-2}$emu/f.u.	≅1.2	≅135	1.5	2.9(1)[c]	146(9)		[98N1]
$Sc_2MnRh_5B_2$	T = 1.9 K, AFM, metamagnetic	1.8[a]	130 (T_N)		4.2	217		[01N1]
$Sc_2FeRh_5B_2$	FM	3.3[a]	450 (T_c)					[01N1]
$Sc_2NiIr_5B_2$	Pauli paramagnet, $\chi_0 = 2 \cdot 10^{-2}$eu/f.u.							[98N1]
$ScFeRu_5B_2$	AFM; C-W: 500 K ≤ T ≤ 700 K	2.10[b]	13 (T_N)	2.4	3.9	−995		[95N1, 07F1]
$ScFeRu_{4.05}Ir_{0.95}B_2$	C-W type 100 K ≤ T ≤ 350 K	0.07[b]		1.039(3)	2.9	−80(2)	0.17	[14H1]
$ScFeRu_{3.2}Ir_{1.8}B_2$	χ vs T: non-linear	0.09[b]			3.6[d]		0.65	[14H1]

(continued)

Table 2.3 (continued)

Compound	Magnetic structure and magnetic moments	M_s (μ_B/f.u.)	T_c (T_N) (K)	C (emuK/f.u.)	M_{eff} (μ_B/f.u.)	θ (K)	H_c at T = 5 K (T)	Reference
$ScFeRu_{2.04}Ir_{2.96}B_2$	FM, C-W: 100 K ≤ T ≤ 300 K	0.45[b]	85	2.845(5)	4.25	44.3(4)	0.66	[14H1]
$ScFeRu_{1.2}Ir_{3.8}B_2$	FM	2.19[b]	150	3.59(4)	4.78	15.5	0.005	[14H1]
$ScFeRu_4RhB_2$	AFM, T > T_c no C-W behavior[d]	0.08[b]	≅ 10 (T_N)		3.2			[07F1]
$ScFeRu_3Rh_2B_2$	AFM, C-W: 100 K ≤ T ≤ 400 K				4.0	−90		[00L1, 07F1]
$ScFeRu_2Rh_3B_2$	FM C-W: 500 K ≤ T ≤ 800 K	3[e]	≅ 300 (T_c)		4.2	375		[07F1]
$ScFeRuRh_4B_2$	FM, T > T_c: no C-W behavior[d]	3.1[e]	≅ 350					[07F1]
$ScFeRh_5B_2$	FM, T > T_c: no C-W behavior[d]	3.3[e]	450 (T_c)					[02D1, 07F1]

[a] T = 1.7 K, μ_0H = 5 T
[b] T = 4 K, μ_0H = 5 T;
[c] M_{eff} determined from Curie constants are higher than those reported [98N1], at T = 350 K;
[d] probably there are magnetic ordered impurities;
[e] T = 4 K, μ_0H = 2 T; 1 emu/g = 1 Am^2/kg

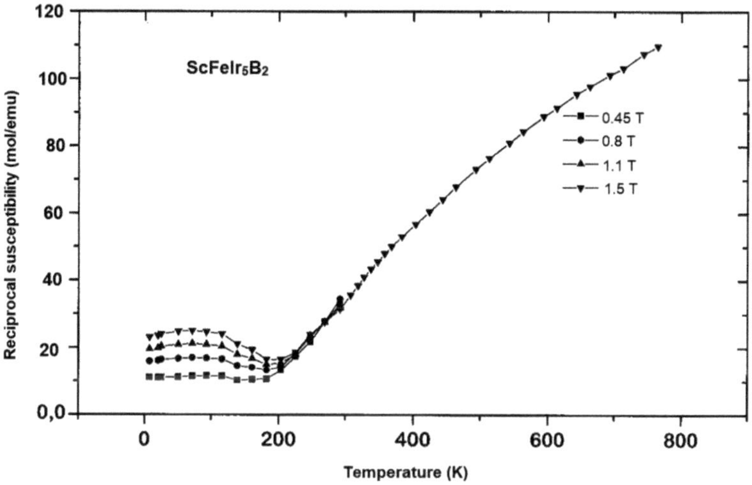

Fig. 2.3 $Sc_2FeIr_5B_2$: temperature dependence of the reciprocal susceptibilities, at the mentioned external fields [98N1]

per formula unit, together with a narrowing of the 4d band. There is a correlation between the character of chemical bonding and the resulting exchange couplings. In the $Sc_2MRu_{5-x}Rh_xB_2$ series, there is a tendency for AFM ordering for nearly the half-filled bands cases (M = Cr, Mn) and FM ordering for almost empty or almost filled bands (M = Fe, Co, Ni) [07S1].

The $Sc_{2-x}Ni_{21}B_6$ shows a Pauli-type paramagnetism [08G1, 08G2]. The $ScNi_3B_{0.5}$ perovskites is also Pauli paramagnetic. The superconducting transition temperatures of $Sc_{1-x}Y_xNi_2B_2C$ series decrease little in the $0.1 \leq x \leq 0.3$ composition range from $T_s = 15.1$ K (x = 0.1), 14.8 K (x = 0.2) and 14.5 K (x = 0.3) [04K1].

References

[54B1] P. Blum, F. Bertaut, Acta Crystallogr. **7**, 81 (1954)
[65M1] V.I. Matkovich, J. Economy, R.F. Giese, R. Barrett, Acta Crystallogr. **19**, 1056 (1965)
[67K1] Yu.B. Kuzma, Yu.Y. Voroshilov, Sov. Phys. Cristallogr. **12**, 297 (1967)
[69K1] Yu.B. Kuzma, P.I. Kripyakevich, N.S. Bilonishko, Dopov. Akad. Nauk. Ukr. RSR **A10**, 939 (1969)
[70K1] Yu.B. Kuzma, Sov. Phys. Cristallogr. **15**, 312 (1970)
[70P1] P. Peshev, J. Etourneau, R. Nanlain, Mater. Res. Bull. **5**, 319 (1970)
[73R1] P. Rogl, Monatsh. Chem. **104**, 1623 (1973)
[76V1] J.M. Vandenberg, B.T. Matthias, E. Corenzwit, H. Barz, J. Solid State Chem. **18**, 395 (1976)
[77S1] G.F. Stepanchikova, Lvov Gos. Univ. Conf. p. 197, 1977
[79S1] G.F. Stepanchikova, Vestn. Lviv Politekn. Inst. **130**, 58 (1979)
[81G1] J.C. Gianduzzo, R. Georges, B. Chevalier, J. Etourneau, P. Hagenmüller, G. Will, W. Schafer, J. less Common Met. **82**, 29 (1981)

[81S1] G.F. Stepanchikova, Yu.B. Kuzma, Vestn. Lvov Univ. Khim. **23**, 48 (1981)
[84P1] E. Parthé, B. Chabot, *Handbook on Physics and Chemistry of Rare Earths*, Chap. 48, Elsevier, 1984
[84R1] P. Rogl, *Handbook of Physics and Chemistry of Rare Earths*, Chap. 49, Elsevier, 1984
[88K1] Yu.B. Kuzma, O.M. Dub, V.A. Bruskov, N.F. Chaban, L.X. Zavalij, Kristallogr. **33**, 841 (1988)
[88Z1] L.V. Zavalij, Yu.B. Kuzma, S.I. Mikhalenko, Inorg. Mater. **24**, 1549 (1988)
[90K1] Yu.B. Kuzma, N.F. Chaban, *Binary and Ternary Systems Containing Boron* (Metallurgiya, Moscow, 1990)
[92G1] I.B. Gubich, Yu.B. Kuzma, Poroshk. Metall. (12), 57 (1992)
[95N1] E.A. Nagelschmitz, PhD Thesis, Univ. Köln, Germany (1995)
[98M1] S.I. Mikhalenko, N.F. Chaban, Yu.B. Kuzma, Poroshk. Metall. (1–2), 116 (1998)
[98N1] E.A. Nagelschmitz, W. Jung, Chem. Mater. **10**, 3189 (1998)
[00L1] G.A. Landrum, R. Dronskowski, Angew. Chem. **112**, 1598 (2000)
[01N1] E.A. Nagelschmitz, W. Jung, P. Feiten, P. Müler, H. Luekin, Z. Anorg, Allg. Chem **627**, 523 (2001)
[01T1] Zh.M. Tomilo, P.V. Molchan, A.S. Shestak, V.M. Finskaya, N.A. Prytkova, S.N. Ustinovich, Physica C **361**, 95 (2001)
[02D1] R. Dronskowski, K. Korczak, H. Lueken, W. Jung, Angew. Chem. **114**, 2638 (2002)
[02R1] V. Raghavan, J. Phase Equil. **23**, 165 (2002)
[03S1] T. Shishido, K. Kudou, T. Sasaki, J. Okada, J. Ye, K. Izumi, A. Nomura, T. Sugawara, K. Obara, M. Tanaka, J. Kohiki, Y. Kawazoe, K. Nagajima, M. Oku, Jpn. J. Appl. Phys. **42**, 7464 (2003)
[04K1] G.V.M. Kiruthika, G. Behr, R. Kulkarni, S.K. Dhar, L.C. Gupta, Physica C **405**, 245 (2004)
[04M1] S. Mykhalenko, V. Babizhetskyj, Yu.B. Kuzma, J. Solid State Chem. **177**, 439 (2004)
[04S1] T. Shishido, K. Kudou, T. Sasaki, S. Okada, J. Ye, K. Izumi, A. Nomura, T. Sugawara, K. Obara, M. Tanaka, J. Kohiki, Y. Kawazoe, K. Nagajima, M. Oku, J. Alloys Comp. **383**, 294 (2004)
[05M1] S.I. Mikhalenko, V.S. Babizhetskii, Yu.B. Kuzma, Powder Metall. Metal. Ceram. **44**, 567 (2005)
[06F1] B.P.T. Fokwa, B. Eck, R. Dronskowski, Z. Krystallogr. **221**, 445 (2006)
[06L1] G. Levchenko, A. Lyashchenko, V. Baumer, A.V. Evdokimova, V. Filipov, Yu Paderno, N. Shitsevalova, J. Solid State Chem. **179**, 2949 (2006)
[06V1] P. Villars, K. Cenzual, J. Daams, Landolt Bornstein vol. III/43A3, Springer Vorlag (2006)
[07F1] B.P.T. Fokwa, H. Lueken, R. Dronskowski, Chem. Eur. J. **13**, 6040 (2007)
[07S1] G.D. Samolyuk, B.P.T. Fokwa, R. Dronskowski, G.J. Miller, Phys. Rev. B**75**, 094404 (2007)
[08G1] R. Gumeniuk, W. Schnelle, H. Rosner, Y. Prots, S. Veremchuk, A. Leithe-Jasper, Y. Grin, Verh. Deutch Phys. Soc. **43**, 1 (2008)
[08G2] R. Gumeniuk, Y. Prots, W. Schnelle, A. Leithe-Jasper, Z. Kristallogr. **223**, 327 (2008)
[10V1] P. Villars, L.B. Handbook, vol. 43A9, Springer Verlag (2010)
[13E1] G. Eickerling, W. Scherer, T. Fickenscher, U.C. Rodewald, R. Pöttgen, Zeit. Anorg. Allgem. Chem. **639**, 2071 (2013)
[14H1] M. Hermus, M. Yang, D. Gruner, F.Y. DiSalvo, B.P.T. Fokwa, Chem. Mater. **26**, 1967 (2014)
[15H1] M. Hermus, J.P. Scheifers, R. Touzani, B.P.T. Fokwa, Inorg. Chem. **54**, 4056 (2015)
[20M1] Materials Research Project, *Lawrence Berkeley Nat.* (Lab. Berkeley, USA, 2020)
[20S1] J.P. Scheifers, R.T.D. Nguyen, Y. Zhang, B.P.T. Fokwa, J. Phys. Chem. C **124**, 26062 (2020)
[22S1] G. Shwetha, S. Chandra, N.V. Chandra Shekar, S. Kalavathi, Physica **B624**, 413369 (2022)
[22Z1] X. Zao, Q. Wang, H. Yu, F. Han, H. Liu, S. Zhang, Phys. Rev. **B105**, 094106 (2022)

Chapter 3
Titanium-Transition Metals-Boron Compounds

The investigations of the R-Ti-B phase diagrams with R = Gd [78C1, 87O1] and R = Ho [93G1] evidenced that both ternary compounds as well as binary solid solutions are not formed. Taking the above into account, the removal of the Ti impurity from liquid Gd metal has been realized in the presence of boron [21Y1]. The TiB_2 phase having high melting temperature and low density was formed and floated up to surface of liquid metal, allowing their removal.

The borides derived from ternary Ti-M-B systems, where M is a 3d transition metal atom, show interesting physical properties. These compounds can be classified in five groups [10F1]: (1) ternary, quaternary and quinary phases derived from $Ti_3Co_5B_2$-tye structure; (2) borides containing ladders of 3d transition metals, as for example $Ti_9M_2Ru_{18}B_8$; (3) borides containing B_4 units; (4) titanium borides having perovskite type structure and (5) derived from $Cr_{23}C_6$-type lattice.

3.1 Crystal Structures

Borides derived from $Ti_3Co_5B_2$—type structure

The crystal structures of ternary phases with general formula $A_3M_5B_2$ are ordered substitutional variants of $Ti_3Co_5B_2$ lattice-type. The $Ti_3Co_5B_2$ compound crystallizes in a tetragonal structure, *P4/mbm* space group [71K2, 12V1, 17Z1]—Fig. 3.1 and Table 3.1a. This can be viewed as constituted by face-connected tetragonal, pentagonal and edge sharing double trigonal prisms of cobalt atoms. The prisms are stacked along the **c**-direction, thereby building channels. The Ti atoms occupy both the tetragonal and pentagonal prism centers. The B atoms reside in trigonal prism, being trigonal-prismatically coordinated by the Co atoms. The prisms are face-connected, building a three-dimensional network. The structure can be viewed as a stacking of planar layers, one containing Co atoms and the other Ti and B atoms, alternating along **c** direction. There are no B-B contacts. Later on, a significant mixed

E. Burzo, *Rare Earths-Transition Metals-Boron Compounds*,
https://doi.org/10.1007/978-3-030-99245-3_3

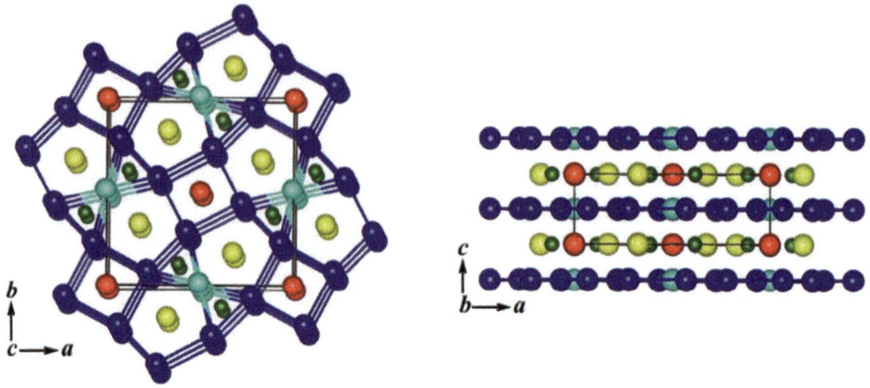

Fig. 3.1 $Ti_3Co_5B_2$: crystal structure [17Z1]. Notations: Ti1(2*a*)-red, Ti2(4*g*)-yellow, Co1(2c)-cyan, Co2(8*j*) -blue and B(4*g*)-green

occupancy of Ti and Co on the 2*a* site was shown, leading to its reformulation of composition as $Ti_{3-x}Co_{5+x}B_2$ with $0 \leq x \leq 0.5$ [17S1].

In the $A_3M_5B_2$ series, the largest A (Ti, Sc…) atoms are pentagonal- and tetragonal-prismatically coordinated by smaller M = Co, Ru, Rh, Ir atoms, but electron richer. Starting from the above ternary borides, quaternary or quinary phases were obtained [90J1, 98N1, 01N1, 08V1, 10F1]. The ordered quaternary substitutional variants of $Ti_3Co_5B_2$, with general formula $A_2MT_5B_2$ where A = Ti, M = Fe, Co, Ni and T = Ru, Rh, Ir, crystallize in a tetragonal structure, *P4/mbm* space group. There are well-separated chains of magnetic M(3d) atoms with intrachain and interchain separations of about 0.3 nm and 0.66 nm, respectively. The ordered quaternary substitutional variants can be also obtained by replacing the tetragonal-prismatically coordinated Ti atoms, by smaller elements, as for example in $Ti_{3-x}Ru_{5-y}Ir_yB_{2+x}$ complex solid solutions. When x = 1, the B atom was found not only in the usual trigonal prisms, but also in the tetragonal prisms [11F1, 11H1]. In the above system, expressed as $Ti_2(Ti_{1-x}B_x)Ru_{5-y}Ir_yB_2$, with $0 \leq x \leq 1$ and $1 < y < 3$, the tetragonal prisms, occupied by Ru/Ir atoms, are also filled with Ti, in the boron poorest phase, $Ti_3Ru_{2.9}Ir_{2.1}B_2$. The gradual substitution of Ti by B results in a successive filling of this site by a Ti/B mixture, leading to the boron richest phase, $Ti_2Ru_{2.8}Ir_{2.2}B_3$. Both Ru and Ir share two sites, but a clear Ru/Ir site preference is found. The DFT calculations confirmed the stability of these phases and showed that the Ru/Ir-B and Ru/Ir-Ti heteroatomic interactions are mainly responsible for their structure stability [11F1].

Quinary $A_2MT_{5-x}T_x'B_2$ phases are obtained by substitution of T atoms in quaternary $A_2MT_5B_2$ borides, by electron richer T' atoms, as for example $Ti_2FeRu_{5-x}T'_xB_2$ ($1 \leq x \leq 5$) with T' = Rh, Ir [11F2, 17Z1]. These compounds, as function of composition, have valence electrons counts from VEC = 63 (x = 1) to VEC = 67 (x = 5) [10F1]. The lattice parameters a and c vary as function of composition, in opposite directions, leading to little volume change—Table 3.2. In the above series, the Ru

Table 3.1 Atomic sites and their coordinates

(a) $Ti_3Co_5B_2$ having tetragonal structure, *P4/mbm* space group [71K2, 12V1][a]

Atom	Sites	Coordinates			Atomic environment
		x	y	z	
Ti	2*a*	0	0	0	18-vertex polyhedron $Ti_6Co_8B_4$
Ti	4*g*	0.173	0.673	0	7-capped pentagonal prisms $Ti_4Co_{10}B_3$
Co	2*c*	0	1/2	1/2	14-vertex Frank-Kasper $Ti_4Co_6B_4$
Co	8*j*	0.068	0.222	1/2	14-vertex Frank-Kasper $Ti_6Co_6B_2$
B	4*g*	0.615	0.115	0	trigonal prism Co_6

(b) $Ti_2FeRu_{2.85}Ir_{2.15}B_2$ having tetragonal structure, *P4/mbm* space group [11H1]

Atom	Sites	Occ	Coordinates		
			x	y	z
Ti1	4*g*		0.6754(2)	0.1754(2)	0
Fe1	2*a*		0	0	0
Ru1/Ir1	8*j*	50/50	0.07261(6)	0.21418(5)	1/2
Ru2/Ir2	2*c*	84/16	0	1/2	1/2
B	4*g*		0.875(2)	0.375(2)	0

(c) $Ti_9Fe_2Ru_{18}B_8$ having tetragonal structure, *P4/mbm* space group [08F2, 12V1][b]

Atom	Sites	Coordinates			Atomic environment
		x	y	z	
Ti	2*a*	0	0	0	square prism (cube) Ru_8
Ti	4*g*	0.69594	0.19594	0	7-capped pentagonal prism $Ru_{10}B_4Ti_3$
Ti	4*g*	0.17256	0.67256	0	square prism (cube) Ru_8
Ti	8*i*	0.1871	0.03574	0	single atom B
Fe	4*g*	0.05022	0.55022	0	14-vertex Frank-Kasper $Fe_3B_2Ru_8Ti$
Ru1	8*j*	0.06896	0.09803	1/2	non-colinear B_2
Ru2	8*j*	0.08043	0.2534	1/2	coplanar square B_4
Ru3	8*j*	0.21082	0.16877	1/2	non-colinear B_2
Ru4	8*j*	0.326	0.05251	1/2	non-colinear B_2
Ru5	4*h*	0.58469	0.08469	1/2	non-coplanar square B_4
B1	8*i*	0.0375	0.3348	0	trigonal prisms Ru_6
B2	8*i*	0.1213	0.1693	0	trigonal prisms Ru_6

(d) $TiCrIr_2B_2$ having hexagonal structure, $P\bar{6}2m$ space group [16K1]

Atom	Sites	Coordinates		
		x	y	z
Ir	6*k*	0.4626(1)	0.2766(2)	1/2
Cr	3*f*	0.1783(6)	0.1783(6)	0
Ti	3*f*	0.4139(8)	0	0

(continued)

Table 3.1 (continued)

(d) $TiCrIr_2B_2$ having hexagonal structure, $P\bar{6}2m$ space group [16K1]

Atom	Sites	Coordinates		
		x	y	z
B1	1*b*	0	0	1/2
B2	3*g*	0.206(4)	0	1/2
B3	2*c*	2/3	1/3	0

(e) $Ti_{2.3}Co_{20.7}B_6$ having cubic structure, $Fm\bar{3}m$ space group [09K2]

Atom	Sites	Occ	Coordinates		
			x	y	z
Co1	32*f*		0.38276(5)	0.38276(5)	0.38271(5)
Co2	48*h*		0	0.17022(4)	0.17022(4)
Ti3	8*c*		1/4	1/4	1/4
Ti4/Co4	4*a*	71.6% Ti 28.4% Co	0	0	0
B	24*e*		0	0	0.2730(5)

[a]transformation from data [71K2], with origin shift 1/2 1/2 0 [12V1]
[b]transformation from published data [08F2]: origin shift 0 0 1/2

and Ru atoms are assumed to be statistically distributed on 8*j* and 2*c* sites. Substituting Rh in the above system by electron richer congener, Ir, a site preference was observed—Table 3.1b and Fig. 3.2. Although both elements are mixed in each site, Ru prefers the 2*c* site, whereas Ir is preferentially located in 8*j* sites [10H1, 11H1]. Their structures consist of layers of Ru and Ir atoms, building trigonal, tetragonal and pentagonal prisms. The trigonal prisms are filled with boron and the pentagonal prisms accommodate titanium. The tetragonal prisms are filled by iron—Fig. 3.3a.

In the quaternary $Ti_{3-x}Si_xRu_5B_2$ system, with $x = 0.89$, having space group *P4/mbm*, the Ru atoms are located at 2*c* and 8*j* sites, corresponding to Co sites in $Ti_3Co_5B_2$ and B atoms occupy 4*g* sites [15X1]. The Ti atoms fully occupy another 4*g* site. The 2*a* sites are nearly fully occupied by Si, with a minor amount of Ti admixture. The layered metal-rich boride $Ti_{2.11}Si_{0.89}Ru_5B_2$ have planes containing only Ru atoms, interleaved with Ti-Si-B planes, alternatively, along **c**-axis. The distance between the Ru layers is ≅ 0.295 nm. These Ru layers form the vertices of trigonal, tetragonal and pentagonal coordination prisms. B atoms fill the trigonal prisms, Ti atoms the pentagonal prisms and the Si/Ti mixture fills the tetragonal prisms.

The $Ti_3(Fe,Co)_5B_2$ system, has been also investigated [16Z1, 17P1]. Two series of nanocrystalline Ti-Fe-Co-B alloys, $Ti_{11+x}Fe_{37.5-0.5x}Co_{37.5-0.5x}B_{14}$ with $x = 0$ and 4, as well as $Ti_{11}Fe_{26}Co_{26}Ni_{10}Al_{11}Cu_2B_{14}$ were prepared by melt spinning and subsequent annealing [16Z1]. In the first system, the presence of $Ti_3(Fe,Co)_5B_2$ and (FeCo)-rich bcc phases were evidenced, the volume fraction of $Ti_3(Fe,Co)_5B_2$ component increasing with x. In the second system, in addition to $Ti_3(Fe,Co)_5B_2$ boride, the (FeCo)-rich bcc and NiAl-rich $L2_1$ phases were present. The grain sizes, of the two

Table 3.2 Crystal structures and lattice parameters

Compound	T (K)	Space group	Lattice parameters (nm)		Reference
			a	c	
$Ti_3Co_5B_2$	RT	*P4/mbm*	0.8489(5)	0.3038(3)	[71K2, 12V1, 17P1]
$Ti_3Co_4FeB_2$	RT	*P4/mbm*	0.8469	0.3018	[17P1]
$Ti_3CoFe_4B_2$	RT	*P4/mbm*	0.8618	0.2899	[17P1]
$Ti_3Fe_5B_2$	RT	*P4/mbm*	0.8633	0.2896	[17P1]
$Ti_2FeRu_4RhB_2$	RT	*P4/mbm*	0.91507(11)	0.29682(6)	[11F2]
$Ti_2FeRu_3Rh_2B_2$	RT	*P4/mbm*	0.91590(8)	0.29586(5)	[11F2]
$Ti_2FeRu_2Rh_3B_2$	RT	*P4/mbm*	0.91790(9)	0.29526(5)	[11F2]
$Ti_2FeRuRh_4B_2$	RT	*P4/mbm*	0.91703(13)	0.29521(7)	[11F2]
$Ti_2FeRh_5B_2$	RT	*P4/mbm*	0.90833(8)	0.30127(5)	[11F2]
$Ti_2FeRu_{2.85}Ir_{2.15}B_2$	RT	*P4/mbm*	0.92176(8)	0.29639(4)	[11H1]
$Ti_2FeRu_{2.34}Ir_{2.66}B_2$	RT	*P4/mbm*	0.92239(7)	0.29514(3)	[11H1]
$Ti_3Ru_{2.9}Ir_{2.1}B_2$	RT	*P4/mbm*	0.9236(5)	0.2972(2)	[11F1]
$Ti_{2.63}Ru_{3.4}Ir_{1.6}B_{2.37}$	RT	*P4/mbm*	0.9180(3)	0.2944(2)	[11F1]
$Ti_{2.27}Ru_{2.5}Ir_{2.5}B_{2.73}$	RT	*P4/mbm*	0.9149(4)	0.2928(2)	[11F1]
$Ti_2Ru_{2.8}Ir_{2.2}B_3$	RT	*P4/mbm*	0.9030(2)	0.2855(2)	[11F1]
$Ti_{2.11}Si_{0.89}Ru_5B_2$	RT	*P4/mbm*	0.91081(7)	0.29522(2)	[15X1]
$Ti_9Cr_2Ru_{18}B_8$	RT	*P4/mbm*	1.7529(6)	0.29620(7)	[10F2]
$Ti_9Mn_2Ru_{18}B_8$	RT	*P4/mbm*	1.7547(3)	0.29683(8)	[10F2]
$Ti_9Fe_2Ru_{18}B_8$	RT	*P4/mbm*	1.7525(3)	0.29678(5)	[08F2, 10F2]
$Ti_9Co_2Ru_{18}B_8$	RT	*P4/mbm*	1.7533(4)	0.29689(11)	[10F2]
$Ti_9Ni_2Ru_{18}B_8$	RT	*P4/mbm*	1.7540(3)	0.29695(6)	[10F2]
$Ti_9Cu_2Ru_{18}B_8$	RT	*P4/mbm*	1.7543(3)	0.29692(7)	[10F2]
$Ti_9Zn_2Ru_{18}B_8$	RT	*P4/mbm*	1.7552(3)	0.29699(5)	[10F2]
$Ti_8Cr_3Ru_{18}B_8$	RT	*P4/mbm*	1.7524(1)	0.29611(1)	[13G1]
$Ti_8Mn_3Ru_{18}B_8$	RT	*P4/mbm*	1.7535(1)	0.29647(1)	[13G1]
$Ti_8Fe_3Ru_{18}B_8$	RT	*P4/mbm*	1.7519(2)	0.29670(4)	[11G1, 13G1]
$Ti_7Fe_4Ru_{18}B_8$	RT	*P4/mbm*	1.7487(2)	0.29621(3)	[11G1]
$Ti_8Co_3Ru_{18}B_8$	RT	*P4/mbm*	1.7531(2)	0.29642(1)	[13G1]
$Ti_8Ni_3Ru_{18}B_8$	RT	*P4/mbm*	1.7527(2)	0.29645(3)	[13G1]
$Ti_{10}Ru_{18.95}B_8$	RT	*P4/mbm*	1.7591(5)	0.29645(13)	[09F1]
$Ti_{1.64}Os_{2.36}B_2$	RT	$P\bar{6}2m$	0.88332(13)	0.30299(4)	[08F1]
$Ti_{0.75}Fe_{0.25}Os_2RhB_2$	RT	$P\bar{6}2m$	0.88292(13)	0.30301(4)	[08F1]
$Ti_{0.67}Fe_{0.33}Os_2RhB_2$	RT	$P\bar{6}2m$	0.88150(8)	0.30280(4)	[08F1]
$Ti_{1.6}Os_{1.4}RuB_2$	RT	$P\bar{6}2m$	0.88554(14)	0.30336(7)	[06F2]
$Ti_{1-x}V_xIr_{1.94}B_2$	RT	$P\bar{6}2m$	0.8601(2)	0.31870(6)	[20S2]

(continued)

Table 3.2 (continued)

Compound	T (K)	Space group	Lattice parameters (nm)		Reference
			a	c	
$TiMn_{0.92}Ir_{2.08}B_2$	RT	$P\bar{6}2m$	0.8610(2)	0.31880(6)	[20S2]
$TiCrIr_2B_2$	RT	$P\bar{6}2m$	08559(2)	0.31817(6)	[16K1]
Ti_2Rh_6B	RT	$Fm\bar{3}m$	0.78191(5)		[06F3]
$TiCo_{22}B_6$	RT	$Fm\bar{3}m$	1.0532(54)		[09K2]
$Ti_2Co_{21}B_6$	RT	$Fm\bar{3}m$	1.0537(5)		[09K2]
$Ti_2Co_{21}B_6$	RT	$Fm\bar{3}m$	1.0510		[67K1]
$Ti_{2.3}Co_{20.7}B_6$	RT	$Fm\bar{3}m$	1.0516(3)		[09K1, 09K2]
$Ti_3Co_{20}B_6$	RT	$Fm\bar{3}m$	1.0526		[71K1]
$Ti_3Co_{20}B_6$	RT	$Fm\bar{3}m$	1.0542		[65G1]
$Ti_3Co_{20}B_6$	RT	$Fm\bar{3}m$	1.0529(6)		[09K2]
$Ti_3Co_{20}B_6$	RT	$Fm\bar{3}m$	1.0549		[83B1]
$Ti_3Ni_{20}B_6$	RT	$Fm\bar{3}m$	1.0577		[65G1]

Compound	T (K)	Space group	Lattice parameters (nm)			Reference
			a	b	c	
$Ti_2Mn_{0.83}Ir_{3.17}B_3$	RT	*Pbam*	0.8439(2)	1.5004(4)	0.3228(7)	[20S2]
$Ti_2Fe_{0.91}Ir_{3.09}B_3$	RT	*Pbam*	0.8618(1)	1.4975(7)	0.3221(1)	[20S2]
$Ti_2Cr_{0.88}Ir_{3.12}B_3$	RT	*Pbam*	0.860(2)	1.500(3)	0.3225(5)	[20S2]
$Ti_2Ni_{0.88}Ir_{3.12}B_3$	RT	*Pbam*	0.8598(2)	1.4970(4)	0.3238(7)	[20S2]
$Ti_{1.68}Rh_{1.62}Ir_{1.94}B_3$	RT	*Pbam*	0.8620(1)	1.4995(2)	0.3234(1)	[12G1]
$Ti_{1.34}Fe_{0.73}Os_{6.93}B_6$	RT	*Cmcm*	0.2982(1)	1.3057(6)	2.117(1)	[20S1]

compositions, were 40 nm and 30 nm, respectively. The electronic structures of $Ti_3Co_{5-x}Fe_xB_2$ bulk alloys and thin films, were investigated by first-principle density functional calculations [17P1].

Borides containing ladders of 3d transition metals

The complex intermetallic borides, containing magnetically active 3d atoms, in close proximity to each other, have been synthesized. These have similar chains, composed of magnetically active elements, as in the $Ti_3M_5B_2$ borides, being variants of the $Zn_{11}Rh_{18}B_8$-type structure, which crystallizes in the *P4/mbm* type lattice [98E1]. The $Zn_{11}Rh_{18}B_8$-type structure, written as $Zn_{2.75}Rh_{4.5}B_2$, is more complex, with a 2 × 2 superlattice of the $Ti_3Co_5B_2$-type, in the tetragonal (ab) plane. Substitutions of zinc by both titanium and iron, along with replacing Rh with Ru, lead to $Ti_9Fe_2Ru_{18}B_8$ composition and their structure analogous [08F2]. Then, the polycrystalline and single crystals of $Ti_9M_2Ru_{18}B_8$ with M = Cr, Mn, Co, Ni, Cu and Zn were investigated [08F2, 09F1, 10F2]. The phases were all isotypic and crystallize in a tetragonal system, substitutional variants of $Zn_{11}Rh_{18}B_8$—type structure [98E1], space group

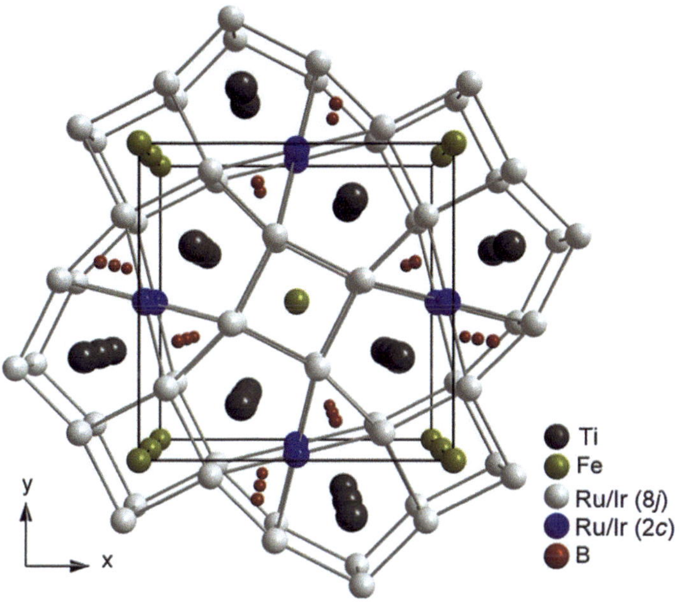

Fig. 3.2 $Ti_2Fe(Ru_{0.8}Ir_{0.2})_5B_2$: crystal structure viewed along the [001] direction [12B1]

P4/mbm. The structure of $Ti_{9+x}Fe_2Ru_{18-x}B_8$ single crystal, contains trigonal, tetragonal, pentagonal and elongated hexagonal prisms of Ru atoms stacked on top of each other to form channels along the [001] direction [10F2]. While the trigonal prisms are centered by boron atoms, the tetragonal and pentagonal prisms accommodate the Ti atoms and the elongated hexagonal prisms contain dumbbells of iron atoms—Table 3.1c. The dumbbells are arranged in chains along the [001] direction with intrachain Fe_2-Fe_2 distance of $\cong$ 0.297 nm, forming ladders along the **c**-directions—Fig. 3.3b. These ladders are separated from each other by intrachain distances of at least 1.12 nm. The crystal chemistry of the Ti-based phases is different from that of Zn-based ones. In the $Zn_{11}Rh_{18}B_8$-type structure, Fe atoms exhibit different local environments after replacing either Zn in the above lattice or Ti in hypothetical "$Ti_{11}Ru_{18}B_8$" compound. In these structures, the Zn and Ti atoms occupy tetragonal, pentagonal and elongated hexagonal prisms of Rh or Ru atoms, respectively with the elongated hexagonal prisms accommodating two-atom dumbbells [08F2]. One half of the elongated hexagonal prism (i.e. the volume accommodating only one atom of the dumbbell), is smaller than a single pentagonal prism, but larger than a tetragonal prism. Consequently, the atomic size factors influence their stability. In $Zn_{10}FeRh_{18}B_8$, Fe replace Zn in the smallest volume site (tetragonal prism), whereas in $Ti_9Fe_2Ru_{18}B_8$, Fe preferentially replaces Ti in the elongated hexagonal prism, therefore the electronic factors, as opposed to size factors, should play a key role in determining substitutional preference of Ti by Fe, leading to $Ti_9Fe_2Ru_{18}B_8$ composition. The different compositions obtained, as compared to Zn-Rh-B system,

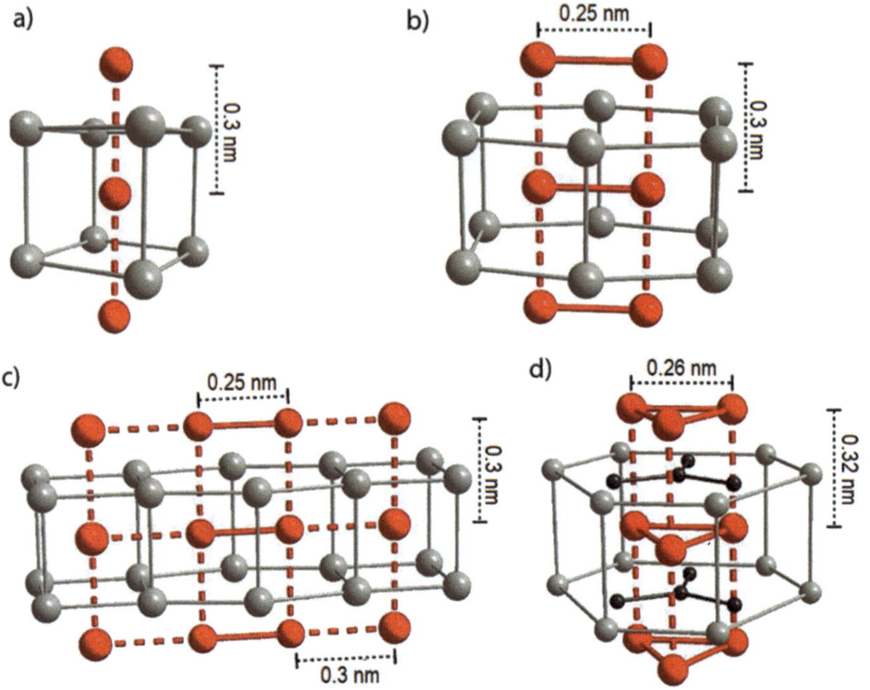

Fig. 3.3 Low dimensional substructures of magnetic atoms (red), 4d/5d metals (gray), boron fragments (dark-gray). (**a**) Chain of M atoms in $Ti_2MT_{5-x}T'_xB_2$ (T/T' = Ru, Rh, Ir); (**b**) Ladders of Fe atoms in $Ti_9Fe_2Ru_{18}B_8$; (**c**) Scaffolds of M atoms in $Ti_{9-x}M_{2+x}Ru_{18}B_8$ with M = Fe, Mn; (**d**) Trimer chains of Cr atoms in $TiCrIr_2B_2$ [16K1]

as already mentioned, were correlated to the strong valence electron count (VEC) dependence of the dumbbells site (4*h*) present in $Ti_9Fe_2Ru_{18}B_8$—related phases [09F1].

Theoretical investigations on $Ti_9M_2Ru_{18}B_8$ system revealed that the structure type is electronically stable across a range of valence electron (VEC) counts [10F2]. The phases $Ti_9M_2Ru_{18}B_8$ with M = Cr (216 VEC), Mn (218 VEC), M = Fe (220 VEC) or R = Zn (228 VEC) were prepared. When M = Ti (212 VEC) titanium was found to mix statistically with Ru in the ladder site, the structure $Ti_9(TiRu)Ru_{18}B_8$ being characterized by a value VEC = 216, confirming that this series exist only between 216 and 228 VEC counts [13G1]. In order to increase the VEC count in these phases, Ti was substituted by more electron rich elements, following the general composition $Ti_{9-x}M_{2+x}Ru_{18}B_8$. Phases as $Ti_8Fe_3Ru_{18}B_8$ (224 VEC) and $Ti_7Fe_4Ru_{18}B_8$ (228 VEC) were prepared [11G1]. The crystal structures are characterized by the simultaneous presence of dumbbells which form ladders of magnetically iron atoms along the [001] direction and two additional mixed iron/titanium chains occupying sites 4*h* and 2*b*. The ladder substructure is at ≅ 0.3 nm from the two chains at the 4*h*, which creates the sequence chain-ladder-chain and thus forms a structural and magnetic motif, the *scaffold*. The structural unit, *scaffold*, is mainly responsible for their magnetic properties—Fig. 3.3c. The electronic structure calculations on a hypothetical structural model of "$Ti_{8.5}Fe_{2.5}Ru_{18}B_8$" (VEC = 222), revealed that both Ti-Ru interactions, as well as ligand field splitting of the Fe3d orbitals, regulated the observed VEC counts between 220 and 228 electrons per formula unit [11B1].

Polycrystalline samples of the boride series $Ti_{9-x}M_{2+x}Ru_{18}B_8$ (M = Cr, Co, Mn, Ni) are also substitutional variants of the $Zn_{11}Rh_{18}B_8$ structure type, space group *P4/mbm* [13G1]. These contain a *scaffold* structural unit (M-ladders interacting with M/Ti-chains), as well as isolated M/Ti chains. According to DFT calculations the Ru-X (X = B, Ti, Ti/M) bonding interactions are nearly constant throughout series and responsible for the structural stability of these phases, whereas the M-M and Ru-M interactions vary significantly with valence electron count.

Titanium borides having B_4 units

Depending on the connection between the boron centered trigonal prisms, several boron clusters have been discovered in borides [17S2]. Five configurations for the clusters have been reported the B_4 cluster is the most versatile. For Ti-borides, only the trigonal planar B_4 fragments and zig-zag B_4 unit are present and of interest.

The presence of trigonal-planar B_4 units has been first evidenced in $Ti_{1.6}Os_{1.4}RuB_2$ compound [06F2, 08F1, 10F1]. The boride crystallizes in hexagonal structure having space group $P\bar{6}2m$ and is composed of two alternating layers, stacked along the **c**-axis. The first layer contains the M1 site (69% Os + 31% Ti) and the trigonal planar B_4 fragments (involving B1 and B2 atoms). The second layer is filled with Ru atoms, M2 site (having 96% Ti and 4% Os) and isolated boron atoms, B3. The Os atoms mainly prefer the layer located at z = 1/2, whereas Ti mostly the layer situated at z = 0. The two layers are interconnected mainly by M1-M1, M1-Ru and Ru–Ru bonds. The Os-B, Ru–B and Ti-B interactions are the strongest in the structure. The crystal structures of $Ti_{1-x}Fe_xOs_2RhB_2$ with $0 \leq x \leq 0.5$ and $Ti_{1.6}Os_{2.4}B_2$ borides are isotypic with that of $Ti_{1.6}Os_{1.4}RuB_2$, although the three types of boron-centered trigonal prisms are build up from different element mixtures [10F1]. In $Ti_{1.6}Os_{1.4}RuB_2$, the three boron centered trigonal prisms are $(Os1/Ti1)_6B$, $(Os1/Ti1)_2Ru_4B$ and Ru_6B, the corresponding prisms in $Ti_{1.6}Os_{2.4}B_2$ being $(Os1)_6B$, $(Os1)_2(Os2/Ti2)_4B$ and $(Os2/Ti2)_6B$. In $Ti_{1-x}Fe_xOs_2RhB_2$ series, the $(Os1)_6B$, $(Os1)_2Rh_4B$ and Rh_6B trigonal prisms are present. In addition, the observed Ti chains in $Ti_{1.6}Os_{2.4}B_2$ are free from osmium, whereas they accommodate 4% Os in $Ti_{1.6}Os_{1.4}RhB_2$ and mixed with iron in $Ti_{1-x}Fe_xOs_2RhB_2$ borides [08F1].

The $TiCrIr_2B_2$, as $Ti_{1+x}Os_{2-x}RuB_2$, crystallizes in a hexagonal structure, space group $P\bar{6}2m$ [16K1] —Fig. 3.3d. As in $Ti_{1+x}Os_{2-x}RuB_2$ borides, the unit cell is built of two different types of layers stacked along the **c**-axis. The z = 1/2 layer contains trigonal planar B_4 fragments (involving B1 and B2 atoms), as well as Ir atoms. The layer, at z = 0, contains isolated boron atoms (B3), Cr_3 triangles and Ti atoms. In this boride, Ir was found only on the z = 1/2 layer, whereas the 3d metals (Cr) are solely located on the z = 0 layer [16K1]—Table 3.1d.

The $Ti_{2-x}M_{1+x-y}Ir_{3+y}B_3$ borides with x = 0.5 for M = V to Mn and x = 0 for the Mn to Ni and y < 0.2 show a structural change by successive substitutions of the 3d transition metals (M = V, Cr, Mn, Fe, Co and Ni) [20S2]. The change of structure from $P\overline{6}2m$ to *Pbam* leads to a change of B_4 shape from trigonal planar B_4 (M = V to Mn), to zigzag B_4 fragments (M = Mn to Ni).

Titanium borides with double perovskite-type structure

A class of metal rich borides with the same metal to boron ratio (M/B = 4), as $Ti_3Co_5B_2$-type is the (anti-) perovskite-type, with the general formula AT_3B. These compounds can be also described as having Cu_3Au-type structure (AT_3) with interstitial boron atoms occupying the centers of T_6 octahedra [06Y1, 17S1]. In most cases a solubility range for boron can be observed, if the binary Cu_3Au exist. There are also ternary boron-deficient perovskites such as $ScNi_3B_{0.5}$, for which the binary Cu_3Au phase does not exist (see Chap. 2). The stability and boron solubility limits, in this class of compounds, depends on valence electron counts (VEC) [87T1]. The stable VEC range was correlated with the location of Fermi level in a deep pseudo-gap in the DOS. Vacancy ordering was observed for several AT_3B-perovskites, resulting in the space group change from $Pm\overline{3}m$ to $Fm\overline{3}m$ [06F3, 10H2]. The presence of a deep pseudo-gap observed in DOS of boron perovskites, is similar with that observed in $Ti_3Co_5B_2$-type borides [14H1].

The Ti_2Rh_6B boride crystallizes in a vacancy containing double perovskite type structure, space group $Fm\overline{3}m$, when the B ordering was observed [06F3]. The Ti occupies the 8*c* sites, being surrounded by 12 Rh atoms to form a cubo-octahedron. Rhodium is located in 24*e* site and its environment is formed by four Ti atoms and eight Rh atoms which form also a cubo-octahedron. One quadratic face of the cubo-octahedron is centered by a B atom. Boron lies at the position 4*e*, octahedrally surrounded by six rhodium atoms. Thus, the crystal structure can be described as a double perovskite $A_2BB'O_6$, where the A site is occupied by Ti, the B site by boron and O site by rhodium, the B' site being vacant.

Borides with $Cr_{23}C_6$-type structure

The $M_{23-x}M'_xB_6$ with M = Co, M' = Ti, Nb, V borides, crystallize in a $Cr_{23}C_6$-type structure, having space group $Fm\overline{3}m$ [65G1, 83B1]—Table 3.1e. Depending on the size of substituting elements M', different sites are affected. When the size of M' atom is larger than that of M elements, it occupies predominantly the 8*c* atomic sites having the highest coordination number, CN = 16, thereby leading to the composition $M'_2M_{21}B_6$. When increasing M' content, these atoms may additionally enter site 4*a* with CN = 12, to give the formula $M'_3M_{20}B_6$. Mixture of two metals are often observed. With decreasing differences in atomic radii, the statistical distribution of M' (Cr, Fe, Mn, Co) and M (Ir, Re) on atomic sites 48*h* and 32*f* is observed, leading to general formula $(M_{1-x}M'_x)_{23}B_6$. In some borides, additional boron atoms have been observed leading to $(M_{1-x}M'_x)_{21}M_{2-y}B_{6+4y}$ (y = 1–2) compositions [10F1]. The crystal structures of borides formed in $Ti_xCo_{23-x}B_6$ series as well as of others Ti-containing borides are listed in Table 3.2.

3.2 Physical Properties

The $Ti_3Co_5B_2$ compound has been reported to be Pauli paramagnetic [71K2]. In some Ti-Fe-Co-B nanocrystalline alloys, the presence of $Ti_3(Fe,Co)_5B_2$ borides has been evidenced, this phase being ferromagnetically ordered [16Z1]. Band structure calculations on $Ti_3Co_{5-x}Fe_xB_2$ bulk alloys suggested that the above borides are paramagnetic, except the "ideal" end series compound "$Ti_3Fe_5B_2$" [17P1]. The half metallicity reflects the hybridization of the Ti, Fe and Co3d orbitals, which causes a band gap in minority spin channels [17P1]. In the form of thin films these compounds are ferromagnetically ordered. The computed anisotropy is uniaxial and decreases when decreasing Co content. The $Ti_3(Fe,Co)_5B_2$ and $Ti_3(Fe,Co,Ni,Al,Cu)_5B_2$ melt spun alloys, are multiphase systems and magnetically ordered [16Z1]. The coercivity increases when increasing the iron content in the mixed phase alloys—Table 3.3.

The $Ti_{2.11}Si_{0.89}Ru_5B_2$ boride was reported to be diamagnetic. The paramagnetic contribution of the conduction electrons to the magnetic susceptibility is small [15X1]. A metallic type behavior, has been evidenced, by resistivity measurements.

The magnetic properties of $Ti_2FeRu_{5-x}Rh_xB_2$ compounds, having valence electron count between VEC = 63 (x = 1) and 67 (x = 5), were investigated [10F1, 11F2]. When changing composition, leading to an increase of VEC values from 63 to 66, the Curie temperatures increase from 220 to 390 K, the saturation magnetizations changing also from 0.6 μ_B to 3.1 μ_B/f.u. The sample with x = 5 (VE C = 67) shows a reversed trend, both T_c and M_s decreasing as compared to sample having VEC = 66. An evolution from soft to semihard magnetism was evidenced as the Ru content increases, decreasing VEC values, respectively. Above the Curie points, the magnetic susceptibilities follow Curie–Weiss-type dependences—Table 3.3.

The saturation magnetizations in $Ti_2M'Ru_{5-n}Rh_nB_2$ series with M' = Fe, Co, as function of valence electron counts, as determined from band structure calculations, follow a Néel-Slater-Pauling-type dependence, the greatest moment occurring typically at VEC = 66 [09B1]. The main contributions to the magnetizations are due to iron and cobalt with minor ones resulting from 4d band polarization of Ru/Rh atoms. The cobalt moments increase as function of electron counts, Rh content respectively—Table 3.3.

The magnetic properties of soft $Ti_2FeRh_5B_2$ and semi-hard $Ti_2FeRu_4RhB_2$ borides were calculated [17Z1]. The spin–orbit coupling effect, magnetic dipole–dipole interactions and spin exchange couplings were investigated. An easy axis of magnetization, parallel to the **c**-direction of the tetragonal structure was suggested. The small coercivity of $Ti_2FeRh_5B_2$ has been attributed to the weak Fe–Fe inter-chain interactions, which result in magnetically isolated iron chains. The $Ti_2FeRu_4RhB_2$ boride has a relative high coercive field, the AFM interactions being present. The existence of new phases as $Ti_2T'Co_5B_2$ with T' = Mn, Fe has been predicted and their magnetic properties calculated [17Z1].

The magnetic properties of $Ti_2Fe(Ru_{0.8}T_{0.2})_5B_2$ with T = Ru, Rh, Ir were investigated [12B1]. In these borides, the valence-electron-rich Rh and Ir atoms, prefer to occupy the 8*j* sites. An increase of the computed iron moment, correlated with

Table 3.3 Magnetic properties

Compound	Magnetic structure and magnetic moments	M_s (μ_B/f.u.)	T_c (T_N) (K)	C (emuK/f.u.)	M_{eff} (μ_B/f.u.)	θ (K)	μ_0H_c(T) or K_1 (MJ/m^3)	Reference
$Ti_3Co_5B_2$ (bulk)	PM							[17P1]
$Ti_3Co_5B_2$ (thin film)	FM	0.635[a1]					1.768(K_1)[a2]	[17P1]
$Ti_3Co_4FeB_2$ (thin film)	FM	0.619[a1]					0.92(K_1)[a2]	[17P1]
$Ti_3CoFe_4B_2$ (thin film)	FM	0.408[a1]					1.024(K_1)[a2]	[17P1]
$Ti_3Fe_5B_2$ (thin film)	FM	0.251[a1]					0.022(K_1)[a2]	[17P1]
$Ti_2FeCo_5B_2$ (BS)	M_{Fe} = 2.71 μ_B, M_{Co}(2c) = 0.08 μ_B, M_{Co}(8j) = 0.44 μ_B M_{Ti} = -0.23 μ_B, M_B(4g) = -0.01 μ_B	4.08						[17Z1]
$Ti_2MnCo_5B_2$ (BS)	M_{Mn} = 2.83 μ_B, M_{Co}(2c) = 0.07 μ_B, M_{Co}(8j) = 0.36 μ_B M_{Ti} = -0.16 μ_B, M_B(4g) = 0 μ_B	4.03						[17Z1]
$Ti_{15}Fe_{35.5}Co_{35.5}B_{14}$[b)]	FM	5.2[c]	743				0.0223	[16Z1]
$Ti_{11}Fe_{26}Co_{26}Ni_{10}\cdot Al_{11}Cu_2B_{14}$[b]	FM	5.04[c]	597				0.0206	[16Z1]

(continued)

Table 3.3 (continued)

Compound	Magnetic structure and magnetic moments	M_s (μ_B/f.u.)	T_c (T_N) (K)	C (emuK/f.u.)	M_{eff} (μ_B/f.u.)	θ (K)	$\mu_0 H_c$(T) or K_1 (MJ/m^3)	Reference
$Ti_{2.11}Si_{0.89}Ru_5B_2$	Diamagnetic, $\chi_0 =$ -7.6·10^{-4} emu/f.u							[15X1]
$Ti_2FeRu_4Rh_1B_2$ VEC = 63	T ≥ 250 K, χ = C/(T-θ)	0.6^d 3.09^f	≅ 220	1.7	3.7	175	0.0300^e	[10F1, 11F2]
$Ti_2FeRu_3Rh_2B_2$ VEC = 64	T ≥ 300 K, χ = C/(T-θ)	1.7^d 3.76^f	≅ 285	3.1	5.0	230	0.0130^e	[10F1, 11F2]
$Ti_2FeRu_2Rh_3B_2$ VEC = 65	T ≥ 400 K, χ = C/(T-θ)	2.2^d 4.34^f	≅ 390				0.002^e	[10F1, 11F2]
$Ti_2FeRuRh_4B_2$ VEC = 66	T ≥ 400 K, χ = C/(T-θ)	3.1^d 4.38^f	≥ 350				0.0030^e	[09B1, 10F1, 11F2]
$Ti_2FeRuRh_4B_2$ (BS)	M_{Fe}(2a) = 2.70 μ_B,M_{Ti}(4g) = -0.07 μ_B,M_{Ru}(2c) = -0.01 μ_B M_{Rh}(8j) = 0.16 μ_B, M_{Ru}(8j) = 0.08 μ_B, M_B(4g) = 0.01 μ_B	2.97						[17Z1]
$Ti_2FeRh_5B_2$ VEC = 67	T > 250 K	2.3^d 4.02^f	≅ 208	3.26	5.1	200	0.0011	[10F1, 11F2]

(continued)

Table 3.3 (continued)

Compound	Magnetic structure and magnetic moments	M_s (μ_B/f.u.)	T_c (T_N) (K)	C (emuK/f.u.)	M_{eff} (μ_B/f.u.)	θ (K)	μ_0H_c(T) or K_1 (MJ/m^3)	Reference
$Ti_2FeRh_5B_2$ (BS)	$M_{Fe}(2a) = 3.10$ μ_B,$M_{Ti}(4\ g) = 0$ μ_B,$M_{Rh}(2c) = 0.12$ μ_B $M_{Rh}(8j) = 0.24\ \mu_B$, $M_B(4\ g) = 0.01\ \mu_B$	4.20						[17Z1]
$Ti_2Fe(Ru_{0.8}Rh_{0.2})_5B_2$ (BS)	FM: $M_{Fe} = 2.51\ \mu_B{}^g$, $M_{Fe} = 2.58\ \mu_B{}^h$, $M_{Fe} = 2.56\ \mu_B{}^i$ AFM: $M_{Fe} = 2.38\ \mu_B{}^g$, $M_{Fe} = 2.49\ \mu_B{}^h$, $M_{Fe} = 2.51\ \mu_B{}^i$							[12B1]
$Ti_2Fe(Ru_{0.8}Ir_{0.2})_5B_2$	FM: $M_{Fe} = 2.53\ \mu_B{}^j$, $M_{Fe} = 2.53\ \mu_B{}^k$, $M_{Fe} = 2.53\ \mu_B{}^l$ AFM: $M_{Fe} = 2.41\ \mu_B{}^j$, $M_{Fe} = 2.45\ \mu_B{}^k$, $M_{Fe} = 2.45\ \mu_B{}^l$							[12B1]
$Ti_{2.11}Si_{0.89}Ru_5B_2$	Diamagnetic $\chi_0 = -7.6 \cdot 10^{-4}$ emu/mol							[15X1]
$Ti_2CoRu_4RhB_2$ (BS)	FM	1.9						[10F1]

(continued)

Table 3.3 (continued)

Compound	Magnetic structure and magnetic moments	M_s (μ_B/f.u.)	T_c (T_N) (K)	C (emuK/f.u.)	M_{eff} (μ_B/f.u.)	θ (K)	μ_0H_c(T) or K_1 (MJ/m^3)	Reference
$Ti_2CoRu_3Rh_2B_2$ (BS)	FM	2.3						[10F1]
$Ti_2CoRu_2Rh_3B_2$ (BS)	FM	2.8						[10F1]
$Ti_2CoRu_2Rh_3B_2$ (BS)	FM	3.02						[09B1]
$Ti_2CoRuRh_4B_2$ (BS)	FM	3.1						[10F1]
$Ti_9Fe_2Ru_{18}B_8$	FM T > 350 K, $\chi = C/(T-\theta)$	1.2[d] 5.1 [m]	200			290		[08F2]
$Ti_{8.5}Cr_{2.5}Ru_{18}B_8$ (BS)		0.16[n]						[13G1]
$Ti_{8.5}Mn_{2.5}Ru_{18}B_8$ (BS)		2.05[n]						[13G1]
$Ti_{8.5}Fe_{2.5}Ru_{18}B_8$ (BS)		2.0[n]						[13G1]
$Ti_{8.5}Co_{2.5}Ru_{18}B_8$ (BS)		0[n]						[13G1]
$Ti_{8.5}Ni_{2.5}Ru_{18}B_8$ (BS)		0.71[n]						[13G1]
$Ti_8Mn_3Ru_{18}B_8$	FIM	0.56[d]	120			−300		[13G1]
$Ti_8Co_3Ru_{18}B_8$	PM, 4 K ≤ T ≤ 300 K, $\chi = C/(T-\theta)$							[13G1]

(continued)

Table 3.3 (continued)

Compound	Magnetic structure and magnetic moments	M_s (μ_B/f.u.)	T_c (T_N) (K)	C (emuK/f.u.)	M_{eff} (μ_B/f.u.)	θ (K)	μ_0H_c(T) or K_1 (MJ/m^3)	Reference
$Ti_8Fe_3Ru_{18}B_8$	FIM, 200 K ≤ T ≤ 300 K, χ = C/(T-θ)	0.868[o]	210	$3.59 \cdot 10^{-4}$ [p]		95	0.0129[r]	[11G1]
$Ti_8Fe_3Ru_{18}B_8$	FIM		225					[06F1]
$Ti_7Fe_4Ru_{18}B_8$	200 K ≤ T ≤ 300 K, χ = C/(T-θ)	1.814[o]	220	$3.35 \cdot 10^{-4}$ [p]		169	0.0358[r]	[11G1]
$TiCrIr_2B_2$	350 K ≤ T ≤ 600 K, χ = C/(T-θ)	0.05 [s]	275			−755		[16K1]
$TiCrIr_2B_2$ (BS)	FIM (mostly canted) M_{Cr}: M_1 = -1.57 μ_B, $M_2 = M_3$ = 1.22 μ_B Ti: $M_1 = M_2$ = 0.03 μ_B, M_3 = -0.14 μ_B Ir: M_1 = 0.02 μ_B							[16K1]
$Ti_3Co_{20}B_6$	FM	13.6[t]	478					[83B1]

[a]Computed; [a1]μ_B/surfaces Ti atom, [a2]K_1 = MJ/m^3; in bulk state the $Ti_3Co_{5-x}Fe_xB_6$ are paramagnetic [17P1]; [b]mixed phase ribbons [c]from figure, μ_0H = 0.1 T; [d]T = 5 K, μ_0H = 7 T; [e]at T = 5 K; [f]from band structure [09B1]; [g]Fe with 0Rh,1Fe n.n.; [h]Fe with 2Rh,1Fe n.n.; [i]Fe with 1Rh,1Fe n.n.; [j]Fe with 0Ir,1Fe n.n.; [k]Fe with 2Ir,1Fe n.n.; [l]Fe with 1Ir,1Fe n.n.; [m]computed [08F2, 10F1]; [n]from figure; [o]T = 5 K, μ_0H = 5.8 T; [p]given in m^3 K mol^{-1}; [r]at T = 10 K; [s]T = 2 K, μ_0H = 5 T; [t]at T = 4.2 K, μ_0H = 1.8 T. 1 emu/g = 1 Am^2/kg

larger effective exchange parameters [07S1] was shown. The iron local environments influence their magnetic moments. Placing the Rh/Ir atoms in the nearest neighbour sites to an iron one, increases the energy difference between FM and AFM-type in favor of AFM ordering.

The $Ti_9Fe_2Ru_{18}B_8$ compound is ferromagnetically ordered, strong Fe–Fe exchange interactions being present [08F2, 10F2, 11B1]. The crystal orbital Hamiltonian populations and DOS curves evidenced the occupation of the Fe–Fe antibonding states and a local maximum in the nonmagnetic DOS, at E_F, both of which point toward the electronic instability of the system [00L1]. Allowing the structure to relax through spin polarization, resulted in the removal of both Fe–Fe antibonding states and the peak in the DOS and effects ferromagnetic ordering along the rungs of the "ladders". The ferromagnetic ordering was shown to be preferred. The "rungs" of the ladder were treated as FM stemming from a triplet spin state of neutral iron dimers in a D_{2h} crystal field.

As already mentioned, in the $Ti_{9-x}Fe_{2+x}Ru_{18}B_8$ series, the Fe atoms partially substitute for Ti in pseudo-cubic prism, at sites 2*b* and 4*h* [01N1, 09F1, 10B1, 11B1]. The 2*b* sites form chains along the [001] direction, well separated from other sites occupied by 3d metals, whereas the 4*h* sites are at $\cong$ 0.3 nm from the Fe "ladders". The proximity of the 4*h* chains to the ladders can also be described as a tetramer of atoms, that along the [001] direction forms a *scaffold* structure. Thus, direct magnetic exchange interactions between the 4*h*-chains and the "ladder" can also be expected. The iron atoms occupying the 2*b*-chains are located at $\cong$ 0.65 nm from the 4*h*-chain and 0.79 nm from the ladder. There are weaker indirect (through bond) exchange interactions, between iron atoms [11B1]. The computed magnetic properties, of $Ti_{9-x}Fe_{2+x}Ru_{18}B_8$ series, evidenced that their magnetizations arise primarily from the iron sites with small contributions from polarized bands of nearest neighbor Ru sites [11B1]. The calculated local moment for the isolated Fe1 ladder, in $Ti_9Fe_2Ru_{18}B_8$, was of 4.48 μ_B/Fe_2-dimer, with Ti atoms in the two "chain" sites, carrying negligible moments. Substituting Fe atoms for Ti, in the "isolated" 2*b*–chain (M3) sites, does not affect the computed moment at the ladder site, while the additional Fe atoms at 2*b* sites exhibit a local moment of 2.23 μ_B/atom. The incorporation of Fe atoms at the 4*h*–chain (M2) sites, which forms the *scaffold* structure with the ladder, creates a calculated moment, of 0.1 μ_B/Fe atom larger than in the 2*b* chains. The substitution of Fe atoms at sites with near neighbor Fe atoms leads to a larger magnetic moments than when iron atoms occupy isolated sites. The correlation between magnetic properties of $Ti_{9-x}Fe_{2+x}Ru_{18}B_8$ series and the valence electron count (VEC) has been also analysed [11B1]. At VEC counts (220–222) corresponding to low Fe content, the most favorable is the ferromagnetic ordering. A transition to a ferrimagnetic arrangement occurs at VEC = 224, flipping the moments of the chain site atoms (M2, M3), from parallel to antiparallel alignment with respect to the Fe1 ladder. The most favorable model, at $224 \leq$ VEC ≤ 228, can be described by a ferromagnetic ordering between the Fe1 ladders, but antiferromagnetic between the Fe1 ladder and the 4*h* chain (M2) sites, making a ferrimagnetic *scaffold* [11B1]. Over this VEC counts range, the total moment is low and nearly constant, the iron magnetic moments at the both M2 and M3 sites being antiparallel oriented to that of the Fe1 ladders. The Ru polarization, at sites closest to the Fe1 ladders, is antiparallel oriented to iron moments of ladders sites. The experimentally determined moment in $Ti_8Fe_3Ru_{18}B_8$ boride (VEC = 224) is smaller than that in $Ti_9Fe_2Ru_{18}B_2$ sample [08F2], which stems for ferrimagnetic ordering. In the composition range $0 \leq n_{Fe} \leq 2$, the theoretically predictions of the magnetic structures and trends in the total magnetic moments coincide with experimental observations. For higher VEC values, as in $Ti_7Fe_4Ru_{18}B_2$ (VEC = 228), do begin to differ from experimental findings, probably due to the differences in mixed Ti/Fe occupancies, at the M2 and M3 sites, between real and assumed values.

The magnetic properties of $Ti_{8.5}M_{2.5}Ru_{18}B_8$ with M = Cr, Mn, Fe, Co and Ni were calculated [13G1]—Table 3.3. It was predicted that the phases with Mn, Fe and Ni develop a magnetic ordering but not for M = Co boride. Magnetic measurements confirmed that $Ti_8Co_3Ru_{18}B_8$ sample, at T > 4 K, is paramagnetic, while the $Ti_8M'_3Ru_{18}B_8$ with M' = Mn or Fe are ferrimagnetically ordered.

The $Ti_{4.34}Fe_{0.73}Os_{6.93}B_6$ boride crystallizes in an orthorhombic type structure, space group *Cmcm* [20S1, 20S2]. The boride contains both isolated boron atoms and infinite zig-zag boron chains. In addition, there are isolated 1D iron chains running parallel to the B-chains. The DFT calculations predict strongly preferred FM ordering within each chain, but only small energy difference with magnetic interactions between chains.

The "likely ferromagnetic" $Ti_{1-x}Fe_xOs_2RhB_2$ borides with $0 \leq x \leq 0.5$ were reported [08F1]. A magnetic frustration, due to indirect antiferromagnetic interactions within Cr_3 triangles and a canted and non-linear magnetic ground state ordering was proposed for $TiCrIr_2B_2$ boride [16K1]. The above magnetic structure is consistent with the small overall magnetic moment, of 0.05 μ_B/f.u., in a field $\mu_0H = 5$ T. The ^{10}B NMR study on $TiCrIr_2B_2$ evidenced a dominant p-d hybridization state of the hyperfine interaction in the vicinity of the boron environment [18K1]. The main contribution to the magnetic ordering is the strong mixture of Bp- and Crd-states. The persistence of AFM, up to T_N, close to ambient temperature is an evidence of low dimensionality of the competing magnetic correlations within the Cr_3 triangles. The magnetic spin fluctuations driven by competitive weak ferromagnetic and strong ferromagnetic interactions support an itinerant d-based character of $TiCrIr_2B_2$.

The band structure calculations, performed on Ti_2Rh_6B, predicted a metallic state with main contributions from the Rh-centered 4d states. [06F3].

The crystal structures of transition metal-rich borides contain low dimensional subunits of 3d elements—Fig. 3.3: chains ($Sc_2FeRh_5B_2$), ladders ($Ti_9Fe_2Ru_{18}B_8$), *scaffolds* ($Ti_8Fe_3Ru_{18}B_8$) and chains of triangles ($TiCrIr_2B_2$). In this way, long range magnetic ordering may be obtained. A diversity of magnetic behavior has been obtained by substituting in some series, the non-magnetic elements, with the magnetically ones, from 3d series. In this way, the electronic structures were changed and the magnetic properties tuned, in correlation with the VEC count. As seen from Table 3.3, a variety of magnetic behavior was experimentally observed or suggested by band structure calculations. The further substitutions to the basic compounds are of interest, these influencing in a larger extent their physical properties.

References

[65G1] E. Ganglberger, H. Nowotny, E. Benesovsky, Monatsh. Chem. **96**, 1144 (1965)
[67K1] Y.B. Kuzma, Y.V. Voroshilov, Sov. Phys. Crystallogr. **12**, 297 (1967)
[71K1] Y.B. Kuzma, Inorg. Mater. **7**, 452 (1971)
[71K2] Y.B. Kuzma, Y.P. Yarmolyuk, Zh. Strukt. Khim. 12, 458 (1971)
[78C1] N.F. Chaban, Yu.B. Kuzma, I.D. Gerasim, Poroshk. Metall. **18**(8), 24 (1978)
[83B1] K.H.J. Buschow, P.G. van Engen, R. Jongebreur, J. Magn. Magn. Mater. **38**, 1 (1983)
[87O1] S.S. Ordanyan, E.E. Nikolaeva, Sov. Powder Metall Metal. Ceram. **26**, 51 (1987)
[87T1] H. Takeya, T. Shishido, J. Less Common Met. **134**, 263 (1987)
[90J1] W. Jung, J. Schiffer, Z. Anorg, Allg. Chem. **581**, 135 (1990)
[93G1] I.B. Gubich, Y.B. Kuzma, Powder Metall. Metal. **32**, 156 (1993)
[98E1] U. Eibenstein, W.Z. Jung, Z. Anorg, Allg. Chem. **624**, 802 (1998)
[98N1] E.A. Nagelschmitz, W. Jung, Chem. Mater. **10**, 3189 (1998)
[00L1] G.A. Landrum, R. Dronskowski, Angew. Chem. Int. Ed. **39**, 1560 (2000)
[01N1] E.A. Nagelschmitz, W. Jung, R. Feiten, P. Müller, H. Lueken, Z. Anorg, Allg. Chem. **627**, 523 (2001)
[06F1] B.P.T. Fokwa, H. Lueken, R. Dronskowski, Z. Anorg. Allg. Chem. 2125 (2006)
[06F2] B.P.T. Fokwa, J. von Appen, R. Dronskowski, Chem. Commun., 4419 (2006)
[06F3] B.P.T. Fokwa, R. Dronskowski, Z. Kristallogr. **221**, 445 (2006)
[06Y1] K. Yubuta, H. Nomura, K. Nakajima, T. Shishido, J. Alloys Comp. **426**, 308 (2006)
[08F1] B.P.T. Fokwa, R. Dronskowski, Z. Anorg, Allg. Chem. **634**, 1955 (2008)
[08F2] B.P.T. Fokwa, G.D. Samolyuk, G.J. Miller, R. Dronskowski, Inorg. Chem. **47**, 2113 (2008)
[08V1] P. Villars, K. Cenzual, Pearson's crystal structure database for inorganic compounds, Materials Park, OH (2008)
[09B1] J. Burghaus, R. Dronskowski, G.J. Miller, J. Solid State Chem. **182**, 2613 (2009)
[09F1] B.P.T. Fokwa, Z. Anorg, Allg. Chem. **635**, 2258 (2009)
[09K1] D. Kotzott, M. Ade, H. Hillebrecht, J. Solid State Chem. **182**, 538 (2009)
[09K2] D. Kotzott, Ph. D. Thesis, Univ. Freiburg (2009)
[10B1] J. Brgoch, S. Yeninas, R. Prozorov, G.J. Miller, J. Solid State Chem. **183**, 2917 (2010)
[10F1] B.P.T. Fokwa, Eur. J. Inorg. Chem. 3075 (2010)
[10F2] B.P.T. Fokwa, C. Goerens, M.Z. Gillsaren, Z. Kristallogr.-Cryst. Mater. **225**, 180 (2010)
[10H1] M. Hermus, B.P.T. Fokwa, Z. Anorg. Allg. Chem., 2047 (2010)
[10H2] M. Hermus, B.P.T. Fokwa, J. Solid State Chem. **183**, 784 (2010)
[11B1] J. Brgoch, C. Goerens, B.P.T. Fokwa, G.J. Miller, J. Am. Chem. Soc. **133**, 6832 (2011)
[11F1] B.P.T. Fokwa, M. Hermus, Inorg. Chem. **50**, 3332 (2011)
[11F2] B.P.T. Fokwa, H. Lueken, R. Dronskowski, Eur. J. Inorg. Chem. 3926 (2011)
[11H1] M. Hermus, B.P.T. Fokwa, Z. Anorg, Allg. Chem. **637**, 947 (2011)
[11G1] C. Goerens, J. Brgoch, G.J. Miller, B.P.T. Fokwa, Inorg. Chem. **50**, 6289 (2011)
[12B1] J. Brgoch, Y.A. Mahmoud. G.J. Miller, J. Solid State Chem. **196**, 168 (2012)
[12G1] C. Goerens, B.P.T. Fokwa, J. Solid State Chem. **192**, 113 (2012)
[12V1] P. Villars et al., Landolt Börnstein Handbook, vol. III 143 A, Springer Verlag (2012)
[13G1] C. Goerens, J. Brgoch, G.J. Miller, B.P.T. Fokwa, J. Solid State Chem. **204**, 283 (2013)
[14H1] M. Hermus, M. Yang, D. Grüner, F.J. DiSalvo, B.P.T. Fokwa, Chem. Mater. **26**, 1967 (2014)
[15X1] W. Xie, M.K. Fuccillo, B.F. Phelan, H. Luo, R.J. Cava, J. Solid State Chem. **227**, 92 (2015)
[16K1] M. Küpers, L.L. Kappelman, Y. Zhang, G.J. Miller, B.P.T. Fokwa, Inorg. Chem. **55**, 5640 (2016)
[16Z1] W.Y. Zhang, R. Skomski, A. Kashyap, S. Vallopilly, X.Z. Li, J.E. Shield, D.J. Sellmyer, AIP Adv. **6**, 056001 (2016)
[17P1] R. Pathak, I. Ahamed, W.Y. Zhang, S. Vallopilly, D.J. Sellmyer, R. Skomski, A. Kashyap, AIP Adv. **7**, 055713 (2017)
[17S1] P. Shankhar, J.P. Scheifers, M. Hermus, K. Yubutu, B.P.T. Fokwa, Z. Anorg Allg. Chem. **643**, 1551 (2017)

[17S2] J.P. Scheifers, Y. Zhang, B.P.T. Fokwa, Acc. Chem. Res. **50**, 2317 (2017)
[17Z1] Y. Zhang, G.J. Miller, B.P.T. Fokwa, Chem. Mater. **29**, 2535 (2017)
[18K1] D. Koumoulis, M. Kupers, R. Touzani, Y. Zhang, B.P.T. Fokwa, L.S. Bouchard, Mater. Res. Bull. **100**, 91 (2018)
[20S1] J.P. Scheifers, Ph. D. Thesis, Univ. California, Riverside (2020)
[20S2] J.P. Scheifers, M. Küpers, Y. Zhang, L. Lutz-Kappelman, G.J. Miller, B.P.T. Fokwa, Solid State Science **107**, 106294 (2020)
[21Y1] Z.Y. Yang, S.M. Pang, S.H. Yan, H.B. Yang, X.W. Zhang, Q. Zhu, D.H. Chen, D.G. Wu, Rare Met. **40**, 2320 (2021)

Chapter 4
Rare-Earths-Vanadium-Boron Compounds

The phase diagrams of R-V-B ternary systems with R = Ce [79M1], R = Gd [84C1], R = Ho [93G1], R = Er [93C1] and R = Lu [02C1] as well as Ce-V-Si [81S1] were reported. The RVB_4 borides, having $YCrB_4$-type structure, are present, when R = Gd, Tb, Dy, Ho, Er, Y, while in Ce-V-B system no ternary compounds are formed [79K1, 90K1, 96C1]. The Lu-V-B phase diagram evidence a different behavior. A ternary $Lu_{1.34}V_{1.66}B_6$ boride was formed and in addition a limited solubility of VB_2 in LuB_2 has been also shown. The above trend was correlated with the smaller lutetium atomic radius in the rare-earths sequence [90K1].

The RVB_4 borides crystallize in $YCrB_4$-type structure—Table 4.1a. In this lattice, all boron atoms are connected in two-dimensional network of condensed five- and seven-membered rings. The pentagons fuse two by two and are surrounded by heptagons [84P1, 84R1, 02C2, 03K1]—Fig. 4.1. All prismatically coordinated centers between two slabs are filled by metallic ions. According to atomic radii, the largest R and the smallest V atoms occupy the centers of heptagonal prisms and pentagonal prisms, respectively.

The $Lu_{1.84}V_{1.66}B_6$ boride [02C1], crystallizes in the Y_2ReB_6-type structure [90K1]—Table 4.1b. The structure is built from two layers of atoms, perpendicular to the **z**-axis. One layer consists of metal atoms and the other of boron ones, which are linked together by five-, six- and seven-fold rings. There are two positions for R atoms. The R atoms located above and below the seven-fold boron rings have coordination number CN = 23, while those that lie above and below six-fold rings have CN = 20. The metal atoms, which are smaller in size are located above and below the five-fold rings and have CN = 15. The position with CN = 20 in R_2MB_6 is fully occupied by rare-earths. The partial substitution of V by Ta, changes the $YCrB_4$-type structure to Y_2ReB_6-type one [03K1]. In $Er(V_{0.77}Ta_{0.23})VB_6$ boride, the hexagonal prisms contain a statistical distribution of Ta and V atoms. There are three different coordination polyhedra around Er, (V, Ta) and V atoms, with CN = 23, 20 and 17 vertices, respectively [72K1, 03K1]. Each boron atom is trigonal prismatically coordinated by metal atoms. The environment is completed by three additional boron neighbors, which is situated in the same plane as the central atom.

E. Burzo, *Rare Earths-Transition Metals-Boron Compounds*,
https://doi.org/10.1007/978-3-030-99245-3_4

Table 4.1 Lattice sites and their coordinates

(a) $ErVB_4$ having orthorhombic structure, *Pbma* space group [02C2]

Atom	Sites	Atomic coordinates		
		x	y	z
Er	*4h*	0.37551(5)	0.34868(2)	1/2
V	*4h*	0.3714(2)	0.08333(8)	1/2
B1	*4g*	0.024(1)	0.3072(6)	0
B2	*4g*	0.115(2)	0.4519(6)	0
B3	*4g*	0.135(2)	0.0321(5)	0
B4	*4g*	0.212(1)	0.1855(6)	0

(b) $Lu_{1.34}V_{1.66}B_6$ with orthorhombic structure having *Pbma* space group [02C1]

Atom	Sites	Atomic coordinates		
		x	y	z
Lu	*4g*	0.8214(1)	0.0851(1)	0
V/Lu[a]	*4g*	0.4450(3)	0.1282(2)	0
V	*4g*	0.1440(6)	0.1834(5)	0
B1	*4h*	0.055(4)	0.067(3)	1/2
B2	*4h*	0.254(4)	0.079(3)	1/2
B3	*4h*	0.300(4)	0.237(3)	1/2
B4	*4h*	0.125(4)	0.316(3)	1/2
B5	*4h*	0.478(4)	0.284(3)	1/2
B6	*4h*	0.100(1)	0.474(3)	1/2

[a] 34 at % Lu and 66 at % V

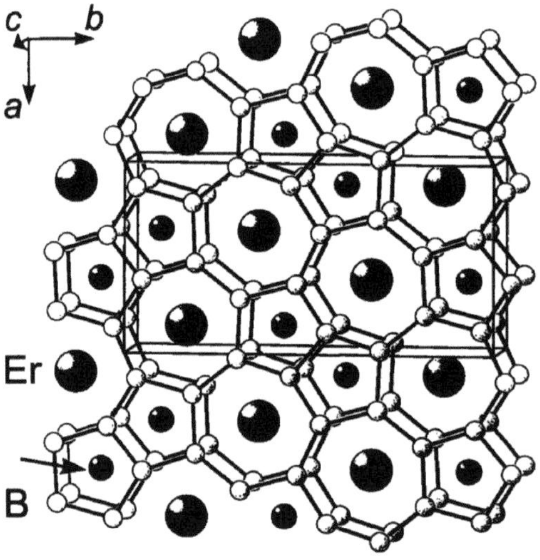

Fig. 4.1 $ErVB_4$: crystal structure [02C2]

Table 4.2 Crystal structures and lattice parameters

Boride	T (K)	Space group	Lattice parameters (nm) a	b	c	Reference
$GdVB_4$	RT	*Pbam*	0.6022(5)	1.166(1)	0.3504(5)	[70K1, 93C1]
$GdVB_4$	RT	*Pbam*	0.3473	0.6025	1.1669	[20M2]
$TbVB_4$	RT	*Pbam*	0.3455	0.6022	1.1659	[20M2]
$TbVB_4$	RT	*Pbam*	0.5970(5)	1.160(1)	0.3475(5)	[70K1]
$DyVB_4$	RT	*Pbam*	0.5966(5)	1.158(1)	0.3475(5)	[70K1]
$DyVB_4$	RT	*Pbam*	0.3452	0.6006	1.1630	[20M2]
$HoVB_4$	RT	*Pbam*	0.3436	0.5990	1.1595	[20M2]
$HoVB_4$	RT	*Pbam*	0.5968(5)	1.156(1)	0.3462(5)	[70K1]
$HoVB_4$	RT	*Pbam*	0.5068(4)	1.1563(6)	0.3438(1)	[93G1]
$ErVB_4$	RT	*Pbam*	0.5961(5)	1.155(1)	0.3448(5)	[70K1]
$ErVB_4$	RT	*Pbam*	0.5957(8)	1.148(2)	0.3460(3)	[93C1]
$ErVB_4$	RT	*Pbam*	0.5979(1)	1.1573(2)	0.34340(6)	[02C2]
$ErVB_4$	RT	*Pbam*	0.3413	0.5992	1.1593	[20M2]
YVB_4	RT	*Pbam*	0.5975(5)	1.160(1)	0.3485(5)	[70K1]
$Lu_{1.34}V_{1.66}B_6$	RT	*Pbam*	0.8947(8)	1.1273(7)	0.3381(2)	[02C1]
$ErV_{0.77}Ta_{0.23}VB_6$	293	*Pbam*	1.1280(2)	0.8940(2)	0.3390(1)	[03K1]
$Lu_{0.95}V_{0.05}B_2$	RT	*P6/mmm*	0.3238(1)		0.3689(1)	[02C1]

The lattice parameters of some R-V-B compounds are listed in Table 4.2.

The bonding and structural arrangements of YMB_4, Y_2MB_6 and Y_3MB_7-type structures were investigated. Band structure calculations were performed and the stability of the above structures has been evaluated [10L1, 11L1, 11O1].

The hard and super hard materials from R-V-B system were predicted and analyzed [20M1]. The calculated Vickers hardness H_V were 31GPa (YVB_4), 25 GPa ($DyVB_4$) and 46 GPa ($TiVB_4$).

The band structure calculations on RVB_4 suggested that the compound with R = Gd are ferromagnetic and those with R = Tb, Dy, Ho and Er are not magnetically ordered [20M2].

References

[70K1] Yu.B. Kuzma, Dopov. Akad. Nauk. Ukr. RSR, A32, 756 (1970)
[72K1] Yu.B. Kuzma, S.L. Svarichevskaya, Sov. Phys. Crystallogr. 17, 569 (1972)
[79K1] Yu.B. Kuzma, N.S. Bilonizhko, S.I. Mikhalenko, G.T. Stepanchikova, N.F. Chaban, J. Less Common Met. 67, 51 (1979)
[79M1] S.I. Mikhalenko, O.I. Bilens, Vestn. Lvov Univ., Ser. Khim. 21, 42 (1979)
[81S1] K.E. Spear, J.H. Blanks, M.S. Wang, J. less Common Met. 82, 237 (1981)
[84C1] N.F. Chaban, Powder Metall. Met. Ceram. 23, 796 (1984)

[84P1] E. Parthé, B. Chabot, Handbook on the Physics and Chemistry of Rare-Earths, Chap. 48, Elsevier Sci. Publ. (1984)
[84R1] P. Rogl, Handbook on the physics and chemistry of rare-earths, chap. 49, Elsevier Sci. Publ. (1984)
[90K1] Yu.B. Kuzma, N.F. Chaban, Binary and Ternary Systems Containing Boron, Metallurgiya, Moscow (1990)
[93C1] N.F. Chaban, O.P. Nechai, Powder Metall. Met. Ceram. 32, 539 (1993)
[93G1] I.B. Gubich, Yu.B. Kuzma, Powder Metall. Met. Ceram. 32, 156 (1993)
[96C1] N.F. Chaban, S.I. Mikhalenko, Yu.B. Kuzma, Neorg. Mater. 32, 44 (1996)
[02C1] N.F. Chaban, S.I. Mikhalenko, V.M. Davidov, Yu.B. Kuzma, Powder Metall. Met. Ceram. 41, 162 (2002)
[02C2] N.F. Chaban, Yu. Prots, Yu.B. Kuzma, Yu. Grin, Z. Kristallogr. 217, 315 (2002)
[03K1] Yu.B. Kuzma, Yu. Prots, Yu. Grin, Z. Kristallogr. 218, 159 (2003)
[10L1] S. Lassoued, R. Gautier, A. Boutarfaia, J.F. Halet, J. Organomet. Chem. 695, 983 (2010)
[11L1] S. Lassoued, R. Gautier, J.F. Halet, in Boron Rich Solids, Springer Science, p. 95, (2011)
[11O1] H. Orsini-Rosenberg, W. Steurer, Phil. Mag. 91, 2567 (2011)
[20M1] E. Mazhnik, A.R. Oganov, J. Appl. Phys. 128, 075102 (2020)
[20M2] Materials Research Project, Lawrence Berkeley Nat. Lab. Berkeley, USA (2020)

Chapter 5
Rare-Earths-Chromium-Boron Compounds

The phase diagrams of ternary R-Cr-B systems with R = La [73K2, 98M1], Ce [73K2], Pr [77M1], Nd [77M1], Sm [77M1], Gd [82C1, 98M1], Tb [93G1], Dy [93G1], Ho [93G1], Er [86C1], Tm [97C1], Yb [96C1], Lu [99C1], and Y [70K2] were investigated and also reviewed [84R1, 90K1]. The R-Cr-B systems, have three series of compounds, with respect to the nature of component interaction and the number of ternary borides. Compounds having $YCrB_4$-type structure, are present in R-Cr-B systems except for R = La, Eu. The RCr_2B_6 borides are formed only with light rare-earths (R = Ce, Pr, Nd, Sm), while those with R_3CrB_7-type structure are present when R = Tb, Dy, Ho, Er, Tm and Y. The constancy of the lattice parameters in bi-phases or tri-phases specimens indicates that there are no solid solutions or regions of homogeneity.

The $YCrB_4$ structure-type consists of vertically stacked planar 3-connected boron nets, containing 5- and 7-membered boron rings [70K1, 72K1, 74R1, 07S1]—Fig. 5.1a. The metal atoms are located halfway between the nets. Covalent bonding between metal and boron atoms plays a significant role in determining their properties. Geometrically, the structure can be seen as build from heptagonal bipyramids decorating the vertices of a tiling of squashed hexagons. The heptagonal bipyramids consist of a heptagonal boron ring capped by large rare-earth ions. The voids left by the heptagonal bipyramids correspond to pentagonal bipyramids capped by the smaller transition metal ions.

The $ThMoB_4$—type structure, having *Cmmm* space group, is related to that of $YCrB_4$-type. In both structure-types, the boron atoms form pentagonal and hexagonal rings. The pentagons fuse two by two and are surrounded by heptagons. Having the same stoichiometry, the two structures are polymorphic in some ways since they differ by the way the fused pentagons are positioned each other [74R1, 12O1].

The structure of Y_2CrB_6-type [73K1] has, in addition to heptagons and pentagons, hexagons forming boron type layers—Fig. 5.1b. The structure can be seen as intermediate between that of YB_2, having only B_6 rings and $YCrB_4$ type containing B_5 and B_7 rings.

E. Burzo, *Rare Earths-Transition Metals-Boron Compounds*,
https://doi.org/10.1007/978-3-030-99245-3_5

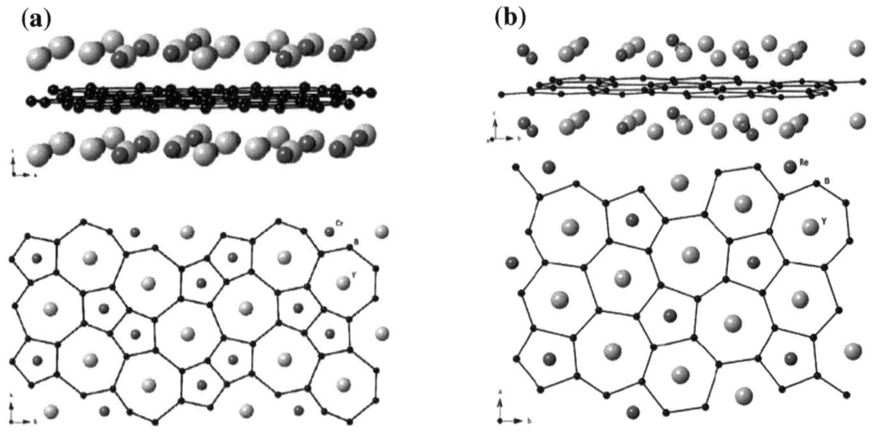

Fig. 5.1 $YCrB_4$ (**a**), Y_2CrB_6 (**b**): crystal structures [10L1]. Large, medium and small spheres represent rare-earth, chromium and boron atoms, respectively

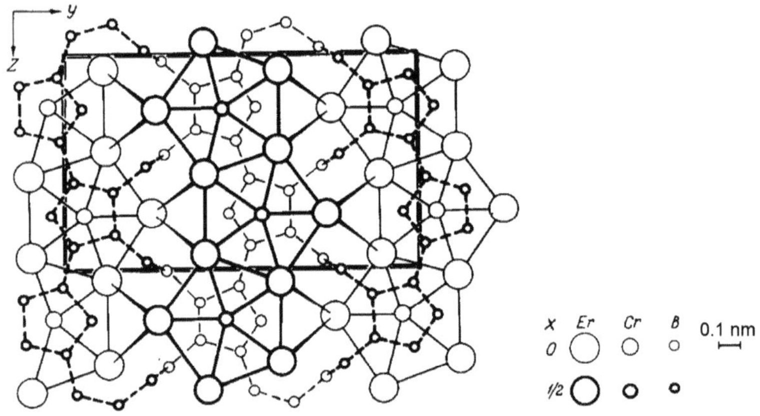

Fig. 5.2 Er_3CrB_7: projection of the crystal structure along the **x**-axis. Stacking of trigonal prisms of metal atoms is shown by solid lines, broken lined indicate the connection of boron atoms [86K1]

A projection of the Er_3CrB_7 structure along the **x**-axis is shown in Fig. 5.2 [86K1]. The metal atoms form trigonal prisms stacked to the slightly kinked walls in rows along [010] direction and displaced relative to one another by a/2. The stacks of filled trigonal prisms [BEr_6] and the empty four faced pyramids [BEr_5] are between the walls. The environment of chromium atoms is like that in the structure of $YCrB_4$ [83K1, 86K1]. The boron atoms form five member rings, connected to form bands along **z**-axis. They are placed on two levels (x = 0 and 1/2)—Table 5.1c. A connection between the rings is realized by zig zag bonds of the B1 atoms. All the boron atoms form a framework which consists of five-membered and thirteen-membered rings [86K1].

Table 5.1 Lattice sites and their coordinates

(a) $ThCrB_4$, having *Pbam*-type structure [96K1]

Atom	Site	Occupancy	x	y	z
Th	4*h*	1.000(1)	0.37715(6)	0.34884(2)	1/2
Cr	4*h*	0.996(5)	0.3762(3)	0.0832(1)	1/2
B1	4*g*	0.92(4)	0.139(2)	0.0329(1)	0
B2	4*g*	0.87(4)	0.022(2)	0.3124(9)	0
B3	4*g*	0.76(3)	0.108(2)	0.4562(8)	0
B4	4*g*	0.87(4)	0.219(2)	0.1830(9)	0

(b) $CeCr_2B_6$, having *Immm* type structure [96K1]

Atom	Site	Occupancy	x	y	z
Ce	2*a*	1.0042(9)	0	0	0
Cr	4*j*	0.969(2)	1/2	0	0.34567(3)
B1	4*h*	0.95(1)	0	0.1438(6)	1/2
B2	8*l*	0.936(9)	0	0.2373(4)	0.3035(3)

(c) Er_3CrB_7, having *Cmcm* space group [83K1, 86C1]

Atom	Site	x	y	z
Er1	8*f*	0	0.1056(1)	0.0597(2)
Er2	4*c*	0	0.7572(2)	1/4
Cr	4*c*	0	0.9462(6)	1/4
B1	8*f*	0	0.274(3)	0.023(5)
B2	8*f*	0	0.367(2)	0.352(4)
B3	4*f*	0	0.481(3)	0.408(5)
B4	4*c*	0	0.544(5)	1/4

The atomic sites and their coordinates of the above mentioned compounds are given in Table 5.1 and the lattice parameters are listed in Table 5.2.

The electronic band structures and the stability of some ternary R-Cr-B compounds have been analysed [07S1, 10L1, 11L1]. The stability of theoretical heptagonal approximants has been also investigated [07S1, 11O1]. Two approximant structures were constructed from supertiles taken for the $ThMoB_4$ and YCr_4B structure types. The total energy convex hulls were calculated for ternary systems R-Cr-B with R = Dy, Er, Ho, Y and R-Mn-B with R = Dy, Y. In all ternary systems, the theoretical approximants were found to be mechanically stable, but chemically unstable, by an amount of energy larger than the margin error. A tendency towards the stabilization of structures showing five- or tenfold local environments was observed.

The crystal structures of pseudoternary compounds crystallizing in *Pbam* type lattice were studied. The $Y_{1-x}Sc_xCrB_4$ series form solid solutions in all the composition range [18A1], while $Y_{1-x}Ce_xCrB_4$ have a narrow stability range, being present only for $x \leq 0.05$ or $x \geq 0.95$ [21F1].

Table 5.2 Crystal structure and lattice parameters

Compound	T (K)	Space group	Lattice parameters (nm)			Reference
			A	B	C	
$CeCrB_4$	RT	*Pbam*	0.5974(5)	1.153(1)	0.3536(1)	[73K2]
$CeCrB_4$	RT	*Pbam*	0.3549	0.5961	1.1524	[20M1]
$PrCrB_4$	RT	*Pbam*	0.6058(5)	1.172(1)	0.5563(4)	[77M1]
$NdCrB_4$	RT	*Pbam*	0.6025(5)	1.169(1)	0.3547(4)	[75M1, 77M1]
$SmCrB_4$	RT	*Pbam*	0.5993(5)	1.161(1)	0.3511(4)	[77M1]
$GdCrB_4$	RT	*Pbam*	0.5872(5)	1.135(1)	0.3485(4)	[70K1]
$GdCrB_4$	RT	*Pbam*	0.5953(3)	1.1527(6)	0.3494(2)	[79S1]
$TbCrB_4$	RT	*Pbam*	0.5832(5)	1.151(1)	0.3463(4)	[70K1, 93G1]
$DyCrB_4$	RT	*Pbam*	0.5792(5)	1.148(1)	0.3451(4)	[70K1, 93G1]
$HoCrB_4$	RT	*Pbam*	0.5774(5)	1.148(1)	0.3444(1)	[70K1]
$HoCrB_4$	RT	*Pbam*	0.3434	0.5914	1.1430	[20M1]
$ErCrB_4$	RT	*Pbam*	0.5889	1.1379	0.3411	[20M1]
$ErCrB_4$	RT	*Pbam*	0.5972	1.146	0.3461	[70K1, 86C1]
$TmCrB_4$	RT	*Pbam*	0.5893(4)	1.1400(5)	0.3412(2)	[97C1]
$TmCrB_4$	RT	*Pbam*	0.5972(5)	1.146(1)	0.3461(4)	[70K2]
$TmCrB_4$	RT	*Pbam*	0.3407	0.5887	1.1372	[20M1]
$YbCrB_4$	RT	*Pbam*	0.5861(2)	1.1359(4)	0.3423(1)	[96C1]
$LuCrB_4$	RT	*Pbam*	0.5726(5)	1.138(1)	0.3404(1)	[70K1]]
$LuCrB_4$	RT	*Pbam*	0.58829(7)	1.1374(1)	0.34068(6)	[99C1]
$LuCrB_4$	RT	*Pbam*	0.3394	0.5870	1.1345	[20M1]
$YCrB_4$	RT	*Pbam*	0.3450	0.5934	1.1462	[20M1]
$YCrB_4$	RT	*Pbam*	0.5972(5)	1.146(1)	0.3461(4)	[70K2, 79K1]
$YCrB_4$	RT	*Pbam*	0.5940(4)	1.1475(9)	0.3463(1)	[18A1]
$Y_{0.25}Sc_{0.75}CrB_4$	RT	*Pbam*	0.5891(6)	1.1314(4)	0.3366(7)	[18A1]
$Y_{0.5}Sc_{0.5}CrB_4$	RT	*Pbam*	0.5919(2)	1.1392(8)	0.3415(8)	[18A1]
$Y_{0.75}Sc_{0.25}CrB_4$	RT	*Pbam*	0.5918(1)	1.1428(9)	0.3438(2)	[18A1]
$Y_{0.96}Ce_{0.93}CrB_{4.11}$	RT	*Pbam*	0.59368(4)	1.14695(8)	0.34606(2)	[21F1]
$Y_{0.05}Ce_{0.95}Cr_{0.96}B_{4.04}$	RT	*Pbam*	0.59684(3)	1.15443(7)	0.35364(2)	[21F1]
$YCr_{0.75}Re_{0.25}B_4$	RT	*Pbam*	0.5943(2)	1.1488(2)	0.3496(3)	[18A1]
$YCr_{0.5}Re_{0.5}B_4$	RT	*Pbam*	0.5949(1)	1.1499(9)	0.3532(4)	[18A1]
$YCr_{0.25}Re_{0.75}B_4$	RT	*Pbam*	0.5959(2)	1.1521(4)	0.3566(3)	[18A1]
$YReB_4$	RT	*Pbam*	0.5961(4)	1.1532(9)	0.3596(1)	[18A1]
Y_2ReB_6	RT	*Pbam*	0.9175(1)	1.1545(9)	0.3670(8)	[18A1]
$YRe_{0.95}Cr_{0.5}B_6$	RT	*Pbam*	0.9173(1)	1.1538(7)	0.3670(6)	[18A1]
$YRe_{0.75}Cr_{0.25}B_6$	RT	*Pbam*	0.9167(1)	1.1543(1)	0.3670(8)	[18A1]

(continued)

Table 5.2 (continued)

Compound	T (K)	Space group	Lattice parameters (nm) A	B	C	Reference
$YRe_{0.50}Cr_{0.50}B_6$	RT	*Pbam*	0.9167(1)	1.1539(8)	0.3663(1)	[18A1]
$CeCr_2B_6$	RT	*Immm*	0.6560(4)	0.8318(5)	0.3102(3)	[73K1, 79K1, 98M1]
$CeCr_2B_6$	RT	*Immm*	0.3105(1)	0.6562(2)	0.8341(1)	[96K1]
$CeCr_2B_6$	RT	*Immm*	0.5510 α = 147.622°	0.5510 β = 107.243°	0.5510 γ = 81.697°	[17M1]
$PrCr_2B_6$	RT	*Immm*	0.6611(4)	0.8325(5)	0.3136(3)	[75M1]
$NdCr_2B_6$	RT	*Immm*	0.6597(4)	0.8319(5)	0.3113(3)	[75M1]
$SmCr_2B_6$	RT	*Immm*	0.6571(4)	0.8306(5)	0.3102(3)	[75M1]
Ho_3CrB_7	RT	*Cmcm*	0.3422(1)	1.5714(5)	0.9303(4)	[93G1]
Ho_3CrB_7	RT	*Gmcm*	0.8064 α = 90°	0.8064 β = 90°	0.9322 γ = 155.457°	[20M1]
Er_3CrB_7	RT	*Cmcm*	0.8027 α = 90°	0.8027 β = 90°	0.9290 γ = 155.438°	[20M1]
Er_3CrB_7	RT	*Cmcm*	0.34201(7)	1.5678(4)	0.9307(2)	[85C1, 86C1, 98M1]
Tm_3CrB_7	RT	*Cmcm*	0.3396(2)	1.5557(8)	0.9183(6)	[97C1]
Y_3ReB_6	RT	*Cmcm*	0.3513(1)	1.5761(2)	0.9337(6)	[18A1]
Y_3MoB_6	RT	*Cmcm*	0.3489(1)	1.5968(6)	0.9461(1)	[18A1]

The $Y(Al_{1-x}M_x)B_4$ series form solid solutions for $x \leq 0.10$ when M = Fe and $x \leq 0.015$ if M = Cr [17Y1]. The presence of *Pbam* type structure was also shown for some compositions as the $R(Al_{1-x-y}Fe_xCr_y)B_4$ series with R = Ho or Er [17Y1].

The $R(Cr_xAl_{1-x})B_4$ with R = Tm and $x \leq 0.1$ [18K1], R = Yb when $x \leq 0.015$ [19K2] or R = Lu for $x \leq 0.01$ [19K1], form solid solutions having orthorhombic type structure, space group *Pbam.*

The boron rich compounds containing rare-earth (yttrium) having $YCrB_4$, $ThMoO_4$ and Y_2CrB_6 type structures are refractory and chemical stable and thus of interest for technical uses. According to their hardness, H_v, these were classified as superhard ($H_r > 40$ GPa) or hard materials. Some mechanical properties of these borides are listed in Table 5.3 [12O1, 18A1, 19C1, 19K1]. Starting from band structure calculations, was shown that the $YMnB_4$ has the highest bulk modulus and $YCrB_4$ the highest shear and Young's moduli, these decreasing from M = Cr to Co sequence. The experimental investigations [18A1, 19K1] and theoretical studies showed that $YRe_{1-x}Cr_xB_4$ solid solutions are superhard materials, while the YMB_4 (M = Mn, Fe, Co), $R(Al_{1-x}M_x)B_4$ with M = Cr, Fe series as well as the Y_3MoB_6 and Y_3ReB_6 borides are hard materials.

The magnetic properties of $GdCrB_4$ [79S1] and $CeCr_2B_6$ [96K1] borides were studied in the paramagnetic region—Table 5.4. The effective moment of $GdCrB_4$ is

Table 5.3 Elastic properties

Compound	T (K)	Elastic modulus (GPa)			Hardness, H (GPa)	Reference
		Bulk B	Shear G	Young E		
$YCrB_4$ (BS)		210.95	208.83	471.04	42.0	[19C1]
$YCrB_4$	RT				37.5(2.29)[a]	[18A1]
$YReB_4$	RT				34.25(1.67)[a]	[18A1]
$YRe_{0.5}Cr_{0.5}B_4$	RT				42.48(2.13)[a]	[18A1]
$YMnB_4$(BS)		214.59	199.80	457.43	37.79	[19C1]
$YFeB_4$ (BS)		210.73	190.47	439.10	35.31	[19C1]
$YCoB_4$ (BS)		190.13	170.64	394.04	32.63	[19C1]
$YbAl_{0.99}Fe_{0.01}B_4$	RT				16.8(1.5)	[19K2]
$YbAl_{0.99}Cr_{0.01}B_4$	RT				14.8(5)	[19K2]
$LuAlB_4$	RT				14.0(6)	[12O1]
$LuAl_{1-x}Cr_xB_4$ $0.05 \leq x \leq 0.10$	RT				13(2)–16(3)	[19K1]
$LuAl_{1-x}Fe_xB_4$ $0.35 \leq x \leq 0.5$	RT				16(2)–20(3)	[19K1]
Y_3MoB_6	RT				29.29	[18A1]
Y_3ReB_6	RT				34.31	[18A1]

BS-band structure calculations; [a]load 0.49 N

Table 5.4 Magnetic properties

Compound	Magnetic behaviour	$C\cdot10^3$ (K emu/f.u.)	$\chi_0\cdot10^3$ (emu/f.u.)	M_{eff} (μ_B/f.u.)	θ (K)	Reference
$GdCrB_4$	$T > 77$ K, $\chi = C/(T-\theta)$			7.70	20	[79S1]
$CeCrB_4$	$1.8\ K \leq T \leq 400$ K $\chi = \chi_0 + C/T$		$5.6\cdot10^{-4}$	0.60(4)		[21F1]
$Ce_{0.05}Y_{0.95}CrB_4$	$1.8\ K \leq T \leq 400$ K $\chi = \chi_0 + C/T$		$6.7\cdot10^{-5}$	0.35(3)		[21F1]
$YCrB_4$	Pauli paramagnet		$3.8\cdot10^{-5}$			[21F1]
$CeCr_2B_6$	$T > 77$ K, $\chi = \chi_0 + C/(T-\theta)$	0.732[a]	$5.0\cdot10^{-3}$	1.71/Cr atom[b]	−30	[96K1]

[a]Determined from Fig. 1 [96K1];
[b]if the cerium is considered nonmagnetic

little smaller than that of Gd^{3+} free ion. In $CeCr_2B_6$, if the valence state of cerium is supposed to be 4+, an effective chromium moment close to that of d^1 ions has been evidenced.

The $YCrB_4$ compound is a Pauli-type paramagnet [21F1]. In $Y_{1-x}Ce_xCrB_4$ series, when $x = 0.05$, the magnetic susceptibilities follow a modified Curie–Weiss law. The small effective moment (0.35 μ_B) suggest that cerium has predominant +4 valence state. A broad maximum at T = 250 K, was evidenced in the temperature dependence of the magnetic susceptibility in $Y_{0.05}Ce_{0.95}CrB_4$, suggesting that cerium is in mixed valence state. Their valence decreases from +3.75 at T = 1.8 K up to +3.5 at T = 400 K [21F1].

The $Er(Cr_{0.62}Fe_{0.18}Al_{0.20})B_{3.8}$ boride is antiferromagnetically ordered at $T < T_N = 6.5$ K [17Y1]. Metamagnetic transitions are present in higher fields than $\mu_0H = 1$ T.

The LMTO calculations performed on $YCrB_4$ showed that this is a semiconductor with a narrow band gap of about 0.05 eV, with the states near the gap being of predominantly Cr3d character and with significant d-orbital hybridization [02M1]. The *mno* rule [01J1] was applied to $YCrB_4$-type alloys, in order to obtain information on their electronic structure [10S1]. The rule postulates that the number of occupied molecular orbitals in a cluster with deltahedral geometry, is given by the sum of *m*, *n* and *o*, where *m* is the number of polyhedra in the cluster, *n* is the number of vertices and *o* is the number of single-vertex-sharing condensation. The onset of d-d orbital hybridization was considered to be likely the root cause of the semiconducting properties of the $YCrB_4$-type compounds. The YMB_4 compounds with M = Mo and W are also semiconductors [10S1, 19B1].

The high hardness, chemical inertness and refractory thermal properties, in combination with light weight make the metal borides of interest for thermoelectric (TE) applications [01N1, 10S1, 16A1, 19M1, 21F1]. The materials used for such applications should perform with a high dimensionless thermoelectric figure of merit $ZT = S^2T/\rho\kappa$, in which S, ρ, κ and T are Seebeck coefficient, electrical resistivity, thermal conductivity and temperature, respectively. Materials with ZT parameter greater than one are desired [01N1]. There is a general difficulty in making ZT large, since typically S and $\sigma(1/\rho)$ are in trade off relationship and a large σ with a small κ is also paradoxical [19M1]. A way to overcome those difficulties for the former problem is the band engineering [18M1, 21F1] or magnetism [19T1]. For the latter paradox, nanostructuring and using the difference between mean free paths of charge carriers and phonons seem to be effective [18L1]. Potential related materials for TE applications are the RMB_4 (M = Cr, Mo, W) [02M1, 10S1, 19B1, 19C1], $Y_{1-x}Ce_xCrB_4$ [21F1] or the Y_3MB_7 series with M = Cr, Mo, Fe [10S1].

The dependences of electrical resistivity, in the temperature range $300\text{ K} \leq T \leq 1000\text{ K}$, of $RCrB_4$ borides with R = Gd, Ho and Y confirmed that these are narrow gap semiconductors [10S1], as predicted from band structure calculations [02M1] - Fig. 5.3. In the temperature range $300\text{ K} \leq T \leq 1100\text{ K}$, the thermal dependences of the electrical resistivity, ρ, of $YCrB_4$ can be described by the relation $\rho \propto \exp(E_g/2k_BT)$ with a band gap $E_g = 0.17$ eV, somewhat higher than theoretically computed value of ≅ 0.05 eV [02M1]. The thermopower was shown to be of moderately high n-type character, which may be indicative of the slightly electron excess predicted by the generalized *mno* counting rules. The increase of the metallic radius from Cr to Mo and W leads to a systematic widening influence on the size of the gap, this being of 0.28 eV in $YMoB_4$. This effect may be due to increased d-d orbital overlap and stronger orbital hybridization. The power factors $PF = S^2\sigma$, of the YMB_4 compounds are modest, and correlated to the moderate thermopower, but comparatively with high resistivity. The

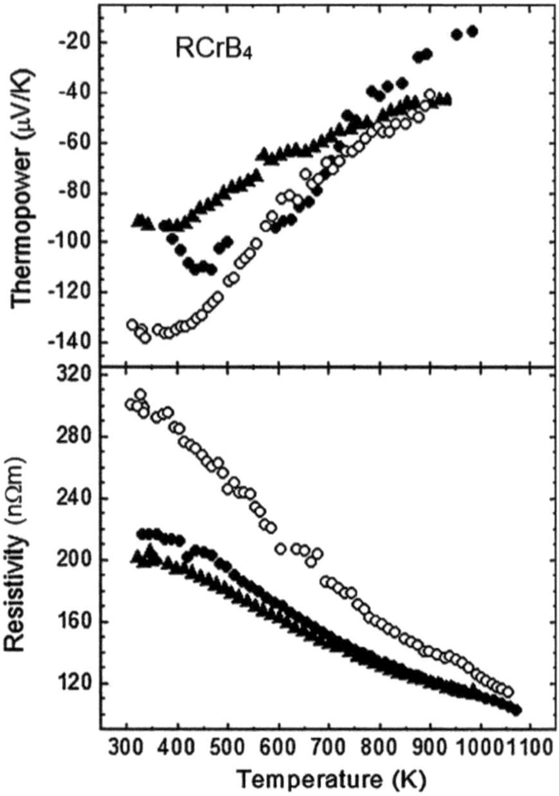

Fig. 5.3 $RCrB_4$ with R = Gd, Ho, Y: temperature dependences of the electrical resistivities and thermopowers [10S1] ●-$YCrB_4$, ○-$GdCrB_4$, ▲-$HoCrB_4$

power factor of $YCrB_4$ reaches $6.0 \cdot 10^{-4}$ W/mK^2, at T = 500 K [10S1]. This seems to be too low for a reasonably efficient TE material. Both the electrical resistivity and thermopower of $RCrB_4$ compounds with R = Gd, Ho, Y, approach convergence at elevated temperature—Fig. 5.3 [10S1]. Either the size difference between R is too small to produce noticeable effect at these temperatures or the decrease in R element size has little to do with the electronic structure, near the gap. The electrical resistivity of $CeCrB_4$ increases with temperature, behavior characteristic for a metallic system [96K1].

The systematic changes of the unit cell parameters with ionic radius of the rare-earth component in RMB_4 series, as well as magnetic measurements showed that the Gd to Er and Yb have the configuration $4f^n$, being in +3 valence state, while cerium configuration is close to $4f^0$ one [16A1]. The electrical resistivity, thermal conductivity and thermopower of RMB_4 with R = Ce, Tb, Ho borides are reminiscent of those of typical metals with low ZT values. These have low thermopowers and relatively high thermal conductivities. Within the rigid-band approach, the substraction of one electron by replacement of M = Re by W or B by Be should lead to semiconducting behavior and thus the thermoelectric properties can be improved.

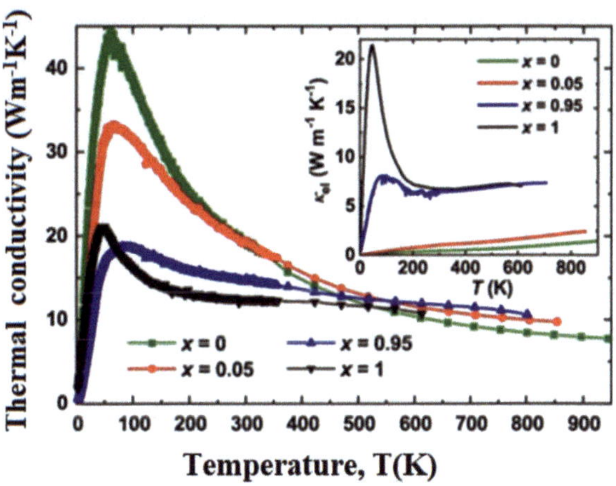

Fig. 5.4 Thermal variation of thermal conductivities in $Y_{1-x}Ce_xCrB_4$ series. In inset is shown the electronic thermal conductivity [21F1]

The effect of dopants on the $YMo_{1-x}Fe_xB_4$ power factor has been investigated in sample with x = 0.2 [10S1]. There was shown to be an increase of the power factor as compared to undoped compound. The TE performances in RMB_4 borides are little improved by simple manipulation of charge carrier concentration, by electron or hole doping.

Another potentially method, in improving the TE properties, might be the introduction of correlations effects by cerium substitution, as in $Y_{1-x}Ce_xCrB_4$ series [21F1]. The band structure calculations of $YCrB_4$ confirmed this as a narrow gap semiconductor. A pronounced Van Hove singularities very close to E_F, originating from nearly dispersionless $Cr3d_{z^2-r^2}$ derived bands, in large part of the Brillouin zone was shown. The insertion of cerium in $YCrB_4$ was expected to lead to a strongly enhanced Seebeck coefficient and a sizable figure of merit. There was shown a strong reduction of resistivity for sample with x = 0.05. In Ce-rich region (x ≥ 0.95), where the Ce is in a mixed valence state, the samples have metallic character. Thus, although the termopower factors for Y-rich compounds are rather large, for the Ce-rich samples it is small and little increased with temperature. A rather good thermal conductivity was shown in $Y_{1-x}Ce_xCrB_4$ series, having as characteristic feature, maxima in low temperature region—Fig. 5.4. The mechanism of thermal conductivity is predominantly electronic in Ce-rich borides and phononic for the Y-rich samples. Taking into account all the data, a figure of merit rather low ZT = 0.06 was obtained at T = 800 K. The electron correlation effects for Ce-rich compounds do not influence significantly the transport properties. The scientific field of TE materials is now still developing.

The temperature dependences of resistivity and thermopower for Y_3MB_7 with M = Cr, Ho, Fe are given in Fig. 5.5 [10S1]. These are low thermopower metallic alloys.

The band structure calculations on $RCrB_4$ with R = Ce, Ho, Er, Tm, Lu and Y, on R_3CrB_7 with R = Ho and Er [20M1], as well as of $CeCr_2B_6$ [17M1] showed that these are not magnetically ordered.

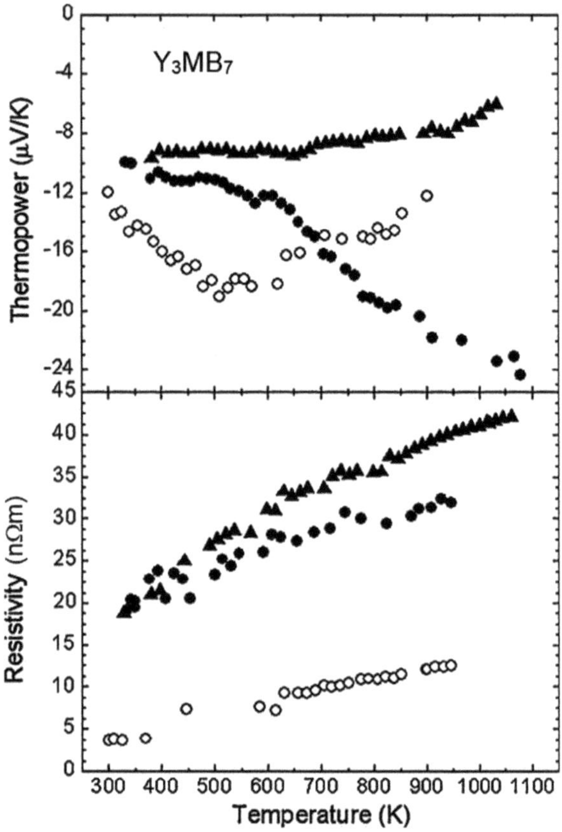

Fig. 5.5 Y_3MB_7 with M = Cr, Mo, Fe: temperature dependences of the electrical resistivities and thermopowers [10S1], ●-Y_3CrB_7, ○-Y_3FeB_7, ▲-Y_3MoB_7

References

[70K1] Yu.B. Kuzma, Kristallogr. **15**, 372 (1970)

[70K2] Yu.B. Kuzma, A.S. Sobolev, M.P. Furtak, Izv. Akad. Nauk. SSR, Neorg. Mater. **6**, 2205 (1970)

[72K1] Yu.B. Kuzma, S.I. Svarichevskaya, Dopov. Akad. Nauk Ukr. **A2**, 166 (1972)

[73K1] Yu.B. Kuzma, S.I. Svarichevskaya, Kristallogr. **17**, 939 (1973)

[73K2] Yu.B. Kuzma, S.I. Svarichevskaya, V.N. Fomenko, Izv. Akad. Nauk SSSR, Neorg. Mater. **9**, 1542 (1973)

[74R1] P. Rogl, H. Nowotny, Monatsh. Chem. **105**, 1082 (1974)

[75M1] S.I. Mikhalenko, Yu.B. Kuzma, Dopov. Akad. Nauk Ukr. RSR **A5**, 465 (1975)

[77M1] S. I. Mikhalenko, Yu.B. Kuzma, Dopov. Akad. Nauk Ukr. RSR **A10**, 951 (1977)

[79K1] Yu.B. Kuzma, N.S. Bilonizhko, S.I. Mikhalenko, G.F. Stepanchikova, N.F. Chaban, J. less Common Met. **67**, 51 (1979)

[79S1] R. Sobczak, P. Rogl, J. Solid State Chem. **27**, 343 (1979)

[82C1] N.F. Chaban, Poroshk. Metall. **1**, 61 (1982)

[83K1] Yu.B. Kuzma, Crystal Chemistry of Borides (in Russian), Vyshcha Shkola, Lvov (1983)

[84R1] P. Rogl, Handbook Phys. Chem. Rare Earth, Chap. 49, 335 (1984)

[85C1] N.F. Chaban, L.G. Akselrud, V.A. Bruskov, Yu.B. Kuzma, Kristallogr. **80**, 187 (1985)

[86C1] N.F. Chaban, V.N. Datsyna, Poroshk. Metall. **1**, 62 (1986)

[86K1] Yu.B. Kuzma, S.I. Mikhalenko, L.G. Akselrud, J. less Common Met. **117**, 29 (1986)
[90K1] Yu.B. Kuzma, N.F. Chaban, Binary and Ternary Systems Containing Boron (in Russian), Metallurgiya, Moscow (1990)
[93G1] I.B. Gubich, N.F. Chaban, Poroshk. Metall. **5**, 48 (1993)
[96C1] N.F. Chaban, S.I. Mikhalenko, Yu.B. Kuzma, Neorg. Mater. **32**, 44 (1996)
[96K1] T. Konrad, W. Jeitschko, M. Danenbrock, C.B.H. Evers, J. Alloys. Comp. **234**, 56 (1996)
[97C1] N.F. Chaban, N.P. Kurylo, Yu.B. Kuzma, Powder Metall. Met. Ceram. **36**, 496 (1997)
[98M1] S.I. Mikhalenko, N.F. Chaban, Yu.B. Kuzma, Powder Metall. Met. Ceram. **37**, 116 (1998)
[99C1] N.F. Chaban, Yu.B. Kuzma, Powder Metall. Met. Ceram. **38**, 458 (1999)
[01J1] E.D. Jemmis, M.M. Balakrishnarajan, J. Am. Chem. Soc. **123**, 4325 (2001)
[01N1] G.S. Nolas, J. Sharp, H.J. Schmidt. The Thermoelectric Basic Principles and New Materials Development, Springer, Berlin (2001)
[02M1] N.I. Medvedeva, Yu.E. Medvedeva, A.L. Ivanovskii, Dokl. Phys. Chem. **383**, 75 (2002)
[07S1] W. Steurer, Phil. Mag. **87**, 2707 (2007)
[10L1] S. Lassoued, R. Gautier, A. Boutarfaia, J.F. Hallet, J. Organomet. Chem. **675**, 987 (2010)
[10S1] J.W. Simonson, S.J. Poon, J. Alloys Comp. **504**, 265 (2010)
[11L1] S. Lassoued, R. Gautier, J.F. Halet, in Boron Rich Solids, Springer, p. 95 (2011)
[11O1] H. Orsini-Rosenberg, V. Steurer, Phil. Mag. **91**, 2567 (2011)
[12O1] S. Okada, T. Mori, K. Kudou, K. Yubuta, T. Shishido, J. Flux Growth, **7**, 6 (2012)
[16A1] M. Abramchuk, W. Schnelle, I. Veremchuk, A. Leithe-Jasper, Y. Grim., R. Gumeniuk, Eur. J. Inorg. Chem. **1**, 111 (2010)
[17M1] Materials Research Project, *Lawrence Berkeley Nat* (Lab. Berkeley, USA, 2017)
[17Y1] T. Yamasaki, K. Kouzu, A. Nomura, S. Okada, T. Shishido, K. Yubuta, A. Yosikawa, T. Mori, Bull. Fac. Sci. Techn. Univ. Kokushikan, **11**, 49 (2017)
[18A1] G. Akopov, H. Yin, I. Roh, L.E. Pangilinan, R.B. Kaner, Chem. Mater. **80**, 6494 (2018)
[18K1] K. Kouzu, T. Yamasaki, S. Okada, A. Nomura, K. Yubata, T. Shishido, A. Yoshikawa, P. Mori, P. Rogl. J. Flux Growth, **13**, 2 (2018)
[18L1] Z. H. Liu, J. Mao, L. H. Liu, G. Chen, Z. F. Ren, MRS Bull. **43**, 181 (2018)
[18M1] J. Mao, Z. Liu, J. Zhou, H. Zhu, R. Zhang, G. Chen, Z. Ren, Adv. Phys. **97**, 69 (2018)
[19B1] C. Benndorf, M. De Oliviera, C. Doerenkamp, F. Haarmann, T. Fickenscher, J. Kösters, H. Eckert, R. Pöttgen, Dalton Trans. **48**, 1118 (2019)
[19C1] A. Candan, G. Suruku, A. Gencer, Phys. Scr. **94**, 125510 (2019)
[19K1] K. Kouzu, T. Yamasaki, S. Okada, T. Mori, Q. Guo, T. Shishido, K. Yubuta, A. Nomura, A. Yoshihawa, P. Rogl, Solid State Phen. **289**, 120 (2019)
[19K2] K. Kouzu, T. Yamasaki, S. Okada, T. Mori, Q. Guo, K. Yubuta, A. Nomura, T. Shishido, A. Yosukawa, P. Rogl, J. Jpn. Soc. Powder Metallurgy, **66**, 525 (2019)
[19M1] T. Mori, J. Solid State Chem. **275**, 70 (2019)
[19T1] N. Tsujii, A. Nishide, J. Hayakawa, T. Mori, Sci. Adv. **5**, 2 (2019)
[20M1] Materials Research Project, Lawrence Berkeley Nat. Lab. Berkeley, USA (2020)
[21F1] S. Flipo, H. Rosner, M. Bobnar, K.O. Kvashnina, A. Luthe-Jasper, R. Gumeniuk, Phys. Rev. **B103**, 195121 (2021)

Chapter 6
Rare-Earths-Manganese-Boron Compounds

The R-Mn-B phase diagrams with R = Ce [75K1, 90K1], R = Pr [90M1, 98M1], R = Nd [90K1, 90M1], R = Sm [91M1], R = Gd [83M1, 89O1], R = Tb [70K1], R = Dy [91M2], R = Ho [91M2], R = Er [91M2], R = Tm [96C1, 01C1, 01K1], R = Yb [96C1], R = Lu [96C1] and R = Y [83M1, 90K1] were reported. Three structure types were identified in the R-Mn-B phase diagrams, one being incommensurate. A classification of crystal structures of borides has been also made [77N1, 83K1, 84P1, 84R1].

The $RMnB_4$ borides crystallize in an orthorhombic structure of $YCrB_4$-type, having *Pbam* space group [70K1, 78K1, 79K1]. The analysis of $TmMnB_4$ lattice parameters evidenced the presence of a narrow homogeneity region which is in the direction of higher manganese contents [01K1].

The R_3MnB_7 ternary compounds have Er_3CrB_7 type structure [83M1, 86K1]. In this lattice, the metal atoms form trigonal prisms stacked to the slightly kinked walls in rows along [010], which are displaced relative to one another by a/2. The stacks of filled trigonal prisms [BR_6] and the empty four faced pyramids [BR_5] are between the walls. The boron atoms form five member rings connected to form bands along the **z**-axis—see Chap. 5.

The reported (Pr, Nd)-Mn-B [90M1] and Sm-Mn-B [91M1] phase diagrams evidenced the presence of compounds, having approximate compositions RMn_4B_4, with tetragonal symmetry. The a lattice constants were in the range 0.72 nm and 0.73 nm, while for c parameters large values were obtained, the unit cell being described as $R_m(Mn_4B_4)_n$. The $R_m(Mn_4B_4)_n$ borides with R = Pr, Nd, Sm have incommensurate type structures. The a lattice constants are not changed, while the c ones are different from those of RMn_4B_4 series, where m = n = 1. The crystal structure of $Pr_{41}(Mn_4B_4)_{35}$ sample, belonging to this series, has been refined [94Z1]. This structure consists from two sublattices. One of them is formed by R atoms, having space group *I4/mmm*, and the second, by Mn_4B_4 units with the space group

E. Burzo, *Rare Earths-Transition Metals-Boron Compounds*,
https://doi.org/10.1007/978-3-030-99245-3_6

$P4_2/ncm$. The two sublattices have the same a parameter, but with different c lattice constants, c_{Pr} = 0.3422(2) nm and $c_{Mn_4B_4}$ = 0.4005(5) nm, respectively. The unit sub-cell parameter ratio is $c_{Mn_4B_4}/c_{Pr} \cong 41/35$ and as a result, the lattice constant is c = 1.4024(2) nm, corresponding to $Pr_{41}(Mn_4B_4)_{35}$ composition. The resultant structure unit cell is described by space group *$P4_22_12$*. The Pr atoms are located in channel formed by four columns of Mn_4 tetrahedra along **z**-axis. The Mn_4 tetrahedra are rotated relative to one another around the **z**-axis. The structure types and lattice parameters of R-Mn-B compounds are given in Tables 6.1 and 6.2, respectively.

The stability of the crystal structures, in some ternary R-M-B with R = Dy, Ho, Er, Y and M = Mn, Cr, has been theoretically investigated [11O1]. The computed structures agree generally with experimental evidence. In case of $DyMnB_4$, theoretically was shown that a structure of $ThMoB_4$-type was more stable, differing from

Table 6.1 Atomic coordinates

(a) $GdMnB_4$ compound shows a $YCrB_4$, type structure, space group *Pbam* [70K1]

Atom	Site	x	y	z
Gd	*4g*	0.125	0.150	0
Mn	*4g*	0.125	0.419	0
B1	*4g*	0.280	0.315	1/2
B2	*4g*	0.340	0.465	1/2
B3	*4g*	0.385	0.050	1/2
B4	*4g*	0.485	0.180	1/2

(b) Dy_3MnB_7 compounds having Er_3CrB_7 type structure, space group *Cmcm* [83M1, 91M1]

Atom	Site	x	y	z
Dy1	*8f*	0	0.104(1)	0.060(2)
Dy2	*4c*	0	0.755(1)	1/4
Mn	*4c*	0	0.950(3)	1/4
B1	*8f*	0	0.274	0.023
B2	*8f*	0	0.367	0.352
B3	*8f*	0	0.481	0.408
B4	*4c*	0	0.544	1/4

(c) $Pr_{41}(Mn_4B_4)_{35}$, with incommensurate structure *$P4_22_12$* space group [94Z1][a]

Atom	Coordinates			i
	x	y	z	
Pr1	0	0	0	
Pr2-Pr21	0	0	i/41	1 to 20
Mn1-Mn35	x_1	1/2 + x_1	$(z_1 + i)/35$	0 to 34
B1-B35	1/4-x_2	3/4-x_2	$(z_2 + i)/35$	0 to 34

[a] x_1, z_1 and x_2, z_2 are atomic coordinates of Mn and B in Mn_4B_4 sublattice, respectively. The Mn_4B_4 sublattice has *$P4_2/ncm$* type structure, the atomic positions having coordinates x, x, z. (site position *8j*): site x_1 = 0.1268(3), z_1 = 0.133(3) and x_2 = 0.069(3), z_2 = 0.630(19)

Table 6.2 Space group and lattice parameters

			Lattice parameters (nm)			Reference
Compound	T (K)	Space group	A	B	C	
$CeMnB_4$	RT	*Pbam*	0.5977(5)	1.153(1)	0.3571(4)	[75K1]
$NdMnB_4$	RT	*Pbam*	0.6033	1.1629	0.3515	[90M1]
$SmMnB_4$	RT	*Pbam*	0.3472	0.5977	1.1563	[17M1]
$GdMnB_4$	RT	*Pbam*	0.3429	0.5935	1.1476	[20M1]
$GdMnB_4$	RT	*Pbam*	0.5944(5)	1.151(1)	0.3464(5)	[70K1]
$GdMnB_4$	RT	*Pbam*	0.5922(3)	1.1466(6)	0.3451(2)	[79S1]
$TbMnB_4$	RT	*Pbam*	0.3420	0.5919	1.1448	[20M1]
$TbMnB_4$	RT	*Pbam*	0.5928(5)	1.146(1)	0.3448(5)	[70K1]
$DyMnB_4$	RT	*Pbam*	0.5897(5)	1.139(1)	0.3439(5)	[70K1]
$DyMnB_4$	RT	*Pbam*	0.5898(3)	1.139(3)	0.3437(1)	[01L1]
$HoMnB_4$	RT	*Pbam*	0.5901(5)	1.138(1)	0.3428(2)	[70K1]
$HoMnB_4$	RT	*Pbam*	0.3394	0.5894	1.1396	[20M1]
$ErMnB_4$	RT	*Pbam*	0.3379	0.5880	1.1369	[20M1]
$ErMnB_4$	RT	*Pbam*	0.5868(5)	1.140(1)	0.3424(5)	[70K1]
$TmMnB_4$	RT	*Pbam*	0.3371	0.5878	1.1356	[20M1]
$TmMnB_4$	RT	*Pbam*	0.5874(3)	1.1333(5)	0.3403(1)	[96C1]
$TmMnB_4$	RT	*Pbam*	0.5868–0.5894	1.1320–1.1370	0.3363–0.34018	[01K1]
$YbMnB_4$	RT	*Pbam*	0.5865(1)	1.1344(8)	0.3385(2)	[96C1]
$LuMnB_4$	RT	*Pbam*	0.5858(2)	1.1322(3)	0.3379(1)	[96C1]
$YMnB_4$	RT	*Pbam*	0.5913(5)	1.142(1)	0.3439(5)	[70K1, 83M1]
Gd_3MnB_7	RT	*Cmcm*	0.3488(1)	1.5875(1)	0.9399(1)	[83M1, 86K1]
Gd_3MnB_7	RT	*Cmcm*	0.8139 $\alpha = 90^o$	0.8139 $\beta = 90^o$	0.9371 $\gamma = 24.460^o$	[20M1]
Tb_3MnB_7	RT	*Cmcm*	0.8088 $\alpha = 90^o$	0.8088 $\beta = 90^o$	0.9348 $\gamma = 155.339^o$	[20M1]
Tb_3MnB_7	RT	*Cmcm*	0.3442(3)	1.5818(7)	0.9357(3)	[91M1, 91M2]
Dy_3MnB_7	RT	*Cmcm*	0.3434(1)	1.5745(4)	0.9329(4)	[91M1, 91M2]
Dy_3MnB_7	RT	*Cmcm*	0.8065 $\alpha = 90^o$	0.8065 $\beta = 90^o$	0.9298 $\gamma = 155.364^o$	[20M1]
Ho_3MnB_7	RT	*Cmcm*	0.7999 $\alpha = 90^o$	0.7999 $\beta = 90^o$	0.9259 $\gamma = 155.294^o$	[20M1]
Ho_3MnB_7	RT	*Cmcm*	0.3421(1)	1.5688(4)	0.9292(2)	[91M1, 91M2]
Er_3MnB_7	RT	*Cmcm*	0.7992 $\alpha = 90^o$	0.7992 $\beta = 90^o$	0.9256 $\gamma = 155.279^o$	[20M1]
Er_3MnB_7	RT	*Cmcm*	0.3417(4)	1.557(1)	0.9262(5)	[91M1, 91M2]
Y_3MnB_7	RT	*Cmcm*	0.8062 $\alpha = 90^o$	0.8062 $\beta = 90^o$	0.9325 $\gamma = 155.347^o$	[20M1]
Y_3MnB_7	RT	*Cmcm*	0.3464(1)	1.5696(1)	0.9313(3)	[83M1, 86K1]
$Pr_{41}(Mn_4B_4)_{35}$	RT	$P4_22_12$	0.7183(3)		1.4024(8)	[94Z1]

experiments; the $DyMnB_4$ crystallizes in $YCrB_4$-type, structure. The electronic structure, anisotropic elastic constants and lattice dynamics of $YMnB_4$ were calculated [19C1]. The boride has dominantly covalent bonding and it is a hard material with hardness $H_v = 37.79$ GPa, somewhat smaller than that in $YCrB_4$. The $YMnB_4$ is a semiconductor having a computed band gap of $\cong 0.25$ eV [19C1] (Table 5.3).

The $GdMnB_4$ boride is magnetically ordered, at low temperatures [79S1]—Table 6.3. The effective moment, determined at $T > T_c$, is little smaller than the expected value for Gd^{3+} free ion. The magnetization isotherm of $DyMnB_4$, at T = 4.2 K, suggests a metamagnetic transition in a field $\mu_0H \cong 6$ T—Fig. 6.1. The magnetization is not saturated even in a field of 14 T [01L1]. The thermal variation of the reciprocal susceptibility is linear at T < 200 K, with a significant smaller effective moment than that of Dy^{3+} free ion. Reliable data can be obtained only when measurements are made in a larger temperature range, if compound is ferrimagnetically ordered [17B1] (Table 6.3).

Table 6.3 Magnetic properties

Compound	Magnetic structure	M_s (μ_B/f.u.)	T_c (K)	M_{eff} (μ_B/f.u.)	θ (K)	Reference
$GdMnB_4$	T > 77 K $\chi = C/(T-\theta)$			7.5	90	[79S1]
$DyMnB_4$	Metamagnetic (?)	6.18[a]	≅ 22	7.5	27 (3)	[01L1]

[a]T = 4.2 K, $\mu_0H = 14$ T

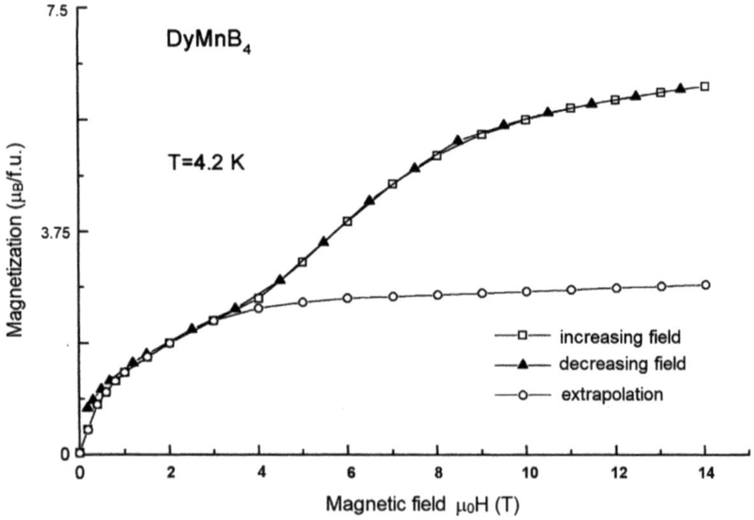

Fig. 6.1 $DyMnB_4$: magnetization isotherm at T = 4.2 K [01L1]

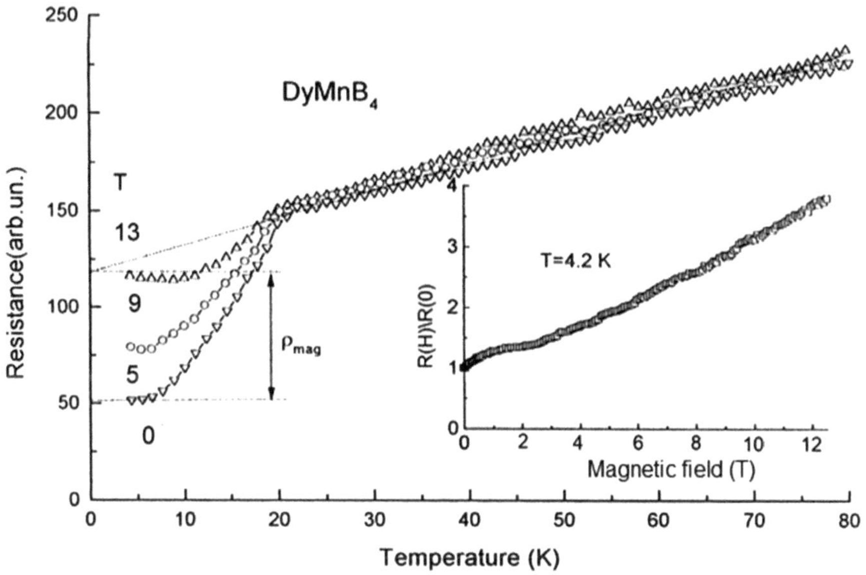

Fig. 6.2 $DyMnB_4$: temperature dependences of the electrical resistance in fields up to 13 T. The inset shows the field dependence of the relative electrical resistance, R(H)/R(0) where R(0) is the resistance at T = 4.2 K, in zero field [01L1]

The temperature dependences of the electrical resistance of $DyMnB_4$, at different external fields, are given in Fig. 6.2. A metallic type conductivity is evidenced [01L1]. At T = 4.2 K, in field of $\mu_0H = 13$ T, the magnetoresistance ratio is of ≅290%.

Band structure calculations suggest that $GdMnB_4$ compound is ferrimagnetic, while those with R = Sm, Tb, Ho and Tm are not magnetically ordered [20M1]. Ferromagnetic ordering of Mn was seen in $ErMnB_4$ compound. The presence of ferrimagnetic ordering was also seen in Gd_3MnB_7 while the R_3MnB_7 compounds with R = Tb. Dy, Ho and Y are not magnetically ordered [20M1].

References

[70K1] Yu.B. Kuzma, Dopov. Akad. Nauk. Ukr. RSR, **A32**, 756 (1970)
[75K1] Yu.B. Kuzma, V.E. Romashov, Visn. Lviv. Derzh. Univ., Khim. **17**, 26 (1975)
[77N1] H. Nowotny, P. Rogl, in Boron and Refractory Borides, p. 413–438, Springer Verlag (1977)
[78K1] Yu.B. Kuzma, Dopov Akad. Nauk. Ukr. RSR A (8) 756 (1978)
[79K1] Yu.B. Kuzma, N.S. Bilonizhko, S.I. Mykhalenko, G.F. Stepanchikova, N.F. Chaban, J. Less Common Met. **67**, 51 (1979)
[79S1] R. Sobczak, P. Rogl, J. Solid State Chem. **27**, 343 (1979)
[83M1] S.I. Mikhalenko, L.P. Gumenna, Yu.B. Kuzma, Poroshk. Metall. (12), 73 (1983)
[83K1] Yu.B. Kuzma, N.S. Bilonizhko, N.F. Chaban, G.V. Chernyak, J. Less Common Met. **90**, 217 (1983)

[84P1] E. Parthé, B. Chabot, Handbook on Physics and Chemistry of Rare Earths, Chap. 48, Elsevier (1984)
[84R1] P. Rogl, Handbook on the Physics and Chemistry of Rare Earths, Chap. 49, Elsevier (1984)
[86K1] Yu.B. Kuzma, S.I. Mykhalenko, L.G. Akselrud, J. less Common Met. **117**, 29 (1986)
[89O1] H. Oesterreicher, K. Oesterreicher, Landolt Börnstein Handbook vol. 19e2, Springer Verlag (1989)
[90K1] Yu.B. Kuzma, N.F. Chaban, *Binary and ternary systems containing boron* (Metallurgiya, Moscow, 1990)
[90M1] S.I. Mykhalenko, Yu.B. Kuzma, T.D. Chuchman, Inorg. Mater. **26**, 1968 (1990)
[91M1] S.I. Mykhalenko, Yu.B. Kuzma, Izv. Akad. Nauk. SSSR, Neorg. Mater. **27**, 2109 (1991)
[91M2] S.I. Mykhalenko, N.F. Chaban, Yu.B. Kuzma, Neorg. Mater. **27**, 2298 (1991)
[94Z1] P.Yu. Zavalij, S.I. Mykhalenko, Yu.B. Kuzma, J. Alloys Comp. **203**, 55 (1994)
[96C1] N.F. Chaban, S.I. Mykhalenko, Yu.B. Kuzma, Neorg. Mater. **32**, 44 (1996)
[98M1] S.I. Mykhalenko, N.F. Chaban, Yu.B. Kuzma, Powder Metall. Met. Ceram. **17**, 116 (1998)
[01C1] N.F. Chaban, S.I. Mikhalenko, Yu.Z. Kernitska, Yu.B. Kuzma, Poroshk. Metall. (5–6), 69 (2001)
[01K1] Yu.B. Kuzma, Powder Metall. Met. Ceram. **40**, 258 (2001)
[01L1] E.M. Levin, T. Palewski, T. Mydlarz, B.S. Kuzhel, Yu.B. Kuzma, S.I. Mykhalenko, J. Alloys Comp. **315**, 95 (2001)
[11O1] H. Orsini-Rosenberg, W. Steurer, Phil. Mag. **91**, 2567 (2011)
[17B1] E. Burzo, P. Vlaic, D.P. Kozlenko, S.E. Kichanov, A.V. Rutkauscas, B.N. Savenko, J. Alloys Comp. **729**, 1184 (2017)
[17M1] Materials Research Project, *Lawrence Berkeley Nat* (Lab. Berkeley, USA, 2017)
[19C1] A. Candan, G. Surucu, A. Gencer, Phys. Scripta **94**, 125710 (2019)
[20M1] Materials Research Project, *Lawrence Berkeley Nat* (Lab. Berkeley, USA, 2020)

Chapter 7
Rare-Earths-Iron-Boron Compounds

7.1 General Problems

The phase diagrams and the crystal structures of R–Fe–B ternary systems, where R is a rare-earth or yttrium, were investigated starting with 1972 year [72B1]. Since the discovery of the high performance permanent magnets, based on $R_2Fe_{14}B$ series [84S1], numerous studies concerning the phase relations in the ternary R–Fe–B systems, particularly when R = Nd, were reported. In this context most studies were focused on the compositions region, near the technologically interesting $R_2Fe_{14}B$ compounds, resulting in an accurate description of the phase relations, in the iron-rich corner of the ternary phase diagrams. The investigations improved the already reported phase diagrams, evidencing the existence of new compounds or metastable phases, or correcting the compositions of some previously reported compounds. A critical review of the R–Fe–B phase diagrams reported in early stages was also presented [83K1, 84P1, 84R1].

The investigated phase diagrams of ternary R–Fe–B system cover all R systems (except R = Eu), namely R = La [77K1, 90U1, 20W1, 21W1], R = Ce [72B1, 90U1, 14Y1], Ce-(Fe,Co)-B [17W1], R = Pr [87T2, 90Z1, 95N1, 03C5, 20C2], R = Nd [79C1, 80C1, 84S6, 85B10, 85M1, 86B11, 86C1, 86D3, 86S2, 87G2, 87Z1, 88R2, 89L3, 89S2, 90U1, 92G2, 92K2, 94M2, 95H2, 99L2, 03R1, 07K1, 13R1, 16F1, 19C2], R = Sm [79C1, 80C1, 02C2, 16X1], R = Eu [90Z1], R = Gd [79C1, 80C1, 17X1], R = Tb [86D3, 87D5, 87G2, 89S2], R = Dy [83C3, 87G2, 89S2], R = Ho [77S1, 78S1, 80S1, 91G1, 20W2], R = Er [77S1, 83C3, 12R1], R = Yb [85D3, 86D4, 05V1, 07R1], R = Tm [77S1, 78S1, 83K2], R = Lu [77S1, 78S1, 86C1, 86D3, 87D5], R = Y [80S1].

A list of the ternary R–Fe–B compounds know up to date are listed in Table 7.1. The larger number of compounds can be evidenced in $R_2Fe_{14}B$ and $R_{1+\varepsilon}$ Fe_4B_4 series. The R_2FeB_6 phase exists only when R = Lu. The number of compounds in ternary compounds, including also metastable phases or in mixture of phases, depends on rare-earth partner: R = Dy, Y (8), R = Pr, Nd, Gd, Tb (7), R = Sm, Er, Lu (6), R =

E. Burzo, *Rare Earths-Transition Metals-Boron Compounds*,
https://doi.org/10.1007/978-3-030-99245-3_7

Table 7.1 Stable and metastable compounds in R–Fe–B ternary systems

Compound	Space group	La	Ce	Pr	Nd	Sm	Eu	Gd	Tb	Dy	Ho[c)]	Er	Tm	Yb	Lu	Y
$RFeB_4$	*Pbam*		○			○		○	○	○	○	○	○	○	○	○
R_2FeB_6	*Pbam*														○	
R_3FeB_7	*Cmcm*								○	○	○	○				○
$R_{5-x}Fe_{2+x}B_6$	$R\overline{3}m$		○	○	○	○	○	○	○	○				○		○
$R_{1+\varepsilon}Fe_4B_4$	*Pccm*	○	○	○	○	○		○	○	○	○	○			○	○
RFe_2B_2	*I4/mmm*								○	○	○	○	○		○	○
$RFe_{12}B_6$[a]	$R\overline{3}m$	○		○	○											
RFe_4B	*P6/mmm*	○[b]						○[b]				○	○		○	○[b]
$R_2Fe_{23}B_3$[a]	$I\overline{4}3d$		○	○	○	○		○		○						
$R_2Fe_{17}B_x$[a]	$R\overline{3}m$ or *P6/mmm*			○	○	○			○							
$R_3Fe_{62}B_{14}$[a]	$Im\overline{3}m$			○	○			○		○						○
$R_2Fe_{14}B$	$P4_2/mnm$	○	○	○	○	○		○	○	○	○	○	○		○	○

[a] Metastable phases, except $LaFe_{12}B_6$; [b] In mixture of phases; [c] The presence of $HoFeB_3$ phase was also mentioned [20W2]

Ce, Ho (5), R = La, Tm (4), R = Yb (2) and R = Eu (1)—Table 7.1. The metastable compounds are mainly formed with light rare-earths.

Some ternary compounds, reported already in R–Fe–B series, with R = La [77K1], R = Y [80S1] or R = Sm [80C1] were not confirmed. As example, the compositions $La_3Fe_{13}B$ and $LaFe_2B_2$ were shown to be really $La_2Fe_{14}B$ and $La_{1+\varepsilon}Fe_4B_4$, respectively [84R1]. The presence of $HoFeB_3$ has been mentioned, without indicating the crystal structure [91G1, 20W2].

7.2 $RFeB_4$ and Lu_2FeB_6 Compounds

The $RFeB_4$ compounds with R = Ce, Sm, Gd, Tb, Dy, Ho, Er, Tm, Yb, Lu, and Y crystallize in a rhombohedral structure of $YCrB_4$-type, having space group *Pbam* [77S1, 79S1, 80B1, 83C3, 85D3, 05V1, 05V2, 16V1]. The crystal structure contains boron nets perpendicular to [001], composed of five and seven membered rings. The metal atoms are located between the boron nets. The rare-earth atoms are situated between the seven-membered rings and the iron atoms between the five membered rings—Fig. 7.1 and Table 7.2. The lattice parameters are listed in Table 7.3.

The stability of the $YCrB_4$ type structure has been analysed comparatively with that of the polymorph, $ThMoB_4$ [10L1, 11L1]. Full-geometry optimizations carried out, at the DFT level of theory, for various R and T (Fe) metals with the $YCrB_4$ and $ThMoO_4$ arrangements, indicate that, regardless of the metals, the $YCrB_4$ arrangement is the most stable.

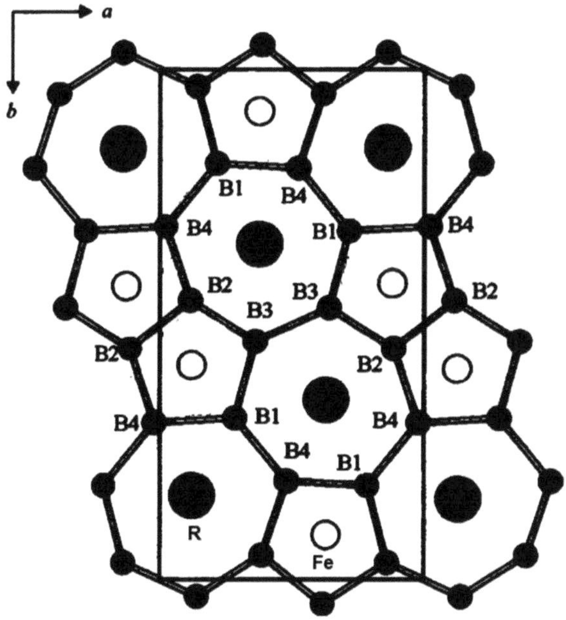

Fig. 7.1 $RFeB_4$: crystal structure [85D3, 05V2]

Table 7.2 Atomic sites in $YbFeB_4$ having orthorhombic-type structure, space group *Pbam* [05V2]

Atom	Site	x	y	z	Atomic environment
Sm	4*g*	0.12736(8)	0.15001(15)	0	5Yb4Fe14B
Fe	4*g*	0.13315(11)	0.41198(6)	0	4Yb1Fe10B
B1	4*h*	0.2848(9)	0.3156(4)	1/2	4Yb2Fe3B
B2	4*h*	0.3624(8)	0.4682(4)	1/2	4Yb2Fe3B
B3	4*h*	0.3868(9)	0.0470(5)	1/2	2Yb4Fe3B
B4	4*h*	0.4756(9)	0.1928(5)	1/2	4Yb2Fe3B

Table 7.3 Crystal structure and lattice parameters of RFeBy and R_2FeB_6 compounds

Compound	T (K)	Space group	Lattice parameters (nm)			References
			a	b	c	
$CeFeB_4$	RT	*Pbam*	0.5934(10)	1.150(2)	0.3511(5)	[77S1]
$SmFeB_4$	RT	*Pbam*	0.5958(5)	1.153(1)	0.3465(4)	[80B1]
$GdFeB_4$	RT	*Pbam*	0.5911(10)	1.150(2)	0.3436(5)	[77S1]
$GdFeB_4$	RT	*Pbam*	0.5918(8)	1.1456(6)	0.3457(2)	[79S1]
$TbFeB_4$	RT	*Pbam*	0.5900(10)	1.141(2)	0.3418(5)	[77S1]
$DyFeB_4$	RT	*Pbam*	0.5885(10)	1.138(2)	0.3403(5)	[77S1, 83C3]
$HoFeB_4$	RT	*Pbam*	0.5871(10)	1.136(2)	0.3391(5)	[77S1]
$ErFeB_4$	RT	*Pbam*	0.5861(10)	1.134(2)	0.3377(5)	[77S1, 83C3]
$TmFeB_4$	RT	*Pbam*	0.5850(10)	1.132(2)	0.3365(5)	[77S1]
$YbFeB_4$	RT	*Pbam*	0.5847	1.132	0.365	[85D3]
$YbFeB_4$	RT	*Pbam*	0.58437(4)	1.1304(1)	0.33520(4)	[05V1]
$LuFeB_4$	RT	*Pbam*	0.5848(1)	1.1316(2)	0.3353(1)	[87D5]
$LuFeB_4$	RT	*Pbam*	0.5832(10)	1.129(2)	0.3353(5)	[77S1]
$YFeB_4$	RT	*Pbam*	0.5906	1.1398	0.3407	[77S1, 80S1]
Lu_2FeB_6	RT	*Pbam*	0.8969(4)	1.1340(5)	0.3490(2)	[87D5]
Lu_2FeB_6	RT	*Pbam*	0.3474	0.8968	1.1317	[20M1]

The physical properties such as structural, electronic, anisotropic elastic and lattice dynamics of $YFeB_4$ have been theoretically investigated [19C1]. The compound has dominantly covalent bonding. The boride shows isotropic elastic properties in (xy} plane. Values 439 GPa for Young modulus and 35.31 GPa for hardness were obtained. The $YFeB_4$ is paramagnetic and shows a metallic behavior.

The ^{155}Gd Mössbauer spectrum of $GdFeB_4$ at $T = 4.2$ K, showed that iron has a vanishing moment [90O1]. The angle of V_{zz} with direction of the hyperfine field is $\theta = 66(5)°$. Thus, the contribution to the Gd hyperfine field from neighbouring Sd atoms is negative. Under an external field of $\mu_0H = 8$ T, the hyperfine field is reduced by $\cong 5$ T, suggesting that stronger external field is necessary to overcome the exchange interactions and to align the Gd moment in the direction of applied field. The reciprocal susceptibility of $GdFeB_4$, at T > 77 K, follows a Curie–Weiss temperature dependence, having paramagnetic Curie temperature $\theta = -46$ K [79S1]. The determined effective moment of $\cong 10\ \mu_B$/f.u. is sensitive higher than that of Gd^{3+} ion, suggesting the possible presence of magnetic ordered impurities in the sample, or contribution from iron atoms.

Only the Lu_2FeB_6 compounds in the R_2FeB_6 series is formed [87D5]. The boride crystallizes also in an orthorhombic Y_2ReB_6 type structure [84P1] space group *Pbam* [87D5]. The structure can be described as an alternation along **c**-axis of distorted close-packed layers of metal atoms and planar boron nets, formed by pentagonal, hexagonal and heptagonal rings [07A1]. The structure stability of this boride has been also investigated [10L1].

7.3 R_3FeB_7 Compounds

The ternary R_3FeB_7 compounds, with R = Tb, Dy, Ho, Er and Y, crystallize in an orthorhombic-type structure having space group *Cmcm* [80S1, 83C3, 83K2, 84P1, 84R1, 86K5, 87D5]—Table 7.4. The metal atoms, in the Y_3FeB_7-type structure, form trigonal prisms, stacked to the slightly kinked walls, in rows along [010], displaced relative to one another by a/2. The stacks of filled trigonal prisms [BY_6] and the empty four faced pyramids [BY_5] are between the walls [86K5]. The environment of the iron atoms is like that in the $YCrB_4$ type structure [83C3]. All the boron atoms form a framework which consists of five-membered and thirteen-membered rings [86K5]. The five-member rings (as in $YCrB_4$ structure) are connected to form bands along the **z**-axis. They are placed on two levels (x = 0 and 1/2). A connection between the rings is accomplished by zig-zag bonds of B1 atoms. The lattice parameters are listed in Table 7.5.

The physical properties of these compounds were little investigated. Band structure calculations were made on R_3FeB_7 compounds with R = Tb and Dy [20M1]. When R = Tb the boride is FM ordered, while the Dy_3FeB_7 is not magnetically ordered.

Table 7.4 Atomic sites in Y_3FeB_7 compound, having space group *Cmcm* [86K5]

Atom	Site	x	y	z
Y1	8*f*	0	0.1061(14)	0.0552(20)
Y2	4*c*	0	0.7580(18)	1/4
Fe	4*c*	0	0.9468(22)	1/4
B1	8*f*	0	0.275(13)	0.024(18)
B2	8*f*	0	0.366(13)	0.353(13)
B3	8*f*	0	0.480(14)	0.406(23)
B4	4*c*	0	0.543(17)	1/4

Table 7.5 Crystal structure and lattice parameters of R_3FeB_7 compounds

Compound	T (K)	Space group	Lattice parameters (nm)			References
			a	b	c	
Tb_3FeB_7	RT	*Cmcm*	0.3397	1.5640	0.9413	[80S1, 87D5]
Tb_3FeB_7	RT	*Cmcm*	0.3974(2)	1.5640(4)	0.9413(2)	[86K5]
Tb_3FeB_7	RT	*Cmcm*	0.4022 $\alpha = 90°$	0.4022 $\beta = 90°$	0.2958 $\gamma = 137.865°$	[20M1]
Dy_3FeB_7	RT	*Cmcm*	0.3375	1.5540	0.9403	[80S1, 83C3]
Dy_3FeB_7	RT	*Cmcm*	0.3375(2)	1.5540(6)	0.9403(3)	[86K5]
Dy_3FeB_7	RT	*Cmcm*	0.8035 $\alpha = 90°$	0.8035 $\beta = 90°$	0.9309 $\gamma = 155.407°$	[20M1]
Ho_3FeB_7	RT	*Cmcm*	0.3369	1.5500	0.9358	[80S1]
Ho_3FeB_7	RT	*Cmcm*	0.3369(1)	1.5504(5)	0.9358(2)	[86K5, 91G1]
Er_3FeB_7	RT	*Cmcm*	0.3363	1.534	0.9350	[80S1, 83C3]
Er_3FeB_7	RT	*Cmcm*	0.3363(2)	1.5341(5)	0.9350(2)	[86K5]
Y_3FeB_7	RT	*Cmcm*	0.3422	1.5640	0.9327	[80S1]
Y_3FeB_7	RT	*Cmcm*	0.3423(1)	1.5658(6)	0.9295(5)	[86K5]

7.4 $R_5Fe_2B_6$ Compounds

The ternary $R_5T_2B_6$ compounds with T = Co [83K3], Fe [86D4] and Mn [90M5] crystallizes in a rhombohedral-type structure having $R\bar{3}m$ space group. The existence of the homogeneity range, described by $R_{5\text{-}x}T_{2+x}B_6$, has been stated in some reports [83K3, 86D4].

The $R_5Fe_2B_6$ compounds are formed with R = Ce, Pr, Nd, Sm, Eu, Gd, Tb, Dy, Tm, Yb and Y [86D4]. In the rhombohedral structure, planar hexagon-mesh B layers, alternate with two puckered triangle-mesh R_2Fe layers with additional R atoms between the iron atoms, having a stacking sequence—B_6-R_2Fe-R-R_2Fe. The structure can be also described by stacking of alternated [R_3] and [$R_2Fe_2B_6$] blocks, along the **c**-axis, yielding $R_5Fe_2B_6$ as the final composition of the stoichiometric boride [92Y1, 97Y1, 17T1]—Fig. 7.2. The [$R_2Fe_2B_6$] layer, contains a flat net of boron atoms, which is surrounded by the iron atoms as nearest neighbors. The R atoms are distributed in three inequivalent sites, having coordination CN = 14 (3*a*), 16 (6*c*) and 15 (6*c*), while Fe and B are each located only in one site, the Fe (6*c*) site having CN = 14. The existence of an homogeneity range, in the presence of constant boron content of 46.2 at %, was related to a statistical occupancy of R3*a* site by R and Fe atoms, as reported in $Tb_{5\text{-}x}Fe_{2+x}B_6$ boride [87D5]. The analysis of ^{57}Fe Mössbauer spectra of compounds with R = Ce, Sm, Tb, Yb confirmed the distribution of iron atoms in 6*c* and 2*a* sites [98Y2], while for R = Nd, only the stoichiometric phase was reported [87D3, 17T1]. The lattice sites and lattice parameters of $R_5Fe_2B_6$ borides,

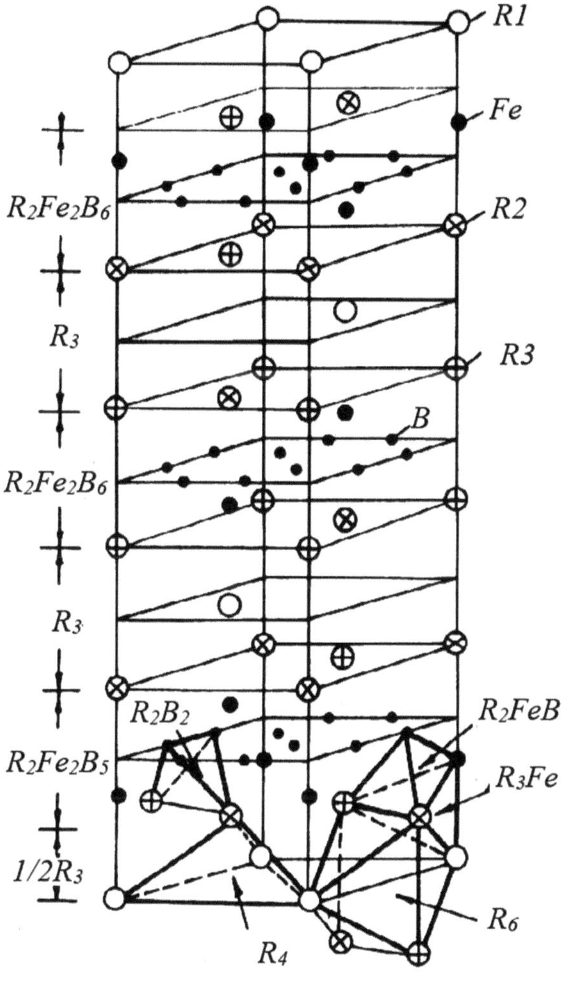

Fig. 7.2 $R_5Fe_2B_6$: crystal structure [92Y1, 97Y1]. The interstices suitable for hydrogen occupation are also indicated

are listed in Tables 7.6 and 7.7, respectively. It is to be noted that in an earlier study the composition of this series was assumed to be $R_2Fe_2B_3$ [86B11]. The following studies evidenced that really this is a mixture of two phases.

Table 7.6 Atomic sites in $Nd_5Fe_2B_6$ compound (a) and their deuteride $Nd_5Fe_2B_6D_{4.1}$ (b), having space group $R\bar{3}m$ [17T1]

(a) $Nd_5Fe_2B_6$

Atom	Site	Symmetry	x	y	z	CN	Atomic environment [07V1]
Nd1	3a	$\bar{3}m$	0	0	0	14	Colinear Fe_2
Nd2	6c	$3m$	0	0	0.2498(2)	16	Non-coplanar hexagon B_6
Nd3	6c	$3m$	0	0	0.4181(2)	15	Non-coplanar hexagon B_6
Fe	6c	$3m$	0	0	0.1171(2)	14	Non-coplanar hexagon B_6
B	18g	.2	0.332	0	0.5	9	Trigonal bipyramid B_3Fe_2

(b) $Nd_5Fe_2B_6D_{4.1}$

Atom	Site	x	y	z	Occ
Nd1	3*a*	0	0	0	1
Nd2	6*c*	0	0	0.2476(2)	1
Nd3	6*c*	0	0	0.4178(2)	1
Fe	6*c*	0	0	0.1170(1)	1
B	18*g*	0.332	0	0.5	1
D1	9*e*	0.5	0	0	0.85(1)
D2	36*i*	0.203(2)	0.026(4)	0.0664(6)	0.13(1)

A very fast, near room temperature hydrogenation of $R_5Fe_2B_6$ borides, has been reported [92Y1, 93P1, 97Y1, 97Y2, 99Y1, 17T1]. Three kinds of sites in the $Nd_5Fe_2B_6$ structure, favorable for H insertion were shown, octahedral Nd_6 and tetrahedral Nd_4 and Nd_3Fe interstices. Their filling leads to a stoichiometry $Nd_5Fe_2B_6H_6$. The H insertion in Nd_2FeB and Nd_2B_2 interstices is also possible for higher hydrogen content. The neutron diffraction study, on $Nd_5Fe_2B_6D_{4.1}$, showed that the deuterium atoms are located in octahedral D1 [Nd_6] and one tetrahedral D2 [Nd_3Fe] interstices, both sites avoiding boron as near neighbor [17T1] (Fig. 7.2). The presence of hydrogen induces an anisotropic cell expansion, mainly along **c**-axis, as evidenced in Table 7.7. The thermal stability of $R_5Fe_2B_6$ compounds and their hydrides increases with increasing the atomic number of the R component [97Y1].

Table 7.7 Lattice parameters and space groups of $R_{5\text{-}x}Fe_{2+x}B_6$ compounds

Compound	T (K)	Space group	Lattice parameters (nm)		Δa/a (%)	Δc/c (%)	References
			a	c			
$Ce_{5\text{-}x}Fe_{2+x}B_6$	RT	$R\bar{3}m$	0.5482(2)	2.443(1)			[86D4]
$Ce_5Fe_2B_6H_{7.6}$	RT	$R\bar{3}m$	0.5472(2)	2.542(2)	−0.18	4.05	[97Y1]
$Pr_{5\text{-}x}Fe_{2+x}B_6$	RT	$R\bar{3}m$	0.5481(1)	2.433(1)			[86D4]
$Pr_5Fe_2B_6$	RT	$R\bar{3}m$	0.5454(6)	2.404(7)			[97Y1]
$Pr_5Fe_2B_6H_{9.3}$	RT	$R\bar{3}m$	0.5447(8)	2.532(5)	−0.13	5.32	[97Y1]
$Nd_{5\text{-}x}Fe_{2+x}B_6$	RT	$R\bar{3}m$	0.5469(1)	2.402(2)			[86D4]
$Nd_5Fe_2B_6$	RT	$R\bar{3}m$	0.5438(3)	2.407(3)			[93P1]
$Nd_5Fe_2B_6H_{6.3}$	RT	$R\bar{3}m$	0.5456(2)	2.494(2)	0.33	3.61	[93P1]
$Nd_5Fe_2B_6$	RT	$R\bar{3}m$	0.5464(5)	2.417(4)			[97Y1]
$Nd_5Fe_2B_6H_{8.6}$	RT	$R\bar{3}m$	0.5455(8)	2.497(8)	−0.16	3.31	[97Y1]
$Nd_5Fe_2B_6$	RT	$R\bar{3}m$	0.5464	2.4272			[16F1]
$Nd_5Fe_2B_6$	100	$R\bar{3}m$	0.54430(4)	2.4133(3)			[17T1]
$Nd_5Fe_2B_6D_{4.1}$	100	$R\bar{3}m$	0.54463(2)	2.4813(2)	0.06	2.74	[17T1]
Nd_2FeB_3 [a)]	RT	$R\bar{3}m$	0.5463	2.4281			[86B11]
$Sm_{5\text{-}x}Fe_{2+x}B_6$	RT	$R\bar{3}m$	0.5459(2)	2.368(1)			[86D4]
$Sm_5Fe_2B_6$	RT	$R\bar{3}m$	0.5437(4)	2.376(3)			[97Y1]
$Sm_5Fe_2B_6H_{8.0}$	RT	$R\bar{3}m$	0.5438(3)	2.446(2)	0.02	2.95	[97Y1]
$Eu_{5\text{-}x}Fe_{2+x}B_6$	RT	$R\bar{3}m$	0.5435(1)	2.336(1)			[86D4]
$Gd_{5\text{-}x}Fe_{2+x}B_6$	RT	$R\bar{3}m$	0.5429(2)	2.330(2)			[86D4]
$Gd_5Fe_2B_6$	RT	$R\bar{3}m$	0.5424(4)	2.344(3)			[97Y1]
$Gd_5Fe_2B_6H_{6.5}$	RT	$R\bar{3}m$	0.5425(4)	2.451(3)	0.02	4.56	[97Y1]
Gd_2FeB_3 [a)]	RT	$R\bar{3}m$	0.5440	2.3524			[86B11]
$Tb_{5\text{-}x}Fe_{2+x}B_6$	RT	$R\bar{3}m$	0.5420(1)	2.323(1)			[86D4]
$Tb_{5\text{-}x}Fe_{2+x}B_6$	RT	$R\bar{3}m$	0.5412(2)	2.321(1)			[86D4, 97Y1]
$Tb_5Fe_2B_6$	RT	$R\bar{3}m$	0.5410(4)	2.307(4)			[97Y1]
$Tb_5Fe_2B_6H_{8.9}$	RT	$R\bar{3}m$	0.5419(3)	2.419(7)	0.17	4.85	[97Y1]
$Dy_{5\text{-}x}Fe_{2+x}B_6$	RT	$R\bar{3}m$	0.5427(1)	2.313(1)			[86D4]
$Yb_{5\text{-}x}Fe_{2+x}B_6$	RT	$R\bar{3}m$	0.5427(2)	2.222(1)			[86D4, 05V1]
$Y_{5\text{-}x}Fe_{2+x}B_5$	RT	$R\bar{3}m$	0.5426(2)	2.328(2)			[86D4]

[a] $R_5Fe_2B_6$ as a main phase

A decomposition of the $R_5Fe_2B_6H_y$ hydrides, with R = Pr, Nd was shown, when heated in hydrogen, at pressures up to $5{\bullet}10^5$ Pa and T > 700 °C, resulting in the formation of rare-earth hydrides $RH_{2\pm y}$, as well as of $R_{1.1}Fe_4B_4$ and R_2B_5 phases [97Y1, 97Y2]. Under the same conditions, such disproportionate reaction did not occur in $R_5Fe_2B_6H_x$ compounds with R = Sm, Gd, Tb [97Y1].

The $Nd_5Fe_2B_6$ compound is feromagnetically ordered [86B11, 17T1]—Table 7.8. The temperature dependences of the magnetization in ZFC and FC sample (in a field $\mu_0H = 0.05$ T) evidenced a magnetic transition, at $T_c = 64.5(5)$ K [17T1]. The ZFC magnetization decreases at T < 40 K, being almost zero at T < 10 K. The magnetization isotherm, at T = 1.8 K, increases linearly with field, up to $\mu_0H = 1.5$ T and then shows a tendency to saturate at H > 7 T, suggesting the presence of a strong pining of domain walls—Fig. 7.3. The Curie temperature, as well as the saturation magnetization, determined in hydrogenated sample, decrease comparatively with those of the parent $Nd_5Fe_2B_6$ boride—Table 7.8. The above trend is evidenced also by specific heat measurements, which exhibit a lambda peak at T ≅ 64 K in boride and at T ≅ 24 K in their deuterated counterpart. An additional small hump, in deuterated sample, is also evidenced at T ≅ 8 K.

The magnetic structures of $Nd_5Fe_2B_6$ and of the deuterated $Nd_5Fe_2B_6D_{4.1}$ samples, have been investigated by neutron diffraction [17T1]—Fig. 7.4 and Table 7.8. The study confirmed that these borides are ferromagnetically ordered, the easy direction of magnetization being **c**-axis. The Nd magnetic moments are dependent on site location; at T = 1.8 K decrease in the sequence $Nd_1 3a$ < $Nd_2 6c$ < $Nd_3$6c. The mean magnetic moment of iron is of ≅0.8 μ_B in boride, this being lost in the deuterated counterpart. The temperature dependence of the neodymium moment at site 3*a*, in deuterated sample, tends to saturate rapidly below T_c, followed by an increase at T < 10 K. This trend can be correlated with the anomaly observed in the specific heat at T ≅ 8 K, and that in the temperature evolution of the ZFC and FC magnetizations, at the same temperature.

The ^{57}Fe Mössbauer spectra, at T = 5 K, in $Nd_5Fe_2B_6$ boride, were fitted assuming the presence of only one iron site [87D3, 17T1]. The small iron moment, as well as the ^{57}Fe hyperfine field, can be correlated with that of Fe6*c* site, where six boron atoms are located in their environment. Starting from the ^{57}Fe hyperfine field and using the proportionality constant between ^{57}Fe hyperfine field and iron moment, in rare-earth-iron compounds [79B1], the iron moment of 0.18 μ_B/atom was estimated. In the $Nd_5Fe_2B_6D_{4.1}$ sample, two iron sites were present [17T1]—Table 7.9. The main component with a small quadrupole splitting value (Fe2), can be related to iron atoms located in low distorted crystallographic sites, having deuterium atoms in their vicinity. The second site (Fe1) was associated with a "non-deuterated" site, similar as the iron one in the parent sample. The ^{57}Fe spectra at the Fe1 site was attributed as due to a sum of quadrupole doublets with Lorentzian shape, having the same isomer shifts, but different quadrupole splittings, ΔQ. This distribution of ΔQ values reflects slight variations of the EFG, at the Fe positions, due to some local disorder or heterogeneity in the sites distortions. It was concluded that the homogeneity range of any significance is absent. For a sample with composition Nd_2FeB_3, two lines corresponding to $Nd_5Fe_2B_6$ and $NdFe_4B_4$ phases were evidenced, the compound being a mixture of two phases [87D3].

The ^{57}Fe Mössbauer studies in $R_5Fe_2B_6$ borides with R = Ce, Sm, Tb and Yb evidenced the presence of iron atoms in sites having coordination number 14 and 15 [98Y2]. Relative high ^{57}Fe hyperfine field values were shown, at T = 80 K, in $Yb_5Fe_2B_6$ compound, but not when R = Ce, Sm, Tb. This behavior was correlated with a smaller *c* lattice parameter comparatively to other borides of this series [98Y2].

Table 7.8 Magnetic properties of $R_5Fe_2B_6$ compounds

Compound	Magnetic structure	M_s (μ_B/f.u.)	T_c (K)	$M_{eff}(R)$ (μ_B/atom)	M_{eff} (Fe) (μ_B/atom)	θ (K)	References
$Nd_5Fe_2B_6$	FM e.a.m ∥ **c**, **q** = (0,0,0) T = 1.8 K Nd:M_1 = 2.36(8) μ_B, M_2 = 2.71(6) μ_B, M_3 = 2.87(5) μ_B Fe: M = 0.8(2) μ_B T > 150 K, $\chi = C/(T-\theta)$[b]	10.8[a]	64.5(5)	3.62	3.1		[17T1]
$Nd_5Fe_2B_6D_{4.1}$	FM T = 1.8 K Nd:M_1 = 1.92(8) μ_B, M_2 = 1.47(5) μ_B, M_3 = 1.56(5) μ_B Fe: M = 0 μ_B	10[a]	21(1)	3.63		9	[17T1]
Nd_2FeB_3[d]	FM	2.83[c]	64	3.51		60	[86B11]
Gd_2FeB_3[d]	FIM	(10.7[c])	57	7.98		60	[86B11]

[a] T = 5 K, μ_0H = 4.6 T; [b] The fit was also made with $\chi = \chi_0 + C(T - \theta)$; gives $\chi_0 = 6\cdot10^{-3}$ emu/mol, M_{eff} = 3.48 μ_B/Fe atom, θ = 56 K [c] T = 4.2 K, μ_0H = 1.8 T; [d] Actually $R_5Fe_2B_6$ and a second phase

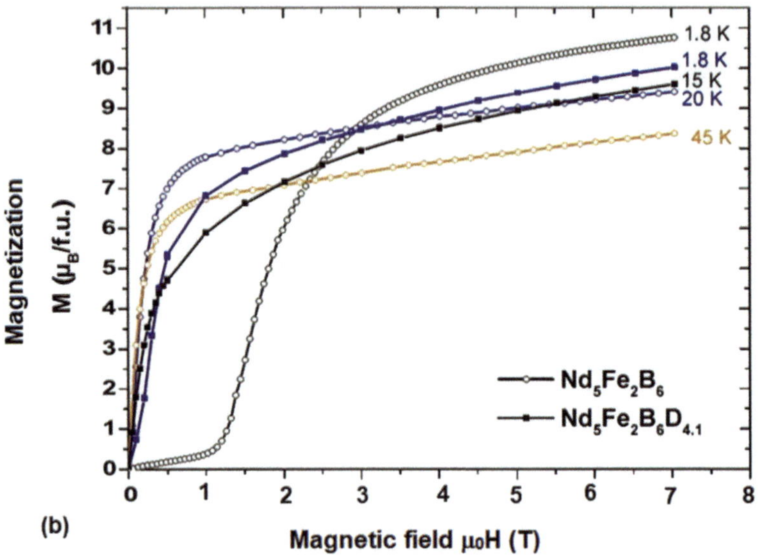

Fig. 7.3 Magnetization isotherms at 1.8 K ≤ T ≤ 45 K of the $Nd_5Fe_2B_6$ and $Nd_5Fe_2B_6D_{4.1}$ borides [17T1]

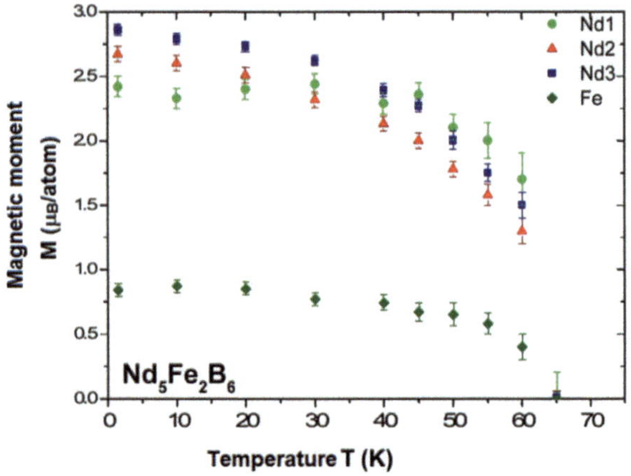

Fig. 7.4 Nd_5FeB_6: thermal variations of the neodymium and iron moments, as determined by neutron diffraction study [17T1]

The different behavior concerning the presence of the homogeneity range, in this system, reported in literature can be associated with the way in which the samples were prepared and thermally treated and/or sample homogeneity, respectively.

Table 7.9 Data obtained by ^{57}Fe Mössbauer spectroscopy on $R_5Fe_2B_6$ compounds

Compound	T (K)	Site	δ^a (mm/s)	ΔQ (mm/s)	DH (mm/s)	B_{eff} (T)	η	A (%)	References
$Ce_5Fe_2B_6$	80	Fe1(6*c*)	0.61(2)	0.83(2)				43(2)	[98Y1]
		Fe2(R3*a*)	0.45(2)	0.45(2)				45(2)	
		Fe3(R6*c*)	0.63(2)					12(2)	
$Nd_5Fe_2B_6$	4.2	Fe	0.35(1)	0.74(3)	0.33(2)	2.4	0.6(1)	100	[17T1]
$Nd_5Fe_2B_6D_{4.1}$	4.2	Fe1	0.30(2)	0.69	0.30			25(3)	[17T1]
		Fe2	0.38(3)	0.24(5)	0.34(2)			75(3)	
$Nd_5Fe_2B_6$	5	Fe(6c)	0.34	0.72	0.28	2.4			[87D3]
	295	Fe(6c)	0.19	0.68	0.28	–			
Nd_2FeB_3[b,c]	5	Fe1	0.34	0.67	0.27	2.6			[87D3]
		Fe2	0.14	0.54	0.30				
$Sm_5Fe_2B_6$	80	Fe1(6*c*)	0.60(2)	0.88(2)				57(2)	[98Y1]
		Fe2(R2*a*)	0.53(2)	0.52(2)				24(2)	
		Fe3(R6*c*)	0.64(2)					19(2)	
$Tb_5Fe_2B_6$	80	Fe1(6*c*)	0.60(2)	0.87(2)				67(2)	[98Y1]
		Fe2(R3*a*)	0.63(2)	0.76(2)				25(2)	
		Fe3(R6*c*)	0.77(2)					8(2)	
$Yb_5Fe_2B_6$	80	Fe1(6*c*)	0.59(2)	0.12(2)		12.2(1)		56(2)	[98Y1]
		Fe2(R3*a*)	0.58(2)	0.21(2)		10.3(1)		20(2)	
		Fe3(R6*c*)	0.62(2)	0.04(2)		2.8(1)		24(2)	

[a] Relative to α-Fe; [b] Actually $Nd_5Fe_2B_6$; [c] Mixture of $Nd_5Fe_2B_6$ and $NdFe_4B_4$ phases [87D3]

7.5 $R_{1+\varepsilon}Fe_4B_4$ Compounds

A large number of investigations were devoted to the analysis of crystal structures of $R_{1+\varepsilon}Fe_4B_4$ compounds [78K1, 79C1, 82B1, 84P1, 84R1, 85B1, 85G1, 86G1, 86G2, 86N1, 87B1, 87O1, 88B2, 88T5, 89Z3–89Z6, 92Z2, 94K1, 96Y2, 97B1, 04L2, 05W1]. The $R_{1+\varepsilon}Fe_4B_4$ structure, with M = Fe, Co, Mn, is build with a tetragonal columnar arrangement. The structural framework consists of M_4 tetrahedra sharing edges in the **c**-direction, with boron pairs connecting the tetrahedral columns to each other. The M and B atoms form tetracapped tetrahedra or straight stellae quadrangulae, commonly labelled *tetraedersterns* [04L2]. In this manner, octagonal channels are formed by the M_4B_4 net. The rare-earths are situated in the channels, ideally one R atom per Fe_4B_4 unit—Fig. 7.5a. The $R_{1+\varepsilon}Fe_4B_4$ structures are incommensurably modulated in the channel direction. The diffraction patterns are indexed as due to two separate substructures, having space groups $P4_2/ncm$ for Fe_4B_4 and $I4/mmm$ for rare-earths, respectively [85G1]. The incommensurability of the structure has been attributed to the modulation of Fe_4 tetrahedra, due to their rotation in basal plane determined by the interaction with rare-earth atoms—Fig. 7.5b. The R atoms move only slightly in the **c**-direction, but have no degrees of freedom in terms of moving around in the channels [04L2]. In the Fe_4B_4 subcell, the Fe and B atoms occupy the 8*i* positions (x,x,z) and in R subcell, the atoms are located on 2*a* position (0,0,0)—Table 7.10. Either of the above substructures can be used as a starting point for symmetry lift into a (3 + 1) dimensional description. If the Fe_4B_4 cell is chosen as the basic structure, the superspace group will emerge from the three-dimensional space group $P4_2/ncm$. If the R sublattice is chosen as the basic set, the starting point would be the three-dimensional space group *I*4/*mmm*.

The crystal structures of $R_{1+\varepsilon}Fe_4B_4$ borides were also analyzed in a commensurate model, the uniaxial periodicity being characterized by two related translation vectors $m\mathbf{c}_R = n\mathbf{c}_{Fe}$ and characterized as Vernier structure or chimney-ladder structure [89H2, 93Y1]. As the ratio c_R/c_{Fe} = m/n, can be represented by two relatively small integers, the structures can be either incommensurate or commensurate with an unusual long period along the **c**-direction. The space groups of the superstructures, were reported to be $P4_2/n$ for R = Ce, Pr, Nd, Gd, Tb [85B1], *Pccn* when R = Nd, Gd [86G2] or $P\bar{4}c2$ for R = Ho [93G3]. The possible commensurate superstructures and their space groups were theoretically analyzed [92Z2]. The symmetries of the $R_m(Fe_4B_4)_n$ superstructures can be either $P4_2/n$ or *Pccn* (*Ccca*), depending of the parity combinations of integers m and n. In describing the superspace approach, it was not stated that m and n must be strict integers. The symmetry of each member of the infinitely adaptive structure (truly incommensurate or unusual long period superstructures) can be described by a four-dimensional superspace group. In both approaches, the interactions between the two substructures are not taken into account. By ignoring the twist modulation of the iron tetrahedra, leads to results in agreement

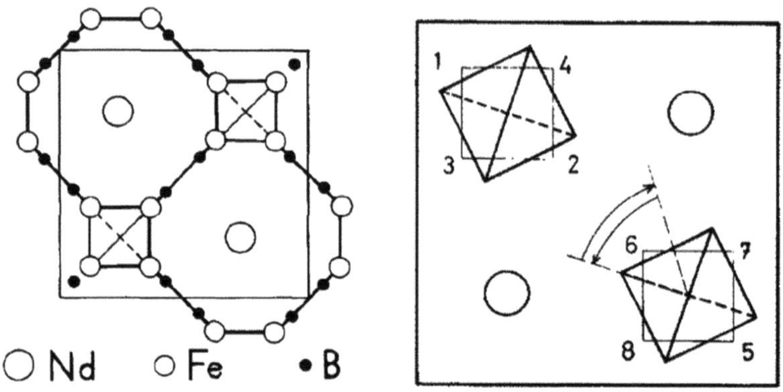

(a) (b)

Fig. 7.5 $R_{1+\varepsilon}Fe_4B_4$: **a** projection of the structure on the (001) plane; **b** rotational modulation of Fe_4 tetrahedra. The broken lines represent the two extreme positions of the iron atom, denoted by 6 [85G1, 86G1]

Table 7.10 Atomic parameters of Nd atoms in space group *I4/mmm* (origin at center *2/m*) and of Fe and B atoms in the space group *P4₂/ncm* (origin at center *4/mmm*) [85G1]

Space group	Atom	Site	x	y	z
I4/mmm	Nd	2*a*	0	0	0
$P4_2/ncm$	Fe	8*i*	0.127	0.127	0.1349
$P4_2/ncm$	B	8*i*	0.068	0.068	0.639

with the experimental space group assignments, in the commensurate approximation, [92Z2].

The structures with a single $R_m(Fe_4B_4)_n$ layer ratio was not strictly observed, on the macroscopic length scale, on a given sample. A very local probe, however yield observations consistent with a definite m/n layer ratio commensurate superstructure, as evidenced in a very thin $Nd_8Fe_{28}B_{28}$ sample (m/n = 8/7) [05W1]. The significant compositional variations, in samples with rather great dimensions, evidenced that there is no reason to believe the presence of a single superstructure through the volume [87B1, 89Z3]. As example, the electron diffraction study, on $Nd_{1+\varepsilon}Fe_4B_4$ sample, revealed a quasi-continuous series of compositions near the $Nd_{1.1}Fe_4B_4$ one [89Z3]. Their presence was related to the local fluctuations in the composition and explained in terms of small energy differences between these phases. The lattice parameters of $R_{1+\varepsilon}Fe_4B_4$ borides are listed in Table 7.11, where the reported superspace groups are also mentioned.

A composite description was used to analyse the incommensurate structures by a pair of superspace groups [92Y1, 93Y1]. This approach considers the two subsets intact and separate, without forcing them into a single unit. The entire set of main and satellite reflections may be indexed by either of the superspace groups. The main reflections, that indicate the dimensions of one of the subcells, will become satellite in the other cell description [04L2].

The $NdCo_4B_4$ boride, has a commensurate type structure with $P4_2/n$ space group [78K1]—Sect. 8.6. In the pseudoternary $Nd_{1+\varepsilon}(Fe_{1-x}Co_x)_4B_4$, series, as the cobalt content increases, a gradual reduction of the difference between the lattice constants c_{Nd} and $c_{(FeCo)_4B_4}$ was found [89Z4], evidencing the transition to a commensurate type structure. There is a large single phase (23/21) and a two phase region with a second ladder type (L2, m/n = 21/19) [87O1].

The disproportionation of $Nd_{1+\varepsilon}Fe_4B_4$ boride, in hydrogen at T > 700 °C, results in the formation of a "sandwich" type structure consisting of $NdH_{2\pm x}$ and Fe_2B phases [96Y2]. No significant recombination was found on heating the hydrogenated sample in vacuum, which was attributed to an increased oxygen content in the decomposed layer.

Since the $R_{1+\varepsilon}Fe_4B_4$ borides are present as a secondary phase in the compositions of the R–Fe–B permanent magnets, their magnetic properties were investigated together with those of hard magnetic components, $R_2Fe_{14}B$ [98B2]. The reported data on $R_{1+\varepsilon}Fe_4B_4$ magnetism include those with R = Pr [86N1, 86R1], R = Nd [85B1, 85G1, 86B11, 86N1, 86R1, 88T1, 17C1], R = Sm [86R1, 87R5] R = Gd

Table 7.11 Space group and lattice parameters of $R_{1+\varepsilon}Fe_4B_4$ compounds

Compound	T (K)	Space group	Lattice parameters (nm)				m/n	References
			a	c_{Fe}	c_R	c		
$La_{1.061}Fe_4B_4$	RT		0.720	0.386	0.3640			[86N1]
$La_{1.06}Fe_4B_4$	RT		0.7221	0.3873	0.3640	12.781	35/33	[87O1]
$Ce_{1.1208}Fe_4B_4$	RT	*P*4$_2$/*n*	0.7090(1)	0.39102(3)	0.34889(4)	12.904	37/33	[85B1]
$Ce_{1.12}Fe_4B_4$	RT		0.7090	0.3910	0.34889	12.904	35/33	[87O1]
$Ce_{1.1}Fe_4B_4$	RT		0.708	0.391	0.351360			[89L2]
$Ce_{1.117}Fe_4B_4$	RT		0.706	0.389	0.349			[86N1]
$Ce_{1.121}Fe_4B_4$	RT		0.7702	0.3898	0.3477		19/17	[86G1]
$Pr_{1.106}Fe_4B_4$	RT	$P4_2/n$	0.7158(1)	0.39042(5)	0.35301(2)	7.418	21/19	[85B1]
$Pr_{1.11}Fe_4B_4$	RT		0.7154	0.3906	0.3529	7.421	21/19	[87O1]
$Pr_{1.1}Fe_4B_4$	RT	$P4_2/n$	0.7158(1)	0.39042(5)	0.35301(2)	7.418	21/19	[82B1]
$Pr_{1.112}Fe_4B_4$	RT		0.7150	0.3890	0.3499		10/9	[86G1]
$Pr_{1.105}Fe_4B_4$	RT		0.713	0.388	0.351			[86N1]
$Nd_{1.1088}Fe_4B_4$	RT	$P4_2/n$	0.7141(3)	0.39073(7)	0.35241(7)	14.457	41/31	[85B1]
$Nd_{1.11}Fe_4B_4$	RT	*Pccn*	0.7117	0.3896	0.3502	3.507	10/9	[86G2]
$Nd_{1.10}Fe_4B_4$	RT	$P4_2/n$	0.7141(3)	0.39073(7)	0.35241(7)	14.457	41/31	[82B1]
$Nd_{1.107}Fe_4B_4$	RT		0.711	0.389	0.352			[86N1]
$Nd_{1.112}Fe_4B_4$	RT		0.7117	0.3896	0.3502		10/9	[85G1, 86G1]
$Nd_{1.1217}Fe_4B_4$	RT		0.71226(1)	0.39025(1)	0.34792(1)	3.2076(2)		[87B1]
$Nd_{1.1313}Fe_4B_4$	RT		0.71233(1)	0.39042(1)	0.34512(1)	2.9744(2)		[87B1]
$Nd_{1.1417}Fe_4B_4$	RT		0.71244(1)	0.39077(1)	0.34227(1)	2.7577(2)		[87B1]
$Nd_{1.11}Fe_4B_4$	RT		0.7111	0.3892	0.3505	14.400	41/37	[87O1]
$NdFe_4B_4$	RT		0.7128	0.3906	0.3478			[86B11]

(continued)

Table 7.11 (continued)

Compound	T (K)	Space group	Lattice parameters (nm)				m/n	References
			a	c_{Fe}	c_R	c		
$Nd_{1.093}Fe_4B_4$	RT		0.7117(1)	0.3891(1)	0.3559(2)			[89Z5]
$Nd_{1.1098}Fe_4B_4$	RT		0.7122(1)	0.3893(1)	0.3546(3)			[89Z5]
$Nd_{1.109}Fe_4B_4$	RT		0.7124(1)	0.3896(1)	0.3513(2)			[89Z5]
$Nd_{1.110}Fe_4B_4$	RT		0.7109(1)	0.3892(3)	0.3505(3)			[89Z5]
$Nd_{1.114}Fe_4B_4$	RT		0.7114(2)	0.3894(4)	0.3496(4)			[89Z5]
$Nd_{1.118}Fe_4B_4$	RT		0.7109(3)	0.3893(2)	0.3482(6)			[89Z5]
$Nd_{1.125}Fe_4B_4$	RT		0.7116(1)	0.3900(1)	0.3446(1)			[89Z5]
$Nd_{1.128}Fe_4B_4$	RT		0.7119(5)	0.3888(4)	0.34446(5)			[89Z5]
$Nd_{1.143}Fe_4B_4$	RT		0.7100(5)	0.392	0.343	2.74(5)	8/7	[05W1]
$Nd_{1+\varepsilon}Fe_3CoB_4$	RT		0.7095	0.3852	0.3527	8.089	23/21	[87O1]
$Nd_{1+\varepsilon}Fe_2Co_2B_4$	RT		0.7084	0.3856	0.3572	7.3188	21/19	[87O1]
$Nd_{1+\varepsilon}Fe_{1.5}Co_{2.5}B_4$	RT		0.7074	0.3833	0.3617	13.4155	37/35	[87O1]
$Nd_{1.1}Fe_4B_4$	RT		0.7116(4)	0.3894(6)	0.3523(5)			[89Z4]
$Nd_{1.1}Fe_3CoB_4$	RT		0.7104(1)	0.3873(2)	0.3548(5)			[89Z4]
$Nd_{1.1}Fe_2Co_2B_4$	RT		0.7094(2)	0.3843(9)	0.3590(7)			[89Z4]
$Nd_{1.1}FeCo_3B_4$	RT		0.7072(2)	0.3820(1)	0.3617(8)			[89Z4]
$NdCo_4B_4$	RT		0.7050(6)	0.3816(1)	0.3816(1)			[89Z4]
$NdCo_4B_4$	RT		0.7070		0.3822			[78K1]
$NdCo_4B_4$	RT		0.7060		0.3795		1/1	[87O1]
$Sm_{1.1316}Fe_4B_4$	RT	$P4_2/n$	0.798(1)	0.39124(5)	0.34574(2)	5.869	17/15	[85B1]
$Sm_{1.133}Fe_4B_4$	RT		0.78	0.390	0.344			[86N1]

(continued)

Table 7.11 (continued)

Compound	T (K)	Space group	Lattice parameters (nm)				m/n	References
			a	c_{Fe}	c_R	c		
$Gd_{1.14}Fe_4B_4$	RT	$P4_2/n$	0.7073(3)	0.39217(4)	0.3442(1)	11.373	33/29	[82B1]
$Gd_{1.14}Fe_4B_4$	RT		0.7064	0.3920	0.3438	11.368	33/29	[87O1]
$Gd_{1.1}Fe_4B_4$	RT	$P4_2/n$	0.7073(3)	0.39217(4)	0.3442(1)	11.373	33/29	[85B1]
$Gd_{1.148}Fe_4B_4$	RT	*Pccn*	0.7051	0.3913	0.3408	2.7327	8/7	[86G1, 86G2]
$Gd_{1.141}Fe_4B_4$	RT		0.704	0.391	0.342			[86N1]
$Tb_{1.19}Fe_4B_4$	RT	$P4_2/n$	0.7049(1)	0.3919(1)	0.34109(1)	10.581	31/27	[85B1]
$Tb_{1.156}Fe_4B_4$	RT		0.700	0.391	0.339			[86N1]
$Dy_{1.152}Fe_4B_4$	RT		0.700	0.390	0.339			[86N1]
$Ho_{1.18}Fe_4B_4$	RT	$P\bar{4}c2$	0.6983(2)	0.3922(1)	0.3337(1)	6.67(2)	19/17	[93G3]
$Ho_{1.162}Fe_4B_4$	RT		0.698	0.391	0.336			[86N1]
$Er_{1.17}Fe_4B_4$	RT		0.6989	0.3917	0.3344	13.710	41/35	[87O1]
$Er_{1.181}Fe_4B_4$	RT		0.697	0.392	0.332			[86N1]
$Tm_{1.197}Fe_4B_4$	RT		0.696	0.390	0.331			[86N1]
$Lu_{1.18}Fe_4B_4$	RT		0.6950	0.3936	0.3344	13.776	41/35	[87O1]
$Y_{1.15}Fe_4B$	RT		0.7013	0.3916	0.3399	5.874	15/13	[87O1]

[86N1, 86R1], R = Tb [86N1], R = Dy [86N1, 86R1] and R = Ho [86N1, 86R1, 93G3]—Table 7.12.

No magnetic ordering was shown, at T > 4.6 K, in $R_{1+\varepsilon}Fe_4B_4$ borides, when R = Ce, Er, Tm [86N1]. The magnetization isotherms, at T = 4.2 K, of $Nd_{1.11}Fe_4B_4$ ($Nd_5Fe_{18}B_{18}$) single crystal, as well as the thermal variation of reciprocal susceptibility showed that the compound is ferromagnetically ordered [85G1]—Fig. 7.6. The Curie temperatures of $Nd_{1+\varepsilon}Fe_4B_4$ compounds were located at $T_c \cong 14$ K [85B1, 85G1, 86R1], $T_c = 10$ K [86B11] or at $T_c \cong 7$ K [17C1]. The latter value was also obtained from specific heat measurements [17C1] and seems to be the most correct. In the $R_{1+\varepsilon}Fe_4B_4$ borides the iron atoms seem to be not magnetic, the magnetic ordering of this series, resulting from the interactions between rare-earths.

The reduced value of Nd moment, in polycrystalline $Nd_{1+\varepsilon}Fe_4B_4$, as compared to that of pure metal, suggests that the higher order CEF terms are important in determining their magnetic properties. The easy direction of magnetization corresponds to [110] axis [85G1, 17C1]. The magnetization, M(θ), of $Nd_{1.143}Fe_4B_4$ single crystal, at T = 4.2 K, has been measured while rotating the sample in (bc) plane from [010] to [001] direction [17C1]. The M(θ) values follow the same trend as the absolute value of a cosine function and approach a node at 90°, when the basal plane moment has no projection along the field direction. At 45°, the magnetization is $M/\sqrt{2}$ (where M is the maximum value of the magnetization), expected for uniaxial ferromagnet rotating in a plane containing the two easy directions or for an easy (ab) plane magnet, rotated in a plane containing the **c**-axis. The evolution of the magnetization, when rotating the sample in (ab)-plane, from [010] to [100] direction, has been also investigated. For a $Nd_{1+\varepsilon}Fe_4B_4$ single crystal, having four easy directions, in the basal plane [85G1, 17C1], the M(θ) looks as a trigonometric function, with the maxima in magnetization located at multiple of 90° and the minima at [100], [010], [$\bar{1}$00] and [0$\bar{1}$0] directions. As the applied field decreases, the form of M(θ) dependences changes. In a field $\mu_0H = 0.7$ T, the M(θ) appears almost flat, as expected if there is no longer in-plane anisotropy. At $\mu_0H = 0.4$ T, the symmetry is different, with the [100] direction now appearing as an easy direction of fourfold symmetry. More complicated behavior is evidenced at lower fields. The above results suggest that the magnetic anisotropy in $Nd_{1+\varepsilon}Fe_4B_4$ boride is more complex than that already described [85G1].

The Curie temperatures of $R_{1+\varepsilon}Fe_4B_4$ borides with light rare-earths are higher than those expected according to De Gennes rule [86N1]. This behavior can be connected with a higher induced polarization on R5d bands, by 4f-5d exchange than in heavy rare-earth borides, as evidenced also $R_2Fe_{14}B$ compounds [21B1]—Sect. 7.10.

The ^{57}Fe Mössbauer spectroscopy has been used to investigate the magnetic behavior of iron in $R_{1+\varepsilon}Fe_4B_4$ borides [85R1, 86N1, 86R1, 87R5, 89Z4, 90O1]. According to [86N1], the ^{57}Fe Mössbauer spectra, at T = 4 K, consist of one quadrupole doublet, evidencing that the iron atoms carry not magnetic moment—Table 7.13. In the presence of an external field, $\mu_0H = 0.65$ T, oriented perpendicular to γ-rays, at T = 4.2 K, an effective field, $\mu_0H_{eff} = 0.7(2)$ T was determined, near the same as the applied field. The presence of magnetic ordered impurities was also shown in some samples [86N1]. Small ^{57}Fe hyperfine fields were evidenced in compounds with R = Sm and Dy [87R5] and their origin has been analysed.

Table 7.12 Magnetic properties of $R_{1+\varepsilon}Fe_4B_4$ compounds

Compound	Magnetic structure	M_s (μ_B/f.u.)	T_c (K)	M_{eff} (μ_B/f.u.)	θ (K)	K_1 (MJ/m^3)	References
$Pr_{1.105}Fe_4B_4$		1.50[a]	7.5(5)				[86N1]
$Pr_{1.11}Fe_4B_4$			3.5				[86R1]
$Nd_{1.11}Fe_4B_4$	FM e.a.m ‖ [110] $M_{Fe} = 0\ \mu_B$	2.44 ‖ [110][b] 1.89 ‖ [100]	15	3.8		6.9	[85G1]
$Nd_{1.11}Fe_4B_4$			14 7 (T_{sR})				[86R1]
$Nd_{1.1}Fe_4B_4$	FM	3.0(1)[c]	13	5.9(6)	≅13		[85B1]
$NdFe_4B_4$	FM	1.7[a]	10				[86B11]
$Nd_{1.107}Fe_4B_4$		2.40[a]	16.0(5)				[86N1]
$Nd_{1.143}Fe_4B_4$	FM, **M** ‖[110]		7	4.5	7		[17C1]
$Sm_{1.11}Fe_4B_4$			36				[86R1]
$Sm_{1+\varepsilon}Fe_4B_4$			37(2)				[87R5]
$Gd_{1.141}Fe_4B_4$		6.02[a]	24.0(5)				[86N1]
$Gd_{1.11}Fe_4B_4$	e.a.m ‖ [110] or [1 $\bar{1}$ 0]		19				[86R1]
$Tb_{1.156}Fe_4B_4$		3.50[a]	16.0(5)				[86N1]
$Dy_{1.152}Fe_4B_4$		5.70[a]	11.5(5)				[86N1]
$Dy_{1.111}Fe_4B_4$	e.a.m ‖ [100] or [110]		9				[86R1]
$Ho_{1.162}Fe_4B_4$		5.50[a]	6.5(5)				[86N1]
$Ho_{1.11}Fe_4B_4$			7.5				[86R1]

[a] $T = 4.2$ K, $\mu_0H = 1.8$ T; [b] Composition $Nd_9Fe_{18}B_{18}$, $M = 11\ \mu_B$/f.u. ‖ [110] and $M = 8.5\ \mu_B$/f.u. ‖ [100] or [010], $T = 4.2$ K, $\mu_0H = 7$ T; [c] $T = 4.2$ K, $\mu_0H = 8$ T

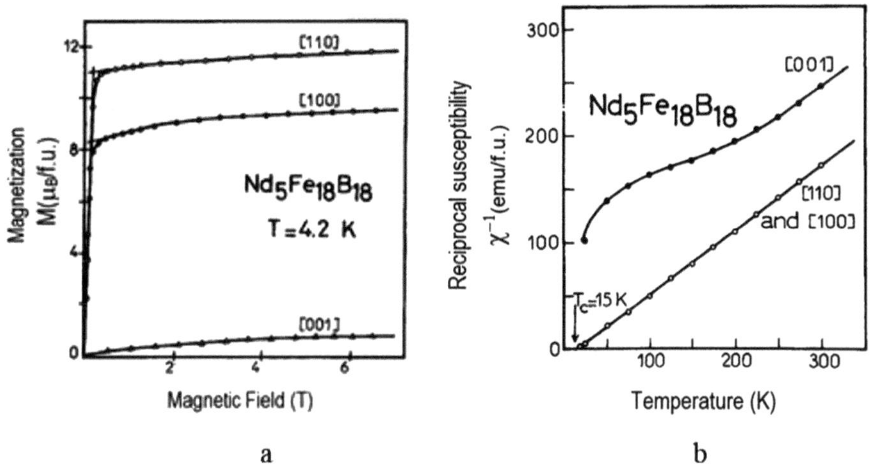

Fig. 7.6 $Nd_{1.1}Fe_4B_4$ ($Nd_5Fe_{18}B_{18}$) single crystal: **a** field dependences of the magnetization at T = 4.2 K; **b** temperature dependences of the magnetic susceptibilities [85G1]

The ^{57}Fe Mössbauer spectra of $Gd_{1.143}Fe_4B_4$ ($Gd_4Fe_7B_7$) were fitted also considering the presence of seven inequivalent 8*i* iron sites, as expected to be present in commensurate *Pccn*—type structure [87R5]. For each iron site, the principal EFG tensor component, V_{zz}, the asymmetry parameter, η and the polar angles θ, Φ of the EFG major axis with respect to the crystallographic axis, were calculated. The principal EFG axis was on the average, situated in the [110] or $[1\bar{1}0]$ directions, with a small angular spread, mostly above and below the basal plane. The ^{57}Fe Mössbauer spectra in $Sm_{1+\varepsilon}Fe_4B_4$ were interpreted in terms of magnetization parallel to the **c**-axis and in case of $Dy_{1+\varepsilon}Fe_4B_4$, consistent with a magnetization lying in the basal plane, either along [100] or [110] axis. Analysing the ^{57}Fe spectra of $R_{1+\varepsilon}Fe_4B_4$ borides, large linewidths were evidenced [85R1, 86R1]. These have been also associated with the multiplicity of iron environments connected with (quasi)incommensurable rare-earth and iron substructures. No magnetic ordered iron moments were seen at $T \geq 1.5$ K [86R1].

The ^{57}Fe Mössbauer spectra of $(Gd_xR_{1-x})_{1+\varepsilon}Fe_4B_4$ series, with R = Nd or Sm, at RT, were fitted assuming the presence of two quadrupole doublets [90O1] and for $Nd_{1+\varepsilon}(Fe_{1+x}Co_x)_4B_4$ borides considering only one doublet [89Z4]—Table 7.13. The addition of cobalt leads to a more negative isomer shift with respect to that of α-Fe, whereas the quadrupole splitting decreases continuously, with the diminution of the mismatch between the two substructures.

The analyses of the ^{155}Gd Mössbauer spectra of $(Gd_xR_{1-x})_{1+\varepsilon}Fe_4B$ pseudoternary systems with R = Nd, Sm, evidenced a nearly constant e^2qQ value [90O1]. The width of the Gaussian, distribution $P(e^2qQ)$, significantly change, when Gd is substituted by Nd. The V_{zz} values spread around the value $2.8 \cdot 10^{22}$ V/m^2, attributed to the multiplicity of the environments, due to different periodicities of R and Fe_4B_4 sublattices as well as the ladder's modulation. These data confirmed also a basal

Table 7.13 Data obtained by Mössbauer spectroscopy on $R_{1+\varepsilon}Fe_4B_4$ compounds

(a) ^{57}Fe nucleus

Compound	T (K)	Nucleus	δ^a (mm/s)	ΔQ (mm/s)	DH (mm/s)	B_{eff} (T)	η	θ°	φ°	References
$La_{1.001}Fe_4B_4$	300	^{57}Fe	0.034(5)	0.632(5)						[86N1]
$La_{13.5}Fe_{44.5}B_{42}$	4.2		0.146(5)	0.60(1)						[86R1]
$Ce_{1.117}Fe_4B_4$	300		0.010(5)	0.530(5)						[86N1]
$Ce_{13.5}Fe_{14.5}B_{4.2}$	4.2		0.120(5)	0.53(1)						[86R1]
$Pr_{1.105}Fe_4B_4$	300		0.032(5)	0.584(5)						[86N1]
$Pr_{13.5}Fe_{44.5}B_{42}$	4.2		0.136(5)	0.55(1)						[86R1]
$Nd_{1.107}Fe_4B_4$	300		0.026(5)	0.582(5)						[86N1]
$Nd_{13.5}Fe_{44.5}B_{42}$	4.2		0.137(5)	0.54(1)						[86R1]
$Sm_{1.133}Fe_4B_4$	300		0.023(5)	0.566(5)						[86N1]
$Sm_{13.5}Fe_{14.5}B_{42}$	4.2		0.125(7)	0.52(1)						[86R1]
$Sm_{1+\varepsilon}Fe_4B_4$	4.2			0.54	0.29	0.78[b]	0	90		[87R5]
$Gd_{1.141}Fe_4B_4$	300		0.022(5)	0.556(5)						[86N1]
$Gd_{13.5}Fe_{44.5}B_{42}$	4.2		0.127(5)	0.54(1)						[86R1]
$Gd_{1.143}Fe_4B_4$[c)]	4.2						0.08(4)	90(7.1)	45.0(8)	[87R5]
$Tb_{1.156}Fe_4B_4$	300		0.018(5)	0.528(5)						[86N1]
$Tb_{13.5}Fe_{44.5}B_{42}$	4.2		0.127(5)	0.52(1)						[86R1]
$Dy_{1.152}Fe_4B_4$	300		0.014(5)	0.532(5)						[86N1]
$Dy_{13.5}Fe_{14.5}B_{42}$	4.2		0.191(5)	0.52(2)						[86R1]
$Dy_{1+\varepsilon}Fe_4B_4$	4.2			0.54	0.30	0.88[b]	0	90		[87R5]
$Ho_{1.162}Fe_4B_4$	300		0.037(5)	0.544(5)						[86N1]

(continued)

Table 7.13 (continued)

(a) ^{57}Fe nucleus

Compound	T (K)	Nucleus	$δ^a$ (mm/s)	ΔQ (mm/s)	DH (mm/s)	B_{eff} (T)	η	θ°	φ°	References
$Ho_{13.5}Fe_{44.5}B_{42}$	4.2		0.150(7)	0.54(1)						[86R1]
$Er_{1.181}Fe_4B_4$	300		0.024(5)	0.548(5)						[86N1]
$Er_{13.5}Fe_{44.5}B_{42}$	4.2		0.180(10)	0.53(1)						[86R1]
$Tm_{1.179}Fe_4B_4$	300		0.024(5)	0.546(5)						[86N1]
$Lu_{13.5}Fe_{44.5}B_{42}$	4.2		0.180(10)	0.61(1)						[86R1]
$Y_{13.5}Fe_{44.5}B_{42}$	4.2		0.140(5)	0.54(1)						[86R1]
$Nd_{1+ε}Fe_4B_4$	300		0.025(1)	0.573(2)	0.159(5)					[89Z4]
$Nd_{1+ε}Fe_3CoB_4$	300		−0.032(1)	0.538(2)	0.157(2)					[89Z4]
$Nd_{1+ε}Fe_2Co_2B_4$	300		−0.040(1)	0.527(2)	0.152(2)					[89Z4]
$Nd_{1+ε}FeCo_3B_4$	300		−0.027(2)	0.504(4)	0.146(3)					[89Z4]

(b) ^{155}Gd and ^{161}Dy nuclei

Compound	T (K)	Nucleus	δ (mm/s)	ΔQ (mm/s)	$σ^e$ (mm/s)	B_{hf} (T)	$θ^g$ (K)	References
$Gd_{1.14}Fe_4B_4$	4.2	$^{155}Gd^d$	0.32(1)	12.67(5)	0.42(8)	23.4(3) $(18.8(5))^f$	isotropic	[87R6, 90O1]
$(Gd_{0.8}Nd_{0.2})_{1+ε}Fe_4B_4$	4.2		0.33(1)	12.60(5)	0.62(4)	24.9(3)	81(2)°	[90O1]
$(Gd_{0.6}Nd_{0.4})_{1+ε}Fe_4B_4$	4.2		0.32(1)	12.68(5)	0.73(4)	24.8(2)	82(2)°	[90O1]
$(Gd_{0.5}Nd_{0.5})_{1+ε}Fe_4B_4$	4.2		0.35(2)	12.67(5)	0.72(4)	25.9(5)	81(5)°	[90O1]

(continued)

Table 7.13 (continued)

(b) ^{155}Gd and ^{161}Dy nuclei

Compound	T (K)	Nucleus	δ (mm/s)	ΔQ (mm/s)	σ^e (mm/s)	B_{hf} (T)	θ^g (K)	References
$(Gd_{0.4}Nd_{0.6})_{1+\varepsilon}Fe_4B_4$	4.2		0.32(1)	12.70(5)	0.88(6)	25.6(5) (18.7(5))^f	90(1)°	[90O1]
$(Gd_{0.2}Nd_{0.8})_{1+\varepsilon}Fe_4B_4$	4.2		0.33(1)	12.83(5)	0.88(6)	25.6(5)	90(1)°	[90O1]
$(Gd_{0.8}Sm_{0.2})_{1+\varepsilon}Fe_4B_4$	4.2		0.32(1)	12.6(1)	0.72(6)	21.8(5) (16.5(5))^f	3(4)°	[90O1]
$Dy_{1.13}Fe_4B_4$	4.2	$^{161}Dy^h$	1.9(5)	74(2)	8.1(5)	556(1)		[87R6]

[a] Relative to α-Fe; [b] In field of 0.108 T; [c] $V_{zz} = 3.88(14)\cdot 10^{20}$ V/m^2 mean values for 7 sites as evidenced from the structure analysis [85B1]
[d] Source ^{155}Eu/$SmPd_3$; [e] Average value and width of the Gaussian distribution of quadrupole interaction; [f] Values in brackets refer to experiments under $\mu_0 H_{ext}$ = 8 T; [g] Angle between V_{zz} and B_{hf}; [h] $^{160}Gd_{0.5}{}^{162}Dy_{0.5}F_3$ source

plane magnetization orientation for Nd-rich alloys and the nearly basal ($\theta = 80°$) one near the Gd-rich end. The magnetization was shown to be parallel to [001] axis for $(Gd_xSm_{1-x})_{1+\varepsilon}Fe_4B_4$ series, excepted for x = 1.0 [90O1]. For $Gd_{1+\varepsilon}Fe_4B_4$ boride, no easy axis was shown, implying a noncollinear spin arrangement, the partial substitution of Gd, by non-S rare-earths, imposing an anisotropy induced by crystal field effects determined by the sign of the B_2^0 coefficient of the substituting rare-earths [90O1]. The ^{155}Gd Mössbauer study of $Gd_{1+\varepsilon}Fe_4B_4$ boride, evidenced a large quadrupolar interaction. A similar study on ^{151}Dy in $Dy_{1+\varepsilon}Fe_4B_4$, showed that this interaction is rather small [87R6]. For both compounds, a strong lattice contribution to the electric field gradient was shown. These interactions, in this series, are much stronger than exchange interactions.

The temperature dependences of the heat capacity, $C_p(T)$ and resistivity, $\rho(T)$, were analysed in $Nd_{1.143}Fe_4B_4$, as well as in non-magnetic $(Nd_{0.5}Ce_{0.5})_{1.143}Fe_4B_4$ borides [17C1]. The cerium substituted sample, has a significant reduced heat capacity compared to the $Nd_{1.143}Fe_4B_4$ one. The differences between the two $C_p(T)$ dependences, persists up to T = 100 K. The entropy per Nd atom, $S/k_B = 0.63$, below T_c, is nearly *ln*2. An entropy of 2.1, which is nearly *ln*10 has been determined, over the entire investigated temperature range (T < 80 K). This result, is expected for Nd^{3+} ion with a $^4I_{9/2}$ ground state and a relatively low T_c value. The peak at T = 7 K evidences the presence of a magnetic transition. An additional broad maximum was shown, at $T \cong 40$ K, reminiscent of a Schottky feature. The Debye temperature is $\theta_D = 540$ K.

The temperature dependence of the electrical resistivity of $Nd_{1.143}Fe_4B_4$, along [001] direction, shows a peak at T_c, the resistivity decreasing below this temperature [17C1]. There is a broad maximum at $T \cong 200$ K, attributed to magnetic correlations among the iron moments. At T > 100 K, the magnetoresistance is positive, 2–3%, in a field of 8 T which is suggestive of magnetic fluctuations.

7.6 RFe_2B_2 Compounds

The RFe_2B_2 compounds with (R = Tb, Dy, Ho, Er, Tm, Lu, Y), crystallize in a tetragonal structure, *I4/mmm* space group [65B1, 78S1, 80S1, 84P1, 84R1, 87D5, 91G1]—Fig. 7.7 and Table 7.14. The presence of $CeFe_2B_2$ was also reported [72B1]. Latter on was shown that really it is $Ce_{1.1}Fe_4B_4$ [85B1]. In the structure, the FeB_4 tetrahedra share edges to form infinite slabs. The R atoms are situated in large voids ($R(Fe_8B_8)$ polyhedron) between the slabs. Pairs of face-linked $B(R_4Fe_4)$ square antiprisms, ($B(Fe_4B)$ square pyramids) share edges to form a 3D-framework [11V1]. The lattice parameters are listed in Table 7.15.

The $R(Co_{1-x}Fe_x)_2B_2$ solid solutions, having *I4/mmm* type structure, are formed with R = Gd ($x \leq 0.4$) [11L2] and R = La ($x \leq 0.27$) [12N1]—see Sect. 8.2.

The RFe_2B_2 compounds were little investigated. The band structure calculations, performed on YFe_2B_2, showed that their ground state is paramagnetic [10T1]. The Fe3d states are hybridized with the Y4d band.

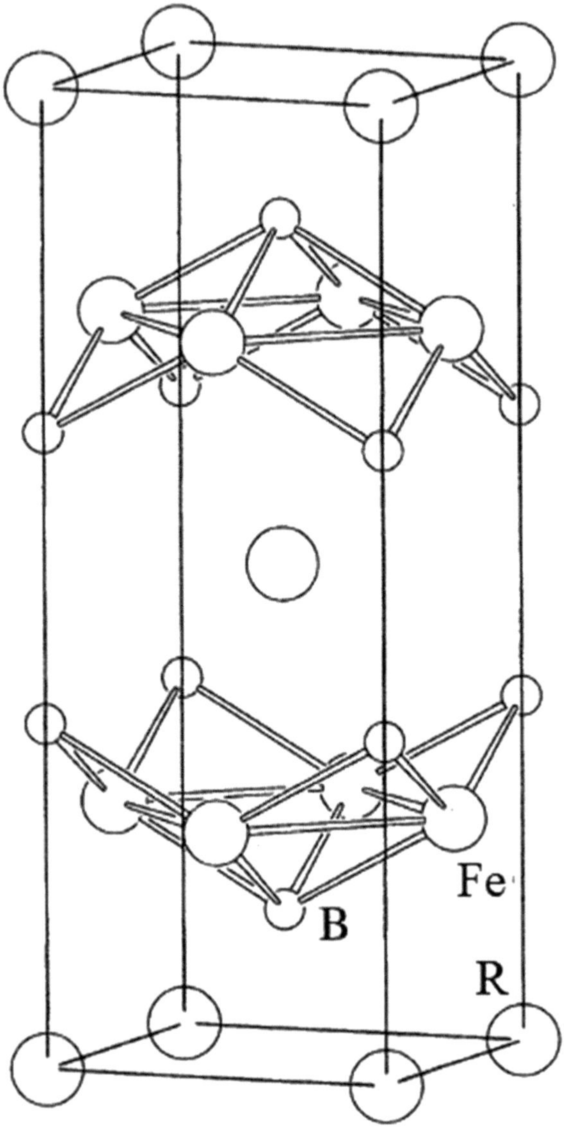

Fig. 7.7 Crystal structure of RFe_2B_2 compounds

7.7 $RFe_{12}B_6$ Compounds

The $RM_{12}B_6$ compounds, where R is a rare-earth or yttrium were first identified in M = Co series [72N1, 89M3]. In the homologous iron series, stable compounds are formed only with R = La [92R3, 97L3]. Metastable structures with R = Pr [10W1] and R = Nd [86B12, 06Z1, 08H1] are also present.

Table 7.14 RFe_2B_2 compounds, having $ThCr_2Si_2$ type structure, space group *I4/mmm* [65B1, 78S1]

Atom	Sites	Symmetry	x	y	z	Atomic environment [11V1]
R	2*a*	4/*mmm*	0	0	0	22-vertex polyhedron $B_{10}Fe_8R_4$
Fe	4*d*	$\bar{4}m2$	0	1/2	1/4	Cuboctahedron $B_4Fe_4R_4$
B	4*e*	4*mm*	0	0	0.387	Tricapped trigonal prism BFe_4R_4

Table 7.15 Space group and lattice parameters of RFe_2B_2 compounds

Compound	T (K)	Space group	Lattice parameters (nm)		References
			a	c	
$TbFe_2B_2$	RT	*I4/mmm*	0.3544(5)	0.9473(10)	[78S1, 87D5]
$DyFe_2B_2$	RT	*I4/mmm*	0.3537(5)	0.9441(10)	[78S1, 83C3]
$HoFe_2B_2$	RT	*I4/mmm*	0.3527(5)	0.9425(10)	[78S1, 91G1]
$ErFe_2B_2$	RT	*I4/mmm*	0.3515(5)	0.9387(10)	[78S1, 83C3]
$TmFe_2B_2$	RT	*I4/mmm*	0.3507(5)	0.9342(10)	[78S1]
$LuFe_2B_2$	RT	*I4/mmm*	0.3499(5)	0.9288(10)	[78S1, 87D5]
YFe_2B_2	RT	I4/mmm	0.3546(5)	9.555(10)	[78S1, 80S1]

Table 7.16 Atomic sites of $LaFe_{12}B_6$ having $R\bar{3}m$ space group[a,b,c]

Atom	Sites	Sym	x	z	Atomic environment [07V1]
La	3*a*	$\bar{3}m$	0	0	Pseudo Frank-Kasper $Fe_{18}B_6$
Fe	18*g*	.2	0.367(2)	1/2	Pseudo Frank-Kasper $B_4Fe_7La_2$
Fe	18*h*	.*m*	0.424(1)	0.032(1)	14-vertex Frank-Kasper B_4Fe_9La
B	18*h*	.*m*	0.483(1)	0.284(1)	Monocapped trigonal prism Fe_7

[a] Crystallographic parameters at T = 50 K; [b] Local environments: La3*a*(12Fe18*g*, 6Fe18*h*, 6B); Fe18*g* (3Fe18*g*, 4Fe18*h*, 2La, 4B), Fe18*h*(4Fe18*g*, 5Fe18*h*, 4B, 1La), B(4Fe18*g*, 3Fe18*h*, 1La) [17D1]; [c] For the atomic sites and their coordinates in $NdFe_{12}B_6$ see [08H1]

The $RFe_{12}B_6$ compounds crystallize in the rhombohedral $SrNi_{12}B_6$-type structure, space group $R\bar{3}m$ [72N1, 80J1, 81K1]. In this lattice, the iron atoms are located in two inequivalent crystallographic positions (18*g*, 18*h*), the rare-earth and boron atoms occupying the 3*a* and 18*h* sites, respectively—Table 7.16.

The $LaFe_{12}B_6$ boride has a stable crystal structure, which do not changes as effect of pressure [17D1]. The a and c lattice parameters decrease linearly with pressure ($p \leq 14$ GPa) with the rates $da/dp = -1.44.10^{-3}$ $nmGPa^{-1}$ and $dc/dp = -1.15.10^{-3}$ $nmGPa^{-1}$. There is an anisotropic shrinkage of the unit cell, with a larger compression in the basal plane. The bulk modulus of $LaFe_{12}B_6$ is $K_0 = 204$ GPa [17D1].

The metastable $RFe_{12}B_6$ compounds with R = Pr and Nd are generally obtained after controlled crystallization of iron-rich R–Fe–B amorphous alloys obtained by direct quenching from molten state or by crystallization from the amorphous state

Table 7.17 Space group and lattice parameters of $RFe_{12}B_6$ compounds

Compound	T (K)	Space group	Lattice parameters (nm)		References
			a	c	
$LaFe_{12}B_6$	1.5	$R\bar{3}m$	0.9583(1)	0.7588(2)	[16D4]
	50	$R\bar{3}m$	0.9596(1)	0.7600(1)	[16D4]
	RT	$R\bar{3}m$	0.9631(5)	0.7612(1)	[16D2]
$LaFe_{12}B_6$	RT	$R\bar{3}m$	0.9618	0.7607	[17F1]
$LaFe_{12}B_6$[a]	RT	$R\bar{3}m$	0.9622(4)	0.7603(3)	[17D1]
$LaFe_{12}B_6$	RT	$R\bar{3}m$	0.9610	0.7604	[92R3]
$La_{0.9}Ce_{0.1}Fe_{12}B_6$	RT	$R\bar{3}m$	0.9630(3)	0.7610(1)	[18D1]
$La_{0.85}Ce_{0.15}Fe_{12}B_6$	RT	$R\bar{3}m$	0.9629(2)	0.7607(1)	[18D1]
$La_{0.825}Ce_{0.175}Fe_{12}B_6$	RT	$R\bar{3}m$	0.9629(1)	0.7606(1)	[18D1]
$PrFe_{12}B_6$[b, c]	200	$R\bar{3}m$	0.9604(2)	0.7557(8)	[10W1]
$PrFe_{12}B_6$[b, c]	201	$R\bar{3}m$	0.9602(0)	0.7554(6)	[10W1]
$NdFe_{12}B_6$[b, d]	RT	$R\bar{3}m$	0.9605	0.7549	[86B12]
$NdFe_{12}B_6$[b, e]	RT	$R\bar{3}m$	0.95859(3)	0.75505(3)	[08H1]
$Nd_{0.1}La_{0.9}Fe_{12}B_6$	RT	$R\bar{3}m$	0.96082(4)	0.75964(8)	[21C1]
$GdCo_{12}B_6$	RT	$R\bar{3}m$	0.9454(1)	0.7449(1)	[13D1]
$GdCo_{11.5}Fe_{0.5}B$	RT	$R\bar{3}m$	0.9462(3)	0.7451(1)	[13D1]
$GdCo_9Fe_3B$	RT	$R\bar{3}m$	0.9482(1)	0.7466(1)	[13D1]

[a] At RT and p = 1 GPa; [b] Metastable compound; [c] α-Fe impurity; [d] $Nd_2Fe_{23}B_3$ phase as impurity; [e] α-Fe,$Nd_5Fe_{18}B_{18}$ impurities

[86B12, 88A1, 89L3]. The single $RFe_{12}B_6$ phases with R = Pr, Nd are hard to be formed. During the crystallization of amorphous samples, other phases precipitate as α-Fe [08H1, 10W1], $Nd_2Fe_{23}B$ [86B12] or $NdFe_4B_4$ [08H1]. Their content is rather high, even of 23% [08H1]. The $NdFe_{12}B_6$ boride transforms to equilibrium phases, at T ≥ 1000 K [86B12, 88B9].

The rhombohedral structure, having space group $R\bar{3}m$, has been also reported in pseudoternary 1/12/6 compounds, in a limited composition range, as in $La_{1-x}Gd_xFe_{12}B_6$ system, for $x \leq 0.5$ [97L3], $La_{1-x}Ce_xFe_{12}B_6$ when $x \leq 0.175$ [18D1], $La_{1-x}Nd_xFe_{12}B_6$ with $x < 0.1$[21C1], $RFe_{12-x}Co_xB_6$ with R = La for $x \leq 10$ [92R3], R = Gd, Y and x = 11.8 [92R3], R = Gd for $x \leq 3$ [13D1] and in $Nd(Co_{1-x}Fe_x)_{12}B_6$ series, with $x \leq 1.8$ [11Z1]. The lattice parameters of $RFe_{12}B_6$ compounds are listed in Table 7.17.

The magnetic properties of $RFe_{12}B_6$ borides with R = La [97L3, 16D1–16D4, 17D1, 17F1, 18D1], R = Pr [10W1] and R = Nd [86B12, 88B9, 08H1, 11Z1] are very interesting.

The stable $LaFe_{12}B_6$ compound, at $T < T_N = 35$ K, has an antiferromagnetic type structure, described by the propagation vector **q** = (1/4, 1/4, 1/4) [16D4]. In this amplitude modulated structure, the iron moments are confined in (ab) plane, their values, at T = 1.5 K, being M_{Fe} = 0.43 μ_B/atom, at both sites—Table 7.18. The temperature dependence of the $LaFe_{12}B_6$ magnetization in ZFC and μ_0H = 3 T shows a broad peak, at $T_N \cong 36$ K, as expected for an antiferromagnetic type ordering. At higher fields (4.5 T $\leq \mu_0H \leq$ 10 T), both ZFC and FC samples, show transitions from antiferromagnetic (AFM) to ferromagnetic (FM) ordering. In ZFC sample, when increasing temperature, a plateau in magnetization develops in fields of 5–6 T, around T $\cong$ 29 K, whose magnitude and width increases when increasing the external field. The saturation magnetization of 18.55 μ_B/f.u. was obtained in a field μ_0H = 10.5 T, corresponding to a mean iron moment of 1.546 μ_B/Fe atom. When increasing temperature, a transition from FM to PM state is evidenced. The magnetic phase diagrams in the field-temperature plane, evidence the presence of a multi critical point, at μ_0H = 4.5 T and T = 33 K, at the crossover of the AFM, FM and PM phases—Fig. 7.8. The magnetic field enhances the FM state. The Curie temperatures, T_c, increase when increasing external field (5 T $\leq \mu_0H \leq$ 10.5 T), with a rate of 5.7 and 6.4 K/T for heating and cooling the sample, respectively [16D3, 16D4]—Fig. 7.8. The magnetization isotherms of ZFC $LaFe_{12}B_6$ boride, at low temperatures, display abrupt jumps followed by plateaus, giving rise to a staircase-like behavior. The M(H) plots begin with a linear dependence, consistent with the AFM ground state. At T = 2 K and 4 K, there are two sharp step transitions, in the initial magnetization, which are also observed in the corresponding field decreasing branch. At T = 6 K only one step is observed. The magnetization jumps were correlated with the transition of a fraction of the sample, from AFM to FM state. After the remove of the magnetic field, a fraction of the sample remains in the FM state, at T < T_c, the magnetization thus depending on the magnetic history. At finite temperatures, both AFM and PM states can be transformed into FM state via a field-induced metamagnetic transition, accompanied with a huge magnetic hysteresis. Above T_N, the field induced transition becomes reversible [16D2].

Under certain combinations of temperature and magnetic field, the time dependence of the magnetization, in $LaFe_{12}B_6$, displays an unusual step-like feature—Fig. 7.9 [16D1]. At lower external fields than μ_0H = 7.1 T and T = 2 K, the time dependences of the magnetizations can be described by a relaxation law, having an exponential form, $M(H,t) \propto \exp(-t/\tau H)$, where τ is the relaxation time, related to the magnitude of the energy barrier between two metastable states and having values in the range 3800 s $\leq \tau \leq$ 7500 s [16D1]—Fig. 7.9. An abrupt magnetization jump occurs over a time interval, in a field μ_0H = 7.1 T, when the magnetization suddenly changes from 1.12 μ_B/f.u. to 15.86 μ_B/f.u. This type of behavior, is described as an "incubation time" effect. The step like transitions at low temperatures, are qualitatively consistent with the martensitic mechanism [48K1, 04H1]. In this scenario, the steep magnetization jumps correspond to a burnstike growth of ferromagnetic fraction, at the expense of antiferromagnetic component, driven by the evolution of the strains at the interfaces between the two kinds of domains. The incubation time, significantly differs from run to run, showing that the characteristic time of the jump is not a material constant. At μ_0H > 7.1 T, the normalized magnetizations at t = 0, for each applied field, yield flat curves without detectable time dependence.

Table 7.18 Magnetic properties of $RFe_{12}B_6$ compounds

Compound	Magnetic structure and magnetic moments	M_s (μ_B/f.u.)	$T_N(T_c)$ (K)	References
$LaFe_{12}B_6$ (ND)	T = 1.5 K; AFM: **q** = (1/4,1/4,1/4); eam ⊥ **c** axis Fe: $M_g = 0.43(3)\ \mu_B$, $M_h = 0.43(3)\ \mu_B$	5.16 18.6^a	36	[16D4]
$LaFe_{12}B_6$ (BS)	Low magnetic state (AFM) Fe: $M_g = 0.71\ \mu_B$, $M_h = 0.28\ \mu_B$ La: $M = -0.02\ \mu_B$, B: $M = -0.03\ \mu_B$ High magnetic state (FM) Fe: $M_g = 1.34\ \mu_B$, $M_h = 1.69\ \mu_B$ La: $M = -0.05\ \mu_B$, B: $M = -0.10\ \mu_B$	5.72 17.43		[11M1]
$LaFe_{12}B_6$	Metamagnetic	4.32 19.2^b	35	[17F1]
$LaFe_{12}B_6$	Metamagnetic	4.6^c 20^b	40(5)	[92R3]
$La_{0.9}Ce_{0.1}Fe_{12}B_6$	AFM	18.22^b	35	[21D5]
$La_{0.85}Ce_{0.15}Fe_{12}B_6$	AFM	17.9^b	22	[21D4]
$La_{0.9}Nd_{0.10}Fe_{12}B_6$	FM	19.26^b	58	[21C1]
$PrFe_{12}B_6{}^f$	FM	23.5^c	200	[10W1]
$NdFe_{12}B_6{}^f$	FM	19.7^d	230	[86B12, 88B9]
$NdFe_{12}B_6{}^f$	FM, Fe: $M_g = 1.1\ \mu_B$, $M_h = 1.3\mu_B{}^e$	14.4^e	205	[08H1]
$GdCo_{12}B_6$ (ND)	FIM, T = 4 K, eam 38(8)° with **c**-axis Gd: $M = 6.9(5)\ \mu_B$; Co: $M_g = 0.41(3)\ \mu_B$, $M_h = 0.50(3)\ \mu_B$	1.68(4)	158 50 (T_{comp})	[13D1]

(continued)

Table 7.18 (continued)

Compound	Magnetic structure and magnetic moments	M_s (μ_B/f.u.)	$T_N(T_c)$ (K)	References
$GdCo_{11.5}Fe_{0.5}B_6$ (ND)	FIM, T = 4 K, eam ⊥ **c**-axis Gd: M = 6.9 μ_B; Co: M_g = 0.22(5) μ_B, M_h = 0.71(7) μ_B	1.30(5)	155 45 (T_{comp})	[13D1]

[a] After metamagnetic transition, T = 2 K, μ_0H = 10 T; [b] After metamagnetic transition, T = 2 K, μ_0H = 5 T; [c] T = 5 K, μ_0H = 5 T; [d] T = 4.2 K, μ_0H = 1.8 T; [e] Determined from ^{57}Fe Mossbauer spectroscopy assuming B_{eff}/M_{Fe} = 14.7 T/μ_B [79B1]; [f] No single phases were obtained

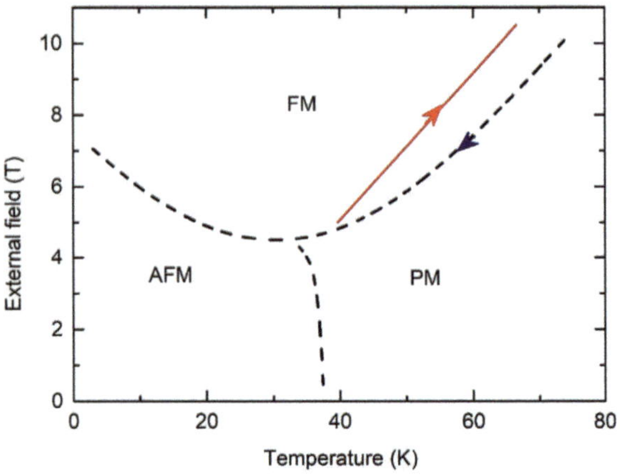

Fig. 7.8 $LaFe_{12}B_6$: magnetic phase diagram where the evolution with temperature and external field of magnetic transition temperatures are given [16D4]. The field dependences of the Curie temperatures when heating (red) and cooling (black) the sample [16D3, 16D4]

The Néel temperature of $LaFe_{12}B_6$ boride decreases, as effect of pressure, with a rate $dT_N/dp = -4.5\ KGPa^{-1}$, evidencing important volume effect on the itinerant electron metamagnetic transition [17D1]. The disappearance of the AFM order was estimated, by extrapolation of experimental data at $T_N = 0$, to a pressure $p = 8$ GPa.

The specific heat, C_p, of $LaFe_{12}B_6$, in the form C_p/T, follows a linear dependence on T^2, at $T \leq 14$ K [17F1]. The Sommerfeld coefficient, obtained by a linear extrapolation, in a field of 6.0 T, is much smaller than that when no external field is present. Such a large reduction observed after the metamagnetic transition, suggests that the internal field resulting from larger iron moment, suppresses the effect of spin

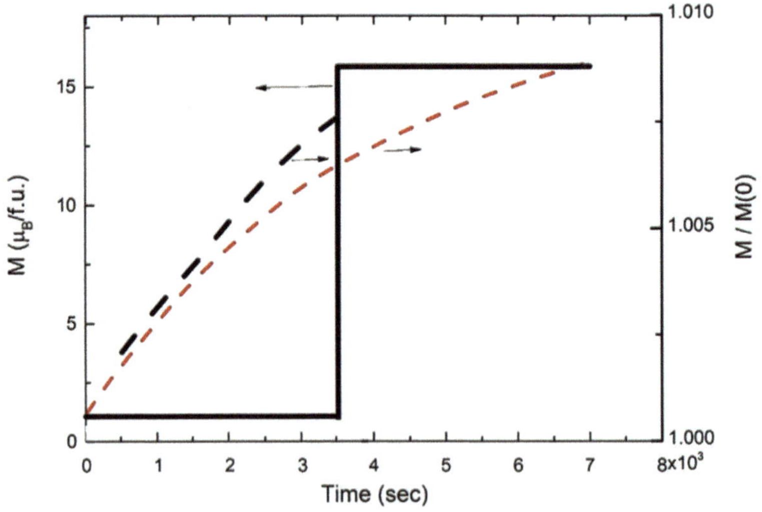

Fig. 7.9 $LaFe_{12}B_6$: time dependences of relative magnetizations, M(t)/M(0), at T = 2 K in external fields of 7 T (black dashed and solid lines) and at 6.8 T (dashed red line) [16D1]

fluctuations, present in the weak ferromagnetic state, as already predicted [68B1] or observed in RCo_2 compounds [92B3].

The magnetocaloric effect (MCE) was studied in $LaFe_{12}B_6$ compound [16D2, 17F1]. The entropy change, ΔS_M, was shown to be negative around T_c and it is changed to positive across the AFM-FM transition for $\mu_0H = 7$ T, leading to normal and inverse values of -6.8 Jkg^{-1} K^{-1} and 19 Jkg^{-1} K^{-1} around T = 38 K and T = 8 K, respectively [16D2]. According to [17F1], at the metamagnetic transition, isothermal magnetic entropy change $\Delta S_M = -16$ Jkg^{-1} K^{-1} as well as large value of the adiabatic temperature change, $\Delta T_{ad} = -8.7$ K at $\mu_0H = 2$ T, were evidenced.

The electronic structure of $LaFe_{12}B_6$ has been theoretically calculated, by using TB-LMTO method in atomic sphere approximation (ASA) [07M3, 11M1]. The total iron spin moment of 5.72 μ_B/f.u. was obtained for the state with low moment and 17.43 μ_B/f.u. in the ferromagnetic state, little smaller than the experimentally determined values [11M1].

The effects of cerium substitutions on the structural, magnetic and transport properties of the $La_{1-x}Ce_xFe_{12}B_6$ series with $x \leq 0.175$ were investigated [18D1, 21D1–21D6, 22D1]. Both AFM and PM states can be transformed into FM state irreversible or reversibly depending on the magnitude of the applied field, the temperature and the direction (increase or decrease) of their changes. The T_N temperature is independent on alloy composition, while the FM → PM transition temperatures T_c increase when La is replaced by Ce with a rate ≅32 K/Ce atom (when heating). The Ce substitution has an effect similar as the application of magnetic field in the end series $LaFe_{12}B_6$. The substitution of 0.35 at% of Ce is equivalent to an applied magnetic field of 2 T.

The most investigated sample was the $La_{1-x}Ce_xFe_{12}B_6$ with x = 0.1 [21D2, 21D5, 21D6. 22D1]. As the end series compound, the specimen is an antiferromagnet below $T_N = 35$ K and presents two consecutive magnetic transitions, AFM-FM and FM-PM, when increasing temperature under the action of magnetic field. The AFM and PM phases may be converted into the FM phase irreversibly or reversible, respectively via a first order metamagnetic transitions accompanied by a large magnetic hysteresis. At T < 8 K, the magnetic field-driven AFM-FM metamagnetic transformation is discontinuous and proceeds by multiple abrupt steps in magnetization and magnetoresistance [21D2]. In addition, a large negative magnetoresistance effect (–77%) was observed. At the phase transition there is a field induced crystallographic phase transition from rhombohedral (AFM, PM) to a monoclinic structure, space group *C2/m* (FM). The transition is driven by magnetoelastic effects and is accompanied by a large forced volume magnetostriction, which at T = 25 K and field of 6 T, it is $\Delta v/v = 1.15\%$ [21D5]. An anisotropic lattice expansion as well as a giant magnetothermal expansion with a volumetric thermal expansion $\alpha_v = -1.93.10^{-4}$ were also shown. The application of a magnetic field on $La_{0.9}Ce_{0.1}Fe_{12}B_6$, leads to a nearly linear increase of T_c at a rate of 5.6 and 4.9 K/T upon cooling and warming, respectively [21D2]. The magnetization (as in $LaFe_{12}B_6$) as well as electrical resistivity undergo a spontaneous transition after an incubation period, when both temperature and applied field are keep constant [21D6]. The presence of a critical point at the crossover of PM, FM and AFM, as in end series compound, has been shown.

The staircase-like shape of magnetization curves are sensitive to the thermomagnetic history of the samples [22D1].

The general features in physical properties of $La_{1-x}Ce_xFe_{12}B_6$ with x = 0.1 were also shown in specimens with x = 0.15 [21D3, 21D4] and x = 0.175 [21D1], with some differences in the involved parameters. As example, in $La_{0.85}Ce_{0.15}Fe_{12}B_6$, at T = 20 K, a giant volume magnetostriction, $\Delta v/v = 0.80\%$, was observed across the first order induced AFM-FM transition in field $\mu_0H = 6$ T. The FM-PM transition is accompanied by a large negative thermal expansion $\alpha_l = -34\cdot10^{-4}$ K^{-1}, over the temperature window of $\Delta T \cong 63$ K [21D3]. A negative magnetoresistance, MR = −75% was associated with the magnetic field induced first order magnetic transition in $La_{0.825}Ce_{0.175}Fe_{12}B_6$ [21D1].

The $PrFe_{12}B_6$ metastable compound is ferromagnetically ordered and experiences first order magnetic phase transition at $T_c \cong 200$ K [10W1]. This type of transition is confirmed also by the sharp peak in the specific heat. The peak shifts at higher temperatures, as the intensity of external field increases, with a rate $dT_c/d\mu_0H \cong 1.8$ K/T. The magnetic transition is accompanied by a lattice contraction which in turn results in a large, ΔS_M entropy change. In the field range 0–5 T, a value $\Delta S_M = -16.2$ J/kgK, was reported.

The $NdFe_{12}B_6$ metastable boride is also ferromagnetically ordered [86B12, 88B9, 06Z1, 08H1]. Curie temperatures $T_c = 205$ K [08H1] or 230 K [86B12] were reported—Table 7.18. The above metastable phases were not obtained in pure state, and as a results, the presence of magnetic impurities can influence their experimentally determined properties.

The $La_{0.9}Nd_{0.1}Fe_{12}B_6$ order ferromagnetically at $T_c = 58$ K, termperature which shifts toward higher temperatures with increased magnetic field [21C1]. At T = 2 K, there is a mixed magnetic state with ≅2/3 of sample in FM state and ≅1/3 in AFM state, after cooling in zero field. The AFM state transforms into FM one via a field-induced metamagnetic transition in field $\mu_0H = 3.4$ T, at T = 2 K. Another metamagnetic transition was observed above the low-field T_c value. Both metamagnetic transitions are of first order and accompanied by large thermal and magnetic hysteresis. Large magnetocaloric effect, of 19.4 J/kgK for a field change of 7 T, was observed in the vicinity of the field induced PM–FM transition.

The $RFe_{12}B_6$ borides with R = La [92R3, 17F1] and Nd [92R3, 08H1] were also investigated by ^{57}Fe Mössbauer spectroscopy—Table 7.19. The mean ^{57}Fe hyperfine field values, at low temperatures, in $LaFe_{12}B_6$, are rather small, in agreement with the corresponding value of the iron moments. At RT, two doublets are evidenced, that having larger quadrupole splitting being attributed to iron at 18*g* site [92R3]. The ^{57}Fe hyperfine fields in $NdFe_{12}B_6$ show little variations up to $T_c \cong 205$ K, when an abrupt decrease is shown, as expected for a first order magnetic transition. The estimated iron magnetic moments, from the determined ^{57}Fe hyperfine fields are listed in Table 7.18 [08H1].

The physical properties of pseudoternary compounds $La_{1-x}Gd_xFe_{12}B_6$ [97L3], $LaFe_{12-x}Co_xB_6$ [92R3], $RFe_{12-x}Co_xB_6$ with R = Gd or Y [92R3], $GdCo_{12-x}Fe_xB_6$ [13D1] and $Nd(Co_{1-x}Fe_x)_{12}B_6$ [11Z1] were investigated. The $Gd_xLa_{1-x}Fe_{12}B_6$ borides are ferrimagnetically ordered when La is partially substituted by Gd [97L3].

The average iron moments, before metamagnetic transitions increases, at low gadolinium content and becomes somewhat saturated for $x \geq 0.3$. After the transition, the mean iron moments increase smoothly when the gadolinium content rises—Fig. 7.10.

The stabilization of $SrNi_{12}B_6$ type structure in $Nd(Co_{1-x}Fe_x)_{12}B$ system, by replacing iron with cobalt, has been investigated as well as their effects on the magnetic and magnetocaloric properties [11Z1]. The saturation magnetizations and Curie temperatures decrease as the cobalt is gradually substituted by iron in $GdCo_{12-x}Fe_xB_6$ series [13D1]—Sect. 8.8. The easy axis of magnetization changes from uniaxial to the basal plane upon substitution of iron by cobalt. The transition metal moments, at 18*g* site, decreases, while that at 18*h* site increases when 0.5 Co atoms are replaced by iron—Table 7.18.

The ^{57}Fe Mössbauer spectroscopy was used to analyze the magnetic properties of $RFe_{12-x}Co_xB_6$ with R = La, Y and Gd [92R3] and by ^{155}Gd the $GdCo_{12-x}Fe_xB_6$ system [13D1—Table 7.19. The average ^{57}Fe hyperfine field in $LaFe_{12-x}Co_xB_6$ decreases from 15.8 T (x = 11.8) to ≅ 5 T in the pure iron compound [92R3]. The ^{155}Gd Mössbauer spectra on $GdCo_{12-x}Fe_xB_6$ showed that for end member of the series (x = 0) the Gd moments order close to **c**-axis (θ = 15(2)°), while they are almost perpendicular to **c**-axis for $x \geq 0.5$ [13D1]—see also Sect. 8.8.

7.8 RFe_4B Compounds

According to R–Fe–B phase diagrams, the RFe_4B compounds are formed only with R = Er [83C1, 12R1], R = Tm [78S1] and R = Lu [86J1, 87D4]. The presence of homogeneity range in $LuFe_{4+x}B_{1-x}$ compound has been reported [83K3, 87D5]. Replacement of iron by boron, decreases a lattice parameter and increases the c one. The existence of RFe_4B with R = La, Gd and Y were also mentioned [12P1, 12P2], in contradiction with the R–Fe–B phase diagrams [80S1, 17X1, 21W1]. These

Table 7.19 Data obtained by ^{57}Fe Mössbauer spectroscopy on $RFe_{12}B_6$ compounds

Compound	T (K)	Site	δ^a (mm/s)	ΔQ (mm/s)	μ_0H_{eff} (T)	A (%)	References
$LaFe_{12}B_6$	RT	18*g* 18*h*	0.012 [b] 0.012 [b]	0.67 [b)] 0.36 [b)]		50 50	[92R3]
	4.2				5.6[c]		
$LaFe_{12}B_6$	5				5.4[c]		[17F1]
$NdFe_{12}B_6$	20	18*g* 18*h*	−0.06(3) 0.09(1)	0.51(5) 0.42(4)	16.9(6) 19.3(5)	}75	[08H1]
		α-Fe	0.03(8)		32(5)	14.7[d]	
		$Nd_{1.11}Fe_4B_4$	0.1(9)	0.0(4)	22(7)	10.5[d]	

[a] Relative to α-Fe; [b] From figure; [c] Mean values; [d] Impurities

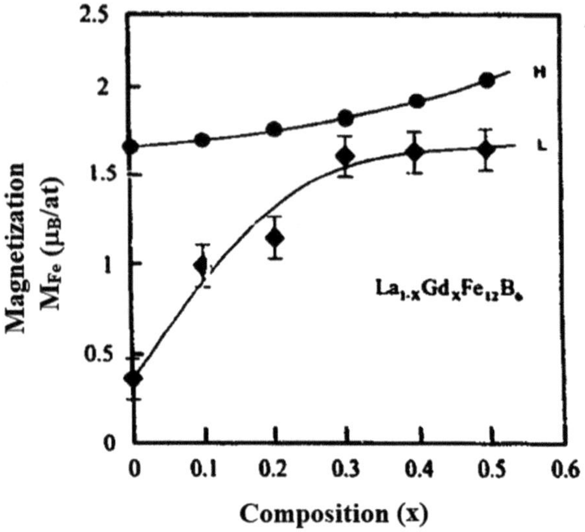

Fig. 7.10 $La_{1-x}Gd_xFe_{12}B_6$: composition dependences of the average iron moments before (L) and after (H) the metamagnetic transition [97L1]

Table 7.20 Atomic sites in RFe_4B compounds, having crystallographic ordered $CeCo_4B$ type structure, space group *P6/mmm*[a] [74B1, 74K1, 11V1]

Atom	Site	Symmetry	x	y	z	Atomic environment
R1	1*a*	6/*mmm*	0	0	0	Pseudo Frank-Kasper $B_6Fe_{12}R_2$
R2	1*b*	6/*mmm*	0	0	1/2	Pseudo Frank-Kasper $Fe_{18}R_2$
Fe	2*c*	$\bar{6}m2$	1/3	2/3	1/2	14-vertex polyhedron $Fe_9R_3B_2$
Fe	6*i*	$\bar{6}m2$	1/2	0	0.213	13-vertex polyhedron $B_2Fe_7R_4$
B	2*d*	$\bar{6}m2$	1/3	2/3	0	Trigonal prism Fe_6

[a] Transformation from published data, origin shift (001/2); The Ce, Co were replaced by R and Fe atoms

compounds were not obtained as single phases. By annealing two months at T = 600 °C and then 10 days at T = 900 °C, the resultant mixtures contained ($LaFe_4B$, Fe_2B, Fe_3B), ($GdFe_4B$, $GdFe_2$, Fe_2B, Fe_3B) and (YFe_4B, $Y_3Fe_{62}B_{14}$, YFe_2), respectively [12P1, 12P2]. The presence of additional phases suggests possible deviations of these compounds from ideal 1/4/1 composition.

The formation of $DyFe_4B$ phase in $(Nd_{53}Dy_{47})_{15}Fe_{77}B_8$ sintered magnet along grain boundaries [87S1, 97S2], as well as in $(Nd,Pr,Dy)_{15}(Fe,Co)_{79}B_6$ alloys [05K1] has been mentioned. The (R,R')$Fe_{4-x}M_xB$ alloys with $CeCo_4B$ type structure it is easily to be formed in multi-components systems and their compositions can be related to the dominant compound [88A2].

The crystal structure of RFe_4B compounds, in earlier reports, has been considered to be of $CeCo_4B$-type, a derivative of $CaCu_5$ one [74B1, 74K1]. In the fully ordered RFe_4B structure, the B atoms occupy 2*d* positions, the Fe the 2*c* and 6*i* sites and the rare-earths 1*a* and 1*b* positions—Table 7.20. In this lattice, the 6*i* sites correspond to the 3*g* positions and (2*c*, 2*d*) to 2*c* sites of $CaCu_5$-type structure. The RFe_5 structure,

Table 7.21 Space group and lattice parameters of RFe_4B compounds

Compound	T (K)	Space group	Lattice parameters (nm)		References
			a	c	
$ErFe_4B$	RT	*P6/mmm*	0.5033(1)	0.6985(3)	[83C2, 83K4]
$ErFe_4B$	RT	*P6/mmm*	0.5046	0.6991	[85V1]
$ErFe_4B$	RT	*P6/mmm*	0.5028	0.6971	[85V3]
$ErFe_4B$	RT	*P6/mmm*	0.5041	0.7002	[86J1]
$ErFe_4B$	RT	*P6/mmm*	0.5047	0.6998	[88Z3]
$TmFe_4B$	RT	*P6/mmm*	0.4973(2)	0.6970(8)	[83C1, 83K4]
$TmFe_4B$	RT	*P6/mmm*	0.5031	0.6977	[85V1]
$TmFe_4B$	RT	*P6/mmm*	0.5023	0.6985	[85V3]
$LuFe_4B$	RT	*P6/mmm*	0.5007(2)	0.6966(6)	[83C1, 83K4]
$LuFe_4B$	RT	*P6/mmm*	0.4994	0.6984	[84O1, 84S5]
$LuFe_4B$	RT	*P6/mmm*	0.4989	0.6927	[85V3]
$LuFe_4B$	RT	*P6/mmm*	0.5046	0.7009	[85V1]
$LuFe_4B$	RT	*P6/mmm*	0.5035	0.6933	[86J1]
$LaFe_4B$ [a)]	RT	*P6/mmm*	0.4896	0.6864	[12P1, 12P2]
$GdFe_4B$ [a)]	RT	*P6/mmm*	0.5089	0.6822	[12P1, 12P2]
YFe_4B [a)]	RT	*P6/mmm*	0.4896	0.6859	[12P1, 12P2]
$SmFe_2Ni_2B$	RT	*P6/mmm*	0.5083	0.6947	[84S5]
$SmFe_2Ni_2B$	RT	*P6/mmm*	0.508	0.691	[87H1]
$SmFeNi_3B$	RT	*P6/mmm*	0.5066	0.6953	[84S5]
$SmFeNi_3B$	RT	*P6/mmm*	0.504	0.696	[87H1]

[a] Mixture of phases

of $CaCu_5$-type, is not formed with R elements, being present only when R = Th [90B1]. The structural features of crystallographic ordered RFe_4B borides involve Kagomé-mesh Fe_3 layers which alternate with RB_2 or RFe_2 layers (B or Fe hexagon mesh, the hexagons of which are centered by an R atom) along [001] direction. The BFe_6 trigonal prisms share edges to form infinite slabs with no B–B contacts [11V1].

The absence of superlattice lines, in XRD patterns of RFe_4B borides, indicates that instead of a fully ordered $CeCo_4B$-type lattice, there is a disordered replacement of iron by boron. Generally speaking, disorder may occur either between 2*c* and 2*d* sites, which correspond to the same site in the parent $CaCu_5$-type structure (2*c*) or can extend to all positions (2*c*, 3*g*) in $CaCu_5$-type, or (2*c*, 2*d* and 6*i*) in $CeCo_4B$-lattice. The above assumptions have been investigated by ^{57}Fe Mössbauer spectroscopy, although the type of disorder assumed, to interpret the experimental data, differed, as will be further discussed [85V1, 85V3, 88Z3]. The reported lattice parameters are given in Table 7.21.

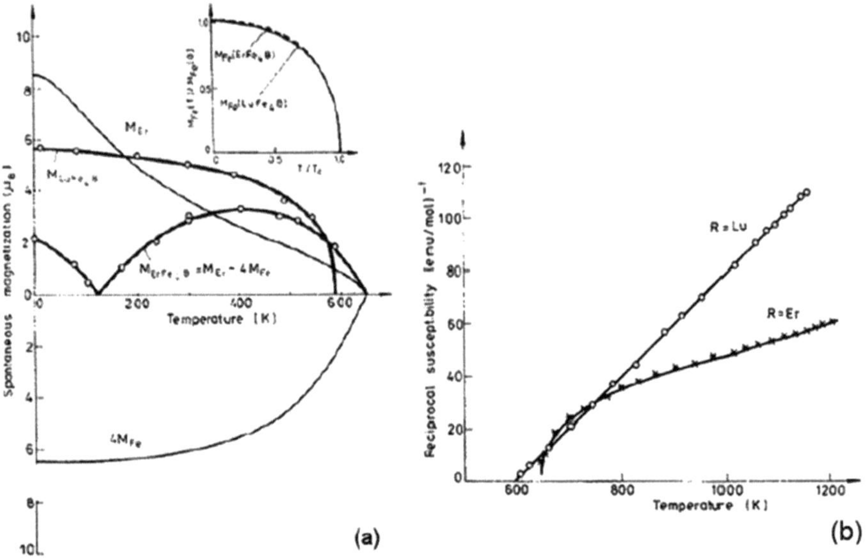

Fig. 7.11 $ErFe_4B$ and $LuFe_4B$ (**a**) thermal variations of spontaneous magnetizations. The computed temperature dependences of erbium and of iron sublattices magnetizations in $ErFe_4B$ are given by solid lines. In inset are shown, in reduced coordinates, the temperature dependences of $LuFe_4B$ and iron sublattice magnetizations in $ErFe_4B$, respectively. The temperature dependences of reciprocal susceptibilities of $ErFe_4B$ and $LuFe_4B$ are given in (**b**) [86J1]. 1 emu/g = 1 Am^2/kg

The crystal structures of a large number of pseudoternary $RFe_{4-x}Co_xB$ borides were investigated as well as the composition ranges where solid solutions are formed-Sect. 8.6. We note that in the $SmFe_{4-x}Ni_xB$ system, solid solutions are formed only for $x \geq 2$ [84O1, 84S5, 87H1].

The RFe_4B amorphous ribbons with R = Pr to Tm, $Er_{0.5}Nd_{0.5}$ and $Er_{0.5}Sm_{0.5}$ were prepared by melt spinning [87A2]. From these ribbons, only $ErFe_4B$ crystallized in hexagonal $CeCo_4B$-type structure. The $RCo_{4-x}Fe_xB$ amorphous ribbons with R = Nd ($x \leq 2$), R = Sm ($x \leq 3$) and R = Er($x \leq 4$) were also obtained by melt spinning and thermally treatment at T = 873 K [88A2]. The maximum coercivity, $\mu_0H_c \cong$ 1.7 T, at RT, was obtained in $SmCo_2Fe_2B$ alloy.

The RFe_4B compounds with R = Er and Tm are ferrimagnetically ordered, the iron and rare earth magnetizations being antiparallel oriented. According to Néel classification [48N1], the temperature dependences of magnetizations of RFe_4B compounds with R = Er and Tm are of V-type. The two sublattices magnetizations compensate at temperatures $T_{comp.}$ = 122 K for R = Er and ≅30 K when R = Tm. The temperature dependences of the R and Fe sublattices as well as of resultant magnetizations, were calculated by using the mean field approximation. A good agreement with the experimental data is obtained—Fig. 7.11a. The iron moments at 6*i* and 2*c* sites have different values, as evidenced by ^{57}Fe Mössbauer spectroscopy. The reciprocal susceptibilities, of $ErFe_4B$ follows a non-linear temperature dependence, characteristic for ferrimagnetic type ordering—Fig. 7.11b [86J1].

The $LuFe_4B$ is a ferromagnet; the magnetic susceptibilities follow a Curie–Weiss type behavior [86J1]—Fig. 7.11b. The magnetizations of $LuFe_4B$, in reduced coordinates M(T)/M(0) vs T/T_c, follows similar dependence as that of iron sublattice magnetization in $ErFe_4B$—Fig. 7.11a inset. The mean ratio $S_p/S_o = 1.1$–1.3, suggests that iron, in this series, has a relative small degree of itineracy.

The iron and rare-earth sublattices in RFe_4B compounds have uniaxial anisotropy, the easy axis of magnetization being oriented parallel to the **c**-axis. The anisotropy fields at RT, are of $\mu_0H_a = 2$–3 T when R = Lu and 8 T as R = Er ($a_J > 0$). High anisotropy fields were evidenced when iron is partially substituted by nickel, in $SmFe_{4-x}Ni_xB$ borides [84O1, 84S5]—Table 7.22.

The locations of iron in RFe_4B compounds were investigated mainly by Mössbauer spectroscopy. The ^{57}Fe spectra of $ErFe_4B$, in earlier studies, have been fitted assuming iron distributed in 2*c* and 6*i* sites, as in fully ordered $CeCo_4B$ lattice [85V1, 85V2]. The following ^{57}Fe Mössbauer studies, on RFe_4B compounds, evidenced a more complex crystallographic arrangement [85V3]. The spectrum of $TmFe_4B$ was fitted starting from a statistical distribution of Fe and B atoms over 6*i*, 2*d* and 2*c* sites, the latter two sites being assumed to be similar. Depending on the number of Fe and B nearest neighbors, different coordinations will exist, designed as $6(i)_n$ and $2(c)_m = 2(d)_m$, where n and m represent the number of nearest neighbors B atoms of a given coordination. The model involves the presence of 8 subspectra—Table 7.23. The B atoms are distributed more or less statistically over the 3*g* site, which correspond to the fully Fe coordinated 6*i* site ($6i_0$), in RFe_4B series. The hyperfine fields decrease in the sequence $6i_1$, $6i_2$ and $6i_3$ sites having respectively, one, two and three iron atoms less than in fully coordinated $6i_0$ sites. Similar decrease of ^{57}Fe hyperfine fields was shown for $(2c_1 + 2d_1)$, $(2c_2 + 2d_2)$ and $(2c_3 + 2d_3)$ sites, when lowering the number of Fe atoms in the first coordination shell of $(2c_0 + 2d_0)$ ones—Table 7.23. These data excluded the possibility that these compounds crystallize in the fully ordered $CeCo_4B$ structure [85V3].

The ^{57}Fe Mössbauer spectra of $ErFe_4B$ have been analysed assuming a binomial distribution of boron atoms [88Z3]. This involved the presence of six subspectra for iron at 6*i* site and five at (2*c*, 2*d*) positions. The effective hyperfine field decreased linearly as the number of boron atoms, in a given site, increases. In this model, an average site occupation of the unit cell, for 6*i* (5.8 Fe + 0.2 B) and (2*c*, 2*d*) sites by (2.6 Fe, 1.4 B) was obtained, indicating a preference for boron to occupy the (2*c*, 2*d*) sites. The above analyses suggest that the disordered structure of RFe_4B compounds may be assumed as derivative either of $CeCo_4B$- or $CaCu_5$-type lattices.

According [88G5, 89G1], the ^{57}Fe Mössbauer spectra of RFe_4B compounds with R = Er, Tm and Lu, at RT, look similar except for slight differences in the relative intensities of the two peaks attributed to Fe2*c* and Fe6*i* sites [89G1]. The ratio of the intensities of the two subspectra attributed to 2*c* and 6*i* sites, of $\cong 1/10$, is different from that expected from composition (1/3)—Table 7.23.

The ^{166}Er Mössbauer spectrum of $ErFe_4B$, at 4.2 K, evidenced near the same hyperfine parameters at Er1*a* and Er1*b* sites, close to the values characteristic of the Er^{3+} free ion. The erbium ground state has a dominant contribution from $J_Z = 15/2$

Table 7.22 Magnetic properties of RFe_4B compounds

Compound	Magnetic structure	M_s (μ_B/f.u.)	M_s (μ_B/Fe atom)	T_c (K)	C (emuK/f.u.)	$M_{eff.}$ (μ_B/f.u.)	θ (K)	μ_0H_a (T)		References
								T = 77 K	T = 295 K	
$ErFe_4B$	FIM eam ‖ **c**	2.10^{a}	1.60	640	17.03	3.32		6.25	8.18	[86J1]
$ErFe_4B$		$2.08^{b)}$	1.61	620^{c}					6.3	[85V3]
$TmFe_4B$	FIM	$1.16^{b)}$		610					4.0	[85V3]
$TmFe_4B$	FIM	$0.75^{d)}$	1.5							[85V1]
$TmFe_4B$	FIM			590						[88Z3]
$LuFe_4B$	FM eam ‖ **c**	5.82^{a}	1.46	586	5.10	3.18	596	2.14	2.14	[86J1]
$LuFe_4B$	FM	$5.47^{b)}$	1.37	580					2.5	[85V3]
$LuFe_4B$	FM	4.97^{e}	1.24	567					3.3^{e}	[84O1, 84S5]
$LuFe_4B$	FM	6^{d}	1.50							[85V1]
$LuFe_4B$	FM eam ‖ **c**								$K_1 = 5.6{\cdot}10^5$ J/m^3	[88T3]
$PrFe_2Ni_2B$	FM	4.76^{f}						0.234^{h1}		[88S3]
$NdFe_2Ni_2B$	FM	5.63^{f}						0.151^{h1}		[88S3]
$SmFe_2Ni_2B$	FM	3.9^{g}		495				49.0^{h2}		[84S5]
$SmFe_2Ni_2B$	FM	2.52^{f}						5.6^{h1}		[88S3]
$SmFe_1Ni_3B$	FM	$2.0^{g,f}$		404				62.8^{h2}		[84S5]
$GdFe_2Ni_2B$	FIM	3.40^{f}						0.023^{h1}		[84O1, 88S3]
$DyFe_2Ni_2B$	FIM	4.00^{f}						0.281^{h1}		[88S3]
$ErFe_2Ni_2B$	FIM	3.88^{f}						3.802^{h1}		[88S3]

[a] T = 4.2 K, μ_0H = 2 T; [b] T = 4.2 K, μ_0H = 1.8 T; [c] T = 620 K [88Z3]; [d] T = 5 K, μ_0H = 1 T; [e] T = 300 K, μ_0H = 1 T, extrapolated at T = 4.2 K; [f] T = 4.2 K, μ_0H = 8 T; [g] T = 4.2 K, μ_0H = 8.5 T; [h1] Coercive field and [h2] Anisotropy field, T = 4.2 K

Table 7.23 Data obtained by Mössbauer spectroscopy on RFe_4B compounds

Compound	T (K)	Nucleus	Site	δ (mm/s)	ΔQ (mm/s)	B_{eff} (T)	References
$ErFe_4B^c$	77	$^{57}Fe^a$	2*c*	−0.04(5)	−0.55(5)	24.6(1.0)	[85V1]
			6*i*	−0.04(5)	0.03(5)	18.5(1.0)	
$ErFe_4B$	RT		2*c*			22.88	[07M1]
			6*i*			17.13	
$TmFe_4B^c$	77		2*c*	−0.07(5)	−0.55(5)	24.2(1.0)	[85V1]
			6*i*	−0.06(5)	0.03(5)	18.1(1.0)	
$TmFe_4B$	300		$6i_0$ $6i_1$ $6i_2$ $6i_3$ $2d_0+2c_0$ $2d_1+2c_1$ $2d_2+2c_2$ $2d_3+2c_3$	−0.2 −0.2 −0.2 −0.2 −0.3 −0.2 −0.2 −0.3	−0.5 0.1 0.0 −0.2 0.0 0.0 −0.3 0.0	22.8 18.3 15.5 10.5 19.0 17.3 16.9 14.5	[85V3]
$LuFe_4B^c$	77		2*c* 6*i*	−0.09(5) −0.02(5)	−0.66(5) 0.03(5)	20.4 14.5	[85V1]
$ErFe_4B^d$	4.2	$^{166}Er^b$		0.37(21)	10.70(96)	854.5(13.2)	[85V1]

[a] Relative to α-Fe; [b] ^{166}Ho source in $Y_{0.6}Ho_{0.4}H_2$; [c] The relative intensities of 2*c*/6*i* population was 1/3; the B atoms replace Fe randomly the 2*c* sites; [d] The hyperfine parameters for occupancy of the 2*c* site with B/Fe = 3/7, instead of 1/3

level. Such a situation occurs when the exchange interaction is much larger than the crystal field interaction [85V1].

The substitutions of Co by Fe in $RCo_{4-x}Fe_xB$ compounds has detrimental effect on the 3d anisotropy, the e.a.m, respectively. Assuming the magnetization parallel to the hexagonal axis in $ErCo_{1-x}Fe_xB$ compounds, the ^{57}Fe Mössbauer spectra could be satisfactorily interpreted in terms of subspectra associated with axial site 2*c* and orthorhombic site 6*i* [89G1]. The $PrCo_{1-x}Fe_xB$ alloys have planar anisotropy [86J2].

The Curie temperatures of RFe_4B [87A2], as well as of pseudoternary $RCo_{4-x}Fe_xB$ with R = Nd, Sm and Er [88A2] amorphous ribbons, are smaller than those of the crystalline samples. The saturation magnetization of amorphous $ErFe_4B$ (4.9 μ_B/f.u.), is more than two times higher than of crystalline sample.

7.9 Metastable R–Fe–B Compounds

The $R_2Fe_{14}B$ compounds are the most important magnetic materials used in manufacturing high energy permanent magnets [98B2]. These magnets are produced either by rapid solidification techniques or by sintering method [84C2, 84S1]. According to R–Fe–B phase diagrams, the $R_2Fe_{14}B$ compounds are formed via a peritectic reaction between the pro-peritectic γ-Fe and residual liquid, if the alloy composition is close to the stoichiometric 2/14/1 composition [86S3]. When casting the corresponding alloy, the α-Fe, which was transformed from γ-Fe phase, remains as a soft magnetic component. In order to suppress the peritectic reaction, two methods can be used: (1) rapid cooling of the melt, by melt spinning, or gas atomizing, with a rate higher than that of the nucleation of the pro-peritectic phase; (2) undercooling the melt deeply below the peritectic temperature achieved when using electro-magnetic levitation and free falling in a drop tube [01G2, 02O1]. From the above amorphous materials, metastable compounds are obtained by annealing, before formation of crystalline stable phases [86B12, 88B9].

The liquid undercooling can induce crystallization of metastable compounds as a primary phase, or as an intermediate peritectic phase following primary γ-Fe formation. The underlying principles for obtaining metastable ternary phases, were summarized [88B9, 91B6]: (1) a free enthalpy which lies between those of amorphous alloy and the stable crystalline phase; (2) an activation energy, for crystallization, that is smaller than the activation energy for crystallization into the stable crystalline structure; (3) the composition of amorphous alloys should fall into the easy glass-forming region.

Metastable crystalline compounds were evidenced in the course of crystallizing the ternary amorphous and microcrystalline alloys, to final $R_2Fe_{14}B$ hard magnetic phases. One of the major factors in determining the type and nature of the metastable transition phases seems to be the R/B atomic ratio [99L2]. A high compositional ratio, $B/R = 1$ to 4, in Nd-Fe-B system, generally promotes the initial formation of Fe_3B and complex boron-rich intermetallic phases such as the cubic $Nd_2Fe_{23}B_3$, the hexagonal $NdFe_{12}B_6$ and the cubic $Y_3Fe_{62}B_{14}$—type compounds, in quenched material. Thus, in $Nd_xFe_{93-x}NbB_6$ series, the $Nd_2Fe_{23}B_3$ and $Nd_3Fe_{62}B_{14}$ metastable intermediate phases are produced during the initial crystallization process, when $4 < x < 7$ [96W1]. The $Nd_2Fe_{23}B_3$ is formed also by annealing the $Nd_8Fe_{80-x}B_{12-x}$ melt spin ribbons with $x > 0$ [99P1]. In $Pr_9Fe_{91-x}B_x$ after crystallization, the $Pr_2Fe_{23}B_3$ phase it is present for a boron content of 8–10.5 at % [06C3]. The low B/R ratio (<0.5) favours the formation of α-Fe and 2/17-type phases or its closely related substitutional RFe_7 analogous. Generally, no pure metastable compounds are formed by annealing the as prepared samples, other phases, particularly α-Fe or $R_{1+x}Fe_4B_4$ being also present [00Z1]. The α-Fe impurity cannot be avoided in obtaining $Nd_2Fe_{23}B_3$ compound [07M2]. The iron formation results primarily from the decomposition of the amorphous $Nd_2Fe_{23}B_3$ and occurs even below the crystallization temperature of the $Nd_2Fe_{23}B_3$ phase.

The formation of metastable phases in Nd–Fe–B system was first suggested [85I1]; later on, as above mentioned, three metastable phases: $Nd_2Fe_{23}B_3$, $NdFe_{12}B_6$ [86B12, 86D2, 88N1, 89C1] and $Nd_6Fe_{124-132}B_{16-18}$(H) [88A1] were found. The $R_2Fe_{23}B_3$ metastable phases were found also when R = Pr [87A3], R = Pr, Sm, Gd [88B9] and R = Ce, Nd, Sm, Gd [91K1]. The presence of metastable compound $Y_3B_{62}B_{14}$ was also reported [87D2]. A metastable phase $Nd_6Fe_{77}B_{17}$ has been obtained by crystallization from an amorphous alloy [94G1]. This phase has been shown to have the same structure as the $Nd_2Fe_{23}B_3$ compound. The so called U-phase, having $Im\bar{3}m$ space group, was also identified in R–Fe–B alloys with R = Tb, Dy, Er [91K1, 91K2].

The metastable phase $Nd_6Fe_{124\text{-}132}B_{16\text{-}18}$ was shown to have the same structure as the cubic $R_3Fe_{62}B_{14}$ compounds. Metastable phases with $TbCu_7$ hexagonal-type crystal structure were produced directly from the drop tube processed samples [00G1, 01G2, 01G3] or by electromagnetic levitation technique, under protective atmosphere [93H2, 02G1, 02G2, 02V1].

A physically-based model was proposed for the competitive precipitation of multiple phases (α-Fe, Fe_3B, $Nd_2Fe_{14}B$, $Nd_2Fe_{23}B_3$, $NdFe_4B_4$ and Fe_2B) from amorphous Nd-Fe-B matrix during crystallization [10L2]. The nucleation and growth of the different phases were calculated. The phase formation sequence during crystallization is sensitive to the heating rate. The possible role of heterogeneous nucleation was also discussed.

A general series $Nd_{1,2}Fe_{7m+n}B_{n+1}$ was proposed to describe some stable and metastable compounds, including $Nd_2Fe_{23}B_3$ (m = 3, n = 2), $Nd_2Fe_{14}B$ (m = 2, n = 0), $Nd_2Fe_8B_2$ ($NdFe_4B$) (m = 1, n = 1) and ($NdFe_{12}B_6$) (m = 1, n = 5) [91H3].

The crystal structures of metastable R–Fe–B compounds will be described in the following, except $RFe_{12}B_6$ ones, whose physical properties are presented in Sect. 7.7, together with their stable $LaFe_{12}B_6$ counterpart.

7.9.1 $R_2Fe_{23}B_3$ Compounds

The $R_2Fe_{23}B_3$ metastable compounds with R = Ce, Pr, Nd, Sm and Gd crystallize in a cubic structure having space group $I\bar{4}3d$ [87D2]—Table 7.24a. The $Nd_2Fe_{23}B_3$ has 224 atoms in a volume of 2.842(2) nm^3, 1.421(1) nm^3 for the primitive lattice, respectively. There are five non-equivalent iron sites and only one for each R and B atoms, respectively. The coordination polyhedron, of the R atom, involves all crystallographic sites (2Fe*c*, 3Fe*d*, 3Fee_1, 6Fee_2, 6Fee_3, 3B). From the above sites only Fee_1 has only one Nd neighbour, the others having two Nd in their first coordination shell. The local environments of the atoms in this lattice are given in Table 7.24a. The crystal structure of $Nd_2Fe_{23}B_3$, where the environments of four Nd atoms are given, is shown in Fig. 7.12a.

The lattice parameters of $R_2Fe_{23}B_3$ metastable compounds with R = Ce [91K1, 92S1], Pr [86B12, 86D2, 91B6, 91K1, 95H1, 02G1], R = Nd [86B12, 86D2, 91B6, 91K1, 92S1, 93G2, 06S1, 08M1], R = Sm [86B12, 86D2, 88B9, 91B6, 91K1] and R = Gd [86B12, 86D2, 88B9, 91B6, 91K1] were reported—Table 7.25. The lattice parameters of $Nd_2Fe_{23}B_3$ increase linearly up to T = 175 K, with a rate of ≅3.5 fm/K followed by a nearly three times higher rate up to T = 300 K [08M1].

During the crystallization of amorphous ribbons, obtained by melt spinning, as already mentioned, the presence of α-Fe cannot be avoided. The iron formation results primarily from the decomposition of amorphous phase, and occurs even below the $R_2Fe_{23}B_3$ crystallization temperature. The very short duration of annealing, required to crystallize $Nd_2Fe_{23}B_3$ phase, indicates that the amorphous alloys could have a local scale structure, close to that of crystalline sample [07M2]. The $R_2Fe_{23}B_3$

Table 7.24 Atomic sites in metastable R-Fe-B compounds

(a) $Nd_2Fe_{23}B_3$ having cubic structure, space group $I\bar{4}3d$ [93G2]

Atom	Site	Symmetry	x	y	z	Atomic environment [04V1]
Nd	16*c*	.3.	0.0579	0.0579	0.0579	23-vertex polyhedron $Fe_{20}B_3$
Fe1	16*c*	.3.	0.1808	0.1808	0.1808	14-vertex Frank-Kasper $Fe_{12}Nd_2$
Fe2	24*d*	2..	0.1657	0	1/4	Pseudo Frank-Kasper $Fe_{10}Nd_2B$
Fe3	$48e_1$	1	0.5623	0.5174	0.3423	11-vertex polyhedron $Fe_{11}Nd$
Fe4	$48e_2$	1	0.6237	0.6825	0.3722	14-vertex Frank-Kasper $B_2Fe_{10}Nd_2$
Fe5	$48e_3$	1	0.7273	0.5339	0.3967	Single atom B, $Nd_2Fe_{10}B$
B	24*d*	2..	1/2	3/4	0.3079	Monocapped trigonal prism BFe_6

(b) $Y_3Fe_{62}B_{14}$ having cubic structure, space group $Im\bar{3}m$ [87D2, 04V1]

Atom	Site	Symmetry	x	y	z	Atomic environment [04V1]
Y	6*b*	4/*mm.m*	0	1/2	1/2	22-vertex polyhedron $Fe_{20}B_2$
Fe1	12*d*	−4 *m*.2	1/4	0	1/2	14-vertex Frank-Kasper $Fe_{12}Y_2$
Fe2	16*f*	.3 *m*	0.191	0.191	0.191	14-vertex Frank-Kasper BFe_{13}
Fe3	48*j*	*m*.	0	0.148	0.291	14-vertex Frank-Kasper $B_3Fe_{10}Y$
Fe4	48*k*	..*m*	0.156	0.156	0.4	icosahedron $Fe_{11}Y$
B1	12*e*	4*m.m*	0.21	0	0	square antiprism Fe_4B_4
B2	16*f*	.3*m*	0.083	0.083	0.083	7-vertex polyhedron B_6Fe

(c) Atomic sites in basic and modified $R\bar{3}m$ and *P6/mmm* type structures [98T1]

Structure	Atom	Site	Occ	x	y	z
Sm_2Fe_{17} $R\bar{3}m$	Sm	6*c*	1	0	0	0.343
	Fe	6*c*	1	0	0	0.096
	Fe	18*f*	1	0.289	0	0
	Fe	18*h*	1	0.501	0.501	0.156
	Fe	9*d*	1	1/2	0	1/2
Sm_2Fe_{17} $R\bar{3}m$ Modified	Sm	6*c*	0.650	0	0	0.340
	Sm	6*c*	0.650	0	0	0
	Fe	6*c*	0.350	0	0	$1/3 - z^a$
	Fe	6*c*	0.350	0	0	$1/3 + z^a$
	Fe	18*f*	1	0.291	0	0
	Fe	18*i*	1	1/2	1/2	0.161
	Fe	9*d*	1	1/2	0	1/2
$TbCu_7$ *P6/mmm*	Sm	1*a*	0.78	0	0	0
	Fe	2*c*	1	1/3	2/3	0
	Fe	2*e*	0.22	0	0	0.306
	Fe	3*g*	1	1/2	0	1/2

(continued)

Table 7.24 (continued)

(c) Atomic sites in basic and modified $R\bar{3}m$ and *P6/mmm* type structures [98T1]						
Structure	Atom	Site	Occ	x	y	z
$TbCu_7$ *P6/mmm* Modified	Sm	1*a*	0.65–0.78	0	0	0
	Fe	2*e*	0.35–0.22	0	0	0.306
	3Fe	6*l*	1/3	0.292	0.584	0
	Fe	3*g*	1	1/2	0	1/2

[a] z = 0.096

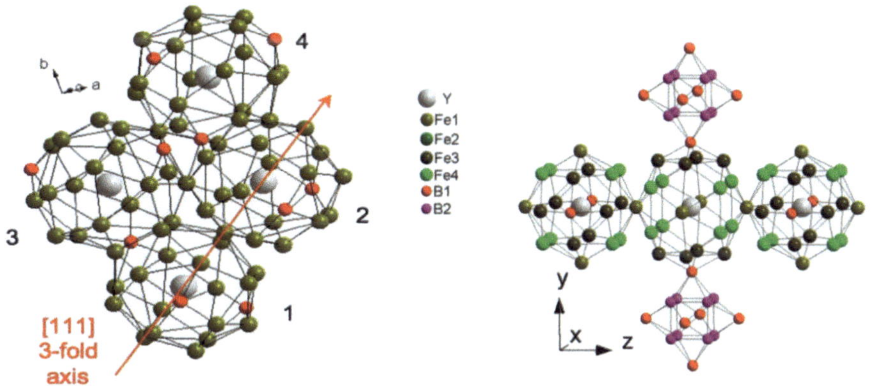

(a) (b)

Fig. 7.12 Crystal structures of **a** $Nd_2Fe_{23}B_3$: four Nd atoms and their environments, Nd-blue, Fe-green, B-red; **b** $Y_3Fe_{62}B_{14}$: Y environment as well as two boron assembles and two from four adjacent environments to Y. The sites notations are also given [08M1]

phase, during crystallization, is located at the periphery of the easy glass-forming region, glass-formation becoming easier when there is a high boron concentration [88B9]. The as-quenched structure of the ribbons is varied by changing the quenching rate during the melt spinning process [00W1]. In as-quenched ribbons, the volume fraction of crystalline phases, which may act as nucleation centres, during the crystallization of amorphous phase, decreases on increasing the initial quenching rate. The paucity of nucleation sites during the crystallization of highly overquenced ribbons could lead to course-grained microstructure. Thus, the occurrence of $R_2Fe_{23}B_3$ metastable phase, during annealing, in the ribbons melt spin at higher speeds, enhances the formation temperature of $R_2Fe_{14}B$ and α-Fe, final mixture and consequently leads to grain coarsening. A sharp concentration gradient of Nd atoms, developed by the initial nucleation and growth of α-Fe during crystallization, reduced or eliminates the thermodynamic driving force for compositional nucleation [00W1] and thus metastable $R_2Fe_{23}B_3$ phase, with a small curvature of the free energy, is more suitable to form prior to the $R_2Fe_{14}B$ hard magnetic phase.

Table 7.25 Space group and lattice parameters of R-Fe-B metastable compounds

Compound	T (K)	Space group	Lattice parameters (nm)		References
			a	c	
$Ce_2Fe_{23}B_3$	RT	$I\bar{4}3d$	1.416		[92S1]
$Ce_2Fe_{23}B_3$	RT	$I\bar{4}3d$	1.414		[91K1]
$Pr_2Fe_{23}B_3$	RT	$I\bar{4}3d$	1.418		[86D2]
$Pr_2Fe_{23}B_3$	RT	$I\bar{4}3d$	1.418		[91B6]
$Pr_2Fe_{23}B_3$	RT	$I\bar{4}3d$	1.417		[92S1]
$Nd_2Fe_{23}B_3$	RT	$I\bar{4}3d$	1.419		[86B12, 86D2, 91B6]
$Nd_2Fe_{23}B_3$	RT	$I\bar{4}3d$	1.41612		[93G2]
$Nd_2Fe_{23}B_3$	RT	$I\bar{4}3d$	1.418		[91K1]
$Nd_2Fe_{23}B_3$	RT	$I\bar{4}3d$	1.416		[92S1]
$Nd_2Fe_{23}B_3$	RT	$I\bar{4}3d$	1.4165(2)		[08M1]
$Nd_2Fe_{23}B_3$	2	$I\bar{4}3d$	1.4141(2)		[08M1]
$Sm_2Fe_{23}B_3$	RT	$I\bar{4}3d$	1.406		[86B12, 86D2, 88B9, 91B6]
$Sm_2Fe_{23}B_3$	RT	$I\bar{4}3d$	1.413		[91K1]
$Gd_2Fe_{23}B_3$	RT	$I\bar{4}3d$	1.411		[86B2, 86D2, 88B9, 91B6]
$Gd_2Fe_{23}B_3$	RT	$I\bar{4}3d$	1.412		[91K1]
$Pr_3Fe_{62}B_{14}$	RT	$Im\bar{3}m$	1.234		[95H1]
$Pr_6(Fe,Nb)_{88}B_6$	RT	$Im\bar{3}m$	1.234		[95W1]
$Nd_3Fe_{62}B_{14}$	RT	$Im\bar{3}m$	1.235		[94W1]
$Nd_6(Fe,Nb)_{88}B_6$	RT	$Im\bar{3}m$	1.231		[95W1]
$Nd_{4.5\pm1}Fe_{82.5\pm5}B_{12.5\pm3.5}$	RT	$Im\bar{3}m$	1.238		[88A1]
$Gd_3Fe_{62}B_{14}$	RT	$Im\bar{3}m$	1.2192(2)		[97C4]
$Dy_3Fe_{62}B_{14}$	RT	$Im\bar{3}m$	1.235		[95H1, 95W1]
$Dy_3Fe_{62}B_{14}$	RT	$Im\bar{3}m$	1.234(1)		[00R1]
$Dy_6(Fe,Nb)_{88}B_{14}$	RT	$Im\bar{3}m$	1.235		[95W1]
$Y_3Fe_{62}B_{14}$	RT	$Im\bar{3}m$	1.235		[87D1, 88B9, 89C1, 91B6]
$Y_3Fe_{62}B_{14}$	RT	$Im\bar{3}m$	1.231		[95H1]
$Y_3Fe_{62}B_{14}$	RT	$Im\bar{3}m$	1.2357(2)		[08M1]
Tb-U phase[a]	RT	*Im*3*m*	1.238		[91K1]
Dy-U phase[a]	RT	*Im*3*m*	1.236		[91K1]
Er-U phase[a]	RT	*Im*3*m*	1.234		[91K1]
$Nd_{16}Fe_{76}B_3$[b]	RT		0.8567	1.247	[05G1]
$Nd_{14}Fe_{79}B_7$[b]	RT		0.8578	1.256	[04O1]
$Nd_{11.8}Fe_{82.3}B_{5.9}$[b]	RT		0.8561	1.264	[02G1]
$Nd_{11.8}Fe_{82.3}B_{5.9}$[b]	RT		0.8548	1.264	[05S3]

(continued)

Table 7.25 (continued)

Compound	T (K)	Space group	Lattice parameters (nm)		References
			a	c	
$Nd_{10}Fe_{85}B_5$	RT		0.8558	1.265	[07G1]
$Nd_{10}Fe_{85}B_5$	RT		0.8552	1.268	[07G1]
Nd_2Fe_{17}	RT	$R\bar{3}m$	0.8578	1.2462	[90B1]
$Nd_{14}Fe_{79}B_7$	RT	$R\bar{3}m$	0.8647	1.2555	[04V2]
$Tb_6Fe_{87}NbB_6$	RT	$TbCu_7$	0.480	0.423	[95H1]
$Tb_2Fe_{17}B_x$	RT	$TbCu_7$	0.4801	0.4225	[95W1]

[a] Composition $R_6Fe_{74}B_{20}$; [b] $Nd_2Fe_{17}B_x$ phase

The activation energies of different phases, during the crystallization process of $Pr_8Fe_{86-x}Zr_xB_6$ amorphous alloys with $0 \leq x \leq 2$, were investigated [02L1]. When the crystallization temperature was below 900 °C, the activation energy of α-Fe remained relatively constant of ≅300 kJ/mol, if the α-Fe content was below 8%. At the beginning of the crystallization, the activation energy of $Pr_2Fe_{23}B_3$ and $Pr_2Fe_{14}B$ phases were ≅510 kJ/mol and ≅726 kJ/mol, respectively. The activation energies decline when increasing the crystalline fraction.

As already mentioned, there it is difficult or even impossible to obtain, by crystallization of melt spin ribbons, a single $R_2Fe_{23}B_3$ metastable phase. During the annealing of $Nd_xFe_{81.5-x}B_{18.5}$ ribbons with $7 \leq x \leq 9$, at T = 670 °C, 10 min, metastable $Nd_2Fe_{23}B_3$ phase is formed, in addition to stable $Nd_2Fe_{14}B$ and $Nd_{1.1}Fe_4B_4$ compounds [97C3]. The same behavior was shown in the multi-component annealed $(Nd_{0.95}La_{0.05})_{9.5}(Fe_{1-x}Cr_x)_{81.5}B_9$ amorphous samples, where in addition to metastable 2/23/3 and 3/62/14 compounds, the $(Nd,La)_2Fe_{14}B$/α-Fe/Fe_3B phases were also shown [97C2]. The crystallization sequence, of the above phases, in the $Nd_{4.5}Fe_{77.5}B_{18}$ amorphous ribbons, depends on the heating rate [06S1]. At low heating rate, the $Nd_2Fe_{23}B_3$ phase is formed, whereas the faster heating leads to the crystallization of the hard magnetic phase. After annealing 5 min at T = 790 °C of the amorphous $Nd_{9.2}Fe_{61.65}B_{21.15}Nb_8$ ribbon, the content of $Nd_2Fe_{23}B_3$ phase was 11.7% [13S2]. Under slow heating rate of 10 °C/min, up to annealing temperature, T_a, the metastable $Nd_2Fe_{23}B_3$ and $Nd_3Fe_{62}B_{14}$ compounds are present, in the initial crystallization stage of the $Nd_7Fe_{86}Nb_1B_6$ ribbon [00G2]. When the heating rate increased to 60 °C/min, the above phases decompose at T_a temperature. The ND study of $Nd_5Fe_{74}Cr_2B_{18}$ melt-spin amorphous alloys, upon heating with a rate of 0.10 K/min, revealed the intermediate $Nd_2Fe_{23}B_3$ and Fe_3B phases, while for a rate of 78 K/min the $Nd_2Fe_{14}B$ and Fe_3B compounds were formed [09S1]. The $R_2Fe_{23}R_3$ phases with R = Ce, Pr and Nd were obtained by melt spinning and heat treated at T = 610–630 °C [92S1].

The non-conventional methods were also used in order to obtain metastable phases in Nd–Fe–B system. The $Nd_2Fe_{23}B_5$ compound was obtained by crystallization of $(Nd_{0.1}Fe_{0.9})_3B$ amorphous alloy at T = 923 K, during 5 min, in the presence of electric field of 300 kV/m. This compound was also the main phase obtained by annealing at T = 1023 K (higher than decomposition temperature), for 5 min, under the action of the above mentioned electric field [04L1, 05L1]. The $Nd_2Fe_{23}B_3$ phase was also synthesized by pulsed-plasma stream-interaction with the surface of $Nd_8Fe_{87}B_7$ crystalline alloy [06G1].

The phase evolution of melt spun $Pr_2Fe_{23}B_3$ and $Pr_2(Fe_{0.975}Ti_{0.025})_{23}B_3$ ribbons, annealed at T > 700 °C evidenced the presence of α-Fe, $Pr_2Fe_{23}B_3$ and Fe_3B, while the samples heated at T = 800 °C, of $Pr_2Fe_{14}B$ and α-Fe phases [03C3]. The formation of $Pr_2Fe_{23}B_3$ phase, in annealed $Pr_{4.5}Fe_{77}B_{18.5}$ amorphous ribbon, occurs immediately after Fe_3B formation [01J1]. In $Pr_xFe_{90-x}B_{10}$ melt spun ribbons after annealing, $Pr_2Fe_{23}B_2$ compound is formed, when x = 10, in addition to $Pr_2Fe_{14}B$ and Fe_3B phases [03C4]. The $Pr_2Fe_{23}B_3$ metastable phase, was formed in melt spun $Pr_9Fe_{91-x}B_x$ alloys, for the composition range $8 \leq x \leq 12$ [03C5]. A fine mixture of $Pr_2Fe_{23}B_3$ compound, with the 2/14/1 and bcc-(Fe,Co,Ni) phases, was shown in $Pr_{10}(Fe_{1-2x}Co_xNi_x)_{84}B_6$ melt spun and crystallized ribbons, when $0.10 \leq x \leq 0.15$ [12D2]. Generally, in amorphous ribbons, crystallization of $R_2Fe_{23}B_3$ phases occurs immediately after Fe_3B formation.

The presence of $R_2Fe_{23}B_3$ phases with R = Pr, Nd, is suppressed by slight substitutions of iron by Ti [03C3, 06C2, 06C4], Mo [01C2], Cr, Nb, V, Ti, Zr [04C1–04C3, 06C1, 07Z1], Zr [07W1], or Cr [97C2, 97H1, 98C2, 99L2]. The addition of M = Ti and Cr, promotes the crystallization of $Nd_2Fe_{14}B$ compound, in $Nd_4Fe_{77.5-x}M_xB_{18.5}$ alloys [04H2]. In the Ti/C co-substituted sample, the C prefers to conjunct with Ti, consuming a part of titanium contained in sample [06C4]. As a results there is a retarding effect of Ti in suppressing the formation of $Pr_2Fe_{23}B_3$ phase. In $Nd_{4.5}Fe_{73}B_{18.5}Co_3Ga$ alloy, in the early stage of crystallization, Co and Ga are rejected from primarily particles of soft magnetic Fe_3B phase and partitioned in the amorphous matrix. In the fully crystallized sample, by annealing at T = 580 °C, for 10 min, the Co enters in $Nd_2Fe_{23}B_3$ and $Nd_2Fe_{14}B$ phases, while Ga substitutes Fe only in $Nd_2Fe_{14}B$ compound [98P1]. The crystallization of amorphous Nd–Fe–B alloys can be modified by alloying with Nb. Such alloy produces numerous dispersed nuclei of crystallization and formation of uniform nanocrystalline structure [06Y1]. The slight substitution of Fe by M = Nb, Zr in $Pr_2(Fe_{1-x}M_x)_{23}B_3$ can effectively supress the metastable 2/23/3 and Fe_3B phases and thus the formation of $Pr_2Fe_{14}B$ + α-Fe mixture [04C2, 04C3]. In $Pr_{8.5}Fe_{81.5}M_2B_{10}$ ribbons, when M = V, Cr, the annealed samples are composed of $Pr_2Fe_{23}B_3$ phase, while the presence of M = Nb, Zr, Ti supresses this phase [07Z1] due to their substitution at R site, as well as to the formation of M-borides which consumes some part of B, which hinders the generation of $Pr_2Fe_{23}B_3$ rich boron phase.

The high-pressure torsion deformation on $Nd_9Fe_{85}B_6$ melt spin alloy, was found to suppress the formation of metastable $Nd_2Fe_{23}B_3$ or $NdFe_7$ phases, precipitated upon annealing the melt-spun amorphous alloy [07P1].

The magnetic properties of $R_2Fe_{23}B_3$ metastable compounds with R = Ce [92S1], R = Pr [86D2, 87A3, 88B9, 91B6, 92S1, 02J1], R = Nd [86D2, 87D2, 88B9, 89C1, 91B6, 92S1, 07M2, 08M1], R = Sm [86D2], R = Gd [86D2] and R = Dy [91H3] were reported—Table 7.26. The compounds with R = Ce, Pr, Nd and Sm are

Table 7.26 Magnetic properties of R-Fe-B metastable compounds

Compound	Magnetic structure	M_s (μ_B/f.u.)	T_c (K)	$\mu_o H_c$ (T)	References
$Ce_2Fe_{23}B_3$	FM	49.8[a]	613		[92S1]
$Pr_2Fe_{23}B_3$	FM		644		[86D2, 88B9, 91B6]
$Pr_2Fe_{23}B_3$	FM	52.7[a]	642		[92S1]
$Pr_2Fe_{23}B_3$	FM		573		[02J1]
$Nd_2Fe_{23}B_3$ (ND)	FM, RT Nd: $M_c = 1.08\ \mu_B$ Fe: $M_c = 1.96\ \mu_B$, $M_d = 1.55\ \mu_B$, $M_{e_1} = 2.26\ \mu_B$, $M_{e_2} = 2.15\ \mu_B$, $M_{e_3} = 1.93\ \mu_B$	48.79[b]	659		[93G2]
$Nd_2Fe_{23}B_3$	FM	52.9[a]	659	0.093[a]	[92S1]
$Nd_2Fe_{23}B_3$	FM	49.1	655		[86D2, 88B9, 89C1, 91B6]
$Nd_2Fe_{23}B_3$ (amorphous)	FM	51.8[c]	517		[07M2]
$Nd_2Fe_{23}B_3$ (crystalline) (ND)	FM, T = 2 K Nd: $M_c = 2.7\ \mu_B$ Fe: $M_c =$ M_{e_2}=2.04(80)μ_B, $M_{e_1} = M_d =$ 2.14(40)μ_B, $M_{e_3} = 1.10(65)\ \mu_B$	54.0[c]	660		[07M2, 08M1]
$Nd_2Fe_{23}B_3$		36[d]			[91H3]
$Sm_2Fe_{23}B_3$	FM		663		[86D2]
$Gd_2Fe_{23}B_3$	FM (?)		689		[86D2]
$Dy_2Fe_{23}B_3$ (MS))	FIM; $\sum M_{Fe}$= 41.17 μ_B[e]	29.4[a]	433		[91H3]
$Pr_3Fe_{62}B_{14}$ ($Pr_6Fe_{87}NbB_6$)	FM		473		[95H1, 95W1]
$Nd_3Fe_{62}B_{14}$ ($Nd_6Fe_{87}NbB_6$)	FM		473		[94W1, 95H1]
$Nd_3Fe_{62}B_{14}$	FM		481		[02M1]
$Nd_3(Fe,Co)_{62}B_{14}$	FM		503		[02M1]
$Nd_5DyFe_{87}NbB_6$	FM		473		[01P1]
$Dy_3Fe_{62}B_{14}$ ($Dy_6R_{87}NbB_6$)	FIM		483		[95W1]

(continued)

Table 7.26 (continued)

Compound	Magnetic structure	M_s (μ_B/f.u.)	T_c (K)	$\mu_o H_c$ (T)	References
$Y_3Fe_{62}B_{14}$ [h]	FM, T = 4 K $M_{Fe} = 1.91(5)\ \mu_B$, crystalline $M_{Fe} = 2.05(5)\ \mu_B$, amorphous		521		[08M1]
$Y_3Fe_{62}B_{14}$	FM $M_{Fe} \cong 1.80\ \mu_B$		510		[87D2]
$Pr_{12}Fe_{17}B_x$	FM		483		[02J1]
$Nd_2Fe_{17}B_x$	FM		440		[96G1]
$Nd_{14}Fe_{79}B_7$ [f]	FM		327–338		[04O1]
$Nd_{10}Fe_{85}B_5$ [f]	FM		363		[07G1]
$Nd_{10}Fe_{85}B_6$ [f]	FM		353		[07G1]
$Nd_{14}Fe_{79}B_7$ [f]	FM	86.7emu/g [g]		0.014	04O2
Nd_2Fe_{17}	FM	30.0 [i]	325		[90B1]
$Sm_2Fe_{17}B_{0.2}$	FM		388 [j]	2.9	[95H3]
$SmFe_{12}B$ ($TbCu_7$-type)	FM	16.4 [k]			[16Z1]
$Tb_2Fe_{17}B_x$	FIM		513		[95H1, 95W1]
$Y_2Fe_{17}B_3$ (BS)	Fe: $M_c = 2.73\mu_B$, $M_d = 2.42\mu_B$, $M_h = 2.09\mu_B$, $M_f = 1.57\ \mu_B$, B: $M = -0.12\mu_B$				[97A2]

[a] $T = 1.5$ K $\mu_0 H = 5$ T; [b] Neutron diffraction data; [c] $T = 4$ K, $\mu_0 H = 10$ T; [d] RT, $\mu_0 H = 0.8$ T; [e] From ^{57}Fe Mössbauer spectroscopy; [f] $R_2Fe_{17}B_x$ phase; [g] 10.1 μ_B/f.u. or 14.4 μ_B per $Nd_2Fe_{11.3}B$ formula unit; [h] $T > T_c$ C-W behavior, $M_{eff}(Fe) = 4.18\ \mu_B$/atom; [i] $T = 4.2$ K; [j] Determined from figure; [k] At RT, data from figure ($Sm_2Fe_{17}B_x$ structure)

ferromagnetically ordered, while those with R = Gd and Dy show a ferrimagnetic type behavior.

The Curie temperatures of $R_2Fe_{23}B_3$ crystalline compounds are rather low, being of 66% from that of metallic iron, although the iron content is of 82%. The mean number of Fe–Fe bonds by one iron atom is ≅10.44, 72.5% from the distances between neighbouring iron atoms being smaller than 0.264 nm [08M1]. Thus, the exchange interactions between iron atoms are weak. In addition, the distance Fe*d*-Fe*e*$_l$ in $Nd_2Fe_{23}B_3$, of 0.2398 nm, suggests possible antiferromagnetic interactions [36N1], which are not satisfied, the positive ones being dominant. These features can explain the relative low Curie temperatures of $R_2Fe_{23}B_3$ series, as shown also in $R_2Fe_{14}B$ compounds [98B2].

The Curie temperatures and the saturation magnetizations of $Nd_2Fe_{23}B_3$ are higher in crystalline phase than in amorphous one [08M1, 08M2], as evidenced also in other $R_2Fe_{23}R_3$ compounds [86B12, 90S1, 91B6]. Assuming a neodymium moment

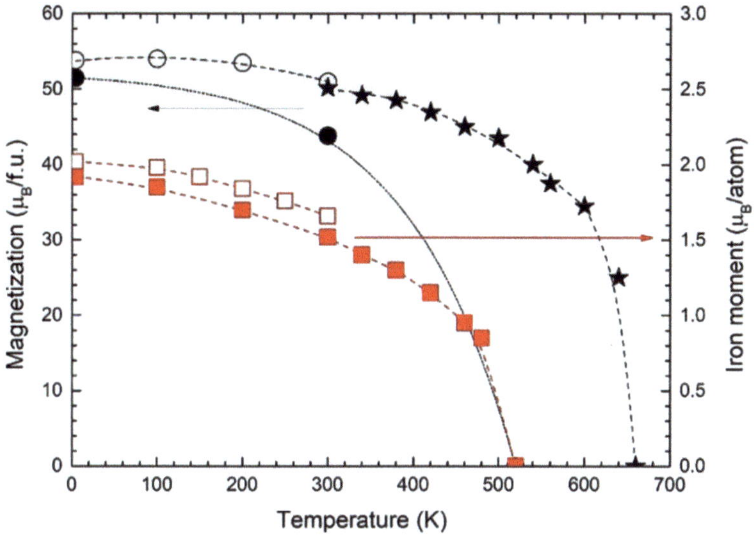

Fig. 7.13 Temperature dependences of magnetizations for amorphous (•) and crystalline (○,⋆) $Nd_2Fe_{23}B_3$ as well as of mean iron moment in amorphous (□) and crystalline (red color) $Y_3Fe_{62}B_{14}$; figure based on data from PhD thesis [08M1]

of $M_{Nd} \cong 2.7$ μ_B/atom, the mean iron moments are of ≅2.11 and 2.01 μ_B/atom, in crystalline and amorphous phase, respectively [08M1, 08M2]—Fig. 7.13. The thermal variations of magnetizations are also different. The above trends can be correlated with different iron environments in crystalline and amorphous samples.

The neutron diffraction measurements on $Nd_2Fe_{23}B_3$ [08M1], evidence that the iron moments, at T = 2 K, decrease in the sequence $M_{Fe}(d) \cong M_{Fe}(e_1) > M_{Fe}(c) \cong M_{Fe}(e_2) > M_{Fe}(e_3)$. The above sequence is somewhat different from that obtained by neutron diffraction, at ambient temperature, on the same compound [93G2]—Table 7.26. Starting from the atomic environments and Voronoi volume of different iron sites, the suggested sequence, is close to that experimentally determined [08M1].

The ^{57}Fe Mössbauer spectroscopy has been used to analyse the phases present in crystalline and amorphous R–Fe–B samples, when metastable $R_2Fe_{23}B_3$ phase is present, as for R = Pr [03C2, 03C4], R = Nd [86B12, 96R1, 97C3], R = Dy [91H3, 96R1] and R = Ho [96R1]. The ^{57}Fe Mössbauer spectrum at RT, of $Nd_2Fe_{23}B_3$, was fitted assuming five non-equivalent iron sites, having the relative intensity ratio 2/3/6/6/6 [86B12]. Although the most reports associated the higher ^{57}Fe hyperfine field with the c and e_1 sites, there are differences in the site sequence of decreasing these values. The different trends can be associated with the presence and amounts of other phases in the analysed samples. Assuming a scaling factor of 14.8 $T\mu_B^{-1}$ [79B1, 85V5], from the determined hyperfine fields, a mean iron moment of ≅2.1 μ_B/atom has been estimated. The ^{57}Fe hyperfine fields in $Dy_2Fe_{23}B_2$ are lower than in the corresponding $Nd_2Fe_{23}B_3$ isomorphous counterpart—Table 7.27.

Table 7.27 Data obtained by ^{57}Fe Mössbauer spectroscopy of R-Fe-B metastable compounds

Compound	T (K)	Site	δ (mm/s)	ΔQ (mm/s)	B_{eff} (T)	A (%)	References
$Nd_2Fe_{23}B_3$ [a]	RT	16*c*	−0.1	0	36.2		[86B12]
		24*d*	0.0	−0.2	33.5		
		48e_1	−0.1	0.0	34.8		
		48e_2	−0.2	0.1	29.8		
		48e_3	−0.1	−0.1	28.2		
$Nd_9Fe_{72.5}B_{18.5}$ [a] ($Nd_2Fe_{23}B_3$)	RT	16*c*	−0.17	0.00	35.6	7.4	[97C3]
		24*d*	−0.09	0.09	32.8	11.1	
		48e_1	−0.08	0.01	35.1	22.1	
		48e_2	0.04	−0.08	29.4	22.1	
		48e_3	−0.01	0.08	27.4	22.1	
		$Nd_{1.1}Fe_4B_4$	0.02	0.559		15.2	
$Dy_2Fe_{23}B_3$ [a,b] (amorphous)	RT	16*c*	0.038	−0.233	28.75		[91H3]
		24*d*	0.094	−0.091	31.86		
		48e_1	−0.088	−0.006	27.03		
		48e_2	−0.128	0.044	25.20		
		48e_3	0.019	−0.064	22.41		
$Nd_2Fe_{23}B_3$ [a,c]	RT	16*c*	0.01	0.40	27.8		[88A1]
		24*d*	−0.05	0.14	34.1		
		48e_1	−0.10	0	35.9		
		48e_2	0.05	−0.24	30.2		
		48e_3	−0.05	0.06	29.0		
$Nd_5DyFe_{37}NbB_6$ $(Nd,Dy)_3Fe_{62}B_{14}$ [a]	17	12*d*	0.06	0.08	37.5	12	[01P1]
		48*j*	0.11	0.01	33.9	36	
		16*f*	0.11	−0.40	36.4	9	
		48*k*	0.07	0.03	25.7	35	
		α-Fe	0.10	0.06	33.9	8	
$Gd_3Fe_{62}B_{14}$ [a,d]	300	Fe1	0.11	0.02	23.2	32	[97C4]
		Fe2	0.11	0.01	21.0	11	
		Fe3	0.04	0.02	17.5	11	
		Fe4	0.05	0.02	15.5	11	
		$Gd_{1.1}Fe_4B_4$	0.06	0.56		5	
		α-Fe(Si)-total				30	
$Nd_8Fe_{78}B_{14}$ [d,e]	RT	$Nd_2Fe_{17}B_x$					[15H1]
		6*c*	−0.086	−0.045	36.72	17.56	
		9*d*	−0.180	−0.048	35.36		
		18*h*	0.018	−0.093	30.44		
		18*f*	0.094	0.342	27.41		
		$Nd_{1.1}Fe_4B_4$	0.033	0.580		7.56	
		α-Fe	0	0	38.0	7.8	
		amorphous			29.06	67.8	

[a] Relative to α-Fe; [b] By using a ratio 14.8 T/μ_B [79B1] values of iron magnetic moments 1.94 μ_B(16*c*), 2.15 μ_B(24*d*), 1.83 μ_B(48e_1), 1.70 μ_B(48e_2) and 1.51 μ_B(48e_3) were determined [91H3]; [c] Fe_3B and amorphous phase was also present; [d] Subspectra were not correlated with Fe sites; [e] Assuming an orthorhombic type structure

7.9.2 $R_3Fe_{62}B_{14}$ Compounds

The $R_3Fe_{62}B_{14}$ metastable compounds with R = Pr, Nd, Gd, Dy and Y crystallize in a cubic structure, space group $Im\overline{3}m$ [87D2]—Table 7.24b and Fig. 7.12b. There are four inequivalent Fe sites, two for boron and only one for R. The R is bonded in a 4-coordinated geometry to twenty Fe atoms. In the structure are located groups of 14 boron atoms, eight (B2) forming a cube and six (B1) disposed above their faces [08M1]. In the $R_3Fe_{62}B_{14}$ structure, for each cubic simple lattice, one corner from four is occupied by a boron group and the remaining three by Fe atoms. The distances between iron atoms are close to those evidenced in α-Fe, excepting that between Fe*j* and Fe*k* sites, which in $Y_3Fe_{62}B_{14}$, is $d_{Fe\text{-}Fe}$ = 0.2354 nm [08M1]. According to Néel-Slater curve [36N1], antiferromagnetic interactions between these atoms are expected. These interactions are not satisfied and thus a magnetic energy is stored. The competition between elastic and magnetic energies, leads to an anomalous thermal expansion, as already evidenced in $R_2Fe_{14}B$ compounds [98B2]. The lattice parameters, of $Y_3Fe_{62}B_{14}$ phase, are not changed from T = 4 K (a = 1.2359 nm) to T = 300 K (a ≅ 1.2357 nm), suggesting, as expected, an Invar type behavior.

The metastable U-phase, proposed to be present in R–Fe–B ribbons with R = Tb, Dy and Er, after crystallization [91K1], as well as the M-phase ($Nd_6Fe_{124\text{-}132}B_{16\text{-}18}$) [88A1] were shown to have cubic structures, space group $Im\overline{3}m$, as the $R_3Fe_{62}B_{14}$ series [91K1, 91K2]—Table 7.25.

The presence of metastable $R_3Fe_{62}B_{14}$-type structure, in addition to α-Fe, was shown after the first crystallization step, T = 580–600 K in $R_6(Fe,Nb)_{88}B_6$ amorphous ribbons with R = Pr, Nd and Dy [95W1, 01P1], Gd-(Fe,Cu,Nb,Si)-B [97C4], (Nd,La)(Fe, Cr)-B [97C2], etc. The $R_3Fe_{62}B_{14}$ metastable phases were also obtained by annealing short time, the amorphous ribbons with R = Nd and Y, at T = 670 °C [86D2, 87D2, 95C3, 95C4]. This metastable phase decomposes upon heating at T = 860 °C [95C4]. The $(Nd,La)_3Fe_{62}B_{14}$ metastable phases were also evidenced by the crystallization of $(Nd_{0.95}La_{0.05})_{9.5}(Fe,Cr)_{81.5}B_9$ ribbons [97C2]. The crystallization temperatures increase as iron is partially substituted by chromium. After annealing the $Gd_5Fe_{68.5}CuNb_3Si_{13.5}B_9$ amorphous ribbons, at T = 600 °C, the content of $Gd_3Fe_{62}B_{14}$ phase was found to be of 65%, in addition to 30% α-Fe(Si) and 5% $Gd_{1.1}Fe_4B_4$ [97C4].

After partial crystallization of $Dy_6Fe_{74}B_{20}$ amorphous ribbon, at T = 450 °C, there was present tetragonal Fe_3B boride; the remaining amorphous material segregates into two amorphous "phases" enriched and impoverished in dysprosium, respectively [00R1]. The tetragonal Fe_3B phase, further transform, at T = 590 °C, in the orthorhombic Fe_3B. Metastable $Dy_3Fe_{62}B_{14}$ compound, then forms from the Dy-impoverished amorphous fraction, at T = 623 °C. At T = 762 °C, the $Dy_3Fe_{62}B_{14}$ phase, decomposes in $Dy_{1+\varepsilon}Fe_4B_4$, o-$Fe_3B$ and bcc iron.

The magnetic properties of $R_3Fe_{62}B_{14}$ metastable compounds with R = Pr [95H1, 95W1, 01P1], R = Nd [94W1, 95H1, 95W1, 01P1, 02M1], R = Dy [95H1, 95W1, 00R1, 01P1], R = Gd [97C4] and R = Y [87D2, 89C1, 01P1, 08M1] were reported. The compounds with R = Pr, Nd, Y are ferromagnetically ordered, while in

$Dy_3Fe_{62}B_{14}$, a ferrimagnetic type ordering is expected. Since pure $R_2Fe_{62}B_{14}$ phases are difficult or impossible to be obtained in pure state, little data on fheir saturation magnetizations were reported until now. As a general feature, the Curie temperatures, T_c, are rather low ($T < 520$ K), although iron content in the above series is of 78.2 at %—Table 7.26. The T_c values, in case of $R_2Fe_{23}B_3$ phases, can be also correlated with the presence of unsatisfied negative exchange interactions between iron atoms situated at distances $d < 0.245$ nm [36N1, 98B2].

The saturation magnetizations of amorphous and crystalline $Y_3Fe_{62}B_{14}$ phases were obtained in external field $\mu_0H > 2$ T [08M1]. The mean iron moments, at T = 4.2 K, determined in the amorphous phase, $M_{Fe}(a) = 2.02(5)$ μ_B/atom, is somewhat higher than that in crystalline sample $M_{Fe}(cryst) = 1.91(5)$ μ_B/atom. The magnetization determined from band structure calculation, of M = 141.187 μ_B/f.u. [20M1], correspond to a mean iron moment of 2.277 μ_B/atom, somewhat higher than the experimentally determined value. The temperature dependences of the saturation magnetizations of crystalline and amorphous samples follow the same trend (T < 300 K)—Fig. 7.13. The reciprocal susceptibility of crystalline $Y_3Fe_{62}B_{14}$, at $T > T_c$, shows a Curie–Weiss type behavior, the mean effective iron moment, $M_{eff} = 4.18$ μ_B/atom, being close to that evidenced in Y_2Fe_{17} compound (3.96 μ_B/atom) [70B1]. The ratio between the number of spins, determined from effective iron moment, S_p and the saturation one, S_0, $r = S_p/S_0 = 1.7$, is an evidence of their itinerant magnetism. The effective iron moment, determined in amorphous $Y_3Fe_{62}B_{14}$ phase, is 3.76 μ_B/atom [08M1].

By magnetic measurements only the mean value of the iron moments can be obtained. The local environments, characterizing the iron sites in $R_3Fe_{62}B_{14}$, is different and as a result the iron moments are expected to follow a sequence determined by the number of magnetic atoms situated in their first coordination shell as well as by boron: $M_{Fe}(d) > F_{Fe}(f) > M_{Fe}(k) > M_{Fe}(j)$ [08M1]. Experimental evidence on this trend can be obtained by ^{57}Fe Mössbauer spectroscopy. Although multiphases alloys are obtained, in some cases [97C4, 01P1], the ^{57}Fe hyperfine parameters of the four iron sites were determined. As seen in Table 7.27, the expected sequence for iron moments is approximately followed by that of the hyperfine fields (B_{eff} (d) > B_{eff} (f) > B_{eff} (j) > B_{eff} (k)) [01P1]. Assuming the same proportionality between hyperfine fields and iron moment, as already mentioned [79B1], a mean iron moment of 2.11 μ_B/atom was estimated in $(Nd,Dy)_3Fe_{62}B_{14}$ series, in agreement with magnetization data [08M1]. In $R_xFe_{80-x}B_{20}$ ribbons with R = Nd, Dy and Ho, a drastic change of the mean hyperfine parameters was shown, when x = 8–9 at % [96R1, 98R1]. The observed discontinuity was attributed to a strong modification of the local atomic order around the iron atoms.

The ^{57}Fe Mössbauer effect studies [95C3, 95C4, 96R1, 97C4, 98R1, 00R1], as well as zero-field spin-echo NMR on ^{11}B, ^{57}Fe and ^{89}Y [95C3, 95C4], were used to analyse the phase compositions in annealed melt spun ribbons, containing $R_3Fe_{62}B_{14}$ metastable phases.

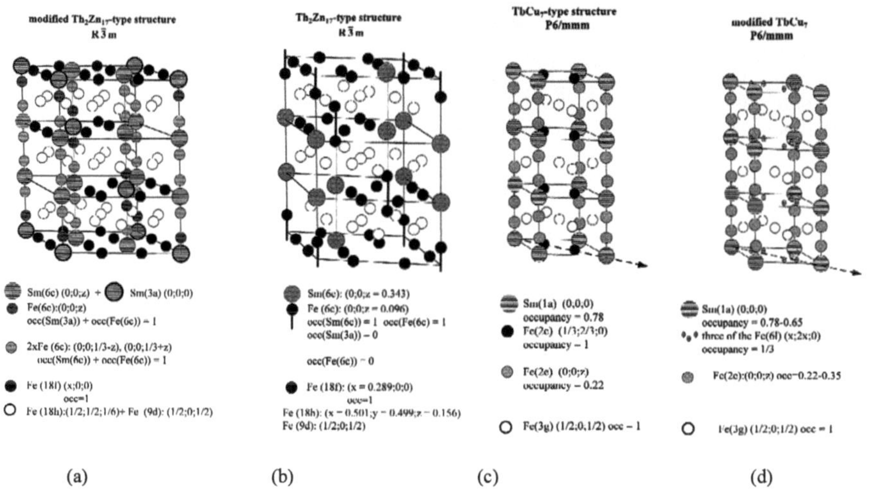

Fig. 7.14 **a** Sm_2Fe_{17} with a modified Th_2Zn_{17} type structure, **b** Th_2Zn_{17} structure, **c** $TbCu_7$ with *P6/mmm* type structure [71B1], **d** $TbCu_7$ with modified *P6/mmm* type structure [98T1]

7.9.3 $R_2Fe_{17}B_x$ Compounds

The stoichiometry of the binary iron-rich rare-earth compounds can be described by the formula $R_{1\text{-}s}Fe_{5+2s}$, as result of rare-earths substitution by dumbbell Fe–Fe pairs [71B1, 03D1]. When one R atom over three is replaced by one dumbbell pair, the resulting composition is R_2Fe_{17} (s = 0.33). The R_2Fe_{17} compounds crystallize in a hexagonal structure, $P6_3/mmc$ space group, when the dumbbell pairs are randomly distributed in lattice and in an orthorhombic $R\overline{3}m$ type structure, for the ordered substitutions. A metastable RFe_9—type structure, having *P6/mmm* space group, was found when s = 0.36 − 0.38 [03D1, 04N1, 18B2]. The last structure, through the literature, is known as of $TbCu_7$-type. In order to analyse the crystal structures of Sm_2Fe_{17} and $Sm_2Fe_{17}B_x$ compounds, prepared by mechanical alloying and subsequent annealing, a modified $SmFe_9$ ($TbCu_7$) type structure was proposed having space group *P6/mmm* [98T1]. In this structure, the Fe*2c* site is replaced by the partially and randomly occupied *6l* site (1/3), maintaining the space group *P6/mmm*—Fig. 7.14. By annealing the sample at 800 °C ≤ T ≤ 900 °C, a disordered modified $R\overline{3}m$ type structure was found, by introducing additional randomly occupied Fe*6c* and Sm*3a* sites. The above crystal structures were used to describe the $R_2Fe_{17}B_x$ metastable phases. The degree of order of Sm atoms and Fe-dumbbells along **c**-direction increases with increasing annealing temperature, T_a, and in addition, a decrease in the Fe concentration, in the unit cell [98T1].

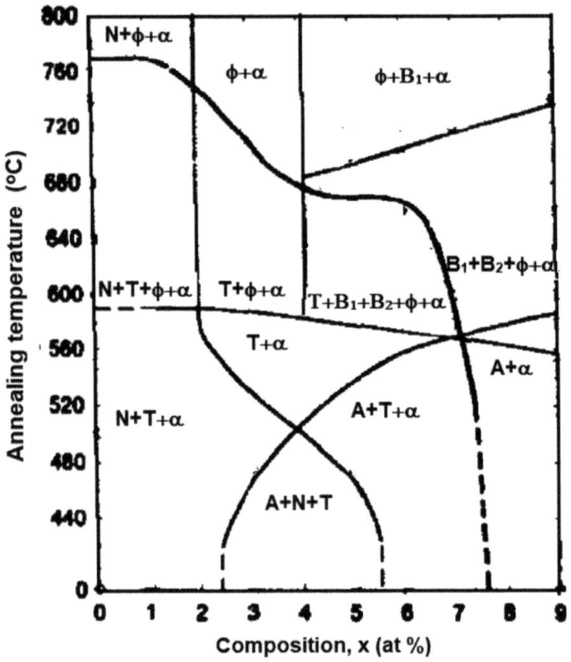

Fig. 7.15 $Nd_9Fe_{91-x}B_x$: metastable phase diagram for rapidly quenched ribbons, annealed for 10 min [96G2]. A-amorphous phase, N-Nd_2Fe_{17}, T-$TbCu_7$, α-α-Fe, ϕ-$Nd_2Fe_{14}B$, B_1-Fe_3B, B_2-$Nd_2Fe_{23}B_3$. The upper bound region, in which the 1/7 phase exists is drawn by heavy lines

The metastable $TbCu_7$ type structure has been identified in R–Fe–B alloys elaborated by melt spinning, electromagnetic levitation, drop tube processing, gas-atomized method or by mechanical milling. Some aspects of their processing will be shortly mentioned.

The $Pr_8Fe_{86}B_6$ amorphous ribbons, during annealing treatment, presents a multi-stage microstructural evolution, where the metastable $TbCu_7$-type phase precipitates firstly and then transforms into metastable $Pr_2Fe_{23}B_3$ one, prior to the formation of $Pr_2Fe_{14}B$ and α-Fe phases [02J1]. The phase diagram of $Nd_9Fe_{91-x}B_x$ melt spun and over quenched ribbons, annealed for 10 min, at different temperatures showed that in the as-quenched ribbons, small amounts of an amorphous phase and α-Fe were shown, in addition to metastable 1/7 phase ($x \leq 7$) and the Nd_2Fe_{17} one ($x \leq 5$) [96G2]—Fig. 7.15. In the composition range $3 \leq x \leq 5$, the amount of 1/7 phase increases simultaneously with the crystallization of amorphous phase. The temperature of complete disappearance of the 1/7 phase decreases monotonously with increasing x, down to 580 °C, for $x = 7$. Both quenching of the $Nd_9(Fe_{1-x}Co_x)_{85}B_6$ alloys and crystallization of the corresponding amorphous phase, are accompanied by the formation of a metastable phase of $TbCu_7$-type [96G1]. There is a stabilizing effect of cobalt on the 1/7 phase.

The Sm–Fe–B melt spun ribbons with single phase $TbCu_7$-type structure were also prepared from the $SmFe_{12}B_x$ alloys with $0 \leq x \leq 0.75$ [16Z1]. The metastable $TbCu_7$-type structure is stabilized because of the increase of the lattice parameters ratio c/a. When the boron content exceeds the limit of solubility ($x > 1.0$) of Sm–Fe alloys, does not impede the appearance of α-Fe and accelerates the formation of $Sm_2Fe_{23}B_6$ metastable phase. The $Tb_6Fe_{87}Nb_1B_6$ melt spun ribbon, crystallizes in two steps [95H1, 95W1]. The intermediate phase is of $TbCu_7$-type, before obtaining a mixture of $Tb_2Fe_{14}B$ and α-Fe phases.

The electromagnetic levitation techniques [93H2, 98H1], has been used to melt the Nd–Fe–B alloys and then to analyse their solidification. The processing of $Nd_xFe_{100\text{-}1.5x}B_{0.5x}$ [01G2, 02G2, 02V1, 03G1] and $Nd_{14}Fe_{79}B_7$ [02G1, 04O2, 04V2] alloys, showed that with increasing the bulk undercooling level, three kinds of primary phase formation were solidified. The γ-Fe solid solution, $Nd_2Fe_{14}B$ and metastable $Nd_2Fe_{17}B_x$ compound were solidified as primary phase in sequence of increasing bulk undercooling level. The critical undercooling for the primary $Nd_2Fe_{17}B_x$ formation was shown to be dependent on the alloy composition [02G1, 02G2, 02V1, 03G1]. The $Nd_2Fe_{17}B_x$ phase decomposes in α-Fe and $Nd_2Fe_{14}B$, during post-solidification thermal process [04O1, 05S3, 07G1].

The metastable phase, with $TbCu_7$-type structure, could be also formed by the drop tube processed samples [99G1, 00G1, 01G2, 01G3]. The solidification path of the droplets depends on the alloy composition and on the droplet size. A large amount of metastable $Nd_2Fe_{17}B_x$ phases was shown, in as dropped $Nd_xFe_yB_z$ alloys with x = 10–14, y = 79–85 and y = 5–7 [05O1, 06O1]. A dual stage phase transformation was induced by heat treatment of as-dropped samples. The first stage of the phase transformation, that is dominant in $Nd_{14}Fe_{79}B_7$ alloy, is the diffusive phase transformation from $Nd_2Fe_{17}B_x$ and Nd-rich phases into the $Nd_2Fe_{14}B$ compound, at T = 950 K. The second stage of phase transformation, dominant in $Nd_{10}Fe_{85}B_5$ alloy, is the decomposition of the metastable phase into $Nd_2Fe_{14}B$ and α-Fe at T = 1100 K. The rate of phase transformation, in either case, is controlled by the diffusion of Nd atoms, the diffusivity being higher, at the first stage due to existence of Nd-rich phase adjacent to the $Nd_2Fe_{17}B_x$ one. The low cooling rate, in alloys processed by electromagnetic levitation, as compared with that of the drop tube method, facilitates the nucleation of equilibrium $Nd_2Fe_{14}B$ phase, at the interface between the metastable $Nd_2Fe_{17}B_x$ and liquid phase. Since in this case, the $Nd_2Fe_{17}B_x$ phase is surrounded by the $Nd_2Fe_{14}B$ one, the first stage of phase transformation, which occurs at the triple junctions with $Nd_2Fe_{17}B_x$ and Nd-rich phases, is blocked and therefore the alternative model of phase transformation, that is, the decomposition into the dual-phases microstructure ($Nd_2Fe_{14}B$ and γ-Fe) is induced [05O1].

Two types of microstructures were also shown in gas-atomized Nd–Fe–B particles of non-peritectic compositions $Nd_xFe_yB_z$ with x = 16–18, y = 73–76 and z = 8–9 [05G1]. The first is featured by primary formation of the metastable $Nd_2Fe_{17}B_x$ dendrites, whereas another is featured by predominant primary formation of stable $Nd_2Fe_{14}B$ grains. They are generally similar to those observed in levitated bulk samples, having the same composition, except for reduced grain sizes and partial suppression of the solid-state decomposition of $Nd_2Fe_{17}B_x$ phase. A tiny amount of α-Fe dendrites, was observed in some grains located near the surface. The containerless solidification, of gas-atomized $Nd_{10}Fe_{85}B_5$ melt droplets by drop tube, evidenced in some samples, a quasi-single phase microstructure consisting of the metastable $Nd_2Fe_{17}B_x$ dendrites [07G1]. The composition of this alloy is very close to that of stoichiometric metastable phase. The boron was assumed to occupy different interstitial sites than those where H, C and N are located in Nd_2Fe_{17} lattice.

The metastable $Nd_2(Fe,Co)_{17}B_x$ phase, having $TbCu_7$ type disordered structure, primarily crystallized, when undercooling small droplets of near-stoichiometric Nd–Fe–B alloys doped with Co, Zr and Ga [00G1]. The presence of this phase was related to large undercooling levels achieved prior to solidification, increased lattice stability due to Co addition and relatively disordered structure.

The presence of metastable $Nd(Fe,Mo)_7B_x$ compound, with volume fractions of 24% was shown in the annealed $Nd_{8.4}Fe_{85.5}Mo_{1.6}B_{4.5}$ alloy, at T = 600 °C for 20 min, prepared by mechanical alloying and mechanical milling, respectively [02C3]. The content of $Nd(Fe,Mo)_7B_x$ phase increases as result of simultaneous substitution of Fe by Mo and Co. The $Nd_2Fe_{17}B_x$ metastable-phase was also obtained by partial decomposition of $Nd_2Fe_{14}B$ compound, when has been washed with distilled water and NH_4Cl, in order to remove the CaO by-product, resulting from R-D process used for their preparation [11J1].

The reported Curie temperatures, T_c, for $Nd_2Fe_{17}B_x$ phase, by different authors, are in a large range of values (327 K $\leq T_c \leq$ 483 K) [02J1, 04O1, 07G1]—Table 7.26. The structure type (*P6/mmm*, $R\overline{3}m$) and B content can influence significantly the T_c values. In addition, the presence of other magnetic ordered phases contributes to the large range of reported T_c values. The Curie temperatures of the $Nd_2(Fe_{1-x}Co_x)_{17}B_x$ series, both in amorphous and crystallized states, increase in the same way with the rise of cobalt content, those of amorphous phase being somewhat smaller [96G1]. The saturation magnetizations are also influenced by the presence of other phases.

The partial replacement of Sm in rapidly quenched $Sm_2Fe_{17}B_x$ with $x < 0.5$, by Nd or Tb improved their permanent magnetic properties [94C1].

The ^{57}Fe Mössbauer spectra of annealed $Nd_8Fe_{78}B_{14}$ alloy, at T = 993 K for 10 min, evidenced the presence of a fraction of ≅18% $Nd_2Fe_{17}B_x$ phase [15H1]. The spectrum of $Nd_2Fe_{17}B_x$ phase was analysed assuming a rhombohedral $R\overline{3}m$ type structure, with iron located in *6c*, *9d*, 18 *h* and *18f* sites. The ^{57}Fe Mössbauer spectrum of $SmFe_{12}B_x$ melt spin ribbons having a modified metastable $TbFe_7$ type structure, was also fitted assuming iron located at 2*e*, 3*g* and 6*l* sites [16Z1]. The ^{57}Fe hyperfine fields, B_{eff}, decrease in a sequence $B_{eff}(2c) > B_{eff}(3g) > B_{eff}(6l)$ [16Z1]. These were correlated with the number of iron atoms situated in their first coordination shell at each site [04N1]. The boron content influences little the hyperfine field values, particularly at 2*e* site.

The $Nd_2Fe_{17}B_x$ has no uniaxial anisotropy, unlike that of 1/7 phase, in Sm-Fe-B system, where a rather high coercive field was reported [16Z1].

7.10 $R_2Fe_{14}B$ Compounds

The $R_2Fe_{14}B$ compounds, where R is a rare-earth or yttrium are the main magnetic phases in manufacturing high energy permanent magnets. The researchers from two different groups, independently, discovered these materials as well as their technical applications. Anisotropic magnets were obtained, by using sintering process [84S1], while from melt quenched alloys, magnets with isotropic properties [84C2] were manufactured. Known as a third generation of rare-earth-based magnets, these are the most powerful and advanced commercialized permanent magnets today. The physical properties of R–Fe–B permanent magnets were intensively investigated in order to improve their technical performances. Reviews on the matter were already published [86B3, 86B6, 86B10, 88B8, 89B2, 90B1, 91H1, 92B2, 98B2, 21C2]. The $R_2Fe_{14}B$ compounds were generally prepared by melting the constituent elements, splat cooling or hydrogen decrepitation [85H1, 98B2]. In the previous chapter, the methods for obtaining $R_2Fe_{14}B$ phases by crystallization of amorphous alloys are described.

In this chapter, the physical properties of the ternary $R_2Fe_{14}B$ compounds as well of the related pseudoternary systems will be reviewed. The crystal structures and elastic properties are presented in Sect. 7.10.1. In the Sect. 7.10.2, the physical properties of $R_2Fe_{14}B$, as well as of $(R_1, R_2)_2Fe_{14}B$ and $R_2(Fe,M)_{14}B$ psudoternary systems, where R_1, R_2 are rare earths and M a magnetic or nonmagnetic elements, are discussed. There is a large number of scientific reports in the field; only a selective reference list, related mainly to their basic properties, has been given. The recent results in developing high energy R–Fe–B permanent magnets are presented in last section.

7.10.1 Crystal Structures, Elastic Properties

The $R_2Fe_{14}B$ compounds have been reported that form with yttrium and all rare-earths, except Eu. In some reports no presence of $La_2Fe_{14}B$ phase was suggested [84C2, 84S2]. Then, the compound has been obtained by annealing long time at 850 °C [87B7]. Latter on [21L1], was shown that this phase is formed at T = 926 °C by peritectic reaction and decomposes into La, Fe and $LaFe_4B_4$, at T = 794 °C via eutectic reaction.

The $R_2Fe_{14}B$ compounds were R is a rare-earth or yttrium crystallize mainly in a tetragonal structure, space group *P4₂/mnm* [84G1, 84H1, 84S3]—Fig. 7.16. The lattice sites and atom coordinates, are listed in Table 7.28. The R atoms are distributed in two crystallographic sites, the Fe atoms on six different positions and B is located in only one site type. The $P4_2/mnm$ crystal structure can be described as an eight-layer arrangement perpendicular to the **z**-axis, related to a sigma phase [54B1]. The R and B atoms and four Fe*c* atoms form alternate arrangements of large and small nearly regular rhombus in the z = 0 and z = 1/2 planes. The iron atoms are arranged in almost planar nets of slightly deformed hexagons and triangles located at z ≅ 1/8, 3/8, 5/8 and 7/8. The planar arrangements in z = 1/8 and 7/8 planes as well as in 3/8 and 5/8 ones, respectively are superposed. The arrangement in z = 1/8 plane, can be obtained by a rotation of 30° from that in z = 3/8 plane. The Fej_2 atoms in planes z = 1/4 and 3/4 are nearly superposed on Nd situated in planes with z = 0 and ½. The Fe*c* atoms are at the center of a double trigonal prism. The boron occupies the

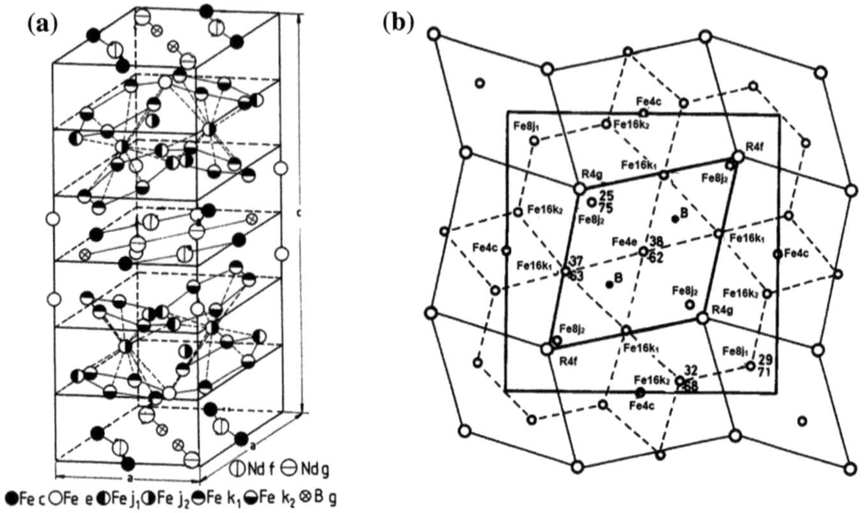

Fig. 7.16 $Nd_2Fe_{14}B$ compound having tetragonal crystal structures: **a** general view [85H2]; **b** schematic representation of the crystal structure; Nd, B and Fe4*c* atoms are at z = 1/2. For all the other Fe atoms, there is one atom above and one other below the plane z = 1/2. The associated couples of z values are indicated in units of c. The square in heavy line represents the unit cell of the structure. The rhombus, in heavy lines, shows the unit cell of $CaCu_5$-type structure [84G1]

Table 7.28 Atomic sites of $Nd_2Fe_{14}B$ with tetragonal structure, space group *P*4$_2$*/mnm* [11V1] [a, b]

Atom	Site	Symmetry	x	y	z	Atomic environment
Nd	4*f*	*m.2m*	0.3585	0.3585	0	Pseudo Frank-Kasper $BFe_{16}Nd_3$
Nd	4*g*	*m.2m*	0.2313	0.7687	0	Pseudo Frank-Kasper $Fe_{16}B_2Nd_2$
Fe	4*e*	2.*mm*	0	0	0.1144	Pseudo Frank-Kasper $B_2Fe_9Nd_2$
Fe	4*c*	2*/m..*	0	1/2	0	14-vertex Frank-Kasper $Fe_8Nd_4B_2$
Fe	$8j_1$	*..m*	0.0979	0.0979	0.2951	Pseudo Frank-Kasper Fe_9Nd_3B
Fe	$8j_2$	*..m*	0.3174	0.3174	0.2535	14-vertex Frank-Kasper $Fe_{12}Nd_2$
Fe	$16k_1$	*1*	0.0671	0.2765	0.1269	Pseudo Frank-Kasper $BFe_{10}Nd_2$
Fe	$16k_2$	*1*	0.0379	0.3587	0.3237	Icosahedron $Fe_{10}Nd_2$
B	4*f*	*m.2m*	0.1243	0.1243	0	Trigonal prism Fe_6

[a] Transformed from published data, at RT; origin shift (1/2,1/2,0); [b] Structure determined by [84G1, 84H1, 84S1], the atomic environments reported by [84G1] are Fe_8Nd_4 for 4*c* site, Fe_9Nd_3 for $8j_1$ site and Fe_9Nd_2B for $16k_1$ site, little diferent from [11V1]

center of trigonal prisms formed by the three nearest iron atoms above and three below the basal (or z = 1/2) planes. These prisms contribute to the stability of the crystal structure. The R atoms are bonded to each boron through the rectangular prism faces. The boron atoms occupy the same lattice sites in the rhombus of R atoms, as in RCo_3B_2 type lattice and related compounds, where they replace Co atoms at the

Table 7.29 Atom parameter correlations used for the crystal structure refinement, at T = 20 K, in *P4₂/mnm* and *Cm*-space groups [96W2]

Atoms	$P4_2/mnm$ sites			Constraints used	Atoms	Coordinates		
	[84S3]	[84G1]	[84H1]					
Nd_1	4*f*	4*g*	4*g*	*Cmmm*[a]	$Nd_1$1	x,	x,	0
					$Nd_1$3	1/2 + x,	1/2−x,	1/2
Nd_2	4*g*	4*f*	4*f*	*Cm*[a]	$Nd_2$1	x,	−y,	0
					Nd_2	1/2 + x,	1/2 + x,	1/2
Fe_1	4*c*	4*e*	4*c*	*Cm*	$Fe_1$1	x,	1/2,	0
					$Fe_1$3	x,	1/2,	1/2
Fe_2	$16k_1$	4*c*	$16k_1$	*C2/m*	$Fe_2$1	x,	y,	z
					$Fe_2$2	x,	y,	−z
					$Fe_2$9	1/2 + x,	1/2−y,	1/2 + z
					$Fe_2$10	1/2 + x,	1/2−y,	1/2−z
Fe_3	$16k_2$	$8j_1$	$16k_2$	*C2/m*	$Fe_3$1	x,	y,	z
					$Fe_3$2	x,	y,	−z
					$Fe_3$9	1/2 + x,	1/2−y,	1/2 + z
					$Fe_3$10	1/2 + x,	1/2−y,	1/2−z
Fe_4	$8j_1$	$8j_2$	$8j_2$	*Cm*	$Fe_4$1	x,	x,	z
					$Fe_4$2	x,	x,	−z
					$Fe_4$3	−x,	−x,	z
					$Fe_4$4	−x,	−x,	−z
					$Fe_4$5	1/2 + x,	1/2 − y,	1/2 + z
					$Fe_4$6	1/2 + y,	1/2 − x,	1/2 − z
Fe_5	$8j_2$	$16k_1$	$8j_1$	*C2/m*	$Fe_5$1	x,	x,	z
					$Fe_5$2	x,	x,	−z
					$Fe_5$5	1/2 + x,	1/2 − y,	1/2 + z
Fe_6	4*e*	$16k_2$	4*e*	*Cmmm*	$Fe_6$1	0,	0,	z
B	4*f*	4*g*	4*g*	*Cm*	B1	x,	x,	0
					B2	−x,	−x,	0
					B3	1/2 + x,	1/2 − y,	1/2

[a] For $Ho_2Fe_{14}B$, the constraint for Ho4*f* was *P4₂/mnm* and for Ho4*g*, *C2/m*. The same constraints as in case of $Nd_2Fe_{14}B$ were used for iron sites in $Ho_2Fe_{14}B$

site 2*c* in RCo_5 type structure [84G1]. There are very short distances between some iron sites, as for example $d_{j_1-k_2} \cong 0.239$ Å and $d_{j_1-j_1} \cong 0.244$ Å, in $Nd_2Fe_{14}B$, at RT.

Latter on [96O1, 96W2], it was shown that the crystal structure of $Nd_2Fe_{14}B$ compound has a lower symmetry. No appreciable deviation from the tetragonal space group was detected and thus, at ambient temperature, the lattice parameters agree with those determined in $P4_2/mnm$ type structure [84G1, 84H1, 84S3]. At temperatures lower than spin reorientation temperature, $T_{sR} \cong 140$ K, the crystal symmetry can be described in a monoclinic space group, *Cm*. The space group constraints used, expressed in the site notations of [84S3], as well as correspondences for other site notations, used in literature, are given in Table 7.29 [84G1, 84H1]. A lowering of the symmetry from space group $P4_2/mnm$ was shown at low temperatures also for others

$R_2Fe_{14}B$ compounds. The crystal structure of $Ho_2Fe_{14}B$, has space group *Cm*, far above the spin reorientation temperature $T_{sR} = 58$ K [01W1]. A coherent sequence of the crystalline and magnetic transformations, when lowering temperature, was proposed as discussed in the following section [01W1]. The crystal structure of $Er_2Fe_{14}B$ is orthorhombic, having *Pnn2* space group, below the spin reorientation temperature, $T_{sR} \cong 330$ K [01W1]. The R moments exhibit a fan structure apart from the (100) plane as result of Er–Fe magnetic coupling, different strength of CEF parameters on the two R sites and because of negative R–R exchange interactions. It was also suggested [01W1] that other so-called planar $R_2Fe_{14}B$ compounds (R = Sm, Tm, Yb), exhibit similar features in their crystal structures, as driven by negative second order CEF parameters. Since the effects of iron axial anisotropy are counterbalanced by those of R-Fe exchange interactions, some deviations from collinearity, within iron magnetic sublattices, can be suspected. A list of lattice parameters of $R_2Fe_{14}B$ compounds are given in Table 7.30.

The thermal variations of lattice parameters of $R_2Fe_{14}B$ compounds with R = Nd, Gd, Tb, Er and Y, in the temperature range $10\text{ K} < T < 1000$ K, are given in Fig. 7.17 [03Y1, 05Y1]. Similar behavior was shown in all $R_2Fe_{14}B$ compounds [84G3, 85G3, 86B9, 87B13]. A large Invar effect with a corresponding large temperature dependence of lattice parameters, at 10–15 K, above T_c, was observed. The **a**-axes show a larger Invar effect than the **c**-axes. The Invar anomalies were attributed to the competition between elastic and magnetic energies [84G3, 85G3]. Such anomalies in the thermal variations of the lattice parameters determine a decrease of the magnetic energy at the expense of the elastic one. The lattice contributions to the thermal expansion, of $Y_2Fe_{14}B$, were calculated, allowing the determination of the magnetic contributions to the anomalous changes in the lattice parameters, $\Delta a/a$ and $\Delta c/c$ [86G3]. Up to $T \cong 500$ K, the deduced volume anomaly was shown to be approximately proportional to the square of spontaneous magnetization, M_s^2, as expected for isotropic exchange interactions.

The shortest Fe–Fe distances occur in the basal plane of $P4_2/mnm$ type structure. Accordingly, the magnetic anomaly is larger for crystallographic parameter a, than for c-one [86G3, 05Y1]. The spontaneous magnetostrains of the lattices and bonds lengths were calculated [03Y1, 05Y1]. The iron sublattice was shown to dominate the volumetric spontaneous magnetostriction of $R_2Fe_{14}B$ compounds, the contribution from the rare-earth sublattice being roughly proportional to the spin magnetic moment of the rare-earths. The different R and Fe sites contribute differently to spontaneous magnetostriction, as determined by the exchange interactions between a given site and its first neighbors. It was shown that Fe*j1* and Fe*j2* sites sustain larger average bonds magnetostrain, which means a greater contribution to the spontaneous magnetostriction. The Fe*k1* and Fe*k2* sites have moderate contributions, whereas the smallest ones are connected with Fe*e* and Fe*c* sites. The average bonds magnetostriction around Fe sites was approximately proportional to their magnetic moments.

The high field magnetostriction studies on polycrystalline $R_2Fe_{14}B$ compounds were made [87B13, 87I1, 90A1, 90I1, 92A1, 99A1, 99A2]. The magnetostrictions of samples with R = Pr, Nd, Dy, Ho, Er, Y, in the temperature range $4.2\text{ K} \leq T \leq 300$ K,

Table 7.30 Space groups and lattice parameters of $R_2Fe_{14}B$ compounds

Compound	T (K)	Space group	Lattice parameters (nm)		γ°	References
			a	c		
$La_2Fe_{14}B$	5	*P4₂/mnm*	0.883	1.237		[95F1]
$La_2Fe_{14}B$	RT	*P4₂/mnm*	0.8822	1.2338		[84S4]
$Ce_2Fe_{14}B$	5	*P4₂/mnm*	0.877	1.215		[95F1]
$Ce_2Fe_{14}B$	RT	*P4₂/mnm*	0.8726	1.2057		[84S4]
$Ce_2Fe_{14}B$	30	*P4₂/mnm*	0.87740	1.2072		[87D1]
$Ce_2Fe_{14}BH_{3.7}$	30.6	*P4₂/mnm*	0.8922	1.2243		[87D1]
$Pr_2Fe_{14}B$	RT	*P4₂/mnm*	0.8838	1.2289		[84S4]
$Pr_2Fe_{14}B$	RT	*P4₂/mnm*	0.881	1.227		[86H4]
$Nd_2Fe_{14}B$	20	*Cm*	0.87118	1.20821	90	[96O1]
$Nd_2Fe_{14}B$	300	*P4₂/mnm*	0.8816(1)	1.2240(2)		[13T1]
$Nd_2Fe_{14}B$	RT	*P4₂/mnm*	0.8804(5)	1.2205(5)		[84S3]
$Nd_2Fe_{14}B$	RT	*P4₂/mnm*	0.8808(1)	1.2207(2)		[00C1]
$Nd_2Fe_{14}B$	RT	*P4₂/mnm*	0.8799(1)	1.2207(1)		[05Y1]
$Nd_2Fe_{14}B$	20	*Cm*	0.8770(6)	1.2016(8)		[96W2]
$Nd_2Fe_{14}B$	RT	*P4₂/mnm*	0.8801	1.2205		[86A4]
$Nd_2Fe_{14}BH_{3.8}$	RT	*P4₂/mnm*	0.8919	1.2345		[86A4]
$Nd_2Fe_{14}BN_{1-\delta}$	300	*P4₂/mnm*	0.8851	1.2238		[92Z1]
$Nd_2Fe_{13}SiB$	RT	*P4₂/mnm*	0.8774(1)	1.2169(2)		[00C1]
$Nd_2Fe_{12}Si_2B$	RT	*P4₂/mnm*	0.8757(1)	1.2101(2)		[00C1]
$Sm_2Fe_{14}B$	RT	*P4₂/mnm*	0.8777	1.2105		[84S4]
$Gd_2Fe_{14}B$	RT	*P4₂/mnm*	0.8777(1)	1.2096(1)		[05Y1]
$Gd_2Fe_{14}B$	RT	*P4₂/mnm*	0.8780	1.2075		[84S4]
$Gd_2Fe_{14}B$	RT	*P4₂/mnm*	0.8772	1.2025		[88Z1]
$Gd_2Fe_{14}BH_{4.4}$	RT	*P4₂/mnm*	0.8862	1.2180		[88Z1]
$Tb_2Fe_{14}B$	RT	*P4₂/mnm*	0.8765(1)	1.2064(1)		[05Y1]
$Tb_2Fe_{14}B$	RT	*P4₂/mnm*	0.8785	1.2070		[84S4]
$Tb_2Fe_{14}B$	RT	*P4₂/mnm*	0.8769	1.2028		[88Z1]
$Tb_2Fe_{14}BH_{4.3}$	RT	*P4₂/mnm*	0.8863	1.2144		[88Z1]
$Dy_2Fe_{14}B$	RT	*P4₂/mnm*	0.8757	1.1990		[84S4]
$Dy_2Fe_{14}B$	RT	*P4₂/mnm*	0.8754	1.1977		[88Z1]
$Dy_2Fe_{14}BH_{4.6}$	RT	*P4₂/mnm*	0.8851	1.2093		[88Z1]
$(Nd_{0.867}Dy_{0.133})Fe_{14}B$	300	*P4₂/mnm*	0.87921(1)	1.21779(1)		[17S1]
$(Nd_{0.5}Dy_{0.5})Fe_{14}B$	300	*P4₂/mnm*	0.87717(1)	1.20943(1)		[17S1]
$Ho_2Fe_{14}B$	4.2	*Cm*	0.87354(4)	1.19450(7)	89.950(4)	[91O1]
$Ho_2Fe_{14}B$	20	*Cm*	0.8770	1.2016	89.913(6)	[90W1, 91O1]
$Ho_2Fe_{14}B$	100	*P4₂/mnm*	0.8755	1.1990		[90W1]

(continued)

Table 7.30 (continued)

Compound	T (K)	Space group	Lattice parameters (nm)		γ°	References
			a	c		
$Ho_2Fe_{14}B$	300	*P4₂/mnm*	0.87485	1.19863		[90W1]
$Ho_2Fe_{14}B$	RT	*P4₂/mnm*	0.8740	1.1966		[88Z1]
$Ho_2Fe_{14}BH_{5.4}$	RT	*P4₂/mnm*	0.8845	1.2082		[88Z1]
$Er_2Fe_{14}B$	300	*P4₂/mnm*	0.8744(1)	1.1968(2)		[13T1]
ErFe14B	77	*P4₂/mnm*	0.87260(6)	1.19185(11)		[86Y5]
$Er_2Fe_{14}B$	RT	*P4₂/mnm*	0.8732(1)	1.1953(3)		[87F4]
$Er_2Fe_{14}B$	6	*P4₂/mnm*	0.87495	1.19508		[87D1]
$Er_2Fe_{14}BH_{2.58}$	25	*P4₂/mnm*	0.8825	1.2069		[87D1]
$Er_2Fe_{14}B$	RT	*P4₂/mnm*	0.8731(1)	1.1974(1)		[05Y1]
$Er_2Fe_{14}B$	RT	*P4₂/mnm*	0.8732	1.1941		[88Z1]
$Er_2Fe_{14}BH_{3.7}$	RT	*P4₂/mnm*	0.8815	1.2028		[88Z1]
$Tm_2Fe_{14}B$	RT	*P4₂/mnm*	0.8728	1.1928		[84S4]
$Tm_2Fe_{14}B$	RT	*P4₂/mnm*	0.874	1.194		[86H4]
$Lu_2Fe_{14}B$	RT	*P4₂/mnm*	0.8712	1.1883		[84S4]
$Lu_2Fe_{14}B$	RT	*P4₂/mnm*	0.8710	1.1878		[05T1]
$Lu_2Fe_{14}BH_{2.5}$	RT	*P4₂/mnm*	0.8766	1.1962		[05T1]
$Lu_2Fe_{14}B$	RT	*P4₂/mnm*	0.8710	1.1876		[85L2]
$Lu_2Fe_{14}BH_{2.3}$	RT	*P4₂/mnm*	0.8764	1.1958		[85L2]
$Y_2Fe_{14}B$	RT	*P4₂/mnm*	0.8757	1.2026		[84S4]
$Y_2Fe_{14}B$	RT	*P4₂/mnm*	0.8751(1)	1.2035(1)		[05Y1]
$Y_2Fe_{14}B$	RT	*P4₂/mnm*	0.8754	1.2024		[86A4]
$Y_2Fe_{14}BH_{4.8}$	RT	*P4₂/mnm*	0.8830	1.2130		[86A4]
$Y_2Fe_{14}B$	10	*P4₂/mnm*	0.8758	1.2005		[87D1]
$Y_2Fe_{14}BH_{3.6}$	10	*P4₂/mnm*	0.8880	1.2114		[87D1]
$Y_2Fe_{14}B$	300	*P4₂/mnm*	0.8761	1.2032		[92Z1]
$Y_2Fe_{14}BN_{0.317}$	300	*P4₂/mnm*	0.8798	1.2075		[92Z1]

were determined both parallel ($\lambda_{\parallel}$) and perpendicular ($\lambda_{\perp}$) to the applied magnetic field [92A1]. Large Invar anomalies were observed at $T < T_c$, where the spontaneous volume striction is connected with both the exchange interactions and strain dependences of the moment within iron sublattice. The forced volume striction was shown to be mainly of exchange origin and comes also from iron sublattice—Table 7.31. The complex temperature dependences of the anisotropic magnetostriction, associated with R sublattice, need recourse, for their analysis, to the high-order crystal field single-ion interactions [92A1]. Linear and volumetric spontaneous magnetostriction as well as the thermal expansion coefficients of some $R_2Fe_{14}B$ compounds, are listed in Table 7.31.

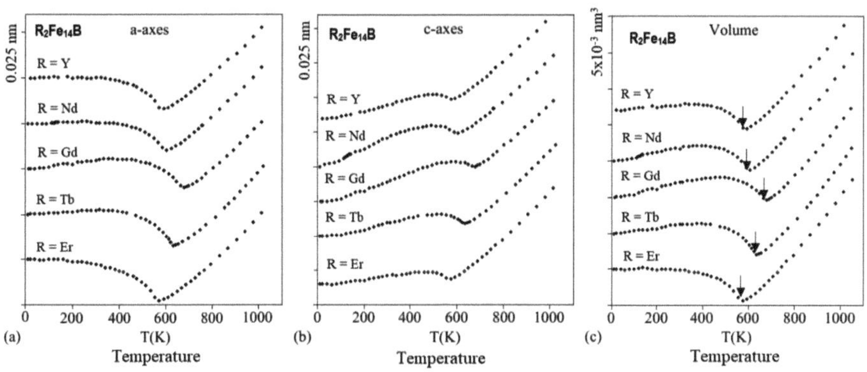

Fig. 7.17 $R_2Fe_{14}B$ compounds with R = Nd, Gd, Tb, Er, Y: temperature dependences of the lattice parameters and volumes [05Y1]. In order to avoid the superposition, the curves were translated. The temperature where changes in volumes, are denoted

The temperature dependences of the elastic properties of $R_2Fe_{14}B$ compounds with R = Pr, Nd, Y [85T2], R = Nd [86D1], R = Nd, Y [89S3], R = Y [19I1], as well as of isotropic hot-pressed, melt spun compounds with R = Ce, Pr, Nd and Er [94F1] were investigated. In case of $Nd_2Fe_{14}B$ and $Er_2Fe_{14}B$ an anomaly in the temperature dependence of the elastic moduli was shown at T_{sR}. The ΔE effect, in $Nd_2Fe_{14}B$ at $T = 77$ K, is of the order of 0.5% [86D1]. Latter on [89S3], the ordinary ΔE effect has been observed only in the shear modulus at $T < 150$ K. A softening in longitudinal modulus was shown for compounds with R = Nd, Y, at $T < T_c$, and explained as an inverse effect of the volume magnetostriction [89S3].

The elastic moduli and the breaking stress were measured in sintered Nd–Fe–B magnets, at $4\ K \leq T \leq 300$ K [91V1]. The c/a ratio of sintered $Nd_2Fe_{14}B$ magnets is greater than that of powdered sample [20K1]. This suggest that the sample texture and hence the internal stress has a determinant effect on the lattice constants. At low temperatures, there is an increase of elastic moduli, by 10–20%, as compared with the value determined at $T = 300$ K. The relatively broad dispersion of the experimental values were correlated with non-homogeneity of sintered sample.

The mechanical properties of La–Ce [16J1] and Dy [16H1] substituted in Nd–Fe–B magnets were also investigated. The optimum compressive and bending strengths of 1137.8 MPa and 339.9 MPa, respectively were achieved at 18 wt % La–Ce substitution. The above substitutions enhance the micro-hardness and Young's modulus of the 2/14/1 phase, gradually [16J1]. In Dy substituted sintered magnet, the bending strength, declined and the Rockwell hardness first declined and then increased with the increase of Dy content [16H1]. The Dy diffused into main phase during sintering.

The Nd self diffusion in nano-crystalline Nd-rich Nd–Fe–B alloys was investigated [05S1]. The effect of the inter-related motion of atoms and changes in the oscillations of the magnetic structure has been revealed in $Nd_2Fe_{14}B$ compound [09G1]. The process was shown to be cyclic with a period of changes in the atomic structure of $\cong$1 month and in the magnetic structure of 8–12 days. The oscillations

Table 7.31 Linear spontaneous magnetostriction λ_a, λ_c, volumetric spontaneous magnetostriction ω_s, thermal expansion and elastic properties of $R_2Fe_{14}B$ compounds

	Temperature range (K)	$R_2Fe_{14}B$												References
		La	Ce	Pr	Nd	Sm	Gd	Tb	Dy	Ho	Er	Lu	Y	
Magnetostriction[a]														
$\lambda_a \cdot 10^{-3}$	10				8.8(3)		10.0(3)	11.6(3)			9.1(3)		8.2(3)	[03Y1, 05Y1]
$\lambda_c \cdot 10^{-3}$	10				5.0(3)		5.9(3)	6.9(3)			5.8(3)		5.4(3)	[03Y1, 05Y1]
$\omega_s \cdot 10^{-3}$	10				22.9(1.0)		27.9(1.0)	31.1(1.0)			24.1(1.0)		22.6(1.0)	[03Y1, 05Y1]
$\omega_s \cdot 10^{-3}$	4	25.2	32.1	27.9	27.9	27.3	33.6	31.8	35.1	35.4	31.2	27.0		[87B13]
$\omega_{sd} \cdot 10^{-3}$	4	25.2	32.1	25.3	25.5	25.8	26.1	26.2	26.4	26.4	26.5	27.0		[87B13]
$\omega_{sf} \cdot 10^{-3}$	4	0	0	0.25	0.24	0.15	0.75	0.56	0.87	0.90	0.46	0		[87B13]
Thermal expansion[b]														
$\alpha_a \cdot 10^{-5}$ (K^{-1})	$T > T_c$				1.29(2)		1.34(2)	1.32(2)			1.27(2)		1.26(2)	[03Y1, 05Y1]
$\alpha_c \cdot 10^{-5}$ (K^{-1})	$T > T_c$				1.18(2)		1.25(2)	1.19(2)			1.24(2)		1.26(2)	[03Y1, 05Y1]
$\alpha_v \cdot 10^{-5}$ (K^{-1})	$T > T_c$				3.77(6)		3.99(6)	3.89(6)			3.75(6)		3.82(6)	[03Y1, 05Y1]
$\alpha \cdot 10^{-5}$ (K^{-1})	$T_c < T < 1000$ K	1.65	2.02	1.62	1.59	1.55	1.56	1.56	1.56	1.55	1.59	1.56		[87B13]
Elastic constant (GPa)[c]														
c_{11}	RT			147	161								146	[85T2]
c_{44}	RT			59.5	70.1								66.9	[85T2]
c_L	300		161.1	179.2	188.5						193.8			[94F1]
c_T	300		56.3	58.4	62.0						65.1			[94F1]

(continued)

Table 7.31 (continued)

	Temperature range (K)	$R_2Fe_{14}B$												References
		La	Ce	Pr	Nd	Sm	Gd	Tb	Dy	Ho	Er	Lu	Y	
Elastic moduli[d]														
E	RT			138	156								144	[85T2]
E[d]	300				144(1)									[91V1]
Breaking stress (MPa)[e]	300				238									[91V1]
Debye temperature,θ_D (K)	RT		390	414	380 [87F6]								429	[85T2]

[a] λ_a, λ_c linear spontaneous magnetostrictions along **a** and **c**-axis, values for $R_2Fe_{14}B$ (R = Nd,Y) at 300 K [03Y1]; ω_s-total spontaneous volume magnetostriction and contribution from 3d (ω_{sd}) and 4f (ω_{sf}) sublattices; [b] α_a, α_c, α_V are the thermal expansion along **a** and **c**-axis and volume expansion, respectively, α is linear thermal expansion coefficient; [c] c_{11}, c_{44} are the elastic constant for isotropic material; c_L, c_T are the longitudinal and transversal elastic constants of melt spun $R_2Fe_{14}B$ alloys; [d] Elastic moduli of sintered magnet; [e] For sintered Nd-Fe-B magnet

Table 7.32 Magnetic properties of $R_2Fe_{14}B$ compounds

Compound	Magnetic Structure	M_s (μ_B/f.u.)	T_c (K)	T_{sR} (K)	C (emuK/f.u.)	θ (K)	M_{eff} (μ_B/Fe atom)	μ_0H_a (T)	References
$La_2Fe_{14}B$	FM	31.0[a]	543					1.97(4.2 K) [88G6]	[95F1]
$La_2Fe_{14}B$	FM	30.6[b]	530					3.0–4.0(4.2 K)	[84S4]
$Ce_2Fe_{14}B$	FM	29.4[c]	422					3.0(4.2 K)	[86H4]
$Ce_2Fe_{14}B$	FM	30.0[a]	425					3.66(77 K)	[95F1]
$Ce_2Fe_{14}B$	FM	30.6[d]	437					3.71(300 K),3.66(77 K)	[85B3]
$Ce_2Fe_{14}B$	FM	30.2 [b]	425					5.0–6.0(4.2 K)	[84S4]
$Ce_2Fe_{14}B$	FM	22.7[d]	424					4.63(RT)	[84Y1]
$LaCeFe_{14}B$	FM	30.8[a]	487						[95F1]
$Pr_2Fe_{14}B$	FM	36.3[d]	566		31.6	575	4.02	7.93 (RT),17.2(77 K)	[85B3, 85B7]
$Pr_2Fe_{14}B$	FM	37.6[c]	569					32.0(4.2 K)	[86H4]
$Pr_2Fe_{14}B$	FM	29.3[d]	564					12.5(293 K)	[84Y1]
$Pr_2Fe_{14}B$	FM	37.3[e]	567					14.5 (4.2 K)	[88Z2]
$Nd_2Fe_{14}B$	FM	35.0(1)[f]	588(3)	140(5)					[00C1]
$Nd_2Fe_{14}B$	FM	38.2[d]	589		32.1	602	4.06	7.07(RT)	[85B3, 85B7]
$Nd_2Fe_{14}B$	FM	37.7[c]	586					6.7(300 K)	[86H4]
$Nd_2Fe_{14}B$	FM	32.1[d]	585					15.0(RT)	[84Y1]
$Nd_2Fe_{14}B$	FM	34.0 [g]	595						[86A4]
$Nd_2Fe_{14}BH_{3.2}$	FM	34.8 [g]	630						[86A4]
$Nd_2Fe_{13}SiB$	FM	31.1(1)[f]	610(3)	107(5)					[00C1]
$Nd_2Fe_{12}Si_2B$	FM	26.8(1)[f]	633(3)	93(5)					[00C1]
$Sm_2Fe_{14}B$	FM	32.4[d]	626						[85B3]
$Sm_2Fe_{14}B$	FM	33.3[c]	620						[86H4]

(continued)

Table 7.32 (continued)

Compound	Magnetic Structure	M_s (μ_B/f.u.)	T_c (K)	T_{sR} (K)	C (emuK/f.u.)	θ (K)	M_{eff} (μ_B/Fe atom)	μ_0H_a (T)	References
$Sm_2Fe_{14}B$	FM	31.3^b	616					40.0(4.2 K)	[84S4]
$Gd_2Fe_{14}B$	FIM	18.8^d	669					2.37(77 K);2.86(RT)	[85B3]
$Gd_2Fe_{14}B$	FIM	17.9^c	659					1.6(4.2 K)	[86H4]
$Gd_2Fe_{14}B$	FIM	18.2^d	655					2.1(77 K);2.7(RT)	[88Z1]
$Gd_2Fe_{14}BH_{4.4}$	FIM	22.3^d	698	355				plane	[88Z1]
$Tb_2Fe_{14}B$	FIM	13.2^c	620					30.6(4.2 K)	[86H4]
$Tb_2Fe_{14}B$	FIM	12.9^b	620						[84S4]
$Tb_2Fe_{14}B$	FIM	12.7^d	639					35.2(RT)	[84Y1]
$Tb_2Fe_{14}B$	FIM	12.5^d	618					32.0(77 K); 21.5(RT)	[88Z1]
$Tb_2Fe_{14}BH_{4.4}$	FIM	13.6^d	657					28.5(77 K); 14.0(RT)	[88Z1]
$Dy_2Fe_{14}B$	FIM	11.9^d	592		57.6		4.09	19.2(77 K), 15.8 (RT)	[85B3, 85B7]
$Dy_2Fe_{14}B$	FIM	11.3^c	598					16.7(4.2 K)	[86H4]
$Dy_2Fe_{14}B$	FIM	11.0^b	585					20.0(4.2 K)	[84S4]
$Dy_2Fe_{14}B$	FIM	12.8^d	602					31.4(RT)	[84Y1]
$Dy_2Fe_{14}B$	FIM	11.4^d	587					14.5(77 K); 16.5(RT)	[88Z1]
$Dy_2Fe_{14}BH_{4.6}$	FIM	13.7^d	638	45				6.5(77 K); 5.6(RT)	[88Z1]
$Ho_2Fe_{14}B$	FIM	11.2^c	573					7.5(300 K)	[86H4]
$Ho_2Fe_{14}B$	FIM	10.9^b	565						[84S4]
$Ho_2Fe_{14}B$	FIM	17.0^d	576					25.0(RT)	[84Y1]
$Ho_2Fe_{14}B$	FIM	11.7^d	568					7.2(77 K); 8.1(RT)	[88Z1]
$Ho_2Fe_{14}BH_{5.4}$	FIM	14.1^d	615					plane(77 K); 4.2(RT)	[88Z1]

(continued)

Table 7.32 (continued)

Compound	Magnetic Structure	M_s (μ_B/f.u.)	T_c (K)	T_{sR} (K)	C (emuK/f.u.)	θ (K)	M_{eff} (μ_B/Fe atom)	μ_0H_a (T)	References
$Er_2Fe_{14}B$	FIM	13.77^{f}	551	336					[87F4]
$Er_2Fe_{14}B$	FIM	14.1^{d}	556		53.8		4.10		[85B3, 85B7]
$Er_2Fe_{14}B$	FIM	12.9^{c}	551						[86H4]
$Er_2Fe_{14}B$	FIM	13.5^{b}	550					26.0(4.2 K)	[84S4]
$Er_2Fe_{14}B$	FIM	13.7^{d}	548	318					[88Z1]
$Er_2Fe_{14}BH_{3.7}$	FIM	16.6^{d}	588	352					[88Z1]
$Tm_2Fe_{14}B$	FIM	18.1^{c}	549						[86H4]
$Tm_2Fe_{14}B$	FIM	17.8^{b}	540					17.0(4.2 K)	[84S4]
$Tm_2Fe_{14}B$	FIM	21.6^{d}	541						[84Y1]
$Yb_2Fe_{14}B$	FIM	23^{h}	524	115					[91H2]
$Yb_2Fe_{14}B$	FIM		524(3)	115(3)					[86B5]
$Lu_2Fe_{14}B$	FM	28.2^{b}	535					4.0–6.0(4.2 K)	[84S4]
$Lu_2Fe_{14}B$	FM	28.1^{i}	549					0.94 (4.2 K)	[05T1]
$Lu_2Fe_{14}BH_{25}$	FM	30.2^{i}	602					0.67 (4.2 K)	[05T1]
$Y_2Fe_{14}B$	FM	29.5^{j}	575					2.35(4.2 K)	[85G2]
$Y_2Fe_{14}B$	FM	31.4^{c}	571					1.2(4.2 K), 2.0(300 K)	[86H4]
$Y_2Fe_{14}B$	FM	30.7^{b}	558					3.0–4.0(4.2 K)	[84S4]
$Y_2Fe_{14}B$	FM	31.6^{e}	565					2.1(77 K)	[85B3]
$Y_2Fe_{14}B$	FM	28.5^{g}	575						[86A4]
$Y_2Fe_{14}BH_{4.8}$	FM	31.7^{g}	610						[86A4]

[a] T = 5 K, μ_0H = 9.0 T; [b] T = 4.2 K, μ_0H = 14.0 T; [c] T = 4.2 K, μ_0H = 1.0 T; [d] T = 4.2 K, μ_0H = 2.0 T; [e] T = 4.2 K, μ_0H = 35.0 T; [f] T = 5 K, μ_0H = 1.0 T; [g] T = 4.2 K, μ_0H = 1.25 T; [h] T = 4.2 K, μ_0H = 9.0 T; [i] T = 4.2 K, μ_0H = 14.0 T; [j] T = 4.2 K, μ_0H = 8.0 T; 1 emu/g = 1 Am^2/kg

are initiated by the migration of the light boron atoms over structural vacancies, under the effect of electrical action.

The lattice constants and atomic coordinates of $Nd_2Fe_{14}B$ change under high pressure ($p \leq 0.5$ GPa) [20T1]. The shift of each atom folded back in the atomic position at ambient pressure.

The physical properties of pseudoternary $R_2Fe_{14}B$ compounds, where substitutions at rare-earth or iron sites are made, were intensively studied, particularly in connection with the technical performance of R–Fe–B permanent magnets. The locations of substituting elements, at the rare earth and iron sites, as well as the compositions ranges in which $R_2Fe_{14-x}M_xB$ solid solutions are formed with magnetic or nonmagnetic M elements are of interest from both scientific and technical poin of view.

The $(R_1,R_2)_2Fe_{14}B$, pseudoternary compounds where R_1 and R_2 are rare earths, yttrium or Th [86P4], form solid solutions in extended composition range. Elements such as Ti, Zr, Hf and Sc have been shown to substitute for rare-earth, up to about 15–20 at % [86J3, 87J1, 93C1]. In the $(R_{1-x}Sc_x)_2Fe_{14}B$ system with R = Nd, Dy, Er, Lu, Y, the solid solutions are formed in the composition range $x \leq 0.2$ [88C4]. The $(Nd_{1-x}Eu_x)_2Fe_{14}B$ series crystallize in *P*4$_2$/*mnm* type structure up to x = 0.4 [03C1].

The R_1 and R_2 elements, in the $(R_1,R_2)_2Fe_{14}B$ pseudoternary compounds, are generally not randomly distributed over the 4*f* and 4*g* sites, as they have different local environmen There is also a slightly larger volume of the 4*g* site, and consequently preferentially occupied by atoms with larger radii [86A2, 89M4]. The preferential substitutions, reported in literature, were not in all cases in agreement. The cerium in $(La_{1-x}Ce_x)_2Fe_{14}B$ was predicted to have a preference to substitute La at 4*f* site [97C1,13A1]. The Ce preference for 4*g* site over the 4*f* one, in $(Nd_{1-x}Ce_x)_2Fe_{14}B$ solid solutions, was of 7/3 [17S2]. In the $(R_{1-x}Ce_x)_2Fe_{14}B$ with R = La, Nd, cerium was found to enter the $R_2Fe_{14}B$ phases over the entire compositional range, and prefers the smaller 4*f* sites [16C1]. In the $(Nd_{1-x}Ce_x)_{25}Fe_{40}Co_{20}Al_4B_{11}$ multiphases system, the Nd prefers to enter in 2/14/1 phase, while cerium enters into the second RM_2 phase [18Z1]. The occupancy of lanthanum on 4*g* site, in $La_{0.1}Y_{0.9}Fe_{14}B$ system, exceeds than at 4*f* site by a factor of four [96K1].

Using the first principles method, the R site preferences were analysed in $(Nd_{1-x}R_x)_2Fe_{14}B$ with R = La, Pr, [16K2]. It was found that the La preferred the 4*f* site, at low doping ratio (x = 0.125), while the 4*g* site was more favourable at higher La content (x = 0.25). An opposite behavior was shown when R = Pr, the site preference being switched. In $(Nd_xR_{1-x})_2Fe_{14}B$, with R = La and Y, the total energy calculation indicates that La and Y prefer the 4*g* and 4*f* sites, respectively [13L1]. The substitution energies of La and Y in 2/14/1 phase are positive and negative, respectively. Thus, Y prefers to enter the 2/14/1 phase, while La tends to be expelled from the 2/14/1 structure to Nd-rich grain boundary phase in Nd–Fe–B magnets [13L1]. By using first-principles density functional calculation, has been analysed the partition of R = Dy and Tb in Nd–Fe–B magnets [12L1]. The substitution energies are negative, as Nd is replaced with 0.25% Dy or Tb. By neutron diffraction studies, on $(Nd_{1-x}Dy_x)_2Fe_{14}B$, was shown that Dy favours the 4*f* site, with the smaller Wigner–Seitz atomic cell [86Y6]. Latter on [13I1], by X-ray energy-dispersive spectroscopy was found that Dy is substituted into a few surface layers of $R_2Fe_{14}B$ grains in (Nd,Dy)-Fe-B hot deformed magnets, involving a grain boundary diffusion process. Thus, Dy diffuses only between $Nd_2Fe_{14}B$ grains. More Dy ions were found on the 4*f* site. The Dy favours the 4*f* site, as resulting from a combination of ab initio and thermodynamic calculations [17S1]. The preference of Dy for 4*f* sites, in $Nd_2Fe_{14}B$, was also evidenced by first principle calculations [12L1, 13H1]. The phase diagram of $Nd_{1-x}Pr_xFe14B$ was constructed [15Y1].

The crystal structures and lattice parameters of $R_2Fe_{14-x}M_xB$ solid solutions, where iron is replaced by a magnetic or non-magnetic M element, have been also investigated [98B2].

In the $Nd_2Fe_{14-x}Sc_xB$ system, the presence of solid solutions was shown up to x = 1.4 [87L1], while in the $R_2Fe_{14-x}Ti_xB$, these are present up to x = 0.7, when R = Nd, Y [89L4] or for $x \leq 1.4$ for R = Nd [87K5]. In vanadium doped samples, a tetragonal structure is shown up to x = 1.4, when R = Nd [87K5] or x = 1.0 for R = Y [91B4]. The $R_2Fe_{14-x}Cr_xB$ solid solutions are formed up to x = 4, for R = Y [87P2], x = 3 when R = Pr [89K4], R = Gd, Y [88K2], R = Y [92B1], x = 2, when R = Nd, Y [86A1], R = Y [90K8], x = 1.4 for R = Nd [87K5] and R = Y [02K1], or x = 1.0 when R = Nd [88M2].

The $R_2Fe_{14-x}Mn_xB$ pseudoternary compounds were investigated in the composition ranges up to x = 9, when R = Er [86M1, 87F3, 87F4], x = 5.6 for R = Pr, Nd, Y [90Y1], R = Pr [90P1], R = Nd [88C3], x = 5 as R = Y [87P2, 89S1, 91M1], x = 3.5 when R = Y [87L2], x = 3 for R = Nd [87B5], R = Nd, Y [85L1, 88M2], R = Er [90V2], x = 2, as R = Pr, Nd, Gd, Y [86H6], R = Nd, Y [86A1], or x = 1.4 when R = Nd [87K5] and Y [02K1].

The $R_2Fe_{14-x}Co_xB$ series form solid solutions in all the composition range when R = La [88G6, 88V2], R = Ce [13S1], R = Pr [85R1, 86P2, 87B6, 87G3], R = Nd [85A1, 85M1, 86A1, 86H3, 86H6, 86H7, 87B4, 87H8, 87T1, 88C2, 88L1, 88Y1, 88Y5, 90G1, 90K2, 01D1, 09B1], R = Gd [85L2], R = Tb [87P7, 98B1], R = Dy [87P4, 01D1], R = Er [86M1], R = Tm [88P8], R = Y [87G5, 87T1, 89H1, 89S1, 06K1]. The pseudoternary $R_2Fe_{14-x}Ni_xB$ compounds, having tetragonal structure, with substitutions up to x = 5 for R = Nd [87C4, 88C6], R = Y [85B5, 86B8, 87P2], x = 3 for R = Nd [87B5, 87C4, 88C6], $x \cong 2$ when R = Nd, Y [86A1, 87L2], $x \cong 1.5$ for R = Ce [18O1] or Nd [87L1, 87Y2] were investigated.

The solid solutions, of $R_2Fe_{14-x}Cu_xB$ type, were formed only for substitutions up to x = 1.5 for R = Nd, Gd, Y [88C2, 88K1, 88K2], R = Nd, Er [90B2], R = Y [87P9] or x = 1.0 when R = Pr [87P9].

The $R_2Fe_{14-x}Al_xB$ solid solutions were investigated in the composition range up to x = 4 for R = Y [86B8], x = 3 when R = Nd [85A1, 86A1, 88H2], x = 2.5–2.0 when R = Ce [18O1], R = Er [88H2, 91K3], x = 1.4–1.7 when R = Y [87L2, 90W2], or x = 1 as R = Nd [88C2, 89K3], R = Gd, Y [89S5]. Solid solutions with aluminium content up to x = 0.7–1.0, for R = Pr [89K4] or R = Gd, Y [89S5], were also analysed. The $R_2Fe_{14-x}Si_xB$ series with $P4_2/mnm$ type structure, were shown in composition ranges up to x = 2.33 for R = Ce [18O1], x = 2 when R = Pr [89K4], R = Pr, Nd, Er [87P8], R = Nd, Y [87J2], R = Nd [93M3], R = Gd, Y [88K2, 89S5] and R = Y [87P6, 88Y4, 90K8, 93M2, 94M1].

The iron was replaced by gallium up to x = 2, in $R_2Fe_{14}B$ compounds with R = Nd [88X1, 89Q1], x = 1.8 for R = Nd [00C2], or up to x = 1.0 when R = Pr [89K4], R = Nd [88G7, 89Q1, 91B4], R = Y [89S5, 91B4]. The $R_2Fe_{14-x}Ge_xB$ pseudoternary compounds, form solid solutions up to x = 0.75 as R = Y [91B4] or x = 0.5 when R = Nd [90K6].

The $Nd_{16}Fe_{76-x}Sn_xB_8$ alloys with $x \leq 4$ were prepared [96M1]. The presence of $Nd_2Fe_{14-x}Be_xB$ pseudoternary system, with $x \leq 1$, was also reported [96L2].

The $Nd_2Fe_{14-x}Nb_xB$ alloys having $x \leq 1.12$ [88H4], $x \leq 0.5$ [89J1] or x = 0.28 [94A1] were investigated. In the last report, the presence of a second phase, in addition to the tetrahedral one, was shown. The $Nd_2Fe_{14-x}Mo_xB$ solid solutions with $x \leq 1.12$ were reported [88H4] and the structure of the $Nd_{16}(Fe_{0.78}Co_{0.18}Mo_{0.04})_{76}B_8$ alloy has been analysed [90D3]. Small quantities of Mo dissolves in $Nd_2Fe_{14}B$ phase [89L5].

The iron was replaced by ruthenium in $Nd_2Fe_{14-x}Ru_xB$, up to x = 2.8 [86K4], x = 2 [86P3], or x = 1.4 [87Y2]. The $Y_2Fe_{14-x}Re_xB$ series, keep the tetragonal structure up to x = 0.5 [89J3]. The $Nd_{16}(Fe_{0.80}Co_{0.12}W_{0.02})_{76}B_8$ alloy was also prepared, as possible permanent magnet material [90D3].

The preferential occupancy of M substitutional elements, at iron sites, in $R_2Fe_{14-x}M_xB$ pseudoternary compounds, influence sensitively their physical properties. Consequently, this matter has been investigated, particularly for improving the performances of R–Fe–B permanent magnets.

The vanadium in $YFe_{14-x}V_xB$ was suggested to be preferentially located on $16k_2$ sites and to a lesser extent in $16k_1$ ones [95M1]. The sites occupancies of M = V, Cr, Mn, Zr and Nb elements in $Nd_2Fe_{14-x}M_xB$ were theoretically investigated [05W2]. All the above elements were shown to prefer the larger $8j_2$ site, the preferential order following the sequence Nb-V-Cr-Mn-Zr. The chromium substitutes iron in $Nd_2Fe_{14-x}Cr_xB$, at the $8j_2$ site [88M2] while in $Y_2Fe_{14-x}Cr_xB$, at the $4e$ positions [92B1].

By using the neutron diffraction method, the manganese in $R_2Fe_{14-x}Mn_xB$ was shown to prefers the $8j_2$ site and the iron atoms the $16k_1$ and $16k_2$ sites, when R = Er [87F3, 87F4, 88F2]. The $8j_2$ sites become essentially fully occupied at x = 9. As the $8j_2$ site saturates, the manganese occupancy of the $8j_1$ site increases. The preference of Mn for $8j_2$ site was shown also in pseudoternary systems with R = Pr [87Y1, 90P1] or R = Y [89S1, 91M1].

The sites preferences in $R_2(Fe,Co)_{14}B$ compounds [85H2, 90R3, 90Z3, 93L1] are presented in the Sect. 8.9 and no more discussed in the present chapter. In $Nd_2Fe_{14-x}Ni_xB$ system, the nickel was reported to be located at $16k_2$ and $8j_2$ sites [88D1], $8j_2$ [88G7] or $16k_2$ sites [89W1].

According [92B1], in a first approximation, copper in $Y_2Fe_{14-x}Cu_xB$, is randomly distributed among iron sites. A site preference of Cu for $16k_2$ site and in lower extent for $16k_1$ site was latter reported [95M1].

In $R_2Fe_{14-x}Al_xB$ series, when R = Nd, the aluminium substitution occurs first at the $8j_2$ site, up to a concentration corresponding to one substituted atom per unit cell and further substitutions takes place at the $16k_2$ sites [88H2]. The $8j_2$ site preference for aluminium in this system was further reported [87Y1, 89G2, 89W1]. In the compounds with R = Gd, Y, a preference of aluminium for $16k_2$ site was suggested [92W1]. The neutron diffraction study, on $Y_2Fe_{14-x}Al_xB$, shows that aluminium atoms occupy the $8j_2$ and $4c$ sites [96H2]. The Al affinity appears to be determined by steric effects ($8j_2$) or by site environment ($4c$).

The silicon location in $R_2Fe_{14-x}Si_xB$ compounds has been more intensively investigated. When R = Nd, a preference of Si at $16k_2$ and $16k_1$ sites and of iron for $8j$ sites was shown [88G7]. By using the ^{57}Fe Mössbauer spectroscopy on pseudoternary systems with R = Nd, Y, a random distribution of silicon over the smaller $16k_1$, $16k_2$, $8j_1$, $4c$ and $4e$ sites was reported [89P2]. Apparently, the silicon is at least in part excluded from the $8j_2$ sites. According [90L4], when R = Nd, the most probably occupancy of silicon is $16k_2$ site, followed by $16k_1$ and $4e$ ones. While in [92W1], reported that Si preferably is located at $16k_2$ sites, EXAFS studies on $Nd_2Fe_{14-x}Si_xB$ evidenced that silicon occupies the $4c$ site [88B6, 89B1]. The neutron diffraction studies on $R_2Fe_{14-x}Si_xB$, with R = Nd, Y, showed that silicon is located preferentially at the $4c$ site and, to a lesser extent, at the $8j_1$ site, it is almost excluded from the $16k_2$ site and avoids the $16k_1$, $8j_2$ and $4e$ positions [93M2, 93M3, 94M1]. The silicon site occupancy was correlated with its preference to have rare-earth (yttrium) atoms in their local environment. The same conclusion was obtained by neutron diffraction study on $Nd_2Fe_{14-x}Si_xBH_y$ hydrides, where an important silicon content, at $4c$ site was found [00C1].

A preference of gallium for $16k_2$ sites, in $Nd_2Fe_{14-x}Ga_xB$ system, was shown [88G7, 92W1]. A neutron diffraction study, on this system, evidenced a preferential location of Ga at Fe$4c$, Fe$8j_1$ and $16k_2$ sites [00C2]. The Sn in $Nd_2Fe_{14-x}Sn_xB$ pseudoternary compounds was shown to be located at $16k_2$ site [96M1].

By ^{57}Fe Mössbauer spectroscopy, on $Nd_{16}(Fe_{0.78}Co_{0.18}Mo_{0.04})_{76}B_8$ alloy, was concluded that molybdenum is located at $16k_2\{4e,4c\}$ sites and in $Nd_{16}(Fe_{0.80}Co_{0.18}W_{0.02})_{76}B_8$, the W atoms are distributed on $8j_1$, $16k_2$ and $4e$ sites [90D3].

From the above data, a reasonably agreements concerning the location of substituting M elements, in $R_2Fe_{14-x}M_xB$ compounds, reported by different methods or authors, are shown. The differences in case of Si substitution can be, in some extent, correlated with the methods used in analysing experimental results. The atomic size considerations, on their own seem to be not sufficient to explain the experimental results. The energy effects, as the relative bond strength between neighbour atoms seems to be also important. Analysis based on the correlation of site occupancy with the Wigner–Seitz atomic cell was shown also to be useful [96L3]. The content of substituting element play also a role.

The $R_2Fe_{14}B$ compounds adsorb significant amounts of hydrogen, forming stable hydrides [84L1, 84O2, 85D2, 85H2, 86F1, 87C5, 88Z1, 95I1, 00C1, 00P1, 05T1]. The hydrogen content, in $R_2Fe_{14}BH_y$ hydrides can rise up to y = 5 and is accompanied by an increase of the unit cell volume; the expansion of the lattice occurs mainly along the **a**-axis (basal plane). The increase of lattice parameters in hydrides, is followed by an increase of the distances between iron atoms. Thus, the distance $d_{k_2-j_1}$ = 0.239 nm, in $Nd_2Fe_{14}BH_y$, for y = 0, increases at y = 4, up to 0.243 nm [95I1]. Also the $d_{j_1-j_1}$ interatomic distances in the same hydrogen composition range increases from 0.244 nm to 0.250 nm. The hydrogen insertion in $Nd_2Fe_{14}B$ crystal structure induces a significant reorganization of Nd environment.

In hydrides with light rare-earth elements, as $Nd_2Fe_{14}BH_y$, four non-equivalent interstitial sites have been found to be occupied by hydrogen atoms, all of which being pseudotetrahedral [95I1]: H1 in 8*j* site with three Nd atoms and one iron at the corners, H2 in site $16k_1$, surrounded by two Nd and two iron, H3 in site $16k_2$ with two iron and two Nd atoms at the corners and H4 in site 4*e* with two Nd and two iron as closest neighbours. The hydrogen is initially accommodated in H1 site. The decrease of the H1 site occupancy, when increasing the content of adsorbed hydrogen was correlated with gradual filling of H2 site, since these tetrahedral sites share an edge, and only then the remaining hydrogen sites. The occupancy of the H4 site it is generally low. This filling scheme is characteristic for hydrides with R = Ce [87D1] or Nd [95I1]. For heavy rare-earth hydrides, a different filling scheme was observed. In this case, a saturation of H1 site was observed, the H2 site occupancy being moderate. The reason for different filling schemes can be correlated with the relative sizes of light and heavy rare-earth elements, cell dimensions, respectively.

The N atom in $Nd_2Fe_{14}BN_{0.352}$ has a preference for the 4*e* sites over the 4*f* interstitial sites [20Z1]. The expansion of the unit cell is slight by introducing nitrogen atoms.

7.10.2 Physical Properties

The physical properties of ternary $R_2Fe_{14}B$ were intensively investigated, since of their interesing basic properties as well as in connection with technical uses. These studies were focussed on all the $R_2Fe_{14}B$ compounds, although with preference for those with neodymium, the basic compound in manufacturing permanent magnets: R = La [84S4, 97Z1, 00M1, 01F1]; R = Ce [84S4, 85B3, 85B8, 85H4, 86H1, 86H4, 86Y2, 87N1, 88L2, 88L4, 93C2, 97Z1, 00M1, 16I1]; R = Pr [84S4, 85B3, 85B7, 85B8, 85H4, 86H2, 86H4, 86Y2, 87H4, 88L2, 88V1, 89S4, 90L6, 91L2, 96K2, 97Z1, 00M1, 01F1, 04A1, 16I1]; R = Nd [84G1, 84S4, 85B3, 85B7, 85B8, 85C1, 85G3, 85H4, 85K1, 86G3, 86G4, 86H2, 86H4, 86Y2, 87C2, 87N1, 88C1, 88L1, 88L2, 88L4, 88V1, 89S4, 90L3, 90L6, 91H2, 91L2, 92M1, 93G1, 96G3, 96K2, 96Y1, 97Z1, 00B2, 00G3, 00M1, 01F1, 04A1, 04K1, 05H1, 08B2, 09G1, 13T1, 16I1, 16K1]; R = Sm [84S4, 85B3, 85B8, 85H5, 86H4, 86Y2, 87K3, 88L2, 96K2, 97Z1, 01F1, 04A1]; R = Gd [84S4, 85B2, 85B3, 85B8, 85H4, 86H4, 86Y2, 88L2, 91L2, 91L3, 91Z3, 93Y2, 96K2, 97Z1, 00M1, 01F1, 04A1, 10K2]; R = Tb [84S4, 85H4, 85K1, 86H4, 86Y2, 87G1, 88G3, 88L2, 89Z7, 90L6, 90Z5, 91L2, 93S1, 94Z1, 96K2, 97Z1, 99L2, 00M1, 01F1, 04A1]; R = Dy [84S4, 85B3, 85B7, 85B8, 85H4, 86H4, 86Y2, 88G3, 88L2, 88L4, 88V1, 89Z7, 90L6, 90Z5, 91L2, 93L2, 93Y2, 94Z1, 95K1, 96G3, 97Z1, 00M1, 16I1]; R = Ho [84S4, 85H4, 86H4, 86Y2, 87F5, 88G6, 88L2, 89Z7, 90Z5, 91L2, 94Z1, 96G3, 96K2, 97Z1, 00M1, 01F1, 04A1]; R = Er [84S4, 85B3, 85B7, 85B8, 85H3, 86Y2, 86Y5, 87C1, 87G1, 88G3, 88L4, 89Z7, 90C1, 90Z5, 91L2, 92L2, 93S1, 93Y2, 94Z1, 96K2, 97Z1, 00M1, 01F1, 04A1,

13T1]; R = Tm [84S4, 85D1, 85H3, 85Y1, 86H4, 86P7, 87C1, 88G1, 88G3, 89Z7, 90C1, 90L6, 90Z5, 91L2, 92L1, 94Z1, 95K1, 96K2, 97Z1, 00M1, 01F1, 04A1]; R = Yb [88B5, 89M2, 00M1, 01F1, 04A1]; R = Lu [84S4, 86H1, 97Z1, 00M1, 01F1, 04A1] R = Y [85B2, 85B3, 85B8, 85G3, 85H4, 85K1, 86G3, 87N1, 88L2, 88L4, 91K8, 91Z3, 93G1, 96K2, 97Z1, 10K2], R = Sc [90C2]. The physical properties of $R_2Fe_{14}B$ compounds with two or more rare earths, in addition to Fe substitutions were also reported [91K4, 91P1, 94I1, 96H1, 09P1].

The electronic structures of $R_2Fe_{14}B$ compounds were also investigated, as for R = Pr [93H3, 93L2, 08B1], R = Nd [85S2, 87G4, 88J1, 90J1, 90R1, 90Z6, 91N1, 92H2, 92H3, 93H3, 93L2, 93N1, 96H3, 08B1, 09K1, 16K1, 19S1]; R = Sm [93H3, 93L2, 08B1] R = Gd [87C3, 88C5, 91C2, 92H2, 92H3, 92L3, 93H3, 93L2, 98Y1, 08B1]; R = Tb [92H2, 92H3, 92L3, 93H3, 93L2, 08B1]; R = Dy [92H2, 92H3, 92L3, 93H3, 93L2, 08B1, 09K1]; R = Ho [92H2, 92H3, 93H3, 93L2, 08B1]; R = Er [92H2, 92H3, 93H3, 93L2, 08B1]; R = Tm [93H3, 93L2]; R = Y [86I1, 87G5, 87I2, 88S1, 91C1, 91C3, 98Y1, 16K2].

The compounds with nonmagnetic R elements (R = La, Ce, Lu, Y) are ferromagnetic, the magnetic moments of iron, at different sites being parallelly aligned. In compounds with magnetic light rare earths (R = Pr, Nd, Sm), there is a parallel alignment of R and Fe magnetizations, respectively. When R is a magnetic heavy rare-earth, the compounds are ferrimagnetically ordered. According to Néel classification [48N1], the temperature dependences of magnetizations in $R_2Fe_{14}B$ compounds with R = Tb, Dy, Ho, Er and Tm are of P-type, while for R = Gd is of Q—type. The $R_2Fe_{14}B$ compounds with R = Nd, Ho, Er, Tm, Yb show discontinuities in their temperature dependences of the magnetizations, attributed to spin reorientation effects. Non-collinear magnetic structures were shown in these systems, at low temperatures. In $R_2Fe_{14}B$ compounds the R4f electrons, responsible for the magnetism of rare-earths, are localized and their magnetic properties are described in term of concepts used in atomic theory, whereas, the Fe3d electrons are itinerant and their magnetic behavior can be analysed better when using band models. The main magnetic properties of $R_2Fe_{14}B$ compounds are listed in Table 7.33.

The Curie temperatures, T_c of $R_2Fe_{14}B$ compounds are more than one half smaller than that of iron metal, although the iron content is of 82.4 at %. As mentioned in Sect. 7.10.1, in the *P4*$_2$*/mnm* type structure, there are rather short distances ($d_{Fe\text{-}Fe}$ < 0.250 nm) between some iron sites, as k_2-j_1 or j_1-j_1. The exchange interactions between iron atoms are of short range and dependent on the iron separation [74G1, 81B1, 85G2, 96H2], as usually described in terms of Néel-Slater curve [36N1]. The magnetic interactions between iron atoms situated at distances $d_{Fe\text{-}Fe}$ < 0.250 nm, are negative, while those between iron atoms, located at greater distances than the above ones, are positive and of higher intensity. The negative interactions are not satisfied and considerable magnetic energy is stored, that brings about their Curie temperatures. The T_c values of $R_2Fe_{14}B$ compounds are linearly dependent on De Gennes factor $G = (g_J - 1)^2 J(J + 1)$, with different slopes for light, s_l, and heavy, s_h, rare-earth compounds, their ratio being $s = s_l/s_h \cong 1.70(10)$—Fig. 7.18. When plotting T_c as function of R5d band polarization a linear dependence is shown, with the same slope both for heavy and light rare earths $R_2Fe_{14}B$ compounds, as expected

Table 7.33 Magnetic moments determined by neutron diffraction, ^{57}Fe Mössbauer spectroscopy and band structure calculations ($P4_2/mnm$ space group)

Compound[a)]	Magnetic structure	T (K)	Magnetic moment (μ_B/atom)									References
			R4*f*	R4*g*	4*e*	4*c*	8*j*$_1$	8*j*$_2$	16*k*$_1$	16*k*$_2$	B	
$Ce_2Fe_{14}B$ (ND)	FM	77			2.1(2)	2.4(1)	2.7(2)	3.4(1)	2.7(1)	2.2(1)		[86H1]
$Ce_2Fe_{14}B$ (MS)	FM	4.2			2.19(10)	1.72(10)	2.36(3)	1.95(10)	1.96(3)	2.10(3)		[87F2]
$Ce_2Fe_{14}B$ (ND)	FM	30	≤0.35	≤0.35	1.90	2.15	2.30	2.55	2.15	2.08		[87D1]
$Ce_2Fe_{14}BH_{3.7}$ (ND)	FM	30			1.80	2.25	2.05	2.50	2.05	2.10		[87D1]
$Pr_2Fe_{14}B$ (ND)	FM	77	2.6(1)	2.8(1)	1.7(2)	1.7(2)	3.3(1)	2.4(1)	2.6(1)	2.5(1)		[85H2]
$Pr_2Fe_{14}B$ (MS)	FM	4.2			2.18(10)	1.77(10)	2.37(3)	2.01(10)	2.07(3)	2.15(3)		[87F2]
$Nd_2Fe_{14}B$ (ND)	FM	4.2	2.30(5)	2.25(5)	2.10(6)	2.75(6)	2.85(5)	2.30(5)	2.60(4)	2.60(4)		[85G2, 85G3]
$Nd_2Fe_{14}B$ (ND)	FM	77	2.3(1)	3.2(1)	1.1(2)	2.2(3)	3.5(1)	2.7(1)	2.4(1)	2.4(1)		[84H1]
$Nd_2Fe_{14}B$ (MS)	FM	4.2			2.28(10)	1.97(10)	2.43(3)	2.06(10)	2.08(3)	2.16(3)		[87F2]
$Nd_2Fe_{14}B$ (MS)	FM	4.2			2.00	2.17	2.21	2.55	2.24	2.30		[87O4]
$Nd_2Fe_{14}B$ (BS)	FM		3.01	3.04	1.99	2.97	2.48	3.40	2.15	2.28	−0.20	[87G4]
$Nd_2Fe_{14}B$ (BS)	FM		−0.55[b]	−0.52[b]	2.13	2.59	2.12	2.74	2.15	2.18	−0.20	[90J1]
$Nd_2Fe_{14}B$ (BS)[c)]	FM		−0.57[b)]	−0.53[b]	2.15	2.45	2.10	2.86	2.35	2.23	−0.18	[91N1]
$Nd_2Fe_{14}B$ (BS)[d1)]	FM		2.63	2.72	1.91	2.41	2.04	2.69	2.00	2.08	−0.17	[93M5]
$Nd_2Fe_{14}B$ (BS)[d2)]	FM		−0.38	−0.37	1.97	2.47	2.17	2.72	2.09	2.17	−0.18	[93M5]
$Nd_2Fe_{14}B$ (BS)	FM				2.03	2.46	2.59	2.30	2.18	2.38	−0.19	[96H3]
$Nd_2Fe_{14}BH_4$ (ND)	FM	300	1.8(3)	2.2(3)	2.4(3)	2.7(2)	2.9(1)	3.0(2)	2.2(2)	2.5(3)		[95I1]
$Nd_2Fe_{13}SiB$ (ND)	FM	2	3.2(2)	3.3(2)	1.7(3)	2.0(3)	2.8(2)	2.4(2)	2.3(2)	2.6(2)		[00C1]
$Nd_2Fe_{12}Si_2B$ (ND)	FM	2	3.0(2)	3.3 (2)	1.6(2)	1.9(2)	2.8(2)	2.9(2)	2.2(1)	2.3(1)		[00C1]
$Nd_2Fe_{13}SiBH_{2.8}$ (ND)	FM	2	3.0(2)	2.8(2)	2.0(2)	2.2(2)	3.0(2)	2.9(2)	2.4(1)	2.6(1)		[00C1]
$Nd_2Fe_{12}Si_2BH_{1.9}$ (ND)	FM	2	3.4(3)	3.2(3)	2.3(3)	2.4(3)	3.0(2)	2.9(2)	2.5(2)	2.5(1)		[00C1]

(continued)

Table 7.33 (continued)

Compound[a)]	Magnetic structure	T (K)	Magnetic moment (μ_B/atom)									References
			R4*f*	R4*g*	4*e*	4*c*	8*j*$_1$	8*j*$_2$	16*k*$_1$	16*k*$_2$	B	
$Nd_2Fe_{13.45}Si_{0.55}B$ (ND)	FM	8	1.8(2)	3.2(2)	2.1(3)	0.2(5)	2.8(1)	3.7(2)	1.8(1)	2.5(2)		[93M3]
$Nd_2Fe_{13.08}Si_{0.92}B$ (ND)	FM	8	1.8(2)	3.5(2)	2.1(2)	0	2.7(1)	3.6(2)	1.9(1)	2.4(1)		[93M3]
$Nd_2Fe_{12.69}Si_{1.31}B$ (ND)	FM	8	1.6(2)	3.3(2)	2.0(4)	0	2.5(1)	3.3(2)	1.8(1)	2.3(2)		[93M3]
$Nd_2Fe_{11.96}Si_{2.04}B$ (ND)	FM	8	1.8(2)	2.7(2)	1.8(2)	0	2.4(2)	3.1(2)	1.7(1)	2.0(2)		[93M3]
$Nd_2Fe_{13}GaB$	FM	4	2.5(1)	3.1(1)	2.1(2)	2.0(2)	2.6(2)	2.6(1)	2.3(1)	2.6(1)		[00C2]
$Nd_2Fe_{13.2}Ga_{1.8}B$	FM	4	2.0(2)	2.6(1)	1.6(3)	1.5(2)	2.6(3)	2.7(2)	2.0(1)	2.6(1)		[00C2]
$Gd_2Fe_{14}B$ (MS)	FIM	4.2			2.28	1.74	2.70	2.52	2.10	2.22		[87F2]
$Gd_2Fe_{14}B$ (BS)	FIM		−7.64	−7.67	2.04	2.48	2.33	2.60	2.19	2.39	−0.18	[92H2, 92H3]
$Gd_2Fe_{14}B$ (BS)	FIM		−7.42	−7.40	2.49	1.99	2.59	2.12	2.41	2.13		[98Y1]
$Gd_2Fe_{14}B$ (BS)	FIM		−7.50	−7.44	2.03	2.42	2.24	2.52	2.10	2.33	−0.16	[91C2]
$Tb_2Fe_{14}B$ (ND)	FIM	8	−7.9(1)	−8.5(1)	2.1(1)	2.5(1)	1.7(1)	2.6(1)	2.4(1)	1.7(1)		[93H1]
$Tb_2Fe_{14}B$ (BS)	FIM		−9.60	−9.63	2.03	2.47	2.32	2.60	2.19	2.39	−0.18	[92H2, 92H3]
$Dy_2Fe_{14}B$ (ND)	FIM	77	−8.9(1)	−9.2(1)	2.4(1)	2.5(1)	3.0(1)	2.5(1)	2.6(1)	2.5(1)		[85H2]
$Dy_2Fe_{14}B$ (MS)	FIM	4.2			2.28(10)	1.69(10)	2.49(3)	2.10(10)	2.06(3)	2.17(3)		[87F2]
$Dy_2Fe_{14}B$ (BS)	FIM		−10.55	−10.58	2.02	2.44	2.30	2.59	2.17	2.38	−0.18	[92H2, 92H3]
$Ho_2Fe_{14}B$ (BS)	FIM		−10.52	−10.54	2.02	2.44	2.29	2.59	2.17	2.37	−0.18	[92H2, 92H3]
$Ho_2Fe_{14}B$ (MS)	FIM	4.2			2.06	1.95	2.29	2.63	2.03	2.76		[87F2]

(continued)

Table 7.33 (continued)

Compound[a)]	Magnetic structure	T (K)	Magnetic moment (μ_B/atom)									References
			R4*f*	R4*g*	4*e*	4*c*	8*j*$_1$	8*j*$_2$	16*k*$_1$	16*k*$_2$	B	
$Er_2Fe_{14}B$ (ND)	FIM	77	−7.07(10)	−7.39(10)	1.36(15)	2.91(12)	2.06(10)	2.79(8)	2.50(8)	2.35(9)		[86Y5]
$Er_2Fe_{14}B$ (ND)	FIM	6	−9.5	−9.5	2.00	2.25	1.95	2.90	2.70	2.00		[87D1]
$Er_2Fe_{14}B$ (MS)	FIM	4.2			2.25(16)	1.96(16)	2.25(5)	2.15(16)	2.08(6)	2.21(5)		[87F2]
$Er_2Fe_{14}B$ (BS)	FIM		−9.48	−9.5	2.02	2.42	2.28	2.58	2.16	2.36	−0.18	[92H2, 92H3]
$Er_2Fe_{14}BH_{2.58}$ (ND)	FIM	25	−9.5	−9.5	2.00	2.05	2.35	2.85	2.75	2.20		[87D1]
$Tm_2Fe_{14}B$ (ND)[e)]	FIM	7	−7.00 Ø= 34.1(6.4)[f)]	−7.00 Ø= 14.9(1.8)[f)]	2.15(5)	1.95(5)	2.80(4)	2.40(4)	2.25(3)	2.25(3)		[85Y1]
$Tm_2Fe_{14}B$ (ND)	FIM	294	−2.6(2)	−2.4(2)	0.9(2)	3.0(2)	1.7(2)	1.7(2)	1.9(2)	1.7(2)		[85D1]
$Tm_2Fe_{14}B$ (ND)	FIM	340	−2.1(3)	−1.4(3)	0.4(2)	2.9(2)	1.8(2)	1.9(2)	1.4(2)	1.1(2)		[85D1]
$Lu_2Fe_{14}B$ (ND)	FM	77			1.7(2)	2.2(2)	3.6(2)	2.9(2)	2.8(2)	2.4(2)		[86H1]
$Lu_2Fe_{14}B$ (MS)	FM	4.2			2.06(10)	1.70(10)	2.32(3)	2.00(10)	1.95(3)	1.99(3)		[87F2]
$Y_2Fe_{14}B$ (ND)	FM	4.2			2.15(5)	1.95(5)	2.80(4)	2.40(4)	2.25(3)	2.25(3)		[85G1, 88G2]
$Y_2Fe_{14}B$ (ND)	FM	20			2.71(11)	2.44(10)	3.09(7)	2.33(8)	2.60(7)	2.64(6)		[01W1]
$Y_2Fe_{14}BN_{0.317}$ (ND)	FM	300			1.15	1.31	1.99	2.72	1.31	1.89		[92Z1]
$Y_2Fe_{14}B$ (BS)					2.32	2.28	2.16	2.74	2.41	2.11		[86I1]
$Y_2Fe_{14}B$ (MS)	FM	4.2			2.28(10)	1.90(10)	2.43(3)	2.31(10)	2.07(10)	2.23(3)		[87F2]
$Y_2Fe_{14}B$ (MS)	FM	4.2			2.05	2.01	2.11	2.55	2.21	2.31		[87E1]
$Y_2Fe_{14}B$ (MS)	FM	4.2			2.10	2.25	2.15	2.59	2.25	2.32		[87O4]
$Y_2Fe_{14}B$ (BS)	FM		−0.33	−0.32	1.87	2.32	2.67	2.16	2.15	2.20	−0.12	[98Y1]
$Y_2Fe_{14}B$ (BS)	FM		−0.43	−0.39	2.01	2.40	2.22	2.51	2.1	2.31	−0.15	[91C1]

(continued)

Table 7.33 (continued)

Compound[a)]	Magnetic structure	T (K)	Magnetic moment (μ_B/atom)									References
			R4*f*	R4*g*	4*e*	4*c*	8*j*$_1$	8*j*$_2$	16*k*$_1$	16*k*$_2$	B	
$Y_2Fe_{14}B$ (BS)	FM				2.20	2.55	2.10	2.83	2.14	2.24		[88S1]
$Y_2Fe_{14}B$ (BS)	FM				2.32	2.28	2.16	2.74	2.41	2.11		[86I1]
$Y_2Fe_{14}B$ (BS)	FM				2.11	2.31	2.35	2.61	2.36	2.62		[87I2]
$Y_2Fe_{14}B$ (ND)	FM	4.2			1.95	2.25	2.25	2.80	2.40	2.15		[86G4]
$Y_2Fe_{14}B$ (ND)	FM	10			1.70	2.25	2.45	2.65	2.50	2.50		[87D1]
$Y_2Fe_{14}BH_{3.6}$ (ND)	FM	10			1.80	1.95	2.41	2.75	2.60	2.21		[87D1]
$Y_2Fe_{13.72}Si_{0.28}B$ (ND)	FM	8			2.4(3)	1.8(3)	2.7(2)	3.3(2)	1.7(2)	1.9(2)		[94M1]
$Y_2Fe_{13.38}Si_{0.67}B$ (ND)	FM	8			2.4(1)	1.9(1)	2.7(1)	3.3(1)	2.1(1)	1.3(2)		[94M1]
$Y_2Fe_{12.13}Si_{1.87}B$ (ND)	FM	8			1.9(1)	1.5(1)	2.2(2)	2.6(2)	1.3(2)	2.1(2)		[94M1]
$Y_2Fe_{13.58}Mn_{0.42}B$ (ND)	FM	295			1.4(3)	1.0(3)	1.7(3)	3.2(3)	1.7(2)	2.2(2)		[91M1]
$Y_2Fe_{12.6}Mn_{1.4}B$ (ND)	FM	8			1.5(3)	1.4(3)	1.4(3)	3.4(4)	1.3(2)	2.1(2)		[91M1]
$Y_2Fe_{10.5}Mn_{3.5}B$ (ND)	FM	8			0.8(4)	0.9(4)	0.4(4)	2.9(4)	0.7(2)	1.1(2)		[91M1]

[a] Determined by neutron diffraction (ND), ^{57}Fe Mössbauer spactroscopy (MS) or from band structure (BS); [b] Conduction electron polarization; [c] Calculated with 4f spin moment; [d] Nd4f electrons as valence[1)] or core[2)] [e] Same iron moments as in $Nd_2Fe_{14}B$; [f] angle with **c**-axis;

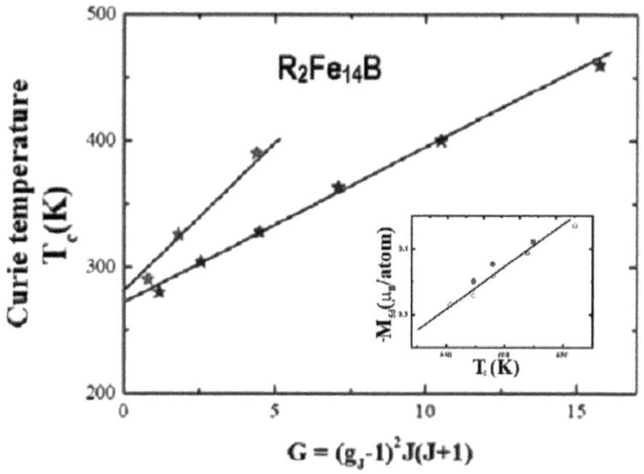

Fig. 7.18 $R_2Fe_{14}B$: Curie temperatures, T_C, as function of De Gennes factor. In inset, the T_C values are ploted as function of the total R5d (Y4d) band polarizations

in the 4f-5d-3d exchange interaction model [20B1, 21B1]. In this model, the exchange interactions are mediated by R5d band polarization.

By magnetization measurements, only a general view on the magnetic behavior $R_2Fe_{14}B$ system can be obtained. Their magnetic structures in some cases, rather complex, were determined mainly by neutron diffraction or Mössbauer spectroscopy—Table 7.33.

The neutron diffraction measurements on $R_2Fe_{14}B$ compounds with non-magnetic R elements, R = Lu [86H1], Ce [86H1] and Y [85G1] showed that iron moments depend on site location, and are parallelly aligned—Table 7.33. The largest iron moment was observed at Fe8j_2 site, having the highest iron coordination number, 12 from 14 neighbouring atoms. The smaller iron moment was found at 4e site, which has in their environment 2R and 2B atoms, in addition to 9Fe. The anisotropy of the iron sublattice was shown to be uniaxial.

Band structure calculations, on $Y_2Fe_{14}B$, evidenced that the Y4d band is negatively polarized, the polarization being induced by Y4d-Fe3d interactions, hybridization effects, respectively [91C1, 98Y1]. The induced polarization, in absolute value, is higher at the Y4*f* site than that at Y4*g* one, behavior connected with their different local environments.

The cerium, in $Ce_2Fe_{14}B$ compound was found to be in α-like state, which is nonmagnetic and strongly mixed valent [86H1, 92C1, 93C2]. The neutron diffraction measurements on $Ce_2Fe_{14}BH_y$ hydride, showed that cerium has a γ-like $4f^1$configuration, on one of the two unequivalent sites [87D1, 87F1]. According to [92C1], the strongly mixed -valent state of parent $Ce_2Fe_{14}B$ phase, do not change on hydriding. Latter on [93C2], reported that, in cerium based compounds, there is a shift towards a γ-like cerium state, as the steric volume of cerium site increases.

The magnetic moments of Pr and Fe, in $Pr_2Fe_{14}B$ are parallelly aligned [85H2]. The magnetic behavior of praseodymium is dependent on their location, in 4*f* or 4*g* sites respectively. The iron moments follow the same trend as function of site location, as in isomorphous compounds with non-magnetic elements. The anisotropy of $Pr_2Fe_{14}B$ is uniaxial, the e.a.m. being along the [001] direction. The magnetization isotherms and their field dependences were described starting from a simplified Hamiltonian, including the crystal electric field parameters and the intersubaltice exchange field [87H4].

The $Sm_2Fe_{14}B$ compound is also ferromagnetically ordered, the easy axis of magnetization being situated in (ab) plane, along [100] direction [85H5, 87K3]. The magnetization isotherms were well described by theoretical calculations, in which the crystalline electric field (CEF) parameters, up to sixth order and molecular field, due to Sm-Fe exchange interactions, at samarium site, were taken into account. No spin reorientation transition was found up to Curie temperature, $T_c \cong 580$ K. The $Sm_2Fe_{14}B$ nanoflakes with planar magnetocrystalline anisotropy and shape anisotropy, exhibit high complex permeability [21H1]. An enhanced microwave absorbtion performance has been achieved.

The magnetic structures of $R_2Fe_{14}B$ compounds with R = Nd, Ho, Er, in earlier studies, at low temperatures, were analysed assuming the presence of $P4_2/mnm$ space group. Latter on, crystal structures with lower symmetries have been determined. At ambient conditions, the tetragonal $P4_2/mnm$ structure, of $Nd_2Fe_{14}B$, was confirmed, as well as the ferromagnetic ordering, collinear to **c**-axis [84S3]. At $T < T_{sR}$, the magnetic structure is not collinear and only the projections of Nd moments along the spontaneous magnetization direction have been determined [85G2]. A deviation of Nd moments from **c**-axis, by ≅30° has been reported [84G2, 84O3, 84S1, 85G3, 85S1]. Fairly complicated models of magnetic structures, based on second-, fourth- and sixth-order crystal field terms acting on the Nd sites and also considering the Fe3d sites anisotropy and a molecular field decription, have been developed [84C1, 85B3, 85G5, 85Y1, 86F1, 86H4, 87O3, 87O4, 88L1]. These models fit well with the data obtained from bulk magnetization on single crystals or by ^{57}Fe Mössbauer spectroscopy. The saturation magnetization of $Nd_2Fe_{14}B$, at 300 K < T < T_c, along [100] direction was found to be 2–4% smaller than for [001] axis [21S2]. The difference was attributed to the reduction of Nd moment when forced to be aligned by external field.

The neutron diffraction experiments, performed on the $R_2Fe_{14}B$ (R = Nd, Ho, Er) single crystals, allowed to determine more accurately their non-collinear magnetic structures at low temperatures [90W1, 92F1, 96O1, 96W2, 00W2, 01W1—Fig. 7.19 and Table 7.34.

At low temperatures ($T < T_{SR}$), the easy direction of magnetization of $Nd_2Fe_{14}B$, deviates from **c**-axis. Consequently, in the refinement of neutron diffraction patterns the space group *Cm* was adopted, whose constraints are given in Table 7.29. The neodymium moments were shown to be not collinear, no longer being confined in the (110) mirror plane [96W2, 01W1]. The resulting magnetization is not parallel to the pseudo-tetragonal axis and at T = 20 K, it deviates from **c**-axis by $\theta = 32.3°$, in agreement with magnetic data and ^{57}Fe Mössbauer spectroscopy [87F1].

The $Ho_2Fe_{14}B$ compound, exhibits a spin reorientation transition at $T_{sR} = 58$ K. At low temperatures, the easy axis of magnetization, deviates, from **c**–direction. In analysing the magnetic structure, constrains imposed by *Cm* space group were considered. As seen in Fig. 7.19 and Table 7.34, the holmium moments, at the two sites, are not parallel to each other and the resulting magnetization is also not along

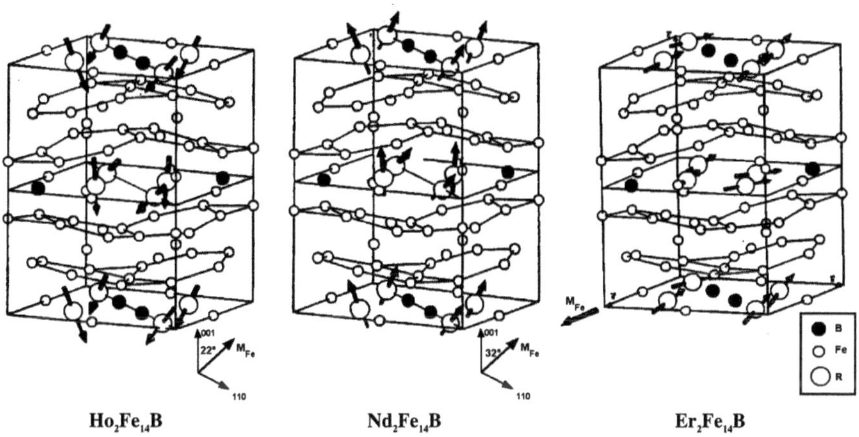

Fig. 7.19 $R_2Fe_{14}B$ with R = Nd, Ho and Er: magnetic structures [01W1]

to the pseudo-tetragonal axis [90W1]. At T = 20 K, the magnetization deviates by θ = 22°, from **c**-axis, in agreement with the magnetic measurements [92F1] and ^{57}Fe Mössbauer spectroscopy [87F5]. At T = 110 K, the deviation from **c**-axis persists, a value θ = 4° being determined.

The $Er_2Fe_{14}B$ compound undergoes a spin reorientation type transition at T_{sR} = 316 K [85H4] or 323 K [86H4]. The low temperature neutron diffraction measurements evidenced that the space group *Pnn*2 was compatible both with the easy magnetization direction [100] and the extinction of the reflections *h0l* and *0kl* [01W1]. No symmetry lowering was detected, by neutron diffraction on $Y_2Fe_{14}B$, at low temperatures [01W1].

The low temperatures crystal and magnetic structures of $R_2Fe_{14}B$ (R = Nd, Ho, Er), exhibiting spin reorientation transitions, can be related to the competing effects, such as exchange and magnetostrictive forces on one site, and also the anisotropies of iron and rare-earth sublattices—Fig. 7.20. In the systems with R = Nd, Ho, at T > T_c, the rare-earth local anisotropy leads to a maximum in magnetic susceptibility, along a direction tilted from the **c**-axis, within the (110) plane. If iron anisotropy is weak enough, a low symmetry should occur. At T < T_c, but far above T_{sR}, the R sublattice is ferromagnetically ordered, via the R–Fe exchange interactions. The resulting direction of the iron anisotropy is perpendicular to the iron layers. The coupling forces, lead to a rotation of the 4f orbitals and small shifts of the neighbouring atoms, but the crystal structure looks as $P4_2/mnm$ one, due to small atom shifts. When the temperature is further decreased, the anisotropy energy of the rare earths increases and compete with that of iron. The rare-earth magnetic moments begin to form a fan arrangement, since the two R sites exhibit different second order CEF parameters. At this point the crystal structure changes from tetragonal one. The negative R–R exchange interactions induce a net distortion of the magnetic structure, with the loss of the symmetry centre. Then, the rotation of easy direction within the (110) plane starts, as a compromise between the CEF parameters set, of the two

Table 7.34 Magnetic moments determined by neutron diffraction, in $R_2Fe_{14}B$ compound having crystal structures with lower symmetry than $P4_2/mnm$

Atom	$Nd_2Fe_{14}B$ [96W2] T = 20 K			$Ho_2Fe_{14}B$ [90W1, 92F1] T = 20 K			$Er_2Fe_{14}B$ [01W1] T = 20 K			$Ho_2Fe_{14}BH_{3.1}$ [91O1] T = 20 K					
	θ_i^0	$\varnothing_i^0$	M (μ_B/atom)	θ_i^0	$\varnothing_i^0$	M (μ_B/atom)	θ_i^0	$\varnothing_i^0$	M (μ_B/atom)	θ^0	$\varnothing^0$	M_x (μ_B/atom	M_y	M_z	M
R_{11} 4*f*	47(6)	45		18(1)	45		94.2(1.8)	−15.9(5)		59	45	5.95	5.95	5.03	9.80
R_{12} 4*f*	12(3)	45	3.20(7)	22(1)	45	−10.00(5)	81.5(1.5)	−1.7(6)	−9	24	45	2.71	2.71	8.45	9.30
R_{13} 4*f*	38(3)	98(3)		21(1)	9(2)		85.8(1.8)	15.9(5)		31	62	2.41	4.61	8.74	10.20
R_{14} 4*f*	38(3)	−8(1)		21(1)	81(2)		98.5(1.5)	1.7(6)							
R_{11} 4*g*	66(3)	69(4)		24(1)	89(2)		97.7(1.5)	22.6(5)		−48	−74	−2.24	8.04	7.54	11.25
R_{12} 4*g*	66(3)	−21(4)	3.21(5)	24(1)	1(1)	−10.00(5)	83.5(1.5)	11.0(5)	−9	46	45	4.84	4.84	6.64	9.55
R_{13} 4*g*	−9(4)	45		29(1)	45		82.3(1.5)	−22.6(5)		44.7	45	4.76	4.76	6.79	9.55
R_{14} 4*g*	50(5)	45		33(1)	45		96.5(4.5)	−11.0(5)							
Fe4*e*			2.22(2)			2.31(4)			2.31(6)	65	45	1.50	1.50	0.97	2.31
Fe4*c*			2.69(1)			2.53(4)			2.41(4)	60	45	1.55	1.55	1.26	2.59
Fe8j_1	33.3(8)	45	2.07(8)	23(1)	45	3.08(5)	90	0	2.15(3)	35	45	1.25	1.25	2.52	3.08
Fe8j_2			2.75(1)			2.21(4)			2.94(3)	35	45	0.90	0.90	1.81	2.21
Fe16k_1			2.28(7)			2.45(3)			2.59(3)	60	45	1.50	1.50	1.22	2.45
Fe16k_2			2.39(6)			2.50(4)			2.63(3)	60	45	1.53	1.53	1.25	2.50
M (μ_B/f.u.)		38.7(8)			15.5(5)			18.5(2)							
θ_M^0		32.6(8)			22.5(5)			90							
$\varnothing_M^0$		45			45			0							

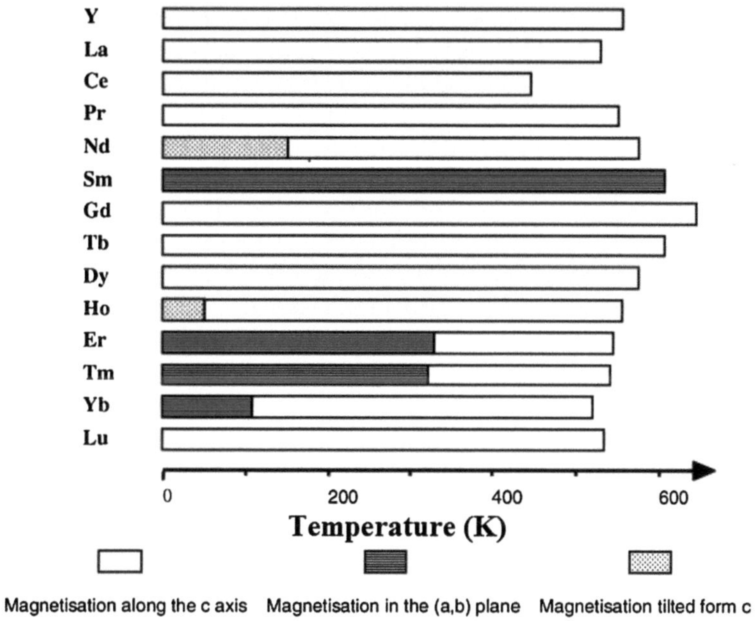

Fig. 7.20 $R_2Fe_{14}B$: temperature dependence of magnetization orientations

different R sites [01W1]. The Er moments in *Pnn*2 orthorhombic type structure of $Er_2Fe_{14}B$, at $T < T_{sR}$, exhibit a fan magnetic structure, apart from the (100) plane, due to Er–Fe exchange interactions, different strength of CEF parameter on the two R sites and negative Er–Er interactions [01W1]. The latter coupling induces the loss of the symmetry centre.

According to [01W1], the $R_2Fe_{14}B$ (R = Sm, Tm, Yb), so called planar compounds, can exhibit similar features, in crystal structures as driven by negative second order CEF parameters. Since the effects of iron axial anisotropy are counter-balanced by the R–Fe exchange interactions, deviations from collinearity within the iron sublattice can be expected, although not reported yet.

The $Tm_2Fe_{14}B$ compound is ferrimagnetically ordered. A spin reorientation transition from the basal plane to **c**-axis, at $T_{sR} \cong 310$–316 K, was shown by ^{57}Fe Mössbauer spectroscopy [86P7], magnetic susceptibility [90L1], resistivity [97S1] and heat capacity studies [87F6, 95L2]. The spin reorientation transition is of first order [01S1].

The magnetic structure of $Tm_2Fe_{14}B$, at low temperatures, has been determined by neutron diffraction [85D1, 85Y1]. In analysing the diffraction patterns, the presence of four Tm sites (f_1, f_2), (g_1, g_2) and 28 iron atoms was assumed [85Y1]. At T = 7 K, the Tm moment at $f_1(g_1)$ site are equal to those at $f_2(g_2)$ ones. Forbidden reflections with finite magnitude have been found. The canting angles of Tm moments, of $\cong 7$ μ_B, at 4*f* and 4*g* sites, were $\phi(4f) = 34.1(6.4)^\circ$ and $\phi(4g) = 14.9(1.8)^\circ$. The magnetic structure has been discussed in terms of a phenomenological model, incorporating both exchange and crystal field interactions [87C1]. At least fourth-order crystal field terms must to be considered in analysing the experimental data. The magnetization curves of $Tm_2Fe_{14}B$, at T = 4.2 K and 78 K, have been calculated for applied magnetic fields up to 200 T, in the (ab) plane [92L1]. It was shown that at T = 4.2 K, there is a field induced planar fan magnetic structure for moderate fields, applied along the [100] direction. In this field range, the iron-sublattice magnetization is not parallel to the field direction. The magnetic behavior of $Tm_2Fe_{14}B$ compound has been also studied by neutron diffraction at temperatures T = 294 K ($T < T_{sR}$) and T = 340 K ($T > T_{sR}$)—Table 7.33. In the limit of experimental accuracy all the magnetic moments lie in the basal plane, at T = 294 K and along **c**-axis at T = 340 K [85D1].

The ^{57}Fe Mössbauer spectroscopy, on $Yb_2Fe_{14}B$, established the existence of magnetization reorientation, at T_{sR} = 115 K [86B5]. From the neutron diffraction measurements, at T = 4.2 K, with and without an applied magnetic field, the easy direction of magnetization was found to be along [100] direction, in the basal plane, as determined by the second order crystal field terms and Yb–Fe exchange interactions. The values of the crystal field parameters and the Yb–Fe exchange interactions have been obtained from a ^{174}Yb Mössbauer spectroscopy [89M2]. The best fit of experimental data, was obtained considering an exchange field close to [001] axis at $4f$ site and close to the (ab) plane, at the $4g$ site. The magnetizations orientations, as function on temperature, in $R_2Fe_{14}B$ compounds are shown in Fig. 7.20.

The $R_2Fe_{14}B$ compounds with R = Gd, Tb and Dy are collinear ferrimagnets. The neutron diffraction study of $Dy_2Fe_{14}B$, at T = 77 K and 395 K, as well as of $Tb_2Fe_{14}B$, at T = 7 K [93H1], showed that R and Fe moments are antiparallel oriented along **c**-axis [85H2]. The Tb moments approach their free ion value and the largest iron moments are located at $8j_2$ sites, as in other $R_2Fe_{14}B$ compounds [91H1, 98B2]. The inelastic neutron spectra, of $Tb_2F_{14}B$, evidenced the presence of two strong peaks, at 23.0 and 27.1 meV, identified as flat modes, originating from $4f$ and $4g$ sites, respectively [90L6]. There is also a small peak at 15.5 meV and an intensity spread between 0 and 15 meV, due to transition between the nearly pure Zeeman states, The molecular field acting on these sites were $\mu_0H_m(4f) = 322$ T and $\mu_0H_m(4g) = 248$ T, their ratio being 1.3 [91L3]. The $Gd_2Fe_{14}B$ compound is a collinear ferrimagnet [84S1, 84S4, 85B3, 86H4]. By means of inelastic neutron scattering experiments, molecular fields $\mu_0H_m(4f) = 358$ T and $\mu_0H_m(4g) = 289$ T were determined [91L3]. Their ratio is 1.24, close to that determined in $Tb_2Fe_{14}B$ compound. There is a significant dependence of molecular field on the rare-earth ion and site location.

The temperature dependences of the *ac* susceptibility, $\chi = \chi' + i\chi''$, were investigated in $R_2Fe_{14}B$ compounds with R = Pr, Nd, Sm, Gd, Tb, Dy, Ho, Er, Tm and Y, in the range $4.2\text{ K} \le T \le 300$ K and frequency (5−1000 Hz) [96K2]. In a small *ac* field, applied perpendicular to e.a.m., the reversible rotation of magnetic moments, is the main contribution to the *ac* susceptibility. The corresponding energy loss is nearly zero. If the data are obtained with field parallel to e.a.m., the irreversible domain wall movements are the main contributions to the *ac* susceptibility (χ'), and the energy loss (χ'') is non-zero.

The paramagnetic behavior of $R_2Fe_{14}B$ compounds was also investigated [85B7, 85B8, 89B2, 93Y2]. According to the type of magnetic ordering, the temperature dependences of the magnetic susceptibilities follow two different trends. In case ferromagnetic $R_2Fe_{14}B$ compounds, the reciprocal susceptibilities show a Curie–Weiss type dependence. The same temperature dependence is characteristic for pseudoternary $Y_2Fe_{14-x}M_xB$ with M = Ga, Ge, V [91B4]. The susceptibilities of ferromagnetic $R_2Fe_{14}B$ compounds, above, but close to the Currie points show nonlinear temperature dependences, described by a relation of the form $\chi \propto (1-T_c/T)^{\gamma}$. In polycrystalline compounds with R = Pr, Nd and Y, values $\gamma = 1.30$–1.40 were evidenced [85B8]. The magnetization isotherms in $Nd_2Fe_{14}B$ single cystal obtained with different field orientations, ψ, from the easy direction, yields a significant dependence of the "critical exponents" as well as of the Curie temperature upon angle ψ [91K9, 92R2]. For $\psi = 0$, the critical exponents of magnetization $\beta \cong 0.43$, magnetic susceptibility $\gamma \cong 1.20$ and critical magnetization isotherm $\delta = 3.79$, do not agree either with the three-dimensional Heisenberg model or the three-dimensional Ising model. The γ and β values were understood by crossover between above models, when the uniaxial magnetocrystalline anisotropy and dipolar interactions are taken into account [92R2]. The difference in the γ value, for polycrystalline and single crystal samples is not clear, but may be due to the presence of small amount of magnetic ordered impurities, which can influence the experimental results [98B2].

The volume effects play a major role in describing the exchange interactions between iron atoms, as already mentioned. These exchange interactions can be modified either by pressure of by the presence of interstitial atoms, the unit cell dimensions, respectively. A large decrease of the Curie temperatures of $R_2Fe_{14}B$ compounds, as effect of pressure, is observed. Values $dT_c/dp = (-46$ to -83 $KGPa^{-1})$ for R = Ce [87N1], -26.5 $KGPa^{-1}$ [87K2] or (-35 to -100 $KGPa^{-1}$) for R = Nd and (-34 to -72 $KGPa^{-1}$) when R = Y [87N1]. These data can be analysed on the basis of Néel-Slater curve [36N1] and correlate with the increase of negative exchange interactions between iron atoms, located below a critical distance ($\cong$0.250 nm). An interesting feature, as effect of pressure, was shown in $Ce_2Fe_{14}B$ compound. When heated up to $T \cong 600$ K, under a pressure $p = 0.55$ GPa, the Curie temperature increased by $\cong$60 K [87N1]. Then, the T_c values decrease with pressure, following a parallel trend to that of the initial sample. The decrease of $Y_2Fe_{14}B$ magnetization, with a rate $d\ln M_s/dp \cong -1.9 \cdot 10^{-2}$ GPa^{-1}, was also reported [88G2].

The transition from ferrimagnetic to paramagnetic state, in $Er_2Fe_{14}B$, at $p = 5.70$ GPa, was shown to be more abrupt and the giant (order of magnitude); softening of the bulk modulus was observed before the transition [94S2]. The spin reorientation temperature shifts to lower temperatures with a rate $dT_{SR}/dp \cong -19$ $KGPa^{-1}$. Values $dT_{SR}/dp \cong -20$ $KGPa^{-1}$, not affected by the substitutions, were shown in $(Er_xR_{1-x})_2Fe_{14}B$ with R = Gd or Y [92I1]. Different behavior was shown when Er was replaced by R = Dy or Nd, the T_{SR} values decreasing when increasing the content of substituting elements. The above trend has been attributed to the strong influence of pressure on the CEF interactions [92I1].

The magnetic properties of $R_2Fe_{14}BH_y$ hydrides with R = La [04B1], R = Ce [86F1, 86P6, 87D1, 87F1], R = Pr [99P6], R = Nd [84O2, 86P6, 03N1, 06T1, 07M4],

R = Sm [86P6], R = Gd [88Z1, 99P2, 06T1], R = Tb [88Z1, 07T1], R = Dy [85L2, 86F1, 87C5, 87R7, 88Z1, 99P2, 07T1], R = Ho [87R7, 88Z1, 91O1, 07T1, 12D1], R = Er [86F1, 87D1, 88Z1, 04T1, 06T1, 07T1], R = Lu [85L2, 05T1, 06T1], R = Y [85L2, 86F1, 86P6, 87D1, 87R7, 04T1] were investigated. The introduction of hydrogen in their structure expands the unit cell volume and increases the saturation magnetization, M_s, Curie temperatures, T_c and reduces the anisotropy fields. The increase of T_c values is related to the reduction of the negative exchange interactions between some closely spaced iron atoms, as the distances between them increased.

Reliable data on the physical properties of $R_2Fe_{14}BH_y$ compounds can be obtained by neutron diffraction. In this context the $Ho_2Fe_{14}BH_y$ samples with $0 \leq y \leq 3.1$, were investigated [91O1]. Because of the influence of the bond length on the exchange interactions, the magnetoelastic couplings, the crystal field parameters, the crystal and magnetic structures are affected by hydrogen adsorption. The most visible features are the continuous rise, upon hydrogen adsorption, of spin reorientation temperature, T_{sR}, from 58 K (y = 0) up to 90 K (y = 3.1). The Curie temperature also increases from T_c = 610 to 670 K, in the above hydrogen concentration range. The neutron diffraction patterns, as for $Ho_2Fe_{14}B$ sample, were analysed, at low temperatures assuming a crystal structure having *Cm* space group. The higher the hydrogen content, the higher is the deviation of the magnetization from **c**-axis, increasing at T = 4.2 K, from 22° (y = 0), up to 50° when y = 3.1 [91O1]. The anisotropy field decreased with temperature, indicating, in addition to a normal decrease due to crystal field effects, that due to the presence of hydrogen [88F1]. In the pseudoternary $Nd_2Fe_{14-x}Si_xBH_y$, the insertion of hydrogen is still possible, but with a dramatic decrease in the maximum adsorbed hydrogen content [00C1]. Hydrogen atoms partially occupy the interstitial sites surrounded by silicon atoms. The effects of Si and H on the spin reorientation temperature of the $Nd_2Fe_{14-x}Si_xBH_y$ was shown to be additive, both reducing the T_{sR} values. The hydrogen induces a spin reorientation in $R_2Fe_{14}BH_y$ hydrides with R = Gd and Dy favoring planar anisotropy at $T < T_{sR}$ [88Z1]. It was postulated that hydrogenation affects differently the anisotropy of the six iron sites.

The X-ray Magnetic Circular Duchroism (XMCD) and magnetization data on $R_2Fe_{14}BH_y$ compounds are consistent with the decrease of rare-earth magnetic moment upon hydriding [98C1]. This result is in agreement with ^{161}Dy and ^{166}Er Mössbauer spectroscopy, which evidence a significant reduction of the hyperfine field, at the rare earth sites [86F1, 86S1]. The reduction of light rare-earths (R = Pr, Nd) moments is of 10%, whereas for heavy rare-earths it is of ≅25% [98C1, 03C1], correlated with different hydrogen filing sequence for the two series [91O1]—Sect. 7.10.1.

The magnetic properties of $R_2Fe_{14}B$ compounds are determined by those of the iron and rare-earths sublattices, respectively. Consequently, a short review on the magnetic behavior of the two sublattices will be firstly presented. Then, the magnetic properties of the $R_2Fe_{14}B$ compounds, determined by the complex interactions between the two sublattices, will be reviewed.

As already mentioned, the neutron diffraction studies on $R_2Fe_{14}B$ compounds, evidenced that the iron moments, M_{Fe}, are dependent on the site location, the largest value corresponding to $8j_2$ site. When substituting a non-magnetic R element, by magnetic ones, the M_{Fe} values increase linearly, along series, with De Gennes factor, the corresponding slopes being dependent on the site location. The rate of increase is the larger for those atoms, having the greater number of rare-earth atoms situated in their first coordination shell [20B1, 21B1].

The magnetic behavior of iron atoms in $R_2Fe_{14}B$ compounds were also investigated by ^{57}Fe Mössbauer spectroscopy, when R = La [86V2, 87E1], R = Ce [85P3, 85V5, 86F1, 86R3, 86V2], R = Pr [85P3, 86V2, 90V2], R = Nd [85K2, 85P3, 85R2, 85V5, 86V2, 87F2, 87O3, 88G4, 88G7, 88G8, 89R2, 90L2, 93M3], R = Sm [86V2], R = Gd [85P3, 85V5, 86R3 86V2, 87F2, 89R2, 90G2, 90L2], R = Tb [86V2, 90L2], R = Dy [85P3, 86F1, 86V2], R = Ho [85P3, 86V2, 87F2, 87F5, 87O3, 93M1], R = Er [86F1, 86V2, 87F2, 89R2], R = Tm [86P7, 87F2, 93M3], R = Yb [86B5], R = Lu [85G5, 86V2, 87F2], R = Y [85P3, 86F1, 86R3, 86V2, 87E1, 87F2, 87F5, 87Y1, 90L2], R = Th [87E1] or in pseudoternary compounds $Nd_2(Fe,Co)_{14}B$ [87M1, 89R2], and $Y_2Fe_{14-x}M_xB$ with M = Mn [88P9], M = Mn, Co, Ni [88B1]. The complex ^{57}Fe Mössbauer spectra, resulting from the superposition of six iron sextets, made their analysis difficult. A consensus on the assignment of the major components was shown, but a controversary over the assignment of the subspectra associated with $4c$ and $4e$ site still existed [89R2, 90G2, 92G1, 92L5, 93F1, 93L3]. The iron moments were also determined assuming that the hyperfine fields, H_{ef}, are proportional to iron moments ($H_{hf}/M_{Fe} = 14.8$ T/μ_B) [79B1]. The above assumption is not entirely valid [90E1, 92H3], the iron moments, thus obtained, being generally smaller than those determined by neutron diffraction. Really, only the core contribution to the hyperfine field is proportional to the local iron moment, whereas the conduction band contribution is not. The spread in the ^{57}Fe hyperfine fields, experimentally determined, are typically between 0.5 and 1.0 T—Table 7.35.

The iron magnetic moments and the ^{57}Fe hyperfine fields were theoretically computed. Some of the reported values are listed in Tables 7.33 and 7.35. The magnetic moments, of constituent atoms, in $Nd_2Fe_{14}B$ were also determined starting from the spin-polarized band structure calculation [85S2, 93N1] or using OLCAO method [87G4, 90Z6], including the spin–orbit interaction. The self-consistent band structure calculations were used to determine theoretically the magnetic moments in the $Nd_2Fe_{14}B$ compound [91N1]. The recursion method was used to compute the electronic density of states in $Y_2Fe_{14}B$ [86I1] or $Nd_2F_{14}B$ [85S2]. Ab initio self-consistent band structure calculations, using the linearized muffin tin orbital method and LSDA, were used to determine the local iron magnetic moments and ^{57}Fe hyperfine fields in $R_2Fe_{14}B$ compounds with R = Nd, Gd, Tb, Dy, Ho, Er [92H3]. The iron moments and the ^{57}Fe hyperfine fields in $Y_2Fe_{14}B$ were calculated by means of first-principles self-consistent band structure calculations, using the augmented plane or spherical wave method [91C1, 91C3, 13A2, 16K1]. Some differences in the computed iron moments, by different groups, were shown [87F2, 91C1, 92H3]. The electronic structures and magnetic moments in $R_2Fe_{14}B$ compounds with R = Nd, Dy, were calculated by using a linear combination of localized pseudo-atomic-orbital (LCPAO) method [09K1]. By including the spin–orbit interaction, the greater iron moments were shown at Fe$8j2$ site and the minimum at Fe$4e$ site, in agreement with

Table 7.35 Hyperfine parameters determined by ^{57}Fe Mössbauer spectroscopy on $R_2Fe_{14}B$ compounds

Compound	T (K)		Iron sites						Reference
			$4e$	$4c$	$8j_1$	$8j_2$	$16k_1$	$16k_2$	
$La_2Fe_{14}B$[a]	RT	δ (mm/s)[a]	−0.21(3)	−0.27(3)	−0.24(3)	−0.10(3)	−0.20(3)	−0.27(3)	[86V2]
		ΔQ (mm/s)	−0.32(3)	0.09(3)	0.15(3)	0.29(3)	0.12(3)	0.09(3)	
		μ_0H_{eff} (T)	28.2(2)	25.8(2)	27.6(2)	33.4(2)	28.6(2)	30.0(2)	
$Ce_2Fe_{14}B$[a]	RT	δ (mm/s)[a]	−0.27(3)	−0.24(3)	−0.26(3)	−0.13(3)	−0.26(3)	−0.31(3)	[86V2]
		ΔQ (mm/s)	−0.34(3)	0.10(3)	0.10(3)	0.32(3)	0.11(3)	0.07(3)	
		μ_0H_{eff} (T)	23.2(2)	21.2(2)	22.4(2)	28.5(2)	23.4(2)	24.4(2)	
$Pr_2Fe_{14}B$[a]	RT	δ (mm/s)[a]	−0.23(3)	−0.28(3)	−0.23(3)	−0.09(3)	−0.22(3)	−0.28(3)	[86V2]
		ΔQ (mm/s)	−0.42(3)	0.11(3)	0.15(3)	0.32(3)	0.13(3)	0.09(3)	
		μ_0H_{eff} (T)	28.0(2)	25.5(2)	27.5(2)	33.8(2)	28.3(2)	30.2(2)	
$Pr_2Fe_{14}B$[c]	RT	δ (mm/s)[a]	−0.03(4)	−0.06(4)	0.04(4)	0.08(4)	−0.03(4)	−0.19(4)	[85P3]
		ΔQ (mm/s)	0.24(10)	−0.83(10)	0.28(10)	0.67(10)	0.36(10)	0.03(10)	
		μ_0H_{eff} (T)	25.7(4)	28.0(4)	29.1(4)	33.6(4)	28.2(4)	29.5(4)	
$Nd_2Fe_{14}B$	RT	δ (mm/s)[a]	−0.23(3)	−0.28(3)	−0.24(3)	−0.09(3)	−0.21(3)	−0.29(3)	[86V2]
		ΔQ (mm/s)	−0.43(3)	0.11(3)	0.15(3)	0.33(3)	0.12(3)	0.09(3)	
		μ_0H_{eff} (T)	28.3(2)	26.1(2)	28.0(2)	34.3(2)	28.9(2)	30.6(2)	
$Nd_2Fe_{14}B$	RT	δ (mm/s)[a1]	−0.04	−0.07	−0.01	0.09	−0.02	−0.14	[87O3]
		ΔQ (mm/s)	0.12	−0.44	0.10	0.32	0.25	−0.01	
		μ_0H_{eff} (T)	25.9	28.5	28.8	34.0	29.1	29.7	
$Nd_2Fe_{14}B$ (BS)		μ_0H_{eff} (T)	26.4	28.6	28.8	28.2	28.6	30.7	[92H3]
$Nd_2Fe_{14}B$	85	δ (mm/s)[b]	0.070	0.131	0.027	0.197	0.076	0.016	[93M3]
		ΔQ (mm/s)	−0.630	−0.298	0.277	0.656	0.299	0.273	
		μ_0H_{eff} (T)	33.5	29.0	30.2	37.5	32.3	34.3	
$Nd_2Fe_{11.96}Si_{2.04}B$[d]	85	δ (mm/s)[b]	0.013	0.216	0.220	0.282	0.149	0.054	[93M3]
		ΔQ (mm/s)	−0.147	−0.897	0.391	0.541	0.333	0.296	
		μ_0H_{eff} (T)	21.1	17.9	21.0	31.5	24.5	26.4	
		θ^o	0	90	2.5	11.5	17.8	32.9	

(continued)

Table 7.35 (continued)

Compound	T (K)		Iron sites						Reference
			4*e*	4*c*	8j_1	8j_2	16k_1	16k_2	
$Nd_2Fe_{14}B$	RT	δ (mm/s)[b]	−0.07	−0.07	−0.01	0.08	−0.05	−0.14	[90L2]
		ΔQ (mm/s)	0.22	−0.25	0.03	0.65	0.24	0.10	
		μ_0H_{eff} (T)	25.4	29.0	28.3	33.8	28.6	30.1	
$Nd_2Fe_{14}B$	RT	δ (mm/s)[b]	−0.042	−0.109	−0.080	0.073	−0.040	−0.128	[07C1]
		ΔQ (mm/s)	−0.747	−0.173	0.335	0.692	0.287	0.314	
		μ_0H_{eff} (T)	28.9	25.6	28.1	34.5	28.9	30.8	
$Nd_2Fe_{14}B$	RT	δ (mm/s)[b]			0.08		0.20	0.03	[87F2]
		ΔQ (mm/s)			0.61		0.06	0.31	
		μ_0H_{eff} (T)	32.1	28.2	34.8	25.7	30.1	29.2	
$Sm_2Fe_{14}B$	RT	δ (mm/s)[b]			0.12		0.13	0.25	[87F2]
		ΔQ (mm/s)			0.18		0.15	0.14	
		μ_0H_{eff} (T)	28.6	30.1	31.5	30.5	32.4	30.5	
$Gd_2Fe_{14}B$	RT	δ (mm/s)[b]	−0.05	−0.03	0.03	0.06	−0.04	−0.14	[90L2]
		ΔQ (mm/s)	0.24	−0.29	−0.07	0.58	0.22	0.10	
		μ_0H_{eff} (T)	26.2	28.6	30.6	36.0	28.9	30.4	

(continued)

Table 7.35 (continued)

Compound	T (K)		Iron sites						Reference
			4*e*	4*c*	$8j_1$	$8j_2$	$16k_1$	$16k_2$	
$Gd_2Fe_{14}B$	85	δ (mm/s)[b]	0.010	0.070	−0.030	−0.230	0.130	0.020	[90G2]
		ΔQ (mm/s)	−0.762	−0.460	0.320	0.642	0.367	0.300	
		μ_0H_{eff} (T)	31.20	31.12	33.17	39.50	33.25	35.41	
$Gd_2Fe_{14}B$	RT	δ (mm/s)[a]	−0.23(3)	−0.24(3)	−0.10(3)	−0.23(3)	−0.22(3)	−0.30(3)	[86V2]
		ΔQ (mm/s)	−0.37(3)	0.13(3)	0.14(3)	0.31(3)	0.13(3)	0.07(3)	
		μ_0H_{eff} (T)	27.9(2)	26.2(2)	27.9(2)	34.5(2)	28.9(2)	30.8(2)	
$Gd_2Fe_{14}B$[c)]	RT	δ (mm/s)[a]	−0.08(4)	−0.11(4)	0.05(4)	0.08(4)	0	−0.22	[85P3]
		ΔQ (mm/s)	−0.73(10)	0.28(10)	0.23(10)	0.58(10)	0.37(10)	0.07(10)	
		μ_0H_{eff} (T)	27.8(4)	28.6(4)	31.3(4)	36.3(4)	30.3(4)	31.4(4)	
$Gd_2Fe_{14}B$ (BS)		μ_0H_{eff} (T)	26.7	27.8	28.3	28.5	28.5	30.6	[92H3]
$Tb_2Fe_{14}B$	RT	δ (mm/s)[a]	−0.23(3)	−0.31(3)	−0.23(3)	−0.10(3)	−0.22(3)	−0.29(3)	[86V2]
		ΔQ (mm/s)	−0.42(3)	0.12(3)	0.16(3)	0.30(3)	0.13(3)	0.10(3)	
		μ_0H_{eff} (T)	28.9(2)	27.7(2)	28.8(2)	36.0(2)	30.0(2)	31.6(2)	
$Tb_2Fe_{14}B$	RT	δ (mm/s)[b]	−0.04	−0.06	−0.03	0.05	−0.07	−0.12	[90L2]
		ΔQ (mm/s)	0.23	−0.28	0.06	0.60	0.24	0.11	
		μ_0H_{eff} (T)	25.9	28.4	30.8	34.8	28.7	31.1	
$Tb_2Fe_{14}B$ (MS)	4.2	μ_0H_{eff} (T)	26.6	30.6	36.0	32.3	31.4	29.2	[87F2]
$Tb_2Fe_{14}B$ (BS)		μ_0H_{eff} (T)	26.7	28.1	28.6	28.7	28.5	30.5	[92H3]
$Dy_2Fe_{14}B$	RT	δ (mm/s)[a)]	−0.26(3)	−0.31(3)	−0.24(3)	−0.10(3)	−0.23(3)	−0.30(3)	[86V2]
		ΔQ (mm/s)	−0.40(3)	0.14(3)	0.14(3)	0.29(3)	0.13(3)	0.10(3)	
		μ_0H_{eff} (T)	28.5(2)	27.1(2)	28.0(2)	35.3(2)	29.3(2)	30.8(2)	
$Dy_2Fe_{14}B$ (MS)	4.2	μ_0H_{eff} (T)	27.3	26.0	35.3	29.4	31.0	29.2	[87F2]
$Dy_2Fe_{14}B$ (BS)		μ_0H_{eff} (T)	26.7	28.3	28.8	28.8	28.4	30.5	[92H3]

(continued)

Table 7.35 (continued)

Compound	T (K)		Iron sites						Reference
			4*e*	4*c*	$8j_1$	$8j_2$	$16k_1$	$16k_2$	
$Ho_2Fe_{14}B$	RT	δ (mm/s)[a]	−0.24(3)	−0.26(3)	−0.26(3)	−0.10(3)	−0.23(3)	−0.31(3)	[86V2]
		ΔQ (mm/s)	−0.39(3)	0.10(3)	0.16(3)	0.29(3)	0.11(3)	0.09(3)	
		μ_0H_{eff} (T)	28.0(2)	26.4(2)	27.0(2)	34.2(2)	28.5(2)	29.7(2)	
$Ho_2Fe_{14}B$	RT	δ (mm/s)[b]	−0.09	−0.15	−0.08	0.09	−0.04	−0.17	[87O3]
		ΔQ (mm/s)	0.14	−0.46	0.14	0.34	0.13	0.08	
		μ_0H_{eff} (T)	26.2	28.5	27.2	33.8	28.8	29.5	
$Ho_2Fe_{14}B$	4.2	δ (mm/s)[b]	0.06	0.07	0.257	−0.012	0.104	−0.008	[93M1]
		ΔQ (mm/s)	0.15	0.24	0.499	0.432	0.143	0.026	
		μ_0H_{eff} (T)	30.3	31.5	39.2	34.8	32.9	34.8	
$Ho_2Fe_{14}B$	RT	δ (mm/s)[b]	−0.10(4)	−0.07(4)	−0.01(4)	0.07(4)	0.00(4)	−0.23(4)	[85P3]
		ΔQ (mm/s)	0.02(10)	−0.70(10)	0.10(10)	0.63(10)	0.35(10)	0.06(10)	
		H_{eff} (kOe)	24.9(4)	28.2(4)	28.7(4)	34.0(4)	27.9(4)	28.8(4)	
$Ho_2Fe_{14}B$	RT	δ (mm/s)[b]	−0.030	−0.138	−0.080	0.067	−0.060	−0.129	[07C1]
		ΔQ (mm/s)	−0.655	−0.194	0.289	0.673	0.270	0.320	
		μ_0H_{eff} (T)	28.7	25.6	27.1	34.2	28.5	29.9	
$Ho_2Fe_{14}B$ (MS)	RT	μ_0H_{eff} (T)	31.3	27.2	34.2	26.9	29.1	28.1	[87F2]
$Ho_2Fe_{14}B$ (MS)	4.2	μ_0H_{eff} (T)	30.6	32.8	32.6	38.7	33.6	33.9	[87F5]
$Ho_2Fe_{14}B$ (BS)		μ_0H_{eff} (T)	26.8	28.5	29.0	28.8	28.5	30.6	[92H3]
$Er_2Fe_{14}B$ (MS)	4.2	μ_0H_{eff} (T)	28.6	27.2	30.5	29.6	29.8	28.0	[87F2]
$Er_2Fe_{14}B$ (BS)		μ_0H_{eff} (T)	26.7	28.7	29.2	28.9	28.4	30.6	[92H3]
$Tm_2Fe_{14}B$	294	δ (mm/s)[b)]	0.15(4)	0.22(4)	−0.08(1)	0.07(1)	−0.18(1)	−0.16(1)	[86P7]
		ΔQ (mm/s)	−0.14(7)	−0.09(8)	0.03(1)	−0.16(1)	−0.07(1)	−0.03(1)	
		μ_0H_{eff} (T)	28.2(7)	28.7(6)	25.8(10)	31.0(2)	29.4(10)	28.0(10)	
$Tm_2Fe_{14}B$ (MS)	RT	μ_0H_{eff} (T)	27.8	25.7	30.5	27.3	29.2	27.5	[87F2]

(continued)

Table 7.35 (continued)

Compound	T (K)		Iron sites						Reference
			4*e*	4*c*	$8j_1$	$8j_2$	$16k_1$	$16k_2$	
$Lu_2Fe_{14}B$[a)]	RT	δ (mm/s)[a)]	−0.28(3)	−0.26(3)	−0.26(3)	−0.11(3)	−0.23(3)	−0.32(3)	[86V2]
		ΔQ (mm/s)	−0.32(3)	0.12(3)	0.11(3)	0.31(3)	0.10(3)	0.08(3)	
		μ_0H_{eff} (T)	27.1(2)	23.7(2)	24.0(2)	30.7(2)	26.4(2)	26.8(2)	
$Lu_2Fe_{14}B$ (MS)	RT	μ_0H_{eff} (T)	28.2	24.1	30.6	25.8	26.4	24.6	[87F2]
$Y_2Fe_{14}B$[a)]	RT	δ (mm/s)[a)]	−0.26(3)	−0.28(3)	−0.26(3)	−0.11(3)	−0.25(3)	−0.29(3)	[86V2]
		ΔQ (mm/s)	−0.28(3)	0.13(3)	0.11(3)	0.29(3)	0.11(3)	0.08(3)	
		μ_0H_{eff} (T)	29.0(2)	26.8(2)	25.9(2)	32.8(2)	28.0(2)	29.3(2)	
$Y_2Fe_{14}B$	RT	δ (mm/s)[b)]	−0.09	−0.04	−0.06	0.05	−0.05	−0.13	[87O3]
		ΔQ (mm/s)	0.14	−0.28	0.12	0.26	0.14	0.06	
		μ_0H_{eff} (T)	25.8	28.4	26.2	32.7	28.3	28.7	
$Y_2Fe_{14}B$	4.2	δ (mm/s)[a)]	0.04	0.04	0.09	0.22	0.08	−0.01	[87O4]
		ΔQ (mm/s)	0.19	−0.33	0.15	0.28	0.15	0.08	
		μ_0H_{eff} (T)	30.8	33.0	31.7	38.0	33.1	34.1	
$Y_2Fe_{14}B$	RT	δ (mm/s)[b)]	−0.08	−0.06	−0.05	0.03	−0.03	0.13	[90L2]
		ΔQ (mm/s)	0.24	−0.48	0.18	0.59	0.20	0.11	
		μ_0H_{eff} (T)	25.8	28.2	26.4	32.4	28.3	29.0	
$Y_2Fe_{14}B$ (MS)	RT	μ_0H_{eff} (T)	29.4	28.7	32.7	28.0	28.6	26.1	[87F2]
$Y_2Fe_{14}B$	4.2	μ_0H_{eff} (T)	30.8	33.0	31.7	38.0	33.1	34.1	[87F5]

[a] ^{57}Co in palladium source; [a1] ^{57}Co in Rh matrix [b] Relative with α−Fe; [c] No distinction of the hyperfine parameters at 4*e* and 4*c* site was shown; [d] θ angle between the hyperfine filed and the principal axis of the electric field gradient for $Nd_2Fe_{14-x}Si_xB$ with $0 \leq x \leq 2.04$

neutron diffraction data [85G2]. Calculated magnetic moments, at the iron sites in $Dy_2Fe_{14}B$ were less than those determined in $Nd_2Fe_{14}B$ compound—Table 7.33.

The $R_2Fe_{14}B$ compounds with R = La, Ce, Sm, Er [88E1] and R = Nd [85R2] were studied by NMR on ^{57}Fe, ^{10}B and ^{11}B, at T = 4.2 K. The hyperfine fields at Fe and B sites were determined and the magnetic moments of iron sites were estimated. The transferred hyperfine field, at boron sites, is negative and of the order of ≅4.0 T, being determined by the six next-nearest Fe atoms in the corner of trigonal prism in which B atoms are located.

The electrical resistivity, ρ, measurements on $R_2F_{14}B$ compounds with R = Y [91S1], R = Nd, Ho, Er, Tm, Yb [90L1] and R = Ce, Pr, Nd, Sm, Gd, Tb, Dy, Ho, Er, Tm [97S1] evidenced that at low temperatures, ρ values, increase with temperature as T^2. The scattering of electrons by modes involving all spins with a dispersion, given by the R–Fe exchange interaction, contributes to this behavior. The high temperature anomaly in ρ(T) was observed just below T_c and considered to be produced by electron–phonon scattering, arising from a pronounced lattice softening. Saturation values of the hight-temperature magnetic resistivity are in agreement with the prediction for spin-disorder scattering. The low temperature resistivity behavior, of $R_2Fe_{14}B$ compounds, was also shown to be consistent with the spin-mixing model for electron scattering, with an intrinsic thermopower contribution related to iron and a spin-dependent impurity scattering term, related to the R magnetic ions [94P1]. The electrical resistivity and Hall efect measurements on $R_2Fe_{14}B$ compounds with R = Nd, Gd, Tm, Y also evidenced that the overall behavior of the resistivity is determined by iron atoms, the contributions from rare-earth atoms, being observed mainly at low temperatures [99S1, 05S2]. Large variations of the Hall resistivity, found near T_{sR} and T_c, were attributed to critical magnetization fluctuations which enhance skew scattering in these regions.

The heat capacity investigations on $R_2Fe_{14}B$, when R = Y, show that the experimental data can be described with a simplified theory of spin fluctuations. In compounds with magnetic R elements the anomalous entropy follows the trend of the De Gennes factor [95L2]. By adiabatic heat capacity study, a first order phase transition has been observed at T_{sR} ≅ 323 K for R = Er and a second order one at T_{sR} ≅ 135 K for R = Nd and T_{sR} ≅ 60 K when R = Ho [90L1, 92P1, 96P1]. The enthalpy of the transition for R = Er and the anomalous heat capacity for R = Nd and Ho were obtained from free energy calculations. The analysis of thermodynamic properties of $Nd_2Fe_{14}B$ indicated that Nd4f component is frozen in the core, supporting open core approach [20M2].

The magnetocaloric effects were investigated in $R_2Fe_{14}B$ compounds with R = Nd [19T1, 20M2], R = Nd, Er and their hydrides [17B1], R = Er [16S1], R = Y [08B1], $(Nd_{1-x}Pr_x)_2Fe_{14}B$ [19P1]. Modest values of entropy changes were obtained.

An increase of the iron magnetization and ^{57}Fe hyperfine fields, was reported in $R_2Fe_{14}BH_y$ hydrides [86C2, 86P6, 87W3, 88P2], as compared with the parent samples. According [93M1], in $Ho_2Fe_{14}BH_y$, the iron moments deviate more and more from the **c**-axis between 295 K and 4.2 K, when the hydrogen content increases and that these deviations differ from site to site, giving rise to a non-collinear structure, as determined by ^{57}Fe Mössbauer spectroscopy and ND studies.

Additional information concerning the magnetic behavior of iron in $R_2Fe_{14}B$ compounds, was obtained in pseudoternary systems, by substitutions at Fe and Fe/R sites. Only a limited fraction of iron can be replaced, with the cobalt exception, as discussed in Sect. 7.10.1. The effects of substitution of iron by more than 20 elements on the physical properties of $R_2Fe_{14-x}M_xB$ pseudoternary systems, were analysed: $R_2Fe_{14-x}Sc_xB$ with R = Nd [87K5]; $R_2Fe_{14-x}Ti_xB$ with R = Y [87K5, 87L2, 89L4]; $R_2Fe_{14-x}V_xB$ with R = Nd [87K5, 89K1, 05W2], R = Y [91B4]; $R_2Fe_{14-x}Cr_xB$ with R = La [97C2], R = Pr [89K4], R = Nd [86A1, 87H2, 87K5, 88K3, 89K3, 89L4, 05W2], R = Gd [88K2, 88K3, 03G2], R = Y [86A1, 87L2, 87L3, 87P2, 88K2, 88K3, 90K7, 90K8, 92B1, 02K1]; $R_2Fe_{14-x}Mn_xB$ with R = Pr [86H7, 90P1, 90Y1], R = Nd [86A1, 86H7, 87B5, 87H2, 87K5, 88C3, 89L4, 90D2, 90I2, 90Y1, 05W2], R = Gd [86H7, 90V1, 90V3], R = Tb, Dy, Ho, Tm [90V1, 90V3], R = Er [87F3, 87F4, 87P1, 88F2, 88P5, 88P6, 88P9, 89S1, 89Y1, 90I2, 90V1, 90V2, 90V3], R = Y [88B1, 88P9, 89S1, 02K1]; $R_2Fe_{14-x}Co_xB$ with R = La [85B9, 88G6, 88V2], R = Ce [13S1, 18W1], R = Pr [86P2, 87B6, 87G3, 87P3, 87S2, 87W2, 88G1], R = Nd [85A1, 85B11, 85M2, 86A1, 86B3, 86F2, 86G6, 86H3, 86H7, 86P1, 87B11, 87C2, 87H2, 87H3, 87K5, 87L3, 87P5, 87R2, 87R3, 87S2, 87T1, 87W2, 88G2, 88G6, 88G7, 88L1, 88V2, 89L4, 89N1, 89W1, 90G1, 90K2, 92J1, 95S3, 96L3, 01C1, 01D1], R = Gd [85L2, 86H7, 87P5, 87S2, 87W2, 90K2, 92H1], R = Tb [87P6, 87P7, 87S2, 87W2], R = Dy [87P4, 87W1, 01D1], R = Er [87P3–87P5, 87W1, 92R1], R = Tm [87P11, 88P8], R = Y [85B6, 85B8, 86A1, 86B8, 86H7, 87B4, 87L3, 87P2, 87P5, 87S2, 87T1, 87W1, 88G2, 88T3, 88T4, 89S1, 90K2, 02K1, 06K1]; $R_2Fe_{14-x}Ni_xB$ with R = Ce [18O1], R = Nd [86A1, 87B5, 87K5, 88C6, 88D1, 88G2, 88G7, 89K3, 89L4, 89W1, 95S3], R = Gd [89K3], R = Y [85B5, 85B8, 86A1, 86B8, 87L2, 87P2, 88B1]; $R_2Fe_{14-x}Cu_xB$ with R = Pr [89K4, 89L4], R = Nd [88C2, 88K1, 90B2, 90K7], R = Gd [88K1, 88K2], R = Er [90B2], R = Y [87P9, 88K1, 88K2, 90K8, 92B1]; $R_2Fe_{14-x}Nb_xB$ with R = Pr [89J1], R = Nd [88H4, 05W2], R = Y [89J3]; $R_2Fe_{14-x}Mo_xB$ with R = Nd [88H4]; $R_2Fe_{14-x}Ru_xB$ with R = Nd [86K4, 86P1, 86P3]; $R_2Fe_{14-x}Re_xB$ with R = Nd [89J2]; $R_2Fe_{14-x}Al_xB$ with R = Ce [18O1], R = Nd [85A1, 86A1, 86K1, 87H2, 87H7, 88G7, 88H2, 89K2, 89L4, 89W1, 90D2, 08B1], R = Gd [86B7, 89S5, 90K5, 92W1], R = Er [87Q1, 91K3], R = Y [86A1, 86B8, 86Y4, 87L2, 87Y1, 89S5, 90K5, 90W2, 92W1]; $R_2Fe_{14-x}Si_xB$ with R = Ce [18O1], R = Pr [86P1, 87P6, 88X2, 89K4], R = Nd [86P1, 86V1, 87B2, 87J1, 87P6, 89B1, 88B6, 88G7, 88X2, 89K2, 89P2, 89W1, 90K7, 90L4, 93M3, 94M1, 99C1, 00C1], R = Gd [88K2, 88X2, 89S5, 92W1], R = Dy [88X2], R = Er [86P1, 87P8, 92R1, 02G3], R = Y [86P1, 87P9, 88B6, 88K2, 88X2, 88Y4, 89P2, 89S5, 90K5, 90K8, 92W1, 93M2, 94M1]; $R_2Fe_{14-x}Ga_xB$ with R = Pr [88P4, 89K4], R = Nd [86P1, 88G7, 88H4, 88P4, 88X1, 89K2, 89Q1, 89W1, 90Q1, 00C2], R = Gd [89S5], R = Er [92R1], R = Y [89S5, 91B4, 92W1]; $R_2Fe_{14-x}Ge_xB$ with R = Nd [90K6], R = Y [91B4]; $R_2Fe_{14-x}Sn_xB$ with R = Nd [96M1]; $R_2Fe_{14-x}Be_xB$ with R = Nd [96L2, 97L2]; $R_2Fe_{14-x}Nb_xB$ with R = Nd [05W2, 06L1] and $R_2Fe_{14-x}Zr_xB$ [05W2] series.

$R_2Fe_{14}B$ pseudoternary compounds with multiple substitutions at Fe sites were investigated as $R_2(Fe,Co,Al)_{14}B$ with R = Pr [87B10, 87B11, 89B2], R = Nd [87B9, 87S3, 88J3, 89B2], $R_2Fe_{12-x}Mn_xCo_2B$ with R = Pr, Nd, Gd [86J4].

The pseudoternary systems, where substitutions at both R and Fe sites were made, have been also investigated. Some such systems are below mentioned: $(Pr,Gd)_2(Fe,Co)_{14}B$ [91Z2], $Pr_{2-x-y}R_xNd_yFe_{11.6}Co_2Al_{0.4}B$ with R = Tb, Dy [88P7], $(Pr,Dy)_2(Fe,Nb,Co)_{14}B$ [89J1], $(R_1,R_2)_2Fe_{14}Co_2B$ with R_1 = Pr, Nd and R_2 = Tb, Dy [86F3]; $(Nd,R)_2(Fe,Co,Re)_{14}B$ with R = Tb, Dy [90J3], $(Nd,Gd)_2(Fe,Co)_{14}B$ [91Z2], $(Nd,Ce)_2(Fe,Co)_{14}B$ [18Z1], $(Nd,Dy)_2(Fe,Co)_{14}B$ [01C1, 01D1], $(Nd,Y)_2(Fe,Cr)_{14}B$ [91K7]. At equilibrium $(Nd_{1-y}Dy_y)_2(Fe_{1-x}Co_x)_{14}B$ had been assumed to be single phase, when $1 > x > 0$ and $1 > y > 0$. Three-phase microstructures were observed, when $x > 0.3$ and $y > 0.5$ [01D1]. The $(Nd,Er)_2(Fe,M)_{14}B$ with M = Fe, Al [90M1], $(Nd,Y)_2(Fe,Cr)_4B$ [91K7], $(Nd,Y)_2(Fe,Cu)_{14}B$ [93K2] were also investigated.

The composition dependences of Curie temperatures in pseudoternary $R_2Fe_{14-x}M_xB$ compounds with M = Co, Ni, Ga, Si, Cu, Be, Cr, Al, Ru, V and Mn [86F2, 86K4, 86P3, 87J3, 87K5, 87P8, 87S1, 88C6, 88K1, 88X1, 89B2, 90B2] showed an increase when Fe is substituted by Co, Ni, Si, Cu or Ga and decrease when iron is replaced by V, Cr, Mn, Ru, Al, Be. The effect of iron substitution on the Curie temperature is a rather complex matter. It has been suggested, in earlier reports [85B6, 85B8, 87P8], that when substituting iron by a non-magnetic or a weak magnetic element, there are present two opposite effects: (1) a decrease of both positive and negative interactions between some iron atoms, by breaking their exchange path. When the contribution of negative exchange interactions between iron atoms, situated below a critical distance, is diminished more than that of positive ones, an increase of T_c values can be expected; (2) changes of iron moments as result of substitutions and magnetic dilution, respectively and consequently of the involved exchange interactions.

The increase of T_c values, when Fe was substituted by Si, in earlier studies, was correlated with possible substitutions at j_1 and k_2 sites, diminishing in this way, the negative exchange interactions [87P8]. Latter studies on $R_2Fe_{14-x}Si_xB$ compounds, with R = Nd or Y, have shown that silicon preferentially occupies the 4*c* site and to a lesser extent, the $8j_1$ site, it is almost excluded from $16k_2$ sites and avoids the $16k_1$, $8j_2$ and 4*e* ones [88Y4, 89B1, 93M2, 93M3]. In addition, the lattice parameters decrease, particularly along **c**-direction, as the silicon content increases. The only apparent pattern, to the occupancy, as already mentiomed, was the preference of silicon for sites coordinated by the rare-earths [93M2, 93M3]. The ND and ^{57}Fe Mössbauer spectroscopy studies on $Y_2Fe_{14-x}Si_xB$ suggested an increase in the covalency of bonding on silicon, which is consistent with the preference of silicon to bond by yttrium [94M1]. The substitution of the silicon in the near neighbour environment of an iron atom, primarily influences the long range contributions to the hyperfine field experienced by the iron [93M3]. In $R_2Fe_{14-x}Si_xB$ solid solutions, only a small fraction of the iron-iron bonds is shorter than 0.245 nm and none of those short bonds involve the 4*c* site. Some silicon does occupy the $8j_1$ and $16k_2$ sites and bonds between these sites are shorter than 0.245 nm. In the $Nd_2Fe_{14-x}Si_xB$ compound with x = 2.04, only 16%, of the silicon atoms break bonds shorter than 0.245 nm and the remainder break longer iron-iron bonds, most of them involving iron at 4*c* site, having relative small magnetic moment [00C1]. It was concluded that the breaking of a few short Fe–Fe bonds by silicon, is more effective at increasing T_c, outweighing the negative effects of breaking longer iron–iron bonds. The same site distribution, as silicon in 2/14/1 compounds, was also shown for gallium. In $Nd_2Fe_{14-x}Ga_xB$ system, gallium has also a preference for iron sites having the larger number of Nd in their environment as 4*c*, $8j_1$ and $16k_2$ sites [00C1, 00C2]. Unlike by Si substitution, the presence of gallium increases the lattice parameters, Fe–Fe distances, respectively. Thus, there is a decrease of the number of iron atoms involved in negative exchange interactions, and an enhancement of positive ones by increasing the sample volumes, leading to an increase of T_c with a rate higher than that involving silicon substitutions. The saturation magnetizations, of iron sublattices, decrease when Fe is gradually replaced by Si or Ga.

The saturation magnetizations and the Curie temperatures, of $Y_2Fe_{14-x}Mn_xB$ pseudoternary compounds, decrease rapidly when increasing Mn content [89S1]. In this system, a preferential Mn occupancy of $8j_2$ site was shown [88P5, 91M1]. Two models, in addition to that involving antiferromagnetic Mn ordering, was used to describe their magnetic behavior. The approach assuming that the Mn couple ferromagnetically with the iron moments was first proposed [88M2]. By neutron diffraction study was found no evidence, in the weak magnetic scattering, for either ferromagnetic or antiferromagnetic coupling of Mn moments with those of iron [91M1]. It was concluded that small amounts of manganese located in $8j_2$ site are very effective in disrupting the long-range ferromagnetic coupling. A binomial distribution model of atoms was used for fitting the ^{57}Fe Mössbauer spectra [91M1].

The neutron diffraction studies and magnetic measurements were used to analyse the magnetic behavior of $Er_2Fe_{14-x}Mn_xB$ system [87F4, 87P10, 88F2]. Both the Curie and spin reorientation temperatures decrease when increasing Mn content. Compensation points were observed in the temperature dependences of magnetizations, for intermediate Mn substitutions [87F4, 88F2]. The neutron diffraction measurements showed that the simple collinear moment model, of the magnetic structure, valid at high temperatures, can not explain the existence of reflections forbidden in the space group $P4_2/mnm$. A five-sublattice model was used, in which each of the two rare-earth sites splits into two distinct sublattices, whose moments are oppositely canted from the direction antiparallel to the transition metal moment [88F2]. The canting angle increases with increasing the manganese fraction. According [89Y1], in $Er_2(Fe_{1-x}Mn_x)_{14}B$ series, ferromagnetic ordering of Fe atoms at high x was observed only on the $16k_1$ and 4*c* sites consistent with extrapolation from lower concentration. The $8j_2$ site, which contains a high concentration of Mn atoms, orders with an antiferromagnetic arrangement. This ordering corresponds to strong coupling along the major disclination line in the system.

The Curie temperatures, of $Y_2Fe_{14-x}Cr_xB$ pseudoternary system, decrease by 51.7 K per substituted chronium atom and the saturation magnetization more rapidly than expected by the simple dilution process [92B1]. The ^{57}Fe Mössbauer spectra were fitted assuming substitution of Cr by Fe, at 4*e* sites. In this situation each of the $16k_1$, $8j_1$, $8j_2$ and 4*c* positions have only one 4*e* site as nearest neighbour, which is occupied either by one iron or chromium atom, with equal probability. Each subspectrum associated with these sites was split in two components, one having strongly reduced hyperfine field as result of Cr substitution [92B1].

The Curie temperatures of $R_2Fe_{14-x}Co_xB$ with R = La [88V2], R = Pr [87B6, 88G1], R = Nd [87T1, 95S3, 01D1], R = Dy [01D1] and R = Y [85B6, 86B8, 87T1, 06K1] increase strongly as the cobalt content is higher.

The saturation magnetizations, at T = 4.2 K, for $R_2Fe_{14-x}M_xB$ series with M = Co, Cu, Si, Mn, Al, Ru [86F2, 86P3, 87P8, 89B2] decrease when iron is gradually substituted by M elements, except when M = Co. The saturation magnetization of $Y_2Fe_{14-x}Co_xB$ system was shown to have a maximum at x = 2.0 [92M2, 95S3], $x \cong 3.5$ [86B8] or x = 1.4 [93M4]. Latter on, by using higher external field, only a little change of magnetization up to x = 2 was shown. Then, the magnetizations decrease as the cobalt content increases [06K1]. The ^{59}Co NMR studies on $R_2Fe_{14-x}Co_xB$ compounds, with R = Y, evidenced an increase of the hyperfine field, at T = 4.2 K, when $9.8 \leq x$ and an almost constant value for $1.4 \leq x \leq 9.8$ [92M2, 93M4]. When R = Nd [88I2, 88J2, 89N1, 89W2, 90J2], R = Sm [90M4] and R = Gd [92H1], the ^{57}Co hyperfine fields increase as the iron content increases. A change in the easy magnetization direction from plane to axis was shown in $Y_2Fe_{14-x}Co_xB$ compounds at x = 9.8 [92M2]. The orbital moments, at 4*c* site, are almost constant for $7 \leq x \leq 14$, with a small maximum at x = 9.8. The orbital moments at other sites are almost constant for $9.8 \leq x \leq 14$ and $1.4 \leq x \leq 8.4$. For $8.4 \leq x \leq 9.8$, the orbital moments at $8j_1$ and $8j_2$ sites increase and those at 16*k* ($16k_1 + 16k_2$) and 4*c* sites decrease. The above trends were correlated with the values of the hyperfine coupling constants. The hyperfine fields, in the composition range $1.4 \leq x \leq 9.8$, depend little on the configuration of surrounding atoms. The largest contribution to the anisotropy of the 3d sublattice, in $Y_2Fe_{14-x}Co_xB$ compounds, is given by cobalt atoms located in $8j_1$ and $8j_2$ sites [92M2].

The $Nd_2Fe_{14-x}Co_xB$ compounds, at T = 4.2 K, have a canted magnetic structure with the magnetization tilted away from the tetragonal **c**-axis towards the [110] direction by $\theta = 12°$ (x = 14), up to $\theta = 32°$ (x = 0) [88H2, 88J2, 92F1]. By increasing temperature, a spin reorientation and uniaxial configuration along **c**-axis is reached, at T = 32–34 K (x = 14) or T = 134–136 K (x = 0) [87P3]. In the canted magnetic state, the in-plane component of magnetization introduces magnetically inequivalently positions among the same crystallographic sites. A doubletlike structure for some ^{59}Co NMR lines has been observed [89P1]. The doublet separation was attributed to the in-plane hyperfine field anisotropy and a large splitting was found at 4*c*, $16k_2$ and $16k_1$ sites [89P1, 89W2, 90J2]. When increasing temperature, and the uniaxial state is reached, the doublets converge into singlets and six NMR lines are observed at $T > T_{SR}$. A change of the hyperfine field anisotropy, is shown when spin-reorientation from **c**-axis to **c**-plane takes place. As a result, the $8j_1$ and 4*e* lines are shifted upon reaching uniaxial state [89P1, 90J2]. The 4*e* site shows a large orbital contribution which is decreased by the addition of iron [89W2]. The $8j_2$ lines were not affected by spin reorientation [89P1].

The ^{59}Co hyperfine fields in $R_2Co_{14-x}Mn_xB$, with R = Nd, Sm, decrease when increasing Mn concentration [88I2, 90M4], while for R = Y an increase is observed, at $x \leq 1.4$, attributed to the change of easy direction to easy plane. In $R_2Co_{14-x}Ni_xB$, the ^{59}Co hyperfine field, for all sites, decrease when cobalt is replaced by nickel, in compounds with R = Y [93M4] and R = Sm [90M4].

The substitution of iron by nickel in $R_2Fe_{14-x}Ni_xB$, with R = Y [85B6, 85B8] or R = Nd [95S3], decrease the saturation magnetization, although the Curie temperatures increase little [85B6]. The ^{143}Nd NMR study showed that the Nd magnetic moments, averaged over the *f* and *g* sites diminishes from 3 μ_B/atom (x = 0) to 2.9 μ_B/atom (x = 1.4) [95S3].

The Curie temperatures of $R_2Fe_{14-x}Cu_xB$ with R = Nd, Er, increase with a rate of ≅9 K per substituted iron atom, while the saturation magnetization decreases with a higher slope than expected for simple dilution trend [90B2]. The ^{57}Fe Mössbauer study on $Y_2Fe_{14-x}Cu_xB$ system, suggested, in a first approximation, a random distribution of Cu atoms among the iron sites [92B1]. The substitution of copper for iron has only a minute influence on the ^{57}Fe hyperfine field values, while that of chromium decreases the hyperfine fields.

A high decrease of the Curie temperatures of $Y_2Fe_{14-x}Al_xB$ compounds, with a rate of ≅56 K per substituted atom, in the composition range $x \leq 3$, while the magnetization decreases with a rate of 2.3 μ_B per substituted Al atom [86B6]. The presence of local environment effects, as well as a decrease of the ^{57}Fe hyperfine field in $Nd_2Fe_{14-x}Al_xB$ system was also found as Al content increases [88H2]. Field induced transitions from ferrimagnetic to ferromagnetic ordering through the intermediate canted magnetic structure, have been observed in magnetically aligned $Er_2Fe_{14-x}Al_xB$ powdered samples [91K3].

The substitutions of Fe by M = V or Ge, in $Y_2Fe_{14-x}M_xB$ compounds, decrease their Curie temperatures, as well as the saturation magnetizations [91B4]. It was also concluded that the T_c temperatures are mainly determined by site distribution, of substituting elements, while the saturation magnetization by their molar fraction.

The magnetocrystalline anisotropy in $R_2Fe_{14}B$ compounds were analysed for R = La [85G4, 86G5], R = Ce [85G4, 86G5, 87R1, 98K1], R = Pr [85G4, 86G5, 87R1, 98K1, 00K2, 14A1], R = Nd [84Y1, 85G4, 86G5, 87R1, 90G3, 00Y2, 05H1, 10K1, 11T1] R = Sm [85H5, 87R1, 89L6], R = Gd [84Y1, 85B12, 85G4, 86G5, 87R1, 88H3, 90G3, 98K1, 10K1], R = Tb [84Y1, 87R1, 90G3, 98K1], R = Dy [84Y1, 87R1, 88H3, 90G3, 98K1, 11T1, 14A1, 18Y1], R = Ho [84Y1, 85G4, 86G5, 87R1, 88H4, 90G3, 18Y1], R = Er [84Y1, 87R1, 89L6, 90G3, 95K2, 00K1], R = Tm [87R1, 89L6, 90G3, 00K1], R = Yb [89L6], R = Lu [85G4, 86G5, 90G3], R = Y [84Y1, 85G4, 86G5, 10K1, 14A1, 14M1, 14T1]—Table 7.37.

The anisotropy energy of the iron sublattice, in $R_2Fe_{14}B$ series, can be determined in compounds with non-magnetic (R = La, Lu, Y, Ce) or S-state (R = Gd) rare-earths. The anisotropy energy, E_a, of tetrahedral compounds was described by the relation:

$$E_a = K_1 \sin^2\theta + K_2 \sin^4\theta + K_3 \sin^4\theta \sin 2\varphi \qquad (7.1)$$

By K_i are denoted the anisotropy constants and θ and φ are the polar angles of the magnetization. The K_3 anisotropy constant is the one that determines the basal plane anisotropy and it is of the fourth order term.

The anisotropy constants K_i, were determined from magnetization isotherms, particularly for single crystals, from torque curves or by using SPD method [80A1]. From these values, the anisotropy fields, H_a, were determined by using only K_1 value,

$H_a = 2K_1/\mu_0M_s$ [85B3], or higher order terms in E_a as $H_a = 2(K_1 + 2K_2)/\mu_0M_s$. According to used relation, some differences in H_a values of $R_2Fe_{14}B$ compounds are reported in literature.

The orbital moments, at different iron sites, in $Nd_2Fe_{14}B$ were calculated [87S4, 90Z6]. The largest orbital moment is located on the $8j_2$ sites. These sites are pulled out from the centre of the hexagon of other iron atoms [84H1]. There is a strongly preferred direction of magnetization along **c**-axis, $8j_2$ sites playing an important role in determining the magnetic anisotropy of iron sublattice. By the XMCD study on $Nd_2Fe_{14}B$ single crystal, through the spin reorientation transition, was shown that the average Fe orbital moment is proportional to the macroscopic iron anisotropy constant K_1 and diverges 15 K below the spin reorientation temperature. The divergence was indicative to a critical behavior and was related to tetragonal distortion [00G3, 01G1].

The anisotropy fields, $H_a(T)$, in $R_2Fe_{14}B$ compounds with R = Y, Gd, increase when rising temperature, leading to a maximum below the Curie temperatures. The above unusual behavior can be correlated with the presence of two mechanisms: (1) six different iron sites give contributions to the bulk anisotropy, which have opposite signs and different temperature dependences; (2) The increase with temperature of the anisotropy is reminiscent of the magneto-volume anomaly, at $T < T_c$, i.e., the Invar effect influences the magnetic anisotropy [85A2, 87B7, 91K5]. In this model, the temperature dependence of the anisotropy constant K_1 is given by $K_1 \sim M^3[1-0.52(c/a)^2]$ [87B7]. The presence of the second mechanism has been evidenced, by a comparative analysis of the physical properties of $R_2Fe_{14}B$ and $R_2Fe_{14}C$ compounds [90G3, 90K4]. Thus, both mechanisms can contribute to the anisotropy of iron sublattice. The temperature dependences of the anisotropy fields in $R_2Fe_{13}MB$ compounds, with R = Gd, Y and M = Al, Si and Ga show differences, as compared to those of parent compounds. The anisotropy fields, at low temperatures in Y-based compounds are different from those of the gadolinium ones. In $Y_2Fe_{13}MB$, the H_a values rise on the substitution of M = Ga and particularly of Si. When R = Gd, the addition of gallium leads to a substantial loss in H_a values. A preference of the substituting elements for 16*k* sites (corresponding to a preference of iron for 8*j* sites, particularly $8j_2$) seems to favour an increase of the anisotropy. Thus, to gain H_a value, this peculiar site has to be occupied by iron as much as possible. According to [92W1], the Ga occupation of $8j_2$ site in $Gd_2Fe_{13}GaB$ is supposed to be responsible for the reduced anisotropy.

The anisotropy of a large number pseudoternary $R_2Fe_{14-x}M_xB$ compounds were investigated, particularly when M = Co and R = Y [87B4, 88T4] or La [88G6]. The composition dependence of the anisotropy constants of $Y_2Fe_{14-x}Co_xB$ series, at T = 4.2 K, has a maximum at $x \cong 10.5$ and change in sign at $x \cong 4.0$ [88T4, 93M4]—Fig. 7.21. The data were analysed in a model considering the individual site anisotropies of the 3d sublattices, the contributions of iron and cobalt to anisotropy having different sign [88T3]:

Fig. 7.21 $Y_2Fe_{14-x}Co_xB$: composition dependence of the anisotropy constant K_1, at T = 4.2 K experimentally determined and computed value (solid line) [88T4]

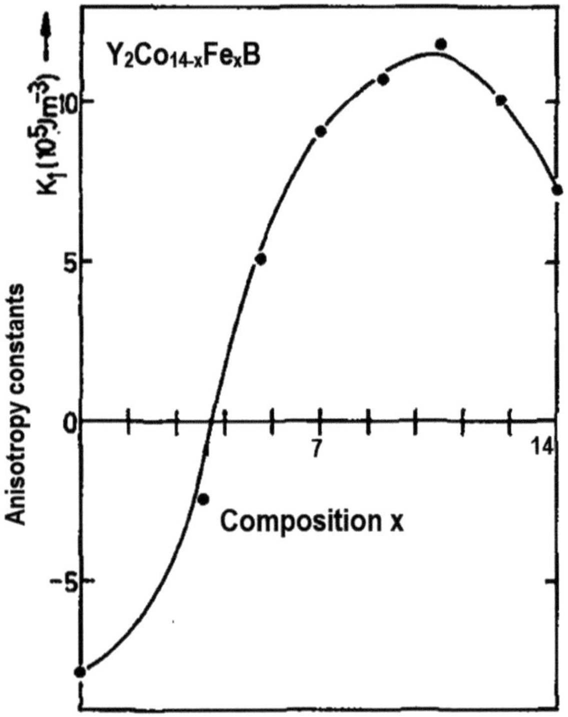

$$K_1 = K_1(0) + \sum_{i=1}^{m} n_i f_i(x) \Delta K_1^i \tag{7.2}$$

By $K_1(0)$ is denoted the anisotropy constant, at x = 0 and $\Delta K_1^i = K_1^i(Fe) - K_1^i(Co)$ is the difference between the contributions to anisotropy of iron and cobalt at the site i. Values ΔK_1^i (10^5 J/m^3) = −0.11(k_1), 0.009(k_2), −6.25(j_1), 4.41(j_2), 9.41(e) and −1.92(c) were obtained. The calculated curve, according relation (7.2) and the above ΔK_1^i values, describes well the experimental data. The anisotropy energy in the $Y_2Fe_{14-x}Co_xB$ system was also estimated from local anisotropy energy per atom, at each site and preferential substitution of Fe for Co at 8j_2 site [93M4].

The temperature and composition dependences of the anisotropy fields were studied in a large number of $R_2Fe_{14-x}M_xB$ compounds, particularly with R = Y and M = Cr [90K8], M = Co [86H7, 87B4], M = Cu [87P9, 88K1], M = Si [87J3, 87P9, 88Y4, 90K8] or R = Nd with M = Mn [87B5], M = Co [86G6, 86H7, 87B4], M = Ni [87B5], M = Cu [88K1], M = Nb, Mo [88H4], M = Si [87P8, 87J3, 87P6, 88Y4], M = Ga [88H4, 88X1] or M = Be [96L2]. Different composition dependences, as function of substituting elements, were shown at ambient temperature. In $Y_2Fe_{14-x}M_xB$ compounds, an increase of H_a values was observed for M = Si or Cr at $x \leq 1$, a continuous decrease when M = Co and a decrease followed by an increase, for M = Cu. In $Nd_2Fe_{14-x}M_xB$ system, for low Ni, Si, Ga, Be and Al contents, the anisotropy fields have maxima [98B2]. A continuous decrease of H_a values, occurs in the compounds when Fe is substituded by M = Mn, Co, Mo. The above trends are related also to changes in Curie temperatures and associated thermal variations of magnetizations as well as due to the effect of site occupancies of substituting elements. The spin reorientation temperature decreases by Co and/or Al substitution for Fe in $Nd_2Fe_{12\text{-}x}Co_2Al_xB$ compounds [89K3].

The composition dependences of the anisotropy field of $Nd_2Fe_{14\text{-}x}M_xB$ system, at $T/T_c = 0.45$ (T = 293 K), evidenced a linear decrease, when Fe is substituted by M = Ni, while when M = Mn, an increase is observed [87B5]. At low temperatures, the anisotropy constants of Mn substituted compounds, linearly diminished by Mn addition, the observed behaviour, at $T/T_c = 0.45$, being correlated with the decrease of magnetization due to thermal effects. The Nd sublattice anisotropy seems to be not affected by Mn substitution, while Ni acts to decrease it. The reduced anisotropy field, as function of reduced temperatures T/T_c, for T = 300 K, in $Nd_2Fe_{14\text{-}x}Co_xB$, decreases as the cobalt content increases [86G6]. The room temperature anisotropy fields in $Pr_2Fe_{14\text{-}x}Co_xB$ decreased up to a minimum located at $x \cong 9.8$ and then strongly increased [87G3].

In $Y_2Fe_{14}BH_y$, there is a sharp fall in the anisotropy constant K_1, when increasing hydrogen content, which in partly offset, at T = 4.2 K, by a rise in K_2 [86C2]. The sharp decrease of the uniaxial anisotropy constant K_1, with y, was correlated with the iron sites which form the tetrahedra occupied by hydrogen, i.e. $16k_2$, $8j_2$ and $4c$. According to [95I1], the dramatic reorganization of Nd environment is probably the main reason for the drop of the uniaxial magnetic anisotropy of $Nd_2Fe_{14}BH_y$ samples.

The $Y_2Fe_{14}BN_y$, has an easy axis of magnetization [92Z1]. As compared to $R_2Fe_{14}B$ (R = Pr, Nd), the anisotropy field and the critical field of the first order magnetic transition (FOMT) of $R_2Fe_{14}BN_y$ compounds are also reduced, whereas spin reorientation temperatures T_{sR} of $R_2Fe_{14}BN_y$ are only slightly decreased [92K3, 92Z1]. The magnetic properties of $R_2Fe_{14}B_{1\text{-}x}C_x$ were also investigated [90K3].

The magnetic state of the rare-earths, in $R_2Fe_{14}B$ ternary system, depends on the crystal electric field (CEF) and of the exchange field acting on these ions. The low temperature rare-earth moments, determined by magnetization measurements, are close to those of free ion g_JJ values [86Y6]. Differences in the R4f and R4g moments were reported by Mössbauer studies on ^{145}Nd [90N2], ^{155}Gd [85B2, 98B3], ^{161}Dy [85F1, 85F2, 85G6, 85V6, 86F1, 86G7], ^{166}Er [86S1, 87F2] and ^{169}Tm [87G6] in $R_2Fe_{14}B$ compounds. The Mössbauer spectra consist generally of two subspectra corresponding to the two R sites. Some hyperfine parameters, determined by Mössbauer spectroscopy and NMR method are listed in Table 7.36. It is to be noted that the hyperfine fields, at light rare-earths, are less than the value characteristic for free ion, whereas for heavy rare-earths (R = Gd, Dy, Er, Tm) are greater. This behavior was attributed to contribution of iron sublattice magnetization, which is parallelly aligned with that of R sublattice in light rare-earth compounds and antiparallel in heavy rare-earth ones. Similar results were obtained by NMR studies at ^{139}La in $(La,R)_2Fe_{14}B$ with R = Nd, Y [96K1], ^{141}Pr in $Pr_2Fe_{14}B$ [95S1], ^{143}Nd, ^{145}Nd in $Nd_2Fe_{14}B$ [89B3, 93K1], $Nd_2(Fe,Co)_{14}B$ [85P4, 87F7, 89N1, 92J1, 93K1, 95S3],

Table 7.36 Hyperfine parameters determined by NMR and MS at rare-earths nuclei in $R_2Fe_{14}B$ compounds

Compound	T (K)	Nucleus	Site	δ (mm/s)	ΔQ (mm/s)	$\mu_0 H_{hf}$ (T)	η	θ°, φ[a]	$V_{zz} \cdot 10^{17} (Vm^{-2})$	References
$La_2Fe_{14}B$ (NMR)	4.2	^{139}La	*4f* *4g*			36.1 44.1			-3.8 -7.9	[96K1]
$Pr_2Fe_{14}B$ (NMR)	4.2	^{141}Pr	*4f* *4g*			257.4(4) 280.3(4)			13.8 -8.9	[95S1]
$Nd_2Fe_{14}B$ (NMR)	4.2	^{145}Nd	*4f* *4g*				0.40 0.22		-2.5 [g)] -6.7 [g)]	[93K1]
$Sm_2Fe_{14}B$ (NMR)	4.2	^{147}Sm	*4f* *4g*				0.40 0.22		-2.5 [g)] -6.7 [g)]	[93K1]
$Gd_2Fe_{14}B$[b,c] (MS)	4.2	^{155}Gd	*4f* *4g*	0.41(1) 0.35(2)	−0.51(1) −2.58(8)	42.3(1) 21.2(5)		90°	2.18 −8.46	[85B12]
$Gd_2Fe_{14}B$[b,c] (MS)	4.2		*4f* *4g*	0.157(3) 0.69(2)	0.60(3) −2.33(14)	35.4(6) 31.0(20)		0°		[85B12]
$Gd_2Fe_{14}B$[b] (MS)	4.2		*4f* *4g*	0.249(5) 0.191(7)	−3.60(8) 4.96(4)	14.7(4) 27.9(5)	0.44(14) 0.34(5)	0°,0° 90°,90°		[85B2]
$Er_2Fe_{14}B$[e] (NMR)	4.2	^{167}Er	*4f* *4g*			804.2(8) 797.7(8)	0.35 0.61	90°,13.7° 90°,7°	14.7 −8.63	[93S1]
$Dy_2Fe_{14}B$[d] (MS)	4.2	^{161}Dy	*4f* *4g*	0(2) 2(2)	57(5) 60(5)	606.0(40) 611.0(40)				[85G6, 86G7]
$Dy_2Fe_{14}B$[d] (MS)	4.2	^{161}Dy	*4f* *4g*			620.0 637.7				[86B1]
$Tb_2Fe_{14}B$(NMR)	4.2	^{159} Tb	*4f* *4g*			367.4(4) 356.5(4)	0.35 0.61	90°,90° 90°,90°	15.6 −10.3	[93S1]
$Tm_2Fe_{14}B$[c)] (MS)	4.2	^{196}Tm	*4f* *4g*		136(3) 167(3)	770.0(30) 760.0(30)		$\varphi^f = \pm 34.1(6.4)°$ $\varphi^f = \pm 14.9(1.8)°$		[87G4]

[a] θ and φ polar angles defining the direction of hyperfine field with respect to EFG axis; [b] $^{154}SmPd_3$ source; [c] Data in the upper and lower parts were fitted under constraints θ = 90° and θ = 0°, respectively; [d] ^{161}TbF source; [e] $^{168}ErAl_3$ in Al matrix source; [f] φ is angle with [100] axis; [g] V_{zz}(lattice)

$Nd_2(Fe,Ni)_{14}B$[95S3]; ^{197}Sm, ^{149}Sm in $Sm_2Fe_{14}B$ [93K1], ^{159}Tb in $Tb_2Fe_{14}B$ [93S1], $(Tb,Er)_2Fe_{14}B$ [95S2] and ^{167}Er on $Er_2Fe_{14}B$, [93S1]. These measurements allowed to determine the electric quadrupole splitting at the two R sites and crystal field parameters—Table 7.38. The ^{141}Pr NMR study, on $Pr_{2-x}La_xFe_{14}B$, showed that the magnetic moments at 4*f* and 4*g* sites are of 2.9 μ_B/atom and 3 μ_B/atom, respectively [95S1]. The ^{145}Nd NMR investigations, performed on $Nd_2Fe_{14-x}Co_xB$ [85P4, 87F7, 89N1, 92J1, 93K1], showed the presence of two components associated with Nd4*f* and Nd4*g* sites. It is to be mentioned that one of the Nd positions was subdivided in two components, depending on the angle between Nd magnetic moments and the electric field gradient [89N1]. Well resolved quadrupole subspectra, assigned to 4*f* and 4*g* sites, were evidenced by NMR on ^{147}Sm and ^{149}Sm in $Sm_2Fe_{14}B$, at T = 4.2 K [93K1] or by ^{159}Tb NMR in $Er_{2-x}Tb_xFe_{14}B$, at T = 1.4 K [95S2]. The hyperfine fields at Tb4*f* and Tb4*g* sites decrease when Tb is replaced by Er.

The R4*f* and R4*g* sites, having local symmetry *mm*, in the tetragonal $P4_2/mnm$ structure, can be further subdivided magnetically into f_1, f_2, g_1 and g_2 sites. A splitting of the ground state multiplet into (2 J + 1)/2 doublet (for J half integer) and (2 J + 1) singlets (for J integer) is expected. By using inelastic neutron scattering technique was possible to determine the CF level scheme of the ground state multiplet of R ions, as the available neutron energies are of the order of this splitting [90L6, 91L2, 91L3, 94Z1]. The presence of molecular field, produces further splittings of the doublets and shifts the positions of all levels. With neutron experiments, performed at low temperatures, one measure only transitions from the ground state to the excited levels, with intensities determined by the corresponding dipole matrix elements.

The Hamiltonian describing the magnetic behavior of R elements in $R_2Fe_{14}B$ compounds, assumed crystal field approximations and in addition, interaction with the molecular field, H_m [94Z1]. The R–R interactions are neglected and the constant term due to Fe–Fe interactions ommitted. In this way nine crystal field parameters $A_n^m(i)$ and four molecular fields $H_m(i)$, for positions i = (f_1,f_2,g_1,g_2) can be determined. Some approximations were made in analysing this matter [88C1, 88G3, 88Y1, 89R1, 91R1]. Some authors [88C1, 88G3, 89R1], assumed that the molecular fields acting on R ions does not depend on the ion position, $H_m(i) = H_m$. The model [88Y1], and partly the model [89R1, 91R1], assumed additionally the same $A_n^m(i)$ parameters for the 4*f* and 4*g* sites, which means that all R ions in the crystal, are subjected to the same crystal field. The model [88C1, 88G3] used different crystal field parameters $A_n^m(f)$ and $A_n^m(g)$ for the two R sites. Many models assumed that $A_4^2 = A_6^{-2} = A_6^{-6} = 0$. The models [88Y1, 21S2], adopted finite values, particularly for Pr and Nd ions. In addition, $A_4^4 = 0$ was assumed [88Y1], while in other models [88C1, 88G3, 89R1, 91R1] considered $A_4^4 \neq 0$, in the model [88C1, 88G3], A_4^4 being different for *f* and *g* sites. Using these approximations, the crystal field parameters are calculated—Table 7.38.

The CEF parameters and R–Fe exchange fields were determined also from the analysis of the magnetization isotherms of $R_2Fe_{14}B$ single crystals [88C1, 88G3, 88R1, 88T2, 88V1, 88W1, 88Y1, 88Y5, 89R1, 89R3, 89Z7, 89Z9, 90N1, 91R1, 92K1, 98K1, 00Y1]. Although the magnetization curves are well reproduced, when

Table 7.37 Anisotropy constants of $R_2Fe_{14}B$ compounds

(a) Anisotropy constants

Compound	T (K)	Anisotropy constant (MJ/m^3)			K_{calc} (MJ/m^3)		References[b]
		K_1	K_2	K_3	K_{1R}[a)]	K_1[a)]	
$La_2Fe_{14}B$	RT	1.2					[88G6]
$Ce_2Fe_{14}B$	4.2	1.65					[85H4, 86H4]
$Ce_2Fe_{14}B$	4.2	3.4			10.8	11.5	[84S4, 87R1]
$Ce_2Fe_{14}B$	300	1.5			1.8	2.8	[86H5, 87R1]
$Pr_2Fe_{14}B$	4.2	23.5					[85H4, 86H4, 86Y2]
$Pr_2Fe_{14}B$	4.2	13	12				[86Y3]
$Pr_2Fe_{14}B$	4.2	9.5	10.5				[86L1]
$Pr_2Fe_{14}B$	4.2	14.2	11.5	−13.5			[87H9]
$Pr_2Fe_{14}B$	4.2	22	11				[00K2]
$Pr_2Fe_{14}B$	4.2	18.4			20.6	21.3	[84S4, 87R1]
$Pr_2Fe_{14}B$	300	5.4			6.1	7.1	[86H5, 87R1]
$Pr_2Fe_{14}B$	280	6.1	1				[00K2]
$Nd_2Fe_{14}B$	4.2	−16	28				[86H4, 86Y2]
$Nd_2Fe_{14}B$	4.2	−8	19.5				[86Y3]
$Nd_2Fe_{14}B$	4.2	−9	24				[86D5]
$Nd_2Fe_{14}B$	4.2	−12.8	45.9	−22.4			[87H9]
$Nd_2Fe_{14}B$	4.2	−6.5	14	22			[87O5]
$Nd_2Fe_{14}B$	4.2	cone			12.3	13.0	[87R1]
$Nd_2Fe_{14}B$	77	−12					[87O2]
$Nd_2Fe_{14}B$	300	4.9					[86A4]
$Nd_2Fe_{14}BH_{2.8}$	300	0.79					[86A4]
$Nd_2Fe_{14}B$	300	4.8			4.3	5.3	[86G5, 87R1]
$Nd_2Fe_{14}B$	300	5.0					[85K1]
$Nd_2Fe_{14}B$	300	4.5	0.66				[85S1]
$Nd_2Fe_{14}B$ $Nd_2Fe_{14}B$[b]	300	4.2	0.7				[86D5, 87I2]
$Sm_2Fe_{14}B$	4.2	−24.9			−20.8	−20.1	[84S4, 87R1]
$Sm_2Fe_{14}B$	4.2	−26.0					[85H4, 86Y3]
$Sm_2Fe_{14}B$	4.2	−21.0					[85H5, 87R1]
$Sm_2Fe_{14}B$	RT	−12	0.29	−0.29			[85S1]
$Sm_2Fe_{14}B$	300	−11.8			−12.5	−11.5	[85H5, 87R1]
$Gd_2Fe_{14}B$	4.2	0.55			0		[85B2]
$Gd_2Fe_{14}B$	4.2	9.0					[90K4]
$Gd_2Fe_{14}B$	RT	0.72					[85H4]

(continued)

Table 7.37 (continued)

(a) Anisotropy constants

Compound	T (K)	Anisotropy constant (MJ/m^3)			K_{calc} (MJ/m^3)		References[b]
		K_1	K_2	K_3	K_{1R}[a)]	K_1[a)]	
Gd2$Fe_{14}B$	300	1.0					[86A2]
$Tb_2Fe_{14}B$	4.2	6.9			27.6	28.3	[86Y2, 87R1]
$Tb_2Fe_{14}B$	300	8.2					[85H4]
$Tb_2Fe_{14}B$	300	15			25.1	26.1	[85K1, 87R1]
$Tb_2Fe_{14}B$	300	3.4					[86A2]
$Dy_2Fe_{14}B$	4.2	4.4			26.6	27.3	[84S4, 87R1]
$Dy_2Fe_{14}B$	4.2	3.8					[85H4, 86Y3]
$Dy_2Fe_{14}B$	300	4.8			20.0	21.0	[86H5, 87R1]
$Ho_2Fe_{14}B$	4.2				10.3	11.0	[87R1]
$Ho_2Fe_{14}B$	4.2	−3.3					[86Y2]
$Ho_2Fe_{14}B$	77	−0.25					[87O2]
$Ho_2Fe_{14}B$	300				6.0	7.00	[87R1]
$Ho_2Fe_{14}B$	300	2.0					[86A2]
$Er_2Fe_{14}B$	4.2	−7.1			−10.2	−9.5	[84S4, 87R1]
$Er_2Fe_{14}B$	4.2	−1.4					[86Y2]
$Er_2Fe_{14}B$	300				−3.7	−2.7	[87R1]
$Er_2Fe_{14}B$	300	−1.36					[86A2]
$Tm_2Fe_{14}B$	4.2	−6.2			−21.9	−21.1	[84S4, 87R1]
$Tm_2Fe_{14}B$	4.2	−3.6					[86Y2]
$Tm_2Fe_{14}B$	300				−4.8	−3.8	[87R1]
$Yb_2Fe_{14}B$	4.2				4.3		[89L3]
$Lu_2Fe_{14}B$	4.2	0.94					[05T1]
$Lu_2Fe_{14}BH_{2.5}$	4.2	−0.67					[05T1]
$Y_2Fe_{14}B$	4.2	0.70					[85B2]
$Y_2Fe_{14}B$	4.2	0.77					[85H4]
$Y_2Fe_{14}B$	RT	1.1	≅ 0	≅ 0			[85S1]
$Y_2Fe_{14}B$	300	0.99					[86A1]
$Y_2Fe_{14}B$					$K_1 = -3.73$ $K_2 = 1.31$ $K_3 = 0.42$		[87I2]

(continued)

Table 7.37 (continued)

(a) Anisotropy constants

Compound	T (K)	Anisotropy constant (MJ/m^3)			K_{calc} (MJ/m^3)		References[b]
		K_1	K_2	K_3	K_{1R} [a)]	K_1 [a)]	
$Y_2Fe_{14}BH_{1.8}$	300	0.25					[86A1]

(b) Anisotropy constants at R sites and a total anisotropy constant

Compound	T (K)	Anisotropy constant of R sublattice (J/kg)					K_1		References
		K_{1R}	K_{2R}	K_{3R}	K_{4R}	K_{5R}	ǁ [100]	ǁ [101]	
$Tb_2Fe_{14}B$	4.2	4300							[98K1]
$Dy_2Fe_{14}B$	4.2	1700	6600						[98K1]
$Er_2Fe_{14}B$	4.2	1300	−4500	−380	2200	400	−1320	20	[88G3, 90K1]
$Er_2Fe_{14}B$	4.2	440	−2100	−980	710	1000	−1180	20	[90V1]
$Er_2Fe_{14}B$	4.2	−1255	0	−30			−2630	−30	[00K1]
$Tm_2Fe_{14}B$	4.2	−2700	1600	630	−1500	−710	−9740	−80	[88G3, 90K1]
$Tm_2Fe_{14}B$	4.2	−1500	−560	230	−370	−330	−8520	−100	[90V1]
$Tm_2Fe_{14}B$	4.2	−6830	0	−260			−14700	−260	[00K1]

[a] Anisotropy constants were calculated by [87R1] from high field measurements [86Y2, 87S2]; [b] For $Nd_2Fe_{14}B$ values $K_{1R} = -14.88$, $K_{2R} = 23.91$ and $K_{3R} = 0.03$ MJ/m^3 and $K_{1d} = -1.12$, $K_2 = 3.09$ and $K_3 = 0.42$ MJ/m^3 were calculated [87I2]

using the determined CEF and molecular field values, there are other sets of parameters which also explain the experimental data [90N1, 94Z1]. This evidence can be correlated with a large number of parameters to be determined, compared with a limited informations than can be obtained from magnetization curves. To test the influence of crystal field parameters, different sets of data were used to analyse the magnetization isotherms. For R = Nd, Tb and Ho, the dependence of the calculated magnetization curves on the crystal field parameters is rather weak. For R = Pr, Dy and Er, quite different values were obtained, depending on the choice of CEF parameters.

If one breaks the ferrimagnetic structure of $R_2Fe_{14}B$ compounds with heavy rare-earth, by applying a high magnetic field, more direct and accurate information on the CEF parameters and exchange fields may be obtained [98B2]. Based on the model [88Y1], the magnetization processes up to $\mu_0H = 110$ T, have been analysed for R = Tb, Dy, Er and Tm [92K1, 95K1]. The high field magnetization isotherms of $R_2Fe_{14}B$ compounds with R = Tb, Dy, Ho, Er, Tm, at T = 4.2 K, were analysed in terms of the exchange and crystal field interactions, including terms up to sixth order, which may differ at R4*f* and R4*g* sites, respectively [87G1, 88G3].

The inelastic neutron scattering INS studies, on polycrystalline $R_2Fe_{14}B$ compounds were performed in order to determine the positions and intensities of the R magnetic excitations. The corresponding spectra, even at low temperatures,

Table 7.38 Crystal field parameters for $R_2Fe_{14}B$ compounds

(a) Crystal field parameters[a, b]

Compound	H_m(T)		A_2^0		A_2^{-2}		A_4^0		A_4^{-2}		A_4^4		A_6^0		A_6^{-2}		A_6^4		A_6^{-6}		References
	$g=f$	$g=f$	f	g	f	g	f	g	f	g	f	g	f	g	f	g	f	g	f	g	
$Pr_2Fe_{14}B$[b]	224	448	295		−454		−12.3		0		0		−6.89		−14.7		−29.8		25.1		[88Y1]
$Pr_{Y12}Fe_{14}B$	230	460	220		0		−15		0		3.9		−6.6		0		−12		0		[89R1, 90R2]
$Pr_2Fe_{14}B$			176	179	−116	351	3	3	0		−9	8	−7	−7					−42	−4	[88G1]
$Nd_2Fe_{14}B$	390	520	295		−454		−12.3		0		0		−1.84		9.8		−15.9				[88Y1]
$Nd_2Fe_{14}B$	342	456	307		−196	600	−14.5	−12.6	0		−43.3	−40.6	−2.2				−33.2	−12.6			[88C1]
$Nd_2Fe_{14}B$	350	470	330		−210	660	−16		0		3.9		−2.9		0		−12		0		[89R1]
$Nd_2Fe_{14}B$[c)]																					
LMTO			470	284	−640	709	−20	−18	30	−31	46	−43	−0.2	−0.3	0	0.4	−1.5	−2.3	0.5	0	[96H3]
valence			497	643	426	1454	−19	−20	19	−10	33	−54	0	0	0.1	0.1	0.2	0.1	−0.1	0.1	[96H3]
lattice			−22	−359	−1066	−745	−0.8	2.2	11	−21	13	−11	−0.2	−0.3	−0.1	0.3	−1.6	−2.3	0.5	−0.1	[96H3]
$Sm_2Fe_{14}B$	230		297		−458		−12.6						−7.95				−20.2				[88Y1]
$Gd_2Fe_{14}B$[d)]																					
band			458	289	−308	1058	−23	−19	24	−30	26	−44	0.2	−0.2	−1.3	1.5	−2.3	−1.8	5.9	1.2	[98Y1]
f–			433	265	−471	804	−19	−16	22	−31	28	−38	−0.3	−0.3	−0.2	0.7	−1.6	−2.4	0.1	−0.4	[98Y1]
core			409	263	−481	806	−20	−17	26	−32	32	−40	−0.3	−0.3	−0.1	0.6	−1.5	−2.5	−0.2	−0.5	[98Y1]
$Tb_2Fe_{14}B$	144	216	300		−462		−12.6		0		0		−0.958		0		−5.17		0		[88Y1]
$Tb_2Fe_{14}B$	157	236	304	308	−200	605	−14.5	−12.7	0		43.4	−40.6	−2.2	−2.2	0		−33.2	−12.6	0		[88C1, 88G3]
$Dy_2Fe_{14}B$	108	216	302		−467		−12.7		0		0		−0.973		0		−5.29		0		[88Y1]
$Dy_2Fe_{14}B$	121	242	292	296	−192	581	−13.9	−12.2	0		41.7	−39.0	−1.3	−1.3	0		−18.9	−7.6	0		[88C1, 88G3]
$Dy_2Fe_{14}B$	130	260	285		0		−14		0		3.9		0.5		0		−12		0		[89R1]
$Ho_2Fe_{14}B$[b)]	86	216	302		−467		−12.7		0		0		−0.973		0		−5.29		0		[88Y1]
$Ho_2Fe_{14}B$	81	202	298	302	−196	593	−8.7	−7.6	0		26.0	−24.4	−1.3	−1.3	0		−18.9	−7.2	0		[88C1, 88G3]
$Ho_2Fe_{14}B$	86	216	334		−210	660	−15		0		3.9		−1.1		0		−12		0		[89R1]
$Er_2Fe_{14}B$[b)]	72	216	303		−471		12.8		0		0		−0.980		0		−5.33		0		[88Y1]
$Er_2Fe_{14}B$	68	204	292	296	−222	689	−16.0	−14.0	0		36.7	−38.6	−1.2	−1.2	0		−5.7	−2.6	0		[88C1, 88G3]

(continued)

Table 7.38 (continued)

(a) Crystal field parameters[a, b]

Compound	H_m(T)		A_2^0		A_2^{-2}		A_4^0		A_4^{-2}		A_4^4		A_6^0		A_6^{-2}		A_6^4		A_6^{-6}		References
	$g=f$	$g=f$	f	g	f	g	f	g	f	g	f	g	f	g	f	g	f	g	f	g	
$Tm_2Fe_{14}B$[b)]	62	216	303		−471		−12.8		0		0		−0.980		0		−5.33		0		[88Y1]
$Tm_2Fe_{14}B$	58	203	258	262	−205	617	−12.3	−10.8	0		34.7	−32.5	−1.7	−1.7	0		−26.5	−10.1	0		[88C1, 88G3]
$Yb_2Fe_{14}B$[b)]	145		303		−471		−12.8		0		0		−0.980		0		−5.33		0		[88Y1]

(b) Crystal field parameters[e]

Compound	H_m(T)		A_2^0		A_2^{-2}		A_4^0		A_4^{-2}		A_4^4		A_6^0		A_6^{-2}		A_6^4		A_6^{-6}		References
$Pr_2Fe_{14}B$[f)]	780		400		210		−150		0		0		−1050		0		−450		0		[00Y1]
$Pr_2Fe_{14}B$[f)]	758		386		207		68		0		0		−1760		0		−516		0		[88G1]
$Pr_2Fe_{14}B$[f)]	620		478		0		−339		0		11		−1660		0		−269		0		[90R2]
$Nd_2Fe_{14}B$[f)]	630		600		320		−260		0		0		−550		0		−400		0		[00Y1]
$Nd_2Fe_{14}B$[f)]	700		591		371		−236		0		0		−365		−190		−281		0		[88Y1]
$Nd_2Fe_{14}B$[f)]	626		682		366		−371		0		0		−529		0		−491		0		[88C1]
$Nd_2Fe_{14}B$[f)]	627		661		380		−307		0		9		−575		0		−212		0		[89R1]
$Nd_2Fe_{14}B$[f)]	627		661		0		−307		0		9		−575		0		−212		0		[88V1]
$Nd_2Fe_{14}B$	269	287	358	463	878	−269	−51	−56	−85	70	−146	100	−3	−2	4	−3	−27	−19	−5	−5	[21S2]
$Tb_2Fe_{14}B$	335		440		±270		−190		0		−70		−60		±30		−115		0		[89Z9, 90Z5]
$Tb_2Fe_{14}B$	430	335	510	410	460	−144	−165	−240	−42	60	84	−32	−90	−50	0	0	−29	−43	0	43	[94Z1]
$Dy_2Fe_{14}B$	320		420		± 260		−170		0		−65		−55		± 28		−100		0		[89R1, 89Z9]
$Dy_2Fe_{14}B$	390	305	480	390	435	−137	−150	−220	−38	55	76	−30	−80	−45	0	0	−25	−38	0	38	[94Z1]
$Dy_2Fe_{14}B$	229	238	351	373	−836	64.5	−40	−40.3	60.9	−44.3	117	54.9	−1.67	−1.38	−3.46	2.43	−43.7	−9.83	4.77	0.102	[18Y1]
$Ho_2Fe_{14}B$	213	221	345	358	−823	45.1	−38.2	−38.3	56.2	−40.6	−111	49.6	−1.46	−1.22	−2.89	2.10	−12.3	−8.64	6.24	0.219	[18Y1]
$Ho_2Fe_{14}B$	310		400		± 250		−160		0		−60		−45		± 25		−90		0		[89Z9, 90Z5]
$Ho_2Fe_{14}B$	345	275	460	370	415	−130	−138	−200	−35	50	70	−27	−70	−40	0	0	−22	−34	0	34	[94Z1]

(continued)

Table 7.38 (continued)

(b) Crystal field parameters[e]																					
$Er_2Fe_{14}B$	300		380		± 240		−155		0		−55		−30		± 22		−75		0		[89Z9, 90Z5]
$Er_2Fe_{14}B$	320	265	440	355	395	−125	−127	−185	−32	46	65	−25	−63	−36	0	0	−20	−30	0	30	[94Z1]
$Tm_2Fe_{14}B$	270		350		± 240		−150		0		−40		−30		± 20		−50		0		[89Z9, 90Z5]
$Tm_2Fe_{14}B$	300	255	420	340	380	−120	−117	−170	−30	43	60	−23	−57	−32	0	0	−18	−27	0	27	[94Z1]

[a] Ka_0^{-1} units; [b] Mean values are also given; [c] Computed LMTO with 4f state of Nd as core electron and 4f contribution to electric field gradients excluded; [d] PLAPW method with 4f state as valence band (band) or as core (core); by f are the data obtained excluding f−f and p−f contributions

[e] Values in K; [f] From magnetization measurements

have complicated shapes. Differences in the molecular field acting on R*f* and R*g* sites and dispersion effects, due to R-R interactions, were considered. The observed peaks in INS spectrum of $R_2Fe_{14}B$ originated not only from the difference in the exchange fields, but also from the CEF parameters of R*f* and R*g* sites. The INS measurements were made on $R_2Fe_{14}B$ compounds with R = Ce, Pr, Nd, Tb, Dy, Ho, Er, Tm [90L6], R = Nd [88Y1], R = Pr, Nd, Gd, Tb, Dy, Ho, Er, Tm [91L2], R = Gd [91L3], R = Dy [95L1], R = Tb, Dy, Ho, Er, Tm [94Z1]. Strongly dispersive (magnonlike) modes give usually a smooth contribution to the spectra, whereas CEF transitions are seen as well pronounced maxima. As example, for R = Dy, at the lowest temperature, besides the elastic line, the INS spectra evidenced a sharp inelastic line at $\Delta = 12$ meV, with shoulder at 11.1 meV and two week lines at 8.8 meV and 5.5 meV [95L1]. The temperature dependence of the average position Δ of the dominant line follows roughly the temperature dependence of the $Y_2Fe_{14}B$ magnetization, indicating that the energy of this mode is substantially fixed by the molecular fields of surrounding iron moments. For Gd^{3+} ions, crystal fields effects are negligible and therefore can be determined, directly, the molecular field, from neutron scattering spectra [91L2]. According [94Z1], the determined exchange fields and CEF parameters vary monotonously across the $R_2F_{14}B$ series. Different values of the R–Fe exchange field at R*f* and R*g* sites, with a ratio of $\cong$1.2 and for the second-order CEF parameters A_2^0 with a ratio of 1.24 have been deduced. The calculations made by using the above parameters, described rather well the first-order magnetization process along the [100] direction in $R_2Fe_{14}B$ compounds, with R = Er, Tm, at low temperatures [94Z1].

The crystal field parameters were determined also by other procedures. The point charge method was used to compute the A_n^m crystal field parameters [88S2, 91B3]. By using a model independent approach, considering four distinct R sites, with respect to CEFs, a relationship between the corresponding crystal field coefficients has been established [86A3]. The crystal field parameters were also obtained by measurements of the lattice contribution to the electric field gradient, at the R nuclei. Thus, the field gradient V_{zz}, at the gadolinium nucleus, where the charge cloud is spherically symmetric, as obtained from ^{155}Gd NMR studies [85B9, 85B12, 86B2, 86R2, 88S2, 89B3, 89C3, 90G4], can be directly used to determined A_2^0. It was concluded that the second order crystal field parameters differ in sign for the two gadolinium sites [85B9, 85B12, 90G4]. According [86B2], the A_2^0 parameters have the same sign for both sites. Two crystallographically unequivalent gadolinium sites were evidenced by the analysis of ^{155}Gd Mössbauer spectrum of $Gd_2Fe_{14}B$ single crystal [89C3]. The electric field gradient, $|V_{zz}|$ decreases by $\cong$5% between 4 and 120 K, for site R4*f* and increases by the same amount for site R4*g*. The total hyperfine field was found to be positive for both Gd sites. Higher hyperfine field was shown at R4*g* site, their temperature dependences being correlated with data obtained from the termal variation of iron sublattice magnetization.

The NMR studies on ^{89}Y of $Y_2Fe_{14}B$ [86B1, 87E1], $Y_2Fe_{14-x}Co_xB$ [92M2] and ^{159}La in $La_2Fe_{14}B$-based compounds [86R2, 96K1], evidenced also differences in the lattice electric field gradient, at R4*f* and R4*g* sites. The ratio of the crystal electric fields gradient at the R4*f* and R4*g* correspond to a difference of crystal electric field parameters A_2^0, by a factor of two, between those sites [96K1].

The field gradients at nuclei ^{143}Nd, ^{145}Nd, ^{147}Sm, ^{149}Sm, ^{159}Tb, ^{161}Dy, ^{165}Er, ^{169}Tm and ^{174}Yb, consist primarily of the contribution V_{zz}^{4f}, due to their asymmetric 4f charge cloud. Subtracting this contribution, from the experimental V_{zz} values, the lattice contribution V_{zz}^{lat} can be obtained (Table 7.36) and as result the A_2^0 parameter [86C3, 86F1, 87F2, 87V1, 89G3]. The ^{169}Tm Mössbauer spectrum consists of two subspectra corresponding to Tm4*f* and Tm4*g* sites. The temperature dependences of the ^{169}Tm hyperfine field in $Tm_2Fe_{14}B$ and the quadrupole splitting were reasonably described, based on the reported magnetic structure [85Y1], taking into account the determined exchange fields and second order crystal fields parameters. A more consistent description, of the above experimental data, requires the inclusion of higher-order crystal field parameters. The ^{174}Yb Mössbauer spectra of $Yb_2Fe_{14}B$ were fitted, assuming, as variable parameters, the hyperfine fields and their directions, at the two Yb sites [89M2]. The best fit was obtained with a field and hence a moment oriented close to [001] direction at the Yb4*f* site and close to the basal plane at the Yb4*g* site [89M2, 98B2]. Different hyperfine fields, at Tb4*f* and Tb4*g* sites, were also reported in $Tb_2Fe_{14}B$ by ^{159}Tb NMR measurements [95S2].

The second order CEF parameters, in $Pr_2Fe_{14}B$, were found to be about 50% smaller, than those for other homologous compounds [88G1]. The above reduction has been attributed initially to an incipient valence instability of the Pr ions. The Pr valence in $Pr_2(Fe_{1-x}Co_x)_{14}B$ series was found latter to be formaly trivalent [92C2]. The hyperfine fields and magnetic moments of Pr ions, in the valence fluctuating state, were initially expected to be smaller than that of Pr^{3+} ion. By ^{141}Pr NMR study, on $Pr_2Fe_{14}B$ compound, the magnetic moment of Pr^{3+} was found to be close to the free ion value [95S1], in disagreement with the valence instability assumption.

The crystal field parameters A_2^0, at R4*f* and R4*g* sites, are different as also evidenced by NMR studies on ^{141}Nd, ^{145}Nd, ^{147}Sm and ^{149}Sm in $Nd_2Fe_{14}B$ and $Sm_2Fe_{14}B$ compounds [93K1]. The coexistence of two different magnetic ground states for Nd has been evidenced by ^{145}Nd NMR study on $Nd_2Fe_{14-x}Co_xB$ system [89N1, 92J1]. Besides the fully polarized state, as in pure $Nd_2Fe_{14}B$, the other ground state was suggested, resulting from the competition between easy orientation, due to crystal field interaction and that resulting from exchange interactions, which are non-collinear, for Nd ions having some cobalt atoms in their neighbours.

The non self-consistent band structure calculations, on $Nd_2Fe_{14}B$, indicate that the crystal field at the R sites originates mainly from the aspherical charge density of valence electrons of the Nd atoms and the contribution from the charge density of the neighbour ions is of secondary importance [89Z8]. Similar qualitative conclusions were obtained from first-principles band structure calculations [90C3, 91C1, 91C2]. These data, stress that the second-order crystal field parameter A_2^0 is determined mainly by the asphericity af 5d and 6p valence electrons, their relative contributions being almost equal for the two Nd sites. This fact restrict the determination of A_2^0, from V_{zz}^{lat} values, although it may still be used for describing the trends of crystal fields parameters [91B5, 98B2]. According [05H1], the **c** axis intrinsic magnetic stability, in

$Nd_2Fe_{14}B$, arises predominantly at Nd4*g* site. The Nd4*f* site undetermines magnetic stability by favouring a magnetic moment orientation in the basal plane.

The magnetic measurements evidence a large Nd anisotropy in $Nd_2Fe_{14}B$ single crystals [84G2, 85K1, 85S1]. The crystal field, at the R sites, in $R_2Fe_{14}B$ compounds, will create a local anisotropy. For R magnetic sublattice, if $A_2^0 > 0$, the sign of B_2^0 will be that of α_J. This is connected with the shape of the spectral distribution of 4f electrons. The R ions in $R_2Fe_{14}B$ compounds, where the electronic density of saturated ions is prolate, exhibits planar anisotropy and the R^{3+} ions with oblate electron density have axial anisotropy (R = Pr, Nd, Tb, Dy, Ho). This means that the 4f electron density tends to be confined in the layers containing R ions. For $R_2Fe_{14}B$ compounds, with R = Nd and Ho, the magnetic structure was also correlated with the decrease of the lattice symmetry, at low temperatures, as already discussed. Considering crystal field effects, in the above compounds, although $\alpha_J < 0$, the higher order terms can give rise to conical low temperature magnetic structures. The CEF is created by the positive charges of R ions in their surroundings. Since the spacing between two consecutive layers containing R atoms is more than 0.6 nm, the CEF sensed by the 4f electrons arises essentially from R^{3+} ions situated in the same layer. The above discussion does not consider the in-plane anisotropy which may be large for the local symmetry of the R^{3+} ions [98B2].

The anisotropy energy of the rare-earth sublattice, as that of the iron one, can be described phenomenologically by the relation [91B5]:

$$E_{aR} = K_{1R}\sin^2\theta + \left(K_{2R} + K'_{2R}\cos 4\varphi\right)\sin^4\theta + \left(K_{3R} + K'_{3R}\cos 4\varphi\right)\sin^6\theta,$$

where θ and φ are the polar and azimuthal angle for the R sublattice magnetization, relative to the crystallographic axis.

The anisotropy constants, K_{iR} (i = 1, 2, 3) can be expressed as a function of crystal field parameters B_n^m, as determined for an hexagonal structure [75L1, 85R1]. The K_{1R} component has the form $K_{1R} = (-3/2)B_2^0\langle O_2^0\rangle$, where $B_2^0 = \alpha_J\langle r^2\rangle A_2^0$, the $\langle r^2\rangle$ being the expected value for the Hartree–Fock R4f radius. The thermal variation of $\langle O_n^m\rangle$ follows that of the K_{iR} (i = 1, 2, 3) anisotropy constants. According to [66C1], this can be described by the temperature variation of reduced magnetization, $m_R = [M_R(T)/M_R(0)]^{n(n+1)/2}$. This gives, at low temperatures, a variation of the anisotropy as m_R^3, m_R^{10} and m_R^{21}, for the second-, fourth- and sixth-order terms of crystal fields. These terms are important, at low temperatures and can lead to a complex magnetic structure [86C3, 98B2]. The contributions of fourth- and sixth-order terms, decrease rapidly with increasing temperature, and at RT, the second order term determines mainly the anisotropy.

The zero-temperature values of the magnetocrystalline energy of trivalent rare-earth ions and their temperature dependences have been calculated for whole $R_2Fe_{14}B$ series [87R1]. The computations, performed assuming a single-ion origin of the anisotropy and within a mean field approximation, revealed a large local anisotropy of the rare earth ions, two order of magnitude larger than that of iron sublattice. The crystal field acting on the R ions is of comparable magnitude, as the mean field. Thus, the magnetic structures are sensitive to applied field, leading to bending angles between R and Fe sublattice magnetizations. The temperature dependences of the anisotropy constants K_{1R} and K_{2R} were also calculated, in $R_2Fe_{14}B$ (R = Pr, Nd, Tb, Dy, Ho) compounds, where the anisotropy is uniaxial [94J1]. Finally, the contribution of iron sublattice anisotropy was added, in order to compare these data with experimentally determined thermal variations of the $R_2Fe_{14}B$ anisotropies. The anisotropy of itinerant iron electrons was shown to be important in some compounds. A rather good agreement with experimental results was shown.

The anisotropy constants, at T = 4.2 K of the R sublattice, in $R_2Fe_{14}B$ compounds with R = Dy, Tb [98K1], R = Er, Tm [00K1] are listed in Table 7.37, together with other reported values. The inequivalence of R ions, located at R4*f* and R4*g* sites appears in the non-collinear magnetic structure of $Ho_2Fe_{14}B$ [18Y1]. A model was used to explain the presence of spin reorientation, when R = Ho and their absence when R = Dy. The inequivalence of R sites in these compounds was shown to be not so important in determining their total magnetocrystalline anisotropy in the thermal equilibrium state.

The spin reorientations in $R_2Fe_{14}B$ compounds can be, from a non-collinear ground state to a **c**-axis structure, on increasing temperature, or from an orientation of net magnetization from the **c**-axis to the basal plane—Fig. 7.20. As already discussed, the iron sublattice magnetization has uniaxial anisotropy, the easy direction of magnetization being along **c**-axis. The contributions of rare-earth sublattice to the anisotropy is dependent on their Stevens factor α_J [52S1, 64H1]. There is faster decrease of the second order R^{3+} crystal fields terms, with increasing temperature, and thus of the rare-earth sublattice anisotropy, as compared to that of iron sublattice. Thus, spin reorientations are expected when increasing temperature, if the R ions have $\alpha_J < 0$. The anisotropy constants and their evolution with temperature, in $R_2Fe_{14}B$ compounds with heavy rare-earths are given in Fig. 7.22 [88M1]. The K_1 value increases monotonically with decreasing temperature in $Tb_2Fe_{14}B$. The K_1 anisotropy constant begins to decrease, at low temperatures in the $Dy_2Fe_{14}B$ compound and changes the sign from positive to negative in $Ho_2Fe_{14}B$, at the spin reorientation temperature [88M1]. In compounds with R = Er and Tm, K_1 is always negative, below T = 300 K in agreement with the [100] easy direction of magnetization; their absolute values increase when decreasing temperature. The K_2 values are positive and increase monotonically with decreasing temperature when R = Tb, Dy, Er and Tm. In $Ho_2Fe_{14}B$, K_2 increases rapidly below T ≅ 100 K, resulting in spin reorientation phenomenon. The absolute values of K_3 across the series, is one order of magnitude smaller than those of K_1 ones.

The anisotropy of $Nd_2Fe_{14}B$ compound, has been investigated in more detail, taking into account their technical interest [84G2, 84S1, 85K1, 86Y1, 87O5, 88C1]. The first anisotropy constant, at T = 4.2 K, is negative and gradually increases with temperature, up to T ≅ 133 K, when it changes sign. The K_2 and K_3 constants are positive and decrease rapidly with increasing temperature, being almost zero at T = 300 K. At T = 20 K, the tilt angle of magnetization is oriented 32.3° from **c**-axes, decreasing with temperature. At T_{sR} = 135 K − 138 K, the e.a.m. is near parallel to the **c**-axis [84G2, 85A2, 85K1, 85T1, 86C3, 86F2, 86H4, 86G5, 86Y1, 86Y6, 87O4, 94F1, 95C2]. Below T_{sR}, the contributions of high-order terms, to the Nd

Fig. 7.22 $R_2Fe_{14}B$: temperature dependences of the anisotropy constants K_1 for (**a**) R = Tb, Dy, Ho, Er, Tm [88M1]

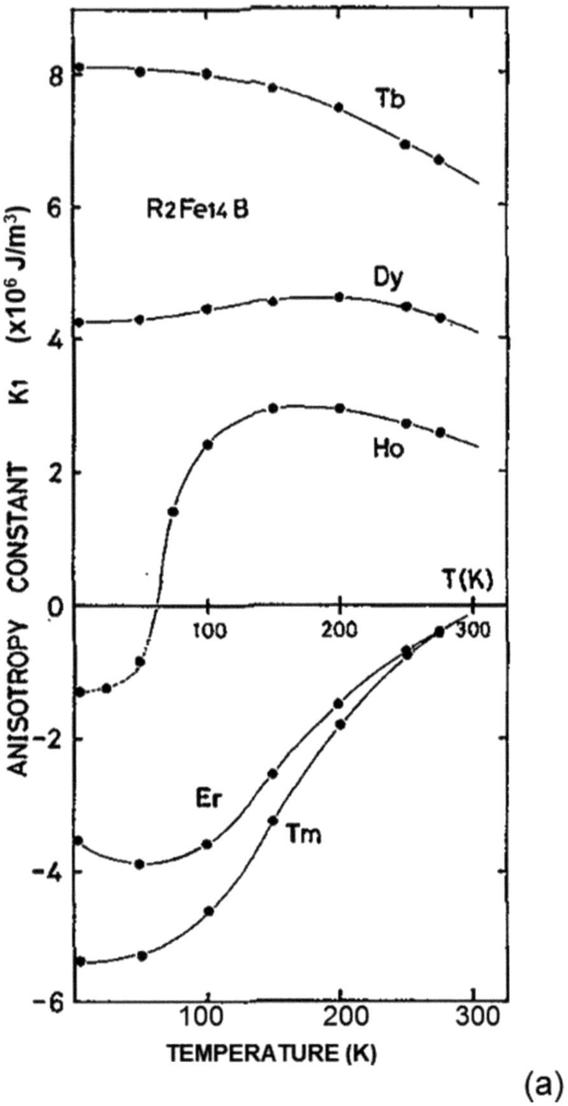

sublattice anisotropy, are important. The magnetic transition, at T_{sR}, is of second order, accompanied by a change in volume [85A2] and an anomaly in the specific heat [87F6]. The magnetic properties of $Nd_2Fe_{14}B$ including the spin reorientation transition was also theoretically investigated [90L3, 91F1, 91Z4, 00Y2, 05H1, 09K1, 10K1, 11T1, 16I1]. The anisotropy constants K_1 and K_2, of $Sm_2Fe_{14}B$, are negative in the temperature range $T < 450$ K and decrease in absolute magnitude when increasing temperature [85H5]. The easy direction of magnetization lies along [100] axis, in the basal plane of the tetragonal structure.

The domain wall energy, related to anisptropy of $R_2Fe_{14}B$ compounds has been analysed when R = Pr, Nd [85S3], R = Nd [97L1, 97P1, 09K2, 20U1, 20X1] and $Nd_2(Fe,Co,Ni)_{14}B$ [87S3] and $Tb_2Fe_{14}B$ [99L1]. For surfaces perpendicular to **c**-axis, domains have been observed with dimensions of 2–5 μm, at T = 4 K [97P1] and about 0.2–0.4 μm at RT [85S3, 97L1, 97P1]. Fine magnetic domains occur at the $Nd_2Fe_{14}B$ surface in uniaxial state [09K2]. At T > T_{sR} only topological trivial magnetic bubbles are observed, while around T_{sR} magnetic skyrmions are present, due to the modulated anisotropy and magnetization, originated from the spin reorientation [20X1]. Below T_{sR} the domain structure is much coarser and the effective volume of domain walls is smaller [09K2]. The domain wall energy, domain wall thickness and critical single domain particle size in $Tb_2Fe_{14}B$ single crystal were determined and compared with those obtained in $R_2Fe_{14}B$ compounds with R = Nd, Gd, Dy and Er [99L1]. The domain wall energy increases and domain thickness decreases with increasing anisotropy.

The anisotropy of $R_2Fe_{14}BH_y$ hydrides were also investigated [85F1, 85L2, 86A4, 86F1, 86S1, 87D1, 88P2, 91O1, 95I1, 95K3, 95S4, 00P1, 02C1, 03N1, 04B1, 04F1, 04T1, 05T1, 06K1, 06T1, 07M4, 07T1, 12D1, 18B1]. The Curie temperature and the average iron magnetic moment, in $Lu_2Fe_{14}BH_y$, increase upon hydrogenation, whereas the first anisotropy constant K_1 decreases [05T1]. As in the parent compound, (y = 0), $K_1(T)$, exhibits a characteristic non-monotonous behavior. The phase diagrams of $R_2Fe_{14}BH_y$ compounds with R = Nd, Gd, Er, Lu were elaborated [95K3]. The hydrogenation reduces the crystal fields acting on R ions, the T_{sR} values being thus dependent on the hydrogen content. The reduction of second order CEF parameter, A_2^0, in $R_2Fe_{14}BH_y$ hydrides, was described by a dimensionless factor $f_2(y) = A_2^0(y)/A_2^0(0)$. When the T_{sR} values are in the ambient temperature range, as for R = Er, Tm, the A_2^0 parameters can be obtained from T_{sR} [95K3, 04B1]. The function $f_2(y)$ was also obtained from electric field gradient V_{zz}, on the ^{161}Dy nucleus in the $Dy_2Fe_{14}BH_y$ hydrides [85F1]. A proportionality between these values has been shown, $A_2^0 = uV_{zz}^{lat}$. Ab intio electronic structure calculations showed that there is no physical reason for the parameter u to be a universal constant. The $f_2(y)$ function was mainly analysed in $R_2Fe_{14}BH_y$ compounds with R = Dy, Er, Tm [85F1, 95K3, 00P1]. This function was also studied by ^{139}La NMR method in $La_2Fe_{14}BH_y$ hydrides [04B1]. An analysis of the published data, demonstrates that the reduction of both the second order crystal field parameter, A_2^0 and the lattice electric field gradient, V_{zz}^{lat}, are well described by the same function $f_2(y)$ in $R_2Fe_{14}BH_y$ compounds. The only distinction that need to be made is between R, as being a light or a heavy rare-earth ion, respectively.

The effects of hydrogen uptake on the spin-reorientation temperatures, of $R_2Fe_{14}BH_y$ compounds, have been analysed [87W3, 88F1, 95B1, 95K3, 03N1]. The compounds with R = Er and Tm, have shallow minima for y $\cong$ 1. The spin reorientation temperatures remain unchanged in $Nd_2Fe_{14}BH_y$ for y $\leq$ 3.3 [03N1]. The T_{sR} values increase monotonically, in $Ho_2Fe_{14}BH_y$ hydrides, as the hydrogen content is greater. In $R_2Fe_{14}BH_y$ compounds with R = Pr and Gd, where no spin reorientations are present in parent samples, after hydrogen uptake, such magnetic transitions can be observed [88F1, 88Z1, 91B2]. In $Gd_2Fe_{14}BH_y$, the spin reorientation appeared for $y \geq 2$, while in $Pr_2Fe_{14}BH_y$ for $y \geq 1$. The spin reorientation temperatures increase with hydrogen content when R = Gd and decrease as R = Pr. The temperatures T_{sR} increase also in $Er_2Fe_{14}BH_y$, while $Tb_2Fe_{14}BH_y$ appears to remain uniaxial when $3.7 \leq y \leq 5.4$ [88Z1].

The magnetic and related properties of a large number of pseudoternary systems were investigated with substitutions at the rare-earth sites or iron sites. As already mentioned (Sect. 7.10.1), when substituting one rare earth R_1, by another R_2, solid solutions of $(R_1, R_2)_2Fe_{14}B$ type are generally formed in large compositional range. A survey of the magnetic properties of these systems is given below: $(La,R)_2Fe_{14}B$ with R = Ce [95C1, 95F1, 97C1, 13A1, 16C1, 16J1]; R = Pr [89M4, 16K2, 18G1], R = Nd [89N2, 90L5, 13L1, 16K2]; $(Ce,R)_2Fe_{14}B$ with R = Nd [86A2, 13A1, 15L1, 16C1, 17H1, 17S2], R = Er [86P1, 96B1, 02P1, 02P2, 05P2, 11K1], R = Zr [86J3]; $(Pr,R)_2Fe_{14}B$ with R = Nd [86Y3, 87H9, 89Z1, 90M2, 90Z2, 91Z1, 93C1, 00Y1, 15Y1, 16J1, 16K2], R = Sm [86L1, 86Y3, 88Y2, 88Y3, 89Z1, 17K1, 17X2], R = Gd [88Z2, 89Z1, 89Z9, 90Z2, 91Z2], R = Er [86P1, 87B3, 88B3, 88B4, 88Y2, 88Y3, 89Z1, 91B1, 06P1, 06W1, 07W2], R = Tm [88B5], R = Y [89M4, 89Z9, 90Z4, 18G1], R = Zr [86I1, 93C1]; $(Nd,R)_2Fe_{14}B$ with R = Sm [86Y3, 90N1], R = Eu [03C1], R = Gd [88Z2, 89Z2, 89Z9, 91Z3], R = Tb [85A1, 87P1, 91Z3, 12L1], R = Dy [86A2, 86Y6, 87S1, 89N2, 91L1, 01K2, 12L2, 13H1, 15W1, 16H1, 17S1, 21S1], R = Ho [85A1, 89Z9, 07C1, 18K1], R = Er [86A2, 88I1, 89D1, 89I1, 90D1, 90I1, 90M3, 91D1, 92D1, 92I1, 07C1], R = Tm [87B12], R = Y [85A1, 85L1, 86B4, 86K2, 88W1, 89N2, 89Z2, 89Z9, 90K1, 90L5, 91Z2, 91Z3, 93K2, 13L1, 18G1], R = Zr [86J3], R = Sc [88C4]; $(Sm,R)_2Fe_{14}B$ with R = Er [89C2, 00B1], R = Y [88L3]; $(Gd,R)_2Fe_{14}B$ with R = Er [86P1, 87V1, 92I1, 01K1, 05P2, 07W2], R = Y [85B4, 85B8], R = Zr [86J3]; $(Tb,R)_2Fe_{14}B$ with R = Dy [88P8], R = Er [88B5, 89C2, 91O2, 92L4, 95S2, 01K1], R = Tm [88B5], R = Y [86K3]; $(Dy,R)_2Fe_{14}B$ with R = Er [86N2, 87R4, 88B5, 88P8, 89A1, 89D1, 89I1, 89I2, 90D1, 92I1], R = Tm [87P11, 88B5, 88P3], R = Th [86P4, 88P3], R = Zr [86J3], R = Sc [88C3]; $(Ho,R)_2Fe_{14}B$ with R = Er [89M1, 89Z9], R = Tm [17K1], R = Y [89Z9]; $(Er,R)_2Fe_{14}B$ with R = Y [86K3, 86P1, 87A1, 87B3, 92I1, 95G1, 02P2, 03P1], R = Th [86P4, 05P1, 05P2, 06W1], R = Sc [88C4]; $(ScR)_2Fe_{14}B$ with R = Nd, Dy, Er, Lu, Y [88C2].

The complicated interplay of exchange and crystal field interactions, in $R_2Fe_{14}B$ based compounds, in addition to spin reorientations, give rise, in high magnetic fields, also to first order magnetization processes (FOMP), characterized by discontinuities or jumps in the magnetization isotherms, M(H), when high external fields are applied along a crystallographic direction in an uniaxial ferro- or ferrimagnet. According [80A1, 88P1], high order anisotropy constants are required in order to describe FOMP, their presence enables one to obtain local minima of free energy and thereby first order transitions. The FOMP process has been also analysed within an anisotropic Heisenberg model, in which only leading-order anisotropy constants of the magnetic sublattices are included [02I1]. The FOMP is characterized by the magnetization values of the initial, M_1, and final, M_2, magnetization states. The type I FOMP describes the situation when $M_2 = M_s$, i.e. the magnetization is saturated. The type II FOMP, characterizes the process, when after the transition, the magnetization is not saturated and still higher fields have to be applied for attaining saturation, M_s. The extreme cases of FOMPs are when the field is applied perpendicularly (P) or parallel (A) to the **c**-axis. Another situation is when the field is applied along the equilibrium conical axis, if it exists (C).

The interplay between magnetocrystalline anisotropies of magnetically ordered sublattices leads to reorientation phenomena in which the easy axis of magnetization changes, at a characteristic temperature T_{SR}. The noncollinearity of R magnetic moments, at low temperatures, was correlated with the decreased symmetry of the crystal structure and crystal field (anisotropy) effects. In $Nd_2Fe_{14}B$, the net magnetization begins to tilt away from **c**-axis, below T_{SR}, where K_1 crosses zero [86Y1], changing from positive to negative and in addition there is $K_2 > 0$. This behavior was correlated with higher-order terms in the Nd sublattice anisotropy. The transition was shown to be of second order. At T_{SR}, there is a change in volume [85A2] and there is a specific heat anomaly [86F2]. The spin reorientation process, in $Nd_2Fe_{14}B$, has been theoretically analysed [90L3, 91F1, 91Z4]. Different micromagnetic models were used to describe the behavior of the *ac* susceptibility, in the temperature range of spin reorientation transition [91F1]. An angular dependent free energy approach was also used [91Z4]. A magnetic Hamiltonian which includes the CEF and exchange terms, and using relativistic band-structure results has been developed. The net magnetic anisotropy was shown to be canted away from the **c**-axis, close to the experimentally reported spin-reorientation one. The method of multi-ions crystal field calculation was also applied to compute the magnetic structure of $Nd_2Fe_{14}B$ [90L3].

The spin reorientations were studied in pseudoternary compounds with substitutions at both R and Fe sites. The larger number of investigations were made on Nd- and Er-based systems: $Pr_{2-x}R_xFe_{14}B$ with R = Nd [90M2] and R = Gd [88Z2], $Nd_{2-x}R_xFe_{14}B$ with R = La [87N2], R = Gd [89Z2, 91Z3], R = Dy [87N2], R = Er [05P2] and R = Y [87N2, 89Z2, 91Z3, 93K2], $Sm_{2-x}R_xFe_{14}B$ with R = Y [88L3], $Er_{2-x}R_xFe_{14}B$ with R = Ce [02P1], R = Gd [05P2, 07W2], R = Dy [86N2] and R = Y [02P2], R = Th [05P1], $R_{2-x}Th_xFe_{14}B$ with R = Dy, Er, Y [86P4] etc.

The spin reorientation temperatures, T_{SR}, in $Nd_{2-x}R_xFe_{14}B$ with R = Sm, Er, Tm substitutions, having $\alpha_J > 0$, opposite to that of Nd ($\alpha_J < 0$), increase since these counter more efficiently the **c**-axis anisotropy of iron sublattice. When $\alpha_J = 0$, as for R = Gd, and Y, the spin reorientation temperatures decrease slowly, while when R = Tb and Ho, having $\alpha_J < 0$, the decrease of T_{SR} values is more pronounced [98B2]. The composition dependences of T_{SR} values in $Nd_{2-x}R_xFe_{14}B$, with R = Gd, Y, were analysed considering the crystal field effects and exchange interactions [89Z2, 91Z3]. The evolutions with composition of T_{SR} values in $Nd_{2-x}Y_xFe_{14}B$ and $Nd_{2-x}Y_xFe_{13}CuB$ were analysed in a combined crystal field-exchange field model [93K2]. By adopting a "simple mixture" model and the characteristic parameters for the end series members, the concentration dependences of the spin reorientation temperatures in $Nd_{2-x}R_xFe_{14}B$ with R = La, Dy, have been described [87N2].

The spin reorientation temperatures, in $Nd_{2-x}Dy_xFe_{14}B$ system, decrease linearly on increasing Dy content, from T = 135 K (x = 0) to 56 K (x = 2), with a rate of −1.11 K/Dy at %, in the composition range $0 \leq x \leq 1.5$ [01K2]. The angle of magnetization with **c**-axis, at T = 4.2 K, in the same composition range, changes from 30.4° to 14.7°. For $x \geq 1.70$, no spin reorientation process is present. When increasing Sm content, in $Y_{2-x}Sm_xFe_{14}B$, up to $x = 0.2$, the anisotropy, at ambient temperature is diminished. A transition to planar anisotropy was shown at $T = 1.5$ K for $x = 0.16$ and 0.2 [88L3].

The magnetic phase diagrams of the pseudoternary $Er_{2-x}R_xFe_{14}B$ compounds, where R elements have different Stevens factor from that of Er, are rather complicated. These reveal, as function of composition and temperature, axial, planar and cone regions [86N2, 87B1, 87F4, 87R4, 88B5, 88I1, 88Y2, 89D1, 89I2, 90M3, 91B3, 91O2]. The critical fields characterizing the first order magnetic transitions have been determined, as function of both temperature and composition in the system $(Er_{1-x}Nd_x)_2Fe_{14}B$ [90M3]. In this system, PI type FOMP with the field applied perpendicular to the **c**-axis have been observed, at low temperatures for $x < 0.7$, whilst type AI (field applied parallel to the **c**-axis) processes were shown when $x =$ 0.8 and 0.9. Such a switch from PI to AI FOMP and the observed inversion of the temperature dependence of the crystal field, with increasing Er content, is consistent with the opposite CEF parameters for the two R elements. The magnetic phase diagram of $(Er_xDy_{1-x})_2Fe_{14}B$ includes axial-planar and axial-cone transitions. The consideration of second- and fourth-order crystal field terms was enough to account for the observed transitions [89I2].

Two spin reorientation temperatures, located at $T_{sR1} = 164(2)$ K and $T_{sR2} =$ 43.4(5) K were shown in $Er_{1.2}Ho_{0.8}Fe_{14}B$ compound [89M1]. The T_{sR1} temperature corresponds to that in which e.a.m. begins to tilt away from **c**-axis, giving rise to a conical magnetic structure. T_{sR2} characterizes the temperature at which higher order CEF terms begin to be important and a cone 1 to cone 2 spin reorientation take place. Transitions from axial to planar, or conical type magnetic structures were shown in $Tm_{2-x}Ho_xFe_{14}B$, as temperature decreases [17K1]. Succesive phase transitions were observed in $(Er_{1-x}Tb_x)_2Fe_{14}B$ system, where Tb and Er have competing anisotropies. First order magnetic transitions were shown in aligned $(Er_{1-x}Tb_x)_2Fe_{14}B$ polycrystalline samples with $0.3 \leq x \leq 0.45$ [91O2, 92L4]. For $H \perp \mathbf{c}$, after an anomalous change, the magnetization does not saturate and continues to increase. The calculated magnetization curves, at $T = 0$ K, exhibit jumps for both [100] and [110] directions. Also was predicted that FOMP transitions can occur in the concentration range $0.26 < x < 0.5$. The FOMP transitions in $(Er_{1-x}Tb_x)_2Fe_{14}B$ system were interpreted to be the result of the cancellation of the second-order CEF potentials, which relatively emphasizes the effect of higher-order CEF terms. The variations with composition of the spin reorientation temperatures in $Pr_xR_{2-x}Fe_{14}B$ with R = Sm, Er [88Y3] and Er [06P1] have been also investigated.

A variation with pressure, p, of spin reorientation temperature was shown. In $Er_2Fe_{14}B$ compound, values $\Delta T_{sR}/\Delta p = -20$ KGPa^{-1} [92I1], or -19 KGPa^{-1} [94S2] were reported. It was concluded [93A1] that the dependence of the crystal electric field interaction on the interatomic distances, is significant in the intermetallics, where the second-order CEF terms dominate their anisotropy. The fourth- and sixth-order CEF parameters do not seem to be sensitive to changes of the interatomic distances. The pressure effect on spin-reorientation temperature, in $Er_xR_{2-x}Fe_{14}B$, is not affected by the Er substitution by R = Gd, Y [92I1]. A different high-pressure behavior has been observed for $Er_xDy_{2-x}Fe_{14}B$ and $Er_xNd_{2-x}Fe_{14}B$ systems, when $\Delta T_{sR}/\Delta p$ rate decreases, as increasing the Nd and Dy content. The different behavior has been attributed to the strong influence of pressure on the CEF interactions. Starting from these data, the variations of CEF parameters, under applied pressure, were determined [92I1, 92I2].

The spin reorientation temperatures characteristic of $R_2Fe_{14-x}M_xB$ pseudoternary compounds are strongly influenced by the M elements. In $R_2Fe_{14-x}Co_xB$ compounds, with R = Gd, Y, ($\alpha_J = 0$), the substitution of Fe by Co, for $x > 9$, changes the anisotropy from axial to planar [87P5]. For R = Pr, Nd, Tb, with $\alpha_J < 0$, the phase diagram is dominated by the axial anisotropy of the iron sublattice, the R elements favouring also this orientation of the magnetization. Only for high cobalt concentration, the plane anisotropy can be shown, induced by its presence, combined with the decrease of R anisotropy, at high temperatures. When R = Er or Tm ($\alpha_J > 0$), the cobalt substitutions enlarge the planar regions thus raising the planar to axial T_{sR} temperatures. By increasing temperature, the $Nd_2Fe_{14-x}Co_xB$ compounds undergo a spin reorientation, and a uniaxial configuration along **c**-axis is reached at T = 34 K for x = 14 and at T = 134 K when x = 0 [87P3].

The magnetic phase diagrams of $R_2Fe_{14-x}Mn_xB$ with R = Nd [86H6, 91I1], R = Pr, Gd, Y [86H6] and R = Er [87F4, 89Y1, 91I1] were elaborated. Manganese decreases the net transition metal moment, in $ErFe_{14-x}Mn_xB$ compounds, and compensation points have been evidenced, the net magnetization being parallel ($x \le 2$) or antiparallel ($x \ge 2$), to the moments of the transition-metal sublattice [87F4]. Although Curie temperature is decreasing, the temperature range of planar alignment is increasing relative to axial orientation. It is expected that for $x \ge 4$, the net magnetization is oriented in the basal plane, at all temperatures.

For a low Al content (x = 1), the spin reorientation temperatures, in $Er_2Fe_{14-x}Al_xB$ system, increase from T_{oR} = 322 K to 348 K, but further substitution depresses T_{sR} to 328 K (x = 2) and 308 K (x = 3) [87Q1]. The presence of aluminium reduces the uniaxial anisotropy of iron sublattice, while the planar anisotropy of Er sublattice is little affected. Thus, there is an increase of the temperature range over which the Er sublattice anisotropy dominates. The anisotropy of Er sublattice, only for high Al substitution, appeared to be weakened. In $Nd_{0.5}Er_{1.5}Fe_{14-x}Al_xB$, system, the aluminium substitution shifts T_{sR} at lower temperatures and for x = 2, an additional spin reorientation is observed, at $T \cong 25$ K [90M1]. The cobalt substitution in $Nd_{0.5}Er_{1.5}Fe_{14-x}Co_xB$, decreases the spin reorientation temperatures.

The spin reorientation temperatures in $Nd_2Fe_{14-x}M_xB$ with M = Mn, Co, Ni Si, Ru [86H6, 86P3, 87B5, 87P3, 88B7] diminish, the rate of decrease, following in the sequence Co, Ni, Si, Mn, Ru. The conical angle, at T = 77 K, is also reduced in the same manner [87B5, 87Y2].

The spin reorientation temperatures in $R_2Fe_{14}B$ compounds, were evidenced also by heat capacity [87F6, 90L1, 96P1], electrical resistivities [91S1, 97A1, 97S1] and Hall efect [99S1] studies. The second-order phase transitions at T_{sR}, were observed from heat capacity measurements for R = Nd and Ho and of first order for R = Er and Tm [96P1]. The experimental results were compared with theoretical predictions, based on a mean field model, where the CEF and exchange parameters, already determined, were used.

Sharp peaks, near the spin-reorientations and Curie temperatures were observed by Hall efect, in $Nd_2Fe_{14}B$ single crystal [99S1]. The effect is very sensitive to the applied magnetic field direction, near the T_{sR} temperature. In $Er_2Fe_{14}B$, there is a steplike decrease in resistivity and an increase in thermopower, over a very narrow temperature interval around T_{sR} = 323 K, indicating a discontinuous spin reorientation transition [97A1]. The same compound, shows a step like decrease in resistivity and an increase in termopower around the spin reorientation temperature [01S1].

The high field magnetization isotherms and the first order magnetization processes were analysed in $R_2Fe_{14}B$ compounds with R = Ce [87H6], R = Pr [85P2, 87H4, 87H5, 88V1, 89Z1, 91K6, 92V1], R = Nd [85P1, 87B8, 87H6, 87K1, 87K4, 88V1, 89Z2, 91K6, 92V1], R = Sm [87K3], R = Gd [87H3, 87H6], R = Tb [87G1, 88G3, 88M1, 94Z1], R = Dy [88G3, 88M1, 88V1, 94Z1, 95K1], R = Ho [87H3, 88G3, 88M1, 94Z1], R = Er [87K4, 88G3, 88M1, 91R1, 92L1, 94Z1], R = Tm [87K4, 88G3, 88M1, 92L1, 94Z1, 95K1]. A large number of studies were also performed on the pseudoternary compounds as $(Nd_{1-x}La_x)_2Fe_{14}B$ [90L5], $(Pr_{1-x}Nd_x)_2Fe_{14}B$ [87H9, 88Z2, 89Z1, 90M2, 91Z1, 00Y1], $(Pr_{1-x}Sm_x)_2Fe_{14}B$ [88Y3, 89Z1] $(Pr_{1-x}Gd_x)_2Fe_{14}B$ [88Z2, 89Z1, 91Z2], $(Pr_{1-x}Gd_x)_2Fe_{10}Co_4B$ [91Z2], $(Pr_{1-x}Er_x)_2Fe_{14}B$ [89Z1, 91Z1], $(Nd_{1-x}Gd_x)_2Fe_{14}B$ [89Z2], $(Nd_{1-x}Dy_x)_2Fe_{14}B$ [91L1], $(Nd_{0.5}Ho_{0.5})_2Fe_{14}B$ [18K1], $(Nd_{1-x}Er_x)_2Fe_{14}B$ [90M3], $(Nd_{1-x}Y_x)_2Fe_{14}B$ [89Z2, 90L5], $(Tb_{1-x}Er_x)_2Fe_{14}B$ [92L4], $Pr_2(Fe,M)_{14}B$ with M = Al, Ga, Si [91Z5], $Er_2Fe_{14-x}Al_xB$ [91K3].

The high fields magnetization isotherms, at T = 1.5(4.2) K, in $R_2Fe_{14}B$ compounds with R = Pr and Nd are given in Fig. 7.23 [88V1]. In $Pr_2Fe_{14}B$, when the field was applied along [100] and [110] directions, jumps are observed at μ_0H = 13 and 16 T [87H4] or at μ_0H = 13.2 and 17.1 T [88V1], respectively. These are characteristic for type II FOMP, feature evidenced up to T = 100 K [87H4]. A FOMP type transition of type I, at T = 4.2 K, is observed for μ_0H = 16.3 T in $Nd_2Fe_{14}B$, when the field was applied along the [100] direction. Such transition type has been

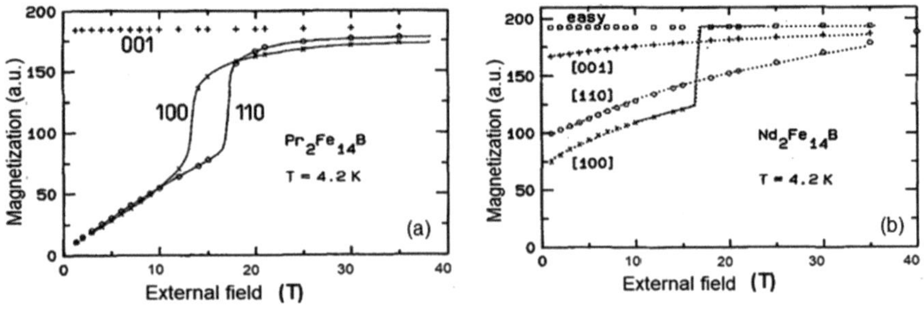

Fig. 7.23 $R_2Fe_{14}B$ with R = Pr (**a**) and Nd (**b**): high field magnetization isotherms at T = 4.2 K [88V1]. In (**a**) full lines and marks represent the experimental data. In (**b**) the dotted lines represent the fit of experimental data

shown in the temperature range T < 220 K [87B8]. The experimental data were well described in a model, where both four and six order CEF parameters (or anisotropy constants), in addition to exchange field acting on R atoms, are taken into account. Acording [92V1], the transition is of first order in $Nd_2Fe_{14}B$, while that in $Pr_2Fe_{14}B$ is not of first order and originates from a fast but nevertheless continuous rotation of magnetic moments.

Magnetic isotherms in fields up to $\mu_0H = 35$ T have been investigated, at T = 4.2 K and 77 K, on magnetically aligned $(Pr_{1-x}R_x)_2Fe_{14}B$ powders with R = Nd, Sm, Gd and Er, with field applied parallel and perpendicular to the alignment direction [89Z1]. In $(Pr_{1-x}Nd_x)_2Fe_{14}B$, the FOMP transitions slightly shift to higher fields, as the Nd content increases. The low field "perpendicular" magnetization is very similar for samples with x = 0 and 0.2, whereas for higher Nd content, a strong increase of the field, reflecting the appearance of the cone structure, was observed. In high fields, the FOMP transition, in $Pr_2Fe_{14}B$ is of PII-type, while for the compound with x = 0.2 it is of type PI, and for other compositions ($0.4 \leq x \leq 1.0$), PIc FOMPs were found. In $(Pr_{1-x}Gd_x)_2Fe_{14}B$ system, except when x = 1, a FOMP transition was found [89Z1]. The anisotropy decreased monotonously with increasing Gd content. In the $(Pr_{1-x}R_x)_2Fe_{14}B$ pseudoternary systems, only for low x values ($x \leq 0.4$ for Sm and $x \leq 0.8$ for Er) FOMP transitions were observed at T = 4.2 K [88Z2, 89Z1].

The high field magnetization isotherms of $Pr_2Fe_{14-x}M_xB$ system with M = Al, Ga, Si, in field up to 35 T, at 4.2 K, showed that the slopes of the magnetization curves at the critical fields are very large, the magnetization exhibiting field dependences which are different from what it is expected for a FOMP transition [91Z5].

The substitution of Nd by R = Y or La in $(Nd_{0.9}R_{0.1})_2Fe_{14}B$ system was found to have significant effect on the jump field, at which the FOMP occurs [90L5]. The magnetization curves of $Nd_2(Fe_{1-x}Co_x)_{14}B$ single crystals, with x = 0.15 and 0.38, show essential features as of the parent compound (x = 0) [89N2]. The FOMP transitions were observed only for the [100] direction, at field intensities, slightly smaller than for the parent compound. At low temperatures, in the $Nd_2Fe_{14-x}M_xB$ system, with M = Ni, Mn and $x \leq 3$, the presence of FOMP processes was evidenced [85P1, 87B5]. The temperature, at which the FOMP transition commences, is decreased for above substitution.

The magnetization curve of $Dy_2Fe_{14}B$ aligned powders, at T = 10 K, show peaks, at $\mu_0H = 105$ T and 101 T, for field increasing or decreasing along [001] direction, respectively, which arise from a first order transition between collinear ferrimagnetism and non-collinear spin-flop-like phase [95K1]. The high field magnetization process, in $(Nd_{1-x}Dy_x)_2Fe_{14}B$ single crystal, has been studied up to $\mu_0H = 40$ T [91L1]. The fields where the FOMP occurred, were found to increase with x, reaching 32 T on the sample with x = 0.33, at T = 4.2 K.

The magnetization process in $(Nd_{0.5}Ho_{0.5})_2Fe_{14}B$ has been investigated in fields up to $\mu_0H = 60$ T [18K1]. The induced phase transitions from ferrimagnetic to the forced ferromagnetic state was found at lower fields in $(Nd,Ho)_2Fe_{14}BH_y$ and $Ho_2Fe_{14}BH_y$ than in parent compounds [21K1]. This behavior was due to a nearly twofold reduction in the R–Fe intersublattice exchange interactions.

A one-step transition with a hysteresis loop was observed at $\mu_0H = 43$ T, in $Er_2Fe_{14}B$ [91K3]. For $Er_2Fe_{13}Al_1B$, a transition was found at $\mu_0H = 28$ T, with hysteresis loop widely deformed, after the transition the magnetization increasimg monotonously up to $\mu_0H = 50$ T. In $Er_2Fe_{12}Al_2B$, two transitions, at $\mu_0H_{c1} = 11$ T and $\mu_0H_{c2} = 28$ T were evidenced. Above the transition at H_{c2}, the magnetization increases monotonously. The observed transitions were explained by considering the change in magnetic ordering, from the ferrimagnetic to ferromagnetic one, through an intermediate canted structure.

The magnetization isotherms, of $Tm_2Fe_{14}B$ compound, were studied in fields up to $\mu_0H = 120$ T and 200 T, at 4.2 K [92L1, 95K1]. First order magnetic transitions, at T = 4.2 K, were shown when the field was applied along the [100] direction at $\mu_0H = 49$ and 70 T, at 60 T along [110] direction [95K1] or at 47.9 and 73.5 T along [100] axis and 59.8 T and 147 T along [110] direction [92L4]. The transitions, when field is applied along [100] direction, were found to arise from the moment reversals, at the $f_1(g_2)$ and the $f_2(g_1)$ sites, respectively. In the intermediate state $H_{c1} \leq H \leq H_{c2}$, the iron moments are canted within the (001) plane.

7.10.3 R–Fe–B Permanent Magnets

As already mentiond, the Nd–Fe–B alloys may be obtained by casting, melt-spinning, reduction-diffusion or mechanical alloying. From these alloys permanent magnets were manufactured mainly by sintering. Other methods were developed as hot working, die upset, extrusion, hot rolling, liquid dynamically compacted or by sputtering [98B2]. The last pocedures are little used, mainly at the level of research laboratories. Magnetic powders for bonded magnets are mainly produced by melt-spinning or hydrogenation-disproportionation-desorbtion-recombination (HDDR) processes [98B2, 11S1].

The weight of the total rare-earth content in Nd–Fe–B magnets is of ≅31 wt %. The Pr is often used to substitute Nd, in order to reduce cost, both in pure form or as a mixture with Nd. For higher temperature applications, beyond 5 wt % Nd is substituted by Dy and sometime by Tb in order to increase the coercive field. Despite that the intrinsic properties of $R_2Fe_{14}B$ with R = La, Ce are lower than those when R = Nd, in manufacturing magnets, these have potential to substitute Nd [14Z1, 20F1, 20H1, 20L1, 20L2, 21L2]. For Ce-containing sintered Nd–Fe–B magnets the maximum energy product $(BH)_{max} = 343$ kJ/m^3 and 225 MJ/m^3 have been obtained with 30 wt% and 40 wt% substitutions by Ce, respectively [14Z1]. Magnets free from Nd has been also obtained. For a composition $Ce_{20.4}La_{5.06}Y_{6.93}Fe_{66.62}B_{0.99}$ the maximum energy product was 60 kJ/m^3 [21L2].

The sintered Nd–Fe–B magnets are produced by powdering the ingot, alignment of powder in a magnetic field, isostating pressing, sintering and finally annealing. The sintered magnets have the greatest energy product when obtained from oriented grains with sizes of several μm and a minimum content of the non-magnetic intergranular binder material. Theoretically, the maximum energy product for oriented sintered

$Nd_2Fe_{14}B$ magnets is $(BH)_{max} = 515$ kJ/m^3, while for random oriented grains the maximum energy product is four times smaller [20C3]. Lower energy product is reported in Nd–Fe–B magnets, due to many reasons, as the presence of intergranular phases, misalignment of grains or connected with grain boundary interface etc.

The $Nd_2Fe_{14}B$ hard magnetic phase, as already discussed, has a relative low Curie temperarure and consequently the permanent magnets can be used no more than 150 °C. The Curie temperature, T_c, can increase by substitutions of iron sites by M = Co, Ni, Ga, Si or Cu, with the greatest rate when M = Co. The high decrease of the anisotropy field overcome the T_c increases of cobalt substituted hard magnetic phase. For small content of M = Cu, Si, Ni or Al the anisotropy field increase little at RT, concomitantly with the decrese of magnetization due magnetic dilution effects.

A large number of investigations were focussed on the increase of coercivity of Nd–Fe–B permanent magnets. As representative study on the relation between grain size refinement and coercive field, that on $Nd_{11.5}Pr_3Fe_{77.7}Co_1Al_{0.5}Zr_{0.1}B_{6.2}$ magnet, will be only mentioned [18N1]. There is a linear increase of coercive field as the grain sizes decrease from $\cong$3.6 to $\cong$2.6 μm, but the remanence decreases since it is difficult to align smaller grains. Sealed system must to be used, during processing, in order to avoid oxidation. Thus, the refinement of hard grains were not considerd usefull in manufacturing permanent magnets with high coercivity.

The coercivity of Nd–Fe–B magnets increases after post-sintering thermal treatements, a large number of investigations being directed on this matter [98B2, 22H1]. This behavior was correlated with microstructure and the grain boundary region.

Changes in compositions were made, in order to obtain high coercive Nd–Fe–B magnets. Thus, the Nd was partially replaced, in $Nd_2Fe_{14}B$ hard magnetic grains, by expensive heavy rare earths as Dy or Tb. These magnets have improved high temperature properties (over 150 °C) even the magnetization is decreased due to antiparallel alignement of Nd and Dy/Tb moments. There is an inhomogeneous distribution of Dy/Tb inside the $Nd_2Fe_{14}B$ grains, as mentioned in Sect. 7.10.1.

The coercivity of Nd–Fe–B magnets are determined by the magnetism of hard magnetic grains as well as that of intergranular phases. Thus, in order to improve the magnetic performances of Nd–Fe–B magnets, the investigations were then directed on the boundary modifications of the grains. The convetional intergranular phases in Nd–Fe–B magnets are mainly Nd-rich ones, their compositions depending on oxygen content [98B2]. As the O content increases their structure changes from double close-packed hexagonal to face-centered cubic and then to close-packed hexagonal [08M3]. The iron content remained virtually unchanged. The resulting Nd-rich oxides, having high melting temperature, mainly distribute at the triple-junction region and exhibit poor wettability with hard magnetic phase. The lack of thin layer for isolation, lead to short-range exchange coupling between grains and do not improve the coercivity [22L1]. Under demagnetizing field, the reversed domains preferentially nucleates in the regions with defects with low anisotropy, or at an adjacent non-magnetic phase.

The interface between grain boundary (GB) phase and hard magnetic grains should be half-coherent or coherent. The covering uniformly the hard magnetic grains by the intergranular phase with nm thickness, effective contributes to their magnetic decoupling. The investigations were then directed in enhancing coercivity without or with less heavy rare-earth content, by modification of the grain boundary phase or grain diffusion process. Two main approaches for GB modification were elaborated: (1) *intergranular addition* and (2) *grain boundary diffusion* [22L1].

The effects of *intergranular additions* were intensively investigated. The addition of oxides of M elements (Al_2O_3, MgO, ZnO, CaO and Cr_2O_3) was shown to enhance the magnetic properties of sintered magnets. The positive effects were attributed to the substitution of the conventional Nd–O–Fe phase by Nd–O–Fe–M phase [22L1]. As example, the particles like Nd–O–Fe–Mg with low iron content decrease the magnetism of soft magnetic phase and hinder the intergranular propagation of reversal domain walls. The intergranular modifications with positive effects on the magnetic properties have been made by adding some elements at grain boundary. The presence of Al, Cu, Zn, Mg, Ni and P or their combinations, improve the wettability of GB. These can form eutectic phases, with Nd and Fe, having low melting point. Consequently, can easily enter in the spaces between the hard grains during the sintering, contributing to their isolation.

The group of refractory elements (Mo, W, Nb, Ta), added at GB, act as pinning sites, inhibiting the growth of hard magnetic grains during sintering. These can react with Fe and B from $Nd_2Fe_{14}B$ grains, with negative effects on remanence and maximum energy product [98B2]. The sulphides WS_2, WS_2/Al and MoS_2 having high melting temperatures were then precipitate at GB, as pinning centers, since these do not react with hard magnetic phase [12L2, 12L3, 16B1]. The grain refinement has been obtained also in the presence of SiO_2 and ZrO_2 at the GB [09C1]. These precipitates can be simultaneous used with the low melting additives.

The method of *grain boundary restructuration* (GBR), was used, in order to improve corrosion resistance of (Pr, Nd)-Fe-B sintered magnets. As example, the intergranular (Pr, Nd)-rich phase has been replaced by $(Pr,Nd)_6Fe_{13}Cu$ one, which reduce significantly the electrochemical potential difference [12N2]. The restructured magnet has a good resistance to corrosion, possessing also outstanding magnetic properties.

The *grain boundary diffusion* (GBD) [10S2, 17L1, 20L3] methods are variants of intergranular addition. The low melting alloys are introduced in magnets by single or dual-alloy method and act as a new type of granular phase, facilitating the densification and modifying the hard magnetic grains. Some variants of method were used, these involving as sources either rare-earths and their alloys as well as non rare-earth grain boundary.

The investigations were directed in enhancing coercivity with less heavy rare-earths, (HR) content, than used in replacing Nd by Tb/Dy in hard magnetic grains. Some variants of GBD method were used in orde to achieve their efficient utilization. Thus, in some cases the diffusion sources containing heavy rare earths, were vapor coated on the surface of the magnet [10S1, 17L1]. The HR elements were infiltrated into the magnet through the grain boundary, increasing the anisotropy of the surface of the hard phase and in addition optimized the grain boundary phase. In another procedure, the intergranular addition of heavy rare-earths or of their alloys were made in Nd–Fe–B firstly bended with hard magnetic powder, before compaction and in situ diffusion during the sintering process. The targeting with Dy/Tb to the grain boundary region decrease the Dy/Tb content replacing Nd in hard magnetic grains during fabrication of Nd–Fe–B magnets. The much negative formation energy of $R_2Fe_{14}B$ with R = Dy and Tb than when R = Nd and the melting GB phase, ensure their diffusion at temperatures lower than sintering one. In this way, Dy/Tb go into grain boundary phase, decreasing the magnetism of the grain boundary phase and weakens the coupling between the hard magnetic grains [10S1]. An in-homogeneous distribution of the HR elements can be achieved at the grain boundary region, by this diffusion process [10S1, 20C1]. The low melting HR-M eutectic alloys with HR = Dy, Tb and M = Al, Mn, Fe, Ni, Co, Cu etc. are also used for GB addition as diffusion sources, by dual alloy method. The M elements in combination with HR ones play important roles in enhancing both coercivity and corrosion resistance.

The light rare-earths (LR)-M eutectic alloys with LR = Pr, Nd, Ce and La and M = Cu, Al, Ni etc. were also used for GBD process [10S2, 22L1]. The formation of continuous GB layer can be facilitated by the diffusion of LR–M alloys and thus to the magnetic decoupling between hard magnetic grains. The enhancement of coercivity was shown when using this method. The non-magnetic alloys such Nd–Cu or Pr–Cu were also used to modify the grain-boundary and interface chemistry of the Nd–Fe–B sintered magnets. Their addition can thicken the grain boundary and reduce the iron content of the grain boundery phase. This lead to an increase of coercivity due to the weaken exchange coupling interactions between adjacent hard magnetic grains. The remanent induction shows no obvious changes [22H1].

The non rare-earths grain boundary modification technique was also developed for improving the properties of Nd–Fe–B magnets. In order to enhance the coercivity of sintered Nd–Fe–B hard magnetic phase, improving the anisotropy field and isolating the magnetic grains, the non-magnetic layers were introduced at GB. The selection of non-rare-earth diffusion sources, M, should follow some conditions [22L1]: (1) if enter in the 2/14/1 structure, should not deteriorate their stability. As evidenced in Sect. 7.10.1 and calculations [18T1], the structures of $R_2(Fe,M)_{14}B$ with M = Mg, Al, Ni, Cu, Zn etc. are more stable than of parent sample. When a Fe atom at 4*c* site is substituted by Mg, Cu, Zn or other elements, the anisotropy constant, K_1, of the adjacent Nd4*f* site shows an increase due Nd5d-Mp interactions, enhancing the anisotropy field; (2) the M elements should not form the soft magnetic phases by reacting with existing intergranular phase, in order to avoid the decrease of magnetization. The M elements can form eutectic phases with Nd and Fe, with low melting point, having great wettability with hard magnetic grains. Thus, the M-based alloys or compounds modify the composition and structure of the GB, by diffusing into the intergranular region, resulting in the enhancement of coercivity and corrosion resistance. As example, the $Al_{62.5}Cr_{37.5}$ coatings deposited on the Nd–Fe–B substrate, followed by annealing, evidenced that Al mainly segregates at GB region. The Al atoms diffuse into the magnet through GBs, enhancing wettability of intergranular phase and formation of a continuous GB layer. There was an increase of both coercive field and energy product [21H2, 22L1].

The melt spinning technique was also used in order to obtain the hard magnetic phases, by crystallization of amorphous ribbons. During the crystallization process, intermediate metastable phases are formed, before obtaining $R_2Fe_{14}B$ compounds—Sect. 7.9.

The principle of nanocomposite magnets are based on the exchange coupling between the hard magnetic phase ($R_2Fe_{14}B$) having high coercivity and a soft magnetic phase (α-Fe, Fe_3B). In this way is possible to reduce the rare-earth content and the cost, respectively. The magnetic properties of these nanocomposites are lower as compared with those of sintered magnet.

The present review has no in view to include the complex matter devoted to manufacturing R–Fe–B permanent magnets. There is a large number of technical data, already presented in review papers, as for example [98B2, 11S1, 18S1, 20C3].

References

[36N1] L. Néel, Ann. Phys. **5**, 232 (1936)
[48K1] G.V. Kurdyumov, O.P. Maksimov, Dokl. Akad. Nauk SSSR **61**, 83 (1948)
[48N1] L. Néel, Ann. Phys. (Paris) **3**, 137 (1948)
[52S1] K. W. H. Stevens, Proc. Phys. Soc. A **65**, 209 (1952)
[54B1] G. Bergman, D. P. Shomaker, Acta Cryst. **7**, 857 (1954)
[64H1] P. T. Hutchings, *Solid State Physics*, vol. 10 (New York, Academic Press, 1964), p. 227
[65B1] Z. Ban, M. Sikirica, Acta Crystallogr. **18**, 594 (1965)
[66C1] H. B. Callen, E. Callen, J. Phys. Chem. Solids **27**, 1271 (1966)
[68B1] W. F. Brinkmam, S. Engelsberg, Phys. Rev. **169**, 417 (1968)
[70B1] E. Burzo, F. Givord, C.R. Acad. Sci. (Paris) B **271**, 1159 (1970)
[71B1] K.H.J. Buschow, A.S. Van der Goot, Acta Cryst. B **27**, 1085 (1971)
[72B1] N. S. Bilonizhko, Yu. B. Kuzma, Izv. Akad. Nauk. SSSR, Neorg. Mater. **8**, 187 (1972) sau 183
[72N1] K. Niihara, S. Yajima, Chem. Lett. **1**, 875 (1972)
[74B1] N.S. Bilonizhko, Y.B. Kuzma, Inorg. Mater. **10**, 227 (1974)
[74G1] D. Givord, R. Lemaire, IEEE Trans. Magn. **10**, 109 (1974)
[74K1] Y.B. Kuzma, N.S. Bilonizhko, Kristallogr. **18**, 710 (1974)
[75L1] P. A. Lindgard, O. Danielsen, Phys. Rev. B **11**, 351 (1975)
[77K1] Yu. B. Kuzma, N. S. Bilonizhko, S. I. Svarichevskaya, in *Refratory Borides and Silicides (in Russian), Naukova Dumka, Kiev* (1977), p. 67
[77S1] G. F. Stepanchikova, Yu. B. Kuzma, Vestn. Lvov. Univ. Ser. Khim. **19**, 37 (1977)
[78K1] Y. B. Kuzma, N.S. Bilonizhko, Dopov. Akad. Nauk. Ukr. RSR A **3**, 275 (1978)
[78S1] G. F. Stepanchikova, Yu. B. Kuzma, B. I. Chernyak, Dokl. Akad. Nauk. Ukr. SSR, A **10**, 950 (1978)
[79B1] E. Burzo, I. Ursu, J. Appl. Phys. **50**, 1471 (1979)
[79C1] N.F. Chaban, Yu.B. Kuzma, N.S. Bilonizhko, O.O. Kachmar, M.P. Petryiv, Dopov. Akad. Nauk. Ukr. RSR, Ser. A **11**, 873 (1979)
[79S1] R.Sobczak, P. Rogl, J. Solid State Chem. **27**, 343 (1979)
[80A1] G. Asti, F. Bolzoni, J. Magn. Magn. Mater. **20**, 29 (1980)
[80B1] H.F. Braun, K. Yvon, Acta Cryst. B **36**, 2400 (1980)
[80C1] N. F. Chaban, Yu. B. Kuzma, N.S. Bilonizhko, O. O. Kachmar, N. V. Petrov, Dopov. Akad. Nauk, Ukr. RSR A **875** (1980)
[80J1] W. Jung, D. Quentmeier, Z. Kristallogr. **151**, 121 (1980)
[80S1] G. F. Stepanchikova, Yu. B. Kuzma, Poroshk. Metall. (10), 44 (1980)
[81B1] E. Burzo, Fizica Fenomenelor Magnetice, Editura Academiei Române (1981), p. 97

[81K1] Yu. B. Kuzma, G. V. Chernyak, N. F. Chaban, Dopov. Akad. Nauk. UKR. RSR A (12), 80 (1981)
[82B1] H.F. Braun, M. Pelizzone, K. Yvon, *Proceedings International Conference on Solid Compounds of Transition Elements, Grenoble* (1982)
[83C1] G.V. Chernyak, Inorg. Mater. **19**, 436 (1983)
[83C2] G. V. Chernyak, N. F. Chaban, Yu. B. Kuzma, Poroshk. Metall. **6**, 477 (1983)
[83C3] G. V. Chernyak, N. F. Chaban, Yu. B. Kuzma, Poroshk. Metall. (6), 65 (1983)
[83K1] Yu. B. Kuzma, Kristallokhimiya Boridov, Lvov Univ. (1983)
[83K2] Yu. B. Kuzma, Kristallokhimiya boridov, Vyshcha Shkola, Lvov (1983), p. 160
[83K3] Yu. B. Kuzma, L. G. Akselrud, N. F. Chaban, V. A. Bruskov, Dopov. Akad. Nauk Ukr. RSR B, 1049 (1983)
[83K4] Y. B. Kuzma, N. S. Bilonizhko, N. F. Chaban, G. V. Chernyak, J. less Common Met. **90**, 217 (1983)
[84C1] J. M. Cadogan, J. M. D. Coey, Phys. Rev. B **30**, 7326 (1984)
[84C2] J. J. Croat, J. F. Herbst, R. W. Lee, F. E. Pinkerton, J. Appl. Phys. **55**, 2078 (1984)
[84G1] D. Givord, H. S. Li, J. M. Moreau, Solid State Cummun. **50**, 497 (1984)
[84G2] D. Givord, H. S. Li, R. Perrier de la Bathie, Solid State Commun. **51**, 857 (1984)
[84G3] R. Grössinger, P. Obitsch, X. K. Sun, R. Eibler, H. R. Kirchmayr, F. Rothwarf, H. Sassik, Mater. Lett. **2**, 539 (1984)
[84H1] J. F. Herbst, J. J. Croat, F. F. Pinkerton, W. B. Yelon, Phys. Rev. B **29**, 4176 (1984)
[84L1] P. L'Héritier, P. Chaudouët, R. Madar, A. Rouault, J. P. Sénateur, R. Fruchart. C R Acad. Sci., Ser II, **299**, 849 (1984)
[84O1] H. Oesterreicher, F. Spada, C. Abache, Mater. Res. Bull. **19**, 1069 (1984)
[84O2] K. Oesterreicher, H. Oesterreicher, Phys. Stat. Solidi (a) **85**, K61 (1984)
[84O3] H. Onodera, Y. Yamaguchi, H. Yamamoto, M. Sagawa, Y. Matsuura, H. Yamamoto, J. Magn. Magn. Mater. **46**, 151 (1984)
[84P1] E. Parthé, B. Chabot, *Handbook of Physics and Chemistry of Rare Earths* 6, 249 (1984)
[84R1] P. Rogl, *Handbook Physics and Chemistry of Rare Earths*, Elsevier Science, 6, 335 (1984)
[84S1] M. Sagawa, S. Fujimura, M. Togawa, Y. Matsuura, J. Appl. Phys. **55**, 2083 (1984)
[84S2] M. Sagawa, S. Fujimura, H. Yamamoto, Y. Matsuura, IEEE. Trans. Magn. **20**, 1584 (1984)
[84S3] C. B. Shoemaker, D. P. Shoemaker, R. Fruchart, Acta Cryst. C **40**, 1665 (1984)
[84S4] S. Sinnema, R. J. Radwanski, J. J. M. Franse, D. B. de Mooij, K. H. J. Buschow, J. Magn. Magn. Mater. **44**, 333 (1984)
[84S5] F. Spada, C. Abache, H. Oesterreicher, J. less Common Met. **99**, L21 (1984)
[84S6] H. H. Stadelmaier, A. El.Masry, N. V. Liu, S. F. Cheung, Mater Lett. **2**, 41 (1984)
[84Y1] H. Yamamoto, S. Matsuura, S. Fujimura, M. Sagawa, Appl. Phys. Lett. **45**, 1141 (1984)
[85A1] C. Abache, H. Oesterrericher, J. Appl. Phys. **57**, 4112 (1985)
[85A2] A. V. Andreev, A. V. Deryagin, S. M. Zadvorkin, S. M. Terentev, Sov. Phys. Solid State **27**, 987 (1985)
[85B1] A. Bezingé, R.F. Braun, J. Miller, K. Yvon, Solid State Commun. **55**, 131 (1985)
[85B2] M. Bogé, J. M. D. Coey, G. Czjzek, D. Givord, C. Jeandey, H. S. Li, J. L. Oddou, Solid State Commun. **55**, 295 (1985)
[85B3] E. B. Boltich, E. Oswald, M. Q. Huang, S. Hirosawa, E. Burzo, J. Appl. Phys. **57**, 4106 (1985)
[85B4] E. Burzo, J. Appl. Phys. **57**, 3232 (1985)
[85B5] E. Burzo, Keynote lecture, in *6th International Symposium on HighPurity Materials in Science and Technology, Dresda* (1985)
[85B6] E. Burzo, L. Stanciu, W. E. Wallace, J. less Common Met. **111**, 83 (1985)
[85B7] E. Burzo, E. Oswald, M. Q. Huang, E. Boltich, W. E. Wallace, J. Appl. Phys. **57**, 4109 (1985)
[85B8] E. Burzo, E. B. Boltich, M. Q. Hunag, W. E. Wallace, in *Proceeding of 8th International Workshop on Rare-Earth Magnets and their Applications, Dayton, Ohio* (1985), p. 771
[85B9] K. H. J. Buschow, J. W. C. De Vries, R. C. Thiel, Physica B **132**, 13 (1985)
[85B10] K.H. J. Buschow, D. B. De Mooij, H. M. Van Noort, Philips Res. **40**, 27 (1985)

[85B11] K. H. J. Buschow, H. M. Van Noort, D. B. De Mooij, J. Less Common Met. **109**, 79 (1985)

[85B12] K. H. J. Buschow, J. W. C. De Vries, R. C. Thiel, J. Phys. F.: Met. Phys. **15**, L93 (1985)

[85C1] J. M. D. Coey, H. S. Li, J. P. Gavigan, J. M. Cadogan, B. P. Hu, *Nd-Fe Permanent Magnets-Their Future Applications* (Elsevier, Amsterdam, 1985), p. 143

[85D1] R. L. Davis, R. K. Day, J. B. Dunlop, Solid State Commun. **56**, 181 (1985)

[85D2] D. De Rotier, D. Fruchart, R. Fruchart, R. Guillen, P. L. Heritier, P. Vuillet, P. Wolfers, A. Yaonac, J. Phys. **46**, 323 (1985)

[85D3] O. M. Dub, N. F. Chaban, Yu. B. Kuzma, Izv. Akad. Nauk SSSR, Neorg. Mater. **21**, 1967 (1985)

[85F1] L. P. Ferreira, R. Guillen, P. Vulliet, A. Yaouanc, D. Fruchart, P. Wolfers, P. L'Heritier, R. Fruchart, J. Magn. Magn. Mater. **53**, 145 (1985)

[85F2] J. M. Friedt, J. P. Sanchez, P. L'Heritier, R. Fruchart, *Nd-Fe Permanent Magnets-Their Present and Future Applications* (Elsevier, 1985), p. 179

[85G1] D. Givord, J.M. Moreau, P. Tenaud, Solid. State Commun. **55**, 303 (1985)

[85G2] D. Givord, H. S. Li, F. Tasset, J. Appl. Phys. **57**, 4100 (1985)

[85G3] D. Givord, H. S. Li, J. M. Moreau, R. Perrier de la Bathie, E du Trémolet de Lacheisserie, Physica B **130**, 323 (1985)

[85G4] R. Grössinger, X. R. Sun, R. Eibler, K. H. J. Buschow, H. R. Kirchmayr, J. Phys. **46**, C6–221 (1985)

[85G5] R. Grössinger, X. R. Sun, R. Eibler, K. H. J. Buschow, H. R. Kirchmayr, J. Magn. Magn. Mater. **58**, 55 (1985)

[85G6] P. C. M. Gubbens, A. Van Der Kraan, K. H. J. Buschow, Phys. Stat. Solidi (b) **130**, 575 (1985)

[85H1] I. R. Harris, C. Noble, T. Bailey, J. less Common Met. **106**, L1 (1985)

[85H2] J. F. Herbst, W. B. Yelon, J. Appl. Phys. **57**, 2343 and 4086 (1985)

[85H3] S. Hirosawa, M. Sagawa, Solid State Commun. **54**, 335 (1985)

[85H4] S. Hirosawa, Y. Matsuura, H. Yamamoto, S. Fujimura, M. Sagawa, H. Yamauchi, Jpn. J. Appl. Phys. **24**, L803 (1985)

[85H5] H. Hiroyoshi, H. Yamauchi, Y. Yamaguchi, H. Yamamoto, Y. Nakagawa, M. Sagawa, Solid State Commun. **54**, 41 (1985)

[85I1] H. Ino, K. Mizushita, I. Homma, M. Tokunaga, Seisan-Kenkyu, **37**, 221 (1985)

[85K1] N. C. Koon, B. N. Das, M. Rubinstein, J. Tyson, J. Appl. Phys. **57**, 4091 (1985)

[85K2] A. Kostikas, V. Papaefthymiou, A. Simopoulos, S. C. Hadjipanayis, J. Phys. F.: Met. Phys. **15**, L129 (1985)

[85L1] F. Leccabue, J. L. Sanchez, L. Pareti, F. Bolzoni, R. Panizzeri, Phys. Stat. Solidi (a) **91**, K 63 (1985)

[85L2] P. L'Héritier, R. Fruchart, J. Phys. **46** C6–319 (1985)

[85M1] Y. Matsuura, S. Hirosawa, H. Yamamoto, S. Fujimura, M. Sagawa, K. Osamura, Jpn. J. Appl. Phys. **24**, L635 (1985)

[85M2] Y. Matsuura, S. Hirosawa, H. Yamamoto, S. Fujimura, M. Sagawa, Appl. Phys. Lett. **46**, 308 (1985)

[85P1] L. Pareti, F. Balzoni, O. Mozé, Phys. Rev. B **32**, 7604 (1985)

[85P2] L. Pareti, H. Szymczak, H. K. Lachowicz, Phys. Satus Solidi (a) **92**, K65 (1985)

[85P3] F. E. Pinkerton, W. R. Durham, J. Appl. Phys. **57**, 4121 (1985)

[85P4] E. Potenziani, J. Appl. Phys. **58**, 2764 (1985)

[85R1] M. Rosenberg, P. Deppe, M. Wojcik, H. Stadelmaier, in *Proceedings of Mössbauer Spectrosopy Conference, Leuven, Belgium* (1985)

[85R2] M. Rosenberg, P. Deppe, M. Wojczik, H. Stadelmeier, J. Appl. Phys. **57**, 4124 (1985)

[85R3] C. Rudowich, J. Phys. C: Solid State Phys. **18**, 1415 (1985)

[85S1] M. Sagawa, G. Fujimura, H. Yamamoto, Y. Matsuura, S. Hirosawa, J. Appl. Phys. **57**, 4094 (1985)

[85S2] B. Szpunar, J. A. Szpunar, J. Appl. Phys. **57**, 4130 (1985)

[85S3] R. Szymczak, E. Burzo, W. E. Wallace, J. Phys. **46**, C6–309 (1985)

[85T1] K. Tokuhara, Y. Ohstu, F. Ono, O. Yamada, M. Sagawa, Y. Matsuura, Solid State Commun. **56**, 333 (1985)
[85T2] G. Turilli, H. Szymczak, H. K. Lachowicz, Phys. Stat. Sol. (a) **90**, K143 (1985)
[85V1] P. P. Vaishnava, C. W. Kimball, A. M. Umarji, S. K. Malek, G. K. Shenoy, J. Magn. Magn. Mater. **49**, 286 (1985)
[85V2] P. P. Vaishnava, C. W. Kimball, A. M. Umarji, S. K. Malik, G. K. Shenoy, J. Appl. Phys. **57**, 3264 (1985)
[85V3] H. M. van Noort, D. B. De Mooij, K. H. J. Buschow, J. Less Common Met. **111**, 87 (1985)
[85V4] H. M. Van Noort, K. H. J. Buschow, J. Less Common Met. **113**, L9 (1985)
[85V5] H. M. Van Noort, B. D. De Mooij, K. H. J. Buschow, J. Appl. Phys. **57** 5414 (1985)
[85V6] A. Vasquez, J. M. Friedt, J. P. Sanchez, P. L'Héritier, R. Fruchart, Solid State Commun. **55**, 783 (1985)
[85Y1] M. Yamada, Y. Yamaguchi, H. Kato, H. Yamamoto, Y. Nakagawa, S. Hirosawa, M. Sagawa, Solid State Commun. **56**, 663 (1985)
[86A1] C. Abache, H. Osterreicher, J. Appl. Phys. **60**, 1114 (1986)
[86A2] C. Abache, H. Osterreicher, J. Appl. Phys. **60**, 3671 (1986)
[86A3] S. Adam, Gh. Adam, E. Burzo, J. Magn. Magn. Mater. **61**, 260 (1985)
[86A4] A. V. Andreev, A. V. Deryagin, N. V. Kudrevatykh, N. V. Mushinov, V. A. Reiner, S. V. Terentiev, Sov. Phys. JETP, **63**, 608 (1986)
[86B1] Y. Berthier, M. Bogé, G. Czjzek, D. Givord, C. Jeandey, H. S. Li, J. L. Oddou, J. Magn. Magn. Mater. **54–57**, 589 (1986)
[86B2] M. Bogé, G. Czjzek, D. Givord, C. Jeandey, H. S. Li, J. L. Oddou, J. Phys. F.: Metal Phys. **16**, L67 (1986)
[86B3] E. B. Boltich, W. E. Wallace, J. less Common Met. **126**, 35 (1986)
[86B4] F. Bolzoni, A. Deriu, F. Leccabue, L. Pareti, J. L. Sanchez, J. Magn. Magn. Mater. **54–57**, 595 (1986)
[86B5] P. Burlet, J. M. D. Coey, J. P. Cavigan, D. Givord, C. Meyer, Solid State Commun. **60**, 723 (1986)
[86B6] E. Burzo, in *Lecture 3nd Int. Conf. Magnetic Materials, Sczyrk,* (World Scientific Press, Poland, 1986), p. 282
[86B7] E. Burzo, I. Lupşa, A. T. Pedziwiatr, J. Phys. F.: Metal Phys. **16**, L83 (1986)
[86B8] E. Burzo, N. Plugaru, V. Pop, L. Stanciu, W. E. Wallace, Solid State Commun. **58**, 803 (1986)
[86B9] K. H. J. Buschow, J. Less Common **118**, 349 (1986)
[86B10] K. H. J. Buschow, Mater Sci. Rep. **1**, 1 (1986)
[86B11] K.H.J. Buschow, D.B. De Mooij, J.L.C. Daams, H.M. Van Noort, J. less Common. Met. **115**, 357 (1986)
[86B12] K.H.J. Buschow, D.B.de Mooij, H.M.van Noort, J. less Common. Met. **125**, 135 (1986)
[86C1] G. Chen, J. Liang, X. Wang, Sci. Sinica, (a) **24**, 1172 (1986)
[86C2] J. M. D. Coey, A. Yaouanc, D. Fruchart, Solid State Commun. **58**, 413 (1986)
[86C3] J. M. D. Coey, J. less Common Met. **126**, 21 (1986)
[86D1] D. Dadon, M. P. Dariel, Y. Gefen, H. Klimker, M. Rosen, Appl. Phys. Lett. **48**, 1444 (1986)
[86D2] D.B. de Mooij, K.H.J. Buschow, Philips J. Res. **41**, 400 (1986)
[86D3] O. M. Dub, Yu. B. Kuzma, M. I. David, Poroshk. Metall. **7**, 56 (1986)
[86D4] O. M. Dub, N. F. Chaban, Yu. B. Kuzma, J. less Common. Met. **117**, 297 (1986)
[86D5] K. D. Durst, H. Kronmüller, J. Magn. Magn. Mater. **59**, 86 (1986)
[86F1] J. M. Friedt, A. Vasquez, J. P. Sanchez, P. L'Héritier, R. Fruchart, J. Phys. F.: Metal Phys. **16**, 651 (1986)
[86F2] C. D. Fuerst, J. F. Herbst, E. A. Alson, J. Magn. Magn. Mater. **54–57**, 567 (1986)
[86F3] H. Fujii, W. E. Wallace, E. B. Boltich, J. Magn. Magn. Mater. **61**, 251 (1986)
[86G1] D. Givord, P. Tenaud, J.M. Moreau, J. less Commun Met. **115**, L7 (1986)
[86G2] D. Givord, P. Tenaud, J.M. Moreau, J. less Commun Met. **123**, 109 (1986)
[86G3] D. Givord, H. S. Li, J. M. Moreau, P. Tenand, J. Magn. Magn. Mater. **54–57**, 445 (1986)

[86G4] D. Givord, H. S. Li, in *Nd-Fe Permanent Magnets. Their Present and Future Applications* (Elsevier, Amseterdam, 1986), p. 131
[86G5] R. Grössinger, X. K. Sun, R. Eibler, K. H. J. Buschow, H. R. Kirchmayr, J. Magn. Magn. Mater. **58**, 55 (1986)
[86G6] R. Grössinger, R. Krewenka, X. K. Sun, R. Eibler, H. R. Kirchmayr, K. H. J. Buschow, J. less Common. Met. **124**, 165 (1986)
[86G7] P. C. M. Gubbens, A. M. Van der Kraan, K. H. J. Buschow, J. Magn. Magn. Mater. **54–57**, 591 (1986)
[86H1] J. F. Herbst, W. B. Yelon, J. Magn. Magn. Mater. **54–57**, 570 (1986)
[86H2] J. F. Herbst, R. W. Lee, F. E. Pinkerton, Ann. Rev. Mater. Sci. **16**, 467 (1986)
[86H3] J. F. Herbst, W. B. Yelon, J. Appl. Phys. **60**, 4224 (1986)
[86H4] S. Hirosawa, Y. Matsuura, H. Yamamoto, S. Fujimura, M. Sagawa, H. Yamauchi, J. Appl. Phys. **59**, 873 (1986)
[86H5] H. Hiroyoshi, N. Saito, G. Kido, Y. Nakagawa, S. Hirosawa, M. Sagawa, J. Magn. Magn. Mater. **54–57**, 583 (1986)
[86H6] M. Q. Huang, E. B. Boltich, W. E.Wallace, E. Oswald, J. less Common. Met. **124**, 55 (1986)
[86H7] M. Q. Huang, E. B. Boltich, W. E.Wallace, E. Oswald, J. Magn. Magn. Mater. **60**, 270 (1986)
[86I1] J. Inoue, M. Shimizu, J. Phys. F.: Metal Phys. **16**, 1051 (1986)
[86J1] S.Y. Jiang, W.E. Wallace, E. Burzo, J. less Common Met. **119**, L5 (1986)
[86J2] S.Y. Jiang, W.E. Wallace, E. Burzo, J. Magn. Magn. Mater. **61**, 257 (1986)
[86J3] M. Jurczyk, W. E. Wallace, J. Magn. Magn. Mater. **59**, L182 (1986)
[86J4] M. Jurczyk, W. E. Wallace, J. less Common Met. **124**, 149 (1986)
[86K1] R. Kamal, M. Rani, M. S. Saini, J. less Common Met. **125**, L7 (1985)
[86K2] Cz. Kapusta, Z. Kakol, H. Figiel, R. J. Radwanski, J. Magn. Magn. Mater. **59**, 169 (1986)
[86K3] N. C. Koon, B. N. Das, C. M. Williams, J. Magn. Magn. Mater. **54–57**, 523 (1986)
[86K4] H. C. Ku, J. C. Chen, L. S. Yen, J. Magn. Magn. Mater. **54–57**, 533 (1986)
[86K5] Yu. B. Kuzma, S. I. Mykhalenko, L. G. Akselrud, J. Less Common Met. **117**, 29 (1986)
[86L1] C. Lin, Z. X. Liu, J. Magn. Magn. Mater. **54–57**, 887 (1986)
[86M1] G. P. Meisner, C. D. Fuerst, I.E.E.Trans. Magn. **22**, 744 (1986)
[86N1] D. Niarchos, G. Zouganelis, A. Kostikas, A. Simopoulos, Solid State Commun. **59**, 389 (1986)
[86N2] D. Niarchos, A. Simopoulos, Solid State Commun. **59**, 669 (1986)
[86P1] A. T. Pedziwiatr, W. E. Wallace, J. less Common Met. **126**, 41 (1986)
[86P2] A. T. Pedziwiatr, S. Y. Jiang, W. E. Wallace, J. Magn. Magn. Mater. **62**, 29 (1986)
[86P3] A. T. Pedziwiatr, W. E. Wallace, E. Burzo, J. Magn. Magn. Mater. **61**, 173 (1986)
[86P4] A. T. Pedziwiatr, W. E. Wallace, E. Burzo, J. Magn. Magn. Mater. **61**, 177 (1986)
[86P5] A. T. Pedziwiatr, W. E. Wallace, E. Burzo, J. Magn. Magn. Mater. **59**, L179 (1986)
[86P6] F. Pourarian, M. Q. Huang, W. E. Wallace, J. less Common Met. **120**, 63 (1986)
[86P7] D. C. Price, R. K. Day, J. B. Dunlop, J. Appl. Phys. **59**, 3585 (1986)
[86R1] H.R. Rechenberg, A. Paduan-Filho, F.P. Missell, P. Deppe, M. Rosenberg, Solid State Commun. **59**, 541 (1986)
[86R2] M. Rosenberg, P. Deppe, K. Erdmann, M. Sostarich, H. Stadelmaier, J. Magn. Magn. Mater. **54–57**, 599 (1986)
[86R3] M. Rosenberg, P. Deppe, K. Erdmann, H. Stadelmaier, Hyp. Inter. **28**, 503 (1986)
[86S1] J. P. Sanchez, J. M. Friedt, A. Vasquez, Ph.L'Héritier, R. Fruchart, Solid State Commun. **57**, 309 (1986)
[86S2] G. Schneider, Z. Metallkunde, **77**, 775 (1986)
[86S3] G. Schneider, E.T. Henig, G. Petzow, H.H. Stadelmaier, Z. Metallkd. **77**, 755 (1986)
[86V1] R. Van Mens, J. Magn. Magn. Mater.**61**, 24 (1986)
[86V2] H. M. Van Noort, D. B. de Mooij, K. H. J. Buschow, J. less Common Met. **115**, 155 (1986)
[86Y1] O. Yamada, M. Tokuhara, F. Ono, M. Sagawa, Y. Matsuura, J. Magn. Magn. Mater. **54–57**, 585 (1986)

[86Y2] H. Yamauchi, M. Yamada, Y. Yamaguchi, H. Yamamoto, S. Hirosawa, J. Magn. Magn. Mater. **54–57**, 575 (1986)
[86Y3] Y. C. Yang, W. J. James, H. Y. Chen, H. Sun, J. Magn. Magn. Mater. **54–57**, 895 (1986)
[86Y4] Y. C. Yang, W. J. James, X. D. Li, H. Y. Chen, L. G. Xu, IEEE Trans. Magn. **22**, 757 (1986)
[86Y5] W. B. Yelon, J. F. Herbst, J. Apppl. Phys. **59**, 93 (1986)
[86Y6] W. B. Yelon, B. Foley, C. Abache, H. Oesterrreicher, J. Appl. Phys. **60**, 2982 (1986)
[87A1] M. Abe, S. H. Liou, C. L. Chien, N. C. Koon, B. N. Das, E. Callen, J. Appl. Phys. **61**, 3568 (1987)
[87A2] S. H. Aly, E. Singleton, G. C. Hadjipanayis, J. Appl. Phys. **61**, 3557 (1987)
[87A3] H. Akiyama, K. Nagayama, K. Takumitsu, H. Imo, K. Yano, Seisan-Kenyu, **39**, 347 (1987)
[87B1] A. Bezingé, K. Yvon, H.F. Braun, J. Muller. Phys. Rev. B **36**, 1406 (1987)
[87B2] E. B. Boltich, A. T. Pedziwiatr, W. E. Wallace, J. Magn. Magn. Mater. **66**, 317 (1987)
[87B3] E. B. Boltich, A. T. Pedziwiatr, W. E. Wallace, Mater. Res. Soc. Symp. Proc. **96**, 119 (1987)
[87B4] F. Bolzoni, F. Leccabue, O. Mozé, L. Pareti, M. Solzi, J. Appl. Phys. **61**, 5369 (1987)
[87B5] F. Bolzoni, F. Leccabue, O. Mozé, L. Pareti, M. Solzi, J. Magn. Magn. Mater. **67**, 373 (1987)
[87B6] F. Bolzoni, J. M. D. Coey, J. Gavigan, D. Givord, O. Mozé, L. Pareti, T. Viadieu, J. Magn. Magn. Mater. **65**, 123 (1987)
[87B7] F. Bolzoni, J. P. Gavigan, D. Givord, H. S. Li, O. Mozé, L. Pareti, J. Magn. Magn. Mater. **66**, 158 (1987)
[87B8] F. Bolzoni, O. Mozé, L. Pareti, J. Appl. Phys. **62**, 615 (1987)
[87B9] E. Burzo, A. T. Pedziwiatr, W. E. Wallace, Solid State Commun. **61**, 57 (1987)
[87B10] E. Burzo, N. Plugaru, V. Pop, Acta Phys. Polonica A **72**, 33, (1987)
[87B11] E. Burzo, N. Plugaru, V. Pop, J. Magn. Magn. Mater. **70**, 343 (1987)
[87B12] K. H. J. Buschow, Proc. 9th Int. Workshop on Rare-Earth Magnets and Their Applications, Bad Soden (1987), p. 453
[87B13] K. H. J. Buschow, R. Grössinger, J. less Common Met. **135**, 39 (1987)
[87C1] J. M. Cadogan, J. less Common. Met. **135**, 269 (1987)
[87C2] B. P. Cheng, Y. C. Yang, W. J. James, J. Appl. Phys. **61**, 3586 (1987)
[87C3] W. Y. Ching, Q. G. Zong, J. Appl. Phys. **61**, 3718 (1987)
[87C4] Y. C. Chuang, C. H. Wu, J. Q. Xie, in *Proceedings of 9th International Workshop on Rare-Earth Magnets and Their Applications, Bad Soden* (1987), p. 617
[87C5] J. M. D. Coey, A. Yaouanc, D. Fruchart, R. Fruchart, Ph L'Hértier, J. less Common Met. **131**, 419 (1987)
[87D1] P. Dalmas de Reotier, D. Fruchart, L. Pontonnier, F. Vaillant, P. Wolfers, A. Yaouanc, J. M. D. Coey, R. Fruchart, Ph L'Héritier, J. less Common Met. **129**, 133 (1987)
[87D2] D.B. De Mooij, J.L.C. Daams, K.H.J. Buschow, Philips J. Res, **42**, 339 (1987)
[87D3] P. Deppe, M. Rosenberg, K.H.J. Buschow, Solid State Commun. **64** 1247 (1987)
[87D4] O. M. Dub, M. I. David, Sov. Powder Metal. Met. Ceramics **26**, 563 (1987)
[87D5] O. M. Dub. Yu. B. Kuzma, M. I. David, Poroshk. Metall. **7**, 56 (1987)
[87E1] K. Erdmann, M. Deppe, M. Rosenberg, K. H. J. Buschow, J. Appl. Phys. **61**, 4340 (1987)
[87F1] D. Fruchart, F. Vaillant, A. Yaouanc, J. M. D. Coey, R. Fruchart, Ph. L'Hértier, T. Riesterer, J. Osterwalder, L. Schlapbach, J. less Common Met. **130**, 97 (1987)
[87F2] R. Fruchart, Ph. L'Hértier, P. Dalmas de Reotier, D. Fruchart, P. Wolfers, J. M. D. Coey, L. P. Ferriera, R. Guillen, P. Vulliet, A. Yaouanc, J. Phys. F: Metal Phys. **17**, 483 (1987)
[87F3] C. D. Fuerst, G. P. Meisner, F. E. Pinkerton, W. B. Yelon, J. less Common Met. **133**, 255 (1987)
[87F4] C. D. Fuerst, G. P. Meisner, F. E. Pinkerton, J. Appl. Phys. **61**, 2314 (1987)
[87F5] A. Fujita, H. Onodera, H. Yamauchi, M. Yamada, H. Yamamoto, S. Hirosawa, M. Sagawa, J. Magn. Magn. Mater. **69**, 267 (1987)
[87F6] H. Fujii, H. Nagata, Y. Uwatoko, T. Okamoto, H. Yamamoto, M, Sagawa, J. Magn. Magn. Mater. **70**, 331 (1987)

[87F7] H. Figiel, D. Fowler, T. Dumelow, P. C. Riedi, Cz. Kapusta, J. Magn. Magn. Mater. **65**, 83 (1987)
[87G1] J. P. Gavigan, D. Givord, H. S. Li, O. Yamada, H. Maruyama, H. Sagawa, H. Hirosawa, J. Magn. Magn. Mater. **70**, 416 (1987)
[87G2] B. Grieb, E. Th. Henig, G. Schneider, G. Petzow, in *Proceedings of 5th International Symposium on Magn. Anisotropy and Coercivity Rare-Earth-Transition Metal Alloys, Bad Soden, Germany* (1985), p. 395
[87G3] R. Grössinger, R. Krewenka, H. R. Kirchmayr, S. Sinnema, F. M. Yang, Y. K. Huang, F. R. de Boer, K. H. Buschow, J. less Common Met. **123**, 265 (1987)
[87G4] Z. Q. Gu, W. Y. Ching, Phys. Rev. B **36**, 8530 (1987)
[87G5] Z. Q. Gu, W. Y. Ching, J. Appl. Phys. **61**, 3977 (1987)
[87G6] P. C. M. Gubbens, A. M. Van der Kraan, R. P. Van Stapele, K. H. J. Buschow, J. Magn. Magn. Mater. **68**, 238 (1987)
[87H1] G.C. Hadjipanayis, A. Nazareth, D. Niarchos, IEEE Trans. Magn. **23**, 3613 (1987)
[87H2] S. Hirosawa, Y. Yamaguchi, K. Tokuhara, H. Yamamoto, S. Fujimura, M. Sagawa, IEEE Trans. Magn. **23**, 2120 (1987)
[87H3] S. Hirosawa, K. Tokuhara, H. Yamamoto, S. Fujimura, M. Sagawa, H. Yamauchi, J. Appl. Phys. **61**, 3571 (1987)
[87H4] S. Hiroyoshi, H. Kato, M. Yamada, N. Saito, Y. Nakagawa, S. Hirosawa, M. Sagawa, Solid State Commun. **62**, 475 (1987)
[87H5] H. Hiroyoshi, M. Yamada, Y. Nakagawa, S. Hirosawa, M. Sagawa, in *Proceedings of ISPMM Conference on Sendai* (1987), p. 41
[87H6] H. Hiroyoshi, M. Yamada, M. Saito, H. Kato, Y. Nakagawa, S. Hirosawa, M. Sagawa, J. Magn. Magn. Mater. **70**, 337 (1987)
[87H7] S. Hoch, H. Kronmüller, in *Proceedings of 5th International Symposium Magnetic Anisotropy and Coercivity in Rare Erath Transition Metal Alloys, Bad Soden* (1987), p. 275
[87H8] H. Honma, H. Ino, IEEE Trans Magn. **23**, 3116 (1987)
[87H9] Y. K. Huang, C. H. Wu, Y. C. Chuang, F. M. Yang, F. R. de Boer, J. less Common Met. **132**, 317 (1987)
[87I1] M. Ibarra, P. A. Algarabel, A. Alberdi, J. Bartolomé, A. del Moral, J. Appl. Phys. **61**, 5451 (1987)
[87I2] T. Itoh, K. Ikosaka, H. Takahashi, T. Ukai. N. Mori, J. Appl. Phys. **61**, 3430 (1987)
[87J1] M. Jurczyk, J. Magn. Magn. Mater. **67**, 187 (1987)
[87J2] M. Jurczyk, A. Kowalczyk, A. Wrzeciono, Phys. Stat. Solidi (a) **101**, K65 (1987)
[87J3] M. Jurczyk, A. Kowalczyk, A. Wrzeciono, in *Proceedings of 9th International Workshop on Rare-Earth Magnets and their Application, Bad Soden* (1987), p. 701
[87K1] S. Kajiwara, G. Kido, Y. Nakagawa, S. Hirosawa, M. Sagawa, J. Phys. Soc. Jpn. **56**, 829 (1987)
[87K2] J. Kamarad, Z. Arnold, J. Schneider, J. Magn. Magn. Mater. **67**, 29 (1987)
[87K3] G. Kido, H. Kato, M. Yamada, Y. Nakagawa, S. Hirosawa, M. Sagawa, J. Phys. Soc. Jpn. **56**, 4635 (1987)
[87K4] G. Kido, S. Kajiwara, Y. Nakagawa, S. Hirosawa, M. Sagawa, IEEE Trans. Magn. **23**, 3107 (1987)
[87K5] H. C. Ku, L. S. Yen, J. less Common Met. **127**, 43 (1987)
[87L1] H. S. Li, Ph.D. Thesis, Univ. J. Fourrier, Grenoble (1987)
[87L2] C. Lin, J. Lan, X. F. Xu, J. Appl. Phys. **61**, 3457 (1987)
[87L3] C. Lin, Z. H. Liu, X. F. Xu, IEEE Trans. Magn. **23**, 2296 (1987)
[87M1] M. Matsui, M. Doi, T. Shimizu, IEEE Trans. Magn. **23**, 3113 (1987)
[87N1] H. Nagata, S. Hirosawa, M. Sagawa, A. Ishibashi, S. Endo, J. Magn. Magn. Mater. **70**, 334 (1987)
[87N2] Y. Nakagawa, G. Kido, H. Hiroyoshi, H. Kato, M. Yamada, S. Hirosawa, in *Proceedings of 5th International Symposium on Magnetic Anisotropy and Coercivity in Rare-Earth Transition Metal Alloys, Bad Soden* (1987), p. 87

[87O1] K. Oesterreicher, H. Oesterreicher, J. Solid State Chem. **70**, 313 (1987)
[87O2] F. Ono, Y. Ohtsu, O. Yamada, J. Magn. Magn.Mater. **70**, 319 (1987)
[87O3] H. Onodera, A. Fujita, H. Yamamoto, M. Sagawa, S. Hirosawa, J. Magn. Magn. Mater. **68**, 6 (1987)
[87O4] H. Onodera, Y. Yamaguchi, M. Yamada, H. Yamamoto, M. Sagawa, S. Hirosawa, J. Magn. Magn. Mater. **68**, 15 (1987)
[87O5] Y. Otani, R. Miyajima, S. Chikazumi, J. Appl. Phys. **61**, 9436 (1987)
[87P1] L. Pareti, F. Bolzoni, M. Solzi, K. H. J. Buschow, J. less Common Met. **132**, L5 (1987)
[87P2] L. Pareti, M. Solzi, F. Bolzoni, M. Mozé, R. Panizzeri, Solid State Commun. **61**, 761 (1987)
[87P3] A. T. Pedziwiatr, W. E. Wallace, J. Magn. Magn. Mater. **65**, 139 (1987)
[87P4] A. T. Pedziwiatr, W. E. Wallace, J. Magn. Magn. Mater. **66**, 63 (1987)
[87P5] A. T. Pedziwiatr, W. E. Wallace, Solid State Commun. **64**, 1017 (1987)
[87P6] A. T. Pedziwiatr, W. E. Wallace, J. Appl. Phys. **61**, 3439 (1987)
[87P7] A. T. Pedziwiatr, H. Y. Chen, W. E. Wallace, J. Magn. Magn. Mater. **67**, 311 (1987)
[87P8] A. T. Pedziwiatr, W. E. Wallace, E. Burzo, IEEE Trans. Magn. **MAG-23**, 1795 (1987)
[87P9] A. T. Pedziwiatr, W. E. Wallace, E. Burzo, V. Pop, Solid State Commun. **61**, 61 (1987)
[87P10] F. E. Pinkerton, C. D. Fuerst, G. P. Meisner, J. Appl. Phys. **61**, 3562 (1987)
[87P11] F. Pourarian, S. Y. Jiang, G. Sankar, W. E. Wallace, Solid State Commun. **63**, 1119 (1987)
[87Q1] P. H. Quang, T. H. Anh, N. H. Luong, L. T. Tai, T. D. Hien, J. J. M. Franse, in *Proceedings of 5th International Symposium on Magnets Anisotropy and Coercivity in Rare-Earth Transition Metal Alloys, Baden Soden* (1987), p. 206
[87R1] R. J. Radwanski, J. J. M. Franse, Phys. Rev. B **36**, 8616 (1987)
[87R2] M. Rani, R. Kamal, J. less Common Met. **128**, 343 (1987)
[87R3] M. Rani, R. Kamal, J. Magn. Magn. Mater. **66**, 379 (1987)
[87R4] H. R. Rechenberg, J. P. Sanchez, P. L'Hértier, R. Fruchart, Phys. Rev. B **36**, 1865 (1987)
[87R5] H.R. Rechenberg, J.P. Sanchez, Solid State Commun. **62**, 461 (1987)
[87R6] H.R. Rechenberg, M. Bogé, C. Jeandey, J.L. Oddou, J.P. Sanchez, P. Tenaud, Solid State Commun. **64**, 277 (1987)
[87R7] J. R. Regnard, A, Yaouanc, J. Appl. Phys. **61**, 3565 (1987)
[87S1] M. Sagawa, S. Hirosawa, K. Tokuhara, H. Yamamoto, S. Fujimura, J. Appl. Phys. **61**, 3559 (1987)
[87S2] M. Shimao, H. Ido, K. Kido, K. Ohashi, IEEE Trans. Magn. **23**, 2722 (1987)
[87S3] R. Szymczak, H. Szymczak, E. Burzo, IEEE Trans Magn. **MAG-23**, 2536 (1987)
[87S4] B. Szpunar, W. E. Wallace, Phys. Rev. B **36**, 3782 (1987)
[87T1] D. E. Tharp, Y. C. Yang, O. A. Pringle, G. J. Long, W. J. James, J. Appl. Phys. **61**, 4334 (1987)
[87T2] J. H. Tian, Y. Y. Hung, J. K. Liang, Sci. Chin. Series Math., Phys Astron; Techn. Sci. **30**, 607 (1987)
[87V1] A. Vasquez, J. P. Sanchez, J. less Common. Met. **127**, 71 (1987)
[87W1] W. E. Wallace, A. T. Pedziwiatr, E. B. Boltich, H. K. Smith, S. Y. Jiang, S. G. Sankar, E. Oswald, Mater. Res. Soc. Symp. Proc. **96**, 17 (1987)
[87W2] W. E. Wallace, A. T. Pedziwiatr, in *Proceedings of 5th International Symposium on Magnetic Anisotropy and Coercivity in Rare-Earth-Transition Metal Alloys, Bad Soden* (1987), p. 31
[87W3] G. Wiesinger, G. Hilscher, R. Grössinger, J. less Common Met. **131**, 409 (1987)
[87Y1] Y. C. Yang, D. E. Tharp, G. J. Long, O. A. Pringle, W. J. James, J. Appl. Phys. **61**, 4343 (1987)
[87Y2] L. S. Yen, J. C. Chen, H. C. Ku, J. Appl. Phys. **61**, 1990 (1987)
[87Z1] N. Zhang, Y. Luo, in *Proceedings of 5th International Symposium on Magnetic Anisotropy and Coercivity in Rare Earth Transition Metal Alloys, Bad Soden, Germany* (1987), p. 453
[88A1] Z. Altounian, D. H. Ryan, J. Appl. Phys. **64**, 5723 (1988)
[88A2] S. H. Aly, E. Singleton, G. C. Hadjipanayis, J. Appl. Phys. **63**, 3704 (1988)
[88B1] J. J. Barra, B. F. Bogacz, A. Szytula, J. Magn. Magn. Mater. **75**, 293 (1988)

[88B2] G. Bocelli, G. Calestani, F. Leccabue, Solid State Commun. **65** 1301 (1988)
[88B3] E. B. Boltich, J. Appl. Phys. **63**, 3124 (1988)
[88B4] E. B. Boltich, J. Appl. Phys. **64**, 4796 (1988)
[88B5] E. B. Boltich, F. Pourarian, R. T. Obermeyer, S. G. Sankar, W. E. Wallace, J. Appl. Phys. **63**, 3964 (1988)
[88B6] S. Brennan, A. Bienenstock, J. E. Keem, J. Appl. Phys. **65**, 697 (1988)
[88B7] E. Burzo, S. Adam, G. Adam, W. E. Wallace, Rev. Roum. Phys. **33**, 457 (1988)
[88B8] K. H. J. Buschow, in *Ferromagnetic Materials*, vol. 4, North Holland (1988), pp. 3–129
[88B9] K.H.J. Buschow, D.B de Mooij, R. Coehoorn, J. less Common Met. **145**, 601 (1988)
[88B10] D. Butler, R. D. Greenough, K. C. Pitman, J. Physique, **49**, C8–1341 (1988)
[88C1] J. M. Cadogan, J. P. Gavigan, D. Givord, H. S. Li, J. Phys. F: Metal Phys. **18**, 779 (1988)
[88C2] S. K. Chen, J. G. Duh, H. C. Ku, J. Appl. Phys. **63**, 2739 (1988)
[88C3] B. P. Chen, S. C. Fu, Y. C. Yang, W. J. James, J. Appl. Phys. **64**, 5543 (1988)
[88C4] S. K. Chen, J. G. Duh, C. H. Ku, J. Appl. Phys. **64**, 5546 (1988)
[88C5] W. Y. Ching, Z. Q. Gu, J. Appl. Phys. **63**, 3716 (1988)
[88C6] Y. C. Chuang, C. H. Wu, J. Q. Xie, J. less Common Met. **144**, 235 (1988)
[88D1] S. Dai, A. H. Morrish, X. Z. Zhou, B. P. Hu, S. G. Zhang, J. Appl. Phys. **63**, 3722 (1988)
[88E1] K. Erdmann, M. Rosenberg, K. H. J. Buschow, J. Phys. **49**, C8–583 (1988)
[88F1] D. Fruchart, L. Pontonnier, P. Vaillant, J. Bartolomé, J. M. Fernandez, J. E. Puertolas, C. Rillo, J. R. Regnard, A. Yaouanc, R. Fruchart, P. L. Héritier, IEEE Trans Magn. **24**, 1641 (1988)
[88F2] C. D. Fuerst, G. P. Meisner, F. E. Pinkerton, W. B. Yelon, J. Appl. Phys. **63**, 3119 (1988)
[88G1] J. P. Gavigan, S. H. Li, J. M. D. Coey, T. Viadieu, L. Pareti, O. Mozé, F. Bolzoni, J. Physique, **49**, C8–577 (1988)
[88G2] J. P. Gavigan, D. Givord, H. S. Li, J. Voiron, Physica B+C**149**, 345 (1988)
[88G3] D. Givord, H. S. Li, J. M. Cadogan, J. M. D. Coey, J. P. Gavigan, O. Yamada, H. Maruyama, M. Sagawa, S. Hirosawa, J. Appl. Phys. **63**, 3713 (1988)
[88G4] F. Grandjean, G. J. Long, D. E. Tharp, O. A. Pringle, W. J. James, J. Phys. **49**, C8–581 (1988)
[88G5] Y. Gros, F. Hartmann-Boutron, C. Meyer, M. A. Fremy, P. Tenaud, P. Auric, J. Phys. **49**, C8–847 (1988)
[88G6] R. Grössinger, H. R. Kirchmayr, K. H. J. Buschow, J. less Common Met. **136**, 367 (1988)
[88G7] R. Grössinger, X. C. Chou, R. Krewenka, G. Wiesinger, R. Eibler, X. K. Sun, Y. C. Chuang, J. Phys. **49**, C8–599 (1988)
[88G8] P. C. M. Gubbens, A. K. Van der Kraan, K. H. J. Buschow, J. Phys. **49**, C8–591 (1988)
[88H1] A. Hauet, D. Lemarchand, J. Teillet, Solid State Commun. **66**, 1191 (1988)
[88H2] A. Hauet, D. Lemarchand, D. Labulle, J. Teillet, P. Vigien, J. Phys. **49**, C8–607 (1988)
[88H3] M. J. Hawton, W. D. Corner, J. Magn. Magn. Mater. **72**, 52 (1988)
[88H4] J. Hu, Y. Wang, X. Li, L. Yin, M. Fen, D. Dai, T. Wang, J. G. Zhao, Z. Wang, J. Phys. **49**, C8–601 (1988)
[88I1] M. R. Ibarra, C. Marquina, P. A. Algarabel, J. I. Arnaudas, A. del Moral, J. Appl. Phys. **64**, 5537 (1988)
[88I2] K. Ichinose, K. Fujiwara, M. Oyasato, H. Nagai, A. Tsujimura, J. Phys. **49**, C8–595 (1988)
[88J1] S. S. Jaswal, M. A. Langell, J. Appl. Phys. **64**, 5577 (1988)
[88J2] E. Jedryka, M. Wojczik, P. Panissad, M. Rosenberg, S. Hieosawa, M. Sagawa, J. Phys. **49**, C8–587 (1988)
[88J3] M. Jurczyk, IEEE Trans. Magn. **24**, 1942 (1988)
[88K1] A. Kowalczyk, A. Wrzeciono, J. Magn. Magn. Mater. **74**, 260 (1988)
[88K2] A. Kowalczyk, A. Wrzeciono, J. Phys. **49**, C6–605 (1988)
[88K3] A. Kowalczyk, A. Wrzeciono, Phys. Stat. Solidi (a) **110**, 241 (1988)
[88L1] H. S. Li, J. P. Gavigan, J. M. Cadogan, D. Givord, J. Magn. Magn. Mater. **72**, L241 (1988)
[88L2] H. S. Li, Z. W. Zhang, Z. M. Dang, J. Magn. Magn. Mater. **71**, 355 (1988)
[88L3] C. Lin, Z. X. Liu, Y. X. Sun, Z. H. Lou, J. Appl. Phys. **63**, 3967 (1988)

[88L4] M. Loewenhaupt, M. Prager, A. P. Murani, H. E. Hoenig, J. Magn. Magn. Mater. **76–77**, 408 (1988)
[88M1] H. Maruyama, H. Yamazaki, D. Givord, J. Gavigan, H. S. Li, M. Sagawa, S. Hirosawa, J. Phys. **49**, C8–563 (1988)
[88M2] O. Mozé, L. Pareti, M. Solzi, F. Bolzoni, W. I. F. David, W. T. A. Harrison, A. W. Hewat, J. less Common Met. **136**, 375 (1988)
[88N1] K. Nagayama, H. Ino, Trans. Jpn. Inst. Metals **29**, 261 (1988)
[88P1] L. Pareti, J. Phys. **49**, C8–551 (1988)
[88P2] L. Pareti, O. Mozé, D. Fruchart, Ph. l'Hèritier, A. Yaouanc, J. less Common Met. **142**, 187 (1988)
[88P3] A. T. Pedziwiatr, Solid State Commun. **85**, 519 (1988)
[88P4] A. T. Pedziwiatr, S. G. Sankar, W. E. Wallace, J. Appl. Phys. **63**, 3710 (1988)
[88P5] F. E. Pinkerton, C. D. Fuerst, G. P. Meisner, J. App. Phys. **64**, 2783 (1988)
[88P6] F. E. Pinkerton, C. D. Fuerst, G. P. Meisner, Appl. Phys. Lett. **52**, 2188 (1988)
[88P7] F. Pourarian, G. Sankar, W. E. Wallace, J. Magn. Magn. Mater. **72**, 177 (1988)
[88P8] F. Pourarian, S. Y. Jiang, L. Y. Zhang, E. B. Boltich, G. Sankar, W. E. Wallace, J. Appl. Phys. **64**, 5540 (1988)
[88P9] O. A. Pringle, G. K. Marasinghe, G. J. Long, W. J. James, W. B. Yelon, D. Xie, F. Grandjean, J. App. Phys. **64**, 5580 (1988)
[88R1] R. J. Radwanski, J. J. M. Franse, J. Magn. Magn. Mater. **74**, 43 (1988)
[88R2] B. Rupp, A. Resnik, D. Shaltiel, J. Mater. Sci. **23**, 2133 (1988)
[88S1] D. J. Sellmeyer, M. A. Engelhardt, S. S. Jaswal, A. J. Arko, Phys. Rev. Lett. **60**, 2077 (1988)
[88S2] H. H. Smit, R. C. Thiel, K. H. J. Buschow, J. Phys. F.: Metal Phys. **18**, 295 (1988)
[88S3] J. Strzeszewski, A. Nazareth, C.C. Hadjipanayis, J. Appl. Phys. **63**, 3978 (1988)
[88T1] L.C. Tai, W.D. Zhong, Solid State Commun. **68**, 407 (1988)
[88T2] Y. Takano, Y. Satoh, K. Sekizawa, J. Phys. **49**, C8–561 (1988)
[88T3] N.P. Thuy, J.J.M. Franse, N.M. Hong, T.D. Hien, J. Phys. **49**, C8–499 (1988)
[88T4] N. P. Thuy, T. D. Hien, N. M. Hong, J. M. Franse, J. Phys. **49**, C8–579 (1988)
[88T5] J. Tian, J. Liang, D. Yang, H. Fan, F. Li, Phys. Stat. Solidi (a) **107**, 48 (1988)
[88V1] R. Verhoef, J. J. M.Franse, A. A. Minovsky, R. J. Radwanski, S. Q. Ji, F. M. Wang, J. P. Gavigan, J. Phys. **49**, C8–565 (1988)
[88V2] V. Villas-Boas, F. P. Missell, S. F. Da Cunha, J. App. Phys. **64**, 5549 (1988)
[88W1] C. M. Williams, N. C. Koon, B. M. Das, J. Phys. **49**, C8–567 (1988)
[88X1] J. Q. Xie, C. H. Wu, Y. C. Chuang, Z. D. Zhang, F. R. de Boer, J. Phys. **49**, C8–603 (1988)
[88X2] F. Xing, W. W. Ho, J. Magn. Magn. Mater. **74**, 271 (1988)
[88Y1] M. Yamada, H. Kato, H. Yamamoto, Y. Nakagawa, Phys. Rev. B **38**, 620 (1988)
[88Y2] F. M. Yang, R. W. Zhao, X. W. Li, Y. K. Huang, F. R. De Boer, R. Radwanski, IEEE Trans Magn. **24**, 1945 (1988)
[88Y3] F. M. Yang, X. W. Li, R. W. Zhao, Y. Zhu, T. S. Zhao, Y. K. Huang, F. R. De Boer, J. Phys. **49**, C8–571 (1988)
[88Y4] Y. C. Yang, F. Xing, L. S. Kong, J. L. Yang, Y. F. Ding, B. S. Zhang, C. T. Ye, L. Lin, H. M. Zhou, J. Phys. **49**, C8–597 (1988)
[88Y5] W. B. Yelon, Mater. Sci. Forum, 27–28, 371 (1988)
[88Z1] L. Y. Zhang, F. Pourarian, W. E. Wallace, J. Magn. Magn. Mater. **71**, 203 (1988)
[88Z2] Z. D. Zhang, X. K. Sun, Y. C. Chuang, F. R. de Boer, R. J. Radwanski, J. Phys. **49**, C8–569 (1988)
[88Z3] G. Zouganelis, A. Kostikas, A. Simopoulos, D. Niarchos, J. Magn. Magn. Mater. **75**, 91 (1988)
[89A1] P. A. Algarabel, M. R. Ibarra, C. Marquina, G.Marusi, O. Mozé, L. Pareti, M. Solzi, J. I. Arnaudas, A. Del Moral, Physica B **155**, 263 (1989)
[89B1] S. Brennan, A. Bienenstock, J. E. Keem, J. Appl. Phys. **65**, 697 (1989)
[89B2] E. Burzo, H. R. Kirchmayr, *Handbook on the Physical Chemistry of Rare-Earths*, Vol. 12, Chap. 82, North Holland (1989)

[89B3] K. H. J. Buschow, J. Magn. Magn. Mater. **80**, 1 (1989)
[89C1] R. Coehoorn, D.D. de Mooij, C. de Ward, J. Magn. Magn. Magn. **80**, 101 (1989)
[89C2] J. M. D. Coey, H. S. Li, J. P. Gavigan, J. M. Cadogan, B. P. Hu, in *Concerted European Action on Magnets*, ed. by I. V. Mitchel (1989), p. 76
[89C3] G. Czjek, H. J. Bornemann, H. S. Li, J. Magn. Magn. Mater. **80**, 23 (1989)
[89D1] A. Del Moral, M. R. Ibarra, C. Marquina, J. I. Arnaudas, P. A. Algarabel, Phys. Rev. B **40**, 7192 (1989)
[89E1] K. Erdmann, T. Sinnemann, M. Rosenberg, J. less Common Met. **146**, 59 (1989)
[89G1] Y. Gros, F. Mardmann-Boutron, C. Meyer, in *Concerted European Action on Magnets* (Springer, 1989), p. 288
[89G2] R. Grössinger, S. Heiss, G. Hilscher, H. R. Kirchmayr, J. Magn. Magn. Mater. **80** ,61 (1989)
[89G3] P. C. M. Gubbens, A. K. Van der Kraan, K. H. J. Buschow, Hyp. Int. **50**, 685 (1989)
[89H1] N. M. Hong, J. J. M. Franse, N. P. Thuy, J. less Common Met. **155**, 151 (1989)
[89H2] B.G. Hyde, S. Anderson, in *Inorganic Crystal Structures* (Wiley, New York, 1989)
[89I1] M. R. Ibarra, P. A. Algarabel, J. I. Arnaudas, A. del Moral, Solid State Commun. **69**, 131 (1989)
[89I2] M. R. Ibarra, P. A. Algarabel, C. Marquina, J. I. Arnaudas, A. del Moral, L. Pareti, O. Mozé, G. Marusi, M. Solzi, Phys. Rev. B **39**, 7081 (1989)
[89J1] M. Jurczyk, O. D. Chistyakov, J. Magn. Magn. Mater. **78**, 279 (1989)
[89J2] M. Jurczyk, O. D. Chistyakov, Phys. Stat. Solidi (a) **114**, K219 (1989)
[89J3] M. Jurczyk, G. S. Burkhanov, O. D. Chistyakov, Phys. Stat. Solidi (a) **112**, K121 (1989)
[89K1] R. Kamal, L. Lovleen, J. less Common Met. **147**, 79 (1989)
[89K2] X. C. Kou, X. K. Sun, Y. C. Chuang, T. S. Zhao, R. Grössinger, H. R. Kirchmayr, J. Magn. Magn. Mater. **82**, 327 (1989)
[89K3] X. C. Kou, X. K. Sun, Y. C. Chuang, R. Grössinger, R. Eibler, H. R. Kirchmayr, Phys. Stat. Solidi (a) **112**, 175 (1989)
[89K4] A. Kowalczyk, J. Magn. Magn. Mater. **82**, L1 (1989)
[89L2] L. Li, F.H. Li, D.Y. Yang, J.H. Tiang, Acta Phys. Sinica, **39**, 788 (1989)
[89L3] L.X. Liao, Z. Altounian, J. Appl. Phys. **66** 768 (1989)
[89L4] C. Lin, Z. X. Liu, Y. X. Sun, C. X. Bai, T. S. Zhao, Phys. Rev. B **39**, 7273 (1989)
[89L5] L. Lovleen, R. Kamal, J. Magn. Magn. Mater. **78**, L9 (1989)
[89L6] N. H. Luong, N. P. Thuy, L. T. Tai, T. D. Hin, Phys. Stat. Solidi (a) **111**, 591 (1989)
[89M1] C. Marquina, M. Solzi, P. A. Alagarabel, M. R. Ibarra, L. Pareti, Solid State Commun. **72**, 1167 (1989)
[89M2] C. Meyer, J. P. Gavigan, G. Czjzek, H. J. Bornemann, Solid State Commun. **89**, 83 (1989)
[89M3] M.Mittag, M. Rosenberg, K. H. J. Buschow, J. Magn. Magn. Mater. **82**, 109 (1989)
[89M4] O. Mozé, L. Pareti, G. Marcusi, M. Solzi, W. J. F. David, Physica B **156–157**, 71 (1989)
[89N1] S. Nadolski, E. Machowski, E. Burzo, Solid State Commun. **71**, 97 (1989)
[89N2] Y. Nakagawa, G. Kido, M. Yamada, H. Kato, H. Hiroyoshi, N. Saito, D. W. Lim, S. Hirosawa, M. Sagawa, Physica B **155**, 279 (1989)
[89P1] P. Panissod, E. Jedryka, M. Wojczik, J. I. Budnick, Phys. Rev. B **40**, 2606 (1989)
[89P2] O. A. Pringle, G. J. Long, G. K. Marasinghe, W. J. James, A. T. Pedziwiatr, W. E. Wallace, F. Grandjean, IEEE Trans. Magn. **23**, 3440 (1989)
[89Q1] C. Quan, Y. Wang, I. Yin, J. Zhao, Solid State Commun. **72**, 955 (1989
[89R1] R. J. Radwanski, J. J. M. Franse, J. Magn. Magn. Mater. **80**, 14 (1989)
[89R2] M. Rosenberg, P. Deppe, S. Sinnemann, Hyp. Int. **45**, 3 (1989)
[89R3] D. H. Ryan, Z. Altounian, Z. Strom-Olsen, W. B. Muir, Phys. Rev. B **39**, 4730 (1989)
[89S1] Y. Sano, H. Onodera, H. Yamauchi, H. Yamamoto, J. Magn. Magn. Mater. **79**, 67 (1989)
[89S2] G. Schneider, E. T. Henig, B, Grieb, G. Knochem, *Concentred European Action on Magnets*, Ed. I. V. Mitchell, Brussels (1989), p. 335
[89S3] M. Shiga, Y. Kusakabe, Y. Nakamura, K. Makita, M. Sagawa, Physica B **161**, 206 (1989)
[89S4] X. K. Sun, X. C. Kou, Y. C. Chuang, R. Grössinger, H. R. Kirchmayr, IEEE Trans. Magn. **25**, 3298 (1989)

[89S5] X. K. Sun, X. C. Kou, Y. C. Chuang, R. W. Zhao, J. Y. Li, F. M. Yang, R. Grössinger, H. R. Kirchmayr, Phys. Stat. Solidi (a) **115**, K219 (1989)
[89W1] G. Wiesinger, R. Grössinger, R. Krewenka, X. C. Chou, X. K. Sun, Y. C. Chuang, Hyp. Interact. **50**, 693 (1989)
[89W2] W. Wojczik, E. Jedryka, P. Panissod, S. Hirosawa, M. Sagawa, J. Magn. Magn. Mater. **80**, 19 (1989)
[89Y1] W. B. Yelon, D. Xie, C. M. Hsueh, C. D. Fuerst, G. P. Meisner, Phys. Rev. B **39**, 9389 (1989)
[89Z1] Y. D. Zhang, Y. K. Huang, X. K. Sun, Y. C. Chuang, F. M. Yang, F. R. de Boer, R. J. Radwanski, Physica B **115**, 269 (1989)
[89Z2] Z. D. Zhang, Y. K. Huang, X. K. Sun, Y. C. Chuang, F. M. Yang, F. R. De Boer, R. J. Radwanski, J. less Common Met. **152**, 67 (1989)
[89Z3] Z. Zhao, R. Ma, B. Miao, Scripta Metall. **23**, 743 (1989)
[89Z4] Z.B. Zhao, B.K. Xia, R.Z. Ma, S.M. Pan, J. less Common Met. **155**, 75 (1989)
[89Z5] Z.B. Zhao, S.K. Xia, C.R. Wang, R.Z. Ma, J. Phys.: Condens. Matter, **1**, 7513 (1989)
[89Z6] Z. Zhao, S. Xia, R. Ma, J. Ping, B. Miao, S. Pan, Solid State Commun. **69**, 1011 (1989)
[89Z7] T. Zhao, H. Jin, X. Zhu, J. Magn. Magn. Mater. **79**, 159 (1989)
[89Z8] X. F. Zhong, W. Y. Ching, Phys. Rev. B **39**, 12018 (1989)
[89Z9] Y. Zhu, T. Zhao, H. Jin, F. Yang, J. Xie, X. Li, R. Zhao, F. R. de Boer, IEEE Trans. Magn. **25**, 3443 (1989))
[90A1] P. A. Algarabel, M. R. Ibarra, C. Marquina, A. del Moral, J. Magn. Magn. Mater. **84**, 109 (1990)
[90B1] E. Burzo, A. Chelkowski, H.R. Kirchmayr, *Landolt Börnstein Handbook*, vol. 19d2 (Springer Verlag, 1990)
[90B2] E. Burzo, N. Plugaru, J. Magn. Magn. Mater. **86**, 97 (1990)
[90C1] J. M. Cadogan, J. less Common Met. **163**, 219 (1990)
[90C2] J. M. Cadogan, J. less Common Met. **158**, L45 (1990)
[90C3] R. Coehoorn, K. H. J. Buschow, M. W. Dirken, R. C. Thiel, Phys. Rev. B **42**, 4645 (1990)
[90D1] A. Del Moral, M. R. Ibarra, C. Marquina, J. I. Arnaudas, P. A. Algarabel, J. Magn. Magn. Mater. **83**, 283 (1990)
[90D2] M. C. D. Deruelle, M. Yamada, H. Yamauchi, Y. Nakagawa, Phys. Rev. B **42**, 10291 (1990)
[90D3] Z. Diao, X. Chen, X. Shen, Y. Wang, J. Appl. Phys. **67**, 4607 (1990)
[90E1] H. Ebert, H. Winter, D. D. Johnson, F. Pinski, J. Phys.: Condens. Matter **2**, 443 (1990)
[90G1] K. Girgis, M. Kraft, U. Weis, P. Fisher, M. Sostarich, J. less Common Met. **162**, 335 (1990)
[90G2] F. Grandjean, G. J. Long, O. A. Pringle, J. Fu, Hyp. Int. **62**, 131 (1990)
[90G3] R. Grössinger, R. Krewenka, X. C. Kou, K. H. J. Buschow, J. Magn. Magn. Mater. **83**, 130 (1990)
[90G4] P. C. M. Gubbens, A. K. Van der Kraan, K. H. J. Buschow, Hyp. Int. **53**, 37 (1990)
[90I1] M. R. Ibarra, A del Moral, J. Magn. Magn. Mater. **83**, 121 (1990)
[90I2] T. Iida, H. Ueki, T. Fukuzawa, T. Saito, K. Shinagawa, T. Tsushima, in *Proceedings of 6th International Symposium on Magnetic Anisotropy and Coercivity in Rare-Earth Transition Metal Alloys, Pittsburgh* (1990), p. 342
[90J1] S. S. Jaswal, Phys. Rev. B **41**, 9697 (1990)
[90J2] E. Jedryka, M. Wojczik, P. Panissod, K. H. J. Buschow, J. Appl. Phys. **67**, 4586 (1990)
[90J3] M. Jurczyk, O. D. Chistyakov, Phys. Stat. Solidi (a) **117**, 299 (1990)
[90K1] Cz. Kapusta, H. Figiel, Z. Kakol, J. Magn. Magn. Mater. **83**, 151 (1990)
[90K2] X. C. Kou, X. K. Sun, Y. C. Chuang, R. Grössinger, H. R. Kirchmayr, J. less Common Met. **160**, 109 (1990)
[90K3] X. C. Kou, X. K. Sun, Y. C. Chuang, T. S. Zhao, R. Grössinger, H. R. Kirchmayr, J. Magn. Magn. Mater. **80**, 31 (1990)
[90K4] X. C. Kou, X. K. Sun, Y. C. Chuang, T. S. Zhao, R. Grössinger, H. R. Kirchmayr, Solid State Commun. **73**, 87 (1990)

[90K5] X. C. Kou, G. Wiesinger, R. Grössinger, in *Proceedings of 6th- International Symposium on Magnetic Anisotropy and Coercivity in Rare Earth-Transition Metal Alloys, Pittsburgh* (1990), p. 262

[90K6] A. Kovalczyk, V. Ivanov, J. Magn. Magn. Mater. **87**, 330 (1990)

[90K7] A. Kovalczyk, P. Stefanski, A. Szlaferek, A. Wrzeciono, J. Magn. Magn. Mater. **83**, 147 (1990)

[90K8] A. Kovalczyk, P. Stefanski, A. Wrzeciono, J. Magn. Magn. Mater. **84**, L5 (1990)

[90L1] F. J. Lazaro, J. Bartolomé, R. Navarro, C. Rillo, F. Lera, L. M. Garcia, J. Chaboy, C. Pique, R. Burriel, D. Fruchart, S. Miraglia, J. Magn. Magn. Mater. **83**, 289 (1990)

[90L2] F. Leccabue, F. Bolzoni, G. Bocelli, G. Calestani, A. Deriu, S. Q. Li, J. G. Zhao, J. L. Sanchez Li, J. Magn. Magn. Mater. **83**, 174 (1990)

[90L3] H. S. Li, B. T. Xie, M. Z. Dang, J. Magn. Magn. Mater. **89**, 355 (1990)

[90L4] Z. W. Li, X. Z. Zhou, A. H. Morrish, Phys. Rev. B **41**, 8617 (1990)

[90L5] D. W. Lim, H. Kato, M. Yamada, G. Kido, Y. Nakagawa, S. Hirosawa, J. Magn. Magn. Mater. **90–91**, 72 (1990)

[90L6] M. Lowenhaupt, I. Sosnowska, B. Frick, Phys. Rev. B **42**, 3866 (1990)

[90M1] S. K. Malik, D. T. Adroja, B. M. Ma, E. B. Boltich, J. G. Sohn, S. G. Sankar, W. E. Wallace, J. Appl. Phys. **67**, 4589 (1990)

[90M2] G. Marusi, N. V. Mushnikov, L. Pareti, M. Solzi, A. E. Ermakov, J. Phys. : Condens. Matter **2**, 7317 (1990)

[90M3] G. Marusi, L. Pareti, M. Solzi, O. Mozé, A. Paoluzi, M. R. Ibarra, J. I. Arnaudas, C. Marquina, P. A. Algarabel, A.del Moral, J. Magn. Magn. Mater. **83**, 133 (1990)

[90M4] F. Maruyama, K. Ichinose, K. Fujiwara, M. Misawa, M. Oyasato, H. Nagai, A. Tsujimura, J. Magn. Magn. Mater. **90–91**, 552 (1990)

[90M5] S.I. Mykhalenko, Yu.B. Kuzma, T.D. Chuchman, Izv. Akad. Nauk. USSR, Neorg. Mater. **26**, 2295 (1990)

[90N1] Y. Nakagawa, H. Kato, G. Kido, M. Yamada, in *Proceedings of 5th International Symposium on Magnetic Anisotropy and Coercivity in Rare-Earth-Transition Metal Alloys, Pittsburg* (1990), p. 12

[90N2] I. Nowik, K. Muraleedharan, G. Wortmann, B. Perscheid,G. Kaindl, Solid State Commun. **76**, 967 (1990)

[90O1] J.L. Oddou, C. Jeandey, H.R. Rechenburg, J. Magn. Magn. Mater. **87**, 135 (1990)

[90P1] O. A. Pringle, J. Fu, G. J. Long, W. J. James, D. Xie, W. B. Yelon, F. Grandjean, J. Appl. Phys. **67**, 4762 (1990)

[90Q1] C. Quan, J. G. Zhao, Y. Z. Wang, L. Yin, B. G. Shen, Z. X. Cheng, S. W. Niu, Phys. Rev. B **42**, 6697 (1990)

[90R1] R. J. Radwasski, J. M. F. Franse, R. Verhoef, J. Magn. Magn. Mater. **83**, 127 (1990)

[90R2] R. J. Radwanski, R. Verhoef, J. J. M. Franse, J. Magn. Magn. Mater. **83**, 141 (1990)

[90R3] D. H. Ryan, Z. Altounian, L. X. Liar, J. O. Strom-Olsen, W. B. Muir, J. Appl. Phys. **67**, 4742 (1990)

[90S1] B.G. Shen, S. Methfessel, Phys. Stat. Solidi (a) **121**, K211 (1990)

[90U1] M. Uma Devi, K. P. Gupta, J. less Common Met. **159**, 13 (1990)

[90V1] R. Verhoef, P. H. Quang, J. J. M. Franse, R. J. Radwanski, J. Magn. Magn. Mater. **83**, 139 (1990)

[90V2] R. Verhoef, R. J. Radwanski, J. J. M. Franse,, J. Magn. Magn. Mater. **89**, 176 (1990)

[90V3] R. Verhoef, P. H. Quang, J. J. M. Franse, R. J. Radwanski, J. Appl. Phys. **67**, 4771 (1990)

[90W1] P. Wolfers, S. Miraglia, D. Fruchart, S. Hirosawa, M. Sagawa, J. Bartolomè, J. Pannetier, J. less Common. Met. **162**, 237 (1990)

[90W2] C. H. Wu, X. P. Zhong, Z. D. Zhang, Y. C. Chuang, F. R. de Boer, J. less Common Met. **159**, 91 (1990)

[90Y1] Y. C. Yang, X. D. Zhang, J. Appl. Phys. **67**, 4765 (1990)

[90Z1] L. V. Zavalij, N. S. Bilonizhko, Yu. B. Kuzma, Izv. Akad. Nauk.: Neorg. Mater. **26**, 2665 (1990)

[90Z2] Z. D. Zhang, X. K. Sun, Y. C. Chuang, F. R. de Boer, R. J. Radwanski, Phys. Stat. Solidi (a) **121**, 617 (1990)
[90Z3] Z. D. Zhang, X. K. Sun, Y. C. Chuang, F. R. de Boer, R. J. Radwanski, J. Magn. Magn. Mater. **89**, 304 (1990)
[90Z4] Z. D. Zhang, X. K. Sun, Z. G. Zhao, Y. C. Chuang, J. Magn. Magn. Mater. **92**, 191 (1990)
[90Z5] T. Zhao, Z. Fan, H. Jin, Phys. Stat. Solidi (b) **157**, 677 (1990)
[90Z6] X. F. Zhong, W. Y. Ching, J. Appl. Phys. **67**, 4768 (1990)
[91B1] J. J. Bara, B. F. Bogacz, A. T. Pedziwiatr, J. Magn. Magn. Mater. **99**, 204 (1991)
[91B2] J. Bartolome, F. Luis, O. Fruchart, O. Isnard, O. Miraglia, S. Obbade, K. H. J. Buschow, J. Magn. Magn. Mater. **101**, 411 (1991)
[91B3] J. Bartolome, Supermagnets, Ed. G. J. Long, F. Grandjean, (Kluwer Academic, 1991), p. 281
[91B4] E. Burzo, N. Plugaru, Phys. Stat. Solidi (a) **126**, K169 (1991)
[91B5] K. H. J. Buschow, Rep. Progr. Phys. **54**, 123 (1991)
[91B6] K.H.J. Buschow, *Supermagnets, Hard Magnetic Materials* (Kluwer Academia Publication, 1991), p. 553
[91C1] R. Coehoorn, J. Magn. Magn. Mater. **99**, 55 (1991)
[91C2] R. Coehoorn, K. H. J. Buschow, J. Appl. Phys. **69**, 5590 (1991)
[91C3] R. Coehoorn, C. J. M. Denissen, R. Eppenga, J. Appl. Phys. **69**, 6222 (1991)
[91D1] A Del Moral, J. L. Arnaudas, P. A. Algarabel, M. R. Ibarra, J. Schweizer, J. Appl. Phys. **69**, 5545 (1991)
[91F1] M. Földeaki, L. Koszegi, R. A. Dunlap, J. Appl. Phys. **69**, 5562 (1991)
[91G1] I. B. Gubich, N. F. Chaban, Neorg.Mater. **27**, 1977 (1991)
[91H1] J. F. Herbst, Rev. Mod. Phys. **63**, 819 (1991)
[91H2] J. F. Herbst, J. J. Croat, J. Magn. Magn. Mater. **100**, 57 (1991)
[91H3] J. Hu, Z. Dong, Z. Liu, Y. Liu, M. Feng, Z. Wang, Solid State Commun. **78**, 953 (1991)
[91I1] T. Iida, T. Fukuzawa, T. Saito, K. Shinagawa, T. Tushima, J. Magn. Soc. Jpn. **15**, 205 (1991)
[91K1] J.K. Kim, H. Saitoh, K. Hirano, J. Non Cryst. Solids **128**, 162 (1991)
[91K2] J.K. Kim, H. Saitoh, K. Tsuda, F. Satou, K.I. Hirano, Mater. Trans. JIM, **32**, 526 (1991)
[91K3] T. Kohashi, F. R. de Boer, R. J. Radwanski, X. P. Zhong, A. Yamagishi, M. Date, J. Phys. Soc. Jpn. **60**, 3543 (1991)
[91K4] A. C. Kou, R. Grössinger, H. R. Kirchmayer, J. Appl. Phys. **70**, 6372 (1991)
[91K5] A. C. Kou, R. Grössinger, H. R. Kirchmayer, J. Magn. Magn. Mater. **101**, 349 (1991)
[91K6] X. C. Kou, R. Grössinger, J. Magn. Magn. Mater. **95**, 184 (1991)
[91K7] A. Kowalczyk, P. Stefanski, A. Wrzeciono, J. Magn. Magn. Mater. **101**, 341 (1991)
[91K8] A. Kowalczyk, A. Szajek, J. Magn. Magn. Mater. **97**, 187 (1991)
[91K9] D. Köhler, H. Kronmüller, J. Magn. Magn. Mater. **92**, 344 (1991)
[91L1] D. W. Lim, H. Kato, M. Yamada, G. Kido, Y. Nakagawa, Phys. Rev. B **44**, 10014 (1991)
[91L2] M. Loewenhaupt, I. Sosnowska, J. Appl. Phys. **70**, 5967 (1991)
[91L3] M. Loewenhaupt, I. Sosnowska, A. Taylor, R. Osborn, J. Appl. Phys. **69**, 5593 (1991)
[91M1] G. K. Marasinghe, O. A. Pringle, G. J. Long, W. J. James, W. B. Yelon, F. Grandjean, J. Appl, Phys. **70**, 6149 (1991)
[91N1] L. Nordström, B. Johanson, M. S. S. Brooks, J. Appl. Phys. **69**, 5708 (1991)
[91O1] S. Obbade, S. Miraglia, P. Wolfers, J. L. Soubeyroux, D. Fruchart, F. Lera, C. Rillo, B. Malaman, G. le Caer, J. less Common Met. **171**, 71 (1991)
[91O2] R. T. Obermeyer, F. Pourariamn, J. Appl. Phys. **69**, 5559 (1991)
[91P1] N. Plugaru, E. Burzo, Mater. Sci. Forum **62–64**, 607 (1990)
[91R1] R. J. Radwanski, F. R. de Boer, X. P. Zheong, F. M. Yang, J. Y. Li, T. Kohashi, M. Ono, M. Date, A. Yamagishi, J. Magn. Magn. Mater. **101**, 392 (1991)
[91S1] J. B. Sousa, M. M. Amado, R. P. Pinto, M. A. Salgueiro, M. E. Braga, K. H. J. Buschow, J. Phys.: Condens. Matter **3**, 4119 (1991)
[91V1] P. Vedrine, P. Tixador, Y. Brunet, J. C. Bodo, A. Fevrier, A. Leriche, Cryogenics **31**, 51 (1991)

[91Z1] Z. D. Zhang, X. K. Sun, F. R. de Boer, Y. C. Chuang, R. J. Radwanski, Phys. Rev. B **43**, 555 (1991)

[91Z2] Z. D. Zhang, X. K. Sun, Z. C. Zhong, Y. C. Chuang, F. R. de Boer, J. Magn. Magn. Mater. **96**, 215 (1991)

[91Z3] Z. W. Zhang, Z. D. Zhang, X. K. Sun, Y. C. Chuang, F. R. de Boer, J. Magn. Magn. Mater. **96**, 206 (1991)

[91Z4] X. F. Zhong, W. Y. Ching, W. Lai, J. Appl. Phys. **70**, 6146 (1991)

[91Z5] X. P. Zhong, Q. Wang, C. H. Win, F. M. Yang, N. Tang, K. H. J. Buschow, F. R. De Boer, J. less. Comon Met. **171**, 213 (1991)

[92A1] P. A. Algarabel, A. del Moral, M. R. Ibarra, C. Marquina, J. Magn. Magn. Mater. **114**, 161 (1992)

[92B1] J. J. Bara, B. F. Bogacz, A. Kowalcyk, A. Wrzeciono, J. Magn. Magn. Mater. **111**, 39 (1992)

[92B2] E. Burzo, H. R. Kirchmayr, *Landold Börnstein Handbook*, vol III/19i2 (Springer Verlag, 1992)

[92B3] E. Burzo, R. Lemaire, Solid State Commun. **84**, 1145 (1992)

[92C1] J. Chaboy, J. Garcia, A. Marcelli, O. Isnard, S. Miraglia, D. Fruchart, J. Magn. Magn. Mater. **104–107**, 1171 (1992)

[92C2] T. W. Capehart, R. K. Mishra, J. F. Herbst, J.Appl. Phys. **72**, 678 (1992)

[92D1] A. Del Moral, P. A. Algarabel, M. R .Ibarra, J. I. Arnaudas, J. Schweizer, C. Marquina, J. Magn. Magn. Mater. **104–107**, 1051 (1992)

[92F1] D. Fruchart, S. Miraglia, S. Obbade, R. Verhoef, P. Wolfers, Physica B **180–181**, 578 (1992)

[92G1] S. H. Ge, Y. D. Zhang, F. S. Li, J. I. Budnick, P. Panissod, J. Magn. Magn. Mater. **116**, 211 (1992)

[92G2] B. Grieb, E.Th. Henig, G. Schneider, G. Knoch, G. Petzow, D. de Mooij, Powder Metallurgy **35**, 221 (1992)

[92H1] M. Hayashi, T. Myojin, Y. Kasamatsu, Y. Imaeda, T. Ushida, A. Tsujimura, T. Hihara, J. Magn. Magn. Mater. **104–107**, 1225 (1992)

[92H2] K. Hummler, M. Fahnle, Phys. Rev. B **45**, 3161 (1992)

[92H3] K. Hummler, T. Beuerle, M. Fahnle, J. Magn. Magn. Mater. **115**, 207 (1992)

[92I1] M. R. Ibarra, Z. Arnold, P. A. Algarabel, L. Morellon, J. Kamarad, J. Phys.: Condens. Matter **4**, 9721 (1992)

[92I2] M. R. Ibarra, L. Morellon, P. A. Algarabel, V. Marquina, Z. Arnold, J. Kamarad, J. Magn. Magn. Mater, **104–107**, 1371 (1992)

[92J1] E. Jedryka, M. Vojczik, P. Panissod, J. Magn. Magn. Mater. **104–107**, 1405 (1992)

[92K1] H. Kato, D. W. Lim, M. Yamada, G. Kido, Y. Nakagawa, H. Aruga Katori, T. Goto, in *Proceedings of 7th International Symposium on Magnetic Anisotropy and Coercivity in Rare-Earth, Transition Metal Alloys, Canberra* (1992), p. 425

[92K2] K. G. Knoch, B. Reinsch, G. Petzow, in *Proceedings of 13th Int Workshop on Rare-Earth Magnets and their Applications* (1992), p. 503

[92K3] X. C. Kou, T. S. Zhao, R. Grössinger, H. R. Kirchmayr, X. Li, F. R. De Boer, Phys. Rev. B **46**, 11204(R) (1992)

[92L1] H. S. Li, J. M. Cadogan, J. Phys. : Condens. Matter **4**, 6629 (1992)

[92L2] H. S. Li, J. M. Cadogan, J. Magn. Magn. Mater. **116**, 361 (1992)

[92L3] M. Liebs, K. Hummler, M. Fahnle, Phys. Rev. B **46**, 11201 (1992)

[92L4] D. W. Lim, H. Kato, M. Yamada, G. Kido, Y. Nakagawa, J. Magn. Magn. Mater. **104–107**, 1429 (1992)

[92L5] G. J. Long, R. Kulasekere, O. A. Pringle, J. Magn. Magn. Mater. **125**, L29 (1992)

[92M1] H. M. Mayer, M. Steiner, N. Stüsser, H. Weinfurter, B. Dorner, P. A. Lindgard, K. N. Clausen, S. Hock, R. Verhoef, J. Magn. Magn. Mater. **104–107**, 1295 (1992)

[92M2] F. Maruyama, K. Ichinose, M. Misawa, A.Tsujimura, H. Nagai, K. Adachi, J. Magn. Magn. Mater. **104–107**, 1165 (1992)

[92P1] C. Piqué, R. Buriel, L. M. Garcia, F. J. Lazaro, J. Bartolomé, J. Magn. Magn. Mater. **104–107**, 1167 (1992)
[92R1] R. J. Radwański, X. P. Zhong, F. R. de Boer, F. M. Yang, J. Y. Li, T. Kohashi, M. Ono, M. Date, A. Yamagishi, J. Magn. Magn. Mater. **104–107**, 1139 (1992)
[92R2] K. Ried, D. Köhler, H. Kronmüller, J. Magn. Magn. Mater. **116**, 259 (1992)
[92R3] M. Rosenberg, T. Sinnemann, M. Mittag, K.H.J. Buschow, J. Alloys Comp. **182**, 145 (1992)
[92S1] B.G. Shen, L.Y. Yang, L. Cao, H.Y. Yang, J.G. Zhao, F.M. Yang, J. Magn. Magn. Mater. **104–107**, 1281 (1992)
[92V1] R.Verhoef, Q. Wang, J. J. M. Franse, Physica B **177**, 21 (1992)
[92W1] G. Wiesinger, P. Grössinger, X. C. Kou, J. Magn. Magn. Mater. **104–107**, 1431 (1992)
[92Y1] V.A. Yartys, A.I. Shtogrin, M.I. Bartashevich, Koord. Khim. **18**, 436 (1992)
[92Z1] X. D. Zhang, Q. Pan, S.L. Ge, Y. C. Yang, J. L. Yang, Y. F. Ding, B. S. Zhang, C. T. Ye, L. Jin, Solid State Commun. **83**, 231 (1992)
[92Z2] Z.B. Zhao, S.Y. Hong, J. Phys.: Condens. Matter **4**, 9221 (1992)
[93A1] Z. Arnold, J. Kamarad, M. R. Ibarra, in *Proceedings of 8th International Symposium on Magnetic Anisotropy and Coercivity in Rare Earth Transition Metal Alloy* (1993), p. 279
[93C1] T. Capehart, R. Mishra, F. Pinkerton, J. Appl. Phys. **73**, 6476 (1993)
[93C2] T. Capehart, R. Mishra, G. P. Meisner, C. D. Fuerst, J. F. Herbst, Appl. Phys. Lett. **63**, 3642 (1993)
[93F1] M. Fähnle, K. Hümmler, T. Beuerle, J. Magn. Magn. Mater. **127**, L278 (1993)
[93G1] D. Givord, H. S. Li, R. Perrier de la Bathie, Solid State Commun. **88**, 907 (1993)
[93G2] C. Gou, Z.X. Cheng, D.F. Chen, S.W. Niu, Q.W. Yan, P.L. Zhang, B.G. Shen, L.Y. Yang, J. Magn. Magn. Mater. **128**, 26 (1993)
[93G3] I.B. Gubich, P.Y. Zavalii, Y.B. Kuzma, Neorg. Mater. **29**, 289 (1993)
[93H1] J. F. Herbst, C. D. Fuerst, W. B. Yelon, J. Appl. Phys. **73**, 5884 (1993)
[93H2] D.M. Herlach, R.F. Cochrane, I. Egry, H.J. Fecht, A.L. Greer, Inter. Mater. Rev. **38**, 3273 (1993)
[93H3] K. Hummler, M. Liebs, T. Beuerle, M. Fahnle, Int. J. Modern Phys. B **07**, 710 (1993)
[93K1] Cz. Kapusta, H. Figiel, G. Stoch, J. S. Lord, P. C. Riedi, IEEE Trans. Magn. **29**, 2893 (1993)
[93K2] A. Kowalczyk, P. Stefanski, V. Ivanov, J. Magn. Magn. Mater. **123**, 299 (1993)
[93L1] L. X. Liao, Z. Altounian, D. H. Ryan, Phys. Rev. B **47**, 11230 (1993)
[93L2] M. Liebs, M. Fahnle, J. Magn. Magn. Mater. **128**, L8 (1993)
[93L3] G. J. Long, F. Grandjean, O. A. Pringle, J. Magn. Magn. Mater. **125**, L29 (1993)
[93M1] B. Malaman, R. Gerardin, G. Le Caer, S. Obbade, S. Miraglia, P. Wolfers, D.Fruchart, Zeit. Phys. Chem. **179**, 421 (1993)
[93M2] G. K. Marasinghe, O. A. Pringle, G. J. Long, F. Grandjean, W. B. Yelon, IEEE Trans. Magn. **29**, 2764 (1993)
[93M3] G. K. Marasinghe, O. A. Pringle, G. J. Long, W. J. James, D. Xie, J. Li, W. B. Yelon, F. Grandjean, J. Appl. Phys. **74**, 6798 (1993)
[93M4] F. Maruyama, H. Nagai, K. Adachi, J. Phys. Soc. Jpn. **62**, 3741 (1993)
[93M5] B. I. Min, J. S. Kang, J. H. Hong, J. I. Jeong, Y. P. Lee, S. D. Choi, W. Y. Lee, C. J. Yang, C. G. Olson, Phys. Rev. B **48**, 6217 (1993)
[93N1] L. Nordström, B. Johanssen, M. S. S. Brooks, J. Phys.: Condens. Matter, **5**, 7859 (1993)
[93P1] V. V. Pananyuk, V. A. Yartys, Zeit. Phys. Chem. **179**, 471 (1993)
[93S1] K. Shimizu, J. Magn. Magn. Mater. **123**, L9 (1993)
[93Y1] A. Yamamoto, Acta Cryst. A **49**, 831 (1993)
[93Y2] Y. Yan, X. H. Zhao, H. M. Jin, J. Magn. Magn. Mater. **127**, L1 (1993)
[94A1] F. M. Ahmed, D. S. Edgley, O. Gutfleish, I. R. Harris, IEEE Trans. Magn. **30**, 616 (1994)
[94C1] W. C. Chang, C. Z. Wu, Mater. Chem. Phys. **39**, 40 (1994)
[94F1] C. D. Fuerst, J. F. Herbst, J. L. Sarrao, A. Migliori, J. Appl. Phys. **75**, 6625 (1994)
[94G1] B.X. Gu, B.G. Shen, H.R. Zai, J. Appl. Phys. **75**, 6955 (1994)
[94I1] O. Isnard, D. Fruchart, J. Alloys Comp. **205**, 1 (1994)

[94J1] S. T. Jiang, H. Li, J. Xu, R. W. Gao, J. Magn. Magn. Mater. **136**, 294 (1994)
[94K1] Y. Khan, E. Kneller, R.J. Wang, Phys. Stat. Solidi (a) **192**, 499 (1994)
[94M1] G. K. Marasinghe, O. A. Pringle, G. J. Long, W. B. Yelon, F. Grandjean, J. Appl. Phys. **76**, 2960 (1994)
[94M2] A. Micski, B. Uhrenius, J. Appl. Phys. **75**, 6265 (1994)
[94P1] R. P. Pinto, M. E. Broga, M. M. Amado, J. B. Sousa, K. H. J. Buschow, J. Appl. Phys. **76**, 6851 (1994)
[94S1] V. A. Sidorov, L. G. Khovostantsev, J. Magn. Magn. Mater. **129**, 256 (1994)
[94W1] L. Withanawasam, G.C. Hadjipanayis, R.F. Krause, J. Appl. Phys. **75**, 6646 (1994)
[94Z1] T. S. Zhao, J. I. Lee, J. Appl. Phys. **75**, 3008 (1994)
[95B1] J. Bartolomé, M. D. Kuzmin, L. M. Garcia, I. Plaza, D. Fruchart, K. H. J. Buschow, J. Magn. Magn. Mater. **140–144**, 1047 (1995)
[95C1] J. Chaboy, A. Marcelli, L. Bozukov, J. Phys. : Condens. Matter **7**, 8197 (1995)
[95C2] J. Chaboy, A. Marcelli, L. M. Garcia, J. Bartolomé, M. D. Kuzmin, N. Maruyama, K. Kobayashi, H. Kawata, T. Iwazumi, J. Magn. Magn. Mater. **140–144**, 1051 (1995)
[95C3] Z.H. Cheng, M.X. Mao, J.J. Sun, B.G. Shen, F.W. Wang, C.L. Yang, Y.D. Zhang, F.S. Li, J. Phys.: Condens. Matter **7**, 2303 (1995)
[95C4] Z.H. Cheng, B.G. Shen, M.X. Mao, J.J. Sun, F.W. Wang, F.S. Li, Y.D. Zhang, J. Phys.: Condens. Matter **7**, 3949 (1995)
[95F1] C. D. Fuerst, T. W. Capehart, F. E. Pinkerton, J. F. Herbst, J. Magn. Magn. Mater **139**, 359 (1995)
[95G1] J. S. Garitaonandia, J. M. Barandiaran, I. Orue, F. Plazaola, M. R. Ibarra, A. del Moral, J. Magn. Magn. Mater. **140–144**, 949 (1995)
[95H1] G.C. Hadjipanajis, I. Withanawasan, R.F. Krause, IEEE Trans. Magn. **31**, 3596 (1995)
[95H2] B. Hallemans, P. Wollants, J. Roos, J. Phase Equilib. **16**, 137 (1995)
[95H3] H. Horiuchi, U. Koike, H. Kaneko, T. Kurino, H. Uchida, J. Alloys Comp. **222**, 131 (1995)
[95I1] O. Isnard, W. B. Yelon, S. Miraglia, D. Fruchart, J. Appl. Phys. **78**, 1892 (1995)
[95K1] H. Kato, D. W. Lim, M. Yamada, Y. Nakagawa, H. Aruga Katori, T. Goto, Physica B **211**, 105 (1995)
[95K2] X. C. Kou, E. H. C. P. Sinnecker, R. Grössinger, J. Magn. Magn. Mater. **146**, L231 (1995)
[95K3] M. D. Kuzmin, L. M. Garcia, I. Plaza, J. Bartolomé, D. Fruchart, K. H. J. Buschow, J. Magn. Magn. Mater. **146**, 77 (1995)
[95L1] M. Loewenhaupt, P. Fabi, I. Sosnowska, B. Frick, R. Eccleston, J. Magn. Magn. Mater. **140–144**, 1053 (1995)
[95L2] L. Luis, P. Infante, J. Bartolomé, R. Burriel, C. Piqué, R. Ibarra, K. H. J. Buschow, J. Magn. Magn. Mater. **140–144**, 1045 (1995)
[95M1] M. Morariu, M. Rogalski, M. Valeanu, N. Plugaru, E. Burzo, J. Magn. Magn. Mater. **140–144**, 1001 (1995)
[95N1] A. C. Neiva, A. P. Tschiptschin, F. P. Missell, J. Alloys Comp. **217**, 273 (1995)
[95S1] K. Shimizu, K. Ichinose, Solid State Commun. **94**, 619 (1995)
[95S2] K. Shimizu, K. Ichinose, J. Magn. Magn. Mater. **140–144**, 953 (1995)
[95S3] K. Shimizu, K. Ichinose, Y. Fukuda, M. Shimotomai, Solid State Commun. **96**, 671 (1995)
[95S4] J. L. Soubeyroux, D. Fruchart, O. Isnard, S. Miraglia, E. Tomey, J. Alloys Comp. **219**, 16 (1995)
[95W1] L. Withanawasam, A.S. Murthy, G.C. Hadjipanayis, K.R. Lawless, R.F. Krause, J. Magn. Magn. Mater. **140–144**, 1057 (1995)
[96B1] J. J. Barra, B. F. Bogacz, A. T. Pedziwiatr, J. Alloys Comp. **232**, 101 (1996)
[96G1] A.M. Gabay, A.G. Popov, V.S. Gaviko, Y.V. Belozerov, A.S. Yermolenko, N.N. Shchegoleva, J. Alloys Comp. **237**, 101 (1996)
[96G2] A.M. Gabay, A.G. Popov, V.S. Gaviko, Ve. V. Belozerov, A.S. Yermolenko, J. Alloys Comp. **245**, 119 (1996)
[96G3] L. M. Garcia, J. Bartolomé, F. J. Lazaro, C de Francisco, J. M. Munoz, Phys. Rev. B **54**, 15238 (1996)
[96H1] N. M. Hong, Physica B **226**, 391 (1996)

[96H2] Z. Hu, W. B. Yelon, G. K. Marasinghe, W. J. James, IEEE Trans. Magn. **32**, 4386 (1996)
[96H3] K. Hummler, M. Fahnle, Phys. Rev. B **53**, 3290 (1996)
[96K1] Cz. Kapusta, J. Magn. Magn. Mater. **157–158**, 71 (1996)
[96K2] X. C. Kou, R. Grossinger, G. Hilscher, H. R. Kirchmayr, F. R. Boer, Phys. Rev. B **54**, 6421 (1996)
[96L1] L. H. Lewis, D. Q. Welch, F. Pourarian, J. Appl. Phys. **79**, 5513 (1996)
[96L2] H. S. Li, R. C. Mohanty, A. Raman, C. G. Greiner, J. Magn. Magn. Mater. **162**, 301 (1996)
[96L3] G. J. Long, F. Grandjean, J. Magn. Magn. Mater. **162**, 162 (1996)
[96M1] M. Manivel Raja, A. Narayanasamy, V. Ravichandran, J. Magn. Magn. Mater. **159**, 345 (1996)
[96O1] S. Obbade, P. Wolfers, D. Fruchart, R. Argoud, J. Müller, E. Palacios, J. Alloys Comp. **242**, 80 (1996)
[96P1] C. Piqué, R. Burriel, J. Bartolomé, J. Magn. Magn. Mater. **154**, 71 (1996)
[96R1] G. Ravach, J. Teillet, A. Fnidiki, J.M. LeBreton, I. Driouch, N. Hassanain, J. Magn. Magn. Mater. **157–158**, 173 (1996)
[96W1] L. Withanawanam, I. Panagiotopoulos, G. C. Hadjipanayis, IEEE Trans. Magn. **32**, 4422 (1996)
[96W2] P. Wolfers, S. Obbade, D. Fruchart, R. Verhoef, J. Alloys. Comp. **242**, 74 (1996)
[96Y1] Y. Yan, H. M. Jin, T. S. Zhao, J. Magn. Magn. Mater. **164**, 201 (1996)
[96Y2] V.A. Yartys, O. Gutfleisch, I.R. Harris, J. Magn. Magn. Mater. **157–158**, 119 (1996)
[97A1] M. M. Amado, R. P. Pinto, M. E. Braga, M. S. Rogalski, J. B. Sousa, J. Appl. Phys. **81**, 5784 (1997)
[97A2] S. Asano, M. Yamaguchi, Physica B **237–238**, 541 (1997)
[97B1] A. Bezingé, K. Yvon, H.F. Braun, J. Muller, H.U. Nissen, Phys. Rev. B **36**, 1406 (1987)
[97C1] T. W. Capehart, R. P. Mishra, C. D. Fuerst, G. P. Meisner, F. E. Pinkerton, J. F. Herbst, Phys. Rev. B **55**, 11496 (1997)
[97C2] W.C. Chang, S.H. Wu, B.M. Ma, C.O. Bounds, J. Appl. Phys. **81**, 4453 (1997)
[97C3] Z.H. Cheng, B.G. Shen, J.X. Zhang, M.X. Mao, J.J. Sun, C.L. Yang, F.S. Li, Y.D. Zhang, Chin. Phys. Lett. **14**, 387 (1997)
[97C4] O. Crişan, J.M. LeBreton, A. Jianu, J. Teillet, G. Filoti, J. Alloys Comp. **262–263**, 381 (1997)
[97H1] T. Hinomura, S. Nasu, H. Kanekiyo, M. Uehara, S. Hirosawa, Mater. Trans. **38**, 1106 (1997)
[97L1] H. Lenke, T. Goldenhenrich, C. Heihen, G. Thomas, IEEE Trans. Magn. **33**, 3865 (1997)
[97L2] H. S. Li, R. C. Mohanty, A. Raman, C. G. Grenier, J. M. Cadogan, IEEE Trans. Magn. **33**, 3868 (1997)
[97L3] Q.A. Li, C.H. de Groot, F.R. de Boer, K.H.J. Buschow, J. Alloys Comp. **250**, 82 (1997)
[97P1] I. G. Pastushenkov, A. Forki, H. Kronmuller, J.Magn. Magn. Mater. **174**, 278 (1997)
[97S1] J. Stankiewicz, J. Bartolome, Phys. Rev. B **55**, 3058 (1997)
[97S2] Y. C. Sui, Z. D. Zhang, Q. F. Xiao, W. Liu, X. G. Zhao, T. Zhao, Y. C. Chuang, J. Phys. Condens. Matter **9**, 9985 (1997)
[97Y1] V.A. Yartys, G. Wiesinger, I.R. Harris, J. Alloys Comp. **252**, 201 (1997)
[97Y2] V.A. Yartys, O. Gutfleisch, I.R. Harris, J. Alloys Comp. **253–254**, 134 (1997)
[97Z1] G. W. Zhang, Y. P. Feng, C. K. Ong, Solid State Commun. **103**, 465 (1997)
[98B1] J. J. Bara, B. F. Bogacz, A. T. Pedziwiatr, P. Stefanski, A. Szlaferek, A. Vrzeciono, Molec. Phys. Repts. **22**, 9 (1998)
[98B2] E. Burzo, Rep. Prog. Phys. **61**, 1099 (1998)
[98B3] K. H. J. Buschow, F. M. Mulder, R. C. Thiel, J. Alloys Comp. **275–277**, 498 (1998)
[98C1] J. Chaboy, L. M. Garcia, F. Bartolomé, H. Maruyama, A. Marcelli, L. Bozukov, Phys. Rev. B **57**, 13386 (1998)
[98C2] W.C. Chang, D.Y. Chiou, S.H. Wu, B.M. Ma, C.O. Bounds, Appl. Phys. Lett. **72**, 121 (1998)
[98H1] R. Hermann, V. Loeser, J. Appl. Phys. **83**, 6399 (1998)
[98K1] Y. B. Kim, H. M. Jin, J. Magn. Magn. Mater. **189**, 251 (1998)

[98P1] D.H. Ping, K. Hono, S. Hirosawa, J. Appl. Phys. **83**, 7769 (1998)
[98R1] G. Ravach, J. Teillet, J. Phys.: Condens. Matter **10**, 7065 (1998)
[98T1] A. Teresiak, M. Kubis, M. Mattern, M. Wolf, K.H.J. Müller, J. Alloys Comp. **274**, 284 (1998)
[98Y1] M. Yamaguchi, S. Asano, Physica B **254**, 73 (1998)
[98Y2] S.O. Yuryev, S.I. Yushchuk, Yu.B. Kuzma, I.E. Lopatynsky, J. Alloys Comp. **270**, 16 (1998)
[99A1] C. Abadia, A del Moral, K. Kulakowski, P. A. Algarabel, J. Phys.: Condens. Matter **11**, 3341 (1999)
[99A2] C. Abadia, A del Moral, K. Kulakowski, P. A. Algarabel, Phys. Rev. B **59**, 486 (999)
[99C1] C. Chacon, O. Isnard, S. Miraglia, J. Alloys Comp. **283**, 320 (1999)
[99G1] J. Gao, B. Wei, J. Alloys Comp. **285**, 229 (1999)
[99L1] S. K. Lee, B. N. Das, V. G. Harris, J. Magn. Magn. Mater. **207**, 137 (1999)
[99L2] L. H. Lewis, V. Panchanathan, J. App. Phys. **85**, 4883 (1999)
[99P1] W. S. Park, H. T. Kim, Y. B. Kim, M. J. Park, J. Korean Phys. Soc. **35**, 5467 (1999)
[99P2] C. Piquer, M. Artigas, J. Bartolomé, J. Magn. Magn. Mater. **196–197**, 757 (1999)
[99S1] J. Stankiewicz, J. Bartolomé, J. Magn. Magn. Mater. **196–197**, 712 (1999)
[99Y1] V.A. Yartys, R.V. Denys, O. Gutfleisch, I.I. Bulyk, Y.B. Kuzma, I.R. Harris, Int. J. Hydr. Energy **24**, 189 (1999)
[00B1] J. J. Bara, B. F. Bogacz, A. T. Pedziwiatr, J. Alloys Comp. **307**, 17 (2000)
[00B2] F. Bartolomé, J. M. Tonnerre, N. Jaouen, D. Roux, J. Chaboy, L. M. García, H. Maruyama, R. Steinmann, J. Appl. Phys. **87**, 4762 (2000)
[00C1] C. Chacon, O. Isnard, J. Appl. Phys. **88**, 2342 (2000)
[00C2] C. Chacon, O. Isnard, J. Appl. Phys. **88**, 3570 (2000)
[00G1] J. Gao, T. Volkmann, D.M. Herlach, J. Alloys Comp. **308**, 296 (2000)
[00G2] Y. D. Gao, S. Q. Zhang, B. C. Liu, J. Magn. Magn. Mater. **208**, 158 (2000)
[00G3] L. M. García, J. Chaboy, F. Bartolomé, J. B. Goedkoop, Phys. Rev. Lett. **85**, 429 (2000)
[00K1] Y. B. Kim, H. M. Jin, J. Magn. Magn. Mater. **222**, 39 (2000)
[00K2] M. J. Kim, Y. B. Kim, T. K. Kim, J. Magn. Magn. Mater. **222**, 86 (2000)
[00M1] J. Miguel-Soriano, J. Chaboy, L. M. García, F. Bartolomé, H. Maruyama, J. Appl. Phys. **87**, 5884 (2000)
[00P1] C. Piquer, J. Bartolomé, M. Artigas, D. Fruchart, Phys. Rev. B **62**, 1004 (2000)
[00R1] G. Ravach, F. Machizaud, J. Teillet, J.M. LeBreton, A. Fnidiki, J. Phys.: Condens. Matter **12**, 3639 (2000)
[00W1] Z.C. Wang, S.Z. Zhou, Y. Qiao, M.C. Zhang, R. Wang, J. Magn. Magn. Mater. **218**, 72 (2000)
[00W2] P. Wolfers, M. Bacmann, D. Fruchart, in *Magnetic and Superconductive Materials* (2000), p. 991
[00Y1] Y. Yan, H. Jin, Y. B. Kim, J. Magn. Magn. Mater. **221**, 382 (2000)
[00Y2] S. Yehia, S. H. Aly, J. Magn. Magn. Mater. **212**, 195 (2000)
[00Z1] Z.D. Zhang, W. Liu, J.P. Liu, D.J. Sellmyer, J. Phys. D: Appl. Phys. **33**, R217 (2000)
[01C1] B. A. Cook, J. L. Harringa, F. C. Laabs, K. W. Denis, A. M. Russel, R. W. McCallum, J. Magn. Magn. Mater. **233**, L136 (2001)
[01C2] B.Z. Cui, X.K. Sun, L.Y. Xiong, Z.D. Zhang, Z.Q. Yang, A.M. Wang, J.N. Deng, J. Mater. Res. **16**, 709 (2001)
[01D1] K. W. Dennis, F. C. Laabs, B. A. Cook, J. L. Harringa, R. W. McCallum, J. Magn. Magn. Mater. **231**, L33 (2001)
[01F1] K. Fukui, H. Ogasawara, A. Kotani, I. Harada, H. Maruyama, N. Kawamura, K. Kobayashi, J. Chaboy, A. Marcelli. Phys. Rev. B **64**, 104405 (2001)
[01G1] L. M. Garcia, J. B. Goedkoop, J. Chaboy, F. Bartolomé, Mater Sci. Forum **373**, 289 (2001)
[01G2] J. Gao, Th. Volkmann, D.M. Herlach, J. Mater. Res. **16**, 2562 (2001)
[01G3] J. Gao, Th. Volkmann, S. Roth, W. Löser, D.M. Herlach, J. Magn. Magn. Mater. **234**, 313 (2001)
[01J1] Y.C. Jung, Y. Ohmari, K. Nagai, S. Hirosawa, H. Kanekiyo, Mater. Trans. **42**, 2102 (2001)

[01K1] H. Kato, T. Ishizaki, T. Miyazaki, IEEE Trans. Magn. **37**, 2702 (2001)
[01K2] M. J. Kim, Y. B. Kim, C. S. Kim, T. K. Kim, J. Magn. Magn. Mater. **224**, 49 (2001)
[01P1] V. Papaefthymiou, A. Moukarika, L. Withanawasam, G.C. Hadjipanayis, IEEE Trans. Magn. **37**, 2618 (2001)
[01S1] J. Stankiewiecz, J. Bartolomé, S. Hirosawa, J. Phys.: Condens. Matter **13**, 303 (2001)
[01W1] P. Wolfers, M. Bacmann, D. Fruchart, J. Alloys Comp. **317–319**, 39 (2001)
[02C1] J. Chaboy, C. Piquer, Phys. Rev. B **66**, 104433 (2002)
[02C2] O. Crisan, J. M. Le Breton, M. Nogues, F. Machizaut, G. Filoti, J. Phys.: Condens. Matter **14**, 12599 (2002)
[02C3] B.Z. Cui, X.K. Sun, L.Y. Xiong, S.T. Car, X.X. Zhang, W. Liu, D.Y. Geng, Z.D. Zhang, J. Alloys Comp. **340**, 242 (2002)
[02G1] J. Gao, T. Volkmann, D.M. Herlach, Acta Mater. **50**, 3003 (2002)
[02G2] J. Gao, T. Volkmann, D.M. Herlach, IEEE Trans. Magn. **38**, 2910 (2002)
[02G3] R. Gargula, A. T. Pedziwiatr, B. F. Bogacz, S. Wrobel, J. Bartolome, J. Stankiewicz, Acta Phys. Polonica, **101**, 289 (2002)
[02I1] V. Yu. Irkhin, J. Phys.: Condens. Matter, **14**, 6865 (2002)
[02J1] Z.Q. Jin, H. Okumura, J.S. Munoz, Y. Zhang, H.L. Wang, G.C. Hadjipanayis, J. Phys. D: Appl. Phys. **35**, 2893 (2002)
[02K1] N. V. Kudrevatykh, S. V. Andreev, M. I. Bartashevich, A. N. Bogatkin, Fiz. Met. Metall. **93**, S45 (2002)
[02L1] J. H. Liu, P. Y. Zhang, J. W. Zhang, H. Li, X. Y. Zhang, J. Iron Steel Res. Internat. **9**, 48 (2002)
[02M1] M. Marinescu, H. Chiriac, J. Opt. Adv. Mater. **4**, 267 (2002)
[02O1] S. Ozawa, H. Sato, T. Motegi, J. Appl. Phys. **91**, 8831 (2002)
[02P1] A. T. Pedziwiatr, B. F. Bogacz, R. Gargula, S. Wrobel, J. Magn. Magn. Mater. **248**, 19 (2002)
[02P2] A. T. Pedziwiatr, B. F. Bogacz, R. Gargula, S. Wrobel, J. Alloys Comp. **336**, 5 (2002)
[02V1] T. Volkmann, J. Gao, D.M. Herlach, Appl. Phys. Letters, **80**, 1915 (2002)
[03C1] J. Chaboy, N. Plugaru, J. Bartolomé, G. Subias, Phys. Rev. B **67**, 014415 (2003)
[03C2] H.W. Chang, W.C. Chang, J.C. Ho, M. Unver, H.H. Hamdeh, J. Appl. Phys. **93**, 4027 (2003)
[03C3] H.W. Chang, C.H. Chiu, W.C. Chang, Appl. Phys. Lett. **82**, 4513 (2003)
[03C4] H.W. Chang, W.C. Chang, J.C. Ho, W.M. Hikal, H.H. Hamdeh, Physica B **327**, 292 (2003)
[03C5] Z. Chen, B.R. Smith, B.M. Ma, M.Q. Huang, Y.Q. Wu, M.J. Cramer, IEEE Trans. Magn. **39**, 2938 (2003)
[03D1] C. Djega-Mariadassou, L. Bessais, A. Nandra, E. Burzo, Phys. Rev. B **68**, 024406 (2003)
[03G1] J. Gao, T. Volkmann, D.M. Herlach, J. Alloys Comp. **350**, 344 (2003)
[03G2] I. Goscinska, T. Tolinski, A. Kowalczyk, A. Szlaferek, Phys. Stat Solidi (a) **196**, 78 (2003)
[03N1] S. A. Nikitin, S. Tereshina, Yu. A. Pankratov, T. Paleweski, H. Drulis, M. V. Makarova, Yu. G. Pastushenko, Phys. Stat. Solidi (a) **196**, 317 (2003)
[03P1] A. T. Pedziwiatr, B. F. Bogacz, R. Gargula, Nucleonica, **48**, S59 (2003)
[03R1] V. Raghavan, J. Phys. Equilib. **24**, 451 (2003)
[03Y1] N. Yang, K. W. Dennis, R. W. McCallum, M. J. Kramer, Y. Zhang, P. L. Lee, J. Appl. Phys. **93**, 7990 (2003)
[04A1] K. Asakura, K. Fukui, H. Ogasawara, I. Harada, A. Kotani, Physica B **345**, 205 (2004)
[04B1] J. Bartolomé, M.D. Kuzmin, Cz. Kapusta, P.C. Riedi, M. Ellouza, Ph. l'Héritier, Solid State Commun. **129**, 331 (2004)
[04C1] H.W. Chang, C.H. Chiu, W.C. Chang, IEEE Trans. Magn. **40**, 2871 (2004)
[04C2] C.H. Chiu, H.W. Chang, W.C. Chang, J. Magn. Magn. Mater. **282**, 186 (2004)
[04C3] C.H. Chiu, H.W. Chang, W.C. Chang, E. Yu, Phys. Stat. Solidi (c) **1**, 3389 (2004)
[04F1] D. Fruchart, S. Miraglia, P. de Rango, P. Wolfers, J. Alloys Comp. **383**, 17 (2004)
[04H1] V. Hardy, S. Majumdar, S. Crowe, M. R. Lees, D. M. Paul, L. Herve, A. Maignan, S. Herbert, C. Martin, C. Yaick, M. Hervieu, B. Raveau, Phys. Rev. B **69**, R020407 (2004)

[04H2] S. Hirosawa, H. Kanekiyo, Y. Shigemoto, T. Miyoshi, J. Jpn. Soc. Powder Metall. **51**, 143 (2004)

[04K1] H. Kato, M. Yamada, T. Miyazaki, J. Magn. Magn. Mater. **272**, 2051 (2004)

[04L1] S. D. Li, J. C. Tang, Z. R. Yuan, B. X. Gu, Y. W. Du, J. Rare Earths **22**, 799 (2004)

[04L2] H. Lind, PhD Thesis, Univ. Stockholm (2004)

[04N1] A. Nandra, E. Burzo, I. Bessais, C. Djega Mariadassou, V. Lalanne, Molec. Cryst. Liquid Cryst. **417**, 97 (2004)

[04O1] S. Ozawa, M. Li, S. Sugiyama, I. Jimbo, S. Hirosawa, K. Kuribayashi, Mat. Sci. Eng. A **382**, 295 (2004)

[04O2] S. Ozawa, M. Li, K. Kuribayashi, S. Sugiyama, H. Jimbo, S. Hirosawa, J. Appl. Phys. **95**, 8478 (2004)

[04T1] I. S. Tereshina, G. S. Burkhanov, O. D. Chistyakov, N. B. Kolchugina, S. A. Nikitin, H. Drulis, NATO Sci. Series II, **172**, 553 (2004)

[04V1] P. Villars et al, *Landolt Börnstein Handbook*, vol. III/43A1 (Springer Verlag, 2004)

[04V2] T. Volkmann, J. Strohmenger, J. Gao, D.M. Herlach, Appl. Phys. Lett. **85**, 2232 (2004)

[05G1] J. Gao, T. Volkmann, S. Reutzel, D.M. Herlach, J. Alloys Comp. **388**, 235 (2005)

[05H1] D. Haskel, J. C. Lang, Z. Islam, A. Cady, G. Srajer, M. van Veenendaal, P. C. Canfield, Phys. Rev. Lett. **95**, 217207 (2005)

[05K1] E. N. Kablov, A. F. Petrakov, V. N. Piskorski, R. A. Valeev, E. B. Chaban, Powder Metal. **47**, 227 (2005)

[05L1] S. D. Li, J. C. Tang, L. Y. Lu, W. Q. Zhou, Z. G. Huang, H. Lai, Y. W. Du, Mater. Sci. Eng. A **405**, 321 (2005)

[05O1] S. Ozawa, S. Sugiyama, M. Li, S. Hirosawa, K. Kuribayashi, Microgravity Sci. Technol. **16**, 89 (2005)

[05P1] A. T. Pedziwiatr, A. Wojciechowska, B. F. Bogacz, S. Wrobel, J. Phys.: Condens. Matter **17**, 6999 (2005)

[05P2] A. T. Pedzwiatr, B. F. Bogacz, A. Wojciechowska, S. Wrobel, J. Alloys Comp. **396**, 54 (2005)

[05S1] W. Sprengel, S. Herth, V. Barbe, H. E. Schaefer, T. Wejczanowski, O. Gutfeisch, R. Wurschum, J. Appl. Phys. **98**, 074314 (2005)

[05S2] J. Stankiewicz, J. Bartolome, J. Magn. Magn. Mater. **290–291**, 1172 (2005)

[05S3] S. Sugiyama, S. Ozawa, I. Jimbo, S. Hirosawa, K. Kuribayashi, J. Crypt. Growth **275**, e2019 (2005)

[05T1] I. S. Tereshina, A. V. Andreev, H. Drulis, E. A. Tereshina, J. Alloys Comp. **404–406**, 212 (2005)

[05V1] I. V. Veremchuk, N. F. Chaban, V. S. Babizhetskyy, O. T. Pilyushchak, Yu. B. Kuzma, Inorg. Mater. **41**, 700 (2005)

[05V2] I. Veremchuk, Yu. Prots, A. Leithe-Jasper, Yu.B. Kuzma, Z. Kristallogr. **220**, 125 (2005)

[05W1] S.C. Wang, Y. Li, J. Mat. Sci. **40**, 3853 (2005)

[05W2] H. Y. Wang, F. A. Zhao, N. X. Chen, G. Liu, J. Magn. Magn. Mater. **295**, 219 (2005)

[05Y1] N. Yang, K. W. Dennis, R. W. McCallum, M. J. Kramer, Y. Zhang, P. L. Lee, J. Magn. Magn. Mater. **295**, 65 (2005)

[06C1] H.W. Chang, C.H. Chu, C.W. Chang, C.H. Chen, W.C. Chang, Y.D. Yao, A.C. Sun, J. Alloys Comp. **407**, 53 (2006)

[06C2] H.W. Chang, C.H. Chin, C.W. Chang, W.C. Chang, A.C. Sun, Y.D. Yao, Scripta Mater. **55**, 529 (2006)

[06C3] Z. M. Chen, D. N. Brown, Y. Q. Wu, M. J. Kramer, J. Iron Steel Res. Internat. **12**, 177 (2006)

[06C4] C. H. Ciu, H. W. Chang, C. W. Chang, W. C. Chang, J. Alloy. Comp. **424**, 376 (2006)

[06G1] I. E. Garkusha, A. M. Bovda, V. A. Bovda, K. V. Chebotarev, V. D. Fedorchenko, N. N. Matyushenko, V. I. Tereshin, A. S. Tortika, AIP Conf. Proc. **812**, 465 (2006)

[06K1] N. V. Kudrevatykh, S. A. Andreev, M. I. Bartashevich, A. N. Bogatkin, O. A. Milyaev, P. E. Markin, I. S. Tereshina, T. Palewski, E. A. Tereshina, J. Magn. Magn. Mater. **300**, e488 (2006)

[06L1] D. O. Lozano, L. E. Zamora, G. A. P. Alcazar, Y. A. Rojas, H. Bustos, J. M. Greneche, Hyp. Int. **169**, 1353 (2006)
[06O1] S. Ozawa, K. Kuribayashi, S. Hirosawa, S. Reutzel, D.M. Herlach, J. Appl. Phys. **100**, 123906 (2006)
[06P1] A. T. Pedziwiatr, A. Wojciechowska, B. F. Bogacz, J. Phys.: Condens. Matter **18**, 8891 (2006)
[06S1] J.C. Shih, A.J. Saldanha, K. Suzuki, T. Shoji, A. Koto, S. Tajima, J. Appl. Phys. **99**, 08B505 (2006)
[06T1] E. A. Tereshina, I. S. Tereshina, Phys. Solid State, **48**, 509 (2006)
[06W1] A. Wojciechowska, B. F. Bogacz, A. T. Pedziwiatr, Acta Phys. Polonica, A **109**, 427 (2006)
[06Y1] C.J. Yang, J.S. Han, E.B. Park, E.C. Kim, J. Magn. Magn. Mater. **301**, 220 (2006)
[06Z1] C.L. Zhang, D.H. Wang, Z.D. Han, S.L. Tang, B.X. Gu, F.W. Du, Appl. Phys. Lett. **89**, 122503 (2006)
[07A1] A. Alekseeva, A. M. Abakumov, P. S. Chishov, A. Leiter-Jasper, W. Schnelle, Yu. Profs, J. Hadermann, E. V. Antipov, Y. Grin, Inorg.Chem. **46**, 7338 (2007)
[07C1] J. Chaboy, C. Piquer, N. Plugaru, F. Bartolomé, M. A. Laguna-Marco, F. Plazaolo, Phys. Rev. B **76**, 134408 (2007)
[07G1] J. Gao, T. Volkmann, S. Yang, S, Reutzel, D.M. Herlach, X.P. Song, J. Alloys Comp. **433**, 356 (2007)
[07K1] A. V. Kutepov, A. K. Kim, S. V. Trihaev, Techn. Metall. **3**, 46 (2007)
[07M1] F. Maruyama, Y. Amako, H. Nagai, J. Alloys Comp. **429**, 72 (2007)
[07M2] H. Mayot, O. Isnard, J.L. Souberyoux, J. Magn. Magn. Mater. **316**, e477 (2007)
[07M3] G.I. Miletic, Z. Blazina, J. Alloys Comp. **430**, 9 (2007)
[07M4] N. V. Mushnikov, P. B. Terentev, E. V. Rosenfeld, Fiz. Met. Metal. **103**, 42 (2007)
[07P1] A.G. Popov, V.S. Gaviko, N.N. Shchegoleva, L.A. Shreder, V.V. Stolyarov, D.V. Gunderov, X.Y. Zhang, W. Li, L.L. Li, Phys. Met. Metallogr. **104**, 238 (2007)
[07R1] V. Raghovan, J. Phase. Equilib. **28**, 382 (2007)
[07T1] I. S. Tereshina, O. D. Chistyakov, G. S. Burkhanov, E. A. Tereshina, S. A. Nikitin, L. Folfik, H. Drulis, in *Hydrogen Materials Science and Chemistry and Carbon Nanomaterials, NATO Series A* (2007), p. 599
[07V1] P. Villars et al, *Landolt Börnstein Handbook*, vol. III/43A6 (Springer Verlag, 2007)
[07W1] C. Wang, M. Yan, Q. Li, J. Phys. D.: Appl. Phys. **40**, 3551 (2007)
[07W2] A. Wojciechowska, A.T. Pedziwiatr, B.F. Bogacz, S. Wrobel, J. Alloys Comp. **442**, 126 (2007)
[07Z1] W.Y. Zhang, C.H. Chu, L.C. Zhang, K. Biswas, H. Ehrenberg, W.C. Chang, J. Eckert, J. Magn. Magn. Mater. **308**, 24 (2007)
[08B1] E. Burzo, Rom. Repts. Phys. **60**, 701 (2008)
[08B2] E. Burzo, P. Vlaic, J. Opt. Adv. Mater. **10**, 750 (2008)
[08H1] Y.J. Huang, Z.L. Shi, C.L. Zhang, S.Z. Li, N. Cheng, Z.D. Han, Y.F. Hsia, Mater. Lett. **62**, 85 (2008)
[08M1] H. Mayot, Ph.D. Thesis Univ. Grenoble (2008)
[08M2] H. Mayot, O. Isnard, J. L. Souberyroux, J. Magn. Magn. Mater. **316**, e477 (2008)
[08M3] W. Mo, L. Zhang, Q. Liu, D.Shan, J. Wu, M. Komura, Scripta Mater. **59**, 179 (2008)
[09B1] E. Burzo, N. M. Codescu, W. Kappel, E. Helerea, J. Opt. Adv. Mater. **11**, 229 (2009)
[09C1] X. C. Cui, M. Yan, T. Y. Ma, W. Luo, S. J. Tu, J. Magn. Magn. Mater. **321**, 392 (2009)
[09G1] S. K. Godovikov, E. S. Lagutina, Phys. Metal. Metall. **108**, 67 (2009)
[09K1] I. Kitagawa, J. Appl. Phys. **105**, 07E502 (2009)
[09K2] A. Kreyssig, R.Prozorov, C. D. Downhurst, P. C. Canfield, R. W. McCallum, A. I. Goldman, Phys. Rev. Lett. **102**, 047204 (2009)
[09P1] C. Piquer, J. Bartolomé, C. de Francisco, J. M. Munoz, Phys. Rev. B **79**, 174430 (2009)
[09S1] K. Suzuki, J. S. Garitaonandia, J. C. Shih, G. J. Cuello, T. Shogi, A. Kato, S. Tajima, J. Appl. Phys. **105**, 07A735 (2009)
[10K1] I. Kitagawa, Y. Asari, Phys. Rev. B **81**, 214408 (2010)

[10K2] M. D. Kuzmin, D. Givord, V. Skumryev, J. Appl. Phys. **107**, 113924 (2010)
[10L1] S. Lassoued, R. Gautier, A. Boutarfaia, J.F. Halet, J. Org. Chem. **695**, 987 (2010)
[10L2] V. Ly, S. Gorsse, K. Suzuki, C. Hutchinson, Mat. Sci. Forum **654–656**, 1166 (2010)
[10S1] H. Sepehri-Amin, T. Ohkubo, K. Hono, J. Appl. Phys. **197**, 09A745 (2010)
[10S2] H. Sepehri-Amin, T. Ohkubo, T. Nishiuchi, S. Hirosawa, K. Hono, Scripta Mater. **63**, 1124 (2010)
[10T1] X. Y. Tang, K.X.Jin, C.L. Cheng, C.C. Zhou, Acta Phys. Sinica, **39**, 3414 (20100
[10W1] F. Wang, B.G. Shen, J. Zhang, J.R. Sun, F.B. Meng, Y.X. Li, Chin. Phys. B **19**, 067501 (2010)
[11J1] T.S. Jang, D.H. Lee, S. Namkung, Rev. Adv. Mater. Sci. **28**, 212 (2011)
[11K1] K. Krawiec, A. T. Pedziwiatr, B. F. Bogacz, J. Przewoznik, J. Zukrowski, Cz. Kapusta, J. Magn. Magn. Mater. **323**, 2968 (2011)
[11L1] S. Lassoued, R. Gautier, J.F. Halet, in *Boron Rich Solids* (Springer Science, 2011), p. 95
[11L2] L. Li, Y. Nishimura, D. Huo, Z. Qian, K. Nishimura, J. Appl. Phys. **110**, 083915 (2011)
[11M1] G.I. Miletic, Z. Blazina, J. Magn. Magn. Mater. **323**, 2340 (2011)
[11S1] S. Sugimoto, J. Phys. D.: Appl. Phys. **44**, 064001 (2011)
[11T1] S. Tanaka, H. Moriya, H. Tsuchiura, A. Sakuma, M. Divis, P. Novak, J. Appl. Phys. **109**, 07A702 (2011)
[11V1] P. Villars et al, *Landolt Börnstein Handbook*, vol. III/43A10 (Springer Verlag, 2011)
[11Z1] C.L. Zhang, J. Chen, T.Z. Wang, G.X. Xie, C. Zhu, D.H. Wang, Physica B **406**, 2840 (2011)
[12D1] R. Daminova, S. A. Nikitin, R. Bezdushnyi, I. S. Tereshina, E. A. Tereshina, G. S. Burkhanov, L. Iliev, Solid State Phenomena **190**, 163 (2012)
[12D2] M. Daniil, L. Minter, M.A. Willard, J. Appl. Phys. **111**, 07B521 (2012)
[12L1] X. B. Liu, Z. Altounian, J. Appl. Phys. **111**, 07A701 (2012)
[12L2] W. F. Li, A. M. Gabay, M. Marinescu-Jasinski, J. F. Liu, C Ni, G. C. Hadjipanays, J. Magn. Magn. Mater. **324**, 1321 (2012)
[12N1] K. K. Nakayama, E. Ieki, Y. Tanaka, T. Sato, T. Takahashi, T. Kuroda, H. Mizaguchi, S. W. Kim, H. Hosona, Phys. Rev. B **86**, 014503 (2012)
[12N2] J. Ni, T. Ma, M. Yan, Mater. Lett. **75**, 1 (2012)
[12P1] L. Pall, Ph.D. Thesis, Univ. Bordeaux (2012)
[12P2] L. Pall, L. Bobet, E. Andronescu, Rev. Romana Mater. **42**, 425 (2012)
[12R1] V. Raghavan, J. Phase. Equilib. **33**, 395 (2012)
[13A1] A. Alam, M. Khan, R. W. McCallum, D. D. Johnson, Appl. Phys. Lett. **102**, 042402 (2013)
[13A2] A, E, Aly, Mater. Trend Phys. Res. (2013), p.155
[13D1] L.V.B. Diop, O. Isnard, N.R. Lee-Hone, D.H. Ryan, J.M. Cadogan, J. Phys. Condens. Matter **25**, 316001 (2013)
[13H1] H. F. Hao, J. Wang, F. Sun, L. T. Zhang, Acta Phys. Sinica **62**, 117501 (2013)
[13I1] M. Itakura, N. Watanabe, M. Nishida, T. Daio, S. Matsumura, Jpn. J. Appl. Phys. **52**, 050201 (2013)
[13L1] X. B. Liu, Z. Altounian, M. Huang, Q. Zhang, J. P.Liu, J. Alloys Comp. **549**, 366 (2013)
[13R1] V. Raghavan, J. Phase Equilib. **34**, 124 (2013)
[13S1] E. J. Skoug, M. S. Meyer, F. E. Pinkerton, M. M. Tessema, D. Haddad, J. F. Herbst, J. Alloys Comp. **574**, 552 (2013)
[13S2] M. Szwaja, K. Pawlik, J.J. Wyslocki, P. Pawlik, P. Gebara, Nucleonika **58**, 159 (2013)
[13T1] A.E. Teplykh, Yu.G. Chukalkin, S. Lee, S. G. Bogdanov, N. V. Kudrevatykh, E. V. Rosenfeld, Yu. N. Skryabin, Y. Choi, A. V. Andreev, A. N. Pirogov, J. Alloys Comp. **581**, 423 (2013)
[14A1] A. Asali, D. Toson, P. Blaha, J. Fiedler, IEEE Trans. Magn. **50**, 7027504 (2014)
[14M1] Y. Miura, H. Tsuchiura, T. Yoshioka, J. Appl. Phys. **115**, 17A765 (2014)
[14T1] Z. Torbatian, T. Ozaki, S. Tsuneyuki, Y. Gohda, Appl. Phys. Lett. **104**, 242403 (2014)
[14Y1] C. J. Yan, S. Guo, R. J. Chen, D. Lee, A. R. Yan, Chin. Phys. B **23**, 107501 (2014)
[14Z1] M. Zhu, W. Li, J. Wang, L. Zheng, Y. Li, K. Zhang, H. Feng, T. Liu, IEEE Trans. Magn. **50**, 1 (2014)

[15H1] M. Hasiak, Nucleonika **60**, 29 (2015)
[15L1] Z. B. Li, B. G. Shen, M. Zhang, F. X. Hu, J. R. Sun, J. Alloys Comp. **628**, 325 (2015)
[15W1] J. Wang, B. Yang, L. Liang, L. M. Sun, L. T. Zhang, S. Hirano, AIP Advances **5**, 097206 (2015)
[15Y1] Q.R. Yao, J.L. Xiong, P. Liu, H.Y. Zhou, G. H. Rao, J. Q. Deng, S. K. Pan, J. Wang, J. Alloys Comp. **633**, 229 (2015)
[16B1] K. H. Bae, S. R. Lee, H. J. Kim, M. W. Lee, T. S. Jang, J. Alloys Comp. **673**, 321 (2016)
[16C1] C. V. Colin, M. Ito, M. Yano, N. M. Dempsey, E. Suard, D. Givord, Appl. Phys. Lett. **108**, 242415 (2016)
[16D1] L.V.B. Diop, O. Isnard, Appl. Phys. Lett. **108**, 132401 (2016)
[16D2] L.V.B. Diop, O. Isnard, J. Appl. Phys. **119**, 213904 (2016)
[16D3] L.V.B. Diop, O. Isnard, J. Alloys Comp. **688**, 953 (2016)
[16D4] L.V.B. Diop, O. Isnard, J. Rodriguez-Carrajal, Phys. Rev. B **93**, 014440 (2016)
[16F1] G. Fu, J. Wang, M. H. Rong, G. H. Rao, H. Y. Zhou, J. Phase Equilib. **37**, 308 (2016)
[16H1] Z. Hu, H. Qu, D. Ma, C. Luo, H. Wang, J. Rare Earths **34**, 689 (2016)
[16I1] M. Ito, M. Yano, N. M. Dempsey, D. Givord, J. Magn. Magn. Mater. **400**, 379 (2016)
[16J1] J. Jin, Y. Zhang, T. Ma, M. Yan, IEEE Trans. Magn. **52**, 2100804 (2016)
[16K1] I. Khan, J. Hong, J. Korean Phys. Soc. **68**, 1409 (2016)
[16K2] I. Khan, J. Hong, J. Korean Phys. Soc. **69**, 1564 (2016)
[16S1] K. P. Skokov, Y. G. Pastushenkov, S. A. Nikitin, M. Fries, O. Gutfleisch, IEEE Trans Magn. **52**, 2500304 (2016)
[16V1] P. Villars, *Landolt Börnstein Handbook* (Springer Verlag, 2016)
[16X1] C. H. F. Xu, Z. F. Chen, Z. F. Gu, D. D. Ma, L. Y. Cheng, G. Cheng, G. H, Rao, J. Alloys. Comp. **667**, 240 (2016)
[16Z1] C. Zheng, D. Yu, K. Li, Y. Luo, J. Jin, S. Lu, H. Li, Y. Mao, N. Quan, J. Magn. Magn. Mater. **412**, 89 (2016)
[17B1] G. S. Burkhanov, I. S. Tereshina, G. A. Politova, I. A. Pelevin, Y. S.Koshkidko, M. A. Paukov, H. Drulis, Doklady Physics **62**, 10 (2017)
[17C1] B.S. Conner, M.A. Susner, P. Lampen-Kelley, A.F. May, M.A. McGuire, J.Q. Yan, R.C. Sales, J. Magn. Magn. Mater. **435**, 100 (2017)
[17D1] L.V.B. Diop, O. Isnard, Z. Arnold, J.P. Itié, J. Kastil, J. Kamarad, Solid State Commun. **252**, 29 (2017)
[17F1] S. Fujieda, K. Fukamichi, S. Suzuki, J. Magn. Magn. Mater. **421**, 403 (2017)
[17H1] J. F. Herbst, L. G. Hector, J. Alloys Comp. **693**, 238 (2017)
[17K1] P. M. Kurzydlo, A.T. Pedziwiatr, B. F. Bogacz, J. Przewoznik, D. Oleszak, Nukleonica **62**, 123 (2017)
[17L1] J. Liu, H. Sepehri-Amin, T. Ohkubo, M. Yano, A. Kato, N. Sakuma, T. Shogi, K. Hono, Scripta Mater. **129**, 44 (2017)
[17S1] K. Saito, S. Doi, T. Abe, K. Ono, J. Alloys Comp. **721**, 476 (2017)
[17S2] M. A. Susner, B. S. Conner, B. I. Saparov, M. A. McGuire, E. J. Crumlin, G. M. Veith, H. B. Cao, K. V. Shanavas, D. S.Parker, B. K. Chakoumakos, J. Magn. Magn. Mater. **434**, 1 (2017)
[17T1] J. Tencé, A. Wattiaux, M. Duttline, R. Decourt, O. Isnard, J. Alloys Comp. **693**, 887 (2017)
[17W1] T. Wang, D. Kevorkov. M. Medraj, Materials **10**, 16 (2017)
[17X1] X. L. Xu, Y. S. Du, G. Cheng, J. J. Huo, L. Li, L. Zhou, G. H. Rao, J. Phase Equilib. Diffus. **38**, 576 (2017)
[17X2] C. F. Xu, K. H. Chen, Z. F. Gu, D. D. Ma, L. Y. Cheng, G. Cheng, G. H. Rao, Phase Transitions **90**, 380 (2017)
[18B1] R. Bezdushnyi, R. Damianova, I.S. Tereshina, N.Yu. Pankratov, S.A. Nikitin, J. Magn. Magn. Mater. **453**, 226 (2018)
[18B2] E. Burzo, J. Surface Invest X-Ray Synchrotron Neutron Tech. **12**, 431 (2018)
[18D1] L. V.B. Diop, O. Isnard, Phys. Rev. B **97**, 014436 (2018)
[18G1] Z. F. Gu, D. D. Ma, C. F. Xu, T. Liu, L. V. Cheng, Y. S. Du, W. F. Zhang, J. Super. Novel Magn. **31**, 271 (2018)

[18K1] N. V. Kostyuchenko, I. S. Tereshina, D. I. Gorbunov, E. A. Tereshina-Chitova, A. V. Andreev, M. Donerr, G. A. Politova, A. K. Zvezdin, Intermetallics **98**, 139 (2018)
[18N1] H. Nakamura, Scripta Mater. **154**, 273 (2018)
[18O1] K. Orimoloye, D. H. Ryan, F. E. Pinkerton, M. Medraj, Appl. Sci.-Basel **8**, 205 (2018)
[18S1] H. Sepehri-Amin, S. Hirosawa, K. Hono, *Handbook of Magnetic Materials*, vol. 27 (2018), p. 269
[18T1] Y. Tatetsu, E. Tsuneyuki, I. Gohda, Materials **4**, 388 (2018)
[18Y1] T. Yoshioka, H. Tsuchiura, Appl. Phys. Lett. **112**, 162405 (2018)
[18W1] T. Wang, M. Medraj, J. Magn. Magn. Mater. **460**, 95 (2018)
[18Z1] L. Z. Zhao, J. S. Zhang, G. Ahmed, X. F. Liao, Z. W. Liu, J. M. Greneche, Sci. Repts. **8**, 6826 (2018)
[19C1] A. Candan, G. Surucu, A. Gencer, Phys. Scripta **14**, 125710 (2019)
[19C2] T. L. Chen, J. Wang, C. P. Guo, C. R. Li, Z. M. Du, G. H. Rao, H. Y. Zhou, Calphad **66**, 101627 (2019)
[19I1] J. I. Inoue, T. Yoshioka, H. Tsuchiura, IEEE Trans Magn. **55**, 2100304 (2019)
[19P1] G. A Politova, I. S. Tereshina, T. P. Kaminskaya, A. A. Viryus, M. A. Paukov, A. A. Lukin, A. A. Andreev, J. Phys.: Conf. Series, **1236**, 012016 (2019)
[19S1] J. Sugiyama, K. Miwa, H. Nozaki, Y. Kaneko, B. Hitti, D. Arseneau, G. D. Morris, E. J. Ansalto, J. H. Brewer, Phys. Rev. Mater. **3**, 064402 (2019)
[19T1] I. Tereshina, T. Kaminskaya, L. Ivanov, G. Politova, H. Drulis, D. Gorbunov, M. Paukov, E. Tereshina-Chitrova, A. Andreev, J. Phys.: Conf. Series, **1389**, 012094 (2019)
[20B1] E. Burzo, Int. J. Modern Phys. **34**, 2050006 (2020)
[20C1] F. Chen, J. Magn. Magn. Mater. **514**, 167227 (2020)
[20C2] T. Chen, C. Guo, C. Li, Z. Du, J. Magn. Magn. Mater. **497**, 165983 (2020)
[20C3] J. M. D. Coey, Engineering **6**, 119 (2020)
[20F1] H. B. Feng, X. S. Chen, A. H. Li, IEEE Trans. Magn. **57**, 2101004 (2020)
[20H1] L. He, J. Li, X. Yu, S. U. Rahman, P. Qu, F. Li, Y. Hu, IEEE Trans. Magn. **57**, 2101407 (2020)
[20K1] S. Kobayashi, A. Martin-Cid, K. Toyoki, H. Okazaki, S. Kawaguchi, S. Hirosawa, T. Nakamura, Phys. Rev. Mater. **4**, 094405 (2020)
[20L1] Z. Li, M. Yue, X. Liu, H. Zhang, M. Liu, Y. Li, Z. Zhang, J. Magn. Magn. Mater. **505**, 166747 (2020)
[20L2] A. H. Li, L. L. Xi, H. B. Feng, N. Zhou, M. Tan, M. G, Zhou, W. Li, J. Iron Steel Res. Int. **27**, 1 (2020)
[20L3] J. Liu, X. Liao, J. He, H. Yu, X. Zhong, Q. Zhou, Z. Lin, J. Magn. Magn. Mater. **504**, 166685 (2020)
[20M1] Materials Research Projet, Lawrence National Lab., Berkeley, CA, USA (2020)
[20M2] M. Morishita, T. Abe, A. Nozaki, I. Ohnuma, K. Kamon, Thermochim. Acta **690**, 178672 (2020)
[20T1] T. Tajiri, M. Mito, J. Magn. Magn. Mater. **498**, 166163 (2020)
[20U1] I. E. Uysal, M. Nishino, S. Miyashita, Phys. Rev. B **101**, 094421 (2020)
[20W1] Q. Wei, J. Wang, Q. Yao, Q. Long, M. Qin, H. Zhou, J. Phase Equi. **41**, 35 (2020)
[20W2] Q. Wei, Q. Yao, E. Rao, Z. Lu, B. Deng, J. Wang, C. Hu, H. Zhou, J. Alloys Comp. **834**, 154560 (2020)
[20X1] Y. Xiao, F. J. Morvan, A. N. He, M. K. Wang, H. B. Luo, R. B. Jiao, W. X. Xia, G. P. Zhao, J. P. Liu, Appl. Phys. Lett. **117**, 132402 (2020)
[20Z1] L. Zha, Z. Lin, F. Wang, Z. Liu, M. Xue, J. Yang, W. Yang, G. Tian, J. Han, C.Wang, H. Du, Y. Yang, J. Yang, J. Magn. Magn. Mater. **514**, 167128 (2020)
[21B1] E. Burzo, P. Vlaic, D.P. Kozlenko, A.V. Ruskauskas, J. Surf. Invest. **15**, 520 (2021)
[21C1] X. Chen, Y. Mudryk, A. K. Patak,V. K. Pecharski, J. Alloys Comp. **884**, 161115 (2021)
[21C2] S. Coey, S. Parkin, (Eds.) *Handbook of Magnetism and Magnetic Materials* (Spirnger Verlag, 2021)
[21D1] L. V. B. Diop, O. Isnard, Appl. Phys. Lett. **119**, 032403 (2021)
[21D2] L. V. B. Diop, Eur. Phys. J. Plus **136**, 840 (2021)

[21D3] L. V. B. Diop, M. Amara, O. Isnard, Appl. Phys. Lett. **118**, 262409 (2021)
[21D4] L. V. B. Diop, T. Faske, O. Isnard, W. Donner, Phys. Rev. Mater. **5**, 104401 (2021)
[21D5] L. V. B. Diop, F. Faske, M. Amara, D. Koch, O. Isnard, W. Donner, Phys. Rev. B **112**, 134412 (2021)
[21D6] L. V. B. Diop, O. Isnard, J. Appl. Phys. **129**, 243902 (2021)
[21H1] R. Han, P. Shen, L. Qiao, H. Chen. Y. Fang, Z. Guo, S. Dong, M. Zhu, D. Zhou, F. Li, W. Li, J. Magn. Magn. Mater. **529**, 167859 (2021)
[21H2] J.He, X. Liao, X.Lan, W. Qiu, H. Yu, J. Zhang, W. Fan, X. Zhong, Z. Liu, J. Alloys Comp. **870**, 159229 (2021)
[21K1] N. V. Kostyuchenko, I. S. Tereshina, E. A. Tereshina-Chitrova, L.A. Ivanov, M. Paukov, D. I. Gorbunov, A. V. Andreev, M. Doerr, G. A. Politova, A. K. Zvezdin, S. V. Veselova, A. P. Pyatakov, A. Miyata, O. Drachenko, O. Portugall, Phys. Rev. Mater. **5**, 074404 (2021)
[21L1] X. Li, Q. Yao, Q. Wei, J. Wang, Y. Du, L. Li, Q. Long, H. Zhou, G. Rao, J. Alloys Comp. **859**, 117780 (2021)
[21L2] X. Liao, J. Zhang, J. He, W. Fan, H. Yu, X. Zhong, Z. Liu, J. Mater. Sci. Tech. **76**, 215 (2021)
[21S1] S. Sakamara, A. Manaf, W. A. Adib, J. Phys. Conf. Series **1571**, 012065 (2021)
[21S2] H. Sato. Y. Kibo, T. Yoshioka, M. Tsuchiura, Y. Mizuno, K. Kolke, K. Takahashi, H. Kato, AIP Advances **11**, 025224 (2021)
[21W1] Q. Wei, Z. Lu, Q. R. Yao, J. Q. Deng, J. Wang, H. Y. Zhou, G. H. Rao, Trans. Nonferrous Met. Sci. China **31**, 1748 (2021)
[22D1] L. V. B. Diop, O. Isnard, Solid State Commun. **341**, 114568 (2022)
[22H1] Q. Huang, Q. Jiang, J. Hu, S. U. Rehman, G. Fu, Q. Quan, J. Huang, D. Xu, D. Chen, Z. Zhong, J. Mater. Sci. Techn. **106**, 236 (2022)
[22L1] Z. Liu, J. He, Q. Zhou, Y. Huang, Q. Jing, J. Mater. Sci. Techn. **98**, 51 (2022)

Chapter 8
Rare–Earths–Cobalt–Boron Compounds

8.1 General Problems

The phase diagrams of R–Co–B compounds have been investigated as La–Co–B [74R1, 76S1], Ce–Co–B [72N1, 74B1, 74G1, 90K2], Pr–Co–B [79K1, 85B2, 87C2, 93C4, 99C1], Nd–Co–B [74K1, 83B1, 83B2, 93C3, 00C6], Sm–Co–B [80B1, 84E1, 00C5, 00C8], Eu–Co–B [83K1, 90K2], Gd–Co–B [77C1, 78S1, 00C4, 07C1], Tb–Co–B [83K1, 88D1], Dy–Co–B [81K3, 83C1], Ho–Co–B [92G1], Er–Co–B [81K3, 83C1], Yb–Co–B [81C1, 04V1], Lu–Co–B [88D2] and Y–Co–B [74K1, 77S1]. The phase diagrams of R–Co–B compounds are also presented in some books or chapters of books [83K1, 84P1, 84R1, 90K2]. For the greatest number of compounds, the mutual solubility of constituting elements were shown to be negligible. This is in contrast with the presence of solid solutions in $R_{5-x}Co_{2+x}B_6$ series [86D1].

The compositions of different structure types, formed in R–Co–B systems, are given in Table 8.1. Not all crystal structures of assumed compositions have been solved. In Ce–Co–B systems, the borides $CeCoB_3$ with unknown structure type, as well as Ce_2Co_4B, could not be obtained in homogeneous form and a peritectic- or peritectoid- type formation was suggested [74B1, 84P1]. The CeCoB compound was shown to be stable at 800 °C, but decomposes at low temperatures in Ce_2CoB_2 and $Ce_2Co_5B_2$ phases [74B1]. In the Nd–Co–B system the crystal structures of compounds with approximate formula Nd_2Co_9B and Nd_2CoB_3 were not determined [83B1]. The later reported Nd–Co–B phase diagrams, did not mentioned the existence of the above compounds. The crystal structures of Sm_2Co_9B, $Sm_5Co_8B_5$ and Sm_2CoB_3 remained also unsolved [80B1]. The crystal structures of $ErCoB_5$ [81K3], Lu_2CoB_6 [88D2] or $Y_4Co_3B_3$ [74K1] were not determined, although their presence was suggested.

The original version of this chapter was revised and the omitted revised abstract now has been replaced. The correction to this chapter is available at https://doi.org/10.1007/978-3-030-99245-3_10

E. Burzo, *Rare Earths-Transition Metals-Boron Compounds*,
https://doi.org/10.1007/978-3-030-99245-3_8

Table 8.1 R–Co–B ternary compounds

Compound	La	Ce	Pr	Nd	Sm	Eu	Gd	Tb	Dy	Ho	Er	Tm	Yb	Lu	Y
R_4CoB_{13}[a]							O	O	O	O	O	O		O	O
R_2CoB_5											b				
$RCoB_4$							O	O	O	O	O	O	O	O	O
R_2CoB_3				O	O[c]		O		O						
$R_{5-x}Co_{2+x}B_6$	O	O	O	O	O		O	O	O	O	O				O
R_2CoB_2	O	e													
$RCoB$		O[d]													
RCo_4B_4	O	O	O	O	O	O	O	O	O	O	O	O	O	O	O
$R_4Co_3B_3$					O										e
RCo_2B_2	O		O	O	O		O	O	O	O	O				O
RCo_3B_2		O	O		O		O	O	O	O	O	O	O	O	O
$R_2Co_5B_3$				O											
$R_2Co_5B_2$		O	O	O	O			O	O	O	O	O			O
$RCo_{12}B_6$	O	O	O	O	O		O	O	O	O	O	O			O
$R_2Co_7B_3$	O	O	O	O	O		O	O	O	O	O	O			O
$R_3Co_{11}B_4$	O	O	O	O	O		O	O		O	O	O		O	O
$R_5Co_{19}B_6$			O	O	O										
R_2Co_9B					c										
$R_5Co_{21}B_4$				O											
RCo_4B	O	O	O	O	O		O	O	O	O	O	O		O	O
$R_3Co_{13}B_2$			O	O	O[f]		O[f]		O[f]	O[f]	O[f]				O
$R_2Co_{14}B$	O		O	O	O		O	O							O
RCo_2B_2C			O				O	O	O	O	O	O			

[a]Identified compounds [91G1, 20M1]
[b]Unidentified
[c]Sm_2Co_9B, $Sm_5Co_8B_5$, and $Sm_2Co_3B_2$ were not identified [80B1]
[d]Stable at 800° C, decomposes at low temperatures in Ce_2CoB_2 and $Ce_2Co_5B_2$ [74B1, 20M1]
[e]Ce_2CoB_2, $CeCoB_3$, Ce_2Co_3B and Ce_2Co_4B structures are not known [74B1]
[f]No lattice parameters were available; a mixture of phases were frequently shown

The phase diagram of Yb–Co–B system, differs from those of other heavy rare-earths-transition metals borides, having a small number of ternary compounds [04V1]. In this respect, this is more close to the Eu–Co–B one [90K2], which contains only one ternary phase. The number of R-Co-B compounds is dependent on the rare-earth, decreasing in the sequence 14 (R = Nd, Sm), 13 (R = Gd, Dy, Y), 12 (R = Pr, Tb, Ho, Er), 9 (R = La, Ce, Tm), 6 (R = Lu), 3 (Yb) and 1 (Eu).

Two homologous series of compounds were shown in R-Co-B systems. The first series is described by the formula $R_{n+1}Co_{5+3n}B_{2n}$ [83K2]. The above compounds are formed by stacking of one layer RCo_5 and n RCo_3B_2 layers, along **c**-axis. The

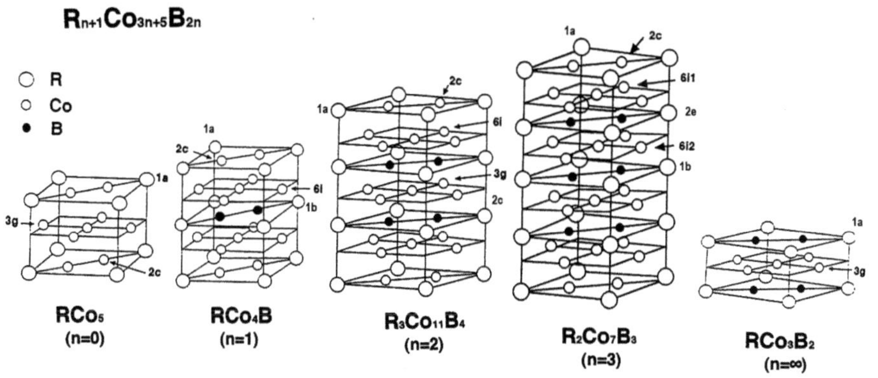

Fig. 8.1 Crystal structures of $R_{n+1}Co_{5+3n}B_{2n}$ compounds

compounds of this series are RCo_5 (n = 0), RCo_4B (n = 1), $R_3Co_{11}B_4$ (n = 2), $R_2Co_7B_3$ (n = 3) and RCo_3B_2 (n = ∞). The crystal structures of these compounds are shown in Fig. 8.1.

A more general series $R_{m+n}Co_{5m+3n}B_{2n}$ was also proposed [00C6], the compounds present in the first series being part of the second one. The crystal structures of compounds belonging on this series are formed by alternating stacking of m RCo_5 layers and n layers of RCo_3R_2 type along **c**-axis. The corresponding compositions are RCo_5 (m = ∞), RCo_4B (m = 1, n = 1), $R_3Co_{13}B_2$ (m = 2, n = 1), $R_5Co_{19}B_6$ (m = 2, n = 3), $R_5Co_{21}B_4$ (m = 3, n = 2) $R_2Co_7B_3$ (m = 1, n = 3), $R_3Co_{11}B_4$ (m = 1, n = 2) and RCo_3B_2 (m = 0, n = 3). The crystal structures of compounds, in addition to those of to first series, are given in Sect. 8.7. The lattice parameters of $R_{m+n}Co_{5m+3n}B_{2n}$ series follow the relations $a \cong a_{RCo_5}$, $c \cong m_{RCo_5} + n_{RCo_5B_2}$. The larger number of compounds are formed in RCo_4B_4, RCo_4B, $R_3Co_{11}B_4$, RCo_3B_2 and $R_2Co_7B_3$ series. Only one compound is present in R_2CoB_2, $R_4Co_3B_3$, $R_2Co_5B_3$ and $R_5Co_{21}B_4$ series.

8.2 R-Co-B Compounds Having Co/B ≤ 1 Ratio and RCo_2B_2C Series

The ternary R-Co-B compounds, having the composition ratio Co/B ≤ 1, include the R_4CoB_{13}, $RCoB_4$, R_2CoB_3, $R_{5-x}Co_{2+x}B_6$, R_2CoB_2, RCo_4B_4, $R_4Co_3B_3$ and RCo_2B_2 series. Since of their interesting magnetic properties, the RCo_2B_2C compounds, having the crystal structure derived from that of RCo_2B_2 series, are also included in the review. The site parameters for representative compounds are given in Table 8.2. The space groups and the lattice parameters are listed in Table 8.3.

Table 8.2 Atomic parameters and temperature isotropic factors, at room temperatures

(a) Tb_4CoB_{13} structure, having P4/mmc space group [88D1]

Atom	Site	x	y	z	$Beq \cdot 10^2$ $(nm)^2$
Tb	*8h*	0.195(2)	0.668(2)	0	0.7(3)
Co	*2a*	0	0	0	0.5(2)
B1	*2b*	0	0	1/2	0.5(2)
B2	*8g*	0.409	0.909	1/4	0.5(2)
B3	*16i*	0.039	0.177	0.233	0.5(2)

(b) $YbCoB_4$ structure, having *Pbam* space group [04V1]

Atom	Site	x	y	z	$B_{eq} \cdot 10^2$ $(nm)^2$
Yb	*4g*	0.12842(6)	0.15001(3)	0	0.325(6)
Co	*4g*	0.1356(2)	0.4105(1)	0	0.38(2)
B1	*4h*	0.288(2)	0.3136(9)	1/2	0.36(13)
B2	*4h*	0.364(2)	0.4700(9)	1/2	0.54(13)
B3	*4h*	0.386(2)	0.0461(10)	1/2	0.66(14)
B4	*4h*	0.476(2)	0.1881(12)	1/2	0.8(2)

(c) $Tb_{5-x}Co_{2+x}B_6$ structure, having $R\bar{3}m$ space group [86DI] [a]

Atom	Site	x	y	z	CN	B $(nm)^2$
$(Tb,Co)_{11}$	*3a*	0	0	0	14	0.9
Tb2	*6c*	0	0	0.2497(7)	16	0.2
Tb3	*6c*	0	0	0.4147(9)	15	1.2
Co	*6c*	0	0	0.112(1)	14	1.9
B	*18g*	0.333	0	0.500	9	2.0

[a] $x = 0.75$

(d) $CeCo_4B_4$ having $P4_2/nmc$ space group [84P1]

Atom	Site	x	y	z
Ce	*2b*	3/4	1/4	1/4
Co	*8g*	1/4	0.503	0.384
B	*8g*	1/4	0.08	0.10

(e) $NdCo_4B_4$ having $P4_2/n$ space group* [84P1]

Atom	Site	x	y	z
Nd	*2a*	1/4	1/4	1/4
Co	*8g*	0.625	0.107	0.362
B	*8g*	0.025	0.600	0.125

* x, y, z changed to y, x, -z and origin shifted by 1/2, 1/2, 1/2

(f) $GdCo_2B_2$ structure having I4/mmm space group [14P2]

Atom	Site	x	y	z
Gd	*2a*	0	0	0
Co	*4d*	0.5	0	0.25
B	*4e*	0	0	0.371

Table 8.3 Space groups ad lattice parameters

Compound	T (K)	Space group	Lattice parameters (nm)			References
			a	b	c	
Gd_4CoB_{13}	RT	*P4/mnc*	0.7378		0.6998	[20M1]
Tb_4CoB_{13}	RT	*P4/mnc*	0.7235(2)		0.7535(2)	[83K3, 88D1]
Tb_4CoB_{13}	RT	*P4/mnc*	0.7338		0.6945	[20M1]
Dy_4CoB_{13}	RT	*P4/mnc*	0.7307		0.6895	[20M1]
Ho_4CoB_{13}	RT	*P4/mnc*	0.7285		0.6872	[20M1]
Er_4CoB_{13}	RT	*P4/mnc*	0.7282		0.6845	[20M1]
Tm_4CoB_{13}	RT	*P4/mnc*	0.7253		0.6813	[20M1]
Lu_4CoB_{13}	RT	*P4/mnc*	0.7203		0.6749	[20M1]
Y_4CoB_{13}	RT	*P4/mnc*	0.7324		0.6920	[20M1]
$GdCoB_4$	RT	*Pbam*	0.5913(9)	1.1451(6)	0.3462(3)	[07C1]
$GdCoB_4$	RT	*Pbam*	0.5887(5)	1.156(1)	0.3406(4)	[77C1]
$GdCoB_4$	RT	*Pbam*	0.5924(3)	1.1472(6)	0.3453(2)	[79S1]
$TbCoB_4$	RT	*Pbam*	0.5899(10)	1.143(2)	0.3421(5)	[83K3]
$DyCoB_4$	RT	*Pbam*	0.5885(10)	1.140(2)	0.3403(5)	[77S1]
$HoCoB_4$	RT	*Pbam*	0.5879(10)	1.136(2)	0.3375(5)	[77S1]
$ErCoB_4$	RT	*Pbam*	0.5869(10)	1.134(2)	0.3353(5)	[77S1]
$TmCoB_4$	RT	*Pbam*	0.5858(10)	1.131(2)	0.3342(5)	[77S1]
$YbCoB_4$	RT	*Pbam*	0.5857(2)	1.1343(3)	0.3360(2)	[04V1]
$YbCoB_4$	RT	*Pbam*	0.5853	1.1299	0.3362	[85D1]
$LuCoB_4$	RT	*Pbam*	0.5827(10)	1.128(2)	0.3327(5)	[77S1]
$YCoB_4$	RT	*Pbam*	0.5878	1.148	0.3410	[77S1]
Nd_2CoB_3	RT	trigonal	0.5423(4)		2.431(4)	[83B2]
Gd_2CoB_3 (d)	RT	$P\bar{6}m2$	0.3130		0.7894	[78S1]
Gd_2CoB_3 (d)	RT	$P\bar{6}m2$	0.3399		0.8193	[20M1]
Dy_2CoB_3	RT	rhomb	0.5403(1)		2.305(1)	[83C1]
$Pr_4(CoB_2)_3$ (d)	RT	$R\bar{3}m$	0.5381		2.300	[20M1]
$Pr_5Co_2B_6$	RT	$R\bar{3}m$	0.5457(2)		2.4718(6)	[83K3]
$Nd_5Co_2B_6$	RT	$R\bar{3}m$	0.5435(1)		2.440(4)	[99Y1]
$Nd_5Co_2B_6H_7$	RT	$R\bar{3}m$	0.5407(1)		2.771(2)	[99Y1]
$La_{5-x}Co_{2+x}B_6$	RT	$R\bar{3}m$	0.5510(2)		2.460(2)	[86D1]
$Ce_{5-x}Co_{2+x}B_6$	RT	$R\bar{3}m$	0.5478(2)		2.447(2)	[86D1]
$Pr_{5-x}Co_{2+x}B_6$	RT	$R\bar{3}m$	0.5457(2)		2.472(1)	[86D1]
$Nd_{5-x}Co_{2+x}B_6$	RT	$R\bar{3}m$	0.5432(2)		2.405(2)	[86D1]
$Sm_{5-x}Co_{2+x}B_6$	RT	$R\bar{3}m$	0.5420(2)		2.382(1)	[86D1]
$Gd_{5-x}Co_{2+x}B_6$	RT	$R\bar{3}m$	0.5419(3)		2.340(1)	[86D1]
$Tb_{5-x}Co_{2+x}B_6$ ($x = 0.75$)	RT	$R\bar{3}m$	0.5401(1)		2.334(1)	[86D1]
$Dy_{5-x}Co_{2+x}B_6$	RT	$R\bar{3}m$	0.5400(1)		2.302(1)	[86D1]

(continued)

Table 8.3 (continued)

Compound	T (K)	Space group	Lattice parameters (nm)			References
			a	b	c	
$Ho_{5-x}Co_{2+x}B_6$	RT	$R\bar{3}m$	0.5390(2)		2.280(2)	[86D1]
$Er_{5-x}Co_{2+x}B_6$	RT	$R\bar{3}m$	0.5388(2)		2.257(2)	[86D1]
$Y_{5-x}Co_{2+x}B_6$	RT	$R\bar{3}m$	0.5422(2)		2.286(1)	[86D1]
La_2CoB_2	RT	trigonal	0.548		2.531	[76S1]
$LaCo_4B_4$	RT	$P4_2/n$	0.7151(4)		0.3811(2)	[78K1]
$CeCo_4B_4$	RT	$P4_2/nmc$	0.5042(3)		0.7081(5)	[73K1, 74B1]
$CeCo_4B_4$	RT	$P4_2/nmc$ [a)]	0.5059(3)		0.7063(5)	[71K1, 72K1, 73K1]
$PrCo_4B_4$	RT	$P4_2/n$	0.7085(4)		0.3815(2)	[78K1]
$NdCo_4B_4$	RT	$P4_2/n$	0.7070(4)		0.3822(2)	[78K1, 79K1, 01C4]
$SmCo_4B_4$	RT	$P4_2/n$	0.7037(4)		0.3824(2)	[78K1]
$SmCo_4B_4$	RT	$P4_2/n$	0.7038		1.1437	[84E1]
$Sm_{1-x}Co_4B_4$	RT	$P4_2/n$	0.7051		0.3871	[90Z1]
$GdCo_4B_4$	RT	$P4_2/nmc$	0.5043(3)		0.7049(5)	[71K2, 77C1]
$TbCo_4B_4$	RT	$P4_2/nmc$	0.5038(2)		0.7030(5)	[72K1, 83K3]
$DyCo_4B_4$	RT	$P4_2/nmc$	0.5026(3)		0.7014(5)	[72K1]
$HoCo_4B_4$	RT	$P4_2/nmc$	0.5020(3)		0.7003(5)	[72K1]
$ErCo_4B_4$	RT	$P4_2/nmc$	0.5016(3)		0.6989(5)	[72K1]
$TmCo_4B_4$	RT	$P4_2/nmc$	0.5009(3)		0.6980(5)	[72K1]
$YbCo_4B_4$	RT	$P4_2/nmc$	0.4999(1)		0.6966(2)	[81C1]
$YbCo_4B_4$	RT	$P4_2/nmc$	0.5003(1)		0.6975(3)	[04V1]
$LuCo_4B_4$	RT	$P4_2/nmc$	0.4998(3)		0.6947(5)	[72K1]
YCo_4B_4	RT	$P4_2/nmc$	0.5028(3)		0.7015(5)	[71K2]
$Sm_{11}(Co_4B_4)_{10}$	RT	$P\bar{4}2_1c$	0.7051(3)		3.8709(13)	[90Z1]
$Sm_4Co_3B_3$	RT	$I4/mmm$	0.704(2)		1.755(3)	[80B1]
CeCoB [(b)]	RT	$P\bar{3}1m$	0.861(5)		0.554(2)	[74B1]
$LaCo_2B_2$	RT	$I4/mmm$	0.3617		1.0225	[73R1]
$LaCo_2B_2$	RT	$I4/mmm$	0.3616(3)		1.0215(5)	[73N2]
$LaCo_2B_2$	RT	$I4/mmm$	0.3641(3)		1.0200(5)	[74K1, 76S1]
$LaCo_2B_2$	RT	$I4/mmm$	0.36186(3)		1.0223(4)	[87R2]
$PrCo_2B_2$	RT	$I4/mmm$	0.3599(3)		0.9932(5)	[73R1]
$PrCo_2B_2$	RT	$I4/mmm$	0.35985(8)		0.9951(2)	[87R2]
$PrCo_2B_2$	RT	$I4/mmm$	0.3600		0.9946	[09L1, 11L1]
$NdCo_2B_2$	RT	$I4/mmm$	0.3586(5)		0.9747(7)	[73N2, 01C4]
$NdCo_2B_2$	RT	$I4/mmm$	0.3592(2)		0.9838(5)	[84F1]
$NdCo_2B_2$	RT	$I4/mmm$	0.35938(2)		0.9850(2)	[87R2]
$NdCo_2B_2$	RT	$I4/mmm$	0.3592		0.9846	[09L2, 11L1]
$SmCo_2B_2$	RT	$I4/mmm$	0.3563(3)		0.9630(7)	[73N2]
$SmCo_2B_2$	RT	$I4/mmm$	0.3580		0.9671	[71N1, 73R1]

(continued)

Table 8.3 (continued)

Compound	T (K)	Space group	Lattice parameters (nm)			References
			a	b	c	
$SmCo_2B_2$	RT	*I4/mmm*	0.35806(2)		0.9673(3)	[87R2]
$GdCo_2B_2$ [c)]	RT	*I4/mmm*	0.3571(2)		0.9529(1)	[14P2]
$GdCo_2B_2$	RT	*I4/mmm*	0.35747(8)		0.9537(2)	[98S1]
$GdCo_2B_2$	RT	*I4/mmm*	0.3573(3)		0.9540(5)	[73N2]
$GdCo_2B_2$	RT	*I4/mmm*	0.3575(2)		0.9561(5)	[73R1, 84F1]
$GdCo_2B_2$	RT	*I4/mmm*	0.35729(8)		0.9541(5)	[87R2]
$GdCo_2B_2$	RT	*I4/mmm*	0.3577		0.9537	[11L1]
$TbCo_2B_2$	RT	*I4/mmm*	0.3557(4)		0.9419(7)	[73N2, 83K3]
$TbCo_{1.8}B_2$	RT	*I4/mmm*	0.3567(2)		0.9489(5)	[84F1]
$TbCo_2B_2$	RT	*I4/mmm*	0.3561		0.9415	[73R1, 12L2]
$TbCo_2B_2$	1.5	*I4/mmm*	0.35483(6)		0.93847(34)	[91A2]
$TbCo_2B_2$	RT	*I4/mmm*	0.35607(11)		0.9423(3)	[87R2]
$DyCo_2B_2$	RT	*I4/mmm*	0.3546(4)		0.9354(6)	[73N2]
$DyCo_2B_2$	RT	*I4/mmm*	0.35548(7)		0.9331(6)	[87R2]
$DyCo_2B_2$	RT	*I4/mmm*	0.3558		0.9308	[12L2]
$HoCo_2B_2$	RT	*I4/mmm*	0.3551		0.9245	[73R1]
$HoCo_2B_2$	RT	*I4/mmm*	0.3556		0.9253	[12L2]
$HoCo_{1.96}B_2$	RT	*I4/mmm*	0.35517(6)		0.9251(5)	[87R2]
$ErCo_2B_2$	RT	*I4/mmm*	0.35450(9)		0.9161(3)	[87R2]
YCo_2B_2	RT	*I4/mmm*	0.3561(3)		0.9358(5)	[73N2, 73R1]
YCo_2B_2	RT	*I4/mmm*	0.3565		0.9338	[76S1]
YCo_2B_2	RT	*I4/mmm*	0.35598(11)		0.9342(3)	[87R2]
YCo_2B_2	RT	*I4/mmm*	0.3562		0.9346	[11L1]
$PrCo_2B_2C$	RT	*Immm*	0.36156(1)		1.03507(6)	[06D1]
$GdCo_2B_2C$	RT	*Immm*	0.35492(7)		1.0491(5)	[98S1]
$TbCo_2B_2C$	RT	*Immm*	0.35246		1.05176	[09E2]
$TbCo_2B_2C$	4.2		0.3535	0.3523	1.0560	[09E2]
$DyCo_2B_2C$	RT		0.35099		1.05268	[09E2]
$HoCo_2B_2C$	RT		0.34997		1.054465	[09E2]
$ErCo_2B_2C$	RT		0.34844		1.05902	[09E2]
$TmCo_2B_2C$	RT	*I4/mmm*	0.3473		1.0647	[09E2]
$TmCo_2B_2C$	30	*I4/mmm*	0.34684(2)		1.06411(6)	[09E3]
$TmCo_2B_2C$	3	*I4/mmm*	0.34681(2)		1.06413(7)	[09E3]
$TmCo_2B_2C$	0.3	*I4/mmm*	0.34689(2)		1.06438(9)	[09E3]

[a]Initial space group *P6/mmm*

[b]Stable at T = 800 °C, decomposes at low temperatures in Ce_2CoB_2 and $Ce_2Co_5B_2$

[c]Single crystal

[d]$\alpha = \beta = 90°$, $\gamma = 120°$

The R_4CoB_{13} compounds crystallize in tetragonal structure, *P4/mnc* space group [88D1, 91G1, 13J1]—Table 8.2a. In the structure, the R atom is bonded in a 6-coodinate geometry to fifteen B atoms and Co to eight boron atoms. Boron is located in three types of sites having in their environments (4R + 8B), (6R + 3B) and (4R + 1Co + 4B) atoms, respectively [13J1, 20M1].

The Gd_2CoB_3 crystallizes in the hexagonal type structure, space group $P\overline{6}m2$. The Gd is bonded in a 9-coordonated geometry to 9 boron sites and Co is bonded in a trigonal planar geometry to three B atoms. The boron is located in three types of sites with different geometries having 6Gd + 3Co in their environments [13J1, 20M1]. The $Pr_4(CoB_2)_3$ or $(Pr_2CoB_3)_2Co$ crystallizes in trigonal structure space group $R\overline{3}m$. The structure is two-dimensional and build from three cobalt molecules and three Pr_2CoB_3 sheets oriented along **z**-direction. There are two Pr sites, six coordinated by boron, having different geometries. The Co and B are located each in only one type site [13J1].

The $RCoB_4$ borides have a bilayer structure, in which one of the layers is made up of seven- and five-membered boron rings with their centers situated directly over and beneath the R and Co atoms, respectively. Accordingly, the R and Co atoms differ in coordination number: 23 and 15, respectively [04V1]. The compounds crystallize in a *Pbam*–type structure—Table 8.2b. The $YCoB_4$ has dominantly covalent bonding and it is a brittle and also hard material [19C1]—Table 5.3. The calculated elastic constants indicated that this boride is mechanically stable but dynamically unstable.

The $R_5Co_2B_6$ compounds with R = La, Ce, Pr, Nd, Sm, Gd, Tb, Dy, Ho and Y, have a trigonal unit cell, space group $R\overline{3}m$ [83K2]. The structure is a derivative of the AlB_2 type, which can be described as the successive alternation of AdAd nets, where A is the net of metal atoms and d the net of boron atoms. In $R_5Co_2B_6$ structure, the nets are arranged in ACdCACdCACdC sequence, perpendicularly to **z**-axis, the A being a defective net of R atoms (2/3 atoms are absent), C is a closely packed crimped net of metal atoms (2/3 R, 1/3 Co) and d is a flat net of boron atoms (analogous to AlB_2). Boron atoms are situated in trigonal prisms $[BR_4Co_2]$. In this structure, the R1 and Co atoms have the same coordination number, CN = 14, while those for R2 and R3 are 16 and 15, respectively—Table 8.2c. The R1(3*a*) site may be statistical occupied by a mixture of rare-earth and cobalt atoms. This means that these borides have regions of homogeneity [86D1]. There is constant boron content (46 at %), and thus, their compositions can be correctly described by $R_{5-x}Co_{2+x}B_6$. Hydride formation provides a strongly anisotropic lattice expansion, as shown in $Nd_5Co_2B_6H_y$, exclusively in the [001] direction, accompanied by a small contraction of the unit cell in the basal plane [99Y1].

The crystal structures of RCo_4B_4 borides consist of two sublattices, one of them being formed by R atoms, the other by Co_4B_4 units—Fig. 8.2 and Table 8.2d, e. The structure contains edge–sharing Co_4 tetrahedra forming one dimensional chains of cobalt. These chains are bonded to adjacent chains by B_2 dimers, forming channels of CoB type. A single atomic chain of R atoms is formed inside each channel. The R atoms form a body-centered tetragonal lattice with neighboring R chains. The c-axis length of R lattice can be equal to c-axis length of the M_4B_4 lattice or the spacing of R atomic chains has a different period than of Co_4B_4 one. In the

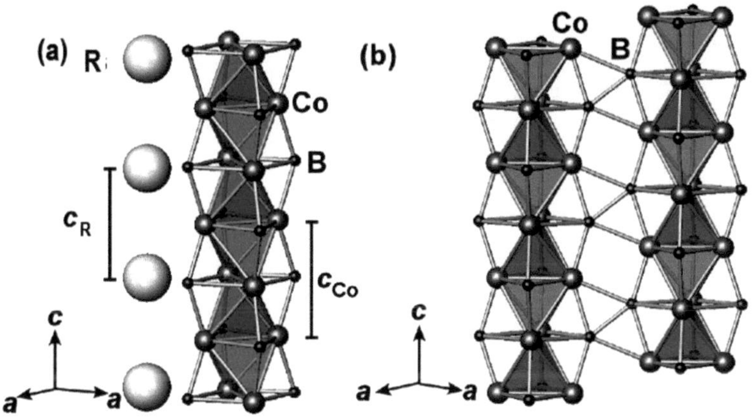

Fig. 8.2 $R_{1+x}Co_4B_4$: crystal structure, view from [001] direction: **a** illustration of two c-axis parameters, c_R and c_{Co}; **b** the connection between two tetrahedral chains [09K1]

incommensurate structures a problem is the space group selection of the complete structure. The sublattice of R atoms is described by *I4/mmm* space group and the other of M_4B_4 units by $P4_2/ncm$ one. The RCo_4B_4 structure can be assumed to have space group $P4_2/n$ [09K1], as in commensurate/normal compounds [85B1, 86G1], or adopts a lower crystal symmetry (*Pccn*) [85G1]. The commensurate structure has been observed in RCo_4B_4 compounds with R = La, Pr, Nd [78K1]. Possible space groups derivatives of RCo_4B_4 structure, considering that R and M_4B_4 (M = Mn, Fe, Co) sublattices are described in space group *I4/mmm* and $P4_2/ncm$, respectively, have been investigated [94Z1]. The crystal structure of the $Sm_{11}(Co_4B_4)_{10}$ boride was analyzed in space group $P\overline{4}2_1c$ [90Z1]. Other incommensurate structure types were also present in $R_{1+\varepsilon}M_4B_4$, particularly when M = Fe [87B1, 93G1]—Sect. 7.5. In $R_{1+\varepsilon}Co_4B_4$ borides, the **c**-axis length of R sublattices and **a**-axis lengths are dependent on the R^{3+} ionic radii. The **c**-axis lengths of M_4B_4 sublattices remain fixed, regardless of R^{3+} ions. Thus, the two sublattices have different **c**-lattice parameters, their ratio defining the superstructure lattice and its common lattice parameter.

A statistical distribution of Rh and Co atoms, at 8*g* sites of $P4_2/nmc$ type structure, was shown in of $R(Co_xRh_{1-x})_4B_4$ series, with R = Lu, Er, Ho [87K1] and R = Y, Er [92D1]. Slight positive deviations from Vegard's law were found in $Y(Rh_{1-x}Co_x)_4B_4$ system. More pronounced "S-shaped" deviations were detected in $Er(Co_xRh_{1-x})_4B_4$ series [92D1].

The RCo_2B_2 compounds crystallize in *I4/mmm* space group [73N2, 91G1]—Fig. 7.7. The R and CoB layers are stacked alternatively along **c**-axis. Each Co atom is coordinated by four B atoms forming a fluorite-type layer. In $LaCo_2B_2$ compound, the length of Co-B bond is 0.20 nm and the B-Co-B angles are 131° and 100°, respectively [11M2]. This indicates that CoB layer is significantly compressed along **c**-axis. In the fluorite-type layer, the Co atom occupies the 4*d* site, resulting in formation of a square lattice net. The B atom occupies 4*e* site on a fourfold rotational axis and resides at the apex of a square pyramid-Table 8.2f.

Fig. 8.3 RCo_2B_2C: crystal structure [98Y2]

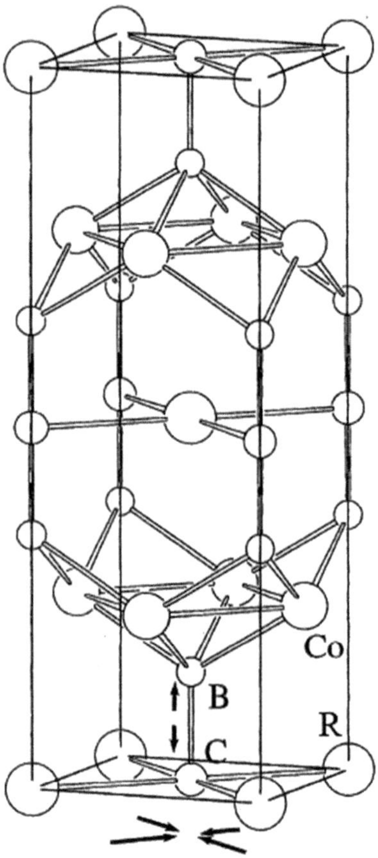

The $Gd_xY_{1-x}Co_2B_2$ forms solid solutions in all the compositional range [93B2]. The $La_{1-x}R_xCo_2B_2$ series with R = Sm ($x \leq 0.3$)[15W1] and R = Y ($x \leq 0.1$) [11M2] were investigated. In the $Gd(Co_{1-x}Fe_x)_2B_2$ system, solid solutions were found in the composition range $x \leq 0.4$ [11L4]. The *a* parameters increase, while the *c* ones decrease when increasing iron content. The $La(Co_{1-x}Fe_x)_2B_2$ with $x \leq 0.3$ and $LaCo_2(B_{1-x}Si_x)_2$ with $x \leq 0.1$ crystallize in a *I4/mmm* space group. In $RCo_2B_{2-x}Si_x$ (R = Ce, Y) series, the presence of two phases were shown in the composition range $0.2 < x < 0.8$ [87R2].

The crystal structure of RCo_2B_2C compounds is derivative from that of RCo_2B_2 one—Fig.8.3. The carbon doping in $GdCo_2B_2C_x$ ($x \leq 1$) [09O1] decreases slightly *a* lattice parameter, while the lattice parameter *c* and unit cell volume increased with increasing x. The C element serves to expand the unit cell in the shape of a square column composed of R atoms, especially along **c**-direction and permit to Co atoms which are too large for the normal lattice to join in combination. The structure can be viewed as an intergrowth of RC (NaCl)-type layers and Co_2B_2 layers [98Y2]. In RCo_2B_2 compounds, the **c**-axis oriented layered structure is established by the

B-B bonding. In the corresponding borocarbides, the two layers are connected, to each other, by B-C-B bonding. The **c**-axis must expand to allow carbon insertion. As example, in $GdCo_2B_2C$, there is an increase of the **c**-axis length by $\cong$ 10%, as compared with that of boride while the **a**-axis parameter shows a small decrease and seems slightly incompressible.

The stability and crystal structure of RCo_2B_2C (R = La, Ce, Pr, Nd, Sm, Gd, Tb, Dy, Ho, Er, Y) was investigated [00E1]. The lattice parameters of this series, are linearly dependent on *r*, the effective R radius, a = 0.2627(29) + 0.0973(31)*r* nm; c = 1.2856(274)−0.2568(290)*r* nm.

The $R(Co_{1-x}Fe_x)_2B_2C$ with R = Tb [12E1, 13E1], R = Dy [18J1] and R = Ho [14E1] form solid solutions in all the composition range. The crystal structures of $R(Co_{1-x}Ni_x)_2B_2C$ series are presented in Sect. 9.4.

The RCo_4B_4 (R = Gd, Tb, Dy, Ho, Er, Tm) compounds are antiferromagnetically ordered [87K2]. The Néel temperatures, T_N, are rather low–Table 8.4. The AF ordering is confirmed by negative paramagnetic Curie temperatures as determined in RCo_4B_4 borides with R = Gd, Dy. The effective moments are only little higher than those of free R^{3+} ions. The zero field ^{11}B and ^{59}Co NMR measurements, at T = 1.3 K, on RCo_4B_4 (R = Gd, Tb, Dy) borides showed that the local symmetry of Co site is not uniaxial [84U1]. The magnitude of the hyperfine fields and NMR patterns of ^{59}Co and ^{11}B, has been explained by the combined effect of both the dipole field from 4f local moment and the quadrupole interaction with asymmetric field gradient.

The RRh_4B_4 borides with R = Er [81O2], Lu [81O1] and Y [20J1] are superconductors with transition temperatures T_s = 8.7 K, 11.45 K and 10.6 K, respectively. The YRh_4B_4 is a conventional superconductor described by BCS theory. $HoRh_4B_4$ is ferromagnetic with T_c = 6.8(2) K [79L1]. The $R(Co_xRh_{1-x})_4B_4$ series with R = Lu and Er are both magnetically ordered and superconductors [87K1]. The superconducting transition temperatures, T_s, in $Lu(Co_xRh_{1-x})_4B_4$ series, decrease from T_s = 11.7 K (x = 0) to 0.70 K (x = 0.4) and then at 0 K (x $\cong$ 0.42). In $Er(Co_xRh_{1-x})_4B_4$, the T_s values changes from T_s = 8.70 K (x = 0) to 0.3 K (x = 0.37) and by extrapolation, at 0 K when x $\cong$ 0.38. No superconductivity was observed in $Ho(Co_xRh_{1-x})_4B_4$ system, due to the persistence of magnetic order in the Rh-rich region and the occurrence of a FM-AFM transition [87K1]. Latter on [92D1], was shown that in $R(Rh_{1-x}Co_x)_4B$ series, dilute cobalt substitutions depress rapidly the superconductivity transition temperatures with rates dT_s/dx = 88 K/mol, when R = Y and 78 K/mol as R = Er. Anomalies in $T_s(x)$ and transitions widths occur in $Y(Rh_{1-x}Co_x)_4B_{4-x}$ system, with x = 0.10.

The $LaCo_2B_2$ is a Pauli type paramagnet [11W1]. The electronic structure indicates a strong hybridization between Co3d and B2p, which removes the magnetic ordering of Co atoms [11W1]. The substitution of La by Y atoms would not change the ground state, while the substitution of La by Sc might induce a magnetic moment on Co atoms. The $LaCo_2B_2$ has a metallic behavior, the resistance at T = 300 K, being of 7.10^{-7} Ω, with a small drop in their temperature dependence, at T = 4 K. The Seebeck coefficient is $\cong$ 7.4 mV/K, indicating that electron carriers are important for conduction [11M2].

Table 8.4 Magnetic properties of R-Co-B compounds with atomic ratio Co/B ≤ 1

Compound	Magnetic structure	M_s (μ_B/f.u.)	T_{ord} (K)	M_{eff} (μ_B/f.u.)	θ (K)	References
$CeCo_4B_4$	PM[a]					[87K2]
$GdCo_4B_4$	AFM		2.48 (T_N)	8.06	−5	[84U1, 87K2]
$TbCo_4B_4$	AFM		3.68 (T_N)			[87K2]
$DyCo_4B_4$	AFM		2.92 (T_N)	11.10	−3	[87K2]
$HoCo_4B_4$	AFM		1.91 (T_N)			[87K2]
$ErCo_4B_4$	AFM		0.32 (T_N)			[87K2]
$TmCo_4B_4$	AFM		0.17 (T_N)			[87K2]
$LuCo_4B_4$	PM[a]					[87K2]
$LaCo_2B_2$	PM, $\chi = 2.45 \cdot 10^{-4}$ emu/mole, at T = 300 K					[87R2]
$PrCo_2B_2$	FM	1.67	18 (T_c) 16(T_c) [11L1]	3.66	21	[09L1]
$PrCo_2B_2$	FM		19.5 (T_c)	3.57	25	[87R2]
$NdCo_2B_2$	FM	1.26[b]	32(1) (T_c)	3.27(10)	0(2)	[84F1]
$NdCo_2B_2$	FM		28 (T_c) 27(T_c) [11L1]	3.59	40	[87R2]
$SmCo_2B_2$[c)]				1.76		[87R2]
$GdCo_2B_2$	FM		26.5 (T_c)	7.80	33	[87R2]
$GdCo_2B_2$	FM	6.42[d]	26 (T_c)	7.22	80(2)	[84F1]
$GdCo_2B_2$	FM	6.97[d]	25 (T_c)	7.96	27.5	[93B2]
$GdCo_2B_2$ (s.c.)	Phase transition at $T_1 = 18.5$ K, $T_2 = 13$ K, $T_3 = 7$ K	7.0	22 (T_c)	‖**a** = ‖**b** = 8.05	‖**a**, 29.5 ‖**b**, 27.5	[14P2]
$GdCo_2B_2$	FM	7.15[d]	21 (T_c)	7.89		[09L3]

(continued)

Table 8.4 (continued)

Compound	Magnetic structure	M_s (μ_B/f.u.)	T_{ord} (K)	M_{eff} (μ_B/f.u.)	θ (K)	References
$TbCo_{1.92}B_2$	AFM	3.50[b]	19 (T_c)	9.60	65	[84F1]
$TbCo_{1.8}B_2$	AFM		13 (T_c)	9.76	76	[87R2]
$TbCo_2B_2$		5.37[e]	15.5 (T_c)			[12L2]
$TbCo_2B_2$	T < 10 K, commensurate **q** = (0.25, 0.25, 0), $\mathbf{M_{Tb}}$ ‖**a** T > 10 K, incommensurate, **q** = (0.244, 0.244,0)	8.5(1.1)[f] 2.5 [g]				[91A2]
$DyCo_2B_2$	AFM		9.3 (T_{ord})	10.87	67	[87R2]
$DyCo_2B_2$		6.77[e]	9.6 (T_{ord})			[12L2]
$HoCo_{1.96}B_2$			8.5 (T_{ord})	10.57	28	[87R2]
$HoCo_2B_2$		6.27[e]	9.2 (T_{ord})			[12L2]
$ErCo_2B_2$	AFM		3.3 (T_{ord})	9.63	22	[87R2]
YCo_2B_2	PM, $\chi = 2.54 \cdot 10^{-4}$ emu/mole, at 300 K					[87R2]
YCo_2B_2	PM, $\chi = 1.9 \cdot 10^{-4}$ emu/mole, at 300 K $\chi = 2.12 \cdot 10^{-4}$ emu/mole, at 2 K $\chi = 3 \cdot 10^{-4}$ emu/mole, at T ≅ 30 K					[93B2]
$CeCo_2B_2C$	PM, $\chi = \chi_0 + C/(T-\theta)$ $\chi_0 = 5.3(1.2) \cdot 10^{-4}$ emu/mole; 13% Ce^{3+}					[95C1]
$PrCo_2B_2C$	AFM	1.7[e)]	10 (T_N)			[04E1]
$PrCo_2B_2C$	AFM		8.5 (T_N)			[06D1]
$NdCo_2B_2C$			3.0 (T_N) 1.1 (T_m)			[00E1]
$SmCo_2B_2C$			5.9 (T_N) 1.6 (T_m)			[00E1]
$GdCo_2B_2C$			8.8 (T_N)			[00E1]

(continued)

Table 8.4 (continued)

Compound	Magnetic structure	M_s (μ_B/f.u.)	T_{ord} (K)	M_{eff} (μ_B/f.u.)	θ (K)	References
$TbCo_2B_2C$ (s.c.)	FM, **q** = (0,0,0), M = 7.6	8.2^{h}	6.3 (T_N) 3.7(2) (T_m)	**H‖c**, 9.7(1) **H‖a**, 9.6(1)	**H‖c**, −29.4(1) **H‖a**, 14.4(1)	[09E1]
$TbCo_2B_2C$	FM, **q** = (0,0,0), M = 7.6‖**a**	7.2^{e}	6.3 (T_N)			[12E1]
$DyCo_2B_2C$	FM, **q** = (0,0,0), M = $5.2(2)^{i}$‖ [110]	$6.2(1)^{j}$	8.0 (T_N) 2.6 (T_m)			[09E2]
$HoCo_2B_2C$	FM, **q** = (0,0,0), M = 7.2(2)‖ [110]	$7.3(1)^{j}$	5.4(2) (T_N)			[09E2]
$HoCo_2B_2C$	AFM, **q** = (1/2,0,1/2)	7.8^{k}	5.4(1) (T_N) 1.46(6) (T_c)	10.5(2)	−15.2	[99R1]
$ErCo_2B_2C$	AFM, **q** = (1/2,0,1/2); M = 6.8(2)‖**b**, e.a.m.‖[010]	7.1^{i}	4.0(1) (T_N) 0.3 (T_m)			[09E2]
$TmCo_2B_2C$	FM, **q** = (0,0,0), e.a.m.‖[001] $\chi = \chi_0 + C/(T-\theta)$, $\chi_0 = 8(1)\cdot 10^{-3}$ emu/mole	$4.5(2)^{l}$	≅ 1.0 K (T_N)	7.6(2)	−4.5(3)	[09E3]

[a] No magnetic order was shown at T > 60 mK
[b] T = 4.1 K, μ_0H 1.7 T
[c] No magnetic order was shown at T ≥ 1.6 K
[d] T = 4.2 K, $\mu_0 H = 7$ T
[e] T = 2 K, $\mu_0 H = 7$ T
[f] 1.5 K
[g] T = 14 K
[h] T = 2 K, $\mu_0 H = 9$ T
[i] T = 1.3 K
[j] T = 1.3 K, extrapolation at $H^{-1} \to 0$
[k] T = 1.8 K, $\mu_0 H = 9$ T
[l] T = 0.65 K, $\mu_0 H = 15$ T; 1 emu/g = 1 Am^2/kg. For magnetic properties of RCo_2B_2 see also [87J4]

The YCo_2B_2 is also a Pauli-type paramagnet [87R2, 93B2]. The magnetic susceptibility increases from $2.12 \cdot 10^{-4}$ emu/mole, at 2 K, up to $\cong 3 \cdot 10^{-4}$ emu/mole at 30 K, then little decreasing up to $1.90 \cdot 10^{-4}$ emu/mole at T = 300 K [93B2].

Bulk superconductivity, with transition temperatures T_s = 4.2 K in $La_{0.9}Y_{0.1}Co_2B_2$, T_s = 5.8 K in $LaCo_{1.7}Fe_{0.3}B_2$ [16M1] and T_s = 3.6 K in $La_{0.8}Sm_{0.2}Co_2B_2$, were reported [15W1]. An anomaly in electrical resistivity around T_s was also shown in the last compound. The angle resolved photoemission spectroscopy in $LaCo_{1.73}Fe_{0.27}B_2$, at T = 30 K, found that E_F is located at a deep in DOS [12N1], in agreement with band structure calculations [11M2]. The reduction in DOS together with the strong Co3d-B2p covalent bonding, removes the magnetic order and may cause the superconductivity [11M2, 12N1]. The site selective NMR experiment revealed the multiband nature of the Fermi surface in $LaCo_{1.7}Fe_{0.3}B_2$ [16M1]. The absence of correlated low energy magnetic spin fluctuations, in the normal state, was also shown. The low upper critical field of $La_{0.8}Sm_{0.2}Co_2B_2$, suggests that the dominant magnetic field suppressing superconducting mechanism is the orbit pair-breaking effect [15W1]. The substitution of Si into B sites in $LaCo_2B_{2-x}Si_x$ raised the resistivity, as for other elements doping and did not induce a superconducting transition above T = 2 K [11M2].

The RCo_2B_2 compounds with R = Pr, Nd and Sm are ferromagnetically ordered—Table 8.4. The magnetic properties of $GdCo_2B_2$ single crystal were conceived in terms of low temperature antiferromagnetism ($T < T_N$ = 22 K), undergoing three order to order magnetic phase transitions, at T_1 = 18.5 K, T_2 = 13 K and T_3 = 7 K, respectively [14P2]. The phase diagrams along **a** and **c** directions are similar, owing the weak magnetocrystalline anisotropy, but along **a**-axis, the values of the critical fields characterizing the magnetic transitions are lower mainly in the high field phase I—Fig.8.4. The phase I, at low temperatures, is stabilized up to ≅ 0.4 T (**H** ||**a**) or 0.8 T (**H**||**c**). The magnetic structures of different phases were not determined. According to [14P2], most likely $GdCo_2B_2$, in zero field, orders first AF, at T_N = 22 K and when cooling it undergoes two consecutive magnetic phase transitions to others AFM structures. Application of magnetic field, leads to the field induced phase I. The phases I, III and IV survive to lowest temperatures. The phase II exists only as a bounded dome between the phases I and III. The low temperatures AF phases III and IV transform by spin flop transitions, up to field induced FM phase I. Above T_N, the reciprocal susceptibility follows a Curie–Weiss type dependence superposed on a Pauli paramagnetic contribution [93B2]. The presence of AFM structure was shown in low magnetic field (<0.2 T) at T < 15 K, whereas a field induced transition from AFM to FM phase has been observed by application of a field $\mu_oH > 0.4$ T.

The effect of pressure on the magnetic properties of polycrystalline $GdCo_2B_2$ has been investigated [12L1, 16H1]. At normal pressure, the magnetic transition temperatures, are close to those reported in $GdCo_2B_2$ single crystal [14P2], but missing the phase transition at T_3 = 7 K, the others being located at T_c = 20.5 K, T_1 = 18 K and T_N = 11.5 K. The low temperature magnetic phase can be suppressed by pressure. The first order-type magnetic phase transitions changed to second one, gradually with increasing pressure [12L1]. The temperatures T_c and T_1 increase with different rates and coincides each other at p > 1.6 GPa. The pressure, decreases the

T_N (T_2) values; these disappear at p > 1.4 GPa, indicating that the low temperature phase can be easily suppressed. A long time relaxation behavior was observed at T = 4 K. In low fields, a spin glass type ordering can be shown. The easy axis of magnetization in $GdCo_2B_2$, as determined from ^{155}Gd Mössbauer study, is located in the basal plane. The quadrupole coupling constant at 4.1 K is 580 MHz [84F1].

The saturation magnetizations of $(Gd_xY_{1-x})Co_2B_2$ series, at 2 K, are close to those of the Gd fraction in the compounds [93B2]. A linear composition dependence of T_c values has been shown. The thermal variations of magnetic susceptibilities can be described by a superposition of a temperature independent term on a Curie–Weiss type behavior. The substitution of cobalt by iron in $Gd(Co_{1-x}Fe_x)_2B_2$ compounds increases the magnetic ordering temperature from 19 K (x = 0) to 32 K (x = 0.4) [11L4]. The saturated magnetic entropy indicates that ground state of gadolinium is 7/2.

Two magnetic phases were reported to exist in $TbCo_2B_2$ below $T_N \cong 19$ K [91A2]. At $T < T_1 \cong 10$ K, the magnetic cell is commensurate with the chemical cell, having propagation vector **q** = (0.25, 0.25, 0). The magnetic moments of Tb atoms are aligned along the **c**-axis and their magnitude varies in sinusoidal way. At $T \cong 10$ K, a first order transition occurs, and the magnetic structure becomes incommensurate with the crystal lattice, the wave vector at T = 14 K being **q** = (0.244, 0.244, 0). Two magnetic phase transitions, located at T_M = 15.5 K and 8.8 K were later reported [12L2]. The magnetic transition at T_M is of first order. Field induced metamagnetic transition from AFM to FM state was shown at low temperatures. A magnetic structure, similar to that of $TbCo_2B_2$, has been theoretically analysed starting from an Ising body centered tetragonal lattice, with competing first-, second- and third-neighbors in plane interactions [92P1]. The stability of the commensurate (1/4, 1/4,

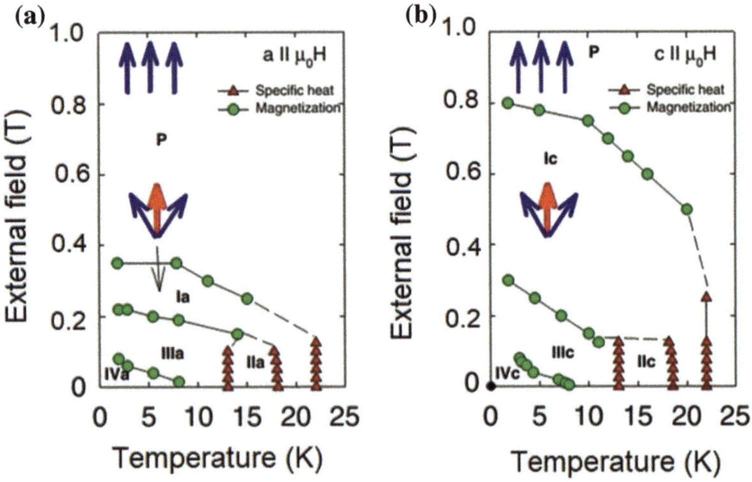

Fig. 8.4 $GdCo_2B_2$ single crystal: magnetic phase diagrams for magnetic field applied along the a-axis (**a**) and c-axis (**b**) [14P2]

0) state, at low temperatures, as well as the (q, q, 0) incommensurate phase, at higher temperatures was shown to be enhanced by the inter-plane interactions. Field induced metamagnetic transition was shown in $TbCo_2B_2$, at 2 K, from AFM to FM ordering [12L2].

The RCo_2B_2 compounds with R = Dy, Ho and Er have rather low Curie temperatures and saturation magnetizations [87J4, 87R2]. The effective moments, determined at $T > T_N$ are close to those of free R^{3+} ions.

The specific heats of RCo_2B_2 compounds with R = La [15W1], R = Pr [09L1], R = Nd [09L2], R = La, Gd [09L1], R = Pr, Nd, Gd, Y [11L1], R = Gd [09L3, 12L1], R = Tb, Dy, Ho [12L2] and $Gd(Co_{1-x}Fe_x)_2B_2$ [11L4] were investigated. The magnetic contributions to the total specific heats were evaluated. The λ-shape peaks at the magnetic transition temperatures are evidenced, except for R = Tb when a broad peak was observed, at $T \cong 15$ K.

The magnetocaloric effects were investigated in RCo_2B_2 compounds with R = Pr [09L1], R = Nd [09L2], R = Pr, Nd, Gd [11L1], R = Gd [09L3, 14P2], R = Tb, Dy, Ho [12L2] and $Gd(Co_{1-x}Fe_x)_2B_2$ with $x \leq 0.4$[11L4] compounds. Entropy changes ($-\Delta S_m$) values, in field $\mu_oH = 5$ T, near magnetic ordering temperatures were 8.1(Pr), 7.1(Nd), 17.1(Gd), 6.2(Tb), 12.1(Dy) and 12.2(Ho) J/kgK [09L1, 09L2, 09L3, 11L1, 12L2]. The corresponding maximum adiabatic temperatures changes, for $PrCo_2B_2$, were evaluated to be 4.3 K and 9.8 K, for field changes of 2 and 7 T, respectively [09L1].

The YCo_2B_2C borocarbiode has an exchange enhanced magnetic susceptibility [00E1, 04E1], as in YCo_2 compound [92B1]. A quadratic temperature dependence of the electrical resistivity was evidenced up to T = 50 K, as in systems which show spin fluctuations. It was assumed that the cobalt sublattice, in RCo_2B_2C borocarbides, as in RCo_2 compounds [92B1], develops a spin fluctuations behavior, at low temperatures [00E1, 04E1].

The local susceptibility measured at the site of ^{140}Ce probe in $CeCo_2B_2C$ indicates the presence of an unstable moment on cerium; fast inter-configurations fluctuation of 4f electron yields strongly temperature dependent average valence of Ce ions [99T2]. The cerium is in mixed valent state, the Ce^{3+} content being of 13%, as determined from temperature dependence of the magnetic susceptibility [95C1].

In $PrCo_2B_2C$, an AFM transition was found at T_N = 8.5–10 K [06D1, 08M6]. As compared to other RM_2B_2C (M = Ni, Pt) borocarbides, the $Pr_2Co_2B_2C$ shows an anomalous contraction in the (ab) plane and volume, which does not correspond to the ionic radii size and/or chemical pressure effects [06D1]. As a result, there is a strong anisotropy, the easy axis of magnetization being situated in [001] direction, instead along the [110] axis, as found in $PrNi_2B_2C$—Sect. 9.4. The presence of a broad peak in magnetization and a tiny anomaly in the specific heat, confirmed the AFM transition, at $T \cong 10$ K, in $PrCo_2B_2C$ [08M6]. The resistivity, as function of temperature, shows a quadratic behavior at T < 5 K, which persists for magnetic fields up to $\mu_0H = 7.8$ T. These characteristics are in agreement with a spin fluctuations scenario. At low applied fields the magnetoresistance (MR) is negative, suggesting that magnetic fluctuations decrease the electron scattering as the magnetic field is increased, when T < 15 K [08M6]. At high magnetic field, the MR values are positive.

In $PrCo_2B_2C$, the onset of magnetic order does not appear as a λ-type event. The magnetic entropy, reaches values 3.43 J/molK at T = 10 K and 5.7 J/molK, at T = 20 K [04E1].

The RCo_2B_2C compounds with R = Nd and Sm are ferromagnetically ordered. Two magnetic transitions were shown for each compound. The $GdCo_2B_2C$ boride is also ferromagnetically ordered. The ^{155}Gd Mössbauer spectroscopy, at T = 4.2 K, showed a large negative value of the hyperfine field (–45.6 T), attributed to interactions inside the Gd sublattice [95M2]. Short range magnetic fluctuations were present at $T > T_c$. The analysis ^{59}Co and ^{11}B NMR spectra, in $Gd(Co_xNi_{1-x})_2B_2C$ compounds revealed the presence of a small hyperfine field (≅6 T), little dependent on composition, assumed to be transferred from Gd. The cobalt has been suggested to be not magnetic [95B2]. Substitution of Ni by Co causes an overall decrease of the magnetic transition temperatures: T_N, from PM to helical AFM state and then to a commensurate state, at T_{sR}. These transitions were attributed to the increase of the distance between GdC layers and changes in the DOS at E_F [95B2].

The magnetic phase diagram of $Y_{1-x}Gd_xCo_2B_2C$ series indicates the presence of the magnetically ordered phase for $0.7 \leq x \leq 1$ [95B3]. The changes of magnetic ordering temperatures, T_N and $T_{ord.}$, were correlated with the magnetic dilution effects as well as the reduction of the distances between $(Gd_xY_{1-x})C$ pairs.

The RCo_2B_2C series with heavy rare earths, show either collinear AFM or are FM as R = Tb [09E1, 12E1, 13E1, 20Z1], R = Dy [01K2, 09E2, 18J1, 20Z1], R = Ho [99R1, 00E1, 09E2, 14E1], R = Er [00E1, 09E2] and R = Tm [09E3]—Table 8.4. In RCo_2B_2C, the CEF induced splitting of the J moment of the R ions, at the magnetic transition temperatures, is substantial and consequently the De Gennes factor scaling is not obeyed [00E1].

The $TbCo_2B_2C$ borocarbide undergoes a magnetic transition, at T_c = 6.3 K, with the easy axis along the [100] direction and concomitantly the unit cell, undergoes a tetragonal to orthorhombic distortion [09E1]. The temperature dependence of magnetization, follows a Brillouin function with J = 6.

The ND patterns of RCo_2B_2C compounds with R = Dy and Ho were indexed on the basis of a FM unit cell, which is of the same dimensions as the crystalline one [09E2]. The magnetic moments are confined in the basal plane. In $HoCo_2B_2C$, a second magnetic transition was reported at T = 1.46(6) K [99R1]. Latter on [09E1], this transition was attributed to the presence of magnetic impurity. Long range ferromagnetic order of Tm sublattice in $TmCo_2B_2C$, develops at $T_c \cong 1$ K [09E3]. The magnetization at T = 0.65 K is only 65% of the value expected for Tm^{3+} ion, due to crystal field effects. The effective moment is in agreement with the expected value for free Tm^{3+} ion.

The $ErCo_2B_2C$ is a collinear antiferromagnet, at $T \leq T_N = 4.0(1)$ K. The magnetic structure is described by the propagation vector $\mathbf{q}$ = (1/2, 0, 1/2). The Er moments are confined within the basal plane. An anomaly in the specific heat and magnetic susceptibility was shown at T_N = 0.37(2) K [00E1]. No magnetic peaks, associated with this "anomaly" was shown by neutron diffraction in the temperature range 50 mK < T < 700 mK [09E2]. The absence of modulated magnetic structures, at T > 50 mK, (different from the corresponding nickel compound) has been attributed to the quenching of the Fermi surface nesting features.

The specific heat of RCo_2B_2C compounds with R = La, Ce, Pr, Nd, Sm, Gd, Tb, Dy Ho, Er, Y) [00E1], R = Pr [04E1, 08M6], R = Nd [09L2], R = Ho [14E1], R = Tb [09E1], R = Tm [09E3] and R = Y [95M1] were investigated. The Sommerfeld constant γ in YCo_2B_2C is rather high, suggesting the presence of spin fluctuations [00E1]. A large γ value was also shown in $PrCo_2B_2C$, while the transition at T_c was not observed [08M6]. The presence of magnon contribution to the specific heat in $TbCo_2B_2C$ was evidenced [09E2]. The power type thermal evolution of magnetic contribution to specific heat, close to 3/2 law, is consistent with would expect from a magnon contribution of ferromagnetic state, wherein the dispersion relation is quadratic and anisotropy field is vanishing small [09E3]. The analysis of specific data in $HoCo_2B_2C$ showed that the ground state it is a doublet [99R1].

The resistivities of RCo_2B_2C borocarbides with R = La, Ce, Pr, Nd, Sm, Gd, Tb, Dy, Ho, Er, Y [00E1], R = Ce [95C1], R = Dy [04E1], R = Ho [99R1] or $PrCo_2B_2C$ [04E1, 06D1, 08M6] were analysed. The RCo_2B_2C compounds show typical metallic behavior, at high temperatures and an incoherent scattering contribution associated with the crystal-electric field (CEF) and hybridization influence at low temperatures [00E1, 04E1]. At high temperatures, the phonon dominates the linear behavior. The resistivity, $\rho(T)$, of $PrCo_2B_2C$ shows a linear variation in the temperature range 7 K$\leq$T$\leq$100 K and at T < 7 K a quadratic shape [08M6]. The magnetoresistance is negative at T < 15 K and low fields and becomes positive at higher fields (>3 T).

A considerable magnetocaloric effect was evidenced in RCo_2B_2C compounds [20Z1]. The isothermal magnetic entropy change (J/kgK) and relative cooling power (J/kg) determined in fields up to 5 T were: R = Gd (10.34, 238), R = Tb (18.09, 438.2), R = Dy (17.79, 480.1)—Table 8.5.

The $Pr_{1-x}Dy_xCo_2B_2C$ system behaves as a normal metallic paramagnet, at high temperatures. At low temperatures, the parent and the low Dy-substituted samples show anomalous features. The T_N temperatures are unexpected high and the magnetic phase diagram manifests a strong non-monotonic composition dependence [04E1]. The CEF effects exercise a strong influence on the low temperature properties, in addition to spin fluctuations type behavior. The temperature range for the manifestation of CEF or spin fluctuations features, depend on the relative strength of the characteristic CEF parameters and spin fluctuation temperature, T_{sf}, respectively. For all x, at T > T_{sf}, the magnetic, resistive and thermal properties are those of paramagnetic metal. For intermediate to lower temperatures, thermal depopulation of the Pr^{3+}-CEF levels gives rise to a deviation from Curie–Weiss behavior and a CEF

Table 8.5 Magnetic entropy change (ΔS_M) and relative cooling power (RCP)

Compound	T_c(K)	$-\Delta S_M$ (J/kgK)	RCP (J/kg)	References
$GdCo_2B_2C$	172	10.34	238.1	[20Z1]
$TbCo_2B_2C$	93	18.09	438.2	[20Z1]
$DyCo_2B_2C$	77	17.79	480.1	[20Z1]

related incoherent resistive contribution. Decreasing temperature up to T = 4 K, the CEF effects lead to a further reduction of magnetization M(x), though T_N for samples with x = 0, is rather high. Alloying with Dy would reduce $T_N(x)$. Beyond a critical concentration (x > 0.17), this character is weakened substantially to the extent that the magnetic properties are determined only by the localized 4f magnetic moments [04E1].

The physical properties of $R(Co_xNi_{1-x})_2B_2C$ borides, as for example with R = Gd [95B2], R = Tb [13E1], R = Dy [18J1], Ho [14E1] R = Dy, Ho, Er, Tm, Lu [97S1], R = Y [95G1, 97I1] or R = Y, Lu [96B1] are reviewed in Sect. 9.4.

8.3 RCo_3B_2 Compounds

The RCo_3B_2 compounds, at normal conditions, crystallize in a hexagonal structure having *P6/mmm* space group. The structure, is a derivative of the RCo_5-type, resulting from ordered substitution of Co by B atoms on the 2*c* site [74K2]—Table 8.6. The R cations form a simple hexagonal lattice along with the B atoms, while the Co atoms form a Kagomé net [77W1, 79K1, 12S1]. Each Co_3 triangle is capped with boron atoms, below and above, to give a trigonal bipyramid. These trigonal bipyramids, are then joined in three dimensions by sharing vertices—Fig. 8.1. The metal–metal distances are close to those in respective transition metals. The R atoms form chains along **c**-axis, these chains being isolated from each other. The RCo_3B_2 compounds with R = La, Pr, Nd and Eu in earlier studies have been not obtained [73N1, 73R1, 79K1]. Later on, the lattice parameters of $PrCo_3B_2$ were reported [84R1, 19V1].

The $R_{1-x}Y_xCo_3B_2$ series with R = Gd [91I1, 95B1] and R = Tb [08W1, 10W1] form solid solutions in all the composition range. At $T < T_c$, the $TbCo_3B_2$ crystallize in an orthorhombic type structure space group *Cmcm* [05D1]. The orthorhombic distortion, reaches at T = 1.5 K a value of $2.8(2)\cdot10^{-4}$. The magnetostriction constant, at this temperature, $\lambda^{\gamma,2} = 3.0(2)\cdot10^{-4}$, is larger than that reported in $TbCo_5$, $\lambda^{\gamma,2} \leq 10^{-4}$ [91A1] and is an evidence for a strong anisotropy. The low temperature crystal structures of $Tb_{1-x}Y_xCo_3B_2$ with $0.05 \leq x \leq 0.25$ were refined in monoclinic type structure of *C2/m* type [10W1]. The maximal subgroup of *Cmmm*, consistent with

Table 8.6 Space groups and atomic sites for three structure types of $TbCo_3B_2$ boride [05D1, 08W1, 10W1]

Compound	*P6/mmm*[a]				*Cmmm*				*C2/m*			
	Site	x	y	z	Site	x	y	z	Site	x	y	z
$TbCo_3B_2$	Tb1*a*	0	0	0	Tb2*a*	0	0	0	Tb2*a*	0	0	0
	Co3*g*	1/2	0	1/2	Co2*c*	1/2	0	1/2	Tb2*d*	0	1/2	1/2
	B2*c*	1/3	2/3	0	Co4*f*	1/4	1/4	1/2	Co4*f*	1/4	1/4	1/2
					B4*g*	≅ 1/3	0	0	B4*d*	x ≅ 1/3	0	z ≅ 0

[a]Atomic environment: Tb1*a* – pseudo Frank-Kasper $B_6Co_{12}Tb_2$; Co3*g* – rhombic dodecahedron $B_4Co_6Tb_4$; B2*c* – trigonal prism Co_6

canted magnetic angles $0 < \theta < 90°$, has the monoclinic *C2/m* space group. For compositions with $x \geq 0.5$, the crystal structure is hexagonal, space group *P6/mmm*.

The $R(Co_{1-x}Ni_x)_3B_2$ series with R = Sm [04M1] and R = Y [06B1, 15B1] crystallize at RT, in *P6/mmm* space group, in all the composition range. The lattice parameters of RCo_3B_2 as well as of some related pseudoternary compounds are listed in Table 8.7.

The physical properties of the RCo_3B_2 compounds are strongly influenced by cobalt atoms, which according to R partner changes their magnetic behavior from Pauli-type paramagnetism to spin fluctuations or as weak ferromagnet. The RCo_3B_2 compounds with R = Y [91I1, 95B1, 96P1, 01C1, 06B1], R = Lu [96P1] and Ce [77O1, 84Y1, 85Y1], at low temperatures, are paramagnetic. The magnetic susceptibility, χ, of YCo_3B_2 increases with temperature up to a maximum located at $T \cong 150$ K and then decreases [95B1]—Fig. 8.5a. Above $T \cong 300$ K the $\chi^{-1}(T)$ follows a Curie–Weiss-type dependence with an effective cobalt moment of 1.34 μ_B/atom. Similar behavior was shown in RCo_2 compounds with R = Y or Lu, although with more flattened maxima, located at T = 260 K (Y) and T = 370 K (Lu), respectively. Such behavior is typical for a spin fluctuations type behavior [79M1, 91M1]. In the systems having rather high Stoner exchange enhanced factor or weak ferromagnets, the wave number dependent magnetic susceptibility, χ_q, has a large enhancement due to electron–electron interactions for small **q** values. The average amplitude of longitudinal spin fluctuations $<S_{loc}> = 3k_BT\Sigma_q\chi_q$ increases with temperature and reaches an upper limit determined by charge neutrality condition at temperature $T > T^*$. Above T^* the magnetic susceptibility follows a Curie–Weiss dependence, the cobalt moments being localized in **q** space. The lower effective cobalt moment in YCo_3B_2 than in YCo_2 compound is due to a strong hybridization of Co3d-B2p states. Band structure calculation performed on YCo_3B_2 boride, did not find a possibility for sustainable ferromagnetic ordering [98Y1, 00K2, 06B1]. Near E_F, the density of states is dominated by the Co3d band. The XPS valence band spectrum was in agreement with theoretical calculations [00K2]. The ^{59}Co NMR study of YCo_3B_2, at 4.2 K, under external field up to 1.8 T, evidenced that there is no hyperfine field at Co site [98I1], in agreement with their paramagnetic response. The same result has been obtained by the analysis of NMR spin echo spectra [92K1].

In order to analyze the part payed by the exchange enhancement factor on the magnetic properties, the $YCo_{3-x}Ni_xB_2$ series, with $x \leq 2$ were investigated [06B1]. The same type of temperature dependence of the magnetic susceptibilities is shown for all compositions, the maxima in $\chi(T)$ being more flattened and shifted to lower temperatures as the Ni content increases. The magnetic susceptibilities at T = 2 K, χ_0, as well as the effective cobalt moments (Curie constants), decrease as Co is gradually replaced by Ni, in parallel way with Stoner enhancement factor. The computed values agree well with the experimentally determined susceptibilities, χ_0—Fig. 8.5d.

The $CeCo_3B_2$ compound, has been reported to be paramagnetic [77O1, 84Y1, 85Y1, 09I1]. The presence of magnetic ordering, in YCo_3B_2 [05S2] and $CeCo_3B_2$ [07H2] compounds, associated with cobalt sublattice, with Curie temperatures $T_c \cong 150$ K and small saturation magnetizations was also suggested. Probably, that

Table 8.7 Space groups and lattice parameters of RCo_3B_2 compounds

Compound	T (K)	Space group	Lattice parameters a	b (β°)	c	References
$CeCo_3B_2$	RT	*P6/mmm*	0.5057(3)		0.3036(2)	[69K1]
$CeCo_3B_2$	RT	*P6/mmm*	0.5058(2)		0.3040(2)	[73N3]
$CeCo_3B_2$	RT	*P6/mmm*	0.5061(4)		0.3038(2)	[73R1]
$PrCo_3B_2$	RT	*P6/mmm*	0.5532(4)		0.3015(2)	[84R1, 19V1]
$SmCo_3B_2$	RT	*P6/mmm*	0.5101(3)		0.2991(2)	[69K1]
$SmCo_3B_2$	RT	*P6/mmm*	0.5079(3)		0.3031(2)	[73N3]
$SmCo_3B_2$	RT	*P6/mmm*	0.5110		0.2995	[94I1]
$GdCo_3B_2$	RT	*P6/mmm*	0.5074		0.3039	[91I1]
$GdCo_3B_2$	RT	*P6/mmm*	0.5056		0.3019	[85M1]
$GdCo_3B_2$	RT	*P6/mmm*	0.5059(3)		0.3019(2)	[69K1]
$GdCo_3B_2$	RT	*P6/mmm*	0.5066(3)		0.3022(2)	[73N3]
$TbCo_3B_2$	RT	*P6*/mmm	0.5048(3)		0.3005(2)	[69K1]
$TbCo_3B_2$	RT	*P6*/mmm	0.5050(3)		0.3009(2)	[73N3]
$TbCo_3B_2$	RT	*P6*/mmm	0.5053(5)		0.3011(3)	[98C1]
$TbCo_3B_2$	220	*P6/mmm*	0.50523(2)		0.300221)	[05D1]
	33.7	*P6/mmm*	0.50509(2)		0.29968(1)	[05D1]
	1.5	*Cmmm*	0.87487(5)	0.50496(3)	0.29971(1)	[05D1]
$Tb_{0.95}Y_{0.05}Co_3B_2$	RT	*P6/mmm*	0.505049(9)		0.301193(7)	[10W1]
	6.5	*C2/m*	0.87345(6)	0.50484(3) β = 90.116(1)°	0.300213(5)	[10W1]
$Tb_{0.90}Y_{0.10}Co_3B_2$	RT	*P6/mmm*	0.50563(1)		0.30127(1)	[10W1]
	6.5	*C2/m*	0.87480(6)	0.50496(3) β = 90.099(1)°	0.300107(8)	[10W1]
$Tb_{75}Y_{0.25}Co_3B_2$	RT	*P6/mmm*	0.504641(8)		0.301640(7)	[10W1]
	7	*C2/m*	0.87334(2)	0.504138(9) β = 90.030(7)°	0.300616(4)	[10W1]
$Tb_{0.5}Y_{0.5}Co_3B_2$	300	*P6/mmm*	0.504148(7)		0.302281(6)	[10W1]
	5.5	*P6/mmm*	0.503761(7)		0.301342(6)	[10W1]
$DyCo_3B_2$	RT	*P6/mmm*	0.5031		0.3018	[94I1]
$DyCo_3B_2$	RT	*P6/mmm*	0.5025		0.3001	[11L3]
$DyCo_3B_2$	RT	*P6/mmm*	0.5028(3)		0.3015(2)	[69K1]
$DyCo_3B_2$	RT	*P6/mmm*	0.5031(3)		0.3021(2)	[73N3]
$DyCo_3B_2$	RT	*P6/mmm*	0.5033(4)		0.3015(1)	[73R1]
$HoCo_3B_2$	RT	*P6*/mmm	0.5026(3)		0.3029(2)	[69K1]
$HoCo_3B_2$	RT	*P6*/mmm	0.5017(5)		0.3024(2)	[73R1]
$HoCo_3B_2$	RT	*P6/mmm*	0.501417(4)		0.302607(4)	[05C1]

(continued)

Table 8.7 (continued)

Compound	T (K)	Space group	Lattice parameters			References
			a	b (β°)	c	
$HoCo_3B_2$	RT	*P6/mmm*	0.5018(3)		0.3023(2)	[73N3]
$ErCo_3B_2$	RT	*P6/mmm*	0.5003(3)		0.3024(2)	[69K1]
$ErCo_3B_2$	RT	*P6/mmm*	0.5006(3)		0.3024(2)	[73N3]
$ErCo_3B_2$	RT	*P6/mmm*	0.5005(5)		0.3029(2)	[73R1]
$TmCo_3B_2$	RT	*P6/mmm*	0.4991(3)		0.3019(2)	[69K1]
$TmCo_3B_2$	RT	*P6/mmm*	0.4999(5)		0.3019(1)	[73R1]
$YbCo_3B_2$	RT	*P6/mmm*	0.4985(5)		0.3020(2)	[73R1]
$LuCo_3B_2$	RT	*P6/mmm*	0.4959(3)		0.3035(2)	[69K1]
YCo_3B_2	RT	*P6/mmm*	0.5051		0.3035	[94I1]
YCo_3B_2	RT	*P6/mmm*	0.50423(4)		0.30304(3)	[05S2]
YCo_3B_2	RT	*P6/mmm*	0.5033(3)		0.3038(2)	[69K1]
YCo_3B_2	RT	*P6/mmm*	0.5020(2)		0.3027(2)	[73N3]

these samples contained magnetic ordered impurities. The specific heat measurements on RCo_3B_2 (R = Y, Lu) [96P1] and Ce [84Y1], as well as resistivity studies on YCo_3B_2 [00K1], suggest also that no magnetic ordered phase is present in the above compounds. According to [84Y1], the electrical resistivity of $CeCo_3B_2$ varies approximately as T^4 law, at $T \leq 20$ K and nearly linear, at $T \geq 200$ K. At low temperatures, the electrical resistivity of YCo_3B_2 follows a T^2 dependence implying that the scattering by spin fluctuations is the dominant process [00K1, 00K2]. At higher temperatures, the evolution of the resistivity has been analyzed [00K2] on the basis of Mott and Jones model [56J1].

The magnetic ordering temperatures, of RCo_3B_2 compounds with magnetic light rare-earths, are rather low. The R sublattice orders ferromagnetically with reduced magnetic moments, as compared with their free ion values. The magnetic behavior of R atoms is strongly controlled by the crystalline electric field (CEF) and exchange fields. The magnetic state of cobalt sublattice and its role in ordering process, in RCo_3B_2 compounds, as already mentioned in YCo_3B_2, were subject of some controversy. The magnetic properties of RCo_3B_2 compounds, reported in literature, are listed in Table 8.8. The critical exponents determined at phase transitions are given in Table 8.9 and data obtained from specific heat measurements in Table 8.10.

The $SmCo_3B_2$ compound is ferromagnetically ordered, with a small spontaneous moment at $T = 4.2$ K [77O1, 94I1, 00I1]. The X-ray absorption and magnetic circular dichroism measurements showed that samarium is in trivalent state, their orbital and spin moments being rather close [07M2]—Table 8.8. The compound undergoes two magnetic transitions [96P1, 00I1, 07M2]. One occurs at T = 43 K, attributed to a change from FM to AFM state [00I1] and another at $T_N = 47$ K. Metamagnetic-type transitions were evidenced in the temperature range $43\ K \leq T \leq T_N$. It was

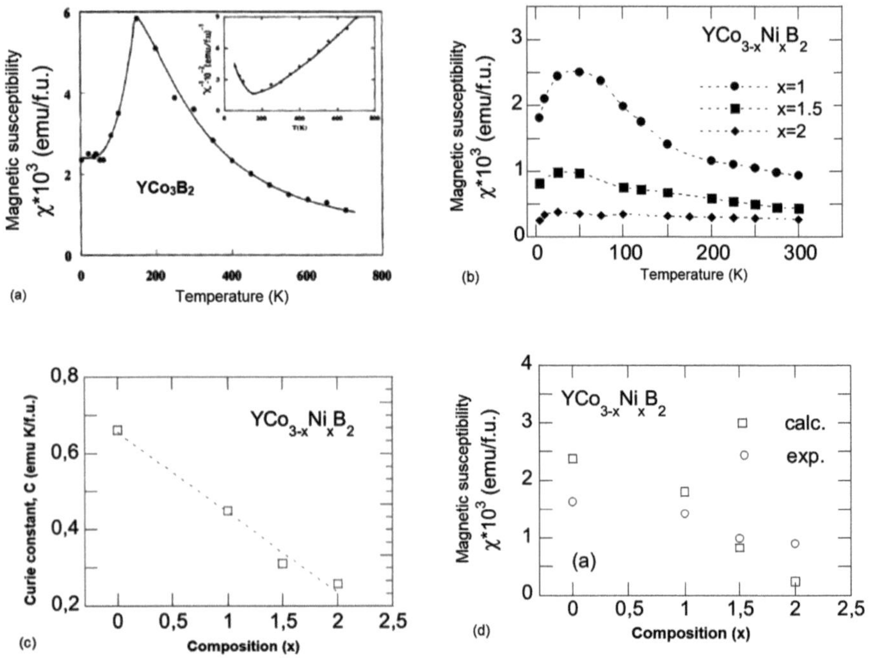

Fig. 8.5 $YCo_{3-x}Ni_xB_2$: thermal variations of magnetic susceptibilities for compounds with x = 0 (**a**), x = 1, 1.5 and 2 (**b**). In inset of (**a**), the temperature dependence of reciprocal susceptibility is given. The composition dependences of the Curie constant (**c**) and of the magnetic susceptibilities at 2(4) K (**d**). In (**d**) magnetic susceptibilities obtained from band structure calculations are also given [06B1]

speculated that this behavior originates from the antiparallel coupled orbital and spin Sm moments [07M2].

The $Sm(Co_{1-x}Ni_x)_3B_2$ compounds are also ferromagnetically ordered. In the composition range $0 \le x \le 0.3$, the induced moments on transition metal sublattice ($Co_{1-x}Ni_x$) by exchange field, decrease from 0.05 μ_B (x = 0) and vanishes at x = 0.1 [04M1]. For $x \ge 0.1$, the magnetic moment, at 4.2 K, $M_s = 0.13\ \mu_B$/f.u., was attributed only to Sm. The ^{59}Co NMR study on $SmCo_3B_2$ evidenced, at T = 4.2 K, the presence of a small hyperfine field (−0.11 T) [98I1]. On this basis, a cobalt moment, of 0.04 μ_B/atom was estimated, close to the value determined by magnetic measurements [04M1].

The cobalt moments in heavy rare-earths compounds are very low and consequently difficult to be evidenced, particularly by using magnetic measurements. Thus, either ferrimagnetic ordering when M_{Co} has been observed, or ferromagnetic one, when cobalt moment was assumed to be nil, were proposed for these compounds. The magnetic ordering temperatures, of RCo_3B_2 compounds, are rather low. Magnetic ordered temperatures of $T_c(Co)$ = 150 K-160 K, for cobalt sublattice were proposed as in compounds with non-magnetic R elements [05D1, 05S2, 07H2, 08R1]. As

Table 8.8 Magnetic properties of RCo_3B_2 compounds

Compound	Magnetic structure	M_s (μ_B/f.u.)	T_N, T_{ord} (K)	M_{eff} (μ_B/f.u.)	θ (K)	References
$CeCo_3B_2$	PM [09I1] FM[07H2] : T = 4.2K, M = 0.008μ_B/f.u. } impurity FM[07H2] : T = 2K, M = 0.01μ_B/f.u. } impurity					[85Y1, 07H2, 09I1]
$SmCo_3B_2$		0.12[a)]	40			[77O1]
$SmCo_3B_2$	FM, T = 43 K, first order magnetic transition 43 K ≤ T ≤ 47 K, metamagnetic transition	0.178[b)]	43 49 (T_N)			[99I1], [00I1]
$SmCo_3B_2$	FM	0.12	40			[94I1]
$SmCo_3B_2$			42 46.5(5) (T_N)			[96P1]
$SmCo_3B_2$	XAS and MCD, Co: $M_s = 0.21\mu_B$, $M_l = 0.04\mu_B$, Sm: $M_{sz} = 1.21\ \mu_B$, $M_{Lz} = 1.16\ \mu_B$		43			[07M2]
$GdCo_3B_2$	FIM	6.9[a)]	58	7.7	75	[77O1]
$GdCo_3B_2$	FIM	6.4[c)]	56	8.05	59	[85M1]
$GdCo_3B_2$	FIM	6.46[b)]	58	7.95		[94I1, 99I1]
$GdCo_3B_2$	FIM	6.2[d)]	54	9.17		[11L5]
$TbCo_3B_2$	FIM $T < T_c$ *Cmmm* collinear canted, $\theta = 74(2)^o$ T = 1.5 K: $M_{Tb} = 5.2(1)\ \mu_B$, $M_{Co} = -0.12(2)\ \mu_B$, $\theta = 74(2)^o$ T = 18.6 K: $M_{Tb} = 4.8(1)\ \mu_B$, $M_{Co} = -0.09(2)\ \mu_B$, $\theta = 75(3)^o$ T = 26.9 K: $M_{Tb} = 2.5(4)\ \mu_B$, $M_{Co} = -0.15(3)\ \mu_B$, $\theta = 49(3)^o$ $T > T_c$ *P6/mmm* T = 33.7 K: $M_{Tb} = 0.9(1)\ \mu_B$, $M_{Co} = -0.21(11)\ \mu_B$, $\theta = 0^o$	6.2(1)[e)]	30(3)			[05D1]
$TbCo_3B_2$	FM (?) T = 18 K: $M_{Tb} = 5.4(2)\ \mu_B$, $M_{Co} = 0.13(7)\ \mu_B$ T > 150 K, $\chi = \chi_0 + C(T\text{-}\theta)^{-1}$, $\chi_0 = 7.1(1)\cdot 10^{-2}$ emu/f.u	6.9[f)]	35(2)	7.7(1)[g]	40(4)	[98C1]

(continued)

Table 8.8 (continued)

Compound	Magnetic structure	M_s (μ_B/f.u.)	T_N, T_{ord} (K)	M_{eff} (μ_B/f.u.)	θ (K)	References
$TbCo_3B_2$	FM, T = 10 K: $M_{Tb} = 4.74\ \mu_B$, $M_{Co} = 0\ \mu_B$		31			[14R1]
$TbCo_3B_2$		5.39^b				[99I1]
$Tb_{1-x}Y_xCo_3B_2$	FIM					[10W1]
x = 0.05	T = 6.5 K; P1 magnetic space group, canted, $\theta = 76(2)^o$ $M_{Tb} = 5.07(5)\ \mu_B$, $M_{Co} = -0.010(7)\ \mu_B$		28.5^h			
x = 0.10	T = 6.5 K; P1 magnetic space group, canted, $\theta = 46(3)^o$ $M_{Tb} = 4.13(9)\ \mu_B$, $M_{Co} = -0.03(3)\ \mu_B$		26.5^h			
x = 0.25 (ND)	T = 7 K; P1 magnetic space group, canted, $\theta = 50(6)^o$ $M_{Tb} = 2.4(1)\ \mu_B$, $M_{Co} = -0.10(6)\ \mu_B$		14^h			
$Tb_{0.75}Y_{0.25}Co_3B_2$ (ND)	FIM, canted, $\theta = 32(9)^o$ from **c**-axis T = 4.6 K: $M_{Tb} = 1.5(2)\ \mu_B$, $M_{Co} = 0.5(2)\ \mu_B$		17			[08W1]
$TbCo_3B_2$ (ND)	T = 10 K, $M_{Tb} = 5.6(3)\ \mu_B$					[08R1]
$Tb_{0.75}Y_{0.25}Co_3B_2$ (ND)	T = 10 K, $M_{Tb} = 3(1)\ \mu_B$					[08R1]
$DyCo_3B_2$	FIM	7.6^c	20.6			[85M1]
$DyCo_3B_2$	FIM	8.64^b	26			[99I1]
$DyCo_3B_2$		8.9^{i2} 8.2^{i1}	27	11.4		[94I1, 99I1]
$DyCo_3B_2$	Dy: $M = 10.13\ \mu_B$, $M_s = 5.13\mu_B$, $M_o = 5\mu_B$,					[98K5, 01J1]
(BS) Dy core state	Co: $M = 0.20\ \mu_B$					
$HoCo_3B_2$	FM(?) T = 4.5 K: $M_{Ho} = 5.08(4)\ \mu_B$; $M_{Co} = 0.11(2)\ \mu_B$		≅ 10			[05C1]
$ErCo_3B_2$		4.8^a	20	10.2	10	[77O1]

(continued)

Table 8.8 (continued)

Compound	Magnetic structure	M_s (μ_B/f.u.)	T_N, T_{ord} (K)	M_{eff} (μ_B/f.u.)	θ (K)	References
$ErCo_3B_2$		8.44[b]	15			[99I1]
$LuCo_3B_2$	PM					[96P1]
YCo_3B_2	PM: spin fluctuations, T > 350 K, C-W: $\chi = \chi_o + C(T\text{-}\theta)^{-1}$ T = 2 K: $\chi_0 = 2.34{\cdot}10^{-3}$, $\chi_{calc} = 1.65{\cdot}10^{-3}$ emu/fu [j]			1.34/Co atom		[95B1, 15B1]
YCo_3B_2	PM, T ≥ 10 K,					[14R1]

[a] T = 4.2 K, $\mu_0H = 0.3$ T
[b] T = 4.2 K, $\mu_0H = 15$ T
[c] T = 5 K, $\mu_0H = 1.0$ T
[d] T = 4 K, $\mu_0H = 7.0$ T
[e] T = 5 K, $\mu_0H = 6.0$ T
[f] Saturated moment, at T = 5 K, $\mu_0H = 5$ T
[g] Determined in limited temperature range T < 275 K
[h] Extrapolated from figure
[i] T = 5 K, $\mu_0H = 4.0$ T[i1] and extrapolated at $\mu_0H = 14$ T[i2]
[j] Band structure calculations

Table 8.9 Critical exponents[a]

Compound or model	Critical exponents				References
	β	γ	δ	n	
$GdCo_3B_2$	0.41(2)	1.32(10)	4.22(20)	0.665(15)	[13L2]
$TbCo_3B_2$	0.265(30)	1.113(150)	5.20(30)	0.467(20)	[13L2]
$DyCo_3B_2$	0.297(30)	1.15(20)	4.90(40)	0.515(30)	[13L2]
$DyCo_3B_2$	0.310(20)	1.15(30)	4.71(40)	0.527(40)	[11L3]
Mean field model	0.5	1.0	3	0.667	[71S1]
3D Heidenberg model	0.365	1.336	4.8	0.627	[71S1]
3D Ising model	0.325	1.24	4.82	0.569	[71S1]
Tricritical field model	0.25	1.0	5.0	0.4	[07F1]

[a] β-associated with spontaneous magnetization, γ with initial magnetic susceptibility, δ-with critical magnetization isotherm at T_c, $\Delta S_m \propto H^n$

Table 8.10 Data obtained from specific heat studies

Compound	Sommerfeld constant, γ (mJ/molK^2)	Debye temperature θ_D (K)	References
$CeCo_3B_2$	9.7(6)	278(6)	[84Y1]
$LuCo_3B_2$	12.4(3)	476(10)	[96P1]
YCo_3B_2	13.9(3)	573(10)	[96P1]

already mentioned, most likely, this behavior can be associated with the presence of magnetic impurities.

The magnetic properties of $GdCo_3B_2$ were analyzed either assuming that cobalt moment is nil, and consequently the compound is ferromagnetically ordered [77O1], or admitting that the cobalt has at low temperatures a small magnetic moment antiparallel oriented to gadolinium one [85M1, 91I1, 94I1, 95B1, 99I1, 09L3, 11L5, 13L2]. According to investigated temperature ranges above T_c, commonly at $T < 300$ K, and assuming that magnetic susceptibility of $GdCo_3B_2$ follows Curie–Weiss dependence, effective moments per f.u., even smaller than that of free Gd^{3+} ion, were reported [77O1, 91I1, 94I1]. Non-linear χ^{-1} Versus T dependence, as expected for a ferrimagnetic type ordering has been evidenced when measurements were made in a larger temperature range [95B1]. According to addition law of magnetic susceptibilities and assuming that the effective Gd moment is given by free ion value, effective cobalt moments of 1.34 μ_B/atom [95B1] or $\cong$ 2.00 μ_B/atom [11L5] were obtained. The ratio $r = S_p/S_o$ is between 4 and 7 indicating a high degree of itinerancy of cobalt magnetism.

The gadolinium hyperfine field, as determined from ^{155}Gd Mössbauer spectra on $GdCo_3B_2$, revealed that this is close to that of Gd^{3+} free ion value and thus cobalt either has a small magnetic moment or is not magnetic [92K1]. The presence of a large crystalline electric field at Gd site was also shown [85M1]. The Curie temperatures and saturation magnetizations decrease nearly linearly when Gd is substituted by Y in $(Gd_xY_{1-x})Co_3B_2$ compounds [91I1, 95B1]. The reciprocal susceptibilities for compounds with $x > 0.2$, follow non-linear temperature dependences. The effective cobalt moments was shown to be little dependent on the samples compositions [95B1].

The magnetic properties of $TbCo_3B_2$ compound have been investigated by magnetic measurements [99I1, 05D1, 11L2, 13L2] and neutron diffraction studies [91A2, 98C1, 05D1, 08R1, 08W1, 10W1, 14R1]. Rather low saturation magnetization as well as Curie temperature, $T_c \cong 30$–31 K, were reported [98C1, 99I1, 05D1]. Neutron diffraction studies [08R1], showed that the CEF-only ground state of Tb ions is a non-magnetic singlet and the magnetic ordered transition is of induced magnetic-type [63B1, 71C1]. At the transition, Tb magnetic moment is induced into the singlet ground state by a component of exchange field, perpendicular to the crystalline electric field, which admixes higher lying states into the singlet ground state via a mechanism similar to that of Van Vleck susceptibility [14R1]. The induced moment causes the rise of exchange field, which in turn leads to the increase of the moment, until magnetic order sets thought a bootstraplike process. A cobalt magnetic moment $M_{Co} = 0.12\ \mu_B$/atom has been determined by ND, at T = 1.5 K [05D1]. The existence of a cobalt moment, at $T > T_c \cong 31$ K, in $TbCo_3B_2$, was suggested by the presence of a neutron count in excess to that contributed by the crystal lattice, which appeared as broadened Bragg reflections [05D1]. Later on [14R1], a revised magnetic structure in which the Tb sublattice is magnetically ordered, at $T < T_c =$ 31 K, with no Co sublattice magnetic ordering was proposed. The magnetic diffuse scattering (MDS), at $T > T_c$, was shown to originate from clusters of magnetically aligned moments which fluctuate over time. There are short-range order with correlation length $\xi_o = 6.5(2.5)$ nm. A similar behavior was shown in $ErCo_2$ compound at $T > T_c$ where the correlation length of short range order was of 0.7(1) nm [07H1]. This behavior can be attributed to coupled R and Co atoms by 4f-5d-3d and 5d-5d exchange interactions still present in paramagnetic region, their intensity being not sufficient to induce magnetic order [17B1, 22B2]. The thermal energy will compensate the exchange interactions only at temperatures of the order of 6–8 T_c, as evidenced also in $TbCo_3B_2$ compound. These results, in addition to those obtained by magnetic measurements [91I1, 95B1, 01C1, 06B1] or NMR [98I1] studies, also excluded the possible presence of a magnetic ordered cobalt sublattice, with $T_c = 150$–160 K, in RCo_3B_2 compounds with magnetic rare-earths.

Neutron powder diffraction on $Tb_{1-x}Y_xCo_3B_2$ series, showed a considerable decrease of the ordered Tb moment, as a result of magnetic dilution, when replacing Tb by Y [08R1, 08W1, 10W1]. As showed already, the Tb magnetic moment is induced into the singlet ground state by a component of exchange field, perpendicular to the crystalline electric field, which admixes higher lying states into the singlet ground state. Diluting Tb sublattice with non-magnetic Y, lowers the exchange interaction strength and a further reduction in Tb moments takes place—Table 8.8. As effect of magnetic dilution by Y, the canting angle,θ, determined by the competition between Tb and Co sublattices anisotropies, diminishes. The Tb sublattice contribution to the magnetocrystalline anisotropy favors the alignment of moments in (ab) plane while the cobalt contribution to the anisotropy, comes from Co3*g* site and is positive [05D1, 10W1]. Magnetic dilution at Tb site with nonmagnetic Y atoms weakens the anisotropy of Tb sublattice and the Tb magnetic moments will align more close to the hexagonal **c**-axis—Table 8.8.

The magnetic properties of $DyCo_3B_2$ compound were also assumed to be either ferromagnetic or ferrimagnetic [85M1, 94I1, 98K5, 99I1, 01J1, 03M2, 11L3, 11T1, 13L2, 14T1]. The presence of a small magnetic Co moment [85M1] induced by exchange field [99I1] and antiparalelly oriented to Dy moment, has been confirmed by magnetic Compton profile [03M2]. Band structure calculations, performed on $DyCo_3B_2$ evidenced also the presence of small cobalt moments, antiparallel oriented to Dy one [98K5, 01J1]. The Dy5d band was found to be polarized. The ^{161}Dy hyperfine field, determined by Mössbauer spectroscopy, was smaller than the free ion value, due to the presence of a large crystalline electric field at Dy site [85M1]. The temperature dependence of the magnetic susceptibility has been analyzed assuming the presence of a Pauli-type contribution superposed on a Curie–Weiss-type behavior [94I1]. Since the compound is ferrimagnetic, the nonlinear temperature dependence of the magnetic susceptibility, at $T > T_c$, can be also analyzed assuming a Néel-type behavior [48N1]. The electrical resistivity of $DyCo_3B_2$, at $4.2\ K \leq T \leq 40\ K$, follows also a T^2 dependence, implying that the electron-spin wave scattering is the dominant mechanism [98K5]. In the high temperature range, the experimental data were analyzed on the basis of s-d scattering mechanism.

The $HoCo_3B_2$ compound shows a second order transition at $T_c = 11.8$ K [18Z1]. Short range correlations exist at temperatures above T_c, as in $TbCo_3B_2$ compound. The neutron diffraction pattern of $HoCo_3B_2$, has been unreasonable analyzed assuming parallel orientation of Ho and Co magnetic moments [05C1]. It seems that this analysis is incorrect, since as a general trend, the magnetic moments of heavy rare-earths, are in all cases antiparallel oriented to those of transition metals [90B1].

The presence of a small cobalt moment, antiparallel oriented to erbium one was also reported in $ErCo_3B_2$ compound [99I1]. This contradict previous data where cobalt was assumed to be not magnetic in this compound [77O1]. The anisotropy was suggested to be low. A very large crystal field parameter A_2^0, was estimated.

The greatest number of studies performed on the RCo_3B_2 compounds with magnetic heavy rare-earths suggest that these are ferrimagnetically ordered, a small cobalt moment being induced by R4f-5d-Co3d exchange path. The cobalt contribution to the resultant magnetization is rather low (<5%), in many cases the Co moment being difficult to be determined and thus assumed to be nil. The reciprocal susceptibilities are believed to follow non- linear temperature dependences. Since of the small cobalt moment, these are close to a Curie–Weiss-type behavior; high temperature measurements are necessary in order to obtain reliable information on the magnetic behavior in paramagnetic range.

The critical exponents of RCo_3B_2 compounds with R = Gd, Tb and Dy, around the magnetic phase transitions, were determined by using the modified Arrott plot method (R = Dy) [11L3], or from field dependence of the magnetic entropy change (R = Gd, Tb) [13L2]—Table 8.9. The critical exponents satisfied the scaling law with values between mean field model and 3D Heisenberg model (R = Gd), close to a tricritical-mean field model (R = Tb), or to a 3D Ising model (R = Dy).

The heat capacity C_p/T of RCo_3B_2 compounds with R = Lu and Y follows, at low temperatures, a T^2 dependence. The heat capacities of RCo_3B_2 R = Sm, Gd, Dy, [96P1], R = Gd [11L5], R = Tb [11L2] and R = Dy [11L3, 11T1], evidenced a λ-shape behavior at the magnetic transition temperatures, as shown in Fig. 8.6 [96P1]. The magnetic contribution to specific heat has been determined. An appreciable amount of short range order occurs above T_c, behavior evidenced also in RCo_2 compounds [18B1, 22B1]. The magnetic specific heats of RCo_3B_2 (R = Gd, Tb, Dy) indicate an important role of crystal field and anisotropy [11T1]. The entropy gain, as well as the ground states of magnetic R elements in RCo_3B_2 compounds, were also evaluated. The Sommerfeld constants and the Debye temperature for some RCo_3B_2 compounds are given in Table 8.10.

The relative large reversible magnetocaloric effects (MCE), in RCo_3B_2 compounds with R = Gd, Tb, Dy and Ho, were related mainly to the R sublattice magnetic phase transition [09L3, 11L2, 11L3, 11L5, 11T1, 14T1, 18Z1]. The observed bulges in the temperature dependence of the magnetic entropy change in $HoCo_3B_2$, at $T > T_c$, were associated with the presence of short range correlations, as already suggested. The determined entropy changes and the adiabatic temperature changes are listed in Table 8.11. The effect of milling $DyCo_3B_2$ boride, on the magnetocaloric effect, was also analyzed [14T1]. Even a long milling time, t_m, does not spoil significantly the MCE parameters. Only when $t_m > 5$ h, the magnetic entropy change is reduced.

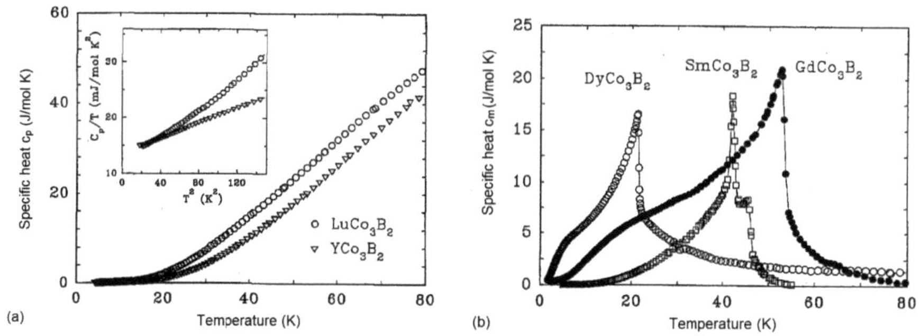

Fig. 8.6 RCo_3B_2: **a** temperature dependences of the heat capacity for RCo_3B_2 (R = Lu, Y) compounds and **b** temperature dependences of magnetic specific heat for RCo_3B_2 compounds with R = Gd, Sm, Dy [96P1]. In inset of (a), the C_p/T versus T^2 plot is given

Table 8.11 Entropy changes, ΔS_m and adiabatic temperature changes, ΔT_{ad}

Compound	T_c (K)	$-\Delta S_m$ (J/kgK)	ΔT_{ad} (K)	$\mu_0 H$ (T)	References
$GdCo_3B_2$	54	11.6	6.4	7.0	[11L5]
$TbCo_3B_2$	28	10.3	8.6	7.0	[11L2]
$DyCo_3B_2$	22	15.4	11.9	7.0	[11T1]
$DyCo_3B_2$	22	15.1	13.8	7.0	[11L3]
$HoCo_3B_2$	11.8	14.1		7.0	[18Z1]

8.4 $R_2Co_7B_3$ Compounds

The crystal structure of $R_2Co_7B_3$ compounds can be viewed as being build up by alternative stacking of one layer of RCo_5 and three layers of RCo_3B_2 unit cells—Sect. 8.1. The compounds crystallize in a hexagonal structure having *P6/mmm* space group [73K1, 74K2]—Fig. 8.1. In this structure, there are three types of R and Co atoms, respectively and two for boron—Table 8.12. Three kinds of layers regularly piled along **c**-axis, where cobalt atoms are located, can be distinguished: (1) Co2*c*, with cobalt in the R-Co layer, sandwiched between two cobalt only layers; (2) Co6i_1, located in the cobalt only layer sandwiched between R-Co and R-B layers; (3) Co6i_2, with the cobalt only layer sandwiched between two R-B layers. The Co6i_1 - B2*d* interatomic distances are very short (<0.201 nm) [01C1, 02Z1, 06N1]. The existence of such a short distance, indicates a break in the regular stacking of the boron containing planes along the **c**-axis. Generally speaking, the shortening of Co-B distances in $R_{n+1}Co_{3n+5}B_{2n}$ compounds may explain the absence of other phases in $3 < n < \infty$ composition range [01C1, 02Z1]. For the same reason, in the $Sm_2Co_7B_3$

Table 8.12 Atomic coordinates of $Ce_2Co_7B_3$ compound, space group *P6/mmm*[a] [74K2, 06V1]

Atom	Site	Symmetry	x	y	z	Atomic environment
Ce	1*a*	6/*mmm*	0	0	0	Pseudo Frank-Kasper $B_6Co_{12}Ce_2$
Ce	1*b*	6/*mmm*	0	0	1/2	Pseudo Frank-Kasper $Co_{18}Ce_2$
Ce	2*e*	6*mm*	0	0	1/4	Pseudo Frank-Kasper $Co_{12}B_6Ce_2$
Co	2*c*(2*d*)	$\bar{6}m2$	1/3	2/3	1/2	14-vertex polyhedron $Co_9Ce_3B_2$
Co	6i_1	2*mm*	1/2	0	0.125	14-vertex Frank-Kasper $B_4Co_6Ce_4$
Co	6i_2	2*mm*	1/2	0	0.35	13-vertex polyhedron $B_2Co_7Ce_4$
B	2*c*	$\bar{6}m2$	1/3	2/3	0	Trigonal prism Co_6
B	4*h*	3*m*	1/3	2/3	0.238	Trigonal prism Co_6

[a]Transformed from published data [74K2], origin shift 0 0 1/2

melt spun and crystallized compound, the B4*h* atoms were not located in the Sm2*e* layer, but between Sm2*e* and Co6i_1 layers [93Y1].

The spontaneous magnetostriction of $Gd_2Co_7B_3$ is larger than that of $Y_2Co_7B_3$ and in both compounds occurs along the **a**-axis. There is a linear thermal expansion of a and c lattice parameters, at $T > T_c$ [98I2]. The elastic properties of $Y_{n+1}Co_{3n+5}B_{2n}$ compounds were also reported [06M3]. The lattice parameters of $R_2Co_7B_3$ compounds are listed in Table 8.13.

Table 8.13 Space groups and lattice parameters of $R_2Co_7B_3$ compounds

Compound	T (K)	Space group	Lattice parameters (nm)		References
			a	c	
$CeCo_3B_3$	RT	*P6/mmm*	0.5053(3)	1.297(2)	[74K2, 79K1]
$Pr_2Co_7B_3$	2	*P6/mmm*	0.5148(1)	1.2710(1)	[02Z1]
$Pr_2Co_7B_3$	RT	*P6/mmm*	0.5150(3)	1.275(2)	[74K2, 87C2]
$Pr_2Co_7B_3$	350	*P6/mmm*	0.5156(1)	1.2766(1)	[02Z1]
$Nd_2Co_7B_3$	2	*P6/mmm*	0.5135(1)	1.2737(1)	[02Z1]
$Nd_2Co_7B_3$	RT	*P6/mmm*	0.5134(3)	1.278(2)	[73K1, 74K2]
$Nd_2Co_7B_3$	RT	*P6/mmm*	0.5159	1.2767	[99C2]
$Sm_2Co_7B_3$	RT	*P6/mmm*	0.5088(3)	1.279(2)	[80B1]
$Sm_2Co_7B_3$	RT	*P6/mmm*	0.5088	1.2793	[00C5]
$Gd_2Co_7B_3$	RT	*P6/mmm*	0.5078(3)	1.285(2)	[74K2]
$Gd_2Co_7B_3$	RT	*P6/mmm*	0.5076	1.2852	[97K2]
$Gd_2Co_7B_3$	RT	*P6/mmm*	0.5067	1.294	[88S2, 00C4]
$Tb_2Co_7B_3$	RT	*P6/mmm*	0.5061(3)	1.283(2)	[74K2, 88D1]
$Tb_2Co_7B_3$	RT	*P6/mmm*	0.5057	1.2846	[97K2]
$Dy_2Co_7B_3$	2	*P6/mmm*	0.5040(1)	1.2796(3)	[02Z1]
$Dy_2Co_7B_3$	RT	*P6/mmm*	0.5046(3)	1.284(2)	[74K2]
$Dy_2Co_7B_3$	RT	*P6/mmm*	0.5049	1.2842	[97K2]
$Ho_2Co_7B_3$	RT	*P6/mmm*	0.5033(3)	1.288(2)	[74K2]
$Ho_2Co_7B_3$	RT	*P6/mmm*	0.5022	1.2881	[94B1]
$Ho_2Co_7B_3$	RT	*P6/mmm*	0.5036	1.2850	[97K2]
$Er_2Co_7B_3$	RT	*P6/mmm*	0.5006(3)	1.287(2)	[74K2]
$Er_2Co_7B_3$	RT	*P6/mmm*	0.5013	1.2874	[94B1]
$Er_2Co_7B_3$	RT	*P6/mmm*	0.5022	1.2843	[97K2]
$Tm_2Co_7B_3$	RT	*P6/mmm*	0.5002(3)	1.285(2)	[74K2]
$Tm_2Co_7B_3$	RT	*P6/mmm*	0.5016	1.2840	[97K2, 02K1]
$Tm_2Co_7B_3$	RT	*P6/mmm*	0.4996	1.2862	[94B1]
$Y_2Co_7B_3$	RT	*P6/mmm*	0.5045(3)	1.288(2)	[74K2]
$Y_2Co_7B_3$	RT	*P6/mmm*	0.5036(2)	1.2899(5)	[01C1]
$Y_2Co_7B_3$	RT	*P6/mmm*	0.50453	1.28872	[93B1, 03M1]

The pseudo-ternary $(R_xY_{1-x})_2Co_7B_3$ systems form solid solutions in all the composition range, when R = Gd [93B1] or R = Er [95Z2]. The $R_2(Co_{1-x}Fe_x)_7B_3$ compounds crystallize in hexagonal structure in a limited compositional range, as $x \leq 0.3$ (R = Ce [04S1] and R = Y [96T2]), $x \leq 0.07$ (R = Pr) [91H2] and $x \leq 0.428$ (R = Nd)[90H1]. The Fe prefers to substitute Co at $2c$ site as can be seen in Ce-based compounds, where an anomaly in lattice parameters was shown at x = 0.14 when this site is populated only by Fe (x = 0.143). The $Pr_2(Co_{1-x}Fe_x)_7B_3$ is a multiphase system when $0.07 < x \leq 0.429$, the main phase being the hexagonal one [91H2]. A major tetragonal phase was also shown in $Y_2(Co_{1-x}Fe_x)_7B_3$ multiphases system for x > 0.3 [90H1]. Solid solutions having hexagonal structure are present in $R_2(Co_{1-x}Ni_x)_7B_3$ series with $x \leq 0.3$(R = Ce) [05S1], $x \leq 0.57$(R = Gd) [97B1] or R = Y [03M1]). The $Y_2(Co_{1-x}M_x)_7B_3$ compounds with M = Cu, V and $x \leq 0.286$ crystallize also in hexagonal *P6/mmm* type structure [05M2].

The effect of Co substitutions by M = Ti, V or Cr in $R_2Co_7B_3$ compounds, with R = Gd, Y, was theoretically analyzed [06Q1]. The M atoms substitute for Co with a strong preference for $6i_1$ site, decreasing the preference in the sequence $6i_1 > 2c > 6i_2$. The computed lattice parameters were in agreement with experimental data.

The $R_2Co_{7-x}Fe_xB_3$ melt spun alloys and crystallized, with x = 1, 2 (R = Sm) and x = 1, 2, 5, 7 (R = Nd), have *P6/mmm* type structure, as a major phase [88A2]. The $Sm_{15}Co_{70-x}Al_xFe_{10}B_5$ ribbons contained in addition to amorphous phase, $Sm_2(Co,M)_7$ and $Sm_2Co_7B_3$ compounds [19D2].

A large number of studies were devoted to $Y_2Co_7B_3$ compound and their pseudoternary alloys. The $Y_2Co_7B_3$ is ferromagnetically ordered [93B1, 93O1, 94B3, 95O1, 96T2, 98M1, 98S2, 01C1, 01V1, 02V1, 03M1, 05M2, 06M1, 06M3, 09B1]—Table 8.14. The reported mean cobalt moments, are situated between 0.33 μ_B and 0.43 μ_B/atom. The reciprocal susceptibility follows a Curie Weiss type dependence, the effective cobalt moment being of $\cong$ 2.10 μ_B/atom [93B1]. The mean S_p/S_o ratio is of 3–4, showing a rather high degree of cobalt magnetism itinerancy. The cobalt moments, as determined by neutron diffraction [01C1] or band structure calculations [98S2, 09B1] are strongly dependent on the number of boron planes around that containing cobalt atoms, decreasing in the sequence $M_{2c} = 1.66\ \mu_B > M_{6i1} = 0.5\ \mu_B > M_{6i2} = 0\ \mu_B$ [01C1]. Taking into account the number, N, of the layers just above and/or just below the cobalt layer, Co_{2c} (N = 0), Co_{6i1}(N = 1) and $Co6_{i2}$ (N = 2), the approximate values of cobalt moments, at different sites, in $R_{n+1}Co_{3n+5}B_{2n}$ (n = 1, 2, 3, ∞) were estimated from cobalt sublattice magnetization, by fitting the experimental data with the relation $M_{Co} = [3(n\text{-}1)\ M_{Co6i2} + 6\ M_{Co6i1} + 2M_{Co2c}]/(3n + 5)$. In all studied series, Co_{6i2} has been assumed to be not magnetic. In this way M_{Co2c} and M_{Co6i1} magnetic moments, obtained in compounds with R = Y (1.55 μ_B, 0.4 μ_B) [95I1], R = Ce (1.2 μ_B, 0 μ_B) [95I2], R = Sm (1.27 μ_B, 0.43 μ_B) [92I2] and R = Gd (1.7 μ_B, 0.77 μ_B) [93O1], were in rather good agreement with those experimentally determined [93O1, 02C2, 02Z1].

The pseudo-ternary $Y_2Co_{7-x}Fe_xB_3$ with $x \leq 2$ [97P2, 06M1, 09B1], $Y_2Co_{7-x}Ni_xB_3$ with $x \leq 4$ [03M1] and $Y_2Co_{7-x}M_xB_3$ with M = Cu,V and $x \leq 2$ [05M2] are ferromagnetically ordered. Both the saturation magnetizations and Curie temperatures increase when cobalt is replaced by iron [97P2, 09B1]. The ^{57}Fe

Table 8.14 Magnetic properties

Compound	Magnetic structure and magnetic moments	M_s (μ_B/f.u.)	T_c (K)	C (emu/f.u.)	M_{eff} (μ_B/Co atom)	θ (K)	References
$Ce_2Co_7B_3$ (NMR)	T = 1.5 K Ce: M = 0 Co: M_{2c}(spin) = 1.24 μ_B, M_{2c}(orb) = 0.13 μ_B M_{6i_1}(spin) = 0.36 μ_B, M_{6i_1}(orb)= 0.01 μ_B, M_{6i_2}= 0 μ_B						[15S1]
$Ce_2Co_7B_3$	Metamagnetic Ce: M = 0 μ_B Co: M_{2c} = 1.2 μ_B, M_{6i_1}= M_{6i_2}= 0 μ_B	0.76[a] 1.30 [95I2]	250				[98M1]
$Ce_2Co_7B_3$	Metamagnetic: transition from 0.9 μ_B/f.u. to 1.8 μ_B/f.u., μ_0 H = 3.0 T, T = 4.2 K	1.8[b]					[95I2]
$Pr_2Co_7B_3$	T = 2 K, FM, e.a.m. in (ab) plane Pr: M_{1a} = 2.0(2) μ_B, M_{1b} = 1.7(2) μ_B, M_{2e} = 2.0(2) μ_B Co: M_{2c} = 1.8(2) μ_B, M_{6i_1}= 0.6(2) μ_B, M_{6i_2}= 0.2(2) μ_B	7.3(1)[c]	340(5)				[02Z1]
$Pr_2Co_7B_3$		7.8[d]	328				[91H2]
$Nd_2Co_7B_3$	T = 2 K, FM, e.a.m. in (ab) plane Nd: M_{1a} = 1.9(2) μ_B, M_{1b} = 1.1(2) μ_B, M_{2e} = 1.6(2) μ_B Co: M_{2c} = 1.5(2) μ_B, M_{6i_1}= 0.6(2) μ_B, M_{6i_2}= 0.1(2) μ_B	7.1(1)[c]	336(5)				[02C2, 02Z1]
$Nd_2Co_7B_3$	FM	8[d]	330				[90H1]

(continued)

Table 8.14 (continued)

Compound	Magnetic structure and magnetic moments	M_s (μ_B/f.u.)	T_c (K)	C (emu/f.u.)	M_{eff} (μ_B/Co atom)	θ (K)	References
$Nd_2Co_7B_3$	FM Co: $M = 0.20\ \mu_B$/atom	9.8[e]	332		2.20		[97K3, 97K5]
$Sm_2Co_7B_3$	T = 2 K, FM, e.a.m. ∥ **c** Co: $M_{2c} = 1.27\ \mu_B$, $M_{6i_1} = 0.43\ \mu_B$, $M_{6i_2} = 0\ \mu_B$	3.52[e,f]	420				[92I2]
$Sm_2Co_7B_3$	FM Sm: $M = 0.54\ \mu_B$	3.68[b]	400				[93O1, 98M1]
$Gd_2Co_7B_3$	FIM $M_{Co} = 0.82\ \mu_B$/atom if $M_{Gd} = 7.0\ \mu_B$	8.26[c]	347	12.62	2.32		[93B1]
$Gd_2Co_7B_3$	FIM $M_{Co} = 0.60\ \mu_B$/atom	9.8	345				[88S2]
$Gd_2Co_7B_3$	FIM Gd: $M = 7\ \mu_B$ Co: $M_{2c} = 1.70\ \mu_B$, $M_{6i_1} = 0.77\ \mu_B$, $M_{6i_2} = 0$	10.4[b]					[93O1, 98M1]
$Gd_2Co_7B_3$	FIM		357	18.23	1.77		[97S2]
$Gd_2Co_7B_3$	FIM	9.6[a]	376				[97K3]
$Gd_2Co_7B_3$	FIM	10.5[a]	370				[04A1]
$Gd_2Co_7B_3$	FIM						[06S3]

(continued)

Table 8.14 (continued)

Compound	Magnetic structure and magnetic moments	M_s (μ_B/f.u.)	T_c (K)	C (emu/f.u.)	M_{eff} (μ_B/Co atom)	θ (K)	References
(BS)	Gd: $M_{1a} = 7.457\ \mu_B$, $M_{1b} = 7.236\ \mu_B$, $M_{2e} = 7.366\ \mu_B$ Co: $M_{2c} = -1.776\ \mu_B$, $M_{6i_1} = -0.795\ \mu_B$, $M_{6i_2} = -0.164\ \mu_B$ B: $M_{2d} = -0.014\ \mu_B$, $M_{4h} = -0.034\ \mu_B$						
$Gd_2Co_7B_3$ (BS)	FIM Gd: $M_{1a} = 7.387\ \mu_B$, $M_{1b} = 7.233\ \mu_B$, $M_{2e} = 7.317\ \mu_B$ Co: $M_{2c} = -1.707\ \mu_B$, $M_{6i_1} = -0.607\ \mu_B$, $M_{6i_2} = -0.140\ \mu_B$ B: $M_{2d} = -0.016\ \mu_B$, $M_{4h} = -0.019\ \mu_B$						[01V1, 02V1]
$Tb_2Co_7B_3$	FIM	12.0[a]	350				[04A1]
$Tb_2Co_7B_3$	FIM	10.4[e]	355				[97K3]
$Tb_2Co_7B_3$ (ND)	FIM, T = 15 K Tb:$M_{1a} = 9(4)\mu_B$,$M_{1b} = 5(4)\mu_B$,$M_{2e} = 9(3)\ \mu_B$; $M_{Tb} = 5.8(8)\ \mu_B$ Co: $M_{2c} = -0.70(25)\ \mu_B$, $M_{6i_1} = 1.50(10)\ \mu_B$, $M_{6i_2} = -0.20(12)\ \mu_B$, $\theta = 56(9)^o$	10.6(2)					[00C1]
$Tb_2Co_7B_3$ (BS)	FIM Tb: $M_{1a} = 9.423\ \mu_B$, $M_{1b} = 9.219\ \mu_B$, $M_{2e} = 9.309\ \mu_B$ Co: $M_{2c} = -1.760\ \mu_B$, $M_{6i_1} = -0.744\ \mu_B$, $M_{6i_2} = -0.184\ \mu_B$ B: $M_{2d} = -0.016\ \mu_B$, $M_{4h} = -0.031$						[06S3]

(continued)

Table 8.14 (continued)

Compound	Magnetic structure and magnetic moments	M_s (μ_B/f.u.)	T_c (K)	C (emu/f.u.)	M_{eff} (μ_B/Co atom)	θ (K)	References
	μ_B						
$Dy_2Co_7B_3$ (ND)	T = 2 K, FIM, e.a.m. in (ab) plane Dy: $M_{1a} = 9.4(2)\ \mu_B$, $M_{1b} = 8.2(2)\ \mu_B$, $M_{2e} = 8.7(2)\ \mu_B$ Co: $M_{2c} = -1.9(2)\ \mu_B$, $M_{6i_1} = -0.7(2)\ \mu_B$, $M_{6i_2} = -0.2(2)\ \mu_B$	12.9(1)[c]	310(5)[e]				[02Z1]
$Dy_2Co_7B_3$	FIM	15.4[a]	350				[04A1]
$Dy_2Co_7B_3$	FIM	13.5[g]	350				[93O1]
$Dy_2Co_7B_3$	FIM	12.2[e]	340				[97K3]
$Ho_2Co_7B_3$	FIM	10.2[i]	350 205(T_{comp})[h]	32.38	2.21		[94B1]
$Ho_2Co_7B_3$	FIM	13.4[a]	333				[04A1]
$Ho_2Co_7B_3$	FIM	11.8[e]	329				[97K3]
$Er_2Co_7B_3$	FIM	7.1[i]	330 200(T_{comp})[h]	28.17	2.42		[94B1]
$Er_2Co_7B_3$	FIM	9.7[e]	320				[97K3]
$Tm_2Co_7B_3$	FIM	5.31[i]	325 120(T_{comp})[h]	19.13	2.31		[94B1]
$Tm_2Co_7B_3$	FIM	6.1[e]	314				[97K3]
$Y_2Co_7B_3$ (ND)	T = 2 K, FM, e.a.m.‖**c** Co: $M_{2c} = 1.5(2)\ \mu_B$, $M_{6i_1} = 0.5(2)\mu_B$, $M_{6i_2} = 0\ \mu_B$	3.0[a]	315				[01C1]

(continued)

Table 8.14 (continued)

Compound	Magnetic structure and magnetic moments	M_s (μ_B/f.u.)	T_c (K)	C (emu/f.u.)	M_{eff} (μ_B/Co atom)	θ (K)	References
$Y_2Co_7B_3$	FM	2.12[c]	310	3.78	2.10	337	[93B1]
$Y_2Co_7B_3$	FM	1.32	257	2.55	1.77	290	[97S2]
$Y_2Co_7B_3$	FM	5.63[b]	259				[93O1]
$Y_2Co_7B_3$ (BS)	FM Y: $M_{1a} = -0.11\ \mu_B$, $M_{1b} = -0.03\ \mu_B$, $M_{2e} = -0.07\ \mu_B$ Co: $M_{2c} = 1.611\ \mu_B$, $M_{6i_1} = 0.474\ \mu_B$, $M_{6i_2} = 0.061\ \mu_B$ B: $M_{2d} = -0.010\ \mu_B$, $M_{4h} = -0.015\ \mu_B$						[01V1, 02V1] [09B1]
$Y_2Co_7B_3$ (BS)	FM Y: $M_{1a} = -0.202\ \mu_B$, $M_{1b} = -0.028\ \mu_B$, $M_{2e} = -0.124\ \mu_B$ Co: $M_{2c} = 1.677\ \mu_B$, $M_{6i_1} = 0.636\ \mu_B$, $M_{6i_2} = 0.120\ \mu_B$ B: $M_{2d} = -0.016\ \mu_B$, $M_{4h} = -0.028\ \mu_B$						[98S2]

[a] $T = 4.2$ K, $\mu_0 H = 8$ T
[b] $T = 4.2$ K, $\mu_0 H = 3$ T
[c] $T = 4$ K, $\mu_0 H = 7$ T
[d] $T = 4$ K, $\mu_0 H = 9$ T
[e] $T = 4$ K, $\mu_0 H = 4$ T
[f] From figure, **H**||**c**, $T = 4.2$ K, extrapolated from 30.0 T
[g] $T = 4$ K, $\mu_0 H = 1.75$ T
[h] Compensation temperature
[i] $T = 2$ K, $\mu_0 H = 7.0$ T

Mössbauer study on $YCo_{7-x}Fe_xB_3$ borides, confirmed that the Fe atoms substitute preferentially the Co2*c* sites and avoid the $Co6_{i2}$ ones [06M1]. The ^{57}Fe hyperfine fields, at each site, increase when increasing the Fe content. Iron moments of 1.33 μ_B (2*c*), 0.87 μ_B ($6i_1$) and 0 μ_B ($6i_2$) were determined in $Y_2Co_6FeB_3$ compound. The weighted average of the isomer shifts decreases when increasing the iron content. The $Y_2Co_{7-x}Ni_xB_3$ compounds, at 4.2 K, are ferromagnetic for $x < 4$ and paramagnetic for higher Ni content [03M1], in agreement with band structure calculations. In $Y_2(Co_{1-x}M_x)_7B_3$ systems with M = Cu or V, there is a higher decrease of magnetizations when Co is substituted by Cu, than for V [05M2]. Thus, in the composition range $0 \leq x \leq 2$, the magnetizations decrease by $\cong$ 1.51 μ_B and 0.49 μ_B, when Co has been substituted by Cu and V, respectively, behavior correlated with the preferential location of V on $Co6i_2$ or $Co6i_1$ sites and Cu, randomly distributed in Co lattice sites.

The Ce has been assumed to be not magnetic in the $Ce_2Co_7B_3$ compound, but so far there is no direct evidence [05S1]. Only Co2*c* atoms were assumed to be magnetic (M_{Co} = 1.19 μ_B), the moments of remainder Co atoms being very small [04I1, 05S1]. The magnetic layers containing Co2*c* atoms are separated by $\cong$ 1.29 nm from other, among non-magnetic layers. The magnetic moments of Co2*c*, at $T < T_N$, order ferromagnetically intra-layer and anti-ferromagnetically inter-layer [04S1, 05S1]. Metamagnetic transitions are evidenced, the critical fields depending on temperature [95I2, 04I1, 04S1, 05S1, 06N1]. At 4.2 K, such a transition takes place at μ_0H = 8.2 T, when increasing the field, accompanied by a hysteresis centered at 5.0 T and having 6.5 T width. The metamagnetic behavior has been considered as a spin-flop transition, where the external magnetic field flops the Co2*c* moments [04I1, 04S1, 05S1]. The magnetic phase diagram of $Ce_2Co_7B_3$ shows that below μ_0H = 3 T, $Ce_2Co_7B_3$ it is in AFM state at $T < T_N$. For 3 T $\leq \mu_0H \leq$ 6 T, depending on temperature, there is a change from AFM to FM ordering at temperature T_{sR}, as a result of spin flop transition [06N1]—Fig. 8.7.

Fig. 8.7 $Ce_2Co_7B_3$: magnetic phase diagram obtained when temperature and external field increase [06N1]

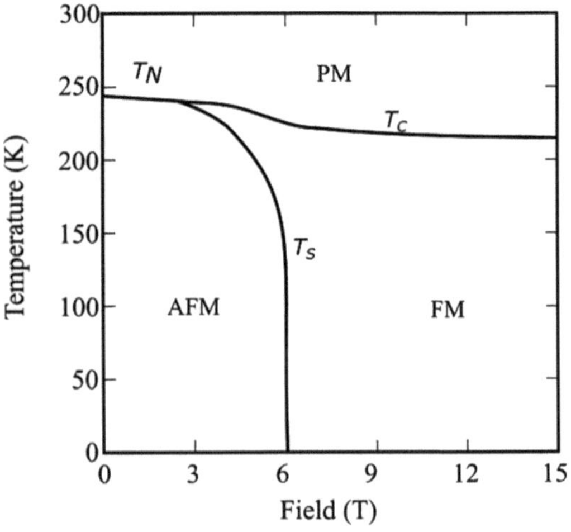

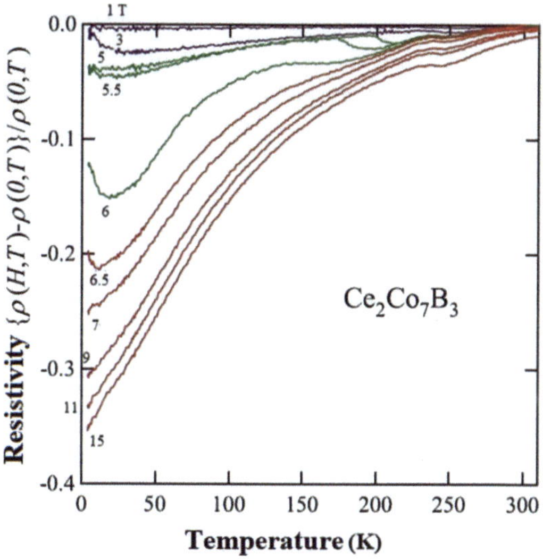

Fig. 8.8 $Ce_2Co_7B_3$: temperature and field dependences of the magnetoresistivities. The field values are given in T [06N1]

The ^{59}Co spin echo NMR study on $Ce_2Co_7B_3$, at T = 1.5 K, showed the presence of resonance frequencies of Co2*c* and Co6i_1 atoms. In addition to zero field spectra, the field-swept spectra for non-magnetic Co6i_2 have been also observed [15S1]. From the determined hyperfine fields, the spin and orbital cobalt moments were determined—Table 8.14.

The resistivity of $Ce_2Co_7B_3$ compound has been studied in external fields up to $\mu_oH = 15$ T [06N1]—Fig. 8.8. Below 3 T a single anomaly is present, at T_N. From 5 to 6 T two anomalies occur. That at lower temperature is due to spin-flip transition, at temperature T_{sR}, from AFM to FM ordering and that at higher one to Curie temperature T_c. The T_{sR} values decrease when increasing magnetic field and disappear at $\mu_0H > 6.5$ T. Thus, the temperature dependence of the $Ce_2Co_7B_3$ magnetoresistivity, can be divided in three regions corresponding to different magnetic states: $\mu_oH < 3$ T (AFM), 5 T < μ_0H < 6 T (AFM with FM divided at T_{sR}) and $\mu_0H > 6.5$ T (FM). A very large negative magnetoresistance was observed (35%) at 4 K and field of 15 T, suggesting the presence of large spin–spin-interaction between s- and d- electrons [06N1].

The $Ce_2(Co_{1-x}Ni_x)_7B_3$ compounds are antiferromagnetically ordered for $x \leq 0.14$ [05S1]. These alloys show metamagnetic transitions when $x < 0.2$, similar with those for sample with $x = 0$ [04S1, 05S1]. The $Ce_2(Co_{1-x}Ni_x)_7B_3$ system at $x > 0.24$, when 2*c* sites are fully occupied by Ni, becomes Pauli paramagnetic. The preferential substitution of Ni atoms for the Co2*c* atoms was shown to be not strictly satisfied.

The easy axes for magnetizations of ferromagnetically ordered $R_2Co_7B_3$ compounds with R = Pr [90Z2, 91H2, 00C3, 02Z1] and R = Nd [90H1, 97K3, 02C2, 02Z1] are situated in the (a,b) plane. There are some common features for these compounds: (1) the magnetic moments of Co2*c* atoms are not affected by the presence of B atoms and are close to 1.7 μ_B, due to the relatively high interatomic distance between Co2c and B atoms; (2) the Co6i_1 atoms are carrying a magnetic moment significantly smaller than the Co2c ones. The distance between the Co6i_1 and B atoms, is short enough (≈0.21 nm), to allow a significant hybridization of Co3d-B2p states, leading to a decrease of cobalt moment, as evidenced also from band structure calculations; (3) no significant local moment was found on Co6i_2 atoms, their environment including two RB_2 layers [02C2, 02Z1]. The atomic neighboring of Nd atoms situated in 1a site is not changed upon the B for Co substitution, as for other Nd sites. Thus, their magnetic moment is little affected as compared to those of Nd atoms situated on the 1b and 2e positions [02C2]. The results of magnetic measurements on the above compounds are listed in Table 8.14 [90H1, 91H2, 97K5, 00C3, 02C2, 02Z1].

The pseudo-ternary $Pr_2Co_{7-x}Fe_xB_3$ [88A2, 91H2] and $Nd_2Co_{7-x}Fe_xB_3$ [88A2, 90H1] series, were investigated particularly in connection with their possible use as permanent magnets materials. For Pr based system and $x \leq 3$, the samples have conical anisotropy, both at T = 77 K and 295 K, which changes to a uniaxial, when $x \geq 3.5$. The saturation magnetizations increase monotonously when increasing iron content, the Curie temperatures having a maximum at x = 3. The $Pr_2Co_2Fe_5B_3$-based sintered magnet has remnant induction $B_r = 0.61$ T, coercive field μ_0H_c = 1.06 T, T_c = 800 K and energy product $(BH)_{max}$ = 72.8 kJ/m^3 [91H2]. The main phase in $Nd_2Co_{7-x}Fe_xB_3$ alloys, is of hexagonal symmetry for $x < 3$, while at $x > 4$, a multiphase system was shown, the major phase having a tetragonal symmetry [90H1]. A conical ($0 < x < 3$) and axial ($x > 4$) magnetocrystalline anisotropy was evidenced at T = 295 K, while at T = 77 K, all the above samples have conical anisotropy. The T_c values have a maximum at x = 3. The $Nd_2Co_2Fe_5B_3$-based magnet, at RT, has an energy product $(BH)_{max}$ = 60 kJ/m^3 and a coercive field μ_0H_c = 0.8 T [90H1]. A higher coercive field, μ_0H_c = 1.38 T, was reported in rapidly quenched $Nd_2Co_2Fe_5B_3$ ribbons [88A2].

The magnetic behavior of $Sm_2Co_7B_3$ has been analyzed in fields up to μ_0H = 35 T [92I2]. The boride is ferromagnetically ordered. The $Sm_2Co_7B_3$ is strongly anisotropic, the anisotropy field, at T = 4.2 K, being estimated at μ_0H_a = 130 T [92I2]. The $Sm_2Co_7B_3$ saturation moments of 3.52 μ_B/f.u [92I2] or 3.68 μ_B/f.u [93O1] were reported. Assuming that the Sm^{3+} moment is 0.71μ_B, a mean cobalt moment of 0.30 μ_B/atom has been determined. When starting from the magnetizations of $Sm_{n+1}Co_{3n+5}B_{2n}$ series, the estimated moments of Co2c, Co6i_1 and Co6i_2 sites, in $Sm_2Co_7B_3$, are 1.27 μ_B, 0.43 μ_B and 0 μ_B, respectively. From the hyperfine field at ^{59}Co, determined by NMR, a value M_{Co6i1} = 0.50 μ_B/atom was obtained [98M1], close to that obtained from magnetic measurements [92I2].

The rate of heating to crystallization, as well as the maximum annealing temperature, of rapidly quenched $Sm_2Co_{7-x}Fe_xB_3$ alloys, were shown to be important in determining the coercive fields [88A2]. Values μ_0H_c = 1.0 T and μ_0H_c = 1.5 T were obtained, for compositions having x = 2 and x = 5, respectively. The $Sm_2Co_6FeB_3$, as cast and heat-treated ingot, has a coercive field of 1.5 T [83E1].

The $Gd_2Co_7B_3$ compound is ferrimagnetically ordered [88S2, 93B1, 97K3, 97S2, 01V1, 02V1, 04A1, 06S3]—Fig. 8.9. The reciprocal susceptibility follows a non-linear temperature dependence, of Néel type [93B1]. The ^{155}Gd Mössbauer study

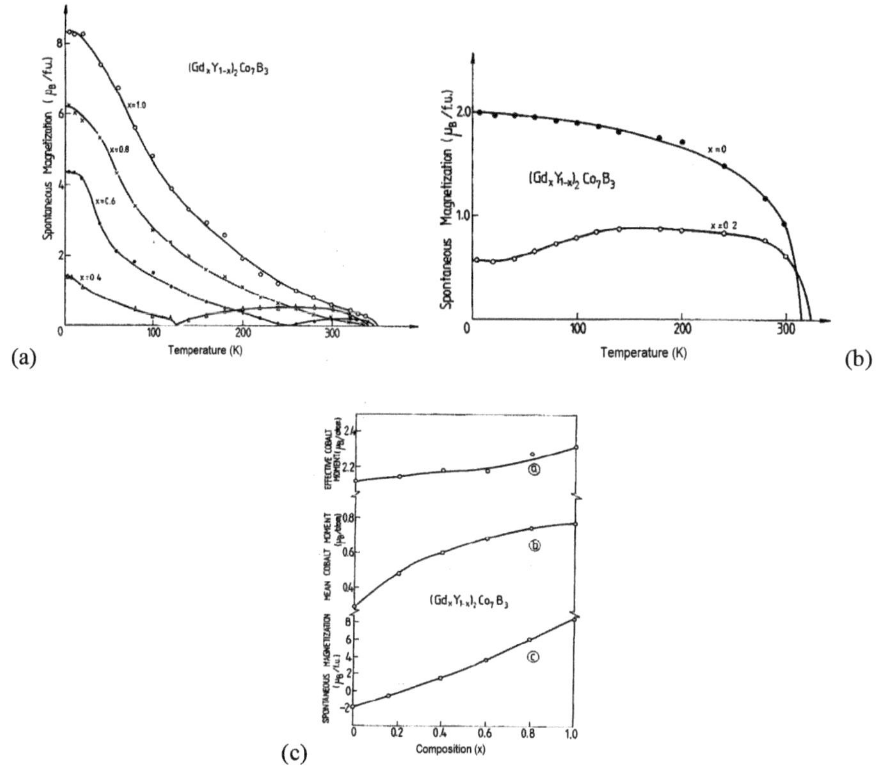

Fig. 8.9 $(Gd_xY_{1-x})_2Co_7B_3$: **a, b** thermal variations of magnetizations; **c** composition dependences of saturation magnetization, mean and effective cobalt moments, respectively [93B1]

of $Gd_2Co_7B_3$ compound, evidenced different types of quadrupole interactions at the Gd non-equivalent crystallographic positions [88S2]—Table 8.15. A semi-empirical method was used to determine the unique set of effective point charges which makes possible to describe crystal field effects.

The thermal variations of the magnetizations of $(Gd_xY_{1-x})_2Co_7B_3$ compounds with $x \geq 0.2$ are typical for ferrimagnetic ordering [93B1, 94B3]—Fig. 8.9a, b. From the Curie constants, determined in the asymptotic region and assuming that the

Table 8.15 Data obtained by ^{155}Gd Mossbauer spectroscopy on $Gd_2Co_7B_3$ compound [88S2]

T (K)	Gd site	Local environment	Isomer shift δ (mm/s)	Hyperfine field μ_0H_h (T)	V^a (10^{19} V/m)	A (%)
4.2	Gd1a	12Co + 6Co	0.25(1)	3.2(3)	9.7(1)	16
	Gd2e	12Co + 6B	0.23(1)	12.9(1)	19.2(1)	59
	Gd1b	12Co + 6B	0.22(1)	7.2(4)	25.8(1)	25

[a] Electric field gradient tensor

Gd^{3+} effective moment is given by its free ion value, the effective cobalt moments were determined. These values increase only little when Gd content is higher—Fig. 8.9c. The magnetic behavior of cobalt has been analyzed in the framework of spin fluctuations model [79M1, 91M1, 95B4]. The S_p/S_o ratio follows a linear dependence on $T^{-2/3}$, as predicted by the model [13B1]. The mean cobalt magnetic moments, at T = 4.2 K, increase as the gadolinium content is higher. The above behavior was correlated with the increase of the exchange, H_{exch}, field, acting on cobalt [93B1, 07B2, 13B1]. The mean induced cobalt moment ΔM_{Co}, seems to be linear dependent on exchange field, $\Delta M_{Co} = A\Delta H_{exch}$, where $A = (3 \cdot 10^2)^{-1}$ μ_B/T. The mean cobalt moment decreases in $Gd_2Co_{7-x}Ni_2B_3$ system, as the nickel content is higher [97B1].

The $R_2Co_7B_3$ compounds with R = Tb [97K3, 97K4, 00C1, 04A1, 06S3], R = Dy [93B1, 94B3] and R = Ho, Er, Tm [94B1, 94B2, 97K2] are ferrimagnetically ordered. A neutron diffraction study on $Tb_2Co_7B_3$, at 1.5 K, attributed the higher cobalt moment to Co6*i* sites instead of Co2*c* one and also a non-collinear arrangement of Tb and Co moments [00C1]—Table 8.14. The ^{59}Co NMR study [04A1] as well as band structure calculations [06S3] evidenced that the general trend for the sequence of decreasing cobalt moments $M_{2c} > M_{6i1} > M_{6i2}$, is valid also in this case, different from that obtained by ND study [00C1]. The neutron diffraction measurements, performed at 2 K, on $Dy_2Co_7B_3$ boride, evidenced smaller Dy magnetic moments than that of free Dy^{3+} ion [02Z1]. This feature was correlated with the crystal field effects [93O1]. The magnetic moment, per formula unit, agree with those obtained by magnetic measurements [97K3, 02Z1, 04A1]. The valence band of $Dy_2Co_7B_3$ is mainly determined by those of Co3d and Dy4f [03K1]. Both the valence band peaks and the core levels peaks were in similar positions as in pure cobalt and dysprosium, respectively. The temperature dependences of magnetizations of $R_2Co_7B_3$ compounds with R = Ho, Er and Tm show compensation of sublattices magnetizations at temperatures T_{comp} = 205 K (R = Ho), 200 K (R = Er) and 120 K (R = Tm) [94B1, 94B2, 97K2]—Fig. 8.10. The effective cobalt moments are little dependent on composition, $M_{eff} \cong 2.30\ (10)$ μ_B/atom [94B1].

The magnetic properties of $Er_{2-x}Y_xCo_7B_3$, with $1.4 \le x \le 2.0$, were studied at low temperatures and fields up to $\mu_oH = 35$ T [95Z2]. The above compounds show metamagnetic transitions from ferrimagnetic state to the ferromagnetic one [95Z2]. The M_{Co} and M_{Er} sublattices rotate discontinuously in low fields ($\mu_0H < 1.3$ T) and then two or three jumps in magnetization were shown. If the erbium magnetization is larger than that of cobalt, hysteresis is present before and after the transition, around $\mu_0H = 12$ T. When $M_{Er} < M_{Co}$, no hysteresis was found. A model was proposed in order to explain the rotation process of Er- and Co- sublattices magnetizations, taken into account the Er and Co sublattices anisotropy energies, Er-Co exchange energy, as well as the Zeeman energy. The coercive field, of $Tm_2Co_7B_3$, decreases faster with temperature up to $T \cong 100$ K and then with a lower rate. This behavior has been analyzed assuming a strongly thermally activated process of domain wall motion [02K1].

The intersublattice coupling constants in $R_2Co_7B_3$ compounds were evaluated in the mean field approximation [94B1, 94L1, 95D1, 97K2, 98M1, 99M1, 01M1]. The

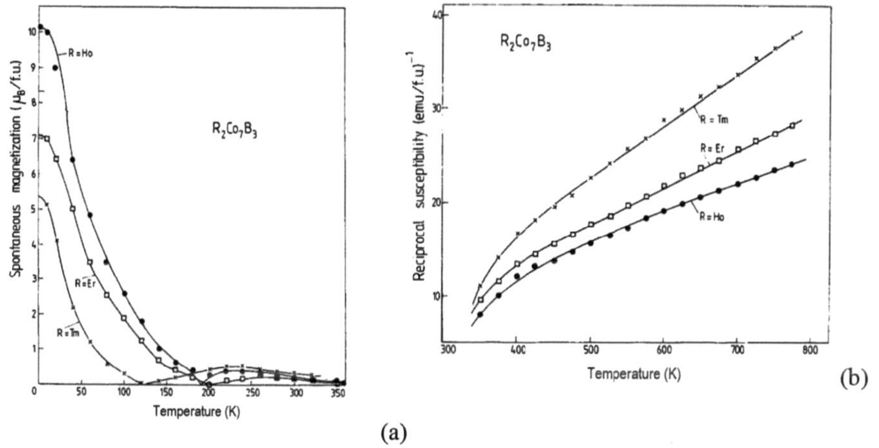

Fig. 8.10 $R_2Co_7B_3$ (R = Ho, Er, Tm): temperatures dependences of the magnetizations (**a**) and of reciprocal susceptibilities (**b**) [94B1]

effects of Co3d-B2p hybridization, on the intersublattice exchange interactions, were also analyzed in $Ce(Co_{1-x}B_x)_5$ [01I1], $Nd(Co_{1-x}B_x)_5$ [97K5] and $Gd(Co_{1-x}B_x)_5$ [95D1] series as well as on end series $R_2Co_7B_3$ compounds.

The NMR studies were also performed on $R_2Co_7B_3$ compounds [92K1, 98M1, 98M2, 04A1, 15S1]. By using zero field spin echo NMR, the variations of the cobalt hyperfine fields, in compounds with R = Y and Gd, were discussed in terms of spin and orbital contributions to the cobalt moments [92K1]—The ^{59}Gd hyperfine fields in $Gd_2Co_7B_3$ are dependent on lattice sites—Table 8.15. The ^{11}B and ^{59}Co NMR spectra were recorded for $R_2Co_7B_3$ compounds with R = Gd, Tb, Dy, Ho and Y, at 4.2 K [04A1]. The ^{59}Co hyperfine fields are strongly dependent on the cobalt local environment of a given site. The highest hyperfine field was shown at 2c site followed by $6i_1$ and $6i_2$ sites. Two kinds of Co signals were shown at each $Co6i_1$ and $Co6i_2$ positions (except for R = Y where only one, at $Co6i_2$ site, is evidenced)—Table 8.16. This behavior was attributed to two kinds of transferred hyperfine fields, likely to the dipole field referred from the Co moments in the **c**-plane. There is a transferred

Table 8.16 The hyperfine field[a] (T) at 4.2 K, in $R_2Co_7B_3$ compounds, determined by ^{59}Co NMR method [04A1]

Site	R = Y	R = Gd	R = Tb	R = Dy	R = Ho
Co2c	−15.6	−12.7	−12.8	−14.3	−13.5
$Co6i_1$	−7.99 −2.83	−7.62 −4.93	−7.76 −4.93	−8.63 −4.97	−8.86 −4.43
$Co6i_2$	0	−1.10 −0.91	−1.09 −0.96	−1.59 0	−1.37 0

[a]Two signals were observed for $Co6i_1$ site and for R = Gd, Tb also at $Co6i_2$ site, attributed to two kinds of transferred hyperfine fields likely to the dipole field referred from the Co moments in the c-plane [04A1]

Table 8.17 Local anisotropy (E_s), anisotropy per unit cell (E_a) anisotropy per unit cell (E_a), anisotropy constant (K_1) and anisotropy field (μ_0H_a)

Compound	E_s(J/atom)$\cdot 10^{22}$			$E_a \cdot 10^{14}$ (J/f.u.)	K_1 (MJ/m^3)	μ_0H_a (T)	References
	Co2*c*	Co6i_1	Co6i_2				
$Y_2Co_7B_3$	4.77	−1.69	0.137	0.226			[94K1]
$Y_2Co_7B_3$	3.10	−1.23	1.6		3.1	16.8[a]	[95O1]
$Y_2Co_7B_3$	1.50		0	0.133	0.8 [93O1]		[97T1, 98M2]
$Pr_2Co_7B_3$	4.77	−1.38	−0.228	−0.143			[90Z2]
$Sm_2Co_7B_3$	4.97	−1.29	−0.203	0.983	3.39	40.0	[93K2]
$Sm_2Co_7B_3$					14.6[a]	130.0[a,b]	[92I2]
$Sm_2Co_7B_3$					10.7[a,b,c]		[91I2]
$Sm_2Co_7B_3$					13.8[a,b,c]		[93O1]

[a]Experimental value
[b]At 4.2 K
[c]K_1/Sm

hyperfine field, in the proportion to the applied field, at ^{11}B nucleus, the B atoms being not magnetic [04A1].

By using the ^{59}Co NMR method, on $R_2Co_7B_3$ compounds with R = Ce, Nd, Sm, Gd, Y [98M1] or R = Nd, Y [98M2], the Co moments at 6i_1 sites were determined, these being of 0.40 μ_B (R = Y), 0.37 μ_B (R = Ce), 0.51 μ_B (R = Nd), 0.50 μ_B (R = Sm) and 0.66 μ_B (R = Gd), consistent with those estimated from magnetic measurements, as already mentioned [92I2, 95I1, 95I2].

The contribution of cobalt sublattice to the anisotropies of $R_2Co_7B_3$ compounds were investigated. According to [78S1, 79S2], the anisotropy of cobalt sublattice, in RCo_5 compounds, could be affected by spin–orbit coupling. A stabilization energy or local anisotropy energy per atom was defined as the difference of the spin–orbit coupling energies in the direction parallel and perpendicular to the hexagonal **c**-axis, $E_s = E_s(\perp) - E_s(\parallel)$. The total anisotropy energy per unit cell, E_{as}, of $R_2Co_7B_3$ compounds was calculated accordingly to the relation $E_{as} = 2E_s(2c) + 6E_s(6i_1) + 6E_s(6i_2)$. The determined E_s and E_a values for $R_2Co_7B_3$ compounds [90Z2, 92H2, 93K2, 94K1, 95O1, 97T1, 98M2] are listed in Table 8.17. The stabilization energies, at 2*c* and 6i_1 sites, are similar to those in RCo_5 compounds, but that at 6i_2 site is much smaller than that at 3*g* site. This suggests that the electronic clouds and the orbital moments of these atoms are significantly altered in $R_2Co_7B_3$ compounds. As a general trend in $R_2Co_7B_3$ compounds, the $E_s(2c)$ is positive, while the $E_s(6i_1)$ and $E_s(6i_2)$ are negative, except when R = Y, when at 6i_2 site it is positive [92H2, 94K1, 98M2]. The Co2*c* atoms contribute to the axial anisotropy, while those at 6i_1 and 6i_2 to planar anisotropy [92H2, 98M2].

The electrical resistivities of $R_2Co_7B_3$ compounds with R = Tb, Er, Tm [97K4], Dy [03K1] and Tm [02K1], at low temperatures, follow a T^2 dependence. At higher temperatures, the resistivities show nonlinear temperature dependences, indicating the presence of electron–phonon interactions, in addition to a small s-d scattering.

Table 8.18 Atomic sites in $Ce_3Co_{11}B_4$ compound having hexagonal structure *P6/mmm* space group [74K2, 06V1]

Atom	Sites	Sym	x	y	z	Atomic environment
Ce	1*a*	*P6/mmm*	0	0	0	Pseudo Frank-Kasper $Co_{18}Ce_2$
Ce	2*e*	6*mm*	0	0	0.333	Pseudo Frank-Kasper $Co_{12}B_6Ce_2$
Co	2*c*	$\bar{6}m2$	1/3	2/3	0	14 vertex polyhedron $Co_9Ce_3B_2$
Co	3*g*	*mmm*	1/2	0	1/2	14 vertex Frank-Kasper $B_4Co_6Ce_4$
Co	6*i*	2*mm*	1/2	0	0.200	13 vertex polyhedron $B_2Co_7Ce_4$
B	4*h*	3*m*.	1/3	2/3	0.350	Trigonal prism Co_6

8.5 $R_3Co_{11}B_4$ Compounds

The $R_3Co_{11}B_4$ series is the member with n = 2 of the $R_{n+1}Co_{5n+1}B_{2n}$ family, where n represent the number of RB_2 type planes in the unit cell. As already mentioned, the crystal structure can be described as alternative stacking along **c**-axis of one layer of RCo_5 and two layers of RCo_3B_2 unit cells. The $R_3Co_{11}B_4$ compounds crystallize in a hexagonal structure having *P6/mmm* space group [74K2, 06V1]—Fig. 8.1. In this lattice, the rare earth (yttrium) atoms occupy two types of sites (1*a*, 2*e*), the cobalt are distributed over three nonequivalent positions (2*c*,3*g*,6*i*) and boron atoms are located in only one site 4*h*. The crystal structures are not distorted below the Curie points, as evidenced by neutron diffraction studies, on $R_3Co_{11}B_4$ compounds with R = Nd, Dy and Y [01C1, 02C2, 02Z1]. The atomic sites and their coordinates, in $Ce_3Co_{11}B_4$ compound, are given in Table 8.18 [74K2]. The lattice parameters of $R_3Co_{11}B_4$ compounds are listed in Table 8.19.

The substitution of rare-earth R by another R' one, leads to the formation of solid solutions having *P6/mmm*-type structure in all the composition range as reported in $(R_xR'_{1-x})_3Co_{11}B_4$ series with R = Gd and R' = Ce [09K2, 09K3], Pr [14K1] and Nd [11K1] or R = Y and R' = Gd [94B3, 98J2], Tb [99T1] and Er [95Z3]. Solid solutions are formed in a limited composition range, when cobalt was replaced by another element. The hexagonal structure is maintained in $R_3Co_{11-x}M_xB_4$ for $x \leq 6$, if Co is substituted by Fe [97P1, 99M1] or Ni [96B2, 96P2, 97P1, 07I1]. When Co is substituted by M = Al [97P1, 01B1] or M = Cu [01B1, 02B2] in $Y_3Co_{11-x}M_xB_4$ series, solid solutions are formed up to x = 3. Preferential occupations of Fe and Al on the Co2*c* sites and Ni at the Co3*g* positions have been experimentally evidenced [97P1]. A theoretical study [13L3], on the $R_3Co_{11-x}Fe_xB_4$ (R = Nd, Gd) series, reported the preference of iron for 3*g* site, the degree of occupation decreasing in the sequence 3*g* > 6*i* > 2*c*. The stability of the corresponding crystal structures was also investigated.

The $Y_3Co_{11}B_{4-x}Si_x$ compounds have hexagonal structure for $x \leq 1.5$ [97B2]. In the $Gd_3Co_{11}B_{4-x}Al_x$ system, solid solutions are formed up to x = 0.5 [97G1]. Further increase of the Al content, leads to the formation of two phase alloys

Table 8.19 Space groups and lattice parameters of $R_3Co_{11}B_2$ compounds

Compound	T (K)	Space group	Lattice parameters (nm)		References
			a	c	
$Ce_3Co_{11}B_4$	RT	*P6/mmm*	0.5045	0.9925	[74K2]
$Ce_3Co_{11}B_4$	RT	*P6/mmm*	0.5076	0.9883	[79K1]
$Pr_3Co_{11}B_4$	RT	*P6/mmm*	0.5147	0.9785	[87C2]
$Pr_3Co_{11}B_4$	2	*P6/mmm*	0.5140(1)	0.9763(1)	[02Z1]
$Pr_3Co_{11}B_4$	RT	*P6/mmm*	0.5145	0.9784	[74K2]
$Pr_3Co_{11}B_4$	400	*P6/mmm*	0.5151(1)	0.9810(1)	[02Z1]
$Nd_3Co_{11}B_4$	2	*P6/mmm*	0.5129(1)	0.9772(1)	[02C2]
$Nd_3Co_{11}B_4$	RT	*P6/mmm*	0.5132	0.9808	[74K2]
$Nd_3Co_{11}B_4$	RT	*P6/mmm*	0.5140	0.9742	[99C2]
$Sm_3Co_{11}B_4$	RT	*P6/mmm*	0.5092	0.9799	[00C5]
$Sm_3Co_{11}B_4$	RT	*P6/mmm*	0.5092	0.9799	[74K2]
$Sm_3Co_{11}B_4$	RT	*P6/mmm*	0.5090	0.9797	[96T1]
$Sm_3Co_{11}B_4$	RT	*P6/mmm*	0.5089	0.9796	[91I2]
$Gd_3Co_{11}B_4$	RT	*P6/mmm*	0.5079	0.9842	[74K2]
$Gd_3Co_{11}B_4$	RT	*P6/mmm*	0.5087(1)	0.9829(1)	[07C1]
$Gd_3Co_{11}B_4$	RT	*P6/mmm*	0.5072	0.9839	[88S2]
$Gd_3Co_{11}B_4$	RT	*P6/mmm*	0.5080	0.9842	[00C4]
$Gd_3Co_{11}B_4$	RT	*P6/mmm*	0.50769(2)	0.98382(6)	[97G1]
$Tb_3Co_{11}B_4$	RT	*P6/mmm*	0.5058	0.9823	[74K2]
$Tb_3Co_{11}B_4$	RT	*P6/mmm*	0.5062	0.9828	[99T1]
$Dy_3Co_{11}B_4$	2	*P6/mmm*	0.5038(1)	0.9794(1)	[02Z1]
$Dy_3Co_{11}B_4$	RT	*P6/mmm*	0.5048	0.9839	[74K2]
$Dy_3Co_{11}B_4$	RT	*P6/mmm*	0.5043	0.9836	[96T1]
$Ho_3Co_{11}B_4$	RT	*P6/mmm*	0.5030	0.9846	[74K2]
$Ho_3Co_{11}B_4$	RT	*P6/mmm*	0.5030	0.9840	[96T1]
$Er_3Co_{11}B_4$	RT	*P6/mmm*	0.5022	0.9845	[74K2]
$Tm_3Co_{11}B_4$	RT	*P6/mmm*	0.5004	0.9853	[74K2]
$Tm_3Co_{11}B_4$	RT	*P6/mmm*	0.5010	0.9850	[97K6]
$Lu_3Co_{11}B_4$	RT	*P6/mmm*	0.4982	0.9867	[74K2]
$Y_3Co_{11}B_4$	2	*P6/mmm*	0.5038(2)	0.9836(4)	[01C1]
$Y_3Co_{11}B_4$	RT	*P6/mmm*	0.5041(2)	0.9853(4)	[01C1]
$Y_3Co_{11}B_4$	RT	*P6/mmm*	0.5086	0.9810	[74K2]

having $Ce_3Co_{11}B_7$ and $CaCu_5$-type structures, respectively. The Al in $Ce_3Co_{11}B_4$—type lattice, shows a preference for 3g site, due to both size and mixed enthalpy considerations.

The thermal expansion coefficients of $Gd_3Co_{11}B_4$ in the temperature range 100 K $\leq T \leq 600$ K were anisotropic with $\alpha_a = 5.4 \cdot 10^{-6}$ K^{-1} and $\alpha_c = 19.2 \cdot 10^{-6}$ K^{-1} [98I2].

The $Y_3Co_{11}B_4$ boride is ferromagnetically ordered [93K3, 93O1, 94K3, 95B5, 96K2, 97S2, 98J2, 98S2, 99T1]. The easy axis of magnetization is along **c**-direction, as evidenced from high field magnetic measurements [93O1, 95O1] or NMR data [05M1]. The site dependence of cobalt moments, determined from neutron diffraction, follows the sequence $M_{2c} > M_{6i} > M_{3g}$ [01C1] in good agreement with the computed values [98S2, 02B2, 02V1]. The reciprocal susceptibility shows a Curie–Weiss-type temperature dependence with an effective cobalt moment of $\cong 2.30$ μ_B/Co atom—Table 8.20.

The magnetizations and the Curie temperatures increase in $Y_3Co_{11-x}Fe_xB_4$ series, when increasing the iron content, correlated with their higher magnetic moment than that of cobalt. [96T2]. The saturation magnetizations and Curie temperatures of $Y_3Co_{11-x}Ni_xB_4$ [96B2], $Y_3Co_{11-x}M_xB_4$ with M = Cu or Al [01B1, 02B2] and $Y_3Co_{11-x}M_xB_4$ where M = Mn, Ti, V [01V2] pseudoternary compounds decrease as cobalt sublattice is magnetically diluted. The mean cobalt moments, at 4.2 K, diminishes when Co is substituted by Ni, from 0.4 μ_B/atom (x = 0) up to 0.1 μ_B/atom (x = 6) [96B2]. A nearly nil cobalt moment has been evidenced at x = 1.5, when Co was replaced by Al. In case of Cu substitution, a value $M_{Co} = 0.04$ μ_B/atom has been shown when x = 3. The $Y_3Co_3Ni_8B_4$ [96B2] and $Y_3Co_9Al_2B_4$ [01B1, 02B2] compounds behave as exchange enhanced paramagnets. The reciprocal susceptibilities for both magnetic and paramagnetic $Y_3Co_{11}B_4$-based compounds follow, at high temperatures, linear dependences. The above magnetic behavior of cobalt can be described in the spin fluctuations model [91M1]. The effective cobalt moments, in nickel substituted compounds, are little dependent on composition, while when Co was replaced by Al and Cu, there is a significantly decrease. A linear dependence of the mean cobalt moment on the exchange field was shown. The degree of itinerancy of transition metal moments, as described by the r values, increases when increasing the Ni, Al, Mn, V and Ti content. The mean cobalt moments, at 4.2 K, decrease when B is gradually substituted by Si, in $Y_3Co_{11}Si_xB_{4-x}$ system, while their effective moments are only little changed [97B2].

The cobalt sublattice anisotropy in $R_3Co_{11}B_4$ has been determined starting from the stabilization energy (local anisotropy energy) per atom, as already defined [78S2, 79S2]—Sect. 8.4. The Co2*c* site in RCo_5 compounds is crystallographically equivalent to the Co2*c* one in $R_3Co_{11}B_4$ boride and Co3*g* site in RCo_5 is equivalent to those of Co3*g* and Co6*i* sites in $R_3Co_{11}B_4$ compounds. Stabilization energies $E_s(2c) = 4.76 \cdot 10^{-22}$ J/atom and $E_s(3g) = -1.39 \cdot 10^{-22}$ J/atom were determined in RCo_5 compounds [78S2]. Since the environment of Co2*c* site was not changed in $R_3Co_{11}B_4$ compounds, the same stabilization energy was assumed, as in RCo_5 compounds. The stabilization energy $E_s(3g) = -1.53(10) \cdot 10^{-22}$ J/atom has been estimated from the anisotropy energy of RCo_4B. The $E_s(6i)$ value has been evaluated from the stabilization energy of cobalt sublattice, E_{as}, in $R_3Co_{11}B_4$ according to the relation $E_{as} = 2E_s(2c) + 3E_s(3g) + 6E_s(6i)$. A negative value was obtained [93K2]. On this basis the stabilization energies in $R_3Co_{11}B_4$ compounds with R = Y, Pr, Sm were

Table 8.20 Magnetic properties of $R_3Co_{11}B_4$ compounds

Compound	Magnetic structure and magnetic moments	M_s (μ_B/f.u.)	T_c (K)	C (emuK/f.u.)[1)]	M_{eff} (μ_B/Co atom)	θ (K)	References
$Ce_3Co_{11}B_4$ (NMR)	T = 1.4 K Co: $M^o_{2c} = 1.21\ \mu_B$, $M^s_{2c} = 0.12\ \mu_B$, $M_{3g} = 0\ \mu_B$, $M^s_{6i} = 0.36\ \mu_B$, $M^o_{6i} = 0.01\ \mu_B$						[15S1]
$Ce_3Co_{11}B_4$ (ND)	FIM, T = 20 K Co: $M_{2c} = 1.4\ \mu_B$, $M_{3g} = 0\ \mu_B$, $M_{6i} = 0.4\ \mu_B$ Ce: $M = -0.11\ \mu_B$	4.88	297 [97I1]				[04I1]
$Ce_3Co_{11}B_4$ (BS)	FIM Ce: $M_{1a} = -0.39\ \mu_B$, $M_{2e} = -0.24\ \mu_B$ Co: $M_{2c} = 1.21\ \mu_B$, $M_{3g} = 0.08\ \mu_B$, $M_{6i} = 0.36\ \mu_B$ B: $M_{4h} = -0.01\ \mu_B$						[09I1]
$Pr_3Co_{11}B_4$ (ND)	FM, T = 2 K Pr: $M_{1a} = 1.7(2)\ \mu_B$, $M_{2e} = 2.2(2)\ \mu_B$ Co: $M_{2c} = 1.7(2)\ \mu_B$, $M_{3g} = 0.2(2)\ \mu_B$, $M_{6i} = 0.9(2)\ \mu_B$	13.8(1)[a]	390(5)				[02Z1]
$Pr_3Co_{11}B_4$	FM	11.57[b]		12.54	2.37		[96T1]
$Pr_3Co_{11}B_4$	FM ⟨Pr⟩:M = 2.4 μ_B ⟨Co⟩:M = 0.42 μ_B	11.8[c]	385				[95K1]

(continued)

Table 8.20 (continued)

Compound	Magnetic structure and magnetic moments	M_s (μ_B/f.u.)	T_c (K)	C (emuK/f.u.)[1)]	M_{eff} (μ_B/Co atom)	θ (K)	References
$Nd_3Co_{11}B_4$ (ND)	FM, T = 2 K Nd: $M_{1a} = 2.1(2)\ \mu_B$, $M_{2e} = 1.7(2)\ \mu_B$ Co: $M_{2c} = 1.7(2)\ \mu_B$, $M_{3g} = 0.2(2)\ \mu_B$, $M_{6i} = 0.7(2)\ \mu_B$	13.4(1)[a]	380(5)				[02C2, 02Z1]
$Nd_3Co_{11}B_4$ (BS)	Nd: $M_{1a} = 2.90\ \mu_B$, $M_{2c} = 2.99\ \mu_B$ Co: $M_{2c} = 1.77\ \mu_B$, $M_{3g} = 0.03\ \mu_B$, $M_{6i} = 0.58\ \mu_B$ B: $M_{4h} = -0.02\ \mu_B$						[98K6, 98J1]
$Nd_3Co_{11}B_4$	FM	11.82[b]		12.52	2.35		[96T1]
$Nd_3Co_{11}B_4$	FM	12.2[c]	392		2.20		[97K5, 98K6]
$Nd_3Co_{11}B_4$	FM	12.4[d]	350 100 (T_{sR})				[00C6]
$Sm_3Co_{11}B_4$	FM	2.4[e]	460				[83G1]
$Sm_3Co_{11}B_4$	Sm: ⟨M> = 0.29 μ_B	6.54[f]	446				[93O1]
$Sm_3Co_{11}B_4$	FM	4.86[g]	458		2.32		[96T1]

(continued)

Table 8.20 (continued)

Compound	Magnetic structure and magnetic moments	M_s (μ_B/f.u.)	T_c (K)	C (emuK/f.u.)[1)]	M_{eff} (μ_B/Co atom)	θ (K)	References
$Sm_3Co_{11}B_4$	FM	1.4	448				[83E2]
$Sm_3Co_{11}B_4$	FM Sm: $\langle M \rangle = 0.71\ \mu_B$ Co: $M_{2c} = 1.27\ \mu_B$, $M_{3g} = 0\ \mu_B$, $M_{6i} = 0.43\ \mu_B$: $\langle M \rangle = 0.465\ \mu_B$	$\cong 6.94^{h}$	446				[92I2]
$Gd_3Co_{11}B_4$	FIM	11.4^{i}	454				[94K3]
$Gd_3Co_{11}B_4$	FIM	10.89^{g}	430	38.40	3.27		[97S2]
$Gd_3Co_{11}B_4$ (BS)	FIM Gd: $M_{1a} = 7.405\ \mu_B$, $M_{2e} = 7.343\ \mu_B$ Co: $M_{2c} = -1.709\ \mu_B$, $M_{3g} = -0.144\ \mu_B$, $M_{6i} = -0.645\ \mu_B$ B: $M_{4h} = 0.02\ \mu_B$						[02V1]
$Gd_3Co_{11}B_4$ (BS)	FIM Gd: $M_{1a} = 7.500\ \mu_B$, $M_{2e} = 7.377\mu_B$ Co: $M_{2c} = -1.648\ \mu_B$, $M_{3g} = -0.171\ \mu_B$, $M_{6i} = -0.792\ \mu_B$ B: $M_{4h} = 0.035\ \mu_B$						[06S3]
$Gd_3Co_{11}B_4$ (BS)	FIM Gd: $M_{1a} = 7.21\ \mu_B$, $M_{2c} = 7.11\ \mu_B$ Co: $M_{2c} = -1.59\ \mu_B$, $M_{3g} = -0.098\ \mu_B$, $M_{6i} = -0.63\ \mu_B$ B: $M_{4h} = 0.02\ \mu_B$						[98K7]

(continued)

Table 8.20 (continued)

Compound	Magnetic structure and magnetic moments	M_s (μ_B/f.u.)	T_c (K)	C (emuK/f.u.)[1)]	M_{eff} (μ_B/Co atom)	θ (K)	References
$Tb_3Co_{11}B_4$ (ND)	FIM, T = 15 K Tb: $M_{1a} = 8.5(7)\ \mu_B$, $M_{2e} = 7.9(7)\ \mu_B$ Co: $M_{2c} = -1.3(5)\ \mu_B$, $M_{3g} = -0.5(4)\ \mu_B$, $M_{6i} = -0.5(2)\ \mu_B$, $\theta = 72^o$	5.2[j]					[00C1]
$Tb_3Co_{11}B_4$ (BS)	FIM Tb: $M_{1a} = 9.439\ \mu_B$, $M_{2e} = 9.349\ \mu_B$ Co: $M_{2c} = -1.749\ \mu_B$, $M_{3g} = -0.126\ \mu_B$, $M_{6i} = -0.787\ \mu_B$ B: $M_{4h} = 0.032\ \mu_B$						[06S3]
$Tb_3Co_{11}B_4$	FIM	15.56[b]	431	32.17	2.60		[96T1]
$Tb_3Co_{11}B_4$	FIM Tb: ⟨M> = 8.40 μ_B, Co: ⟨M> = 0.88 μ_B	15.56[k]	430		2.60		[99T1]
$Dy_3Co_{11}B_4$ (ND)	FIM Dy: $M_{1a} = 10.0(7)\ \mu_B$, $M_{2e} = 9.8(2)\ \mu_B$ Co: $M_{2c} = -1.6(2)\ \mu_B$, $M_{3g} = -0.2(2)\ \mu_B$, $M_{6i} = -0.7(2)\ \mu_B$	19.5[a]	375(5)				[02Z1]
$Dy_3Co_{11}B_4$	FIM	19.1[i]	375				[95K2]
$Dy_3Co_{11}B_4$	FIM	16.63[g]	365 [97S2]	51.44	2.56		[96T1]
$Dy_3Co_{11}B_4$	FIM, e.a.m planar	18.83[f]	375				[93O1]

(continued)

Table 8.20 (continued)

Compound	Magnetic structure and magnetic moments	M_s (μ_B/f.u.)	T_c (K)	C (emuK/f.u.)[1)]	M_{eff} (μ_B/Co atom)	θ (K)	References
$Ho_3Co_{11}B_4$	FIM	13.78[g]		51.06	2.54		[96T1]
$Er_3Co_{11}B_4$	FIM	14.2[i]	350				[95K2]
$ErY_2Co_{11}B_4$	Metamagnetic T = 4.2 K, H < 20 T, FIM $M_{Er} = 9.00(2)\ \mu_B$/f.u., $M_{Co} = -5.27(5)\ \mu_B$/f.u., $\mu_0H > 20$ T, FM, $M \cong 14.3\ \mu_B$/f.u						[95Z3]
$Tm_3Co_{11}B_4$	FIM	8.62[c]	347				[97K6]
$Tm_3Co_{11}B_4$ (BS)	Tm: $M_{1a} = 7.274\ \mu_B$, $M_{2e} = 7.204\ \mu_B$ Co: $M_{2c} = -1.731\ \mu_B$, $M_{3g} = -0.079\ \mu_B$, $M_{6i} = -0.746\ \mu_B$ B: $M_{4\mathbf{h}} = 0.032\ \mu_B$	13.363					[01S1]
$Y_3Co_{11}B_4$ (ND)	FM, T = 2 K, eam ‖ **c** Co: $M_{2c} = 1.7(2)\ \mu_B$, $M_{3g} = 0.1(2)\ \mu_B$, $M_{6i} = 0.7(2)\ \mu_B$	7.9	≅ 350				[01C1]
$Y_3Co_{11}B_4$ (BS)	Y: $M_{1a} = -0.256\ \mu_B$, $M_{2e} = -0.169\ \mu_B$ Co: $M_{2c} = 1.697\ \mu_B$, $M_{3g} = 0.098\ \mu_B$, $M_{6i} = 0.713\ \mu_B$ B: $M_{4h} = -0.035\ \mu_B$	7.37					[98S2]
$Y_3Co_{11}B_4$ (BS)	Y: $M_{1a} = -0.149\ \mu_B$, $M_{2e} = -0.112\ \mu_B$ Co: $M_{2c} = 1.629\ \mu_B$, $M_{3g} = 0.088\ \mu_B$, $M_{6i} = 0.519\ \mu_B$ B: $M_{4h} = -0.019\ \mu_B$	7.18					[02B2, 02V1]

(continued)

Table 8.20 (continued)

Compound	Magnetic structure and magnetic moments	M_s (μ_B/f.u.)	T_c (K)	C (emuK/f.u.)[1)]	M_{eff} (μ_B/Co atom)	θ (K)	References
$Y_3Co_{11}B_4$	FM, $\chi = \chi_0 + C(T\text{-}\theta)^{-1}$, $\chi_0 = 3.26 \cdot 10^{-3}$ emu/f.u.[l]	4.70	351	7.67	2.36	345	[97S2]
$Y_3Co_{11}B_4$	FM	4.5[k]	353		2.31	351	[99T1]
$Y_3Co_{11}B_4$	FM Co: $M_{2c} = 1.7(2)\ \mu_B$, $M_{3g} = 0.1(2)\ \mu_B$, $M_{6i} = 0.5(2)\ \mu_B$						[01C1]
$Y_3Co_{11}B_4$	FM	4.73[i]	345				[93K3, 94K3]
$Y_3Co_{11}B_4$	FM Co: $M_{2c} = 1.66\ \mu_B$, $M_{3g} = 0\ \mu_B$, $M_{6i} = 0.5\ \mu_B$	5.98[f]	345				[93O1]

[a] $T = 2$ K, $\mu_0H = 7$ T
[b] $T = 4.2$ K, $\mu_0H = 8$ T
[c] $T = 4.2$ K, $\mu_0H = 4$ T
[d] $T = 5$ K, $\mu_0H = 6.5$ T
[e] $T = 77$ K
[f] $T = 4$ K, $\mu_0H = 1.5$ T
[g] $T = 4.2$ K, $\mu_0H = 8$ T
[h] $T = 4.2$ K, $\mu_0H = 30$ T
[i] $T = 4.2$ K, $\mu_0H = 14$ T
[j] $T = 5$ K, $\mu_0H = 5$ T
[k] $T = 4.2$ K, $\mu_0H = 10$ T
[l] 1 emu/g = 1 Am^2/kg

estimated—Table 8.21a. In $Y_3Co_{11}B_4$, the $E_s(6i)$ is positive, but smaller than $E_s(3g)$, as determined in YCo_5 [76D1]. In this case, the anisotropy favors the **c**-direction as the easy axis. The environment of Co3g site in $R_3Co_{11}B_4$ is changed as compared to RCo_5 one, due to Co substitution by boron. As a consequence, the corresponding Co orbital moment is modified. The experimentally determined anisotropy constants, anisotropy field and coercive fields, of some $R_3Co_{11}B_4$ compounds are listed in Table 8.21b. The uniaxial anisotropy of end series Y compound, changes from uniaxial to planar for an amount of substitution larger than x = 1 when M = Fe or Al, due to their preference for Co2c site. The composition dependence of the anisotropy constant has been analysed in pseudoternary $Y_3Co_{11-x}Fe_xB_4$ compounds, in the framework of the individual site anisotropy model [96T2].

The $Ce_3Co_{11}B_4$ is magnetically ordered at temperatures by ≅ 100 K smaller than that of isomorphous Y compound, the Ce being in mixed valence state. The cobalt magnetic moments follow the sequence $2c > 6i > 3g$, as evidenced by a neutron diffraction study [04I1]. The NMR study [15S1], as well as band structure

Table 8.21 Anisotropy constants of $R_3Co_{11}B_4$ compounds

(a) Local site anisotropy E_s and total anisotropy energy E_{as}

Compound	$E_s \cdot 10^{-22}$ (J/atom)			$E_{as} \cdot 10^{-22}$ (J/u.c.)	References
	2c	3g	6i		
$Pr_3Co_{11}B_4$	4.76	−1.62	−0.228	−0.875	[90Z2]
$Sm_3Co_{11}B_4$	4.76	−1.53	−0.202	0.159	[93K2]
$Y_3Co_{11}B_4$	4.76	−1.60	0.137	0.226	[94K1]
$Y_3Co_{11}B_4$	3.1	−1.23	2.5	0.61	[95O1]

(b) Anisotropy constants, anisotropy fields and coercive fields

Compound	T (K)	Site	Anisotropy constants (M J/m^3)			Anisotropy field (T)	Coercive field, μ_0H_c (T)	References
			K1	K2	K3			
$Nd_3Co_{11}B_4$	5					49.0		[00C6]
$Nd_3Co_{11}B_4$	298					Plane		[00C6]
$Sm_3Co_{11}B_4$	4		16.9			116.0	21.0	[92I2]
$Sm_3Co_{11}B_4$	RT		0.72			17.5		[93K2]
$Sm_3Co_{11}B_4$	0[a]	Sm	15.3					[93O1]
$Er_3Co_{11}B_4$	4.2						0.14	[95K2]
$ErY_2Co_{11}B_4$	4.2	Er	2.9[b]	15.6[b]	−8.5[b]			[95Z3]
		Co	4.9[b]	0.9[b]	0.7[b]			
$Tm_3Co_{11}B_4$	30						0.8	[97K6]
$Y_3Co_{11}B_4$	4.2		3.1				23.4	[95O1]
$Y_3Co_{11}B_4$	0[a]		1.6					[93O1]

[a]Extrapolated at T = 0 K
[b]J/fu·10^{-18}

calculations [09I1], confirmed this trend—Tables 8.20 and 8.22. The Ce4f spin is parallel to the sample's magnetization [01I1]. The Ce4f orbital angular momentum is larger than the spin one and therefore antiparallelly oriented to cobalt moments. The ^{59}Co spin echo NMR measurements, at 14 K, as well as the field swept spectra of non-magnetic cobalt, were studied in $Ce_3Co_{11}B_4$ compound [15S1]. From the ^{59}Co hyperfine fields, the cobalt moments at 2*c*, 6*i* and 3*g* sites were estimated—Table 8.20. The NMR spin echo study on $R_3Co_{11}B_4$ compounds with R = Gd, Y has been also reported [92K1].

The $R_3Co_{11}B_4$ compounds with R = Pr [95B5, 95K1, 96T1, 02Z1], Nd [96T1, 98J1, 98K6, 00C6, 02C2, 02Z1] and Sm [83E2, 83G1, 91I2, 92I2, 93K2, 93O1, 96T1, 00C5] are ferromagnetically ordered—Fig. 8.11a. The determined cobalt moments follow the general trend, $M_{2c} > M_{6i} > M_{3g}$, that at the 3*g* site is nil or nearly nil, since it is located between two RB_2 planes [02Z1]—Table 8.20. The mean effective cobalt moments, determined at $T > T_c$, of 2.36(1) μ_B/atom, are little dependent on R partner (R = Pr, Nd) [96T1]. The Pr and Nd magnetic moments in $R_3Co_{11}B_4$ compounds are less sensitive to the presence of boron in their environments and imposed the

Table 8.22 Hyperfine fields determined by NMR method in $R_3Co_{11}B_4$ compounds

Compound	T (K)	Hyperfine field (T)					References
		Co2*c*	Co3*g*	Co6*i*	Y1*a*	B4*h*	
$Ce_3Co_{11}B_4$	1.4	−5.88	0	(−)3.34			[15S1]
$Gd_3Co_{11}B_4$[a]	4.2	5.3	0.9	1.8		6.8	[92K1]
$Y_3Co_{11}B_4$[a]	4.2	5.0	2.7	5.3	4.8		[92K1]

[a]From spin echo spectra

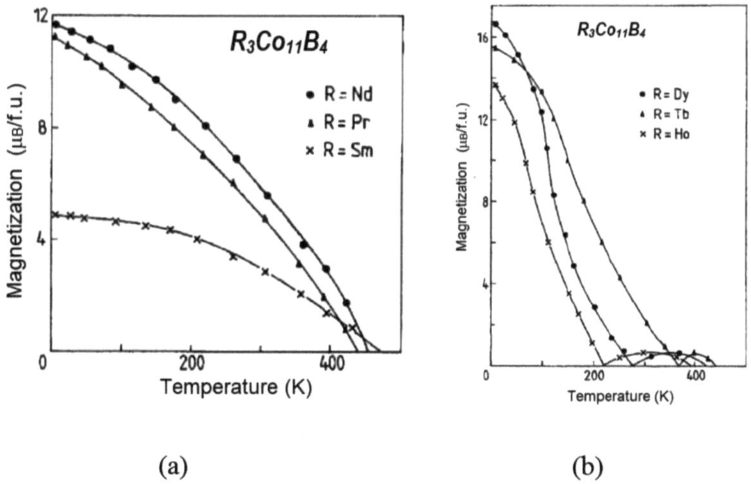

Fig. 8.11 $R_3Co_{11}B_4$: thermal variations of magnetizations for compounds with R = Pr, Nd, Sm (**a**) and R = Tb, Dy, Ho (**b**) [96T1]

alignement of the magnetic moments in the basal plane [02C2]. A planar anisotropy of cobalt sublattice was also suggested in $Pr_3Co_{11}B_4$, where both $E_s(3g)$ and $E_s(6i)$ contributions to Co sublattices anisotropies are negative [90Z2]. The presence of a spin reorientation at $T_{sR} = 100$ K, from easy plane to easy cone in $Nd_3Co_{11}B_4$ [00C6] was not confirmed [02C2, 02Z1]. The XPS spectra, of $Nd_3Co_{11}B_4$ compounds [98K6, 98K8] were correlated with the data obtained from band structure calculations. The $Sm_3Co_{11}B_4$ compound has the easy axis of magnetization along **c**-direction. The anisotropy energy of $Sm_3Co_{11}B_4$ is smaller than that in $SmCo_5$, since both $E_s(3g)$ and $E_s(6i)$ are negative. Assuming that the Sm moment, at 4 K, is $M_{Sm} = 0.71$ μ_B/atom, the magnetic contributions of cobalt atoms to the magnetization, as well as the mean cobalt moment, were estimated [92I2]—Table 8.20. The magnetization reversal mechanism, of $Sm_3Co_{11}B_4$, at T = 295 K, involves nucleation of reverse domain walls, while that at T = 77 K, implies pinning centers such as stacking faults and perhaps even due to lattice itself [83G1].

The $Gd_3Co_{11}B_4$ compound is ferrimagnetically ordered [83E2, 94B3, 94K3, 95B5, 96K2, 97S2, 98K8, 01V1, 02V1, 06S3, 09K2, 09M1, 13L3]. The temperature dependence of magnetization, according to Néel classification [48N1], is of V-type—Fig. 8.12. Assuming $M_{Gd} = 7\mu_B$, the determined mean cobalt moment, at 4.2 K, is $M_{Co} = 0.87$ μ_B/atom. The thermal variation of magnetic susceptibilities, follow a non-linear temperature dependence of Néel-type, this trend being characteristic also for others $R_3Co_{11}B_4$ compounds with magnetic heavy rare-earths (R = Tb, Dy, Ho, Er, Tm). The mean effective cobalt moments, determined in the above compounds, is of ≅ 2.55 μ_B/atom—Table 8.20. The XPS valence band spectra of

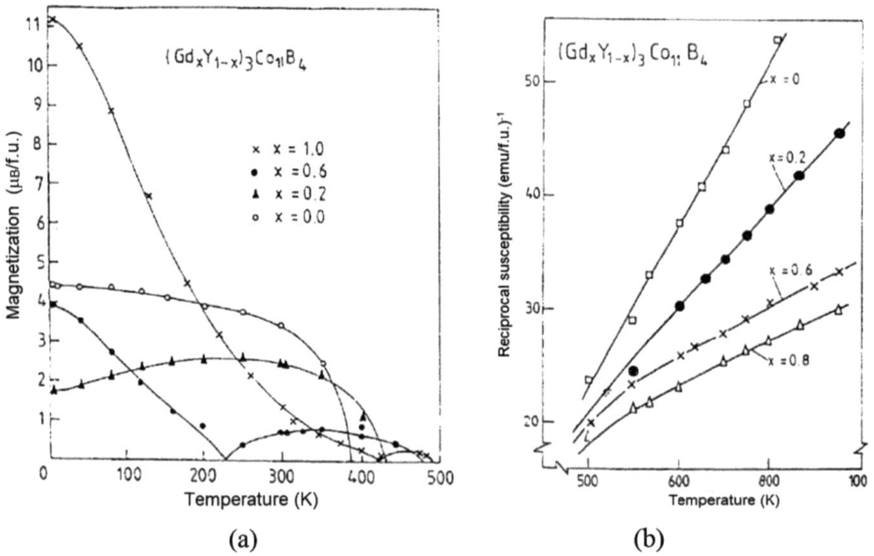

Fig. 8.12 $(Gd_xY_{1-x})_3Co_{11}B_4$: thermal variations of magnetizations (**a**) and of reciprocal susceptibilities (**b**) [94B3]

$Gd_3Co_{11}B_4$ compound has been found to be in good agreement with band structure calculations [98K7].

The $(Gd_xY_{1-x})_3Co_{11}B_4$ compounds are ferrimagnetically ordered, when $x \geq 0.2$ [94B3]—Fig. 8.12. The mean cobalt moments, determined at T = 4.2 K, decrease as the yttrium content is higher, while the effective ones are little dependent on composition. The ratio $r = S_p/S_0$, between the number of spins determined from effective cobalt moment, S_p, and from saturation magnetization, S_0, respectively, follow a $T_c^{-2/3}$ dependence, as predicted by the spin fluctuations model [79M1, 91M1]. Starting from experimental data [94B3], the exchange interactions parameters inside and between magnetic sublattices (Gd, Co) were determined by using mean field approximation [98J2]. By magnetic dilution of Gd sublattice in $(Gd_xY_{1-x})_3Co_{11}B_4$ compounds, the compensation temperatures of the two magnetic sublattices are shifted to lower values [94B3]. The magnetocaloric effect in $(Gd_xY_{1-x})_3Co_{11}B_4$ system has been correlated with temperature dependencies of the resultant magnetization determined mainly by that of gadolinium sublattice as can be shown from the thermal variation of magnetic entropy changes ΔS_M [10B1]—Fig. 8.13. The ΔS_M values show normal and inverse magnetic entropy change below or near the compensation temperature or when magnetization increase with temperature as in sample with x = 0.8 or 0.2, respectively. This trend has been theoretical analysed [20A1].

The pseudoternary $Ce_{3-x}Gd_xCo_{11}B_4$ [09K2], $Pr_{3-x}Gd_xCo_{11}B_4$ [14K1] and $Nd_{3-x}Gd_xCo_{11}B_4$ [11K1] compounds, change the magnetic ordering from ferromagnetic to ferrimagnetic as Gd content increases. Thus, at 4 K, the sublattice magnetizations compensate for compositions x = 1 (R = Ce) or x = 1.4 (R = Pr, Nd).

The increase in the weighted average of the ^{57}Fe hyperfine fields, at RT, in the $Gd_3Co_{11-x}Fe_xB_4$ borides, in composition range $2 \leq x \leq 4$, of 1.9 T, corresponds to

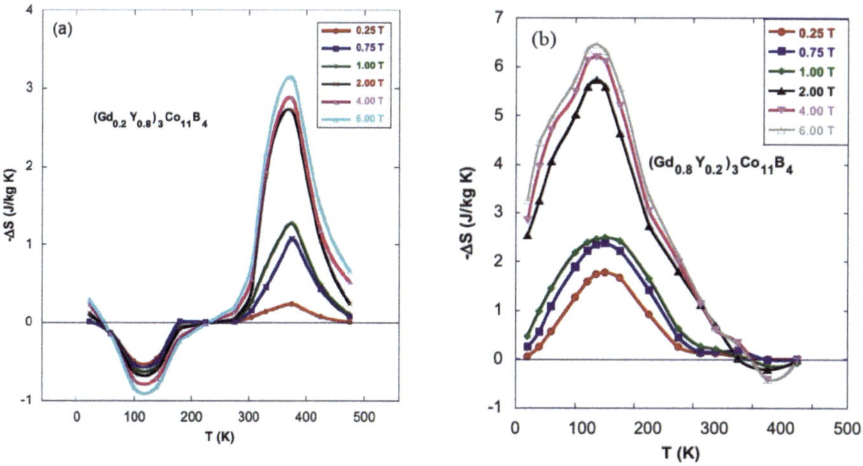

Fig. 8.13 Entropy changes as function of temperature for $(Gd_xY_{1-x})_3Co_{11}B_4$ compounds with x = 0.2 (**a**) and x = 0.8 (**b**) [10B1]

an additional mean iron moment of 0.13 μ_B [09M1]. The iron magnetic moments at RT, estimated from ^{57}Fe hyperfine fields in $Gd_3Co_7Fe_4B_4$ were of 1.4 μ_B/atom (2*c*), 1.1 μ_B/atom (6*i*) and 0 μ_B/atom (3*g*). The weighted average values of the isomer shifts decrease when increasing iron content.

The $R_3Co_{11}B_4$ compounds with R = Tb [95B5, 96T1, 99T1. 00C1], R = Dy [93O1, 95K2, 95K3, 96T1, 97S2, 0Z12], R = Ho [94K2, 96T1], R = Er [95K2, 95K3] and R = Tm [97K6, 01S1] are ferrimagnetically ordered—Fig. 8.11b. The neutron diffraction study on $Tb_3Co_{11}B_4$ [00C1] confirmed the presence of ferrimagnetic type ordering, as suggested by magnetic measurements [95B5, 96T1, 99T1]. A relative high cobalt moment was reported at Co3*g* site [00C1], in disagreement with the general trend evidenced in $R_3Co_{11}B_4$ compounds or computed from band structure [06S1]. When it is assumed that M_{Tb} = 8.4 μ_B/atom, the mean cobalt moment at 4.2 K, is the same as that determined in $Gd_3Co_{11}B_4$ compound, M_{Co} = 0.88 μ_B/atom. The corresponding effective cobalt moments are little changed along the $R_3Co_{11}B_4$ series with heavy rare-earths [96T1]. The Tb and Co sublattice magnetizations in $(Tb_xY_{1-x})_3Co_{11}B_4$ series, at T = 4.2 K compensate at x = 0.23 [99T1]. The mean cobalt moments decrease from 0.88 μ_B/atom (x = 1) up to 0.40 μ_B/atom (x = 0), while their effective moments are little dependent on composition. An increase of the degree of itinerancy of cobalt moment was shown, as the yttrium content increases. Thus, the ratio r = S_p/S_0, increases from r = 2.03 (x = 1) up to r = 3.79 (x = 0).

The dysprosium moments at both 1*a* and 2*e* sites, are close to that of free Dy^{3+} ion, as evidenced by a neutron diffraction study [02Z1]. A very small moment, of the order of experimental errors, has been shown at Co3*g* site. The mean effective cobalt moments in compounds with R = Dy and Ho are of ≅ 2.55(5) μ_B/atom—Table 8.20.

The $ErY_2Co_{11}B_4$ compound shows metamagnetic transitions [95Z3]. At 4 K, in a field μ_0 H ≅ 20 T, oriented along easy direction, the magnetization increases from ≅ 4 μ_B/f.u. to ≅ 14 μ_B/f.u., corresponding to a transition from ferrimagnetic state to a ferromagnetic ordering. The above process has been correlated with the Er and the Co sublattices magnetic anisotropies and the intersublattice interactions.

The calculated $Tm_3Co_{11}B_4$ moment per formula unit, starting from band structure [01S1], is somewhat higher than the experimental value [97K6]—Table 8.20. The reduction of Tm moment from its free ion value was attributed to crystal field effects. The XPS spectrum of the valence band and DOS convoluted by Lorentzians and multiplied by the atomic cross-section, respectively have similar positions of the peaks [01S1]. The coercive field of $Tm_3Co_{11}B_4$ increases from μ_0H_c = 0.01 T, at T = 295 K, up to 0.8 T, at T = 30 K [97K6].

The exchange interactions inside and between magnetic sublattices in $R_3Co_{11}B_4$ series, were evaluated by using the mean field approximation [94K3, 95D1, 95K2, 98J2, 99M1, 99T1, 01M1, 02M1]. The effects of Co3d-B2p hybridization on the exchange interactions in these compounds were also evaluated [95D1].

The study by ^{155}Gd Mössbauer spectroscopy of $Gd_{n+1}Co_{3n+5}B_{2n}$ series, evidenced that the second order crystal field coefficients, A_2^0, increase in the same way as the boron content and magnetocrystalline anisotropy [88S2]. On this basis the hard magnetic properties of some Sm-based compounds were discussed. Averaged over the two Sm sites, A_2^0 reaches a value $A_2^0 = -1200\ Ka_0^{-2}$ in $Sm_3Co_{11}B_4$ compound, higher in absolute magnitude than $A_2^0 = -700$ Ka_0^{-2} in $SmCo_5$. The negative sign of A_2^0 means that the rare-earths having a positive Stevens coefficient, α_J (Sm^{3+}, Er^{3+}, Tm^{3+}, Y^{3+}), show uniaxial anisotropy, as in the corresponding RCo_5 compounds.

The electrical resistivities of $R_3Co_{11}B_4$ compounds with R = Gd, Y [96K2, 98K7], R = Dy, Er [95K3], R = Nd [98K6, 98K8] and R = Tm [97K6] increase, at low temperatures, according to a T^2 law, implying that the electron spin wave scattering is the dominant mechanism. In the high temperature range (T > 100 K), the results were discussed considering the presence of s-d scattering [98K6].

8.6 RCo_4B Compounds

The RCo_4B compounds, where R is a rare-earth or yttrium crystallize in a hexagonal structure of $CeCo_4B$-type, having *P6/mmm* space group [73K1, 74K3]. The structure is derivative from RCo_5-type, by a regular substitution of Co by B in every second layer in 2*c* site—Fig. 8.1. The R atoms are located in 1*a* and *1b* positions. The boron occupies only half of the 2*c* sites in the parent RCo_5-type structure and are located in 2*d* sites in RCo_4B structure. The Co3*g* site, in the RCo_5 structure with two RCo_2 neighbouring planes, become the 6*i* site in RCo_4B type-lattice having one RCo_2 and one RB_2 neighbouring planes. The Co2***c*** site in RCo_4B structure is similar to the Co2*c* site in the parent RCo_5 structure. The Co6*i* layer is slightly displaced toward RB_2 atomic plane (z = 0.287 [85S2] or z = 0.285 [74K2, 06M2])—Table 8.23. The RCo_4B lattices with R = Ce, Y, are stable when some deviations from 1:4:1 composition are present, as evidenced in $YCo_{4\pm x}B_{1\mp x}$ solid solutions with $x \leq 0.2$ [93H2].

The lattice parameters of RCo_4B compounds and some of their pseudoternary alloys are listed in Table 8.24. The c lattice parameter is determined by Co2*c*-Co6*i* and Co6*i*-B2*d* distances and thus remains nearly constant whatever the R element is, while a significant change of the a parameter is observed. The change of R element by another one modifies significantly the R1*a*-Co2*c* and R1*b*-B interatomic distances, in (u,v,0) plane of the structure. Thus, both a lattice parameter, as well as the unit cell volume increase when R metallic radius is greater. There is a near unchanged

Table 8.23 Atomic sites in $CeCo_4B$ compound, having hexagonal structure *P6/mmm* space group[a] [74K3, 06V1]

Atom	Site	Sym	x	y	z	Atomic environment
Ce	1*a*	6/*mmm*	0	0	0	Pseudo Frank-Kasper $B_6Co_{12}Ce_2$
Ce	1*b*	6/*mmm*	0	0	1/2	Pseudo Frank-Kasper $Co_{18}Ce_2$
Co	2*c*	$\bar{6}m2$	1/3	2/3	1/2	14-vertex polyhedron $Co_9Ce_3B_2$
Co	6*i*	*mm*	1/2	0	0.213	13-vertex polyhedron $B_2Co_7Ce_4$
B	2*d*	6*m*2	1/3	2/3	0	Trigonal prism Co_6

[a]Transformation from published data [74K3]: origin shift 001/2

Table 8.24 Space groups and lattice parameters of RCo_4B compounds

Compound	T (K)	Space group	Lattice parameters (nm)		References
			a	c	
$LaCo_4B$	RT	*P6/mmm*	0.5172(3)	0.6860(4)	[73K1, 74K3]
$LaCo_4B$	RT	*P6/mmm*	0.5185	0.6880	[91T1]
$LaCo_4B$	RT	*P6/mmm*	0.5172	0.6863	[85S2]
$LaCo_4BH_{2.8}$	RT	*P6/mmm*	0.5289	0.7140	[85S2]
$CeCo_4B$	RT	*P6/mmm*	0.5005(3)	0.6932(4)	[74K3, 79K1]
$CeCo_4B$	RT	*P6/mmm*	0.5011	0.6944	[89B2]
$PrCo_4B$	2	*P6/mmm*	0.5106(1)	0.6867(1)	[02Z1]
$PrCo_4B$	RT	*P6/mmm*	0.5118(1)	0.6888(1)	[02Z1]
$PrCo_4B$	RT	*P6/mmm*	0.5114	0.6880	[87P8]
$PrCo_4B$	RT	*P6/mmm*	0.5112	0.6875	[85S2]
$PrCo_4BH_{2.5}$	RT	*P6/mmm*	0.5240	0.7098	[85S2]
$PrCo_4BH_{3.8}$	RT	*P6/mmm*	0.5299	0.7249	[85S2]
$NdCo_4B$	2	*P6/mmm*	0.5094(1)	0.6857(1)	[02Z1]
$NdCo_4B$	RT	*P6/mmm*	0.5109(1)	0.6888(1)	[00C3]
$NdCo_4B$	RT	*P6/mmm*	0.5079	0.6879	[87P8]
$NdCo_3NiB$	RT	*P6/mmm*	0.50931(2)	0.69143(3)	[12S2]
$NdCo_3NiBD_{2.99}$	373	*P6/mmm*	0.52786(3)	0.72675(5)	[12S2]
$SmCo_4B$	RT	*P6/mmm*	0.5077(4)	0.6888(3)	[00C3]
$SmCo_4B$	RT	*P6/mmm*	0.5078(3)	0.6871(4)	[73K1]
$SmCo_4B$	RT	*P6/mmm*	0.5084	0.6861	[00C5]
$SmCo_{3.9}B$	RT	*P6/mmm*	0.5077	0.6885	[85S2]
$SmCo_{3.9}BH_{2.6}$	RT	*P6/mmm*	0.5231	0.7044	[85S2]
$SmCo_{3.9}BH_{4.1}$	RT	*P6/mmm*	0.5271	0.7311	[85S2]
$GdCo_4B$	RT	*P6/mmm*	0.5059(1)	0.6901(1)	[07C1]
$GdCo_4B$	RT	*P6/mmm*	0.5058(3)	0.6892(3)	[00C3]
$GdCo_4B$	RT	*P6/mmm*	0.5034	0.6879	[87P8]
$GdCo_2Fe_2B$	RT	*P6/mmm*	0.5104	0.6924	[89D1]
$TbCo_4B$ $TbCo_4B$	RT RT	*P6/mmm*	0.5037(2) 0.4998	0.6886(7) 0.6864	[00C3] [88D1]
$TbCo_4B$ $TbCo_4B$	RT 2	*P6/mmm*	0.4995 0.5035(1)	0.6863 0.6855(2)	[89B2] [09M2]
$DyCo_4B$	RT	*P6/mmm*	0.4987	0.6870	[89B2]
$DyCo_4B$	RT	*P6/mmm*	0.4991(3)	0.6863(4)	[74K3]
$DyCo_4B$	RT	*P6/mmm*	0.5020(2)	0.6885(7)	[08M4]
$DyCo_2Fe_2B$	RT	*P6/mmm*	0.5052	0.6884	[89D1]
$HoCo_4B$	RT	*P6/mmm*	0.4978	0.6873	[89B2]

(continued)

Table 8.24 (continued)

Compound	T (K)	Space group	Lattice parameters (nm)		References
			a	c	
$HoCo_4B$	RT	*P6/mmm*	0.4976(3)	0.6873(4)	[74K3]
$ErCo_4B$	RT	*P6/mmm*	0.4979	0.6869	[87P8]
$ErCo_4B$	RT	*P6/mmm*	0.4968(3)	0.6858(4)	[74K3]
$TmCo_4B$	RT	*P6/mmm*	0.4948(3)	0.6862(4)	[74K3]
$LuCo_4B$	RT	*P6/mmm*	0.4940(3)	0.6867(4)	[73K1, 74K3]
$LuCo_4B$	RT	*P6/mmm*	0.4950	0.6850	[88H1]
YCo_4B	2	*P6/mmm*	0.5012(2)	0.6871(2)	[01C1]
YCo_4B	RT	*P6/mmm*	0.5026(1)	0.6899(1)	[01C1]
YCo_4B	RT	*P6/mmm*	0.5020(2)	0.6891(2)	[00C3]
YCo_3FeB	RT	*P6/mmm*	0.5066(3)	0.6859(4)	[00C2, 01C2]
YCo_2Fe_2B	RT	*P6/mmm*	0.5070(3)	0.6911(4)	[00C2, 01C2]
$YCoFe_3B$	RT	*P6/mmm*	0.5071(3)	0.6973(5)	[00C2, 01C2]

distances between the planes containing Co2*c* and Co6*i* atoms as well as between those containing Co6*i* and B2*d* atoms. No distortions of the RCo_4B structure-type were shown below the Curie temperatures [00C2, 00C3, 01C2, 02Z1], or as effect of pressure as shown in YCo_4B compound ($p \leq 4.5$ GPa) [19B1]. The lattice parameters of the the RCo_4B compounds, with R = Pr [87C2, 93C4, 99C1], R = Nd [99C2, 00C6] and R = Sm [73K1, 00C5] were further investigated.

The RCo_4B compounds adsorb about 4.5 hydrogen atoms per formula unit [85S2]. Deuterium occupies only those interstices that have no boron atoms in their coordination sphere [12S2]. These findings are consistent with repulsive B-D interaction. Multiplateau behavior indicates that at least three distinct hydride phases exist around RT, in the hydrogen pressure range 0–100 atm. An orthorhombic structure was seen only in the α-phase region (RCo_4BH_y with $y < 1$), all higher hydrides reverting to the hexagonal structure.

When an R atom is substituted by another R', solid solutions are formed in all the series. There are limited composition ranges of solid solutions when Co is substituted by Fe, particularly when the RFe_4B phases are not found. Pseudoternary compounds $CeCo_{4-x}Fe_xB$ ($x \leq 0.8$) [95T2, 02Y1], $PrCo_{4-x}Fe_xB$ ($x \leq 1.5$) [86J1], $NdCo_{4-x}Fe_xB$ ($x \leq 2$) [88A1], $SmCo_{4-x}Fe_xB$ ($x \leq 2$) [84O1, 84S3, 94I2, 14J1] or for $x \leq 3$ [88A1], $RCo_{4-x}Fe_xB$ with R = Gd ($x \leq 2$) [89D1, 92D2], R = Dy ($x \leq 2$) [89D1], $ErCo_{4-x}Fe_xB$ ($x \leq 4$) [88A1, 07M1], $YCo_{4-x}Fe_xB$ ($x \leq 1.5$) [89B1, 90B2, 93D2, 94Z2], $x \leq 1.6$ [93O2], $x \leq 3$ [00C2, 01C1] crystallize in $CeCo_4B$-type structure. The Fe substitutes preferentially Co, in RCo_4B compounds, at 2*c* site [93O2, 00C2, 01C2].

The $RCo_{4-x}Ni_xB$ solid solutions are formed in all the composition range when R = Pr [09A1, 14C1], R = Nd [12S2], R = Gd [91B1] or R = Y [90B2, 01C3, 04C1, 07B3]. In $YCo_{4-x}Ni_xB$ system, superlattices are formed, in some compositions ranges, these being related to YNi_4B superstructure—Sect. 9.8. Although YNi_4B was first reported to be isostructural with $CeCo_4B$ [73N1, 75K1], then was shown to crystallize in a $CeCo_4B$ superstructure, characterized by a tripling a and b lattice parameters [83K2, 99B1]. Under an inappropriate thermal treatment, $YCo_{4-x}Ni_xB$ alloys, have an average distorted structure characterized by broad Bragg peaks, which in first approximation, may be considered to be of $CeCo_4B$-type [01C3, 04C1]. When an adequate thermal treatment is followed, the structure is more complex. The presence of the additional lines to those characteristics for a structure with treble a and b lattice parameters, suggested an additive superstructure along **c** axis, either commensurate or incommensurate. The superstructure seems to be connected with the presence of B atoms in the cobalt sites of YCo_4B lattice-type [01C3, 04C1].

Solid solutions, in a limited composition range, were reported when Co was substituted by Cu, Al or when B is replaced by C as in: $YCo_{4-x}Cu_xB$ ($x \leq 1.5$) [04B5, 08B2], $YCo_{4-x}Al_xB$ ($x \leq 2$) [07B3] or $YCo_4B_{1-x}C_x$ ($x \leq 0.3$) [94Z2].

The $NdCo_3NiB$ crystal structure, under hydrogenation, shows interesting features [12S2]. While uncycled $NdCo_3NiB$ crystallizes in the $CeCo_4B$-type structure, hydrogenation and cycling this compound, at T = 373 K, between 0 and 6 MPa hydrogen atmosphere, induced a symmetry decrease to space group *P6mm*. Evidence for partial Co/Ni ordering was found in $NdCo_3NiB$, but not in its deuteride form.

The thermal variations of the lattice parameters were reported in RCo_4B compounds with R = Ce [03A1], R = Y [93K1], R = Nd [90I3], R = Ce, Y [93L2] or in $GdCo_{4-x}Ni_xB$ system [04A2]. Anomalous thermal expansions were observed at $T_c \cong 375$ K and $T_{sR} = 150$ K in YCo_4B, which correspond to the magnetic order–disorder trasition and spin reorientation trasition temperatures, respectively.

The cobalt sublattice has an important role in determining the magnetic properties of RCo_4B compounds. In these systems, the same trend of cobalt moments, M_{Co}, at the two sites is evidenced; only little differences in their values are present when substituting the R partner.

The RCo_4B compounds with R = La [85S2, 91T1, 95O1, 06I1], R = Lu [88H1, 88T1] and R = Y [84O1, 84S3, 87B7, 87P8, 88D3, 88F1, 89B1, 90K1, 91D3, 91I3, 91Z1, 92C1, 92I3, 92K1, 92T1, 92Z4, 93H1, 93H2, 93L2, 93O1, 93O2, 94K1, 94W1, 94Z2, 96T2, 97S2, 97T1, 98I2, 98K1, 98S2, 98Y1, 00C2, 00C3, 01C1, 01C2, 01M1, 01V1, 02C3, 02V1, 03A2, 04B5, 04C1, 06M2, 06M3, 07B3, 08M3, 13B1, 15B2] are ferromagnetically ordered—Table 8.25. The cobalt magnetic moment, at 2*c* site in these compounds is close to that determined for the corresponding site in YCo_5 compound [00C2, 01C1]. The Co6*i* moment is strongly reduced due to Co6*i*-B2p hybridization. The M_{Co} values, determined by neutron diffraction are close to those obtained from band structure calculations. The cobalt moments in YCo_4B were also determined by XMCD study [02C3]. The La5d and Y4d bands are negatively polarized as result of 5d(4d)-Co3d band hybridization [98K1, 98S2, 01V1, 02V1, 06I1, 19B1]—Table 8.26. The total computed spin moment in YCo_4B is 6.95 μ_B/f.u. and the orbital one of 0.89 μ_B/f.u. [98K1]. The saturation magnetizations, reported by various authors, in RCo_4B compounds, are somewhat different, mainly in connection with the external fields used for measurements. The deviations from nominal compositions also influence the magnetic properties, as shown in case of YCo_4B

Table 8.25 Magnetic properties of RCo_4B compounds

Compound	Magnetic structure and magnetic moments	M_s (μ_B/f.u.)	T_c (K)	C (emuK/f.u.)	M_{eff} (μ_B/Co atom)	θ (K)	References
$LaCo_4B$	FM	3.2[a] [95O1]	431				[85S2]
$LaCo_4B$ (BS)	FM La:$M_{1a} = -0.15\ \mu_B$, $M_{1b} = -0.18\ \mu_B$ Co:$M_{2c} = 1.49\ \mu_B$, $M_{6i} = 0.61\ \mu_B$ B:$M_{2d} = -0.04\ \mu_B$	3.1(calc)					[06I1]
$CeCo_4B$	FM, T = 1.4 K Co: $M^s_{2c} = 1.21\ \mu_B$, $M^o_{2c} = 0.083\ \mu_B$, $M^s_{6i} = 0.47\ \mu_B$, $M^o_{6i} = 0.035\ \mu_B$	2.81					[15S1]
$CeCo_4B$	FIM	1.10[b]	297	1.04	1.44	292	[89B2]
$CeCo_4B$	Ce:$M_{1a} = -0.439\ \mu_B$, $M_{1b} = -0.591\ \mu_B$ Co:$M_{2c} = 1.164\ \mu_B$, $M_{6i} = 0.418\ \mu_B$	1.79(exp.)[d] 1.9(theor.)	280				[06T2]
$CeCo_4B$ (BS)	FM Ce:$M_{1a} = -0.53\ \mu_B$, $M_{1b} = -0.42\ \mu_B$ Co:$M_{2c} = 1.35\ \mu_B$, $M_{6i} = 0.44\ \mu_B$	2.67[c] 2.15 (theor.)					[02B1]

(continued)

Table 8.25 (continued)

Compound	Magnetic structure and magnetic moments	M_s (μ_B/f.u.)	T_c (K)	C (emuK/f.u.)	M_{eff} (μ_B/Co atom)	θ (K)	References
$CeCo_4B$ (BS)	FM Ce: $M^s_{1a}= -0.25\ \mu_B$, $M^o_{1a}= 0.20\ \mu_B$, $M^s_{1b}= -0.44\ \mu_B$, $M^o_{1b}= 0.16\ \mu_B$ Co: $M^s_{2c}= 1.31\ \mu_B$, $M^o_{2c}= 0.19\ \mu_B$, $M^s_{6i}= 0.59\ \mu_B$, $M^o_{6i}= 0.04\ \mu_B$	2.69					[09I1]
$CeCo_4B$ (NMR)	FM, T = 4.2 K Co: $M_{2c} = 0.44\ \mu_B$, $M_{6i} = 0.22\ \mu_B$ $M^o_{2c}= 0.11\ \mu_B$, $M^o_{6i}= 0.005\ \mu_B$	1.1[f]					[98Y3]
$Ce(Co_{0.9}Fe_{0.1})_4B$ (NMR)	FM e.a.m ǁ **c** Co: $M_{2c} = 0.84\ \mu_B$, $M_{6i} = 0.42\ \mu_B$ $M^o_{2c}= 0.21\ \mu_B$, $M^o_{6i}= 0.05\ \mu_B$						[02Y1]
$PrCo_4B$	FM e.a.m in (ab) plane	5.10[b]	455	4.04	2.19	462	[87P8]

(continued)

Table 8.25 (continued)

Compound	Magnetic structure and magnetic moments	M_s (μ_B/f.u.)	T_c (K)	C (emuK/f.u.)	M_{eff} (μ_B/Co atom)	θ (K)	References
$PrCo_4B$	FM	5.4[e]	459				[00C3]
$PrCo_4B$	FM	6.8[f]	459				[84O1]
$PrCo_4B$	FM	4.37 [g]	459				[85S2]
$PrCo_4BH_{1.0}$	FM	3.67 [g]	219				[85S2]
$PrCo_4B$ (ND)	FM, T = 2 K Nd:M_{1a} = 2.3(2) μ_B, M_{1b} = 2.2(2) μ_B Co:M_{2c} = 1.5(2) μ_B, M_{6i} = 0.7(2) μ_B	5.5(1)[h]	459(5)				[02Z1]
$PrCo_4B$ (NMR)	FM[j] Co: M^s_{2c}= 0.82 μ_B, M^o_{2c}= 0.26 μ_B M^o_{6i}= 0.43 μ_B, M^s_{6i}= 0.11 μ_B						[97Y1]
$NdCo_4B$	FM Co: M_{2c} = 1.4 μ_B, M_{6i} = 0.1 μ_B	4.99	473				[90I2]
$NdCo_4B$	FM	5.88[i]	473				[91I3]
$NdCo_4B$	FM	5.8[e]	468				[00C3]

(continued)

Table 8.25 (continued)

Compound	Magnetic structure and magnetic moments	M_s (μ_B/f.u.)	T_c (K)	C (emuK/f.u.)	M_{eff} (μ_B/Co atom)	θ (K)	References
$NdCo_4B$	FM e.a.m ‖**c**	5.8^b	458				[87P8]
$NdCo_4B$	FM, T = 2 K Nd:M_{1a} = 2.4(2) μ_B, M_{1b} = 2.0(2) μ_B Co:M_{2c} = 1.6(2) μ_B, M_{6i} = 0.6(3) μ_B	5.8^i	460				[02Z1, 07I1]
$NdCo_3FeB$	FM, T = 2 K Nd:M_{1a} = 2.5(2) μ_B, M_{1b} = 1.8(1) μ_B Co:M_{2c} = 1.4(1) μ_B, M_{6i} = 0.8(2) μ_B Fe:M_{2c} = 2.0(1) μ_B, M_{6i} = 1.1(2) μ_B	6.8^i	688(5)				[07I1]
$NdCo_{2.5}Fe_{1.5}B$	FM, T = 2 K Nd:M_{1a} = 2.3(2) μ_B, M_{1b} = 1.8(1) μ_B Co:M_{2c} = 1.5(1) μ_B, M_{6i} = 0.8(2) μ_B Fe:M_{2c} = 2.1(1) μ_B, M_{6i} = 1.2(2) μ_B	$7.4(1)^i$	709(5)				[07I1]
$SmCo_4B$	FM	3.16^k	500				[93O1]

(continued)

Table 8.25 (continued)

Compound	Magnetic structure and magnetic moments	M_s (μ_B/f.u.)	T_c (K)	C (emuK/f.u.)	M_{eff} (μ_B/Co atom)	θ (K)	References
$SmCo_4B$	FM	3.00[e]	501				[00C3]
$SmCo_4B$	FM	3.22[i]	516				[91I2]
$SmCo_4B$ (BS)	FM Sm:M_{1a} = 1.475 μ_B, M_{1b} = 1.656 μ_B Co:M_{2c} = 1.617 μ_B, M_{6i} = 0.67 μ_B B:M = −0.044 μ_B						[07Y1]
$SmCo_4B$	FM	3.34 [k]	510				[85S2]
$SmCo_4BH_{1.0}$	FM	2.74 [k]	298				[85S2]
$SmCo_4BH_{2.5}$	FM	1.98 [k]	181				[85S2]
$SmCo_4B$	FM	3.8[f]	510				[84O1]
$SmCo_2Fe_2B$	FM	6[f]	768				[84O1]
$SmCo_3NiB$	FM	2.5[f]	424				[84O1]
$GdCo_4B$	FIM, eam ‖ **c**	3.25[b]	505	10.59	2.31		[87P8]
$GdCo_4B$	FIM	3.40 [l]	496 397 (T_{comp})	14.13	3.50		[97S2]
$GdCo_4B$	FIM	2.85 [m]	505 410 (T_{comp})				[90D1]
$GdCo_4B$ (BS)	FIM Gd:M_{1a} = 7.48 μ_B, M_{1b} = 7.38 μ_B Co:M_{2c} = −1.62 μ_B, M_{6i} = −0.75 μ_B	3.57(theor.)					[02B1]

(continued)

Table 8.25 (continued)

Compound	Magnetic structure and magnetic moments	M_s (μ_B/f.u.)	T_c (K)	C (emuK/f.u.)	M_{eff} (μ_B/Co atom)	θ (K)	References
$GdCo_4B$	FIM	3.2[n]	503				[00C3]
$GdCo_4B$	FIM Co:$M_{2c} = 1.7\ \mu_B$, $M_{6i} = 0.8\ \mu_B$	2.85 [m]	517 421 (T_{comp})				[93I2]
$GdCo_4B$ (BS)	FIM Gd:$M_{1a} = -7.442\ \mu_B$, $M_{1b} = -7.556\ \mu_B$ Co:$M_{2c} = 1.521\ \mu_B$, $M_{6i} = 0.778\ \mu_B$ B:$M_{2d} = -0.048\ \mu_B$						[06S3]
$GdCo_4B$ (BS)	FIM Gd:$M_{1a} = -7.386\ \mu_B$, $M_{1b} = -7.496\ \mu_B$ Co:$M_{2c} = 1.625\ \mu_B$, $M_{6i} = 0.766\ \mu_B$ B:$M_{2d} = -0.040\ \mu_B$	3.56					[01V1, 02V1]
$GdCo_2Fe_2B$	FIM	1.2 [m]	770				[89D1]
$GdCo_3NiB$ (BS) Ni in 6*i*	FIM Gd:$M_{1a} = 7.34\ \mu_B$, $M_{1b} = 7.54\ \mu_B$ Co:$M_{2c} = -1.59\ \mu_B$, $M_{6i} = -0.86\ \mu_B$, $M_{Ni}(6i) = -0.13\ \mu_B$ B:$M = 0.07\ \mu_B$	4.56[o]			2.17		[04B5]

(continued)

Table 8.25 (continued)

Compound	Magnetic structure and magnetic moments	M_s (μ_B/f.u.)	T_c (K)	C (emuK/f.u.)	M_{eff} (μ_B/Co atom)	θ (K)	References
$Gd_{0.5}Y_{0.5}Co_3NiB$ (BS) Ni in 6*i*	FIM Gd:$M_{1a} = -7.31\ \mu_B$, Y: $M_{1b} = -0.28\ \mu_B$ Co:$M_{2c} = 1.58$, $M_{6i} = 0.79\ \mu_B$, Ni: $M_{6i} = 0.11\mu_B$	0.60(theor.) 1.29(exp.)[o]	362		210		[04B5]
$TbCo_4B$	FIM	5.17[b]	455	13.92	2.27		[89B2]
$TbCo_4B$	FIM	5.0[e]	458				[00C3]
$TbCo_4B$ (ND)	FIM, T = 4.5 K, Tb:$M_{1a} = 8.1(1)\ \mu_B$, $M_{1b} = 7.4(1)\ \mu_B$ Co:$M_{2c} = -1.56(6)\ \mu_B$, $M_{6i} = -0.77(3)\ \mu_B$		450 360 (T_{comp})				[06C1]
$TbCo_4B$ (ND)	FIM (FM?), T = 15 K Tb:$M_{1a} = 7(1)\ \mu_B$, $M_{1b} = 7.3(7)\ \mu_B$ Co:$M_{2c} = -1.6(3)\ \mu_B$, $M_{6i} = -0.6(1)\ \mu_B$	4.9(1)[n]					[00C1]
$TbCo_4B$ (BS)	FIM, Tb:$M_{1a} = -9.394\ \mu_B$, $M_{1b} = -9.508\ \mu_B$ Co:$M_{2c} = 1.619\ \mu_B$, $M_{6i} = 0.863\ \mu_B$ B:$M_{2a} = 0.056\ \mu_B$						[06S3]

(continued)

Table 8.25 (continued)

Compound	Magnetic structure and magnetic moments	M_s (μ_B/f.u.)	T_c (K)	C (emuK/f.u.)	M_{eff} (μ_B/Co atom)	θ (K)	References
$TbCo_4B$ (ND)	T = 2 K, FIM Tb:$M_{1a} = 8.0(7)\ \mu_B$, $M_{1b} = 7.7(7)\ \mu_B$ Co:$M_{2c} = -1.6(2)\ \mu_B$, $M_{6i} = -0.7(2)\ \mu_B$	5	455 400 (T_{comp})				[09M2]
$TbCo_3FeB$ (ND) M = Co + Fe	T = 2 K, FIM Tb:$M_{1a} = 8.4(2)\ \mu_B$, $M_{1b} = 8.7(3)\ \mu_B$ M:$M_{2c} = -1.8(1)\ \mu_B$ (72 at % Fe), $M_{6i} = 0.8(1)\ \mu_B$ (12 at % Fe)	4.1(2)	699(5) 400(5)(T_{comp})				[09M2]
$TbCo_2Fe_2B$ (ND)	T = 2 K, FIM Tb:$M_{1a} = 7.8(2)\ \mu_B$, $M_{1b} = 8.0(3)\ \mu_B$ Co:$M_{2c} = -1.7(1)\ \mu_B$ (78 at % Fe) $M_{6i} = -1.1(1)\ \mu_B$ (46 at % Fe)	2.9(2)	719(5) 350(5)(T_{comp})				[09M2]
$DyCo_4B$	FIM	6.20[b]	427	16.66	2.39		[89B2]
$DyCo_4B$	FIM	5.90 [m]	420 340 (T_{comp})				[90D1]
$DyCo_4B$	FIM	5.86	428				[93O1]
$DyCo_3FeB$	FIM	5.4(2)[n]	650(5) 320(5)(T_{comp})				[08M4]
$DyCo_2Fe_2B$	FIM	4.0 [m]	650				[89D1]
$DyCo_2Fe_2B$	FIM	3.4(2)[n]	663(5) 300(5)(T_{comp})				[08M4]

(continued)

Table 8.25 (continued)

Compound	Magnetic structure and magnetic moments	M_s (μ_B/f.u.)	T_c (K)	C (emuK/f.u.)	M_{eff} (μ_B/Co atom)	θ (K)	References
$DyCoFe_3B$	FIM	3.2(2)[n]	660(5) 270(5)(T_{comp})				[08M4]
$HoCo_4B$	FIM	6.10[b]	396	16.90	2.38		[89B2]
$HoCo_4B$	FIM	5.67[i]	409 204 (T_{comp})				[91I3]
$HoCo_4B$	FIM Co:M_{2c} = 1.6 μ_B, M_{6i} = 1.0 μ_B	5.47 [m]	409				[90I2]
$HoCo_4B$	FIM	6.40 [m]	398 240 (T_{comp})				[90D1]
$ErCo_4B$	FIM, e.a.m ‖ **c**	5.10[b]	386	14.45	2.42		[87P8]
$ErCo_4B$	FIM	5.45 [m]	387 163 (T_{comp})				[90D1]
$LuCo_4B$	FM	2.14	396				[88H1]
YCo_4B	FM	2.66	380	2.35	2.18	392	[87B6, 87B7]
YCo_4B (ND)	FM, T = 2 K Co:M_{2c} = 1.6(2) μ_B, M_{6i} = 0.6(2) μ_B	3.00[p]	380(5)				[01C1]
YCo_4B	FM	2.53 [l]	380	3.93	2.80	379	[97S2]
YCo_4B	FM e.a.m ‖ **c**	2.65[b]	382				[87P8]
YCo_4B	FM	5.8[f]	383				[84O1]

(continued)

Table 8.25 (continued)

Compound	Magnetic structure and magnetic moments	M_s (μ_B/f.u.)	T_c (K)	C (emuK/f.u.)	M_{eff} (μ_B/Co atom)	θ (K)	References
YCo_4B	FM Co:$M_{2c} = 1.66\ \mu_B$, $M_{6i} = 0.5\ \mu_B$	6.22 [m]	379				[93O1]
YCo_4B (BS)	FM Y:$M_{1a} = -0.16\ \mu_B$, $M_{1b} = -0.17\ \mu_B$ Co:$M_{2c} = 1.63\ \mu_B$, $M_{6i} = 0.67\ \mu_B$	3.475(theor.)					[98K1]
YCo_4B (BS, with spin–orbit coupling)	FM Y: $M^s_{1a} = -0.16\ \mu_B$, $M^o_{1a} = 0.003\ \mu_B$, $M_{1a} = -0.157\ \mu_B$ $M^s_{1b} = -0.17\ \mu_B$, $M^o_{1b} = -0.01\ \mu_B$, $M_{1b} = -0.18\ \mu_B$ Co: $M^s_{2c} = 1.63\ \mu_B$, $M^o_{2c} = 0.15\ \mu_B$, $M_{2c} = 1.78\ \mu_B$ $M^s_{6i} = 0.67\ \mu_B$, $M^o_{6i} = 0.10\ \mu_B$, $M_{6i} = 0.77\ \mu_B$	3.925(theor.)					[98K1]
YCo_4B (BS)	FIM, Y:$M_{1a} = -0.128\ \mu_B$, $M_{1b} = -0.233\ \mu_B$ Co:$M_{2c} = 1.428\ \mu_B$, $M_{6i} = 0.592\ \mu_B$ B:$M_{2d} = -0.034\ \mu_B$						[01V1, 02V1]

(continued)

Table 8.25 (continued)

Compound	Magnetic structure and magnetic moments	M_s (μ_B/f.u.)	T_c (K)	C (emuK/f.u.)	M_{eff} (μ_B/Co atom)	θ (K)	References
YCo_4B (BS)	FIM, Y:$M_{1a} = -0.210\ \mu_B$, $M_{1b} = -0.309\ \mu_B$ Co:$M_{2c} = 1.498\ \mu_B$, $M_{6i} = 0.773\ \mu_B$ B:$M_{2d} = -0.055\ \mu_B$						[98S2]
YCo_4B	FM	3.0[e]	378				[00C2]
YCo_3FeB[r] (ND)	FM[s], T = 300 K Co:$M_{2c} = 1.5(3)\ \mu_B$, $M_{6i} = 0.6(3)\ \mu_B$, 2*c* (74%Fe) Fe:$M_{2c} = 1.9(3)\ \mu_B$, $M_{6i} = 1.1(3)\ \mu_B$, 6*i* (10%Fe)	4.3(1)[e]	635(4)				[00C2, 01C2]
YCo_2Fe_2B[r] (ND)	FM[s], T = 300 K Co:$M_{2c} = 1.6(2)\ \mu_B$, $M_{6i} = 0.6(2)\ \mu_B$, 2*c* (87%Fe) Fe:$M_{2c} = 2.0(2)\ \mu_B$, $M_{6i} = 1.3(2)\ \mu_B$, 6*i* (39%Fe)	5.5(1)[e]	655(4)				[00C2, 01C2]
$YCoFe_3B$[r] (ND)	FM[s)], T = 300 K Co:$M_{2c} = 1.5(2)\ \mu_B$, $M_{6i} = 0.6(2)\ \mu_B$, 2*c* (83%Fe) Fe:$M_{2c} = 2.1(2)\ \mu_B$, $M_{6i} = 1.3(2)\ \mu_B$, 6*i* (72%Fe)	6.1(1)[e]	674(4)				[00C2, 01C2]
YCo_3NiB (BS)	FM, Y:$M_{1a} = -0.19\ \mu_B$, $M_{1b} = -0.28\ \mu_B$ Co:$M_{6i} = 0.80\ \mu_B$, Ni:$M_{2c} = 0.32\ \mu_B$, B: $-0.07\ \mu_B$	2.40(theor.) 1.70(exp.)[o]					[04B5]

(continued)

Table 8.25 (continued)

Compound	Magnetic structure and magnetic moments	M_s (μ_B/f.u.)	T_c (K)	C (emuK/f.u.)	M_{eff} (μ_B/Co atom)	θ (K)	References
$YCo_{3.5}Cu_{0.5}B$ (BS)	FM, Y:$M_{1a} = -0.15\ \mu_B$, $M_{1b} = -0.10\ \mu_B$ Co:$M_{2c} = 1.25\ \mu_B$, $M_{6i} = 0.44\ \mu_B$ B:$M_{2d} = -0.03\ \mu_B$	1.80(theor.) 1.70(exp.)^{o}	265		1.98		[08B2]
YCo_3Cu_1B (BS)	FM Y:$M_{1a} = -0.07\ \mu_B$, $M_{1b} = -0.15\ \mu_B$ Co:$M_{2c} = 1.11\ \mu_B$, $M_{6i} = 0.32\ \mu_B$ B:$M_{2d} = -0.02\ \mu_B$	1.20(theor.) 0.95(exp.)^{o}	210		1.83		[08B2]
$YCo_{2.5}Cu_{1.5}B$	FM	0.35^{o}	164				[08B2]
YCo_3AlB (BS)	FM, Y:$M_{1a} = -0.03\ \mu_B$, $M_{1b} = -0.04\ \mu_B$ Co:$M_{2c} = 1.15\ \mu_B$, $M_{6i} = 0.24\ \mu_B$ B:$M_{2d} = -0.03\ \mu_B$	1.63(theor.) 1.10(exp.)^{o}			1.50		[07B3]

(continued)

Table 8.25 (continued)

Compound	Magnetic structure and magnetic moments	M_s (μ_B/f.u.)	T_c (K)	C (emuK/f.u.)	M_{eff} (μ_B/Co atom)	θ (K)	References
YCo_2Al_2B	PM $\chi_{theor} = 1.95 \cdot 10^{-3}$ emu/mol[t] $\chi_{exp} = 1.50 \cdot 10^{-3}$ emu/mol				1.42		[07B3]

[a] $T = 6$ K, $\mu_0 H = 40$ T
[b] $T = 4.2$ K, $\mu_0 H = 5$ T
[c] $T = 1.5$ K, $\mu_0 H = 7$ T
[d] $T = 1.5$ K, $\mu_0 H = 9$ T
[e] $T = 5$ K, $\mu_0 H = 7$ T
[f] $T = 4.2$ K, $\mu_0 H = 8.5$ T
[g] $T = 77$ K, $\mu_0 H = 1$ T
[h] $T = 2$ K, $\mu_0 H = 7$ T
[i] $T = 4$ K, $\mu_0 H = 10$ T
[j] Two Co 6*i* sites (mean value)
[k] $T = 77$ K, $\mu_0 H = 2$ T
[l] Extrapolated at $T = 0$ K, $\mu_0 H = 2$ T
[m] $T = 4.2$ K, $\mu_0 H = 1.8$ T
[n] $T = 5$ K, $\mu_0 H = 5$ T
[o] $T = 4.2$ K, $\mu_0 H = 9$ T
[p] $T = 4$ K, $\mu_0 H = 7$ T
[r] Distribution of Fe in 2*c* and 6*i* sites
[s] Plane anisotropy at $T = 300$ K
[t] 1 emu/g = 1 Am^2/kg

Table 8.26 Anisotropy energies in RCo_4B-type compounds

Compound	T (K)	$E_s \cdot 10^{-22}$ (J/atom)		$E_a \cdot 10^{-22}$ (J/f.u.)	T_{sR} (K)	T_c (K)	Referencess
		2*c*	6*i*				
$PrCo_4B$	RT	4.76	−1.61	−0.179			[90Z2]
$SmCo_4B$	4.2	4.95	−1.52	0.76			[93K2, 94K2]
YCo_4B	4.2	4.76	−1.59	0.06			[94K1]
YCo_4B	4.2	3.1	−1.23				[95O1]
$YCo_4B^{a)}$	5	2.64	−1.02	−0.41			[97T1]
$YCo_4B^{a)}$	5	2.71	−0.99	−0.25			[97T1]
$YCo_4B^{b)}$		4.77	−4.80		150	380	[14C1]
$YCo_3NiB^{b)}$		4.00	−1.61		No	314	[14C1]
$YCo_2Ni_2B^{b)}$		2.38	−2.40		150	307	[14C1]
$YCoNi_3B^{b)}$		0	−1.60		No	180	[14C1]

[a]Different values of magnetization
[b]Computed values

compound [93H2]. The magnetizations decrease when boron content is higher than one, the spin reorientation temperature is nearly unchanged, the character of spin reorientation process being strongly affected.

The anisotropy of $LaCo_4B$ is uniaxial along **c**-axis [85S2, 91T1, 95O1, 06I1], while that of $LuCo_4B$ is planar [88H1, 88T1]. The positive anisotropy of $LaCo_4B$ was assigned to lattice deformation [91T1]. The e.a.m. in YCo_4B is along **c**-axis, at $T > T_{sR} \cong 150$ K and changes to basal plane at lower temperatures [88D3]. The YCo_4B magnetic anisotropy and their temperature dependence have been intensively investigated [84O1, 87P8, 88D3, 90K1, 91D3, 91Z1, 92C1, 92I3, 92Z4, 93H1, 93O1, 94K1, 95O1, 96T2, 97T1, 00C2]. The experimentally determined trend, has been attributed to the competition between opposite anisotropy of cobalt at 2*c* and 6*i* sites, respectively. Starting from a phenomenological approach, developed for YCo_5 [78S2, 79S2] and extended to RCo_4B compounds [88D3, 88F1], a stabilization energy or a local energy per atom, E_s, has been defined as the difference of the spin–orbit coupling energies in directions parallel and perpendicular to the hexagonal axis [78S2, 79S2]. The overall anisotropy energy was considered to be the sum of contributions arising from those of individual n_i sites, $E_a = \sum n_i E_{s_i}$. In YCo_5 compound the Co2*c* atoms have a large positive contribution to anisotropy, while that at Co3*g* sites is smaller and negative [79S2, 80S1, 91D3]. As already mentioned, the Co2*c* sites, in YCo_5 and YCo_4B, have similar environments, and thus nearly the same E_s as determined in YCo_5 [88D3, 90Z2, 94K1, 94K2, 96D1]—Tables 8.26 and 8.27. From the anisotropy constant of YCo_4B, $K_1 \cong -0.4 \cdot 10^6$ J/m^3 (T = 4 K), a stabilization energy $E_s(6i) \cong -1.5 \cdot 10^{-22}$ J/m^3 was obtained—Table 8.26. The above value is slightly more negative than that determined for Co3*g* site in YCo_5 and can be correlated with changes in their environment when Co is substituted by B. The spin reorientation, at

Table 8.27 Anisotropy constants of RCo_4B-based compounds

Compound	T(K)	Anisotropy constant (MJ/m^3)			Anisotropy constants of Co and R sublattices (J/atom)·10^{22}				Anisotropy field, μ_0H_a(T)	Coercive field μ_0H_c(T)	Referencess
		K_1	K_2	K_3	K_1(Co)	K_2(Co)	K_1(R)	K_2(R)			
$LaCo_4B$	6	0.65							3.5		[95O1]
$LaCo_4B$	4.2	0.58									[92T1]
$CeCo_4B$	4.2								60.0–80.0		[95T2]
$PrCo_4B$	RT	−0.12							3.9		[90Z2]
$NdCo_4B$	77 295								12.2 15.8		[87P8]
$SmCo_4B$	4.2	22.3							110.0		[94I2]
$SmCo_4B$	0	24.7			7.38						[93O1]
$SmCo_4B$	4.2	24.3					−2.7(1a) −3.45(1b)		120.0	8.0	[92I2]
$SmCo_4B$	RT	20.6							90.6		[84O1]
$SmCo_2Fe_2B$	4.2 77 300	22.6 24.2 14.6							74.0 81.0 51.0		[94I2]
$SmCo_2Fe_2B$	RT								60.2		[84O1]
$SmCo_3FeB$	RT								43.9		[84O1]
$SmCo_3NiB$	RT								76.3		[84O1]

(continued)

Table 8.27 (continued)

Compound	T(K)	Anisotropy constant (MJ/m^3)			Anisotropy constants of Co and R sublattices (J/atom)·10^{22}				Anisotropy field, μ_0H_a(T)	Coercive field μ_0H_c(T)	Referencess
		K_1	K_2	K_3	K_1(Co)	K_2(Co)	K_1(R)	K_2(R)			
$GdCo_4B$	77 295								3.2 11.0		[87P8]
$GdCo_4B$	78 147 225	0.33 0.45 0.50					0.58(0.57)[a] 0.39(0.44) 0.29(0.32)		2.0 3.6 7.0		[93I2]
$GdCo_4B$	RT								2.0		[84O1]
$GdCo_4B$	4.2	0.43									[91D2]
$GdCo_3Fe_1B$	4.2	−0.47									[91D2]
$GdCo_2Fe_2B$	4.2	−0.165									[91D2]
$Gd_xDy_{1-x}Co_4B$ $x \leq 0.75$	5				−0.38[d]	0.32[d]	−27.7(Dy)[d] 0.546(Gd)[d]	6.78(Dy)[d]			[95I3]
$Gd_{0.9}Dy_{0.1}Co_4B$	4	−2.3[b]	1.3[b]		−0.287[b]	−0.174[b]	−21(Dy)[c] −0.262(Gd)[c]	13(Dy)[c]			[94I3]
$ErCo_4B$	77								7.8		[87P8]
$LuCo_4B$	4.2	−0.396									[88H1]
YCo_4B	6	−0.764			3.1(2*c*) -1.23(6*i*)				4.0		[95O1]
YCo_4B	4.2	−0.35									[88T1, 92T1]

(continued)

Table 8.27 (continued)

Compound	T(K)	Anisotropy constant (MJ/m^3)			Anisotropy constants of Co and R sublattices (J/atom)·10^{22}				Anisotropy field, μ_0H_a(T)	Coercive field μ_0H_c(T)	Referencess
		K_1	K_2	K_3	K_1(Co)	K_2(Co)	K_1(R)	K_2(R)			
YCo_4B	77 295								1.5 2.15		[87P8]
YCo_4B	138	0.55	0.081								[93O1]
YCo_4B	4	−0.4									[91D3]
YCo_4B[b]	5 100 230	−0.232 −0.135 0.170	0 0 −0.029	−0.073 −0.053 0.0045							[08M3]
$YCo_{2.5}Fe_{1.5}B$	4	−0.955									[91D3]

[a]Calculated values in bracket
[b]$K_i.10^{-22}$ J/fu
[c]$K_i.10^{-22}$ J/atom for both R and Co sublattices
[d]MJ/m^3

the T_{sR} temperature, is determined by the different thermal evolutions of the cobalt anisotropies at 2*c* and 6*i* sites, respectively.

The magnetocrystalline anisotropy energy and the anisotropy of the orbital angular momentum has been calculated in YCo_5 [91D1] and YCo_4B [98K1] compounds. Changes in easy axis of magnetization were evidenced, as function of band filling parameter. In case of YCo_4B, the e.a.m. changes from **c**-axis to the (ab)-plane for a band filling parameter, $q = 83.2$, close to the normal state value, $q = 84$.

The anisotropy constants K_1 and K_2 and their temperature dependences were determined from magnetization isotherms on YCo_4B single crystal [97T1]. These data suggest a transition from a behavior typical for a first order magnetic process (FOMP), to a normal region through and intermediate state characterized by a "quasi-FOMP" type, where the magnetization is not fully saturated after FOMP transition [92Z4, 97T1]. The temperature dependences of K_1 and K_2 constants are different; the K_1 becomes positive around the $T_{sR} = 150$ K while K_2 at $T > 250$ K. Due to above trends, the condition for a FOMP process cannot be satisfied above or below a critical temperature and a "quasi FOMP" intermediate region occurs. In addition to that at T_{sR} temperature, anomalies in the temperature dependence of the ac susceptibility and of the low field magnetization were observed, at T = 80 K [97T1] and T = 50 K [91Z1, 97T1], particularly when the field was applied along **a-** and **b**-axis. No reasons for their presence are given.

The magnetic properties of YCo_4B are affected by pressure [97K1, 03A1, 03A2, 08M3, 15B1, 15B2, 19B1]—Table 8.28. The spontaneous magnetization decreases with a rate $d\ln M_s/dp = -0.09\ GPa^{-1}$, while the Curie temperature with a rate $d\ln T_c/dp = -3.1\ K \cdot 10^{-2}\ GPa^{-1}$ [03A2]. The ND measurements up to p = 4.5 GPa showed that cobalt moments at 2*c* and 6*i* sites decrease with a rate $d\ln M_{Co}/dp = -0.11\ GPa^{-1}$ [19B1]. By extrapolation of experimental data it was suggested that cobalt atoms will be non-magnetic at $p \cong 7$ GPa.

Above the Curie temperature, the reciprocal susceptibility of YCo_4B follows a linear dependence, the effective mean cobalt moment being $M_{eff} \cong 2.20\ \mu_B$/atom [87B6, 87B7, 87P8].

The magnetic properties of pseudo-ternary $Y_{1-x}R_xCo_4B$ compounds with R = La [91T1, 92T1] and R = Th [06M2, 10M1] were also investigated. The Curie temperatures as well as the magnetizations increase gradually as Y is replaced by La. There is a plane to easy axis transition, for $x \leq 0.2$, and then to cone-axis at x

Table 8.28 Pressure effects

Compound	M_{Co} (μ_B/f.u.)	T_c (K)	dT_c/dp (K/GPa)	$dM_s/dp \cdot 10^{-2}$ (μ_B/f.u. GPa)	References
$CeCo_4B$	1.24	282	15(1)	−26	[03A1, 03A2]
YCo_4B	2.74	375	11.5(10)	−90, −0.09[a]	[03A2]
$GdCo_4B$	3.34	515	7.5(10)	27	[03A2]
YCo_4B	2.72	377		−0.11[a]	[19B1]

[a] $d\ln M_{Co}/dp$

= 0.5. The samples with x ≥ 0.8, at T = 4.2 K, show the **c**-axis anisotropy. In high fields, the mean cobalt moment increases, at T = 4.2 K, with a rate ≅ 4(1)·10^{-2} μ_B/T [92T1], similar to that evidenced in RCo_2-based compounds [75B2]. When Y is substituted by Th, both saturation magnetizations and Curie temperatures decrease, mainly determined by the valence state of Th^{4+} [06M2, 10M1]. The spin reorientation phenomenon disappears for x < 0.2.

The presence of carbon enlarges the temperature range where the anisotropy of $YCo_4B_{1-x}C_x$ system is uniaxial; for x ≤ 0.3 both T_c and M_s are little modified [94Z2].

The $YCo_{4-x}Fe_xB$ [84S3, 88H1, 88T1, 89B1, 91D3, 92C1, 93D2, 93O2, 94W1, 94Z2, 00C2, 01C2, 06L1, 09G1] and $LaCo_{4-x}Fe_xB$ [88H1, 88T1] compounds are ferromagnetically ordered. Both the saturation magnetizations and Curie temperatures increase, as the iron content is higher [89B1]. The neutron diffraction [00C2, 01C2] and those by ^{57}Fe Mössbauer spectroscopy [92C1, 93O2, 94W1, 06L1, 09G1], evidenced the presence of a smaller iron moment at 6*i* site—Table 8.25. In the limit of experimental errors, the Co2*c* and Co6*i* moments does not change, while those of Fe2*c* and Fe6*i* increase little, in the same manner as the iron content [00C2, 01C2]. There are opposite contributions to the overall anisotropy, of Co and Fe atoms, on the same crystallographic site. Thus the spin reorientation temperatures were shown to increase with iron content and for x > 0.08 [91D3] or x > 0.20 [94Z2]; no spin reorientations can be shown at T ≤ T_c, due to the preference of iron for 2*c* site. In $YCoFe_3B$, the larger iron content, leads to a significant occupation of the 6*i* site and to a competition between Fe2*c* and Fe6*i* contributions to the magnetocrystalline anisotropy [09G1]. This competition leads to the appearance of spin reorientation at 400 K ≤ T_{sR} ≤ 450 K, with a change from uniaxial, at T < T_{sR}, to plane anisotropy, at T > T_{sR} [00C2, 01C2, 06L1, 09G1].

The composition dependences of the anisotropy constants, K_1, in $YCo_{4-x}Fe_xB$ [88T1, 91D3] and $LuCo_{4-x}Fe_xB$ [88T1] compounds have the same trend. A minimum was observed for $K_1(x)$ at x = 1.2 (R = Y), or at x = 0.38 (R = Lu), followed by a change to positive values at x ≅ 1.2, as determined by site preference of iron atoms.

The increase of the Ni content in $YCo_{4-x}Ni_xB$ series, induces a significantly decrease of the Curie temperatures [95B6, 04B4, 04B5, 04C1, 07B3, 07I1, 08B2, 14C1]. The mean transition metal magnetization changes with a rate of ≅ −1.1 μ_B/Ni atom [07B3]. A metamagnetic transition was shown, in a field μ_oH = 1 T, for a sample with x = 2, where the transition metal sublattice undergoes a change from a low moment, to a high magnetic state [07I1]. The magnetic anisotropy of $YCo_{4-x}Ni_xB$ system has been analysed in a model based on preferential site occupation under Co/Ni substitution. The model predicts the presence of spin reorientations for compounds with x = 0 and 2, in agreement with experimental evidence [14C1].

The substitution of Co by Cu in $YCo_{4-x}Cu_xB$ system, decreases the magnetization with a rate of 1.4 μ_B/Cu atom [07B3, 08B2]. Supposing, as in YCo_5, that Cu atoms are located in 3*g* site [90B1], the data obtained from band structure calculations, agree rather well with the experimentally determined magnetizations—Table 8.25. The cobalt substitution for Al, changes dramatically the magnetic properties of $YCo_{4-x}Al_xB$ system, the sample with x = 2 being paramagnetic [07B3]. The magnetic susceptibility of this compound, has a maximum at $T \cong 20$ K and at $T >$ 100 K follows a Curie–Weiss type behavior, as predicted by spin fluctuation model [91M1, 95B4]. A mean effective cobalt moment, $M_{eff}(Co) = 1.42\ \mu_B$/atom, was determined.

The hybridization of Ce4f and Co3d states play an important role in determining the magnetic properties, particularly the magnetic anisotropy of $CeCo_4B$ compound [89B2, 95H1, 02B1, 08G1, 09I1]. The Ce5d band polarization is antiparallely orientd to cobalt moments. There is an important proportion of Ce4f occupation in the Co3d band and a large orbital moment at Co2*c* sites [95H1, 98Y3]. The cobalt magnetic moments at 2*c* sites, estimated from ^{59}Co NMR measurements [15S1], agree with those obtained from band structure calculations [02B1, 09I1]—Table 8.25. In the $Ce_{n+1}Co_{3n+5}B_{2n}$ series (n = 1 to 3), the Co magnetic moments, at equivalent positions, derived from RCo_5–type structure, are almost unchanged [15S1]. The Curie temperature of $CeCo_4B$ is smaller than those determined in compounds with non-magnetic elements (R = La, Lu, Y) and the magnetic phase transition is of second order. The maximum of the entropy change, of 0.361 J/kgK in a field $\mu_oH = 1$ T, was obtained around T_c [08G1].

An isostructural transition, in $CeCo_4B$, characterized by a sudden shrinkage of the a-parameter, at $T_t \cong 120$ K, has been attributed to a change of 4f-3d hybridization [95H1]. At ambient pressure and low temperatures, $CeCo_4B$, as already mentioned, is in a state with strong 4f-3d hybridization. When increasing temperature, as result of thermal expansion, the Co-Ce separation in the basal plane reach a critical value, and a transition to weakly 4f-3d hybridization state occurs [93H1]. Anomalies in the physical properties of $CeCo_4B$ at the above transirion, were shown also by μSR [06T2] or resistivity measurements [95H1]. The frequency corresponding to the oscillating signal, in μSR experiments, has a maximum at $T \cong 150$ K, and then split into two components at $T < 120$ K [06T2]. The temperature dependence of the electrical resistivity shows a broad maximum at $T_t \cong 115$ K [95H1], associated with the anomalous decrease of the **a**-axis lattice [93L1]. The 4f-3d hybridization degree is changed also as effect of pressure and thus the Curie temperatures, T_c, decrease with a rate $dT_c/dp = -15$ $KGPa^{-1}$ [03A1]—Table 8.28. The increase of T_t temperature with pressure can be described by a parabolic curve, with $T_t = 190$ K, at p = 1.2 GPa [03A1].

The magnetization of partially aligned $CeCo_4B$ powders, along "hard" direction, is far from saturation [95H1]. This can be correlated with a high anisotropy, generated by the 4f-3d hybridization. The large hysteresis (thermal and magnetic) has been correlated with the presence of narrow Bloch domain walls [95H1, 02B1].

The linear magnetostriction coefficients λ_a and λ_c, of $CeCo_4B$ compound, are smaller than those determined in YCo_4B, at $T \leq 130$ K. A negative volume magnetostriction, of the order of 10^{-3} was observed and related to Ce magnetism [93L2]—Table 8.29.

The cobalt substitution by iron, in $CeCo_{4-x}Fe_xB$ system, leads to an increase of Curie temperatures with a rate $\cong$ 12 K/Fe at %, whereas the magnetic anisotropy rapidly decreases, due to changes in 4f-3d bands hybridization and Fe site preference

Table 8.29 Magnetostriction and elastic properties of RCo_4B compounds[a]

Compound	T (K)	Magnetostriction·10^3		Bulk modulus B (GPa)	Elastic constant, c_{44} (GPa)	Compressibility K $GPa^{-1}\cdot 10^{-2}$	References
		λ_a	λ_c				
$CeCo_4B$	0	−0.75	−1.1				[93L2]
$CeCo_4B$[b]						0.74	[03A2]
$NdCo_4B$	>100		−2.2				[90I3]
YCo_4B	0	1.1	−2.1				[93L2]
YCo_4B[c]	145	−0.38	0.63				[93D2]
YCo_4B				187	98		[06M2]

[a]See also [04A2]
[b]Measured value
[c]Spontaneous magnetostriction at spin reorientation temperature

[95T2]. The ^{59}Co NMR measurements made on $CeCo_{4-x}Fe_xB$, with $0 \le x \le 0.8$, confirmed the iron preference for 2*c* site [02Y1]. An additional line to those of Co2*c* and Co6*i*, in the composition range $0 \le x \le 0.6$, has been attributed to Co6*i* atoms having one Fe atom in their nearest neighbor—Table 8.30. A similar study, performed on $CeCo_{4-x}Ni_xB$ (x = 0 and 0.2), evidenced that Ni is preferentially located in 6*i* site [98Y3]. The orbital moments, at Co2*c* and Co6*i* sites, were estimated at 0.11 μ_B and 0.005 μ_B respectively, around 25% from the total magnetic moment at 2*c* site.

The $PrCo_4B$ [84O1, 84S3, 85S2, 86J1, 87P8, 88G6, 90Z2, 97Y1, 99M1, 00C3, 02Z1, 05M1], $NdCo_4B$ [84O1, 84S3, 87P8, 90I2, 90I3, 90K1, 92F1, 92I3, 99M1, 00C3, 00C6, 02C2, 02Z1, 05M1] and $SmCo_4B$ [83E2, 84O1, 84S3, 85S2, 88S2, 88G6, 91I2, 92I2, 93K2, 93O1, 94I2, 94K2, 99M1, 00C3, 01M1, 07Y1, 18C1] compounds are ferromagnetically ordered. The cobalt moments at 2*c* and 6*i* sites are only little higher that in compounds with non-magnetic R elements. The ^{59}Co NMR study, of $NdCo_4B$ [92F1] and $PrCo_4B$ [97Y1], evidenced a split Co6*i* line, due to basal plane anisotropy. A high orbital contribution has been evidenced at Co2*c* site. Both Pr^{3+} and Nd^{3+} ions have negative Stevens coefficients and as result an easy plane magnetization. The presence of boron, in the neighbor of R1*b* (R = Pr, Nd) sites, does not change their moments as compared to that at R1*a* site [02C2, 02Z1]—Table 8.25. Somewhat higher Nd moments than those determined by ND were estimated by ^{145}Nd NMR [92F1].

The anisotropy of $PrCo_4B$, as well as of their cobalt sublattice has been analysed starting from stabilization energies of Co atoms [90Z2]. The $SmCo_4B$ is highly anisotrop, the anisotropy field being $\mu_0H_a = 120$ T, at 4.2 K [92I2, 19S1]—Tables 8.26 and 8.27. The anisotropy constant, K_1, is even higher than that of $SmCo_5$ compound [88S2]. The anisotropy of cobalt sublattice, compared with that of samarium, is relative small [93K2, 94K2] and thus the reported high anisotropy of $SmCo_4B$ can be attributed mainly to Sm sublattice (Sm^{3+} has positive α_J Stevens coefficient). The reciprocal susceptibilities of RCo_4B (R = Pr, Nd) compounds follow

Table 8.30 Data obtained by NMR method

Compound	T (K)	Nucleus	Hyperfine field (T)				Easy axis	References
			2*c*	6*i*	R	B		
$CeCo_4B$	1.5	^{59}Co	−8.32	(−)3.09				[15S1]
$CeCo_{3.2}Ni_{0.8}B^a$	4.2		0.25	−3.20				[98Y3]
$Ce(Co_{1-x}Fe_x)_4B$	4.2							
x = 0			0.25	−3.26			c-axis	[98Y3, 02Y1]
$x = 0.05^b$			0.3(1)	−3.4(1) −1.0(3)			c-axis	[02Y1]
$x = 0.10^b$			0.3(1)	−3.4(1) −1.2(3)			c-axis	[02Y1]
$x = 0.20^b$			0.3(1)	−3.90 −4.88			c-plane	[02Y1]
$PrCo_4B$	4.2		0.7	−7.5 −1.0				[97Y1]
$GdCo_4B$	4.2		7.3	−8.0 −5.4				[95Y1]
$GdCo_4B$	4.2		5.3	5.7				[92K1]
YCo_4B	4.2		4.8	3.0				[92K1]
		^{11}B						
$PrCo_4B$	4.2					1.48		[97Y1]
$GdCo_4B$	4.2					1.02		[92K1]
YCo_4B	4.2					0.73		[92K1]
		^{89}Y						
YCo_4B	4.2				4.8			[92K1]
		^{145}Nd						
$NdCo_4B$	4.2				351.6(1*a*) 360.6(1*b*)			[92F1]

[a] Ni at 6*i* site
[b] Two 6*i* sites, having different environments

Curie–Weiss type dependences [87P8]. An effective mean cobalt moment, $M_{eff} = 2.10\ \mu_B$/atom, has been determined.

Both the Curie temperatures and saturation magnetizations of RCo_4B (R = Pr, Nd) decrease as result of hydrogenation [85S2].

The spin reorientation temperatures, in $Y_{1-x}Nd_xCo_4B$, increases in the same way as Nd content [94Z2]. The easy axis region can be found only in the Y-rich composition range, its temperature widths decreasing considerably when Nd content increases. The saturation magnetizations decrease strongly when Co was substituted by Ni in $PrCo_{4-x}Ni_xB$ system [09A1]. The samples with $x \geq 3$ are paramagnetic, at T = 4.2 K.

A large number of pseudoternary Pr-, Nd- and $SmCo_4B$-based compounds were investigated, both due to their interesting physical properties as well as for possible technical applications. The Curie temperatures and saturation magnetizations increase in a non-linear fashion as function of iron content in $RCo_{4-x}Fe_xB$ series with R = Pr [86J1, 88G6] and R = Nd [88A1, 88G6, 07I1]. In the $NdCo_{4-x}Fe_xB$ system, the iron magnetic moments, at both 2*c* and 6*i* sites, are higher than those of cobalt and little dependent on composition [07I1]—Table 8.25. The ^{57}Fe Mössbauer studies of RCo_3FeB with R = Pr, Nd [88G6], confirmed the easy plane of magnetization as well as the preference of iron for 2*c* sites—Table 8.31. A large anisotropy field, $\mu_0H_a = 8$ T, was shown for the $NdCo_2Fe_2B$ powder [88A1]. The $Ce_xPr_{1-x}Co_2Fe_2B$ melt spin ribbons with x = 0.5–1.0, annealed at T = 973 K, have high coercivity with a peak value of $\mu_0H_c = 2$ T when x = 0.75 [20S1].

Some compositions from $Sm_{1-x}Pr_xCo_{4-y}Fe_yB$ series, were investigated in connection with their permanent magnet properties [93I1]. In the $SmCo_{4-x}Fe_xB$ pseudoternary compounds, although the saturation magnetizations and Curie temperatures increase in the same way as iron content, the anisotropy constants, for $x \leq 2$, are little changed [84O1, 94I2, 96G2]. The nanocrystalline $SmCo_2Fe_2B$ magnet, prepared by ball milling, after thermal treatment, has a coercive field, $\mu_0H_c = 1.35$ T, while in the case of mechanically alloyed sample, this is somewhat lower [96G2]. The pseudoternary $SmCo_{4-x}Fe_xB$ compounds obtained by melt spinning were particularly investigated [84O1, 84S3, 88G6, 88G7, 94I2, 96G2, 14J1, 18C1, 18C2, 19C3, 19S1]. The $SmCo_4B$ compound, obtained by melt spinning and annealing, has a coercive field $\mu_0H_c = 4.4$ T [14J1]. When alloyed with iron, the critical composition of $SmCo_{4-x}Fe_xB$ series, for obtaining a single phase, after crystallization, is x = 1.0. For $x \geq 1$, a two phases system was shown, the second phase, being $Sm_2(Co,Fe)_{17}B_y$ one. The analysis of magnetization isotherms revealed nucleation controlled process, at x < 1 and a combination of nucleation and pinning mechanisms when x = 1 or 2. A comparative study of $SmCo_4B$ and $SmCo_{3.1}Fe_{0.9}B$ ribbons showed that both interatomic distances and the disorder degree increase, when a part of Co is substituted by Fe [18C1]. Such change in atomic structure, increases the effect of atomic mismatch and the glass forming ability, but reduces the magnetocrystalline anisotropy. Annealing, is quite helpful to improve the magnetic properties of $SmCo_{3.1}Fe_{0.9}B$ ribbons.

The Cu addition increased the coercivity of $SmCo_{3.1}Fe_{0.9}B$ ribbons up to 4.45 T, when 1 at% Cu was added, correlated with changes of microstructure [18C2]. The $SmCo_{3.94-x}Fe_xCu_{0.06}B$ melt spin ribbons with $0.7 \leq x \leq 2$, annealed at T = 800 °C, have a short rod shaped microstructure for x = 0.8–0.9, a cellular one at x = 1.0 and equiaxial one at x = 1.5 [19C3]. The maximum coercivity, $\mu_0H_c = 5.01$ T, was obtained when x = 1.0. The coercivity is controlled by the pinning mechanism for sample with x = 1.0 and nucleation one for other compositions.

The ^{57}Fe Mössbauer study on $SmCo_{4-x}Fe_xB$ with x = 1 or 2, evidenced a rather large hyperfine anisotropy [88G6]. The cobalt substitution by Ni, in $SmCo_{4-x}Ni_xB$ series, decreases both T_c and M_s [84O1, 84S3]. The hydrogenation of $SmCo_4B$, as in others RCo_4B compounds, decrease both the magnetizations and Curie temperatures.

Table 8.31 Data obtained by Mössbauer spectroscopy on $RCo_{4-x}Fe_xB$ compounds

Compound	T (K)	Nucleus	Site	δ (mm/s)	ΔQ (mm/s)	B_{eff} (T)	η^a	θ^b (°)	ϕ^c (°)	A^d (%)	References
$CeFe_{0.01}Co_{0.99}B^{e)}$	4.2	^{57}Fe	6*i*	0.02	−0.62	19.7				100	[94W1]
	295		6*i*	−0.16	−0.56	9.4				100	
$PrCo_3FeB^f$	RT		2*c*	−0.194	−1.197	14.45	0	90	0	60.9	[88G6]
			$6i_1$	−0.074	1.23	14.61	1	0	90	13.0	
			$6i_2$	−0.074	1.23	13.32	1	±123.2	90	26.1	
$NdCo_3FeB^f$	RT		2*c*	−0.216	−1.144	14.67	0	90	0	56.7	[88G6]
			$6i_1$	−0.142	1.073	15.44	1	0	90	14.4	
			$6i_2$	−0.142	1.073	14.12	1	123.1	90	28.9	
$SmCo_3FeB^f$	RT		2*c*	−0.219	−1.163	17.29	0	0	0	60	[88G6]
			6*i*	−0.16	1.181	15.98	0.898	90	0	40	
$SmCo_2Fe_2B^f$	RT		2*c*	−0.204	−1.155	19.17	0	0	0	37.7	[88G6]
			6*i*	−0.157	1.146	16.65	0.936	90	0	62.3	
$GdCo_{4-x}Fe_xB^g$	RT		2*c*	0.006(3)	−1.028(6)	10.73(1)	0	0		93.5(6)	[07G1]
x = 0.10						13.73(1)				6.5 imp	
x = 1.0	RT		2*c*	0.002(1)	−0.895(6)	13.65(1)	0	90	0	72.0(6)	[07G1]
			$6i_1$	0.006(6)	0.89(1)	12.6(1)	1	0	90	9.4(6)	
			$6i_2$	−0.006(6)	0.89(1)	12.08(5)	1	120	90	18.6(6)	
x = 2.6	RT		2*c*	0.04(1)	−0.85(1)	18.31(5)	0	0		29.4	[07G1]
			6*i*	0.04(1)	0.78(1)	14.72(3)	1	90	0	70.6	
$TbCo_3Fe_1B^h$	85		2*c*	−0.018(2)	−1.072(7)	17.60(2)	0	90	0	85(3)	[09M2]
			$6i_1$	−0.21(1)	1.23(1)	12.9(2)	1	0	90	5(1)	
			$6i_2$	−0.21(1)	1.23(3)	14.5(1)	1	120	90	10(1)	
$TbCo_2Fe_2B^h$	85		2*c*	−0.007(1)	−1.024(6)	19.61(1)	0	90	0	40(2)	[09M2]
			$6i_1$	0.175(6)	0.90(1)	18.48(7)	1	0	90	20(2)	
			$6i_2$	0.175(6)	0.90(1)	18.86(7)	1	120	90	40(2)	

(continued)

Table 8.31 (continued)

Compound	T (K)	Nucleus	Site	δ (mm/s)	ΔQ (mm/s)	B_{eff} (T)	η^a	θ^b (°)	ϕ^c (°)	A^d (%)	References
$DyCo_3FeB^h$	85		2*c*	−0.007(5)	−1.10(2)	17.44(5)	0	90	0	74(5)	[08M4]
			$6i_1$	−0.071(28)	1.21(7)	15.2(5)	1	0	90	9(1)	
			$6i_2$	−0.071(28)	1.21(7)	15.3(2)	1	120	90	18(2)	
$DyCo_{2.5}Fe_{1.5}B^h$	85		2*c*	−0.021(4)	−1.07(2)	18.55(4)	0	90	0	0(2)	[08M4]
			$6i_1$	0.004(11)	1.03(2)	17.6(2)	1	0	90	13(1)	
			$6i_2$	0.004(11)	1.03(2)	16.83(8)	1	120	90	26(2)	
$DyCo_2Fe_2B^h$	85		2*c*	−0.014(5)	−1.02(2)	19.5(5)	0	90	0	50(3)	[08M4]
			$6i_1$	0.013(11)	1.04(2)	18.3(2)	1	0	90	17(1)	
			$6i_2$	0.013(11)	1.04(2)	17.55(9)	1	120	90	33(3)	
YCo_4B^e (0.5 wt % ^{57}Fe)	4.2		2*c*		−0.17	26.9					[92C1]
			6i		0.54	15.9					
	293		2*c*		−0.34	25.7					
			6*i*		−1.15	13.5					
$Y(Fe_{0.01}Co_{0.99})_4B^{e)}$	4.2		2*c*	0.03(1)	−0.64	19.0		0		83	[94W1]
			$6i_1$	0.02(1)	−0.61	15.0		90		17	
	295		6*i*	−0.11	−0.59	14.6		0		100	
YCo_3FeB^g	78		2*c*	−0.016(2)	−1.08(1)	16.66(2)	0	90	0	73	[06L1]
			$6i_1$	0.030	1.18(3)	16.1(2)	1	0	90	9	
			$6i_2$	0.030	1.18(3)	15.9(1)	1	120	90	18	
$YCo_2Fe_2B^g$	78		2*c*	−0.013(3)	−1.05(1)	18.52(3)	0	90	0	43	[06L1]
			$6i_1$	0.070(4)	0.86(1)	17.7(1)	1	0	0	19	
			$6i_2$	0.070(4)	0.86(1)	18.40(5)	1	120	90	38	
$YCoFe_3B^h$	4.2		2*c*	0.029(5)	−0.87(1)	24.88(5)				22(1)	[09G1]
			6*i*	0.051(1)	0.97(1)	18.8(1)				78(1)	
$YCoFe_3B^g$	78		2*c*	−0.029(2)	−1.035(3)	23.94(1)	0	0	0	21.1(1)	[06L1]
			6*i*	−0.042(1)	0.974(3)	18.48(1)	1	90	0	78.9(1)	

(continued)

Table 8.31 (continued)

Compound	T (K)	Nucleus	Site	δ (mm/s)	ΔQ (mm/s)	B_{eff} (T)	η^a	θ^b (°)	ϕ^c (°)	A^d (%)	References
$GdCo_4B^i$	4.2	^{155}Gd	1*a* 1*b*	0.32(2) 0.28(1)	8.9(1)^j 14.0(1)^j	6.3(5) 14.2(2)					[88S2]

^{a}Asymmetry parameter
^{b}Angle of B_{eff} with Oz at 2*c* site and with OZ at 6*i* site
^{c}Angle of B_{eff} (projection on xoy) and Ox for 2*c* site and of B_{eff} (projection XOY) and OX at site 6*i*. The θ and ϕ imposed values [88G6]
^{d}Fractional content
^{e}Relative to ^{57}CoRh source
^{f}Relative to RhFe
^{g}Relative to α-Fe foil if no other notations are mentioned
^{h}Relative to α-Fe powder
^{i}Source ^{154}SmPd
^{j}Values of the electric field gradient $V \cdot 10^{19}$ V/m

The $GdCo_4B$ compound is ferrimagnetically ordered, the easy direction of magnetization being **c**-axis [84O1, 84S3, 87B6, 87B7, 87P8, 88B2, 88S2, 90D1, 91L1, 93I2, 94B3, 94I3, 95D1, 95I3, 95Y1, 96T2, 96T3, 97S2, 00C3, 01K1, 01M1, 03A2, 04B3, 06S3, 07C1, 08I1]. The thermal variations of magnetization as well as of magnetic susceptibilities are characteristic for this type of magnetic ordering—Figs. 8.14 and 8.15. Band structure calculations show the same trend for Co2*c* and Co6*i* magnetic moments as in other RCo_4B compounds, although little higher [01V1, 02V1, 06S3]—Table 8.25. The Gd5d band is negatively polarized, the polarization being induced both by 4f-5d local exchange and 5d-3d short range exchange interactions [04B3, 11B1]. The presence of two Co6*i* lines, at 4.2 K, as evidenced by ^{59}Co NMR measurements [92K1, 95Y1], were attributed to domain wall edge signals [92K1]—Table 8.30. The quadrupole interactions, at non-equivalent Gd sites, were evaluated on

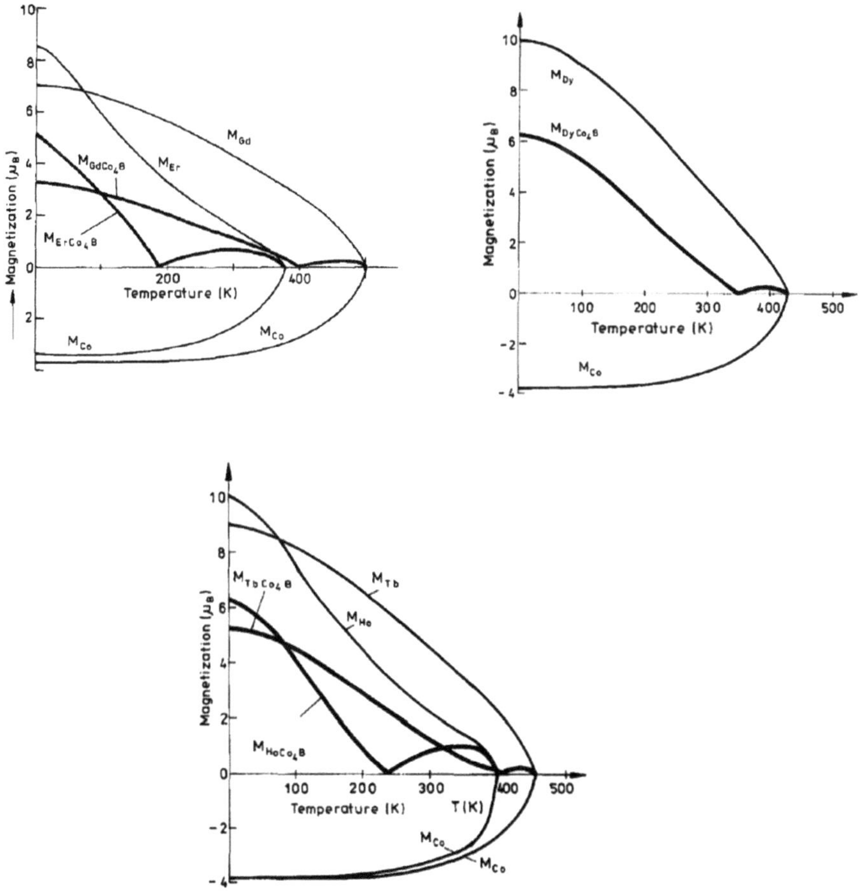

Fig. 8.14 RCo_4B with R = Gd, Tb, Dy, Ho, Er: thermal variations of spontaneous magnetizations as well of the sublattices magnetizations [87P8, 88B2]

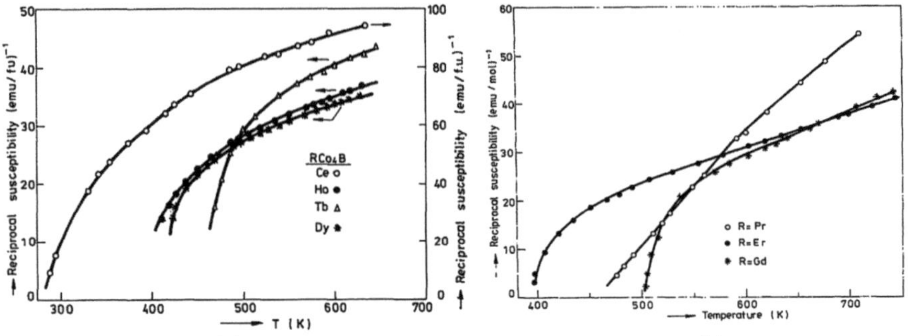

(a) (b)

Fig. 8.15 RCo_4B with R = Ce, Tb, Dy, Ho (**a**) and R = Pr, Gd, Er (**b**): thermal variations of reciprocal susceptibilities [87P8, 88B2]

the basis of ^{155}Gd Mössbauer effect measurements [88S2]. The experimental data allowed a semi-empirical determination of a set of effective point charges which can describe the crystal field.

The temperature dependence of the sublattices contributions to the anisotropy constants K_1 of $GdCo_4B$ was analysed—Fig. 8.16. The cobalt sublattice change from planar to uniaxial anisotropy at $T \cong 110$ K and has a maximum at $T \cong 200$ K. The Gd sublattice anisotropy, due to Gd-Gd and Gd-Co dipole interactions, is positive, at $T < T_c$ [93I2, 96I1].

The magnetic properties of pseudoternary $Gd_xR_{1-x}Co_4B$ alloys with R = Ce [01B2, 02B1], R = Pr [08K1], R = Nd [08K2], R = Dy [94I3, 95I3, 96I1] and R = Y [87B6, 87B7, 88B2, 91L1, 92H2, 94B3, 94L1, 95B4, 04B3, 07B2, 13B1]

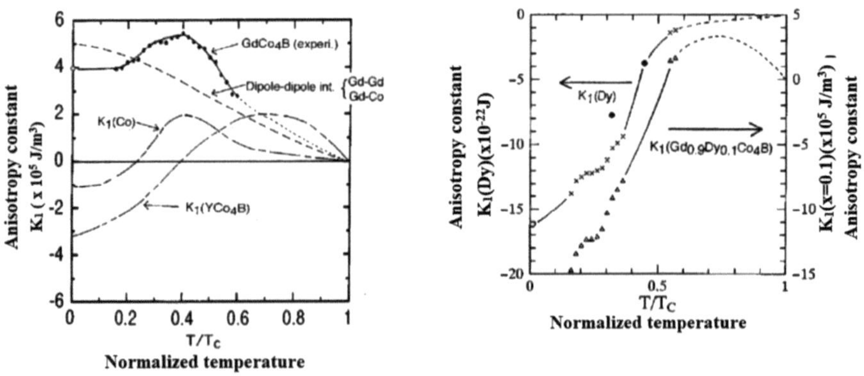

(a) (b)

Fig. 8.16 $Gd_{1-x}Dy_xCo_4B$: **a** temperature dependence of K_1, anisotropy constant for x = 0, of cobalt sublattice and of YCo_4B, **b** anisotropy constants for a sample with x = 0.1 and that of the Dy sublattice [96I1]

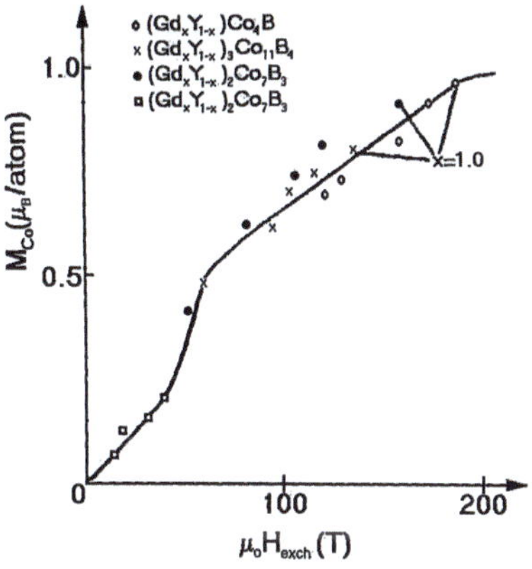

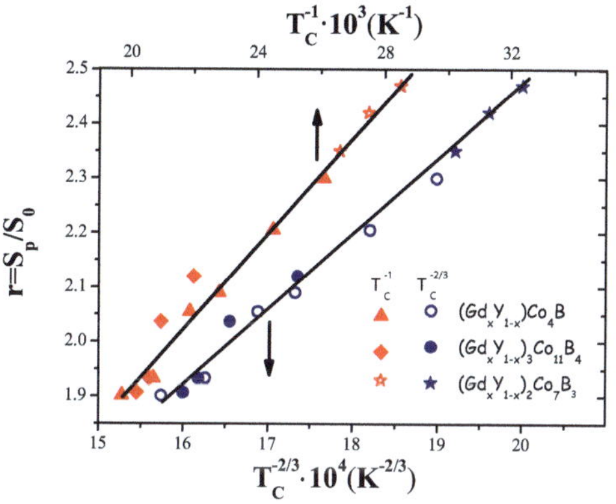

Fig. 8.17 $(Gd_xY_{1-x})_{n+1}Co_{3n+5}B_{2n}$: **a** mean cobalt moments as function of exchange fields [88B2, 04B3, 13B1], **b** ratio $r = S_p/S_o$ as function of Curie temperatures T_c^{-1} and $T_c^{-2/3}$

were investigated. The temperature dependences of spontaneous magnetizations, when $x \geq 0.1$, are typical for ferrimagnetic ordering. When substituting Gd by Y or light rare-earths, at 4.2 K, the rare-earth and cobalt sublattices magnetizations, determined at 4.2 K, compensate at compositions x = 0.32 (R = Ce), x = 0.43 (R = Y) and x = 0.6 (R = Pr or Nd). The Curie temperatures in $Gd_xR_{1-x}Co_4B$ series increase with the rates of 2.2 (R = Ce), 1.28 (R = Y), 0.45 (R = Pr) and 0.55 (R = Nd) K/at % Gd. As results of the increase of the exchange field acting on cobalt, an additional cobalt moment to that determined in compounds with non-magnetic rare-earths, is induced. The dependence of the mean cobalt moments as function of exchange fields, in $(Gd_xY_{1-x})_{n+1}Co_{3n+5}B_{2n}$ series is plotted in Fig. 8.17 [04B3]. A similar trend was shown in RCo_2-based compounds, behavior described in the model of induced moment [75B2, 78B1, 81B1], or considering the metamagnetism of itinerant electrons [62W1, 75B1]. A transition from low to high cobalt moment is evidenced at $\mu_0H_{exch} \cong 60$ T, somewhat smaller value than that reported in RCo_2-based compounds ($\mu_0H_{exch} \cong 70$ T).

The reciprocal susceptibilities of the $(Gd_xY_{1-x})_{n+1}Co_{3n+5}B_{2n}$ series with $x \geq 0.2$ show non-linear temperature dependences, of Néel type [48N1]. Assuming that the effective Gd moment is given by the free ion value, from the Curie constants, the mean effective cobalt moments were determined. The ratio $r = S_p/S_o$ between the number of cobalt spins determined from Curie constants, S_p and saturation magnetizations, S_o, follow better a $T_c^{-2/3}$ dependence than the T^{-1} one [04B1], as predicted by spin fluctuations model [78W1, 79M1, 91M1]—Fig. 8.17b.

The effects of substitutions of Co in $GdCo_{4-x}M_xB$ series by M = Fe [89D1, 90D1, 90D2, 91B1, 91D2, 92D2, 95T1, 07G1, 08M1] or M = Ni [91B1, 04A2, 04B5] were investigated. The Fe substitutions increase near linearly T_c values for $x \leq 1$ and then more slowly. This trend is due mainly due to preferential substitution of Co by Fe on 2*c* site, for low cobalt content. In addition, the transition metal sublattice magnetization also increase. The $GdCo_{4-x}Fe_xB$ compounds, at T = 77 K and 300 K and x ≤ 0.04 exhibit axial magnetic anisotropy, whereas in the composition range 0.08 ≤ x ≤ 0.10, this is of basal plane-type. The compounds with 0.05 ≤ x ≤ 0.07 show a spin reorientation, from basal to axial magnetic anisotropy, at 77 K ≤ T ≤ 300 K [95T1, 07G1]. By using the already mentioned model [88T1, 95T1], a change from axial magnetic anisotropy, in $GdCo_4B$ (x = 0), to basal plane, for x = 0.3 is predicted. A latter study [07G1], showed that for $GdCo_{4-x}Fe_xB$ series, a spin reorientation actually occurs in the 0.15 < x < 0.25 composition range. A second spin reorientation from basal to axial anisotropy for x > 2.3 was shown, connected with the filling of 6*i* sites by iron. The ^{57}Fe hyperfine fields in $GdCo_{4-x}Fe_xB$ series increase with increasing iron content and was found to be larger for axial than for basal plane orientation of the magnetization.

The saturation magnetizations of $GdCo_{4-x}Ni_xB$ system increase when increasing the Ni content, while the Curie temperatures decrease [91B1, 04A2]. This is the result of the diminution of transition metal sublattice magnetization and of exchange interactions, respectively. The two sublattice magnetizations, at 4.2 K, in $Gd_xY_{1-x}Co_3NiB$, compensate at x = 0.275, gadolinium content smaller than x = 0.43, when the two sublattices compensate in $Gd_xY_{1-x}Co_4B$ series.

The RCo_4B compounds with R = Tb [89B2, 91L1, 00C1, 00C3, 06C1, 09M2], R = Dy [89B2, 90D1, 91L1, 93O1, 96I1, 97S2, 02Z1, 03M2], R = Ho [87D2, 89B2, 90D1, 90I3, 91I3, 91L1, 92I3, 96T2] and R = Er [87P8, 88A1, 88G7, 90D1, 91L1, 19R1] are ferrimagnetically ordered—Figs. 8.14 and 8.15. The magnetic properties of pseudoternary $R_xY_{1-x}Co_4B$ with R = Tb, Dy, Ho, Er were also investigated [91L1, 94L1]. The temperature dependences of anisotropy constants, of $Dy_xGd_{1-x}Co_4B_4$ (x = 0.1), as well as that of the dysprosium sublattice were reported [94I3, 95I3, 96I1]—Fig. 8.16b. The spin reorientation temperatures increase from T_{SR} = 162 K (x = 0.05) to 225 K (x = 0.1) [95I3, 96I1].

The substitution of cobalt by iron, in $TbCo_{4-x}Fe_xB$ series, induces both an anisotropic increase of the unit cell volumes as well as of the Curie temperatures [09M2]. The compensation temperatures decrease by ≅ 50 K, in the composition range 0 ≤ x ≤ 2, as result of the increase of the transition metal magnetization—Table 8.25. The $DyCo_{4-x}Fe_xB$ [89D1, 91D2, 08M4], $HoCo_{4-x}Fe_xB$ [91D2] and $ErCo_{4-x}Fe_xB$ [87D2, 88A1, 88G7, 07M1] pseudoternary systems are ferrimagnetically ordered and show close related properties to those of corresponding Gd-based series.

The $Dy_{n+1}Co_{3n+5}B_{2n}$ series with n = 1 and ∞ were investigated by magnetic Compton profile, at T = 10 K [03M2]. As a general trend, the cobalt moments antiparallely oriented to Dy magnetization decrease when increasing B content, while the Dy4f and the delocalized spin moment increased.

The $ErCo_4B$ compound has a planar anisotropy [88G7]. The ^{57}Fe Mössbauer effect measurements on $ErCo_{4-x}Fe_xB$, evidenced the increase of the weighted average of the ^{57}Fe hyperfine field, by 2.34 T, in the composition range 1 ≤ x < 4, which correspond to ≅ 0.16 μ_B/atom [07M1]. This is the result of the increase of the transition metal moments by 0.32 μ_B at 2*c* site and by 0.17 μ_B at 6*i* site.

The $Er_{17}Co_{75}B_8$ ($ErCo_{4.41}B_{0.47}$) alloy, obtained by high energy ball milling and annealing at 1073 K, has $ErCo_{4.5}B_{0.5}$ as the dominant phase, which crystallizes in

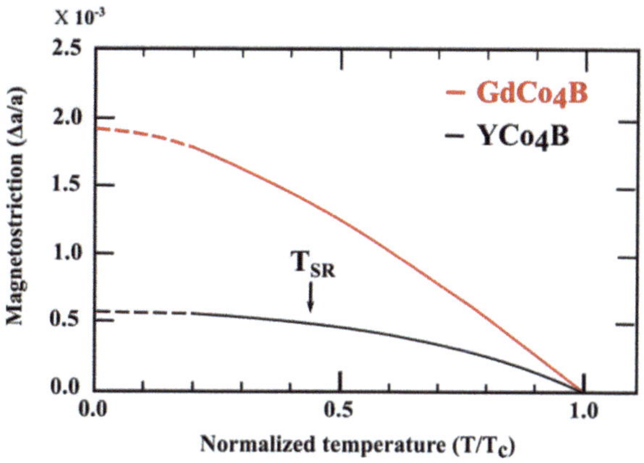

Fig. 8.18 Spontaneous magnetostrictions as functions of reduced temperatures for RCo_4B (R = Gd, Y) compounds [98I2]. The solid lines are the fitted curves through experimental data

hexagonal *P6/mmm* type structure [19R1]. The coercivity exhibits its maximum value, of 0.8 T, at RT, for optimized grain size of 35.7 nm. In the $Sm_xCo_{80-x}B_{20}$ amorphous alloys, after annealing, the coercivity changed, from $\mu_0H_c = 0.13$ T to 1.56 T, as x increased from 5 to 15, while the magnetic induction decreased from 0.51 T to 0.16 T [20L1].

The magnetic properties of $R_{n+1}Co_{3n+5}B_{2n}$ series were analysed in correlation with boron content, as for example the average cobalt moment when R = Pr, Nd [05M1], the effect of Co3d-B2p hybridization, for R = Gd [94B3, 95D1] and the exchange interactions were evaluated in the mean field approximation [88B2, 90D1, 91L1, 95B4, 96T3, 99M1, 01M1, 04B3, 08I1, 13B1].

The thermal expansion of $Gd_{n+1}Co_{3n+5}B_{2n}$ [98I2] and of YCo_4B [93K1, 98I2] were reported. In the first system, the thermal coefficient α_c increases and α_a decreases as the boron content is higher. The temperature dependences of the spontaneous magnetostrictions in RCo_4B (R = Y, Gd) are well described by the relation $\omega_s = A_{Co-Co}M^2_{Co} + B_{Co-Gd}M_{Co}M_{Gd} + C_{Gd-Gd}M^2_{Gd}$, where A, B and C are constants proportional to products of reciprocal elastic constant and the variation of exchange interactions along the *a* lattice parameter [98I2]—Fig. 8.18. Some elastic properties of RCo_4B compounds are given in Table 8.29.

8.7 $R_{m+n}Co_{5m+3n}B_{2n}$ Compounds with (m = 2, n = 1), (m = 2, n = 3) and (m = 3, n = 2)

The $R_3Co_{13}B_2$ (m = 2, n = 1), $R_5Co_{19}B_6$ (m = 2, n = 3), $R_5Co_{21}B_4$ (m = 3, n = 2) compounds are members of $R_{m+n}Co_{5m+3n}B_{2n}$ series, formed by alternating stacking of m layers RCo_5 and n RCo_3B_2 layers along the **c**-axis [99C1, 99C3]—Sect. 8.1 and Fig. 8.19. The above borides are not formed with all rare-earth elements. The $R_3Co_{13}B_2$ series included 8 compounds (R = Pr, Nd, Sm, Gd, Dy, Ho, Er and Y),

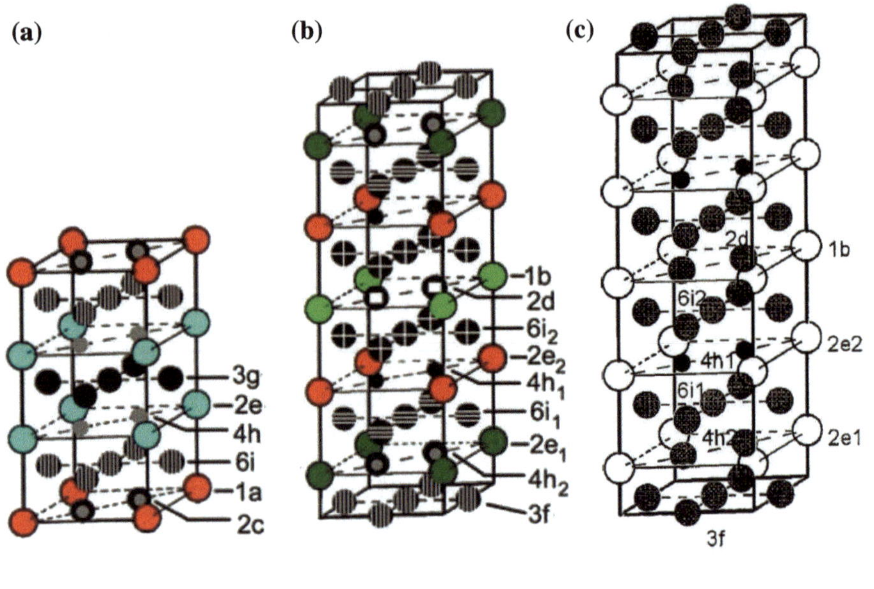

Fig. 8.19 $R_3Co_{13}B_2$ (**a**), $R_5Co_{19}B_6$ (**b**) and $R_5Co_{21}B_4$ (**c**): crystal structures [01C5, 06S1]

the $R_5Co_{19}B_6$ three (R = Pr, Nd, Sm) and only one compound has been reported as having $R_5Co_{21}B_4$ composition (R = Nd) [99C3, 00C7, 01C5].

The above borides are formed by peritectoid reactions, followed by prolonged annealing at relatively low temperatures (T < 1100 K). Even for such treatment, a mixture of phases were shown. Thus, in addition to $R_3Co_{13}B_2$ compounds with R = Sm [00C5] and R = Gd [00C4], always exists a large amount of 1/4/1 ($CeCo_4B$-type) and 1/5 ($CaCu_5$-type) phases. In $Nd_2Co_{13\text{-}x}Ni_xB_2$ series, with $0 \leq x \leq 13$, the content of 1/4/1 phase was between 3.5 and 16.5% and that of 1/5, in the range 1.0 and 5.8% [05P2]. A large number of stacking faults were also shown in $Sm_5Co_{19}B_6$ boride [08M2].

The compounds from the above series crystallize in hexagonal structure, space group *P6/mmm.* The R atoms are distributed on two sites in $R_3Co_{13}B_2$ structure and three sites in those of $R_5Co_{19}B_6$ and $R_5Co_{21}B_4$ series, while the cobalt is located in three, four and five types of sites, respectively. The Co sites are derived from 2*c* and 3*g* ones of the RCo_5-type structure. As example, the 2*c* and 3*g* sites in $CaCu_5$ type structure are related to (2*c*, 4*h*) and (3*g*, 6*i*) sites, respectively of the $R_3Co_{13}B_2$-type lattice, the B atom being located in 2*c* site. The lattice parameters of the above borides, as already mentioned, follow basically the relation $a \cong a_{RCo_5}$ and $c \cong mc_{RCo_5} + nc_{RCo_3B_2}$. The atomic sites are given in Tables 8.32 and 8.33, the lattice parameters are listed.

Table 8.32 Atomic sites of $Nd_{m+n}Co_{5m+3n}B_{2n}$ compounds having *P6/mmm* space group

(a) $Nd_3Co_{13}B_2$ (m = 2, n = 1) [99C4, 06V1]

Atom	Sites	Sym	x	y	z	Atomic environment
Nd	1*a*	6/*mmm*	0	0	0	Pseudo Frank-Kasper $B_6Co_{12}Nd_2$
Nd	2*e*	6*mm*	0	0	0.3292	Pseudo Frank-Kasper $Co_{13}Nd_2$
Co	4*h*	3*m*.	0.3333	0.6667	0.3159	Anticuboctahedron Co_9Nd_3
Co	6*i*	2*mm*	0.5000	0	0.1346	13-vertex polyhedron $B_2Co_7Nd_4$
Co	3*g*	*mmm*	0.5000	0	0.5000	Cuboctahedron Co_8Nd_4
B	2*c*	$\bar{6}m2$	0.3333	0.6667	0	Trigonal prism Co_6

(b) $Nd_5Co_{19}B_6$ (m = 2, n = 3) [99C3, 06V1]

Atom	Sym	Sites	x	y	z	Atomic environment
Nd	6/*mmm*	1*b*	0	0	0.500	Pseudo Frank-Kasper $Co_{12}B_6Nd_2$
Nd	6*mm*	2e_1	0	0	0.0964(1)	Pseudo Frank-Kasper $B_6Co_{12}Nd_2$
Nd	6*mm*	2e_2	0	0	0.2976(1)	Pseudo Frank-Kasper $Co_{18}Nd_2$
Co	3*m*.	4h_1	0.3333	0.6667	0.2955(2)	Icosahedron Co_9Nd_3
Co	2*mm*	6i_1	0.5000	0	0.1742(1)	13-vertex polyhedron $B_2Co_7Nd_4$
Co	2*mm*	6i_2	0.5000	0	0.4114(1)	13-vertex polyhedron $B_2Co_7Nd_4$
Co	*mmm*	3*f*	0.5000	0	0.0000	Rhombic dodecahedron $B_4Co_6Nd_6$
B	$\bar{6}m2$	2*d*	0.3333	0.6667	0.5000	Trigonal prism Co_6
B	3*m*	4h_2	0.3333	0.6667	0.0936(6)	Trigonal prism Co_6

(c) $Nd_5Co_{21}B_4$ (m = 3, n = 2) [01C5]

Atom	Sites	x	y	z
Nd	1*b*	0	0	0.5000
Nd	2e_1	0	0	0.1001
Nd	2e_2	0	0	0.2959
Co	2*d*	0.6667	0.3333	0.5000
Co	4h_2	0.6667	0.3333	0.1084
Co	6i_1	0.5000	0	0.2216
Co	6i_2	0.5000	0	0.3916
Co	3*f*	0.5000	0	0
B	4h_1	0.6667	0.3333	0.2862

The crystal structures of pseudoternary compounds, based on the above end series, were reported: $Nd_3Ni_{13-x}Co_xB_2$ with, $0 \le x \le 5$ [00C8], $0 \le x \le 13$ [05P2, 07B1, 07P1], $Y_3Co_{13-x}Ni_xB_2$ with $0 \le x \le 13$ [05P2, 07B1, 07P1], $Y_3Co_{13-x}Ni_xB_2$ with $0 \le x \le 13$ [05P1, 07B1] or $x \ge 8$ [05P1, 07R1]. In $Y_3Ni_{13-x}Co_xB_2$ series, as the cobalt content increases up to x = 5, the c-lattice constant decreases by 0.2%, while a parameter increases by 0.2%, accounting for the monotonic decrease of the c/a ratio from 2.202 (x = 0) to 2.192 (x = 5) and 2.168 (x = 13) [05P1].

Table 8.33 Crystal structure and lattice parameters of $R_3Co_{13}B_2$, $R_5Co_{19}B_6$ and $R_5Co_{21}B_4$ compounds[a]

Compound	T (K)	Space group	Lattice parameters (nm)		References
			a	c	
$Pr_3Co_{13}B_2$	RT	*P6/mmm*	0.50672(3)	1.06850(6)	[99C5]
$Pr_3Co_{13}B_2$	RT	*P6/mmm*	0.50919(7)	1.0853(4)	[08M2]
$Nd_3Co_{13}B_2$	RT	*P6/mmm*	0.50722(4)	1.07840(5)	[99C2]
$Nd_3Co_{13}B_2$	RT	*P6/mmm*	0.50881(2)	1.0846(1)	[05P2]
$Nd_3Co_{12}NiB_2$	RT	*P6/mmm*	0.50690	1.08060	[00C8]
$Nd_3Co_{10}Ni_3B_2$	RT	*P6/mmm*	0.50630	1.08500	[00C8]
$Nd_3Co_8Ni_5B_2$	RT	*P6/mmm*	0.50575(3)	1.0892(1)	[05P2]
$Nd_3Co_6Ni_7B_2$	RT	*P6/mmm*	0.50476(2)	1.0915(0)	[05P2]
$Nd_3Ni_{13}B_2$	RT	*P6/mmm*	0.50255(1)	1.0959(0)	[05P2]
$Nd_3Co_{13}B_2$	295	*P6/mmm*	0.51024(2)	1.0877(1)	[07P1]
	503	*P6/mmm*	0.51160(2)	1.0905(1)	[07P1]
$Nd_3Co_3Ni_{10}B_2$	1.5	*P6/mmm*	0.50035(2)	1.0865(1)	[07P1]
	100	*P6/mmm*	0.50049(1)	1.0867(1)	[07P1]
$Sm_3Co_{13}B_2$	RT	*P6/mmm*	0.507	1.083	[08M2]
$Gd_3Co_{13}B_2$[b]	RT	*P6/mmm*			[01L1, 00C4]
$Dy_3Co_{13}B_2$[b]	RT	*P6/mmm*			[01L1]
$Ho_3Co_{13}B_2$[b]	RT	*P6/mmm*			[01L1]
$Er_3Co_{13}B_2$[b]	RT	*P6/mmm*			[01L1]
$Y_3Co_{13}B_2$	RT	*P6/mmm*	0.50063(3)	1.0853(1)	[14P1]
$Y_3Co_5Ni_8B_2$	RT	*P6/mmm*	0.49661(3)	1.08867(3)	[05P1]
$Y_3Co_3Ni_{10}B_2$	RT	*P6/mmm*	0.49597(2)	1.08930(3)	[05P1]
$Y_3Co_1Ni_{12}B_2$	RT	*P6/mmm*	0.49585(2)	1.09059(3)	[05P1]
$Y_3Ni_{13}B_2$	RT	*P6/mmm*	0.49579(1)	1.09141(3)	[05P1]
$Pr_5Co_{19}B_6$ $Pr_5Co_{19}B_6$ $Nd_5Co_{19}B_6$	RT RT RT	*P6/mmm* *P6/mmm* *P6/mmm*	0.51264(4) 0.5139 0.5122	1.65602(6) 1.6570 1.6555	[00C7] [20M1] [20M1]
$Nd_5Co_{19}B_6$	RT	*P6/mmm*	0.51158(1)	1.67226(1)	[08M2]
$Nd_5Co_{19}B_6$	RT	*P6/mmm*	0.51328(3)	1.66519(5)	[99C3]
$Sm_5Co_{19}B_6$	RT	*P6/mmm*	0.5101(5)	1.663(2)	[08M2]
$Nd_5Co_{21}B_4$	RT	*P6/mmm*	0.50987	1.76523	[01C5]

[a]See also [81K1, 84P1, 84R1]
[b]Lattice constants were not found [00C4, 01L1], these compounds are hard to be obtained as a single phase and are formed in a mixture of phases

The structural stability of compounds from $R_3Ni_{13-x}Co_xB_2$ series with R = Nd, Sm, Y and the site preferences of cobalt atoms were investigated by using a series of interatomic pair potential [10P1]. The space group remains unchanged upon substitution of Co by Ni and the calculated lattice parameters were found to agree with experimental data. The Co atoms substitute for Ni, with a strong preference for the 3*g* site, the sequence of Co occupation being 3*g* > 4*h* > 6*i*. In $Nd_3Ni_{13-x}Co_xB_2$ series, at intermediate composition range, the Ni preferentially occupies smaller 4*h* sites and strongly avoids the 3*g* sites [05P1, 07P1]. At higher concentrations, Ni maintains its preferential distribution at the 4*h* sites and avoids 3*g* and 6*i* sites. In $Y_3Ni_{13-x}Co_xB_2$, for $x \leq 5$, Co atoms substitutes preferentially Ni at 6*i* and 3*g* sites [05P1, 07R1].

The crystallization of amorphous $Nd_3Co_{13}B_2$ thin films, generates the precipitation of the Co clusters [04G1].

The $Y_3Co_{13}B_2$ compound is ferromagnetically ordered [05P1, 07B1, 14P1]. The cobalt moments are dependent on lattice sites, decreasing in the sequence $M_{4h} > M_{3g} > M_{6i}$, as determined from band structure calculations [14P1]. A large orbital moment at Co4*h* site has been evidenced—Table 8.34. The low magnetic state of Co6*i* site was ascribed to strong Co3d-B2p hybridization. The cobalt sublattice anisotropy has been estimated in the model of individual site contributions [79S2]. The Co4*h* sites give a large positive contribution to the anisotropy, being related to 2*c* position in RCo_5–type structure. The contributions to the anisotropy of cobalt sites, derived from 3*g* sites are relatively small and with reversed anisotropy. The resultant anisotropy of cobalt sublattice, as a balance of the mentioned contributions, is uniaxial.

The magnetic properties of $Y_3Ni_{13-x}Co_xB_2$ series, revealed interesting features [05P1, 07B1, 07R1, 14P1]. The $Y_3Ni_{13}B_2$ compound is a weak itinerant antiferromagnet with $T_N = 68$ K—Sect. 9.8. A weak ferromagnetic component appears at T_N, suggesting an uncompensated antiferromagnetic structure. In the composition range $0 < x \leq 1$, the samples are also antiferromagnetically ordered; the intrinsic weak ferromagnetic component increases with the cobalt content. The temperature dependences of the magnetic susceptibilities, in the composition range $0.5 \leq x \leq 1.0$, show two maxima located at temperatures T_N and also at T_{WF}, where the weak ferromagnetic component sets on [05P1]. Field induced transitions can be shown in this composition range. The critical fields for AFM to FM transition, H_{t1}, decrease when the temperature increases. There is a second transition, at fields H_{t2} (higher than H_{t1}), these fields increasing with temperature. A heterogeneous ferromagnetic state and in plane magnetic anisotropy are present in samples with $2 \leq x \leq 5$, the T_c values increasing nearly linear with cobalt content [07B1]. Assuming that the Ni magnetic moment is negligible, the average cobalt moments, at T = 5 K, are in the range $\langle M_{Co}$ = 0.86–0.73 μ_B/atom. Band structure calculations, predict **c**-axis anisotropy for $Y_3Co_{13}B_6$ and $Y_3Ni_{10}Co_3B_2$ compounds, in agreement with experimental evidence [14P1]. The specific heat measurements on $Y_3Ni_{13-x}Co_xB_2$ series show that both the γ electronic constants and hyperfine constants, α, at low temperatures, increase in the same way as the cobalt content [07R1].

The $Pr_3Co_{13}B_2$ [00C7, 05M1, 06S1] and $Nd_3Co_{13}B_2$ [99C3, 99C4, 00C6, 00C8, 04G1, 04K1, 04S2, 05M1, 05P1, 05P2, 07B1, 07P1, 07R1] compounds are also ferromagnetically ordered—Table 8.34. The neutron diffraction measurements on $Nd_3Co_{13}B_2$ [07P1], evidenced that the Co4*h* site has the higher magnetic moment. The smallest moments reside on Co6*i* and Co3*g* sites derived from 3*g* site of RCo_5 series [04K1, 06S1, 14P1].

The cobalt magnetic moments, determined in Nd-based compounds, are somewhat higher than in Y ones, resulting from increased exchange interactions [05M1]. The

Table 8.34 Magnetic properties of $R_3Co_{13}B_2$, $R_5Co_{19}B_6$ and $R_5Co_{21}B_4$ compounds

Compound	Magnetic structure and magnetic moments	M_s (μ_B/f.u.)	T_c (K)	M_{eff} (μ_B/3d atom)	θ (K)	μ_0H_a (T)	References
$Pr_3Co_{13}B_2$	FM, T = 4 K, eam ‖ **c** Co $M_{4h} = M_{3g} = 1.3\ \mu_B$; $M_{6i} = 0.6\ \mu_B$	20.9[a]	360				[99C3]
$Pr_3Co_{13}B_2$ (BS)	FM Pr: $M_{1a} = 2.88\ \mu_B$, $M_{2e} = 2.76\ \mu_B$ Co: $M_{4h} = 1.60\ \mu_B$, $M_{6i} = 0.66\ \mu_B$, $M_{3g} = 1.64\ \mu_B$ B: $M_{2c} = -0.04\ \mu_B$	23.56					[06S1]
$Nd_3Co_{13}B_2$	FM, basal plane $T < T_{sR} = 370$ K Co: $M_{4h} = 1.2\ \mu_B$, $M_{6i} = 0.6\ \mu_B$, $M_{3g} = 0\ \mu_B$	20.8[a]	710			$\mu_0H_a{}^i = 18$	[00C6]
$Nd_3Co_{13}B_2$ (BS)	FM Nd: $M_{1a} = 1.85\ \mu_B$, $M_{2e} = 2.18\ \mu_B$ Co: $M_{4h} = 1.53\ \mu_B$, $M_{6i} = 0.81\ \mu_B$, $M_{3g} = 1.60\ \mu_B$ B: $M_{2c} = -0.04\ \mu_B$	21.91					[04K1]
$Nd_3Co_{13}B_2$ (ND)	FM, T = 295 K, θ = 61(6)° [c] Nd: $M_{1a} = 1.7(2)\ \mu_B$, $M_{2e} = 1.5(1)\ \mu_B$ Co: $M_{4h} = 1.35(6)\ \mu_B$, $M_{6i} = 0.89(7)\ \mu_B$, $M_{3g} = 0.7(1)\ \mu_B$	17.52[d]				$K_1{}^i = -42.4$ MJ/m^3 $K_2{}^i = 19.6$ MJ/m^3	[07P1]

(continued)

Table 8.34 (continued)

Compound	Magnetic structure and magnetic moments	M_s (μ_B/f.u.)	T_c (K)	M_{eff} (μ_B/3d atom)	θ (K)	μ_0H_a (T)	References
$Nd_3Co_3Ni_{10}B_2$ (ND)	FM, T = 1.5 K, θ = 55(6)° [c] Nd: M_{1a} = 2.0(1) μ_B, M_{2e} = 2.2(1) μ_B Co: M_{4h} = M_{6i} = M_{3g} = 0.5(1) μ_B	12.9[e]					[07P1]
$Nd_3Co_3Ni_{10}B_2$ (ND)	FM, T = 100 K, θ = 36(6)° [c] Nd: M_{1a} = 1.2(2) μ_B, M_{2e} = 1.3(1) μ_B Co: M_{4h} = M_{6i} = M_{3g} = 0.4(1) μ_B	9.0[f]					[07P1]
$Nd_3Co_{13}B_2$	FM	20.8[a]	720 370 (T_{sR})			$\mu_0H_a^i = 18$	[00C8]
$Nd_3Co_{12}NiB_2$	FM	19.7[a]	660 340 (T_{sR})			$\mu_0H_a^i = 25$	[00C8]
$Nd_3Co_{11}Ni_2B_2$	FM	18.3[a]	600 300 (T_{sR})			$\mu_0H_a^i = 34$	[00C8]
$Nd_3Co_{11}Ni_2B_2$	FM	17.2[b]	600 293(T_{sR})				[05P2]
$Nd_3Co_{10}Ni_3B_2$	FM	17.4[a]	520 270(T_{sR})				[00C8]
$Nd_3Co_8Ni_5B_2$	FM	15.6[a]	420 210(T_{sR})				[00C8]
$Nd_3Co_8Ni_5B_2$	FM	14.7[b]	425 205(T_{sR})			$K_1^i = -17.7(3)$ MJ/m^3 $K_2^i = -8.4(2)$ MJ/m^3	[14P1]

(continued)

Table 8.34 (continued)

Compound	Magnetic structure and magnetic moments	M_s (μ_B/f.u.)	T_c (K)	M_{eff} (μ_B/3d atom)	θ (K)	$\mu_0 H_a$ (T)	References
$Nd_3Co_6Ni_7B_2$	FM	13.0[b]	308 182(T_{sR})				[05P2]
$Y_3Co_{13}B_2$ (BS)	FM Y: $M^s_{1a} = -0.12\ \mu_B$, $M^o_{1a} = 0\ \mu_B$, $M^s_{2c} = -0.21\ \mu_B$, $M^o_{2c} = 0.03\ \mu_B$ Co: $M^s_{4h} = 1.36\ \mu_B$, $M^o_{4h} = 0.30$ μ_B, $M^s_{6i} = 0.51\ \mu_B$ $M^o_{6i} = 0.04\ \mu_B$, $M^s_{3g} = 1.33\ \mu_B$, $M^o_{3g} = 0.22\ \mu_B$		636 [07B1]				[14P1]
$Y_3Co_{3.06}Ni_{9.94}B_2$ (BS)	Y: $M^s_{1a} = -0.02\ \mu_B$, $M^o_{1a} = 0\ \mu_B$, $M^s_{2c} = -0.08\ \mu_B$, $M^o_{2c} = 0.01\ \mu_B$ Co: $M^s_{4h} = 1.30\ \mu_B$, $M^o_{4h} = 0.50$ μ_B, $M^s_{6i} = 0.26\ \mu_B$ $M^o_{6i} = 0.03\ \mu_B$, $M^s_{3g} = 1.25\ \mu_B$, $M^o_{3g} = 0.20\ \mu_B$ Ni: $M^s_{4h} = 0.20\ \mu_B$, $M^o_{4h} = 0.02$ μ_B, $M^s_{6i} = 0.04\ \mu_B$ $M^o_{6i} = 0.0\ \mu_B$, $M^s_{3g} = 0.28\ \mu_B$, $M^o_{3g} = 0.03\ \mu_B$						[14P1]
$Y_3Co_5Ni_8B_2$[g]	FM	3.63[b]	241				[05P1]

(continued)

Table 8.34 (continued)

Compound	Magnetic structure and magnetic moments	M_s (μ_B/f.u.)	T_c (K)	M_{eff} (μ_B/3d atom)	θ (K)	μ_0H_a (T)	References
$Y_3Co_3Ni_{10}B_2$[g]	FM	2.39[b]	170				
$Y_3CoNi_{12}B_2$	FM, $\chi = \chi_0 + C/(T-\theta)$, $\chi_0 = 7.8.10^{-6}$ m^3/kg	1.56[b]	106 92(T_{WF})[h]	1.67	76		
$Y_3Co_{0.5}Ni_{12.5}B_2$	FM, $\chi = \chi_0 + C/(T-\theta)$, $\chi_0 = 1.3.10^{-6}$ m^3/kg	1.08[b]	94 86 (T_{WF})[h]	1.16	76		
$Y_3Ni_{13}B_2$	$\chi = \chi_0 + C/(T-\theta)$, $\chi_0 = 3.10^{-6}$ m^3/kg	0.25[b]	68 (T_{WF})[h]	0.71	79		
$Y_3Ni_{13}B_2$ (BS)	FM, Y: $M^s_{1a} = -0.01\ \mu_B$, $M^o_{1a} = 0\ \mu_B$, $M^s_{2c} = -0.02\ \mu_B$, $M^o_{2c} = 0.0\ \mu_B$ Ni: $M^s_{4h} = 0.12\ \mu_B$, $M^0_{4h} = 0.01\ \mu_B$, $M^s_{6i} = 0.01\ \mu_B$ $M^o_{6i} = 0.0\ \mu_B$, $M^s_{3g} = 0.20\ \mu_B$, $M^o_{3g} = 0.02\ \mu_B$						[14P1]
$Pr_5Co_{19}B_6$	FM, eam in plane Pr: $M_b = M_{e1} = M_{e2} = 2.4\ \mu_B$ Co: $M_{h1} = 1.3\ \mu_B$ $M_{i1} = M_{i2} = 0.5\ \mu_B$, $M_f = 0\ \mu_B$	23.7[a]	380			$\mu_0H_a{}^i = 25.0$ $\mu_0H_a = 4.5$ (300 K)	[00C7]

(continued)

Table 8.34 (continued)

Compound	Magnetic structure and magnetic moments	M_s (μ_B/f.u.)	T_c (K)	M_{eff} (μ_B/3d atom)	θ (K)	$\mu_0 H_a$ (T)	References
$Pr_5Co_{19}B_6$ (BS)	FM Pr: $M_b = 2.84\ \mu_B$ $M_{e1} = 3.03\ \mu_B$, $M_{e2} = 2.89\ \mu_B$ Co: $M_{h1} = 1.70\ \mu_B$, $M_{i1} = 0.44\ \mu_B$, $M_{i2} = 0.85\ \mu_B$, $M_f = 0.22\ \mu_B$ B: $M_d = -0.07\ \mu_B$, $M_{h2} = -0.03\ \mu_B$	26.62					[06S1]
$Nd_5Co_{19}B_6$	FM, eam in plane Nd: $M_b = M_e = 2.4\ \mu_B$ Co: $M_{h1} = 1.3\ \mu_B$, $M_i = 0.6\ \mu_B$, $M_f = 0\ \mu_B$	21.5	380			$\mu_0 H_a{}^i = 4.5 \bullet 10^{-4}$	[99C3]
$Nd_5Co_{19}B_6$ (BS)	FM Nd: $M_b = 2.92$, $M_{e1} = 3.07$, $M_{e2} = 2.98\ \mu_B$ Co: $M_{h1} = 1.65\ \mu_B$, $M_{i1} = 0.40\ \mu_B$, $M_{i2} = 0.80\ \mu_B$, $M_f = 0.21\ \mu_B$ B: $M_d = -0.05\ \mu_B$, $M_{h2} = -0.02\ \mu_B$	29.46					[06S2]
$Sm_5Co_{19}B_6$	M ‖ **c**		475				[08M2]
$Nd_5Co_{21}B_4$	FM, eam in plane	31.1[a]	570			$\mu_0 H_a{}^i = 86.1$	[01C5]

(continued)

Table 8.34 (continued)

Compound	Magnetic structure and magnetic moments	M_s (μ_B/f.u.)	T_c (K)	M_{eff} (μ_B/3d atom)	θ (K)	μ_0H_a (T)	References
$Nd_5Co_{21}B_4$ (BS)	FM Nd: $M_b = 2.98\ \mu_B$, $M_{e1} = 2.84\ \mu_B$, $M_{e2} = 2.93\ \mu_B$ Co: $M_{h1} = 1.18\ \mu_B$, $M_{i1} = 0.88\ \mu_B$ $M_{i2} = 0.46\ \mu_B$, $M_f = 0.63\ \mu_B$, $M_d = 1.17\ \mu_B$ B: $M_{h1} = -0.08\ \mu_B$	31.53					[06S2]

[a] $T = 5$ K, $\mu_0H = 6.5$ T
[b] $T = 5$ K, $\mu_0H = 5$ T
[c] Canting angle
[d] $T = 295$ K, $\mu_0H = 5$ T
[e] $T = 1.5$ K, $\mu_0H = 5$ T
[f] $T = 100$ K, $\mu_0H = 5$ T
[g] YCo_5 and YCo_4B impurities
[h] Week ferromagnetism
[i] $T = 5$ K

site dependence of cobalt moments were also evaluated from the mean magnetic moment of cobalt sublattices obtained by subtracting that of R elements from the total magnetizations, in case of $Pr_3Co_{13}B_2$ [99C3] or $Nd_3Co_{13}B_2$ [99C3, 00C6] borides. The magnetic contribution of Co3*g*, that has B layers just above and below containing plane, was assumed to be zero. The nearest neighbor environment of Co4*h*, being not changed as compared to the corresponding site in RCo_5, was considered to keep the same value, of 1.2 μ_B, as in $NdCo_5$ [90B1, 96I1] or 1.3 μ_B in $PrCo_5$ [68V1]. A cobalt magnetic moment, $M_{6i} = 0.6\ \mu_B$/atom, has been obtained—Table 8.34. The Co3d bands dominate the density of states near the Fermi level, E_F. The R and B atoms provide very small contributions to the DOS, at E_F [04K1, 06S1].

The anisotropy of Pr-sublattice in $Pr_3Co_{13}B_2$, is negative, favoring planar orientation of magnetization. The uniaxial Co sublattice anisotropy overcomes that of Pr, the e.a.m. in $Pr_3Co_{13}B_2$ being parallel to the **c**-axis in all the temperature range ($T < T_c$) [99C3]. In $Nd_3Co_{13}B_2$, there is a higher negative contribution of the Nd sublattice to the anisotropy, at low temperatures (that arises from the coupling between Nd ion orbital magnetic moment and the crystal electric field), than that of cobalt. As the temperature increases, the crystal-field induced Nd sublattice anisotropy decreases, due to strong temperature dependence of K_{1Nd} constant, proportional with $O_2^0 = \langle 3J_z^2 - J(J+1)\rangle$ term. As a result, there is a spin reorientation, at temperature $T_{sR} = 370$ K [00C6, 00C8] or 395 K [05P2], from easy plane to **c**-direction. The electrical resistivity, of $Nd_3Co_{13}B_2$ compound, follows a T^2 dependence, at low temperatures [04S2]. At higher temperatures, the resistivity has a non-linear temperature dependence, which indicates an electron–phonon interaction in the presence of small s-d scattering.

In $Nd_3Ni_{13-x}Co_xB_2$ series, both Curie temperatures and saturation magnetizations decrease gradually, as the Ni content increases [00C8, 05P2, 07B1, 07P1]—Fig. 8.20. The spin reorientation temperatures also decrease as a result of the changes in the transition metal sublattice anisotropy, due to Ni substitution at 4*h* site. Upon cooling, the compounds with $x < 7$ order axially at T_c and undergo spin reorientations [05P2].

The neutron diffraction on $Nd_3Ni_{13-x}Co_xB_2$ series with $x = 13$ and $x = 3$, evidenced only small differences between Nd magnetic moments, at sites 1*a* and 2*e* [07P1]—Table 8.34. The refinement of the ND patterns in sample with $x = 3$ has been made assuming equal magnetic moments at the three cobalt sites. According to [07B1], the fits of the data, at $T < T_{sR}$, required a conical type magnetic structure, having a cone angle $\theta = 55°–61°$ with **c**-axis.

The low temperature heat capacity, of $Nd_3Ni_{13-x}Co_xB_2$ ($0 \leq x \leq 13$) series has been analyzed in the temperature range 350 mK $\leq T \leq 300$ K [09A2]. The electronic heat capacity contribution revealed an increase of DOS at E_F as the Co content increased. The substitution of Ni by Co in the composition range $1 \leq x \leq 3$ induced interaction disorder on the Nd sublattice.

The magnetic properties of $R_3Co_{13}B_2$ compounds with R = Sm, Gd, Dy, Ho, Er [01L1, 08M2] or R = Sm [00C5] were little investigated. These samples are difficult to be obtained in pure state. All the reports, mentioned the presence of a large amount of additional phases.

The $R_5Co_{19}B_6$ compounds, with R = Pr [00C7, 06S1] and R = Nd [99C3, 06S2], are ferromagnetically ordered. The Curie temperatures are significantly lower than

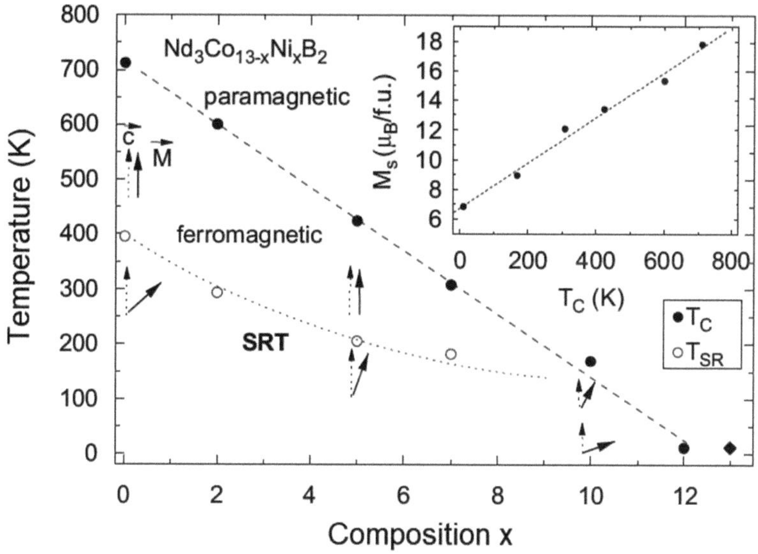

Fig. 8.20 $Nd_3Co_{13-x}Ni_xB$: magnetic phase diagram [05P2]. The plot of the spontaneous magnetizations as function of Curie temperatures is given in inset [07P1]. By SRT is denoted the composition dependence of spin reorientation temperature

those of the $R_3Co_{13}B_2$ series. Both compounds have planar anisotropy, no spin reorientations being detected up to T_c. In a model based on the correspondence between Co sites in RCo_5 series and those in $R_5Co_{19}B_6$ compounds, the magnetic moments at the cobalt sites were estimated [99C3, 00C6]—Table 8.34. These are somewhat different from those obtained from band structure calculations [06S1, 06S2]. The magnetocrystalline anisotropies of cobalt sublattices were analysed in the model of individual site anisotropy [79S2]. At T = 5 K, anisotropy fields of $\mu_0H_a = 25$ T (R = Pr) and $4.5.10^{-4}$ T (R = Nd) were reported [99C3, 00C7].

The $Sm_5Co_{19}B_6$ has been not obtained as single phase [08M2]. The Curie temperature was estimated at $T_c = 475$ K. The easy axis of magnetization is along **c**-direction.

The $Nd_5Co_{21}B_4$ compound has been synthesized by melt spinning [01C5]. The boride is ferromagnetically ordered. The Nd sublattice anisotropy is negative. Substitution of B for Co at sites related to 2*c*, severely weakens the positive contribution at the Co sublattice to the total anisotropy, the compound having only planar anisotropy. The anisotropy field at T = 5 K was estimated at $\mu_0H_a = 86.1$ T. The band structure calculations evidenced, as in other R-Co-B compounds, the importance of local environments in determining the cobalt moments [06S2]—Table 8.34. The largest moments were found at the Co4h_1 and Co2*d* sites (1.18 μ_B and 1.17 μ_B, respectively).

8.8 $RCo_{12}B_6$ Compounds

The compounds belonging to $RCo_{12}B_6$ ternary system have been firstly identified [72N1] and their crystal structures investigated [72N1, 77C1, 80B1, 80J1, 81K2, 85B4]. The compounds crystallize in a rhombohedral type structure, of $SrNi_{12}B_6$-type, having $R\overline{3}m$—space group [80J1]. The atomic sites are given in Table 8.35 and the lattice parameters are listed in Table 8.36. The $RCo_{12}B_6$ compounds are stable for essentially all rare-earths (with exception of europium and probable ytterbium and lutetium [15M2]). The lattice parameters follow the lanthanide contraction [89M1].

The cobalt atoms are located in two inequivalent lattice sites (18*h*, 18*g*), the R atoms at site 3*a* and B at 18*h* sites. The local environment of Co18*g* atoms in $YCo_{12}B_6$, consists of three Co18*g* at ≅ 0.253 nm, four Co18*h* at ≅ 0.253 nm, four B at ≅ 0.204 nm and two R neighbors at ≅ 0.324 nm. The total number of neighbors of the Co18*h* site is also 13, but there one finds only one R atom at ≅ 0.321 nm, four Co18*g* at ≅ 0.250 nm, five Co18*h* at ≅ 0.250 nm and three B around 0.203 nm [88E1].

The R_{1-x} $R'_xCo_{12}B_6$ solid solutions are formed in a limited composition ranges, for $x \leq 0.5$ (R = La, R' = Gd) [97L2], $x \leq 0.9$ (R = Gd, R' = Y), $x \leq 0.8$(R = Tb, R' = Y), $x \leq 0.45$ (R = Dy, Ho, R' = Y) and $x \leq 0.4$ (R = Er, R' = Y) [92Z5]. Substitution of cobalt by iron in $RCo_{12-x}Fe_xB_6$ series allowed the formation of solid solutions in all the composition range when R = La, at $x \leq 2$ (R = Ce) [08M5], $x \leq 1.2$ [11Z1] or $x \leq 2.3$ [18M1] (R = Nd) and $x \leq 3$ [13D1, 13D2, 13L1] (R = Gd).

The experimental [92R1, 93C2] as well as DFT calculation [11M1, 16L1] showed that the Fe atoms prefer to replace Co at 18*h* site. The lattice parameters in $NdCo_{12-x}Fe_xB_6$ series, initial increase along **c**-axis more rapid than in the basal plane, as Co is replaced by Fe [18M1]. This anisotropic expansion of the unit cell seems to be the result of strong preferential location of Fe for 18*h* site [93C2, 02C1], the atomic size of Fe being larger than of Co. Such a structural evolution may be at the origin of limited solubility of Fe in $RCo_{12}B_6$ compounds. The substitution of Co by Ti has been theoretically analyzed in $GdCo_{12-x}Ti_xB_6$ series [15Z1]. The 18*h* site is also preferred by Ti atoms. It was suggested that Ti can stabilize the $R\overline{3}m$ type structure.

The $RCo_{12}B_6$ compounds, with nonmagnetic rare earths, as R = La [89M1, 91N1, 94W2, 97L1, 11M1, 13D2, 14D1] and R = Y [72N1, 87J3, 88E1, 88R1, 89M1, 91N1,

Table 8.35 Atomic sites in $RCo_{12}B_6$ compounds, $R\overline{3}m$ space group [14D1]

Atom	Sites	x	y	z	Atomic environment
R	3*a*	0	0	0	12Co18*g*,6Co18*h*,6B
Co	18*g*	0.36840	0	1/2	3Co18*g*,4Co18*h*,4B,2R
Co	18*h*	0.42380	0.57620	0.03550	4Co18*g*,5Co18*h*,3B,1R
B	18*h*	0.47547	0.52453	0.29123	4Co18*g*,3Co18*h*,1R

Table 8.36 Space groups and lattice parameters of $RCo_{12}B_6$ compounds

Compound	T (K)	Space group	Lattice parameters (nm)		References
			a	c	
$LaCo_{12}B_6$	2	$R\bar{3}m$	0.9490(2)	0.7487(2)	[14D2]
	200	$R\bar{3}m$	0.9499(2)	0.7489(3)	[14D2]
	300	$R\bar{3}m$	0.9530(2)	0.7498(1)	[14D2]
$LaCo_{12}B_6$	RT	$R\bar{3}m$	0.9534	0.7512	[94W2]
$CeCo_{11}B_6$	RT	$R\bar{3}m$	0.9515	0.7458	[87J3]
$CeCo_{12}B_6$	RT	$R\bar{3}m$	0.9469	0.7433	[80J1]
$CeCo_{12}B_6$	RT	$R\bar{3}m$	0.9479	0.7426	[89M1]
$PrCo_{12}B_6$	RT	$R\bar{3}m$	0.9485	0.7468	[89M1]
$PrCo_{12}B_6$	RT	$R\bar{3}m$	0.9484	0.7475	[93C3]
$PrCo_{12}B_6$	RT	$R\bar{3}m$	0.9586	0.7475	[87C2]
$PrCo_{12}B_6$	RT	$R\bar{3}m$	0.9502	0.7483	[87J3]
$NdCo_{12}B_6$	RT	$R\bar{3}m$	0.9489	0.7475	[89M1]
$NdCo_{12}B_6$	RT	$R\bar{3}m$	0.9493	0.7476	[87J3]
$NdCo_{12}B_6$	RT	$R\bar{3}m$	0.9502	0.7471	[83B2, 99C2]
$SmCo_{12}B_6$	RT	$R\bar{3}m$	0.9432(6)	0.7482(5)	[80B1]
$SmCo_{12}B_6$	RT	$R\bar{3}m$	0.9471	0.7458	[87J3]
$GdCo_{12}B_6$	4	$R\bar{3}m$	0.9435(7)	0.7442(5)	[13D2]
	200	$R\bar{3}m$	0.9444(7)	0.7437(5)	[13D2]
$GdCo_{12}B_6$	RT	$R\bar{3}m$	0.9454	0.7453	[89M1]
$GdCo_{11.5}Fe_{0.5}B_6$	4	$R\bar{3}m$	0.9440(7)	0.7443(5)	[13D2]
	200	$R\bar{3}m$	0.9446(7)	0.7436(5)	[13D2]
$GdCo_{12}B_6$	RT	$R\bar{3}m$	0.9454	0.7435	[88R1]
$TbCo_{12}B_6$	RT	$R\bar{3}m$	0.9454	0.7453	[81K2]
$TbCo_{12}B_6$	RT	$R\bar{3}m$	0.9454	0.7440	[89M1]
$DyCo_{12}B_6$	RT	$R\bar{3}m$	0.9444	0.7439	[89M1]
$DyCo_{12}B_6$	RT	$R\bar{3}m$	0.9455	0.7451	[87J3]
$HoCo_{12}B_6$	1.5	$R\bar{3}m$	0.9404(1)	0.7416(2)	[14D1]
	78	$R\bar{3}m$	0.9404(1)	0.7414(2)	
	200	$R\bar{3}m$	0.9410(2)	0.7414(3)	
	RT	$R\bar{3}m$	0.9451(2)	0.7454(3)	
$HoCo_{12}B_6$	RT	$R\bar{3}m$	0.9449	0.7439	[89M1]
$ErCo_{12}B_6$	RT	$R\bar{3}m$	0.9438	0.7431	
$TmCo_{12}B_6$	RT	$R\bar{3}m$	0.9435	0.7417	
$YCo_{12}B_6$	RT	$R\bar{3}m$	0.9443	0.7435	
$YCo_{12}B_6$	RT	$R\bar{3}m$	0.9437(2)	0.7434(1)	[14D2]

(continued)

Table 8.36 (continued)

Compound	T (K)	Space group	Lattice parameters (nm)		References
			a	c	
$YCo_{12}B_6$	RT	$R\bar{3}m$	0.9435	0.7435	[80J1]
$YCo_{12}B_6$	RT	$R\bar{3}m$	0.9453	0.7450	[87J3]
$YCo_{12}B_6$	RT	$R\bar{3}m$	0.9443	0.7435	[88R1, 95N1]

92R1, 93C1, 97L1, 10A1, 11M1, 12A1, 15M2] are ferromagnetically ordered—Table 8.37. The Curie temperatures, as well as the mean cobalt moments, are rather low and little dependent on R partner. The above features were correlated with a relatively small number of cobalt atoms, in the unit cell, with respect to boron ones and to the strong hybridization of Co3d-B2p electron states, due to small distances between these atoms [89M1]. The cerium, in $CeCo_{12}B_6$ compound, is in mixed valence state, close to Ce^{4+} and thus there is a slight diminution of both saturation magnetization, M_s and of Curie temperature, T_c, comparatively to those determined in compounds with nonmagnetic elements as La and Y [87J3, 89M1, 92K2, 08M5, 08S1, 10A1, 11M1, 12A1].

The mean effective cobalt moments, determined in $RCo_{12}B_6$ (R = La, Ce, Y) compounds are higher than those determined from saturation magnetization [89M1]—Table 8.37. The ratio $r = S_p/S_o$ between number of spins determined from effective cobalt moments, S_p and saturation ones, S_o, is in the range 2 to 3, evidencing a relative high degree of itinerancy. The moderate decrease with pressure of the magnetizations and Curie temperatures of $RCo_{12}B_6$ compounds (R = La, Y, Ce) [08M5, 10A1, 14D2], has been also correlated with the high degree of hybridization of Co3d and B2p states. The variation with pressure of the Curie temperatures as well as of saturation magnetizations are higher in $LaCo_{12}B_6$ than in $YCo_{12}B_6$ [14D2] and correlated with a higher volume of $LaCo_{12}B_6$ elementary cell ($\cong$ 2%) than that of $YCo_{12}B_6$ compound [14D2]—Table 8.38. The somewhat higher sensitivity of cobalt moment to pressure in $CeCo_{12}B_6$ can be attributed to the pressure induced shift of cerium valence toward the tetravalent state.

The ^{59}Co NMR studies, performed on $RCo_{12}B_6$ compounds with R = La, Y [91N1] and R = Ce, Pr, Nd [92K2] evidenced the presence of complex spectra corresponding for cobalt located in domain (Co^d) and domain walls (Co^w), respectively. The hyperfine fields at Co18*h* sites (Co^d (H)) are larger than those at 18*g* (Co^d(L)) ones—Table 8.39. A discontinuous shift of the low frequency ^{59}Co domain signal (Co^d (L)) to Co^d (L') was reported in compounds with R = La, Pr and Gd over the 25 K $\leq T \leq$ 40 K temperature range [91N1, 92K2]. Neither a jump in resonance frequency nor unusual behavior of the intensity as a function of temperature was shown when R = Y. The specific heat of $LaCo_{12}B_6$ compound evidenced no anomalies in their temperature dependence [91N1]. In case of $GdCo_{12}B_6$ the anomalous behavior can be correlated with the compensation of magnetization at $T_{comp.}$ = 50 K. No reasonable explanations are now for the unusual behavior of $RCo_{12}B_6$ compounds with R = La or Pr. From the ^{57}Co effective fields, determined in $YCo_{12}B_6$, the magnetic

Table 8.37 Magnetic properties of $RCo_{12}B_6$ compounds

Compound	Magnetic structure and magnetic moments	M_s (μ_B/f.u.)	T_c, T_{comp} (K)	C (emuK/f.u.)	M_{eff} (μ_B/Co atom)	References
$LaCo_{12}B_6$	FM	5.41[a]	160.4	5.57	1.93	[89M1]
$LaCo_{12}B_6$ (ND)	T = 2 K, conical structure, $\mathbf{q} = (0,0,0.0643(15))$ Co: $M_g = 0.55(10)\ \mu_B$, $\theta^{b)} = 60.6(4.0)^o$ $M_h = 0.84(8)\ \mu_B$, $\theta^{b)} = 60.6(4.0)^o$	5.42(5)[c]	160(1)			[14D2]
$LaCo_{12}B_6$ (BS)	FM La: $M = -0.04\ \mu_B$ Co: $M_g = 0.38\ \mu_B$, $M_h = 0.84\ \mu_B$ B: $M = -0.04\ \mu_B$	6.40				[11M1]
$CeCo_{12}B_6$	FM	5.4[a]	154			[87J3]
$CeCo_{12}B_6$	FM	4.56[a]	134.3	3.14	1.45	[89M1]
$CeCo_{12}B_6$ (BS)	FM Ce: $M = -0.21\ \mu_B$ Co: $M_g = 0.44\ \mu_B$, $M_h = 0.90\ \mu_B$ B: $M = -0.04\ \mu_B$	6.51				[11M1]
$PrCo_{12}B_6$	FM	8.4[a]	177			[87J3]
$PrCo_{12}B_6$	FM	8.11[a]	157.7	6.52	1.81	[89M1]
$NdCo_{12}B_6$	FM,	7.68[c]	155 47 (T_{comp})			[18M1]
$NdCo_{12}B_6$	FM	8.21[a]	159	5.12	1.52	[89M1]
$SmCo_{12}B_6$	FM e.a.m. in basal plane	6.8[a]	172			[87J3]
$SmCo_{12}B_6$	FM	7.08[a]	157.5	6.07	2.00	[89M1]
$GdCo_{12}B_6$	FIM,	2.1[a]	169 46 (T_{comp})			[87J3]

(continued)

Table 8.37 (continued)

Compound	Magnetic structure and magnetic moments	M_s (μ_B/f.u.)	T_c, T_{comp} (K)	C (emuK/f.u.)	M_{eff} (μ_B/Co atom)	References
$GdCo_{12}B_6$	FIM,	1.36[a]	162 48.2 (T_{comp})			[89M1]
$GdCo_{12}B_6$	FIM	1.79[d]	161			[88R1]
$GdCo_{12}B_6$ (ND)	FIM, T = 4 K, $\theta \cong 38(8)^o$, Gd: M = 6.9(5)μ_B Co: $M_g = -0.41(3)\ \mu_B$, $M_h = -0.50(3)\ \mu_B$	1.68[e]	158 47.5 (T_{comp})			[13D2]
$GdCo_{12}B_6$ (BS-PAW)	FIM, Gd: M = 7.08μ_B Co: $M_g = -0.34\ \mu_B$, $M_h = -0.79\ \mu_B$	0.30				[16L1]
$GdCo_{12}B_6$ (BS-LMTO)	FIM, Gd: M = 7.44μ_B Co: $M_g = -0.24\ \mu_B$, $M_h = -0.74\ \mu_B$	1.56				[16L1]
$GdCo_{11.5}Fe_{0.5}B_6$ (ND)	FIM, T = 4 K, $\theta = 90°$, Gd: M = 6.9(5) μ_B M: $M_g = -0.22(5)\ \mu_B$, $M_h = -0.71(7)\ \mu_B$	1.30[e]	155[f] 45 (T_{comp})			[13D2]
$GdCo_{11}FeB_6$	Fe: M_h = 1.03 μ_B	1.13(5)	148			[16L1]
$GdCo_{11}FeB_6$ (BS-PAW)	FIM Gd: M = 7.04 μ_B Co: $M_g = -0.27\ \mu_B$, $M_h = -0.65\ \mu_B$ Fe: M = -1.40 μ_B	0.77				[16L1]
$GdCo_{11}FeB_6$ (BS-LMTO)	FIM Gd: M = 7.45 μ_B Co: $M_g = -0.23\ \mu_B$, $M_h = -0.70\ \mu_B$ Fe: M = -1.53 μ_B	1.04				[16L1]

(continued)

Table 8.37 (continued)

Compound	Magnetic structure and magnetic moments	M_s (μ_B/f.u.)	T_c, T_{comp} (K)	C (emuK/f.u.)	M_{eff} (μ_B/Co atom)	References
$GdCo_{12-x}Fe_xB_6$	FIM[f]					[13D1]
x = 0		1.68[e]	165			
			49 (T_{comp})			
x = 0.5		1.30[e]	161			
			47.5 (T_{comp})			
x = 1		1.13[e]	153			
			46.5 (T_{comp})			
x = 2		0.99[e]	128			
			48.6 (T_{comp})			
x = 3		0.89[e]	95			
			53.5 (T_{comp})			
$TbCo_{12}B_6$	FIM, e.a.m. in basal plane [16L1],	3.2[a]	154.4 82.8 (T_{comp})			[89M1]
$TbCo_{12}B_6$	FIM	3.8 [g]				[92Z5]
$DyCo_{12}B_6$	FIM	5.9[a]	165 72 (T_{comp})			[87J3]
$DyCo_{12}B_6$	FIM	4.2[a]	145.8 73 (T_{comp})			[89M1]
$HoCo_{12}B_6$	FIM		146(2) 75(2) (T_{comp})			[02C1]
$HoCo_{12}B_6$	FIM	4.06[a]	146 44.3 (T_{comp})			[89M1]
$HoCo_{12}B_6$ (ND)	FIM, T = 1.5 K, **q** = (0,0,0), e.a.m. ‖ **c**, at $T < T_{SR} = 76$ K Ho: M = 8.89(6) μ_B Co: $M_g = -0.14(2)\ \mu_B$, $M_h = -0.63(2)\ \mu_B$	3.69[e]	147			[14D3]

(continued)

Table 8.37 (continued)

Compound	Magnetic structure and magnetic moments	M_s (μ_B/f.u.)	T_c, T_{comp} (K)	C (emuK/f.u.)	M_{eff} (μ_B/Co atom)	References
$HoCo_{12}B_6$ (ND)	FIM, T = 78 K, **M**⊥**c**, **q** = (0, 0, 0.05373(6)) Ho: M = 2.29(21) μ_B Co: $M_g = 0\ \mu_B$, $M_h = -0.63(8)\ \mu_B$					[14D3]
$ErCo_{12}B_6$	FIM	3.54[a]	145.1 25.4 (T_{comp})			[89M1]
$ErCo_{12}B_6$	FIM	4.2[g]				[92Z5]
$TmCo_{12}B_6$	FIM	1.7[a]	143.9 11.66 (T_{comp})			[89M1]
$YCo_{12}B_6$	FM	6.5[a]	171			[87J3]
$YCo_{12}B_6$	FM	5.16[a]	148.5	5.96	1.99	[89M1]
$YCo_{12}B_6$	FM	5.32[a]	151.5			[88R1]
$YCo_{11}FeB_6$ (BS)	FM Y: M = −0.03 μ_B Co: $M_g = 0.35\ \mu_B$, $M_h = 0.82\ \mu_B$ B: M = −0.04 μ_B	6.22				[11M1]
$YCo_{12}B_6$ (NMR)	Co: $M_g = 0.30(9)\ \mu_B$, $M_h = 0.69\ \mu_B$					[88E1]

[a] T = 4 K, $\mu_0H = 2$ T
[b] θ angle of moments with **c**-axis
[c] T = 5 K, $\mu_0H = 5$ T
[d] T = 4.2 K, $\mu_0H = 5$ T
[e] T = 17 K, $\mu_0H = 10$ T
[f] Little different values for T_c and $T_{comp.}$ even for the same composition [13D1, 13D2]
[g] T = 4.2 K, $\mu_0H = 350$ T

Table 8.38 Pressure effects on the magnetic properties of $RCo_{12}B_6$ compounds

Compound	T_c (K)	dT_c/dp (K/GPa)	$T^{a)}$ (K)	dM/dp (μ_B/GPa·10^2)	dlnM/dlnv	References
$LaCo_{12}B_6$	160(1)	−4.7(2)	5	−7.6(2)	1.8(1)	[14D2]
$CeCo_{12}B_6$	138(1)	−5.2(2)	5	−14.0(4)	3.9(2)	[10A1]
$CeCo_{12}B_6$	135(1)	−5(1)	5	−15(1)		[08M5]
$CeCo_{10}Fe_2B_6$	85(1)	−5	5	−17(1)		[08M5]
$GdCo_{12}B_6$	163(2)	−5.5	5	5		[12A1]
$HoCo_{12}B_6$	147	−3.5(4)	5	−3.3		[19D1]
$YCo_{12}B_6$	150(1)	−3.9(2)	5	−7.0(2)	1.7(1)	[10A1]

[a]Temperature of measurements

Table 8.39 Nuclear magnetic resonance data

Compound	T (K)	Nucleus	Site	ν (MHz)	B_{eff} (T)	$\Delta\nu^a$ (MHz)	References
$LaCo_{12}B_6$[c]	4.2	^{59}Co	$Co^d(H)^b$ $Co^d(L')$ Co^w	73.4 39.2 26.8 24.9	7.27 3.88 2.65 2.46	2.07 1.88	[91N1]
$CeCo_{12}B_6$	4.2	^{59}Co	$Co^d(H)$ $Co^d(L')$ Co^w		6.11 3.69 1.94	1.99 1.56	[92K2]
$PrCo_{12}B_6$	4.2	^{59}Co	$Co^d(H)$ $Co^d(L')$ Co^w		7.07 3.78 2.33 2.20	2.05 1.38	[92K2]
$GdCo_{12}B_6$	4.2	^{59}Co	$Co^d(H)$ $Co^d(L')$ Co^w		7.52 5.73 2.03	1.98 0.94	[92K2]
$GdCo_{12}B_6$[c]	4.2	^{59}Co	Co18*h* Co18*g*	80.5 38 15 20	7.97(7 lines) 3.76 1.49 1.98	1.85	[88E1]
$YCo_{12}B_6$[d]	4.2	^{59}Co	Co18*h* Co18*g*	76.2 26.7 39.4 49.5	7.55(7 lines) 2.64 3.9 4.9	1.85	[88E1]
$LaCo_{12}B_6$	4.2	^{139}La		58.0 17.8	9.64 2.06	0.34	[91N1]

[a]Δν, spacing between the quadrupole-split component lines
[b]The hyperfine fields at 18*g* and 18*h* Co sites were considered to be identical, $Co^d(H)$ and $Co^d(L')$ due to cobalt located in domain and Co^w in center of domain walls
[c]Hyperfine fields for $Co^d(L)$ at T = 77 K are 3.73 T (R = Pr) and 4.05 T (R = Gd)
[d]With average hyperfine field of 3.3(1.0) T and 7.55 T for the two sites, one finds values 0.30(9) μ_B and 0.69 μ_B for magnetic moments at 18*g* and 18*h* sites, respectively

moments at Co18*g* and Co18*h* sites were estimated [88E1], by using the hyperfine coupling constant of 11.0 T/μ_B [86B1]. There is a significant larger moment at 18*h* site by a factor of 1.3, correlated with their different local environments.

The Mössbauer spectra of ^{57}Fe doped $RCo_{12}B_6$ with R = La [94W2] and R = Y [88E1, 92R1, 93C1], in the paramagnetic range, were fitted with two doublets. In the magnetic ordered state (T = 4.2 K), the outer regions of the spectra has been fitted with three sextets of nearly equal intensities, attributed to iron in 18*h* sites (h_1, h_2, h_3), suggesting the presence of a planar anisotropy. An additional sextet, with lower hyperfine field in the central region of the spectra, was attributed to iron located in 18*g* site. Their intensities are of 12–15%—Table 8.40. The presence of planar anisotropy instead of uniaxial one, as determined by other methods, was correlated with the presence of ^{57}Fe. It was shown that the axial to basal plane transition is sensitive to iron content in the sample [13D2]. The boundary for this transition lies below x = 0.1. The ^{57}Fe Mössbauer spectra of $LaCo_{12}B_6$, doped with ^{57}Fe, evidenced no anomalies in their temperature dependences.

In spite of difficulties in performing ND on heavy neutron absorbers like boron compounds, the magnetic structures of $RCo_{12}B_6$ with R = La [14D2], R = Gd [13D2] and Ho [14D3] were determined. The powder neutron diffraction investigation of $LaCo_{12}B_6$, at T = 2 K, evidenced the presence of an incommensurate conical type magnetic structure, with a propagation vector **q** = (0, 0, 0.0643(15)), with a long period along **c**-axis. The M_{Co} moments are tilted 60° away from **c**-axis. The magnetic moment of the cobalt, at 18*h* sites, is greater than at the 18*g* positions, as also suggested by NMR studies. Band structure calculations performed on $RCo_{12}B_6$ with R = La, Ce, Y [11M1] showed the same trend of cobalt moments as those determined by neutron diffraction. The magnetizations per formula unit agree with experimentally determined values, at T = 4 K, for R = La, Y, with some differences in case of $CeCo_{12}B_6$ compound.

In the $LaFe_{12}B_6$ boride, iron moments have a marked instability—Sect. 7.7. In $RCo_{12-x}Fe_xB_6$ series, there is a strong preference for iron to occupy the 18*h* sites, where it carries a significant larger moment, about twice that the moment of iron on the 18*g* site [92R1]. The fact that the 3d sublattice magnetization, in the intermediate concentration range of $LaCo_{12-x}Fe_xB_6$, show a minimum at x = 9, is partly due to preferential site occupancy. The Fe atoms go also into the 18*g* site that is less favorable for the 3d moment formation [92R1]. A similar composition dependence was shown also for the Curie temperatures, with a minimum value at x ≅ 6. The Curie temperatures decrease as the Co is substituted by Fe in $CeCo_{12-x}Fe_xB_6$ series [08M2, 08M5]. As effect of pressure both the saturation magnetization and Curie temperatures decrease—Table 8.38. The specific heat of $CeCo_{12}B_6$ compound exhibits pronounced λ-type anomaly in the vicinity of T_c.

The physical properties of $RCo_{12}B_6$ compounds with R = Y, La and Gd as specific heat [91N1, 95N1], resistivity [97L1, 98K2, 20M2], thermoelectric power [98K2] or spontaneous magneto-independence [15M2] were investigated. A logarithmic divergence, predicted by Ginzburg–Landau theory, satisfactory explains the contribution of short-range spin correlations to the specific heat above the phase transition [95N1]. The critical exponents, α, of the specific heats have almost the same values as that

Table 8.40 Data obtained by Mössbauer spectroscopy on RCo_4B-based compounds

Compound	T (K)	Nucleus	Site	δ^a (mm/s)	ΔQ (mm/s)	B_{hf} (T)	Area (%)	References
$NdCo_{12}B_6$ (0.5 wt % ^{57}Fe)	RT	^{57}Fe	*h*	0.10(1)	0.64(2)		84(5)	[93C1, 93C2]
			g	0.02(1)	0.87(2)		16(5)	
	4.2		*h*			15.3	81(5)	
			g			7.0	19(5)	
$GdCo_{11.8}Fe_{0.2}B_6$	4.2		h_1	0.08 (mean value)		14.8	85	[92R1]
			h_2			15.7		
			h_3			16.7		
			g			7.5(2.0)	15	
$HoCo_{12}B_6$ (^{57}Fe)	4.2		*h*			15.5(5)	88	[02C1]
			g			6.6(7)	12	
$HoCo_{12}B_6$ (^{57}Fe)	90		h_1, h_2, h_3				85	[02C1]
			g				15	
$YCo_{11.8}Fe_{0.2}B_6$	4.2		h_1			14.3	88	[92R1]
			h_2			15.3		
			h_3			16.5		
			g			3.6(2)	12	
$YCo_{12}B_6$ (0.5 wt % ^{57}Fe)	42		h_1	0.20(1)	0.04(2)	14.9	28(5)	[93C1]
			h_2	0.20(1)	−0.31(2)	15.5	22(5)	
			h_3	0.21(1)	−0.45(2)	16.9	32(5)	
			g	0.20(1)	0.04(2)	7.0	18(5)	
$YCo_{12}B_6$ (0.5 wt % ^{57}Fe)	RT		*h*	0.10(1)	0.58(2)		89(5)	[93C1]
			g	−0.13(1)	0.63(2)		11(5)	
$ErCo_{12}B_6$[b)]	4.2	^{166}Er			0.55(5)	78.5		[90G2]

[a]Reference to α-foil
[b]Calibrated by using ^{166}Er spectrum of $ErFe_2$

predicted by a three-dimensional Heisenberg model [97L1]—Table 8.41. The Debye temperatures and the coefficient of electronic specific heat, γ are given in Table 8.42. The thermoelectric power of $RCo_{12}B_6$ (R = Y, Gd) compounds was shown to be non-linear below the magnetic transition temperatures. The mean contribution to the non-linear behavior comes from the scattering mechanism of conduction electrons by magnetic atoms [98K2]. The temperature dependences of magnetizations of $GdCo_{12}B_6$ [15M2] and $YCo_{12}B_6$ [88R1, 15M2] in low field, showed pronounced maxima below T_c, behavior known as Hopkinson effect [889H1]. By using impedance measurements, it was shown that this effect occurs at high enough

Table 8.41 Critical exponents

Compound	Temperature range	t[a] range	α	References
$GdCo_{12}B_6$	$T > T_c = 162$ K $T < T_c = 162$ K	0.001–0.3 −0.20–0.001	−0.10(2) −0.20(2)	[97L1]
$YCo_{12}B_6$	$T > T_c = 148.5$ K $T < T_c = 148.5$ K	0.001–0.3 −0.20–0.001	−0.15(3) −0.19(2)	[97L1]
3D Heisenberg model			−0.115	[80L1]

[a]$t = (T - T_c)/T_c$

Table 8.42 Specific heat data of $RCo_{12}B_6$ compounds

Compound	Temperature range	γ (J/kgK^2)	θ_D (K)	References
$CeCo_{12}B_6$	2.5 K≤T≤300 K	0.126[a]	185	[08S1]
$GdCo_{12}B_6$	80 K≤T≤250 K	3	480	[95N1]
$YCo_{12}B_6$	50 K≤T≤250 K	4	580	[95N1]

[a]115 mJ/molK^2

frequencies near T_c [15M2], for both compounds. This phenomenon was associated with initial permeability.

The $RCo_{12}B_6$ compounds with R = Pr [87J3, 89M1], R = Nd [87J3, 89M1, 93C2, 11Z1] and R = Sm [87J3, 89M1, 96Z1, 98K4] are ferromagnetically ordered. The magnetic susceptibilities follow Curie–Weiss-type dependences, the effective cobalt moments being in the range 1.5 to 2.0 μ_B/atom [89M1]—Table 8.37.

The ^{57}Fe Mössbauer spectra of $NdCo_{12}B_6$ sample, doped with ^{57}Fe, suggested that the magnetic moments, at T = 4.2 K, are oriented along or close to the crystallographic **c**-axis and undergoes a spin reorientation, to basal plane, at $T_{SR} \cong 55$ K [93C1, 93C2]. Thus, at T = 4.2 K, the spectrum was fitted with two sextets, while at T = 90 K and 120 K, required four sextets, the 18*h* line being split into three equal area subspectra—Fig. 8.21. The gradually substitution of Co by Fe up to x = 2.3, in $NdCo_{12-x}Fe_xB_6$ series, lead to a decrease both of magnetizations and Curie temperatures [18M1]. A first order magnetization process (FOMP) was shown in magnetization curve, at T = 2 K, which could be related to the magnetocrystalline anisotropy. The relative cooling power in $NdCo_{12}B_6$ around T_c, is of 36 J/kg, in a field $\mu_0H = 1$ T [11Z1], a value comparable with that reported in $NdFe_{12}B_6$ (39 J/kg) [06Z1].

The $SmCo_{12}B_6$ boride, is ferromagnetically ordered [89M1]. As other R elements in this series, Sm moment is close to the free ion value. The magnetic moments in $SmCo_{12}B_6$, doped with ^{57}Fe, are oriented in the basal plane [96Z1]. The crystallization of $Sm_{16-x}Co_{68+x}B_6$ (x = 0–10.2 at %) and $Sm_6Co_{92-y}B_y$ (y = 10–18.2) melt spun ribbons, revealed the presence of both $SmCo_{12}B$ and $Sm_2Co_{14}B$ phases [98K4].

The $RCo_{12}B_6$ compounds with heavy rare-earths R = Gd [87J3, 88E1, 88R1, 89M1, 92K2, 92Z5, 92Z6, 97L1, 12I1, 13D1, 13D2, 13L1, 14D1, 15M2, 16L1], R

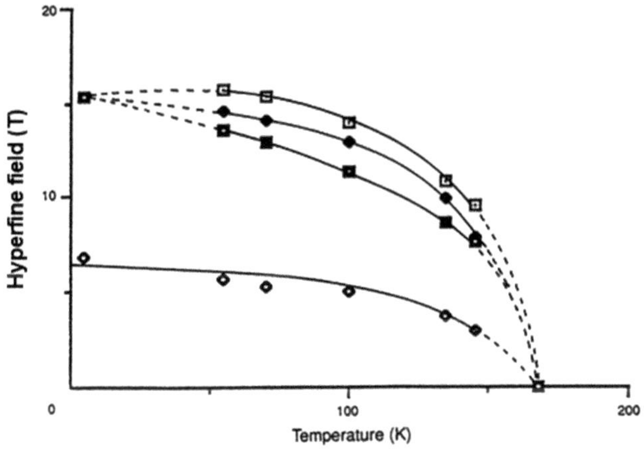

Fig. 8.21 Temperature dependences of the hyperfine fields, as determined by ^{57}Fe Mössbauer spectroscopy at ^{57}Fe doped $NdCo_{12}B_6$ [93C1]

= Tb [89M1, 92Z5, 92Z6], R = Dy [87J3, 89M1, 92Z5, 92Z6, 95F1, 97D1], R = Ho [89M1, 92Z5, 92Z6, 02C1, 14D3], R = Er [89M1, 90G2, 92Z5, 92Z6] and R = Tm [89M1, 92Z5, 92Z6] are ferrimagnetically ordered. At T = 4.2 K, the rare-earth magnetizations are higher than those of cobalt and decrease with temperature with a higher rate than the cobalt one. As a result, the magnetizations of the two sublattices compensate at temperatures T_{comp} = 45 K (Gd) [13D2], 82.8 K (Tb) [89M1], 72 K (Dy) [87J3, 89M1], 46 K (Ho) [14D3], 25.4 K (Er) [89M1] and 11.6 K (Tm) [89M1]. Some small differences, in T_{comp} values, as compared with the above mentioned ones, were sometimes reported—Table 8.37. The exchange interactions between and inside the magnetic sublattices were evaluated in the framework of mean field model [93D1].

The neutron diffraction study showed that the magnetic moments in $GdCo_{12}B_6$, at T = 4 K, are canted away from **c**-axis by θ = 38(3)° [13D2]. The cobalt moment in the 18*h* site is slightly larger than that at the 18*g* site, consistent with NMR studies [88E1, 92K2]. The ^{155}Gd hyperfine field determined by Mössbauer spectroscopy is almost due to local contribution of the Gd moment [13D2, 13L1]. Since B_{eff} is collinear with the Gd moment, their direction will correspond with e.a.m. An angle θ = 15(2)°, between **c**-axis and $\mathbf{B}_{hf}$ was obtained by fitting the experimental data. The reason for obtaining two different tilting angles, by neutron diffraction and ^{155}Gd Mössbauer spectroscopy, is not clear. The crystal field parameter $A_2^0 = 254\ Ka_0^{-2}$ of gadolinium has been determined from the quadrupole splitting [68O1, 03B1]. Since α_J is negative, the easy direction of magnetization is parallel to **c**-axis. In the $GdCo_{12-x}Fe_xB_6$ system, the axial to basal plane boundary lies below x = 0.04. Thus, the easy plane magnetization, previously reported in ^{57}Fe doped $GdCo_{12}B_6$ at T = 4.2 K [92R1], can be attributed to the presence of a small number of iron atoms in 18*h* sites [13D2, 13L1].

The easy axis of magnetizations, in $TbCo_{12}B_6$ [96Z1] and $ErCo_{12}B_6$ [95Z1] compounds, is situated in the basal plane. The presence of an easy plane anisotropy was evidenced also by ^{57}Fe Mössbauer spectroscopy on iron doped samples, although the previous results suggested an easy axis [90G2]—Table 8.40. As already mentioned, by iron doping the easy direction of magnetization can be changed. The $HoCo_{12}B_6$ exhibits axial ordering at $T = 4.2$ K and undergoes spin reorientation, to the plane, at $T_{SR} = 75$ K [02C1]. At $T = 1.5$ K, the magnetic structure of $HoCo_{12}B_6$ boride was indexed with the propagation vector $\mathbf{q} = (0, 0, 0)$, with the moments oriented close to the crystallographic **c**-axis. At $T = 78$ K, above but close to T_{SR}, the Bragg peaks can be indexed with the incommensurate propagation vector $\mathbf{q} = (0, 0, 0573(6))$, revealing a long incommensurate period along **c**-axis [14D3]—Table 8.37. The Curie temperature of $HoCo_{12}B_6$ decreases with pressure with a rate $dT_c/dp = -3.5(4)$ $KGPa^{-1}$ and the spin reorientation temperature by $dT_{SR}/dp = -4.2(4)$ $KGPa^{-1}$ [19D1].

The transitions from ferrimagnetic order to forced ferromagnetism has been analysed in $RCo_{12}B_6$ compounds with R = Gd, Tb, Dy, Ho, Er and Tm in high fields [92Z5, 92Z6, 95F1, 97D1, 12I1]. The lower field ($H < H_1$) part of magnetization isotherms can be described by an antiparallel alignment of R and Co magnetizations $M_s = |M_R - M_{Co}|$ [87J3, 88R1, 89M1, 94L1]. A non-collinear magnetic structure was found in the fields range $H_1 < H < H_2$, the magnetizations begin to turn at H_1, the field H_2 being the field necessary for imposing a parallel alignment of M_R and M_{Co} moments and thus $M_s = M_R + M_{Co}$. The intersublattice coupling constants $J_{R\text{-}Co} = (dM/dH)^{-1}$ were determined from linear part of M vs H ($H_1 < H < H_2$) dependences. In case of $GdCo_{12}B_6$, the magnetization isotherm at $T = 4$ K was found to reaches parallel orientation of the two sublattices in a field of $\mu_0H = 68$ T [12I1]. The determined magnetic moment $M_s = 12.9$ μ_B/f.u. is in good agreement with those estimated starting from ND, of 12.5 μ_B/f.u. [14D1] or from band structure calculations of 13.4(2) μ_B/f.u. [16L1]. An intersublattice exchange parameter $J_{Gd\text{-}Co} = -5.3$ K was determined. Owing that the intersublattice $J_{R\text{-}Co}$ coupling constants are large, not in all cases the parallel orientation of R and Co moments was obtained in the available fields range in ferrimagnetic $RCo_{12}B_6$ compounds. The parallel alignment of R and Co sublattices can be obtained in lower fields, by magnetic dilution of R sublattice, as for example in $Dy_{1-x}Y_xCo_{12}B_6$ system [92Z5]. The magnetic moments bend towards each other, at lower fields, as the difference between the two sublattices magnetizations become smaller. The magnitude of cobalt moment is not influenced, even in the presence of a very high field, behavior different from other borides where induced cobalt moments are present both due to increasing the exchange field or external field (Sect. 8.7). The stability of cobalt moment can be connected with the strong hybridization of Co3d and B2p states. Starting from mean field model, the exchange interactions inside and between magnetic sublattices were determined in $RCo_{12}B_6$ compounds [93D1].

The magnetic properties of $GdCo_{12-x}Fe_xB_6$ are sensitive to iron content [92R1, 13D1, 13D2, 13L1, 16L1]—Fig. 8.22. The T_c values decrease monotonously, from $T_c = 158$ K ($x = 0$) to 93 K ($x = 3$), while the compensating temperatures show a minimum near $x = 1$, then increasing [13D2]. The anisotropy of sample with $x =$

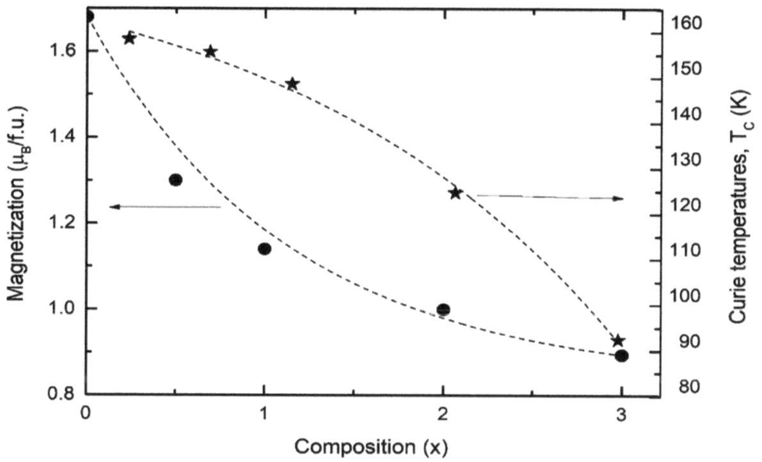

Fig. 8.22 $GdCo_{12-x}Fe_xB_6$ ($x \leq 3$): composition dependences of spontaneous magnetizations at T = 4 K and Curie temperatures [13D2]. The easy axis of magnetization is changed from axial to planar even at low iron content

0.1 is planar. The anisotropies of cobalt and iron in 18*h* and 18*g* sites have opposite sign [13L1]. The axial anisotropy contribution of 18*h* sites is weakened and even the presence of a small iron content changes their direction. The band structure calculations, performed on $GdCo_{12-x}Fe_xB_6$ series, confirmed that iron atoms prefer to substitute for Co at 18*h* site with energy gain of 0.1 eV/f.u. [16L1]. While the $GdCo_{12}B_6$ has a weak easy **c**-axis, the $GdCo_{11}FeB$ displays an easy (ab) plane. The spin reorientation induced by doping with iron, is related to the electronic structure change near E_F and different contributions of Co and the Fe to the anisotropy [16L1].

Interesting results were obtained by analyzing the transport properties of $RCo_{12}B_6$ compounds with R = Y, Gd and Ho [20M2]. The resistivities, ρ(T), were described in terms of the interplay between the electron–electron interaction and magnetic term, due to electron scattering by spin excitations. No signature of the compensation temperature was seen in ρ(T) curves of $RCo_{12}B_6$ compounds with R = Gd and Ho. The magnetoresistance (MR) at low temperatures is positive and show a curvature with increasing field. In $YCo_{12}B_6$ the magnetoresistance is almost isotropic with respect to the relative orientations of current **i** and field **H** directions. In $GdCo_{12}B_6$, the MR is larger when **H** ⊥ **i**. The experimental data were analyzed in two-current model [82C1]. According to this model two spin polarized sub-bands carry the current in parallel and spin mixing is taken into account via pseudo-resistivity term. The spin polarization ratio for the two sub-bands in $RCo_{12}B_6$ (R = Gd, Y) compounds, $\rho(\uparrow)/\rho(\downarrow) < 1$ were determined. It was suggested that a mechanism involving magnetic rare-earths is important in describing the MR data in $GdCo_{12}B_6$ [20M2]. The temperature and field dependences of Hall coefficient in $RCo_{12}B_6$ with R = Y, Gd and Ho were also analyzed in the framework of two bands model. There was present an anomalous term, proportional to the magnetization, in addition to the ordinary effect due to Lorentz curvature of the electron trajectories. In all compounds the Hall coefficient is negative at low temperatures and shows a sign reversal when increasing temperature up to T = 60 K. A sign reversal in Hall resistivity for all compounds, at low temperatures, indicates that the majority spin-up band is hole-like, whereas the minority spin sub-band is electron-like. The anomalous Hall resistivity was separated from the ordinary one. That described by the intrinsic Kappler-Luttinger term [10N1] is the most relevant contribution to the anomalous Hall resistivity while the chiral term has its maximum amplitude near T_c.

The thermoelectric power in $RCo_{12}B_6$ with R = Y, Gd compounds is not linear in the temperature range 4.2 K $\leq T \leq T_c$ [98K2]. The main contribution to the non-linear behavior comes from the scattering mechanism of conduction electrons by the magnetic atoms.

8.9 $R_2Co_{14}B$ Compounds

The $R_2Co_{14}B$ phases constitute an isostructural group of intermetallic compounds which exist for R = La, Pr, Nd, Sm, Gd, Tb and Y [85B1]. The missing compounds may be a consequence of a small size of heavy rare-earth radius, as R = Dy, Ho, Er, Tm, Yb and Lu atoms. There is no evidence that $Ce_2Co_{14}B$ is formed [85B2]. In $Ce_2Fe_{14-x}Co_xB$ series, a single phase is formed up to x = 5 [13S1] and no traces of $Ce_2Co_{14}B$ was shown by annealing Ce-Co-B ingots with various compositions [17H1]. The theoretical analysis offered encouraging potential to form $Ce_2Co_{14}B$, showing that it is mechanically and vibrationaly stable [17H1].

The $R_2Co_{14}B$ compounds crystallize in a tetragonal structure, space group *P4₂/mnm*—Fig. 7.16. The atomic sites are listed in Table 8.43 and lattice parameters in Table 8.44. In the crystal structure, the cobalt atoms are located in six types of sites: $16k_1$, $16k_2$, $8j_1$, $8j_2$, $4c$ and $4e$, the R atoms in $4f$ and $4g$ and B atoms occupy

Table 8.43 Atomic sites in $Nd_2Co_{14}B$ compound, having tetragonal structure $P4_2/mnm$ space group [85L1, 11V1]

Atom	Site	Sym	x	y	z	Atomic environment
Nd	4*f*	*m.2m*	0.14414(4)	x	0	3R16Co1B
Nd	4*g*	*m.2m*	0.72471(4)	x	0	2R16Co2B
Co	$16k_1$	1	0.72375(1)	0.06925(4)	0.37357(3)	2R10Co1B
Co	$16k_2$	1	0.46300(4)	0.13985(4)	0.32192(3)	2R10Co
Co	$8j_1$	*..m*	0.18165(8)	x	0.25336(5)	2R12Co
Co	$8j_2$	*..m*	0.40123(8)	x	0.29495(5)	3R9Co
Co	4*e*	*2.mm*	0	0	0.38420(6)	2R9Co2B
Co	4*c*	*2/m.*	0	1/2	0	4R8Co
B	4*g*	*m.2m*	0.37665(96)	x	0	6Co

Table 8.44 Space groups and lattice parameters of $R_2Co_{14}B$ compounds

Compound	T (K)	Space group	Lattice parameters (nm)		References
			a	c	
$La_2Co_{14}B$	RT	$P4_2/mnm$	0.867	1.201	[85B4]
$La_2Co_{14}B$	RT	$P4_2/mnm$	0.879	1.205	[88Z1]
$La_2Co_{14}BH_{3.8}$	RT	$P4_2/mnm$	0.886	1.219	[88Z1]
$Pr_2Co_{14}B$	RT	$P4_2/mnm$	0.8633	1.1876	[87C2]
$Pr_2Co_{14}B$	RT	$P4_2/mnm$	0.8657	1.1896	[93C4]
$Pr_2Co_{14}B$	RT	$P4_2/mnm$	0.8630	1.1870	[87P4, 87P7]
$Pr_2Co_{14}B$	RT	$P4_2/mnm$	0.8635	1.1890	[87J1]
$Pr_2Co_{14}BH_{3.5}$	RT	$P4_2/mnm$	0.872	1.194	[88Z1]
$Nd_2Co_{14}B$	RT	$P4_2/mnm$	0.8642(2)	1.1859(3)	[86H1]
$Nd_2Co_{14}B$	RT	$P4_2/mnm$	0.8657	1.1878	[86H2]
$Nd_2Co_{14}B$	RT	$P4_2/mnm$	0.8646	1.1865	[00C6]
$Nd_2Co_{14}B$	RT	$P4_2/mnm$	0.8640	1.1830	[88J3]
$Nd_2Co_{14}B$	RT	$P4_2/mnm$	0.8651	1.1871	[87P4, 87P7]
$Nd_2Co_{14}B$	RT	$P4_2/mnm$	0.863	1.182	[88Z1]
$Nd_2Co_{14}BH_{3.4}$	RT	$P4_2/mnm$	0.869	1.188	[88Z1]
$Sm_2Co_{14}B$	RT	$P4_2/mnm$	0.862	1.179	[88Z1]
$Sm_2Co_{14}BH_{3.6}$	RT	$P4_2/mnm$	0.866	1.185	[88Z1]
$Gd_2Co_{14}B$	RT	$P4_2/mnm$	0.8633	1.1779	[86H2, 00C4]
$Gd_2Co_{14}B$	RT	$P4_2/mnm$	0.861	1.176	[88Z1]
$Gd_2Co_{14}BH_{3.1}$	RT	$P4_2/mnm$	0.866	1.183	[88Z1]
$Tb_2Co_{14}B$	RT	$P4_2/mnm$	0.8581	1.1713	[87J1]
$Tb_2Co_{14}B$	RT	$P4_2/mnm$	0.8581	1.1710	[88J3]
$Tb_2Co_{14}B$	RT	$P4_2/mnm$	0.8600	1.1730	[87P6]
$Tb_2Co_{14}B$	RT	$P4_2/mnm$	0.858	1.169	[88Z1]
$Tb_2Co_{14}BH_{2.1}$	RT	$P4_2/mnm$	0.863	1.174	[88Z1]
$Y_2Co_{14}B$	RT	$P4_2/mnm$	0.8595	1.1648	[89H1]
$Y_2Co_{14}B$	RT	$P4_2/mnm$	0.8581	1.1711	[87P4, 87P7]
$Y_2Co_{14}B$	RT	$P4_2/mnm$	0.8616	1.1744	[86H2]
$Y_2Co_{14}B$	RT	$P4_2/mnm$	0.860	1.166	[88Z1]
$Y_2Co_{14}BH_{2.3}$	RT	$P4_2/mnm$	0.865	1.171	[88Z1]

the 4g sites. There are 68 atoms (four $R_2Co_{14}B$ units) per unit cell [86H1]. The crystal structure can be described as an alternate sequence of two different pseudolayers of Co atoms and a layer of R and Co atoms. The Co atoms on sites Co4e, Co8j_1, Co16k_1 and Co16k_2 form a net of distorted hexagons and triangles. The Co8j_2 atoms are at the center of a hexagonal antiprism between two such layers, like in the sigma phase [54B1, 84G1, 84S2]. The R atoms are at the center of a hexagonal prism in the next stacking sequence and form with B atoms in the same layer and Co atoms in adjacent layers, an arrangement which is reminiscent of the RCo_5-type structure. The Co4*c* at the center of a rhombohedral prism of Co atoms, are surrounded by four R atoms in their plane. This phase is in line compound and therefore has no range in stoichiometry.

The $R_2Co_{14}B$ compounds were obtained by melting the constituting elements or by oxides-reduction-diffusion technique, which consists in reduction of rare-earths oxides in a melt of calcium under argon and simultaneous diffusion reaction of just formed rare-earth metal with other elements, as in case of $Y_2Co_{14}B$ compound [97K7]. The $R_2Co_{14}B$ compounds were also obtained by melt-spun and annealing [93C5]. In this way nanocomposite magnets were also elaborated as $Pr_2Co_{14}B$/Co [96W1]. The 2/14/1 compounds can be obtained also by mechanical milling [97M1]. The $Sm_2Co_{14}B$/polyaniline composites with potential for application in the microwave absorption, has been obtained combining the arc-melting process and the in situ oxidative polymerization method [19C2].

The pseudoternary $(R_{1-x}R_x')_2Co_{14}B$ compounds form solid solutions in all the composition range, when end series compounds are formed, as for $(La_{1-x}Tb_x)_2Co_{14}B$ [01G1], $(Pr_{1-x}R_x)_2Co_{14}B$ with R = Nd [87P9, 88W2, 91S1, 92I1], R = Tb [86F2, 87J1, 98K3], R = Y [87P7]; $(Nd_{1-x}R_x)_2Co_{14}B$ with R = Sm [92I1], R = Tb [86F2, 88J3, 88P1], R = Y [87P7, 88K1, 90I1, 91J1, 91J2]; $(R_{1-x}R'_x)_2Co_{14}B$ with R = Gd, Y and R' = Tb series [94M1].

When one of the end series compound of $(R_{1-x}R'_x)_2Co_{14}B$ pseudoternary systems is not formed, solid solutions are present only in a limited composition range, as for example in $(Dy_xR_{1-x})_2Co_{14}B$ with $x \leq 0.8$ (R = La), $x \leq 0.7$ (R = Gd) or $x \leq 0.6$ (R = Y)) [97M1] series. The tetragonal structure of $(Y_{1-x}R_x)_2Co_{14}B$ compounds is maintained up to x = 0.6 (R = Dy) or x = 0.5 (R = Ho) [94M1]. $Pr_{2-x}R_xCo_{14}B$ solid solutions are formed for $x \leq 1.5$ (R = Dy, Er) [88P2], $x \leq 1.5$ (R = Dy, Ho) and $x \leq 1.0$ (R = Tm) [87J1]. The amount of R content, needed to stabilize the *P*4$_2$*/mnm*-type structure is larger, as the smaller the heavy rare-earth radius of R = Dy, Ho. Er and Tm is [87J1]. The rare-earths constituents seem to be not randomly distributed over 4*f* and 4*g* sites, the occupation of the 4*g* site by La exceeds that of 4*f* one, as showed in $(La_{1-x}Y_x)_2Fe_{14}B$ [96K1]. This trend can be correlated with the slightly larger volume of the 4*g* site, which is preferentially occupied by the rare-earth atoms with greater radii—Sect. 7.10.

The B atoms can be substituted by C ones up to x = 0.5 in $R_2Co_{14}B_{1-x}C_x$ series, as reported in compounds with R = Sm, Y, [93Z1] and R = Pr, Nd [92Z2] or all the boron in $Nd_{16}Co_{76}B_{8-x}C_x$ alloys [07C2].

The crystal structures and lattice parameters of $R_2Co_{14-x}M_xB$ compounds, with substitutions at cobalt sites, were also investigated, particularly those with M = Fe, of interest for permanent magnets. These series form solid solutions in all the compositional range when the end series compounds (x = 0 and x = 1.0) are present, as evidenced for R = La [88G5, 88V1], R = Pr [86P3, 87B5, 87G1, 87P2, 88G3, 88J2, 91H1, 92C2, 98B1, 17G1], R = Nd [84S1, 85M2, 86F1, 86G2, 86H1, 86H2, 86M1, 87B4, 87B12, 87C1, 87D1, 87H2, 87P2, 87T1, 88G1, 88G5, 88V1], R = Sm [90M1], R = Gd [86H2, 87P2], R = Tb [87P2, 87P6], or R = Y [86H2, 87B4, 87P2, 88T2,

89S2]. When one of the $R_2M_{14}B$ compound is not formed, the $R_2Co_{14-x}Fe_xB$ solid solutions are present if $x \leq 8$ (R = Dy) [87P3], $x \leq 5$ (R = Er) [87P2, 87P3, 09K4] or $x \leq 4$ (R = Tm) [87P2]. The location of iron when substituting cobalt atoms in the above series were studied by ND and Mössbauer spectroscopy [85V1, 86H1, 87B4, 87D1, 87H1, 87M1, 87Y3, 89R1, 89S2, 90G1, 93L1, 96L1, 17G1, 21E1]. There are discrepancies in the preferential occupancy reported in different works. A large number of studies mentioned the preference of Fe for $8j_2$ site [85V1, 86H1, 87H1, 87Y3, 89R1, 90R2, 93L1] and of Co for $16k_2$ sites [87H1, 87P6, 93L1, 17G1]. The anomalous X-ray synchrotron diffraction measurements in $Pr_2(Fe_{1-x}Co_x)_{14}B$ compounds evidenced that the c parameter, decreased twice much as the a parameter, when increasing cobalt content [17G1]. The above behaviour was correlated with preferential occupation of $16k_2$ site by cobalt. For remaiming sites, randomly occupation was reported [86H1, 89R1, 90R2]. A randomly occupation of cobalt in all lattice sites was also suggested [87M1, 89S2] (Co-rich). The cobalt site preferences in iron-based rare-earth compounds were shown to be strongly controlled by local atomic environment, based on analysis of the Wigner–Seitz cells [93L1]. Site volume was dominant factor controlling preferential cobalt substitutions. As the site volume becomes sufficiently large, the negative enthalpy of mixing between Co and R plays a significantly role in controlling the cobalt site preference in combination with the site volume effect. The ^{57}Co emission Mössbauer spectra of $Nd_2(Fe_{1-x}Co_x)_{14}B$, showed also that the preferential site substitution of cobalt is controlled by the volume of the crystallographic site [96L1]. Latter on, the normalization of the occupancy values to those expected for a random distribution, as obtained by neutron scattering on $R_2Fe_{14-x}Co_xB$ series with R = Ce, Nd and Y, was analysed in a statistical model for: (1) very diluted Co in $R_2Fe_{14}B$ and (2) very diluted Fe in $R_2Co_{14}B$ [21E1]—Fig. 8.23. The $8j_2$ site is over occupied by Fe atoms in Co-rich samples and under occupied by Co in Fe rich ones. The inverse, but to a lesser extent, was found for the $16k_2$ sites. In a first approximation the occupancy of the sites are determined by volume effects (solid lines)—Fig. 8.23. Thus, steric considerations primarily determine atom substitutions. The deviations from this behavior shows that the enthalpy of mixing and entropy considerations also contribute to Co preferential occupancy, particularly in Fe rich compounds.

The tetragonal $R_2(Co_{1-x}M_x)_{14}B$ solid solutions are formed, in a limited composition range, when cobalt was substituted by M elements, as $Sm_2(Co_{1-x}M_x)_{14}B$ for $x \leq 0.4$ (M = Mn) or $x < 0.1$ (M = Ni) [90M1]. In $Y_2(Co_{1-x}M_x)_{14}B$ series, the *P4₂/mnm*-type structure was formed for $x \leq 0.6$ (M = Mn) [87P1, 93M1], $x \leq 0.05$ [93M1] or $x \leq 0.3$ [87P1] for M = Ni and $x \leq 0.3$ (M = Mn, Cr, Ni) [87P1], while in $Nd_2(Co_{1-x}M_x)_{14}B$ for $x \leq 0.20$ (M = Ni), $x \leq 0.08$ (M = Cu), $x \leq 0.10$ (M = Ga, Ge) [89C1], $x \leq 0.2$ (M = Ga) or $x \leq 0.3$ (M = Nb) [88X1]. Solid solutions are formed in $Pr_2Fe_{14-x-y}Co_xCr_yB$ for $x \leq 3$, $y \leq 1$ [88K3]. The multi-components $R_2(Fe,Co,M)_{14}B$ alloys, having tetragonal structure were reported as example when M = Al, V, Cr [88J6], M = Cr, Mn, Fe, Ni [87S1], M = Al [87B8, 87B9, 87B11, 87S2], M = Si, V, Cr, Ta, W [88J5], M = Mo, Al [88J4] or in $Nd_2Fe_{12\text{-}x}Co_xM_xB$ with M = Si for $x \leq 0.6$ [87J2]. The Mn substitutes preferentially Co in $Y_2(Co_{1-x}Mn_x)_{14}B$ series at $8j_2$ site [93M1]. The same site was shown to be occupied by M = Mn or

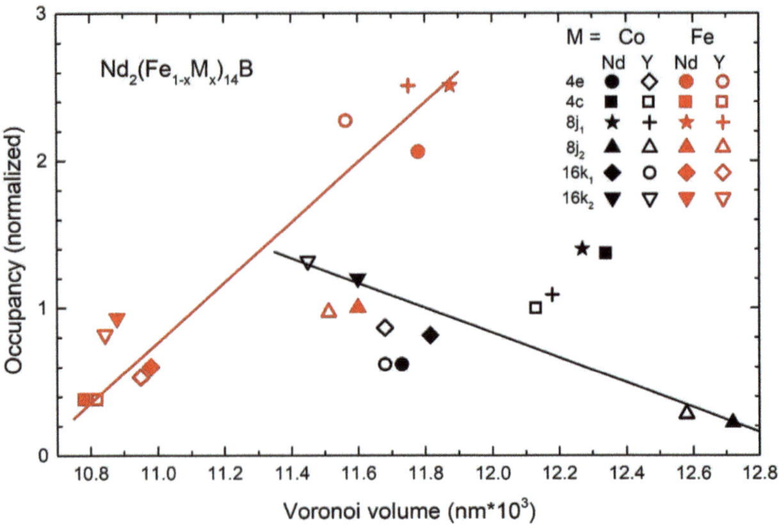

Fig. 8.23 $R_2(Fe_{1-x}Co_x)_{14}B$: normalized occupancies of Co (black) and Fe (red) in Fe-rich and Co-rich compounds, respectively, when R = Y or Nd, using published data [21E1]. Solid line represents linear dependence of occupancy as function of the site volume

Al in $Nd_2(Fe_{1-x}M_x)_{14}B$ system [87H2]. The multiphases alloys, containing mainly $R_2(Co_{1-x}M_x)_{14}B$ components, were also elaborated, particularly in connection with possible technical uses as permanent magnets.

The temperature dependences of lattice parameters, in $Nd_2(Fe_xCo_{1-x})_{14}B$ series, evidenced a discontinuity at the spin reorientation temperature, T_{sR2} [85L1]. The preferred replacement of iron by cobalt in sites $16k_2$, where the distances with nearest neighbours are very short, diminished the unsatisfied negative interactions between iron atoms—Sect. 7.10. As a result, the tendency to expand the lattice parameters in magnetic ordered state, at the expense of elastic energy, is diminished [85L1, 87C1]. Linear temperature dependences of lattice parameters can be seen at $T > T_{sR2}$.

The diffusion coefficients, D, of Fe and Co in $Nd_2(Fe_xCo_{1-x})_{14}B$, are D(Fe) = $3.28.10^{-14}$ m^2/s and D(Co) = $7.63.10^{-14}$ m^2/s [01C6]. The rare-earth diffusivity was found to be significantly lower, with D(Nd) = $2.3.10^{-16}$ m^2/s in $Dy_2(Fe_xCo_{1-x})_{14}B$ and D(Dy) = $2.9.10^{-16}$ m^2/s in $Nd_2(Fe_{1-x}Co_x)_{14}B$ series.

The $R_2Co_{14}BH_y$ hydrides with $2 \leq y \leq 4$ retain the tetragonal type structure of parent compounds [88Z1]. The a and c lattice parameters increase, the volume expansions being in the range 1.5 to 3%, depending on the nature of R atoms and hydrogen content [88Z1].

The $R_2Co_{14}B$ compounds with R = La [85B4, 85S1, 87H1, 87H3, 87S1, 87V1, 87Y2, 88W3, 88Z1] and R = Y [85B4, 85L1, 85S1, 86I1, 86B2, 87G2, 87G3, 87S1, 88B1, 88F1, 88K1, 88Z1, 90W1, 91C1, 92K3, 93Z1, 99M1] are ferromagnetically ordered—Table 8.45. These compounds show a planar spin arrangement, imposed

Table 8.45 Magnetic properties of $R_2Fe_{14}B$-based compounds

Compound	Magnetic structure and magnetic moments	M_s (μ_B/f.u.)	T_c (K)	Anisotropy constant MJ/m^3			Anisotropy field, μ_0H_a (T)	References
				K_1	K_2	$K_3 \cdot 10^{-3}$		
$La_2Co_{14}B$	FM, T = 4.2 K, e.a.m. ‖[100], T = 300 K, e.a.m. ‖ [110]	19.3[a]		-1.4[b]		-4.9[b],2.5[c]		[87Y2]
$La_2Co_{14}B$	FM	19.3[a]		-1.3[b]			2.86[b]	[87H1]
$La_2Co_{14}B$	FM, T = 4.2 K, e.a.m. ⊥ **c**	20.35[d]	955				6.0–7.0[b]	[85B4, 85S1]
$La_2Co_{14}B$	FM, T = 4.2 K, e.a.m. ⊥**c**	20.5[a]	957					[88Z1]
$La_2Co_{14}BH_{3.8}$	FM, T = 4.2 K, e.a.m. ⊥**c**	19.8[a]						[88Z1]
$Pr_2Co_{14}B$	FM, T = 4.2 K and 300 K e.a.m. ‖ **c**	24.82[e]	995				75.0[b]	[85B4, 85S1]
$Pr_2Co_{14}B$	FM, T < T_{sR2} e.a.m. ‖ **c**, T > T_{sR2} e.a.m. ⊥**c**	24.8[a]	986 664(T_{sR2})				34.8[f],13.7[c]	[86P1, 87P7]
$Pr_2Co_{14}B$	FM	25.2[a]	994 680(T_{sR2})				38.0[f],14.6[c]	[88Z1]
$Pr_2Co_{14}BH_{3.5}$	FM	24.5[a]	495(T_{sR2})				26.5[f],6.7[c]	[88Z1]
$Pr_2Co_{14}B_{0.8}C_{0.2}$	FM	24.0[g]	105(T_{sR2})					[92Z2]
$Pr_2Co_{14}B_{0.5}C_{0.5}$	FM	23.2[g]	85(T_{sR2})					[92Z2]
$Nd_2Co_{14}B$	FM, T < T_{sR1}, conical, $T_{sR1} \le T \le T_{sR2}$ axial T > T_{sR2}, planar	25.8[h]	1004 34(T_{sR1}) 546(T_{sR2})				5.5[c], 10.5[f]	[86P1, 87P4, 87P7]
$Nd_2Co_{14}B$	FM	27[i]	993	2.2[b]			5.2[b]	[85L1]
$Nd_2Co_{14}B$	FM, e.a.m. ‖ **c**, T ≤ 300 K	24.83[d] (25.37)	1007				30.0[b]	[85B4, 85S1]

(continued)

Table 8.45 (continued)

Compound	Magnetic structure and magnetic moments	M_s (μ_B/f.u.)	T_c (K)	Anisotropy constant MJ/m^3			Anisotropy field, μ_0H_a (T)	References
				K_1	K_2	$K_3{\cdot}10^{-3}$		
$Nd_2Co_{14}B$	FM	26.3[a]	1002 35(T_{sR1})				5.5[g],10.5[f]	[86H2]
$Nd_2Co_{14}B$	FM	25.7[h]	37(T_{sR1})	-1.4[b]	19[b]			[87H1, 87Y2]
$Nd_2Co_{14}B$	FM	25.8[a]	1006 35(T_{sR1}) 590(T_{sR2})				11.8[f],7.8[c]	[88Z1]
$Nd_2Co_{14}B$	FM	24.7[f]	1002 550(T_{sR2})					[88J3]
$Nd_2Co_{14}BH_{3.4}$	FM	25.5[a]	<4.2(T_{sR1}) 385(T_{sR2})				10.5[f],1.9[c]	[88Z1]
$Nd_2Co_{14}B$	FM[j]	24.9[d]	40(T_{sR1})					[92Z2]
$Nd_2Co_{14}B_{0.8}C_{0.2}$	FM[k]	23.9[d]	20(T_{sR1})					[92Z2]
$Nd_2Co_{14}B_{0.5}C_{0.5}$	FM	22.7[d]						[92Z2]
$Nd_2Co_{14}B$	FM, $M_{Nd} = 3.2\ \mu_B$, $T < T_{sR1}$, $\theta^j = 11°$	25.7[a]	38(T_{sR1})	-1.4	19			[87Y2]

(continued)

Table 8.45 (continued)

Compound	Magnetic structure and magnetic moments	M_s (μ_B/f.u.)	T_c (K)	Anisotropy constant MJ/m^3			Anisotropy field, μ_0H_a (T)	References
				K_1	K_2	$K_3 \cdot 10^{-3}$		
$Nd_2Co_{14}B$ (NMR)	FM, T = 4.2 K hyperfine fields in Table 8.46 $M_{k1} = 1.2\ \mu_B$, $M_{k2} = 1.7\ \mu_B$, $M_{j1} = 1.5\ \mu_B$ $M_{j2} = 1.5(1)\ \mu_B$, $M_e = 0.9(1)\ \mu_B$, $M_c = 1.8(2)\ \mu_B$							[89P1]
$Nd_2Co_{14}B$ (ND)	FM, T = 293 K Nd: $M_f = 1.3(2)\ \mu_B$, $M_g = 2.1(1)\ \mu_B$ Co: $M_{k1} = 1.2(1)\ \mu_B$, $M_{k2} = 1.7(1)\ \mu_B$, $M_{j1} = 1.5(1)\ \mu_B$, $M_{j2} = 1.5(1)\ \mu_B$, $M_e = 0.9(2)\ \mu_B$, $M_c = 1.8(2)\ \mu_B$	23.6	995					[86H1]
$Nd_2Co_{14}B$ (NMR)	FM, T = 4.2 K Co: $M_{k1} = 1.2\ \mu_B$, $M_{k2} = 1.7\ \mu_B$, $M_{j1} = 1.5\ \mu_B$ $M_{j2} = 1.5\ \mu_B$, $M_c = 1.8\ \mu_B$, $M_e = 0.9(1)\ \mu_B$							[89W1]

(continued)

Table 8.45 (continued)

Compound	Magnetic structure and magnetic moments	M_s (μ_B/f.u.)	T_c (K)	Anisotropy constant MJ/m^3			Anisotropy field, μ_0H_a (T)	References
				K_1	K_2	$K_3 \cdot 10^{-3}$		
$Nd_2Co_{12.6}Fe_{1.4}B$ (ND)	FM, T = 293 K Nd: $M_f = 0.7(1)\ \mu_B$, $M_g = 1.8(1)\ \mu_B$ M^k: $M_{k1} = 1.3(1)\ \mu_B$, $M_{k2} = 2.1(1)\ \mu_B$, $M_{j1} = 1.1(2)\ \mu_B$, $M_{j2} = 2.2(1)\ \mu_B$, $M_e = 1.7(1)\ \mu_B$, $M_c = 0.7(3)\ \mu_B$	25.3	994					[86H1]
$Nd_2Co_{9.8}Fe_{4.2}B$ (ND)	FM, T = 293 K Nd: $M_f = 0.9(2)\ \mu_B$, $M_g = 1.8(2)\ \mu_B$ $M^{k)}$: $M_{k1} = 1.7(1)\ \mu_B$, $M_{k2} = 1.8(1)\ \mu_B$, $M_{j1} = 1.3(2)\ \mu_B$, $M_{j2} = 2.7(2)\ \mu_B$, $M_e = 2.1(3)\ \mu_B$, $M_c = 1.9(3)\ \mu_B$	28.7	965					[86H1]

(continued)

Table 8.45 (continued)

Compound	Magnetic structure and magnetic moments	M_s (μ_B/f.u.)	T_c (K)	Anisotropy constant MJ/m^3			Anisotropy field, μ_0H_a (T)	References
				K_1	K_2	$K_3 \cdot 10^{-3}$		
$Nd_2Co_7Fe_7B$ (ND)	FM, T = 293 K Nd: M_f = 1.3(2) μ_B, M_g = 2.1(2) μ_B $M^{k)}$: M_{k1} = 2.1(1) μ_B, M_{k2} = 1.2(1) μ_B, M_{j1} = 1.7(2) μ_B, M_{j2} = 2.8(2) μ_B, M_e = 2.2(3) μ_B, M_c = 3.0(2) μ_B	30.8	918					[86H1]
$NdCo_{4.2}Fe_{9.8}B$ (ND)	FM, T = 293 K Nd: M_f = 1.4(2) μ_B, M_g = 1.3(2) μ_B $M^{k)}$: M_{k1} = 2.2(1) μ_B, M_{k2} = 1.7(1) μ_B, M_{j1} = 1.9(1) μ_B, M_{j2} = 3.2(2) μ_B, M_e = 2.4(3) μ_B, M_c = 2.4(2) μ_B	33.5	818					[86H1]

(continued)

Table 8.45 (continued)

Compound	Magnetic structure and magnetic moments	M_s (μ_B/f.u.)	T_c (K)	Anisotropy constant MJ/m^3			Anisotropy field, μ_0H_a (T)	References
				K_1	K_2	$K_3 \cdot 10^{-3}$		
$Nd_2Co_{1.4}Fe_{12.6}B$ (ND)	FM, T = 293 K Nd: $M_f = 0.3(2)\ \mu_B$, $M_g = 1.4(2)\ \mu_B$ $M^{k)}$:$M_{k1} = 1.3(1)\ \mu_B$, $M_{k2} = 2.9(1)\ \mu_B$, $M_{j1} = 3.0(2)\ \mu_B$, $M_{j2} = 3.7(2)\ \mu_B$, $M_e = 1.2(2)\ \mu_B$, $M_c = 1.3(3)\ \mu_B$	34.4	668					[86H1]
$Nd_2Co_{14}B$ (BS)	FM, Nd: $M_f = 3.03\ \mu_B$, $M_g = 3.04\ \mu_B$ Co: $M_{k1} = 1.18\ \mu_B$, $M_{k2} = 1.29\ \mu_B$, $M_{j1} = 1.59\ \mu_B$, $M_{j2} = 2.44\ \mu_B$, $M_e = 1.16\ \mu_B$, $M_c = 2.27\ \mu_B$ B: $M_g = 0.39\ \mu_B$							[87G2]
$Sm_2Co_{14}B$	FM, e.a.m. ⊥ **c**	18.10	1029					[85B4, 85S1]
$Sm_2Co_{14}B$	FM, e.a.m. ⊥ **c**	18.3[a]	1031					[88Z1]
$Sm_2Co_{14}BH_{3.6}$	FM, e.a.m. ⊥ **c**	18.6[a]						[88Z1]
$Sm_2Co_{14}B$	FM	19.83[1]	122(T_{sR})				60.0[b]	[93Z1]

(continued)

Table 8.45 (continued)

Compound	Magnetic structure and magnetic moments	M_s (μ_B/f.u.)	T_c (K)	Anisotropy constant MJ/m^3			Anisotropy field, μ_0H_a (T)	References
				K_1	K_2	$K_3 \cdot 10^{-3}$		
$Sm_2Co_{14}B_{0.5}C_{0.5}$	FM	19.54[l]	115(T_{sR})				70.0[b]	[93Z1]
$Gd_2Co_{14}B$	FIM, e.a.m. ⊥ **c**	6.90[e]	1050					[85B4]
$Gd_2Co_{14}B$	FIM, T = 4.2 K, e.a.m. ǁ [100], T = 294 K ǁ [110]	5.3[a]		−1.2		−1.3		[87Y2]
$Gd_2Co_{14}B$	FIM, T = 4.2 K, e.a.m. ǁ[100], T = 294 K ǁ[110]	5.36[h]		−1.27		2.54	10.7	[87H1]
$Gd_2Co_{14}B$	FIM, e.a.m. ⊥ **c**	6.2[a]	1057					[86H2]
$Gd_2Co_{14}B$	FIM, e.a.m. ⊥ **c**	6.0[a]	1050					[88Z1]
$Gd_2Co_{14}BH_{3.1}$	FIM, e.a.m. ⊥ **c**	6.2[a]						[88Z1]
$Tb_2Co_{14}B$	FIM, T = 294 K, e.a.m. ǁ **c**		1035					[85B4, 01G1]
$Tb_2Co_{14}B$	FIM, T > T_{sR2} e.a.m. ⊥ **c**	12.2[a]	1031 795 (T_{sR2})					[86P1]
$Tb_2Co_{14}B$	FIM	2.3[f]	1035 799(T_{sR2})					[88J3]
$Tb_2Co_{14}B$	FIM, T ≤ 300 K	3.7[a]	1036 795(T_{sR2})				13.4[c],8.0[f]	[88Z1]
$Tb_2Co_{14}BH_{2.1}$	FIM, T ≤ 300 K	3.3[a]	764(T_{sR2})				11.5[c],3.8[f]	[88Z1]
$Y_2Co_{14}B$	FM, e.a.m. ⊥ **c**	19.8[i]	1003	−1.2[b]			2.8[b]	[85L1]

(continued)

Table 8.45 (continued)

Compound	Magnetic structure and magnetic moments	M_s (μ_B/f.u.)	T_c (K)	Anisotropy constant MJ/m^3			Anisotropy field, μ_0H_a (T)	References
				K_1	K_2	$K_3 \cdot 10^{-3}$		
$Y_2Co_{14}B$	FM, e.a.m. ⊥ **c**	19.7[a]	1012					[86H2]
$Y_2Co_{14}B$	FM, e.a.m. ⊥ **c**	19.5[h]	1015					[87P4, 87P7]
$Y_2Co_{14}B$	FM, e.a.m. ⊥ **c**	19.42[e]	1015				6.0[b]	[85B3, 85S1]
$Y_2Co_{14}B$	FM, e.a.m. ⊥ **c**	19.87[l]					2.8[b]	[93Z1]
$Y_2Co_{14}B_{0.5}C_{0.5}$	FM, e.a.m. ⊥ **c**	18.76[l]					4.6[b]	[93Z1]
$Y_2Co_{14}B$	FM, e.a.m. ⊥ **c**	20.0[a]	1016					[88Z1]
$Y_2Co_{14}BH_{2.3}$	FM, e.a.m. ⊥ **c**	19.6[a]						[88Z1]
$Y_2Co_{14}B$ (NMR)	FM, T = 4.2 K Co: M_{k1} = 1.26 μ_B, M_{k2} = 1.22 μ_B, M_{j1} = 0.93 μ_B, M_{j2} = 0.84 μ_B, M_e = 0.72 μ_B, M_c = 1.43 μ_B							[88B1]
$Y_2Co_{14}B$ (ND)	FM, T = 4.2 K Co:M_{k1} = 1.36(6) μ_B,M_{k2} = 1.17(11) μ_B, M_{j1} = 1.53(8) μ_B, M_{j2} = 1.23(16)μ_B,M_e = 0.66(16) μ_B, M_c = 1.50(14)μ_B							[86L1]

(continued)

Table 8.45 (continued)

Compound	Magnetic structure and magnetic moments	M_s (μ_B/f.u.)	T_c (K)	Anisotropy constant MJ/m^3			Anisotropy field, μ_0H_a (T)	References
				K_1	K_2	$K_3 \cdot 10^{-3}$		
$Y_2Co_{14}B$ (BS)	FM Co: $M_{k1} = 1.35$ μ_B, $M_{k2} = 1.47$ μ_B, $M_{j1} = 1.34$ μ_B, $M_{j2} = 1.48$ μ_B, $M_e = 1.37$ μ_B, $M_c = 1.28$ μ_B							[86I1]
$Y_2Co_{14}B$ (BS)	FM Co: $M_{k1} = 1.17$ μ_B, $M_{k2} = 1.41$ μ_B, $M_{j1} = 1.46$ μ_B, $M_{j2} = 1.43$ μ_B, $M_e = 0.96$ μ_B, $M_c = 1.59$ μ_B Y: $M_f = -0.33$ μ_B, $M_g = -0.32$ μ_B B: $M_g = -0.11$ μ_B							[91C1]

(continued)

Table 8.45 (continued)

Compound	Magnetic structure and magnetic moments	M_s (μ_B/f.u.)	T_c (K)	Anisotropy constant MJ/m^3			Anisotropy field, μ_0H_a (T)	References
				K_1	K_2	$K_3 \cdot 10^{-3}$		
$Y_2Co_{14}B$ (BS)	FM Co: $M_{k1} = 1.21\ \mu_B$, $M_{k2} = 1.44\ \mu_B$, $M_{j1} = 1.51\ \mu_B$, $M_{j2} = 1.38\ \mu_B$, $M_e = 1.09\ \mu_B$, $M_c = 1.75\ \mu_B$ Y: $M_f = -0.42\ \mu_B$, $M_g = -0.42\ \mu_B$ B: $M_g = -0.10\ \mu_B$							[88J1]

[a] $T = 4.2$ K, $\mu_0H = 2$ T
[b] $T = 4.2$ K
[c] at RT
[d] $T = 4.2$ K, $\mu_0H = 35$ T
[e] $T = 4.2$ K, $\mu_0H = 18$–35 T
[f] $T = 77$ K, $\mu_0H = 3$ T
[g] $T = 300$ K, $\mu_0H = 35$ T
[h] $T = 4.2$ K, not mentioned field
[i] $T = 4.2$ K, $\mu_0H = 5$ T
[j] angle from **c**-axis to [110] in **c**-plane at $T = 4.2$ K
[k] $M = (Fe + Co)$
[l] $T = 1.5$ K, $\mu_0H = 6$ T

by cobalt sublattice anisotropy [86P1]. In $La_2Co_{14}B$, at RT, the easy axis of magnetization is along [110] and it changes to [100] direction, at T = 4.2 K [87Y2]. The easy direction of magnetization in $Y_2Co_{14}B$ is also located in the basal plane, where no anisotropy was shown [85L1]. The anisotropy constants, determined from magnetization isotherms, along **c**-axis and plane, were K_1 = -(1.2–1.4) MJ/m^3, at T = 4.2 K—Table 8.45.

The ^{59}Co hyperfine fields in $La_2Co_{14}B$ at various sites. correlated well with their local environements [88W3]. The Co orbital contributions in the easy plane vary strongly from site to site, reaching a maximum value for $8j_1$ position and a minimum for the 4*c* site. The ND [86L1] and NMR [87B12] studies as well as most band structure calculations [87G2, 88J1, 91C1] on $Y_2Co_{14}B$ evidenced that the higher cobalt moment is located on 4*c* site and the smaller one on 4*e* site, this position having two B atoms as first neighbors—Table 8.45. The ^{59}Co NMR studies on $Y_2Co_{14}B$ allowed the assignement of the lines to cobalt atoms located at the six non-equivalent sites [88B1, 92K3, 92M1, 93M1]—Table 8.46.

A comparative analysis of the NMR spectra of $R_2Co_{14}B$ compounds with R = Y and Gd evidenced that the spacing between the quadrupole split components are almost the same, except at $16k_1$ sites. An anisotropy of hyperfine fields, in the **c** plane, for both compounds, was shown. The ^{59}Co NMR study on $Y_2Co_{14}B$, evidenced significant orbital contributions at $8j_1$ and $8j_2$ sites [88B1, 93M1]. The hyperfine fields corresponding to these sites are positive, contrary to other cobalt sites. According to [88B1], the cobalt at $8j_1$ and $8j_2$ sites, where an important orbital component of magnetic moments was detected, are submitted to the larger EFG. This is in agreement with the model [74M1] which explain the induction of orbital cobalt moment through the spin–orbit coupling in the presence of the uniaxial crystalline electric field. In $La_2Co_{14}B$ compound an orbital contribution was found to be significantly in the basal plane, particularly on $8j_1$ site [88W3].

The $Y_2Co_{14}B$ was shown to be a promising microwave soft magnetic material, working in GHz frequency band [15L1].

The $La_{15}(Co_{1-x}Fe_x)_{77}B_8$ alloys obtained by melting and heat treatment, contained as the main phase the $La_2(Co_{1-x}Fe_x)_{14}B$ compounds [88V1]. The Curie temperatures increase as the cobalt content is higher.

The $Y_2(Co_{1-x}Fe_x)_{14}B$ compounds are ferromagnets. The magnetizations, at T = 4.2 K, has been reported either to have a small maximum located at low iron content [85B4, 86B2, 86H2, 87G3, 93M1], similar as in $Co_{1-x}Fe_x$ alloys [84V1], or to decrease gradually when increasing iron content [06K1]. A decrease of the magnetization when increasing the Fe content was also shown in nanocrystalline samples [10T1]. By Mössbauer spectroscopy has been reported that the ^{57}Fe hyperfine fields, at various sites, either increase or decrease in low iron concentration range. This suggest that the composition dependence of magnetization is strongly connected with the preferential site occupation. The ^{59}Co NMR lines at T = 4.2 K, in $Y_2(Co_{1-x}Fe_x)_{14}B$ series, broaden and weaken in intensity and their positions shift rapidly to higher frequency, for $x \leq 0.3$, and are almost constant for $0.3 \leq x \leq 0.9$ [92M1, 93M1]. A change of easy axis of magnetization from plane to axial was shown at x = 0.3. The ^{89}Y hyperfine fields increase for compounds with $x \geq 0.7$.

Table 8.46 Data obtained by NMR on $R_2Fe_{14}B$ compounds

Compound	T(K)	Nucleus	NMR frequency (MHz) or Hyperfine field (T)						NMR frequency ν and Δν (MHz)				References
			$16k_1$	$16k_2$	$8j_1$	$8j_2$	$4c$	$4e$	ν	$\Delta\nu^c$	ν	$\Delta\nu^c$	
									f	f	g	g	
$La_2Co_{14}B^a$ ∥ **c** ⊥ **c** Orbital moment (μ_B)	4.2	^{59}Co	15.65 9.9 0.063	16.12 10.81 0.081	16.03 8.91 0.120	17.05 12.95 0.063	15.62 12.77 0.044	15.20 12.12 0.047					[88W3]
$Nd_2Co_{14}B^a$	4.2		12.95	16.63	18.21	12.57	19.60	5.48					[89P1, 89W1]
$Nd_2Co_{14}B^b$	4.2		130.6/133	184.4	162.6	172.8	127.0	168.1					[88K1]
$Nd_{0.4}Y_{1.6}Co_{14}B^b$	4.2		135.6	189.2	177.0	184.2	160.2	146.6					[88K1]
$Gd_2Co_{14}B^a$	4.2		14.17	16.17	13.11	11.09	17.75	9.91					[92K3]
$Gd_2Co_{14}B^b$	4.2		162	156	128	116							[92M2]
$Y_2Co_{14}B^b$	4.2		166	158	123	108							[92M2, 93M2]
$Y_2Co_{14}B^a$	4.2		16.04	16.44	11.98	10.94	17.38	9.85					[93M1]
$Y_2Co_{14}B^a$	4.2		15.04	16.58	12.40	11.49	18.89	10.66					[92K3]
$Y_2Co_{14}B^a$	4.2		16.43	15.84	12.07	10.9	18.6	9.4					[88B1]
$Y_2Co_{14}B^b$	4.2		159/164.5	167.5	161.5	152.0	170.5	187					[88K1]
$Nd_{15}Fe_{77-x}Co_xB_8{}^b$	1.3	^{57}Fe											[87Z1]
x = 0			44.7	48.5	46.0	52.5	44.0	41.5					
x = 3.9			44.3	48.0	45.5	51.5	43.3	41.5					
x = 7.0			43.5	47.5	45.0	50.5	42.5	41.5					
x = 12.0			43.0	47.0	44.8	49.8	42.2	41.0					
$Nd_2Fe_{13.3}Co_{0.7}B^b$	4.2		44.0	48.0	45.5	51.0	43.3	41.5					[88Z2]

(continued)

Table 8.46 (continued)

Compound	T(K)	Nucleus	NMR frequency (MHz) or Hyperfine field (T)						NMR frequency ν and Δν (MHz)				References
			$16k_1$	$16k_2$	$8j_1$	$8j_2$	$4c$	$4e$	ν	Δν[c]	ν	Δν[c]	
									f	f	g	g	
$Nd_2Fe_{14}B$[a]	4.2		32.5	35.2	33.4	37.8	32.0	30.2					[88Z2]
$Y_2Co_{14}B$[b]	4.2	^{89}Y							22.8	0.17	24.3	0	[87F1]
$Nd_2Co_{14}B$[b]	4.2	^{145}Nd							544	2.58	566	0.3	[89N1]
$Nd_2Co_{14}B$[b]	4.2								545		566		[92J1]
$Nd_2Co_{14}B$[b]	4.2								565.80[d)]	0.36(1)	565.78[d)]	0.31(1)	[87F1]

[a] B_{eff} values are given in T
[b] Resonance frequency
[c] Quadrupole splitting
[d] In original paper was not attributed to a given site

The anisotropy energy of $Y_2(Co_{1-x}Fe_x)_{14}B$ series, at T = 4.2 K, were experimentally determined [88T2] and also computed starting from individual sites anisotropies [79S2], in correlation with sites occupancy [89H1, 93M1]. The anisotropy was also calculated starting from band structures [91U1]. The cobalt contribution to the magnetic anisotropy, at the $8j_2$ site, has opposite sign to that of the iron. Substituting Co by Fe, particularly at $8j_1$ and $8j_2$ sites, increased the uniaxial contribution to the anisotropy and for a critical concentration, the easy direction changes from plane to axis. The computed anisotropy energies, were shown to describe rather well the experimentally determined values, at T = 4.2 K [88T2]—Fig. 8.24 [93M1].

The saturation magnetizations of $Y_2(Co_{1-x}Mn_x)_{14}B$ series have a maximum value around x = 0.1 and then decrease with x. The T_c values decrease nearly linear up to x = 0.5 and the magnetic anisotropy changes at RT from planar to axial, in the composition range $0.05 < x < 0.1$. The saturation magnetizations decrease for Ni substitution, indicating a simple dilution-type behavior [93M1]. The Co at $8j_1$ and $8j_2$ sites give the larger contribution to the anisotropy of the 3d sublattice. The form of ^{59}Co NMR spectra, after doping with Mn, changes abruptly, in the composition range $0.1 \leq x \leq 0.25$, correlated with the spin reorientation, from plane to axis. The ^{59}Co hyperfine fields decrease with the same rate at all sites in $Y_2(Co_{1-x}Ni_x)_{14}B$ series, as Ni content increases.

The wide variety of the $R_2Co_{14}B$ magnetic properties, when R is a magnetic rare-earth, arises from the fact that the R ions are subject not only to the R-Co exchange field, but also to the crystalline electric field, as also evidenced in

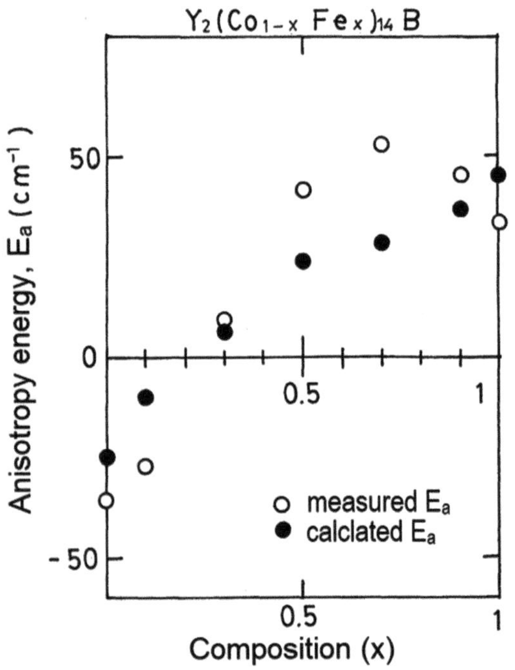

Fig. 8.24 $Y_2(Co_{1-x}Fe_x)_{14}B$: anisotropy energy E_a, experimentally determined from anisotropy constant, at T = 4.2 K [88T2] and the computed values [93M1] $1\ cm^{-1} \cong 2\cdot10^{-23}$ J

$R_2Fe_{14}B$ compounds [88Y1]—Sect. 7.10. The overall (bulk) anisotropy of $R_2Co_{14}B$ compounds with magnetic rare-earths is the result of the complex interplay between competing anisotropies and the exchange fields. The single ion anisotropy of the R ion is determined by the sign of the Steven operator equivalent coefficient, α_J. The rare-earths having $\alpha_J > 0$ (R = Sm), show strong planar tendencies, while those with $\alpha_J < 0$ (R = Pr, Nd, Tb) have axial preference—Table 8.47. The R atoms with $\alpha_J < 0$ and Co sublattices anisotropies have different temperature dependences, that of the R decreasing more rapidly than that of cobalt. Thus, when the anisotropy of R sublattice, at low temperatures, is higher than that of cobalt, spin reorientation type transition occurs [87P2].

The $R_2Co_{14}B$ with R = Pr [85B4, 85S1, 86P1, 87B2, 87G1, 87S1, 88F2, 88G2, 88G3, 88K2, 88Z1, 89C2, 90W1, 91S2, 91Y1, 92Z2, 93C5, 93J1, 99M1] and R = Nd [85B4, 85L1, 85P1, 85S1, 86P1, 87C1, 87G2, 87H1, 87H2, 87S1, 87Y2, 88G1, 88G3, 88K1, 88K2, 88L1, 89N1, 89P1, 90W1, 91Y1, 92Z2, 93J1, 99M1] are ferromagnetically ordered—Table 8.45.

The $Pr_2Co_{14}B$ is a strong uniaxial material with easy axis along **c**-direction, showing a spin reorientation to planar anisotropy, at relative high temperatures, where the cobalt sublattice anisotropy dominates—Fig. 8.25. The anisotropy field, H_a, both at T = 4.2 K and T = 300 K, is two time higher than that in the $Nd_2Co_{14}B$ [87G1]. This behavior has been tentatively explained through the valence instability of Pr [88G3]. An intriguing anomaly reflected in the very rapid rise of $H_a(x)$ in the $Pr_2(Fe_{1-x}Co_x)_{14}B$ series, at RT as x increases from 0.7 to 1.0 was shown; over the $0 \leq x \leq 0.7$ composition range, H_a decreses modestly. In contrast, $H_a(x)$ in $Nd_2(Fe_{1-x}Co_x)_{14}B$ series decline from x = 0 to x = 0.9 and features only a small upturn thereafter. The CEF-exchange model [88C1] was applied to analyse the above features [88G2, 88G3]. It was speculated, on the basis of composition

Table 8.47 Crystal field and exchange parameters (K)

Parameter	$Pr_2Co_{14}B$		$Nd_2Co_{14}B$		
	[91Y1, 93J1]	[88K2]	[91Y1, 93J1]	[88L1]	[88K2]
$2\mu_B H_{exch}$[a]	+480	+440	+440	+410	+370
A_2^0	+890	+980	+820	+780	+900
A_2^2/i	±450 (±445)[b]	±380	±410	±410	±340
A_4^0	430	−170	360	+80	−150
A_4^2/i	0	0	0	0	0
A_4^4	−220	0	−190	0	0
A_6^0	−560	−1700	−460	−420	−260
A_6^2/i	±170	±120	∓ 140	0	±90
A_6^4	−490	−900	−410	−390	−650
A_6^6/i	±330 (±230)[b]	±350	± 270 (±190)[b]	0-	±310

[a] Anisotropic exchange interactions
[b] Values given by [93J1]

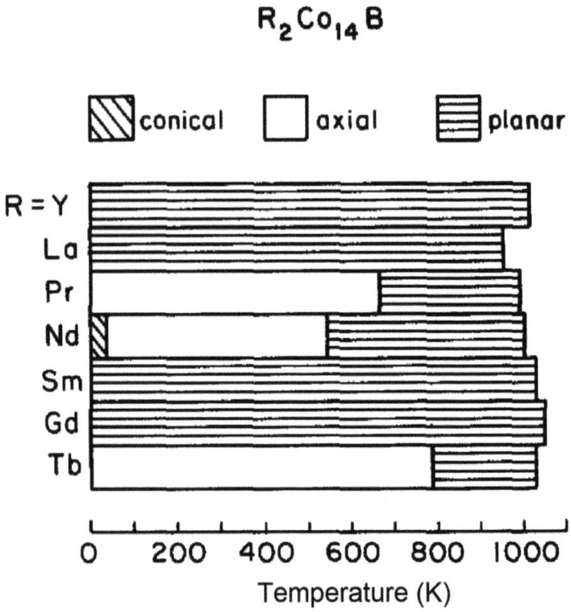

Fig. 8.25 $R_2Co_{14}B$ compounds: evolutions with temperature of the spin arrangements

dependence of the anisotropy field, that the Pr^{3+} configuration is unstable for $x \leq 0.7$ and stable for larger Co content. Latter on [92C2], by using X-ray adsorption near-edge spectroscopy (XANES), found that Pr valence (Pr^{3+}) is constant through $Pr_2(Fe_{1-x}Co_x)_{14}B$ series [92C2]. Thus, the anomalous composition dependence of the magnetocrystalline anisotropy, in $Pr_2(Fe_{1-x}Co_x)_{14}B$ compounds, vis-à-vis their Nd-based counterparts does not arise from a Pr instability. Further analysis of the mater is of interest.

The magnetic properties of $Pr_2(Co_{1-x}M_x)_{14}B$, with M = Cr, Mn, Fe and Ni were investigated at $T \geq 77$ K [87S1]. In Ni doped samples, T_c has a maximum at $x = 0.15$–0.20, while the Co substitutions by M = Cr, Mn and Fe, decrease gradually the Curie temperatures as their content increases. A diminution of the magnetizations was also evidenced, in the above systems, except when M = Fe. The simultaneous substitutions of Fe by Co and Cr, in $Pr_2Fe_{14-x-y}Co_xCr_yB$ ($1 \leq x \leq 3, 0 \leq y \leq 1$) lead to alloys having high T_c and anisotropy fields [88K3]. The saturation magnetizations are lower than those of parent compound, mainly due to chromium presence.

The magnetic properties of $Nd_2Co_{14}B$ compound, are somewhat more complicated as of their counterpart $Nd_2Fe_{14}B$—Sect. 7.10. The easy axis of magnetization is tilted 11°–12° away from the **c**-axis, at T = 4.2 K [87H1, 87Y2]. This behavior is a consequence of the interplay between the exchange interactions and the electric field acting on Nd ion. As already mentioned, the pure crystalline electric field favors an axial anisotropy of Nd sublattice, at low temperatures, but is the exchange field that pulls the moment away from **c**-axis. The result, is a conical type arrangement. When the temperature increased, the exchange field diminish and the e.a.m approaches the **c**-axis and becomes parallel to this axis, at temperature $T_{SR1} = 35$–38 K [86P1, 86Y1,

87H1, 88K1] and remains in this state up to higher spin reorientation temperature, $T_{sR2} = 550$ K [88Z1] or 543 K [85L1, 86P1], due to the decrease of Nd sublattice anisotropy. At T > T_{sR2}, the easy axis of magnetization lies in the **c**—plane. The ^{59}Co NMR study, performed on $Nd_2Co_{14}B$, showed that the low temperature spin orientation, from the conical to axial state, is of second-order phase transition with a canting angle changing smoothly between T = 20 K and 32 K [86K1, 90W1]. While the positions of the lines do not change significantly, at 32 K ≤ T ≤ 77 K, the effect of canting at T < 30 K, in $Nd_2Co_{14}B$ compound, is pronounced as the magnetization rotates away from **c**-axis; its in plane component introduces magnetically non-equivalent sites. The largest line splitting was observed for 4*c*, $16k_2$ and $16k_1$ sites, whereas the lines at *e*, j_1 and j_2 sites do not exhibit significant changes. The splitting measures the in-plane anisotropy, evidencing different contributions to the anisotropy of the cobalt sites. The signals arise from the walls in canted magnetic state and disappear upon reaching uniaxial state. Above T_{sR1} = 32 K–35 K, only six lines are observed. As the samples reach the uniaxial state, signal intensity drops rapidly and thus a reliable spectrum, of $Nd_2Co_{14}B$, can be obtained only after long accumulation.

The ^{145}Nd spin echo spectra on $Nd_2Co_{14}B$ showed, in an earlier study, the presence of a single line, assumed to be a superposition of signals due the two Nd sites [87F1]. Later on, the two lines, with different quadrupole splittings, corresponding to Nd4*f* and N4*g* sites, were resolved [89N1, 92J1]—Table 8.46.

The anisotropy of the $(Nd_{1-x}Y_x)_2Co_{14}B$ system, at T = 4.2 K, is determined by Nd sublattice, up to a threshold concentration x_c = 0.90–0.92 [88K1, 90I1, 91J1, 91J2]. Above the x_c value, the exchange interactions cause the Nd4f sublattice to follow 3d anisotropy (planar), at 4.2 K ≤ T ≤ 300 K. Below threshold concentration, planar structures are reached, when increasing temperature, by a process involving first order transition. The orbital contributions to cobalt moments were evaluated, for samples with x = 0, 0.2 and 1.0 [88K1].

The first order magnetization process (FOMP), characterized by discontinuities or jumps in M(H) curves, was shown along [100] direction, at T = 4.2 K, in $Nd_2Co_{14}B$, at $\mu_0H_{cr} \cong 21$ T [87H3]. The spontaneous magnetization, tilts from [001] towards [110] direction, at an angle $\theta \cong 12°$. The presence of a FOMP was reported also in $Nd_2Fe_{14}B$, along [100] direction, at μ_0H_{cr} = 17 T [98B1]—Sect. 7.10. The quantitative difference between $Nd_2Co_{14}B$ and $Nd_2Fe_{14}B$ can be correlated with different anisotropies of transition metals sublattices, which favor the **c**-axis in case of iron and planar, in case of cobalt. The FOMP transitions in $R_2Co_{14}B$ with R = Nd, Pr were analysed [85S1, 88K2, 91S1, 91Y1]. According to [91S1], in $(Nd_{1-x}Pr_x)_2Co_{14}B$ series, at T = 4.2 K, for x = 0, a FOMP transition of PIc- type occurs at μ_0H_{cr} = 21-22 T, in the [100] hard direction of magnetization Sect. 7.10. When the magnetization direction changes from easy cone (x = 0), to easy axis (x > 0.2), the type of FOMP changes from P1c- to P1-type. Starting from the determined composition dependence of the anisotropy constants, K_i (i = 1, 2, 3), at T = 4.2 K, the critical fields, H_{cr}, for FOMP transitions, were estimated [91S1]. These were in agreement with experimental values, for x < 0.2, when was possible to be experimentaly determined (the H_{cr} is in many cases higher than the available external fields).

A large number of studies were devoted to the analysis of the magnetic behavior of $RCo_{14-x}Fe_xB$ series, with R = Pr [86F2, 86P3, 87B5, 87G1, 87P2, 87S1, 88G3, 88K3, 88P1, 88W2, 91H1, 92C2, 98B1, 98K3, 01Z2, 17G1] and R = Nd [85M2, 86F1, 86F2, 86G2, 86H1, 86H2, 87B4, 87H1, 87H2, 87H4, 87J2, 87M1, 87P2, 87P5, 87R1, 87T1, 87Y1, 88G3, 88J3, 88J6, 88W1, 88X1, 88Z2, 89N1, 96G1]; R = (NdY) [99P1]. The $R_2(Co_xFe_{1-x})_{14}B$ compounds with R = Pr and Nd are ferromagnetically ordered, the T_c values increasing as the cobalt content increases. The saturation magnetizations, at T = 4.2 K, show a similar composition dependence as the Slater-Pauling curve with a maximum at x = (0.15–0.20). In $Pr_2(Co_xFe_{1-x})_{14}B$, at RT, the anisotropy field, first decreases with x, but after passing through a minimum, at x = 0.7, strongly increases [87G1]. The H_a values in $Nd_2(Co_xFe_{1-x})_{14}B$ series, declines from x = 0 up to x = 0.9 and then features a small upturn. The uniaxial anisotropy prevails, at RT, over the entire composition range. At low temperatures, the $Nd_2(Co_xFe_{1-x})_{14}B$ system shows conical anisotropy, the cone angle decreasing from 134 K (x = 0), to ≅ 35 K (x = 1). In the $NdYFe_{14-x}Co_xB$ series, the transition from cone- to axial-type anisotropy (T_{sR1} < 125 K) takes place in all the composition range and from axial to plane at T > 325 K; the T_{sR2} values decrease for x > 8 [99P1]. The composition dependence of spin reorientation temperatures was investigated also in $MM_2Fe_{14-x}Co_xB$, where MM is a Ce-free mishmetal [92Z1].

The ^{145}Nd spin echo NMR spectra, of $Nd_2(Co_{1-x}Fe_x)B$ series, show that the transition between the end series compounds, does not occurs as a smooth and gradual change of the hyperfine field at Nd sites [92J1]. With the iron addition, a new signal was observed to build slightly below 500 MHz. The abrupt change of the hyperfine field was correlated with the existence of two different magnetic ground states for Nd in Fe substituted $Nd_2Co_{14}B$. Besides the fully polarized state, as in $Nd_2Co_{14}B$, the other ground state was attributed to the competition between easy-axis orientation, due to crystal field interaction and that due to exchange interactions, which are non-collinear for Nd ions having some Fe atoms in their near neighbour.

A large number of ^{59}Co NMR studies were performed on $Nd_2(Co_{1-x}Fe_x)_{14}B$ compounds [87F1, 88J2, 88Z2, 89W1]. According to [88J2], the line structure disappears and for x = 0.148 collapsed into two broad lines. The average ^{59}Co hyperfine fields was assumed to be constant up to x = 0.3, then increasing. The strong increase in the ^{59}Co local hyperfine field, at the $16k_2$ site, was correlated with their preferential substitution. Later on [89W1], was shown that for x = 0, the $4e$ site revealed a strong uniaxial hyperfine field anisotropy (3.86 T) due to the large orbital contribution, which rapidly decreased by addition of iron. Large basal plane hyperfine field anisotropy has been found also in $4c$ (1.3 T), $16k_2$ (1.1 T) and $16k_1$ (2.4 T) sites.

The $Nd_2Fe_{14-x}Co_xB$ solid solutions, with x ≤ 0.7, have been studied by ^{10}B, ^{11}B, ^{57}Fe and ^{59}Co spin-echo NMR [88Z2]. The substitution of Co for Fe, decreases the hyperfine fields, at the iron nuclei at j_2, k_2, j_1, k_1 and c sites, while no obvious changes at the Coe site, not at the B nuclei. The spin echo amplitude, corresponding to Fe$16k_2$ site, decreases with increasing cobalt content, confirming the presence of cobalt at this site. A complex ^{59}Co NMR spectrum was found in the frequency range 150–260 MHz, with a strong peak around 250 MHz. A similar study has been made on $Nd_{15}Fe_{77-x}Co_xB$ alloys, with $x < 12$, where the main phase was the $Nd_2(Co,Fe)_{14}B$ [87Z1].

The neutron diffraction studies, as well as those by ^{57}Co NMR [87B12, 89P1] or band structure calculations [86I1, 87G2, 88J1, 91C1, 08B1], showed a correlation between cobalt magnetic moments and site environments [86H1, 86L1], Generally, there was not a common point of view concerning the attribution of cobalt moments at the six sites. The higher cobalt moments, or ^{59}Co hyperfine field, in $Nd_2Co_{14}B$, have been attributed to those located at k_2 and c sites [86H1, 88J1, 89P1, 89W1]. The smaller cobalt moments were located at e and k_1 sites having a larger number of B atoms in their nearest environments. A higher spread of the data was shown in $Y_2Co_{14}B$ [86I1, 86L1, 87G2, 87J1, 88B1, 91C1, 21E1]. In a larger extent, the higher cobalt moments in this compound, were considered to be located in c, j_1, k_2 and j_2 sites and the smaller ones in k_1 and e sites—Table 8.45.

The ^{57}Fe hyperfine parameters were determined by Mössbauer spectroscopy in $R_2(Fe_{1-x}Co_x)_{14}B$ with R = Nd [85V1, 87D1, 87M1, 87R1, 87T1, 88S1] and R = Y [87T1, 89S2]. The above studies confirmed that there is a strong preference of iron atoms to occupy the $8j_2$ site (Co-rich samples), while the opposite holds for the $16k_2$ sites (Fe-rich samples). The spectra were fitted according to site occupancies reported by [86H1]. The hyperfine fields, determined at RT, at ^{57}Fe doped $Nd_2Co_{14}B$ are listed in Table 8.48. The ^{57}Fe hyperfine fields are smaller in the Y series, as compared to Nd one. The effects of spin reorientation in $Nd_2Co_{13.85}Fe_{0.15}B$, on the ^{57}Fe Mössbauer hyperfine parameters [87R1] were compared with the data obtained on $Nd_2Fe_{14}B$ compound [85K1, 88G4]. The temperature coefficients of hyperfine fields, at iron sites, in $Nd_2(Co_{0.87}Fe_{0.13})_{14}B$, follow the sequence $j_2 > j_1 > k_1 > k_2 > e > c$ [87R1]. In $R_2(Fe_{1-x}Co_x)_{14}B$, when R = Nd, the ^{57}Fe hyperfine field at j_1 site changes slightly if increasing cobalt content, while for R = Y, decreases markedly. The above behavior was correlated with the presence of three R atoms in their near neighbor. These atoms are magnetic for R = Nd and non-magnetic when R = Y, influencing in opposite way the cobalt moment.

The magnetizations of tetragonal $Nd_2(Co_{1-x}M_x)_{14}B$ compounds with M = Ni ($x \leq 0.20$), M = Cu ($x \leq 0.08$), Ga ($x \leq 0.10$) and M = Ge ($x \leq 0.10$), decrease

Table 8.48 Data obtained by ^{57}Fe Mössbauer spectroscopy on ^{57}Fe doped $Nd_2Co_{14}B$

Compound	T (K)	Site	δ^b (mm/s)	QS (mm/s)	B_{eff} (T)	A^a (%)	References
$Nd_2Co_{14}B$:^{57}Fe	300 K	$8j_2$	−0.11	0.45	26.9	38(14)	[85V1]
		$16k_1$	−0.30	0.23	25.5	9(29)	
		$16k_2$	−0.28	0.13	24.3	23(29)	
		$8j_1$	−0.21	0.08	23.2	17(14)	
		$4e$	−0.31	−0.43	23.1	9(7)	
		$4c$	−0.06	0.10	19.4	4(7)	

[a]Statistical occupation (parentheses)
[b]Relative to α-Fe

when increasing the content of substituting elements [89C1]. The axial-to-planar spin reorientation temperatures, increase slightly with the addition of Ga and Cu, but decrease in the Ni and Ge systems. The addition of M = Nb and Ga in the $Nd_2(Fe,Co,M)_{14}B$ system change the coercive field of sintered magnets. The $Nd_{0.8}Dy_{0.2}(Fe_{86\text{-}x\text{-}y}Co_{0.06}B_{0.08}Nb_xGa_y)_{5.5}$ sintered magnets with, x = 0.015 and y = 0.01 have remanent induction $B_r = 1.045$ T and energy product $(BH)_{max} = 210$ kJ/m^3 [87T2, 87X1].

The saturation magnetizations and Curie temperatures, of $R_2Co_{14-x}Si_xB$, with R = Pr, Nd, Y compounds, decrease as the silicon content increases up to x = 2 (R = Pr, Nd) or x = 1.6 (R = Y) [87P4]. The spin reorientation temperatures, T_{sR2}, disappear for $x \geq 1.3$ (R = Nd) or $x \geq 1.0$ (R = Pr). The low temperature spin reorientations T_{sR1} vanish at $x > 0.3$ (R = Nd). The above trends can be correlated with weakening of the 3d sublattice magnetization and its anisotropy, due to magnetic dilution effects. The anisotropy field in $Nd_2Co_2Fe_{12\text{-}x}Si_xB$, has a maximum at T = 295 K, for $x \cong 0.5$ [87J2]. Although the saturation magnetizations decrease, the T_c values increase up to x = 0.6, as result of Co and Si preference for some iron sites.

The saturation magnetization of $Sm_2Co_{14}B$ is somewhat lower [85B4, 85S1, 86P1, 87S1, 88Z1, 93Z1, 99M1] than those determined in $R_2Co_{14}B$ compounds with non-magnetic elements. This suggests an antiparallel orientation of Sm moment to Co sublattice magnetization, although Sm^{3+} is an L-S state ion [85B4]. It may arises as a consequence of the combined action of exchange and crystal fields on the 4f electron system of Sm^{3+} ion, for which the energy separation of the ground multiplet J = 5/2 and two excited multiplets J = 7/2 and 9/2 is comparatively low and this may cause a sign reversal of the S_z expectation value, relative to that of L_z+2S_z, as already evidenced in $SmCo_5$ [90B1]. The ^{59}Co hyperfine field in $Sm_2(Co_xM_{1-x})_{14}B$ series increase for M = Fe and decrease when M = Mn and Ni with increasing M content [90M1]. In the $Sm_2(Co_{1-x}M_x)_{14}B$ series, with $x \leq 1.0$ (M = Co), $x \leq 0.4$ (M = Mn) and $x \leq 0.1$ (M = Ni) alloys, the hyperfine fields at $16k_1$, $16k_2$, $8j_1$ and $8j_2$ sites were only obtained, the line intensities for 4c and 4e sites being rather small (<7%) [90M1]. The ^{59}Co hyperfine fields decrease if Co is substituted by Mn or Ni.

The $R_2Co_{14}B$ compounds with R = Gd [85B4, 86P1, 87H1, 87H3, 87S1, 87V1, 87Y2, 88Z1, 90R1, 92K3, 93J1, 93M2, 99M1, 01M1] and R = Tb [85B4, 86P1, 87P6, 87P9, 87S1, 88J3, 88Z1, 99M1] are ferrimagnetically ordered. Since Gd^{3+} is in an S-state, the anisotropy of the compound, essentially determined by that of cobalt sublattice, is planar. The Tb^{3+} ion has negative Stevens factor and thus, at low temperatures, the axial symmetry is preferred. Due to planar tendency of cobalt sublattice magnetization and different temperature dependences of the two sublattices anisotropies, a spin reorientation from uniaxial to planar is shown, at high temperatures—Fig. 8.25.

The largest contribution to Kerr effect, in $R_2Co_{14}B$ compounds with R = La, Gd, is given by cobalt sublattice magnetization [87V1]. The angle of rotation spectra was $\varphi_K = 0.24°$ (R = La) and 0.29° (R = Gd).

The ^{59}Co NMR measurements on $Gd_2(Co_{1-x}Fe_x)_{14}B$, with $x \leq 0.5$, by spin echo method [92H1], at T = 4.2 K, as in corresponding Y system, show that the resonance lines broaden and weaken in intensity and line positions shift gradually to higher frequencies when increasing iron content. For x > 0.3, the resonance lines seem to be disturbed as result of spin reorientation, as in case of $Sm_2(Co_{1-x}Fe_x)_{14}B$ series [90M1]. Only the ^{59}Co hyperfine fields corresponding to $16k_1$, $16k_2$, and $8j_1$ sites were possible to be determined.

The spin reorientations in $(La_{1-x}Tb_x)_2Co_{14}B$ compounds arise from a competition between the Tb and Co sublattices anisotropies [01G1]. The former dominates at low temperatures, the latter favoring an easy direction perpendicular to **c**-axis, as evidenced at high temperatures. Two spin reorientation temperatures were found in the composition range $0.3 \leq x \leq 0.7$, suggesting the occurrence of an easy cone structure, in limited temperature range.

The introduction of hydrogen in $R_2Co_{14}B$ compounds, decreases the cobalt sublattice magnetization [88Z1]. This effect was correlated with electron charge transfer from hydrogen to the Co3d sublattice. The hydrogenation significantly decreased the anisotropy field, H_a and the spin reorientation temperatures T_{sR2} of Pr- and Tb-based compounds.

The exchange interactions in $R_2Co_{14}B$ compounds with R = Pr, Nd, Sm, Gd, Tb [99M1], R = Nd [88L1, 94L1], R = Sm, Gd [01M1], R = Gd, Tb, (DyLa) and (HoLa) [90R1] were evaluated. The exchange fields H_{exch}, and crystal field parameters, A_n^m were analysed in $R_2Co_{14}B$ compounds with R = Pr, Nd, Gd [88K2, 88L1, 91Y1, 93J1]—Table 8.47. The A_n^m values in $R_2Co_{14}B$ are comparable with those reported in $R_2Fe_{14}B$ series—Sect. 7.10. The fact that A_2^0 and A_4^0 values, relative to those calculated in $R_2Co_{14}B$ (R = Pr, Nd) are slightly larger and smaller, respectively than the values determined in iron compounds, implies quantitatively an enhancement of the uniaxial anisotropy in $R_2Co_{14}B$ series, since the sign of A_4^0 is negative in contrast to A_2^0, in both series [88K2]. This sentence can be reanalysed [91Y1]. The exchange fields, H_{exch}, are smaller in $R_2Co_{14}B$ (R = Pr, Nd) compounds than in isomorphous iron series. This may be correlated with the smaller value of averaged cobalt magnetic moment as compared to iron ones [88K2, 88L1]. The anisotropic exchange interactions in $R_2Co_{14}B$ (R = Pr, Nd, Gd) compounds are by two order of magnitude smaller than the isotropic interactions and are of the same order of magnitude as the crystalline electric-field interactions [93J1].

The physical properties of pseudoternary $(R'_{1-x}R''_x)_2Co_{14}B$ compounds as: $(La_{1-x}R_x)_2Co_{14}B$ with R = Tb [01G1], R = Dy, Ho [97M2], R = Ho [90R1], $(Pr_{1-x}R_x)_2Co_{14}B$ with R = Tb, Dy, Ho, Tm [87J1], R = Dy, Er [88P2], R = Y [87P7], R = Nd [87P9, 91S1, 92I1], $(Nd_{1-x}R_x)_2Co_{14}B$ with R = Y [87P7, 88K1, 90I1, 91J1, 91J2, 92I1], R = Sm, Gd, Tb [92I1], R = Tb [88J3], $(R'_{1-x}R''_x)_2Fe_{12}Co_2B$ with R' = Pr, Nd and R" = Tb, Dy [86F2], $(Gd_{1-x}R_x)_2Co_{14}B$ with R = Gd, Dy, Ho [94M1, 96M1], $(Y_{1-x}R_x)_2Co_{14}B$ with R = Tb, Dy, Ho [94M1, 96M1] were investigated. The spin reorientation temperatures, obtained by extrapolation at x = 1, of the values determined in $(La_{1-x}R_x)_2Co_{14}B$ with R = Dy, Ho, up to $x \cong 0.6$, as well as of $Tb_2Co_{14}B$, follow a linear dependence on the absolute value of the Stevens factor [97M2].

In the pseudoternary series involving both light and heavy rare-earths their magnetic moments are antiparallelly coupled. Thus, $Pr_{2-x}R_xCo_{14}B$ compounds with R = Dy or Er are ferrimagnetically ordered, when R sublattice moment is higher than of Pr one. The replacement of Pr by Dy or Er, diminished the spin reorientation temperatures, T_{sR2}, as well as the anisotropy field [88P2]. In $(Pr_xR_{1-x})_2Co_{14}B$ series, increasing Pr content, diminished the spin reorientation temperatures when R = Tb and increased when R = Dy, Ho, Tm [87J1]. The magnetic phase diagrams of $(Nd_{1-x}R_x)_2Co_{14}B$ series with R = Pr, Gd, Tb, Y were determined by NMR method. [92I1]. The critical concentrations of R elements, where the spin reorientation phenomena occurs (axis → plane), is around x = 0.9 (R = Gd, Y) and x = 0.45 (R = Sm). Near these concentrations, the ^{57}Co NMR spectra change drastically. The region of axial spin arrangement is getting larger, by addition of R = Tb and Pr.

A large number of studies were devoted to the analysis of the magnetic behavior of $RCo_{14-x}Fe_xB$ series, R = Gd [86H2, 87P5, 92H1]; R = Tb [87P2, 87P6, 98K3]; R = Dy [87P3]; R = Er [87P2, 87P3, 87P5, 09K4]; R = (Nd,Dy) [90Z3, 91S3]; R = Tm [87P2], R = Y [85B3, 86B2, 86H2, 86S1, 87B4, 87B8, 87C1, 87P2, 87T1, 88F1, 88G1, 88B1, 88T2, 88Z1, 89H1, 89S2, 92M1, 93M1]; R = MM (mishmetal) [88J3, 88J4, 92Z1]. The $R_2Fe_{14-x}Co_xB$ series with R = Tb, Dy, Ho, Er and Tm are ferrimagnetically ordered. A spin reorientation from axis to plane occurs at high temperatures (T_{sR2}) in $Tb_2Fe_{14-x}Co_xB$, for $x \geq 12.5$, due to competing effects of Tb and 3d sublattices anisotropies. The T_{sR2} values decrease, as the cobalt content increases [87P6]. In $NdTbFe_{14-x}Co_xB$ system, the addition of cobalt increases T_c, but reduces the anisotropy fields, at RT [88J3]. In the composition range $x \leq 5$, the spin reorientation temperatures from planar ($T < T_{sR2}$) to axial ($T > T_{sR2}$), in the $Er_2Fe_{14-x}Co_xB$ pseudoternary system, increased from 330 K (x = 0) to 395 K (x = 5) [87P3, 09K4]. The longitudinal magnetostriction, λ_l, undergoes an anomaly around spin reorientation temperatures [09K4]. The longitudinal magnetostriction below the spin reorientation temperature, decreases when cobalt content increases. An invar type behavior was shown in $Er_2Fe_{13}CoB$ at T > 200 K, attributed to the compensation of phonon and magnetic contributions.

The magnetic phase diagrams for $R_2Fe_{14-x}Co_xB$ systems, with R having Stevens coefficients $\alpha_J < 0$ (R = Pr, Nd, Tb) and $\alpha_J > 0$ (R = Er, Tm), were reported [87P2]. The dominant regions in the alloys having rare earths with $\alpha_J < 0$, are those with axial arrangement. For these compounds, the strong axial anisotropies of rare earths, dominate over the 3d sublattices anisotropies. However, due to competing effects of the Co sublattice and also thermal effects, the planar and conical spin arrangements also occur. The compositions for which axis-to-plane transitions are obtained are $x \geq 9.5$ (R = Pr), $x \geq 10$ (R = Nd) and $x \geq 12.6$ (R = Tb). The conical anisotropy can be observed mainly for Nd-based systems. The magnetic phase diagrams of $R_2Fe_{14-x}Co_xB$ with R = Er and Tm ($\alpha_J > 0$) show only a change from planar to axial orientation, as the temperature increases. When the cobalt content is higher, the T_{sR2} values increase, their planar contributions adding to those of Er and Tm sublattices.

The domain wall energy in $Nd_2(Fe,Co,Al)_{14}B$ alloys has been analysed [87S2]. The magnetic properties of multicomponent alloys, having tetragonal $P4_2/mnm$ structure with substitutions both at R and at Co sites, were reported, particularly for their evaluation for technical uses, as permanent magnets: $Pr_2Fe_{14-x-y}Co_xAl_yB$ [87B9, 87B10], $Pr_{2-x-y}R_xR'_yFe_{11.6}Co_2Al_{0.4}B$ with R' = Nd, R = Tb, Dy and $x \leq 0.2$, $y \leq 0.5$

[88P1], $R_2Fe_{12-x}Mn_xCo_2B$ wit R = Pr, Nd [86J2], $R_2(Fe,Co,Nb)_{14}B$ with R = Pr, Nd [89J1], $Nd_2(Fe,Co,Al)_{14}B$ [87B8, 87B9, 87B11], $Nd_2Fe_{12-x}M_xCo_2B$ with M = Al, V, Cr [88J6], $MM_2Fe_{12-x}Co_2M_xB$ with M = Mo, Al [88J6], or with M = Si, V, Cr, Ta and W [88J5]. The substitution of Fe by Mn in $R_2Co_2Fe_{12-x}Mn_xB$ series with R = Pr, Nd decreases the Curie temperatures and saturation magnetizations, while for x ≤ 0.6, the anisotropy field increases [86J2]. In $Pr_{2-x-y}R_xR'_yFe_{11.6}Co_2Al_{0.4}B$ system, partial replacement of Pr by either R = Tb or Dy, significantly enhances the H_a, while the presence of Nd reduces the anisotropy field. The former trend was ascribed to the strong magnetic anisotropy of Tb (Dy) ions, which originates from the strong crystal field interaction [88P1].

Permanent magnets containing multiphases, the tetragonal one being the dominant one, were obtained by sintering mainly multicomponent alloys as $Pr_2Co_{14}B$, (Pr,Dy,Gd)(Fe,Co)B [10P2], (Nd,Pr)(Gd,Tb,Dy)(Fe,Co,Al,Cu)B [13A1], (Nd,Dy)(Fe,Co)B [90Z3], Nd(Fe,Co)B [88W1], Nd(Fe,Co,Ga)B [90F1], (Nd,Dy)(Fe,Co)B with Al, Cr, Nb additions [91S3], (Nd,Dy)(Fe,Co,Al) [07K1], (Nd,Dy)(Fe,Co,Cu,Ga,Nb)B [01P1], by HDDR as Pr(Fe,Co,Nb)B [04B1], by melt spinning $PrCo_{14}B$ [88F2, 93C5], (Pr,Tb)(Co,M)B with M = Fe,Cr,Mn [98K3], Pr(Fe,Co)B ribbons [01Z2], $Nd_9(Fe,Co)_{85}B_6$ [96G1], Nd(Fe,Co)B [87Y1, 96G1], (Nd,Y,Dy)(Fe,Co)B+Ti_2C_2 [06T1], (Nd,Pr)(Fe,Co)B [04L1, 11X1], Nd(Fe,Co,Al)B [04B2, 09K5] or by rapid quenching (Nd,Pr)(Fe,Co)B [88W2]. In addition to the 2/14/1 hard magnetic phase, the above alloys contain a soft one, mainly α-(Fe,Co) or (Fe,Co)-B [96G1, 01Z2, 04L1]. Energy product $(BH)_{max}$ = 115.2 kJ/m^3 and $\mu_0H_c > 2$ T were obtained for $Nd_{13}Dy_2Fe_{77-y}Co_xC_6B_2$ nanocomposites [01Z1] or 94.4 kJ/m^3 in (Nd,Dy,Y)(Co,Fe)B + Ti_2C_2 alloys [06T1]. The magnetic hardness, of rapidly quenched $(Nd, Pr)_2(Fe,Co)_{14}B$ alloys, was correlated with the grain sizes, as well as the existence of secondary phases [09K5]. Coercivities of μ_0H_c = 2.2–2.5 T were reported in the above system. In all the magnets, the temperature coefficient of remanent induction is improved in the presence of cobalt, although their coercivities decrease [85A1, 86M1]. The magnetic properties of multiphase systems as $Nd_{15}(Fe,Co)_{78}B_7$ [85Y2], $Nd_{16}Co_{76-x}Ru_xB_6$ with $x \leq 27$ [06P1] or $Y_{15}Fe_{77-x}Co_xB$ with $x \leq 17$ [86S1], were also investigated in correlation with their basic properties.

The effects of substitutions of B by C, on the magnetic properties of $R_2Co_{14}B$-based compounds, were also analysed. The saturation magnetizations of $R_2Co_{14}B_{1-x}C_x$ with R = Pr, Nd decrease when replacing C for B [92Z2]. The critical fields of the FOMP, increase in the same way with C content in $Nd_2Co_{14}B_{1-x}C_x$ series, while in the corresponding Pr compounds no FOMP was observed. The spin reorientation temperatures in $R_2Co_{14}B_{1-x}C_x$ with R = Pr, Nd decrease, as the carbon content increases. In the $R_2Co_{14}B_{1-x}C_x$ with R = Y, Sm series, the anisotropy increased, whereas saturation magnetizations diminishe when B is substituted by C [93Z1].

The main phase in the $Nd_{16}Co_{76}B_{8-x}C_x$ alloys with $x \leq 7$, is $Nd_2Co_{14}(B,C)$ one, while for x = 8, a $TbCu_7$-type structure ($NdCo_7C_x$), was shown [07C2]. A similar behavior was reported in $Dy_{15}Fe_{77}(B_{1-x}C_x)_6$ system [89S1]. As evidenced in single tetragonal phase, the spin reorientation temperatures, of the above system, increase with C content, due to enhancement of the magneto-crystalline anisotropy of hard magnetic phase. The spin reorientation temperatures, for $Nd_{16}Co_{76}B_{8-x}C_x$ samples, with $x \leq 7$, decrease after sample's hydrogenation [07C2].

References

[889H1] J. Hopkinson, Phil. Trans. Royal. Soc. London A **180**, 443 (1889)
[48N1] L. Néel, Ann. Phys. (Paris) **3**, 137 (1948)
[54B1] G. Bergman, D.P. Shoemaker, Acta Cryst. **7**, 857 (1954)
[56J1] H. Jones, *Handbuch der Physik*, Vol. 19, (Springer Verlag, Berlin, 1956), p. 227
[62W1] E.P. Wohlfahrt, P. Rhodes, Phil. Mag. **7**, 1817 (1962)
[63B1] B. Bleaney, Proc. Royal Soc. A **276**, 19 (1963)
[6801] S. Ofer, I. Nowik, S.G. Cohen, in *Chemical Applications of the Mössbauer Spectroscopy* (Academic Press, New York, 1968)
[68V1] W.A.J.J. Velge, K.H.J. Buschow, J. Appl. Phys. **39**, 1717 (1968)
[69K1] Y.B. Kuzma, P.I. Krypyakevich, N.S. Bilonizhko, Dopov. Akad. Nauk. Ukr. RSR, A **10**, 939 (1969)
[71C1] B.R. Cooper, O. Vogt, J. Physique (Paris) **32**, 958 (1971)
[71K1] Y.B. Kuzma, N.S. Bilonizhko, Kristallografiya **16**, 3211 (1971)
[71K2] Y.B. Kuzma, N.S. Bilonizhko, Kristallografiya, **16**, 1030 (1971)
[71N1] K. Nihara, T. Shishido, S. Yajima, Bull. Chem. Soc. Jpn. **44**, 3214 (1971)
[71S1] H.E. Stanley, *Introduction to Phase Transitions and Critical Phenomena* (Oxford University Press, London, 1971)
[72K1] Y.B. Kuzma, N.S. Bilonizhko, Sov. Phys. Cristallogr. **16**, 897 (1972)
[72N1] K. Niihara, S. Yajima, Chem. Lett. Chem. Soc. Jpn. 875 (1972)
[73K1] Y.B. Kuzma, N.S. Bilonizhko, Kristallogr. **18**, 710 (1973)
[73N1] K. Niihara, Y. Katayama, S. Yajima, Chem. Lett. **6**, 613 (1973)
[73N2] K. Niihara, T. Shishido, S. Yajima, Bull. Chem. Soc. Jpn. **46**, 1137 (1973)
[73N3] K. Niihara, S. Yajima, Bull. Chem. Soc. Jpn. **46**, 770 (1973)
[73R1] P. Rogl, Monatsh. Chemie, B **104**, 1623 (1973)
[74B1] N.S. Bilonizhko, Y.B. Kuzma, Izv. Akad. Nauk., SSSR, Neorg. Mater. **10**, 266 (1974)
[74G1] K.A. Gschneinder, M.E. Verkade, *Selected Cerium Phase Diagrams, IS-RIC-7* (Iowa State University, Ames, 1974), p. 15
[74K1] Y.B. Kuzma, G.F. Stepanchikova, Izv. Acad. Nauk.: Neorg. Mater. **10**, 2223 (1974)
[74K2] Yu B. Kuzma, N.N. Bilonizhko, Sov. Phys. Crystallogr. **18**, 447 (1974)
[74K3] Y.B. Kuzma, N.S.Bilonizhko, Sov. Phys. Cryst. **18**, 710 (1974)
[74M1] N. Mori, T. Ukai, H. Yoshida, J. Phys. Soc. Jpn. **37**, 1272 (1974)
[74R1] A.E. Ray, Cobalt **1**, 13 (1974)
[75B1] D. Bloch, D.M. Edwards, M. Shimizu, J. Voiron, J. Phys.: Metal Phys. **5**, 1217 (1975)
[75B2] E. Burzo, Solid State Commun. **20**, 569 (1975)
[75K1] Y.B. Kuzma, M.P. Khaburskaya, Inorg. Mat. **10**, 1625 (1975)
[76D1] J. Deportes, D. Givord, J. Schweizer, F. Tasset, IEEE Trans. Magn. **12**, 1000 (1976)
[76S1] G.F. Stepanchikova, Y.B. Kuzma, Vest. Lvov. Gos. Univ., Khim. **18**, 16 (1976)
[77B1] N.S. Bilonizhko, Y.B. Kuzma, An. Ukr. SSSR Khim. 194 (1977)
[77C1] N.F. Chaban, Y.B. Kuzma, Izv. Akad. Nauk SSSR, Neorg. Mater. **13**, 923 (1977)
[77O1] H. Oesterreicher, F.T. Parker, M. Misroch, Appl. Phys. **12**, 287 (1977)
[77S1] G.F. Stepanchikova, Y.B. Kuzma, Vestn. Lvov. Univ. Khim. **19**, 37 (1977)
[77W1] A. F. Well, *Three-dimensional nets and polyhedral* (Wiley, New York, 1977)
[78B1] E. Burzo, Rev. Roum. Phys. **23**, 689 (1978)
[78K1] Y.B. Kuzma, N.S. Bilonizhko, Dopov. Akad. Nauk. Ukr. RSR A **276** (1978)
[78S1] H.H. Stadelmaier, H.J. Lee, Z. Metallkd. **69**, 685 (1978)
[78S2] R.L. Streever, Phys. Lett. A **65**, 360 (1978)

[78W1] E.P. Wohlfarth, J. Magn. Magn. Mater. **7**, 113 (1978)
[79K1] Y.B. Kuzma, N.S. Bilonizhko, S.I. Mykhalenko, G.F. Stepanchikova, J. less Common Met. **67**, 51 (1979)
[79L1] G.H, Lander, S.K. Sinha, F.Y. Fradin, J. Appl. Phys. **50**, 1990 (1979)
[79M1] T. Moriya, J. Magn. Magn. Mater. **14**, 1 (1979)
[79S1] R. Sobczak, P. Rogl, J. Solid State Chem. **27**, 343 (1979)
[79S2] R.L. Streever, Phys. Rev. B **19**, 2704 (1979)
[80B1] N.S. Bilonizhko, Y.B. Kuzma, I.D. Polyanskaya, Izv. Acad. Nauk. SSSR, Neorg. Mater. **16**, 832 (1980)
[80J1] W. Jung, D. Quentmeier, Z. Kristallogr. **151**, 121 (1980)
[80L1] J.C. Le Guillou, J. Zinn-Justin, Phys. Rev. B **21**, 3976 (1980)
[80S1] J. Schweizer, F. Tasset, J. Phys. F: Metal Phys. **10**, 1789 (1980)
[81B1] E. Burzo, J. less Common Met. **155**, 251 (1981)
[81C1] N.F. Chaban, Vestn. Lvov Gos. Univ. Khim. **23**, 43 (1981)
[81K1] Y.B. Kuzma, N.S. Bilonizhko, Dopov. Akad. Nauk. Ukr. RSR. Ser. A : Fiz-Mat. Tekn. Nauki, **10**, 87 (1981)
[81K2] Y.B. Kuzma, G.V. Chernyak, N.F. Chaban, Dopov. Acad. Nauk. Ukr., RSR A **12**, 80 (1981)
[81K3] Y.B. Kuzma, N.S. Bilonizhko, N.F. Chaban, G.V. Chernyak, J. less Common Met. **82**, 364 (1981)
[81O1] H.R. Ott, A.M. Campbell, H. Rodriguez, H.C. Hamaker, M.B. Maple, Physica B+C **108**, 751 (1981)
[81O2] H.R.Ott, W.A. Fertig, D.C. Johnson, M.P. Maple, B.T. Mathias, J. Phys. **39**, C6–375 (1981)
[82C1] I. A. Campbell, A. Fert, *Transport Properties of Ferromagnetics* (North Holland, 1982)
[83B1] N.S. Bilonzhko, Y.B. Kuzma, Inorg. Mater. **19**, 438 (1983)
[83B2] N.S. Bilonizhko, Yu. B. Kuzma, Izv. Akad. Nauk. SSSR., Neorg. Mater. **19**, 487 (1983)
[83C1] G.V. Chernyak, N.F. Chaban, Y.B. Kuzma, Inorg. Mater. **19**, 1554 (1983)
[83E1] N.A. El Masry, H.A. Stadelmair, Zeit. Metallk. **74**, 86 (1983)
[83E2] N.A. El Masry, H.H. Stadelmaier, L.Y. Shahwan, L.T. Jordan, Z. Metall. **74**, 33 (1983)
[83G1] H.O. Gupta, S.K. Malik, W.E. Wallace, J. Magn. Magn. Mater. **40**, 32 (1983)
[83K1] Y.B. Kuzma, *The Chemistry of Borides (in Russian)* (Vyshcha Shkola, Lvov, 1983)
[83K2] Yu. B. Kuzma, N.S. Belonizhko, N.F. Chaban, G.V. Chernyak, J. less Common Met. **90**, 217 (1983)
[83K3] Y.B. Kuzma, L.G. Akselrud, N.F. Chaban, V.A. Bruskov, Dopov. Akad. Nauk. Ukr. RSR, B Geol. Khim. Biol. Nauki, **10**, 51(1983)
[84E1] N.A. El Masry, H.H. Stadelmaier, J. less Common Met. **96**, 165 (1984)
[84F1] I.Felner, Solid State Commun. **52**, 191(1984)
[84G1] D. Givord, H.S. Li, J.M. Moreau, Solid State Commun. **50**, 497 (1984)
[84O1] H. Oesterreicher, F. Spada, C. Abache, Mater. Res. Bull. **19**, 1069 (1984)
[84P1] E. Parthé, B. Chabot, Handbook Phys. Chem. Rare Earths, North Holland, **6**, Chap. 48 (1984)
[84R1] P. Rog, Handbook on Physics and Chemistry of Rare Earths, North Holland, **6**, Chap. 49 (1984)
[84S1] M. Sagawa, S. Fujimura, H. Yamamoto, K. Hiraga, IEEE Trans. Magn. MAG **20**, 1584 (1984)
[84S2] C. B. Shoemaker, D. P. Shoemaker, R. Fruchard, Acta Cryst. C **40**, 1665 (1984)
[84S3] F. Spada, C. Abache, H. Oesterreicher, J. less Common Met. **99**, L21 (1984)
[84U1] K. Ueda, Y. Kitaoka, Y. Kohori, T. Kohara, K. Asayama, K. Kumagai, H.C. Ku, F. Acker, J. Phys. Soc. Jpn. **53**, 2674 (1984)
[84V1] A.R. Victora, L.M. Falicov, Phys. Rev. B **30**, 259 (1984)
[84Y1] K.N. Yang, M.S. Torikachvili, M.S. Maple, H.C. Ku, J. Low Temp. Phys. **56**, 601 (1984)
[84A1] S. Arai, T. Shibata, IEEE Trans. Magn. MAG **21**, 1952 (1985)
[85B1] A.Bezinge, N.F.Braun, J.Muller, K.Yvon, Solid State Commun. **55**, 131 (1985)

[85B2] N.S. Bilonizhko, L.V. Leshko, Y.B. Kuzma, Inorg. Matter **21**, 1568 (1985)
[85B3] E. Burzo, L. Stanciu, W.E. Wallace, J. less Common Met. **111**, 83 (1985)
[85B4] K.H.J. Buschow, B.D. de Mooij, S. Sinnema, R.J. Radwanski, J.J.M. Franse, J. Magn. Magn. Mater. **51**, 211 (1985)
[85D1] O.M. Dub, N.F. Chaban, Y.B. Kuzma, Inorg. Mater. **21**, 1718 (1985)
[85G1] D. Givord, J.M. Moreau, P. Tenaud, Solid State Commun. **55**, 303 (1985)
[85K1] R. Kamal, Y. Andersson, Phys. Rev. B **32**, 1756 (1985)
[85L1] D. Le Roux, H. Vincent, P.L'Héritier, R. Fruchart, J. Physique, **46**, C6–243 (1985)
[85M1] S.K. Malik, A.M. Umarji, G.K. Shenoy, J. Appl. Phys. **57**, 3252 (1985)
[85M2] Y. Matsuura, S. Hirosawa, H. Yamamoto, S. Fujimura, M. Sagawa, Appl. Phys. Lett. **46**, 308 (1985)
[85P1] E. Potenziani, J. Appl. Phys. **58**, 2764 (1985)
[85S1] S. Sinnema, J.J.M Franse, R.J. Radwanski, K.H.J. Buschow, J. Physique, **46**, C6–301 (1985)
[85S2] F. Spada, H. Oesterreicher, J. less Common Met. **107**, 301 (1985)
[85V1] H.M. van Noort, K.H.J. Buschow, J. less Common Met. **113**, L9 (1985)
[85Y1] K.N. Yang, M.S. Torikachvili, M.P. Maple, H.C. Kiu, J. Appl. Phys. **57**, 3140 (1985)
[85Y2] Y.C. Yang, W.W. Ho, H.Y. Chen, J. Wang, J. Lan, J. Appl. Phys. **57**, 4118 (1985)
[86B1] W. Boehner, H. Lütgemeier, W. Zinn, J. Magn. Magn. Mater. **62**, 152 (1986)
[86B2] E. Burzo, N. Plugaru, V. Pop, L. Stanciu, W.E. Wallace, Solid State Commun. **58**, 803 (1986)
[86B3] K.H.J. Buschow, B.D.de Mooij, H.M van North, J. less Common Met. **125**, 136 (1986)
[86D1] O.M. Dub, N.F. Chaban, Y.B. Kuzma, J. less Common Met. **117**, 297 (1986)
[86F1] C.D. Fuerst, J.F. Herbst, E.A. Alson, J. Magn. Magn. Mater. **54–57**, 567 (1986)
[86F2] H. Fujii, W.E. Wallace, E.B. Boltich, J. Magn. Magn. Mater. **61**, 251 (1986)
[86G1] D.Givord, J.M.Moreau, P.Tenaud, J. less Common Met. **123**, 109 (1986)
[86G2] R. Grössinger, R. Krewenka, X.K. Sun, R. Eibler, H.R. Kirchmayr. K.H.J. Buschow, J. less Common Met. **124**, 165 (1986)
[86H1] J.F. Herbst, W.B. Yelon, J. Appl. Phys. **60**, 4224 (1986)
[86H2] M.Q. Huang, E.B. Boltich, W.E. Wallace, E. Oswald, J. Magn. Magn. Mater. **60**, 270 (1986)
[86I1] J. Inoue, M. Shimizu, J. Phys. F.: Met. Phys. **16**, L157 (1986)
[86J1] S.Y. Jiang, W.E. Wallace, E. Burzo, J. Magn. Magn. Mater. **61**, 257 (1986)
[86J2] M. Jurczyk, W.E. Wallace, J. less Common Met. **124**, 149 (1986)
86K1] Cz. Kapusta, Z. Kakol, H. Figiel, R.J. Radwanski, J. Magn. Magn. Mater. **59**, 169 (1986)
[86L1] D. Le Roux, PhD Thesis, J. Fourier University, Grenoble, 1986
[86M1] T. Mizoguchi, I. Sakaki, H. Niu, K. Inomata, IEEE Trans. Magn. MAG **22**, 919 (1986)
[86P1] A.T. Pedziwiatr, W.E. Wallace, Solid State Commun. **60**, 653 (1986)
[86P2] A.T. Pedziwiatr, W.E. Wallace, J. less Common Met. **126**, 41 (1986)
[86P3] A.T. Pedziwiatr, S.Y. Jiang, W.E. Wallace, J. Magn. Magn. Mater. **62**, 29 (1986)
[86S1] A. Szytula, J. Drajewicz, Solid State Commun. **57**, 809 (1986)
[86Y1] J.Yamauchi, M. Yamada, Y. Yamaguchi, S. Hirosawa, M. Sagawa, J. Magn. Magn. Mater. **54–57**, 575 (1986)
[87B1] A. Benzinge, K. Yvon, H.F. Braun, J. Muller, Phys. Rev. B **36**, 1406 (1987)
[87B2] E.B. Boltich, A.T. Pedziwiatr, W.E. Wallace, J. Magn. Magn. Mater. **66**, 317 (1987)
[87B3] F. Bolzoni, J.R. Gavigan, D. Givord, H. S. Li, O. Mozé, L. Pareti, J. Magn. Magn. Mater. **66**, 158 (1987)
[87B4] F. Bolzoni, F. Leccabue, O. Mozé, L. Pareti, M. Solzi, A. Deriu, J. Appl. Phys. **61**, 5369 (1987)
[87B5] F. Bolzoni, J.M.D. Coey, J. Gavigan, D. Givord, O. Mozé, L. Pareti, T. Viadieu, J. Magn. Magn. Mater. **65**, 123 (1987)
[87B6] E. Burzo, M. Ursu, J. Magn. Magn. Mater. **70**, 345 (1987)
[87B7] E. Burzo, I. Creanga, M. Ursu, Solid State Commun. **64**, 585 (1987)
[87B8] E. Burzo, A.T. Pedziwiatr, W.E. Wallace, Solid State Commun. **61**, 57 (1987)

[87B9] E. Burzo, N. Plugaru, V. Pop, J. Magn. Magn. Mater. **70**, 343 (1987)
[87B10] E. Burzo, N. Plugaru, V. Pop, Acta Phys. Polonica, A **72**, 33 (1987)
[87B11] E. Burzo, V. Pop, N. Plugaru, L. Stanciu, Acta Phys. Polonica, A **72**, 37 (1987)
[87B12] K. H. J. Buschow, Mat. Res. Soc. Symp. Proc. **96**, 1 (1987)
[87C1] B.P. Cheng, Y.C. Yang, S.C. Fu, W.J. James, J. Appl. Phys. **61**, 3586 (1987)
[87C2] Y.C. Chuang, C.H. Wu, F.H. Wang, J. less Common Met. **135**, 5 (1987)
[87D1] P. Deppe, M. Rosenberg, S. Hirosawa, M. Sagawa, J. Appl. Phys. **61**, 4337 (1987)
[87D2] Z. Drzazga, A. Winiarska, S. Kubiak, Acta Phys. Polonica, A **72**, 45 (1987)
[87F1] H. Figiel, D. Fowler, T. Dumelow, P.C. Riedi, Cz. Kapusta, J. Magn. Magn. Mater. **65**, 83 (1987)
[87G1] R. Grössinger, R. Krewenka, H.R. Kirchmayr, S. Sinnema, F.M. Yang, Y.K. Huang, F.R. de Boer, K.H.J. Buschow, J. less Common Met. **132**, 265 (1987)
[87G2] Z.Q. Gu, W.Y. Ching, Phys. Rev. B **36**, 8530 (1987)
[87G3] Z. Q. Gu, W. Y. Ching, J. Appl. Phys. **61**, 3977 (1987)
[87H1] S. Hirosawa, K. Tokuhara, H. Yamamoto, S. Fujimura, M. Sagawa, H. Yamauchi, J.Appl. Phys. **61**, 3571 (1987)
[87H2] S. Hirosawa, Y. Yamaguchi, K. Tokuhara, H. Yamamoto, S. Fujimura, M. Sagawa, IEEE Trans. Magn. MAG **23**, 2120 (1987)
[87H3] H. Hiroyoshi, M. Yamada, N. Saito, H. Kato, Y. Nakagawa, S. Hirosawa, M. Sagawa, J. Magn. Magn. Mater. **70**, 337 (1987)
[87H4] H. Honma, H. Ino, IEEE Trans. Magn. MAG **23**, 3116 (1987)
[87J1] S.Y. Jiang, F. Pourarian, E.B. Boltich, W.E. Wallace, IEEE Trans. Magn. MAG **23**, 2720 (1987)
[87J2] M. Jurczyk, A. Kowalczyk, J. Magn. Magn. Mater. **68**, 331 (1987)
[87J3] M. Jurczyk, A.T. Pedziwiatr, W.E. Wallace, J. Magn. Magn. Mater. **67**, L1 (1987)
[87J4] M. Jurczyk, A.T. Pedziwiatr, S.G. Sankar, W.E. Wallace, J. Magn. Magn. Mater. **68**, 257 (1987)
[87K1] H.C. Ku, P. Klavins, R.N. Shelton, Physica B **148**, 117 (1987)
[87K2] H.C. Ku, M.F. Tai, P. Klavins, R.N. Shelton, Jpn. J. Appl. Phys. **26**, 827 (1987)
[87M1] M. Matsui, M. Doi, T. Shimizu, IEEE Trans. Magn. MAG **23**, 3113 (1987)
[87P1] L. Pareti, M. Solzi, F. Bolzoni, R. Panizzieri, Solid State Commun. **61**, 761 (1987)
[87P2] A.T. Pedziwiatr, W.E. Wallace, Solid State Commun. **64**, 1017 (1987)
[87P3] A.T. Pedziwiatr, W.E. Wallace, J. Magn. Magn. Mater. **66**, 63 (1987)
[87P4] A.T. Pedziwiatr, W.E. Wallace, J. Appl. Phys. **61**, 3439 (1987)
[87P5] A.T. Pedziwiatr, W.E. Wallace, J. Magn.Magn.Mater. **65**, 139 (1987)
[87P6] A.T. Pedziwiatr, H.Y. Chen, W.E. Wallace, J. Magn. Magn. Mater. **67**, 311 (1987)
[87P7] A.T. Pedziwiatr, H.Y. Chen, W.E. Wallace, S.K. Malik, IEEE Trans. Magn. MAG **23**, 2717 (1987)
[87P8] A.T. Pedziwiatr, S.Y. Jiang, W.E. Wallace, E. Burzo, V. Pop, J. Magn. Magn. Mater. **66**, 69 (1987)
[87P9] F. Pourarian, S.G. Sankar, A.T. Pedziwiatr, E.B. Boltich, W.E. Wallace, Mat. Res. Symp. Proc. **96**, 103 (1987)
[87R1] M. Rani, R. Kamal, J. Less Common Met. **128**, 343 (1987)
[87R2] B. Rupp, P. Rogl, F. Hulliger, J. Less Common Met. **135**, 113 (1987)
[87S1] M. Shimao, H. Ido, G. Kido, K. Ohashi, IEEE Trans. Magn. MAG **23**, 2722 (1987)
[87S2] R. Szymczak, H. Szymczak, E. Burzo, IEEE Trans. Magn. MAG **23**, 2536 (1987)
[87T1] D.E. Tharp, Y.C. Yang, O.A. Pringle, G.J. Long, W.J. James, J. Appl. Phys. **61**, 4334 (1987)
[87T2] M. Tokunaga, H. Kogure, M. Endoh, H. Harada, IEEE Trans Magn. MAG **23**, 2287 (1987)
[87V1] P.P.J. van Engelen, K.H.J. Buschow, J. Magn. Magn. Mater. **66**, 291 (1987)
[87X1] Y. Xiao, K.J. Strnat, H.F. Mildrum, A.E. Ray, in *Proc. 9th International. Workshop on RE Magnets and Their Applications* (1987), p. 467
[87Y1] H. Yamamoto, S. Hirosawa, S. Fujimura, K. Tokuhara, H. Nagata, M. Sagawa, IEEE Trans. Magn. MAG **23**, 2100 (1987)

[87Y2] H. Yamauchi, H. Yamamoto, H. Hirosawa, M. Sagawa, J. Magn. Magn. Mater. **70**, 340 (1987)
[87Y3] Y.C. Yang, S.C. Fu, N.J. James, W.B. Yelon, D. Xie, J. Yang, J. Appl. Phys. **61**, 3589 (1987)
[87Z1] Y.D. Zhang, J.I. Budnick, M. Wojcik, E. Potenziani, A.T. Pedziwiatr, W.E. Wallace, Phys. Rev. B **36**, 8213 (1987)
[88A1] S.H. Aly, E. Singleton, G.C. Hadjipanayis, D.J. Sellmyer, Z.R. Zhao, J. Appl. Phys. **63**, 3704 (1988)
[88A2] S.H. Aly, E. Singleton, G.C. Hadjipanayis, G. Zouganalis, D. Niarchos, J. Magn. Magn. Mater. **75**, 88 (1988)
[88B1] Y. Berthier, N. Nassar, T. Viadeu, J. Phys. **49**, C8–585 (1988)
[88B2] E. Burzo, I. Creanga, N. Plugaru, V. Pop, M. Ursu, Rev. Roum. Phys. **33**, 57 (1988)
[88C1] J.M. Cadogan, J. P. Gavigan, D. Givord, H.S. Li, J. Phys. F: Metal Phys. **18**, 779 (1988)
[88D1] O.M. Dub, N.F. Chaban, Y.B. Kuzma, Porovsk. Metall. **10**, 82 (1988)
[88D2] O.M. Dub, Y.M. Shostak, Y.B. Kuzma, Inorg. Mater. **24**, 1270 (1988)
[88D3] T.T. Dung, N.P. Thuy, N.M. Hong, T.D. Hien, Phys. Stat. Solidi (a) **106**, 201 (1988)
[88E1] K. Erdmann, M. Rosenberg, K.H.J. Buschow, J. Appl. Phys. **63**, 4113 (1988)
[88F1] J.J.M. Franse, N.P. Thuy, N.M. Hong, J. Magn. Magn. Mater **72**, 361 (1988)
[88F2] C.D. Fuerst, J.F. Herbst, F.E. Pinkerton, J. Appl. Phys. **64**, 5556 (1988)
[88G1] J.P. Gavigan, D. Givord, H. S. Li, J. Voiron, Physica B **149**, 345 (1988)
[88G2] J.P. Gavigan, H.S. Li, J.M.D. Coey, J.M. Cadogan, D. Givord, J. Physique, **49**, C8-557 (1988)
[88G3] J.P. Gavigan, H.S. Li, J.M.D. Coey, T. Viadieu, L. Pareti, O. Moze, F. Bolzoni, J. Physique **49**, C8-577 (1988)
[88G4] F. Grandjean, G.J. Long, D.E. Tharp, O.A. Pringle, W.J. James, J. Physique, **49**, C8-581 (1988)
[88G5] R. Gössinger, H.R. Kirchmayr, K.H.J. Buschow, J. less Common Met. **136**, 367 (1988)
[88G6] Y. Gros, F. Hartmann-Boutron, C. Meyer, M.A. Frémy, P. Tenaud, J. Magn. Magn. Mater. **74**, 319 (1988)
[88G7] Y. Gros, F. Hartmann-Boutron, C. Meyer, M.A. Frémy, P. Tenaud, P. Auric, J. Physique, **49**, C8-547 (1988)
[88H1] N.M. Hong, J.J.M. Franse, N. P. Thuy, T.D. Hien, J. Physique, **49**, C8-545 (1988)
[88J1] S.S. Jaswal, M.A. Langell, Y.G. Ren, M.A. Engelhardt, D.J. Sellmyer, J. Appl. Phys. **64**, 5577 (1988)
[88J2] E. Jedryka, M. Wojcik, P. Panissod, M. Rosenberg, S. Hirosawa, M. Sagawa, J. Physique, **49**, C8-587 (1988)
[88J3] S.Y. Jiang, F. Pourarian, J.X. Yan, S.G. Sankar, W.E. Wallace, J. Magn. Magn. Mater. **74**, 181 (1988)
[88J4] M. Jurczyk, J. Magn. Magn. Mater. **73**, 199 (1988)
[88J5] M. Jurczyk, J. Magn. Magn. Mater. **73**, 367 (1988)
[88J6] M. Jurczyk, IEEE Trans. Magn. MAG **24**, 1942 (1988)
[88K1] Cz. Kapusta, H. Figiel, J. Physique, **49**, C8-559 (1988)
[88K2] H. Kato, M. Yamada, G. Kido, Y. Nakagawa, S. Hirosawa, M. Sagawa, J. Physique, **49**, C8-575 (1988)
[88K3] A. Kowalczyk, P. Stefanski, A. Wrzeciono, Phys. Stat. Solidi (a) **107**, K61 (1988)
[88L1] H.S. Li, J.P. Gavigan, J.M. Cadogan, D. Givord, J.M.D. Coey, J. Magn. Magn. Mater. **72**, L241 (1988)
[88P1] F. Pourarian, S.G. Sankar, W.E. Wallace, J. Magn. Magn. Mater. **74**, 177 (1988)
[88P2] F. Pourarian, S.Y. Jiang, S.G. Sankar, W.E. Wallace, J. Appl. Phys. **63**, 3972 (1988)
[88R1] M. Rosenberg, M. Mittag, K.H.J. Buschow, J. Appl. Phys. **63**, 3586 (1988)
[88S1] A. Simopoulos, D. Niarchos, Hyp. Int. **40**, 425 (1988)
[88S2] H.H.A. Smit, R.C. Thiel, K.H.J. Buschow, J. Phys. F.: Metal Phys. **18**, 295 (1988)
[88T1] N.P. Thuy, J.J.M. Franse, N.M. Hong, T.D. Hien, J. Physique, **49**, C8-499 (1988)
[88T2] N.P. Thuy, T.D. Hien, N.M. Hong, J.J.M. Franse, J. Physique, 49, C8–579 (1988)

[88V1] V. Villas-Boas, F.P. Missell, S.F. Da Cunha, J. Appl. Phys. **64**, 5549 (1988)
[88W1] Z. Wang, B. Yu, Y. Wang, J.H. Huang, L.Yin, M. Feng, Mat. Sci. Eng. **99**, 123 (1988)
[88W2] J. Wecker, L. Schultz, J. Physique, **49**, C8-611 (1988)
[88W3] M. Wojcik, E. Jedryka, G. Jasiolek, W. Paskowicz, M. Slepowroriski, J. Magn. Magn. Mater. **72**, 330 (1988)
[88X1] J.Q. Xie, C.H. Wu, Y.C. Chuang, F.M. Yang, J. Magn. Magn. Mater. **75**, 361 (1988)
[88Y1] M. Yamada, H. Kato, H. Yamamoto, Y. Nakagawa, Phys. Rev. B **38**, 620 (1988)
[88Z1] L.Y. Zhang, F. Pourarian, W.E. Wallace, J. Magn. Magn. Mater. **74**, 101 (1988)
[88Z2] Y.D. Zhang, J.I. Budnick, E. Potenziani, A.T. Pedziwiatr, W.E. Wallace, P. Panissod, J. Appl. Phys. **63**, 3719 (1988)
[89B1] E. Burzo, N. Plugaru, V. Pop, I. Creanga, Phys. Stat. Solidi (a) **113**, K253 (1989)
[89B2] E. Burzo, N. Plugaru, I. Creanga, M. Ursu, J. less Common Met. **155**, 281 (1989)
[89C1] S.K. Chen, J.G. Duh, H.C. Ku, W.C. Chang, J. Magn. Magn. Mater. **82**, 12 (1989)
[89C2] C.N. Christodoulou, W.E. Wallace, T.B. Massalski, J. Appl. Phys. **66**, 2749 (1989)
[89D1] Z. Drzazga, A. Winiarska, F. Stein, J. less Common Met. **153**, L21 (1989)
[89H1] N.M. Hong, J.J.M. Franse, N.P. Thuy, J. less Common Met. **155**, 151 (1989)
[89J1] M. Jurczyk, O.D. Chistjakov, J. Magn. Magn. Mater. **78**, 279 (1989)
[89M1] M. Mittag, M. Rosenberg, K.H.J. Buschow, J. Magn. Magn. Mater. **82**, 109 (1989)
[89N1] S. Nadolski, E. Machowska, E. Burzo, Solid State Commun. **71**, 97 (1989)
[89P1] E. Panissod, E. Jedryka, M. Wojcik, J.I. Budnick, Phys. Rev. B **40**, 2606 (1989)
[89R1] D.H. Ryan, Z. Altounian, J.O. Strom-Olsen, W.B. Muir, Phys. Rev. B **39**, 4730 (1989)
[89S1] J.L. Sanchez, S. Garcia, N. Suarez, S. Diaz, F. Leccabue, R. Panizzieri, G. Bocelli, G. Calestani, J. Magn. Magn. Mater. **79**, 249 (1989)
[89S2] Y. Sano, H. Onodera, H. Yamauchi, H. Yamamoto, J. Magn. Magn. Mater. **79**, 67 (1989)
[89W1] M. Wojcik, E. Jedryka, P. Panissod, S. Hirosawa, M. Sagawa, J. Magn. Magn. Mater. **80**, 19 (1989)
[90B1] E. Burzo, A. Chelkowski, H.R. Kirchmayr, *Landolt Börnstein Handbook*, vol. 19d2, (Springer Verlag, 1990)
[90B2] E. Burzo, V. Pop, N. Plugaru, Mat. Sci. Forum, **62–64**, 611 (1990)
[90D1] Z. Drzazga, J. Magn. Magn. Mater. **89**, 97 (1990)
[90D2] Z. Drzazga, A. Winiarska, K. Borgiel, T. Mydlarz, J. Magn. Magn. Mater. **83**, 155 (1990)
[90F1] J. Fidler, C. Groiss, M. Tokunaga, IEEE Trans. Magn. MAG **26**, 1948 (1990)
[90G1] K. Girgis, M. Kraft, U. Weis, P. Fischer, M. Sostarich, J. Less Common Met. **162**, 335 (1990)
[90G2] P.C.M. Gubbens, A.M. Van der Kraan, K.H.J. Buschow, J. Magn. Magn. Mater. **87**, 276 (1990)
[90H1] M.Q. Huang, B.M. Ma, L.Y. Zhang, W.E. Wallace, S.G. Sankar, J. Appl. Phys. **67**, 4981 (1990)
[90I1] K. Ichinose, K. Fujiwara, F. Maruyama, M. Misawa, M. Oyasato, H. Nagai, A. Tsujimura, J. Magn. Magn. Mater. **90–91**, 549 (1990)
[90I2] H. Ido, K. Konno, S.F. Cheng, S.G. Sankar, W.E. Wallace, J. Magn. Magn. Mater. **90–91**, 579 (1990)
[90I3] H. Ido, W.E. Wallace, T. Suzuki, S.F. Cheng, V.K. Sinha, S.G. Sankar, J. Appl. Phys. **67**, 4635 (1990)
[90K1] Cz. Kapusta, N. Spiridis, H. Figiel, J. Magn. Magn. Mater. **83**, 153 (1990)
[90K2] Y.B. Kuzma, N.F. Chaban, *Dvoinye i troine sistemy soderezhashkhie bor* (Moscow, Metallurgiya, 1990)
[90M1] F. Maruyama, K. Ichinose, K. Fujiwara, M. Misawa, M. Oyasato, H. Nagai, A. Tsujimura, J. Magn. Magn. Mater. **90–91**, 552 (1990)
[90R1] R.J. Radwanski, X.P. Zhong, F.R. de Boer, K.H.J. Buschow, Physica B **104**, 131 (1990)
[90R2] D.H. Ryan, Z. Altounian, L.X. Liar, J.O. Strom-Olsen, W.B. Muir, J. Appl. Phys. **67**, 4742 (1990)
[90W1] W.Wojcik, E. Jedryka, P. Panissod, K.H.J. Buschow, J. Magn. Magn. Mater. **83**, 243 (1990)
[90Z1] P.Yu. Zavalij, Y.B. Kuzma, L.V. Zavalij, Russ. Cristallogr. **35**, 638 (1990)

[90Z2] Z.D. Zhang, C.H. Wu, X.K. Sun, Y.C. Chuang, J. Magn. Magn. Mater. **89**, 173 (1990)
[90Z3] G.F. Zhou, Y.C. Chuang, X.K. Sun, R. Grössinger, H.R. Kirchmayr, J. Less Common Met. **158**, 41 (1990)
[91A1] A.V. Andreev, S.M. Zadvorkin, Physica B **172**, 517 (1991)
[91A2] G. André, P. Thuery, M. Pinot, A. Oles, A. Szytula, Solid State Commun. **80**, 239 (1991)
[91B1] E. Burzo, V. Pop, N. Plugaru, J. Magn. Magn. Mater. **97**, 147 (1991)
[91C1] R. Coehoorn, J. Magn. Magn. Mater. **99**, 55 (1991)
[91D1] G.H.O. Daalderop, J.P. Kelly, M.F.H. Schuurmans, Phys. Rev. B **44**, 12054 (1991)
[91D2] Z. Drzazga, Physica, B **168**, 115 (1991)
[91D3] T.T. Dung, N.P. Thuy, N.M. Hong, T.D.Hien, J.J.M. Franse, J. Appl. Phys. **69**, 4633 (1991)
[91G1] I.B. Gubich, Y.B. Kuzma, N.F. Chaban, Inorg. Chem. **27**, 1677 (1991)
[91H1] J.F. Herbst, Rev. Modern Phys. **63**, 819 (1991)
[91H2] M.Q. Huang, B.M. Ma, S.F. Cheng, W.E. Wallace, J. Appl. Phys. **69**, 5599 (1991)
[91I1] H. Ido, H. Yamauchi, S.F. Cheng, S.G. Sanakar, W.E. Wallace, J. Appl. Phys. **79**, 6540 (1991)
[91I2] H. Ido, K. Konno, H. Ogata, K. Sugiyama, H. Hachino, M. Date, K. Maki, J. Appl. Phys. **70**, 6128 (1991)
[91I3] H. Ido, K. Konno, T. Ito, S.F. Cheng, S.G. Sankar, W.E. Wallace, J. Appl. Phys. **69**, 5551 (1991)
[91J1] E. Jedryka, M. Wojcik, P. Panissod, A.T. Pedziwiatr, M. Slepowronski, J. Appl. Phys. **69**, 6043 (1991)
[91J2] E. Jedryka, M. Wojcik, A.T. Pedziwiatr, M. Slepowronski, Proc. Zakopane School on Physics, p. 277 (1991)
[91L1] J.P. Liu, F.R. de Boer, K.H.J. Buschow, J. Less Common Met. **175**, 137 (1991)
[91M1] T. Moriya, J. Magn. Magn. Mater. **100**, 261 (1991)
[91N1] M. Nagahama, S. Satohira, T. Akazawa, F. Nakamura, T. Fujita, M. Kawakami, J. Phys. Soc. Jpn. **60**, 3855 (1991)
[91S1] X.K. Sun, Z.G. Zhao, Q. Wang, X.P. Zhong, Z.D. Zhang, Y.C. Chuang, J. Appl. Phys. **69**, 5548 (1991)
[91S2] X.K. Sun, Z.G. Zhao, Q. Wang, X.P. Zhong, Z.D. Zhang, T. Zhao, Y.Z. Chuang, Chin. Sci. Bull. **36**, 80 (1991)
[91S3] S. Szymura, Yu.M. Rabinovich, V.V. Sergeev, H. Bala, D.V. Pokrovski, J. Physique, III **1**, 1657 (1991)
[91T1] N.P. Thuy, N.M. Hong, N.H. Nghi, D.C. Thanh, Phys. Stat. Solidi (a) **127**, 201 (1991)
[91U1] T. Ukai, K. Yamaki, M. Takahashi, N. Mori, J. Appl. Phys. **69**, 4662 (1991)
[91Y1] Y. Yan, T. Zhao, H. Jin, J. Phys.: Condens. Matter, **3**, 195 (1991)
[91Z1] Z.G. Zhao, J.Y. Wang, Y.P. Ge, X.K. Sun, Y.C. Chuang, J. Magn. Magn. Mater. **98**, L231 (1991)
[92B1] E. Burzo, R. Lemaire, Solid State Commun. **83**, 159 (1992)
[92C1] J.M. Cadogan, H.S. Li, S.J. Campbell, J. Jing, Solid State Commun. **81**, 121 (1992)
[92C2] T.W. Capehart, R.K. Mishra, J. F. Herbst, J. Appl. Phys. **72**, 676 (1992)
[92D1] L.E. DeLong, V. Tondiglia, P. Rogl, J. Alloys Comp. **181**, 311 (1992)
[92D2] Z. Drzazga, E. Popiel, A. Winiarska, J. Magn.Magn. Mater. **104–107**, 1437 (1992)
[92F1] H. Figiel, Cz. Kapusta, N. Spiridis, G. Stoch, P.C. Riedi, M. Rosenberg, J. Magn. Magn. Mater. **104–107**, 1198 (1992)
[92G1] I.B. Gubich, Y.B. Kuzma, P.Y. Zavalii, G.M. Pidgaetskaya, Inorg. Mater. **28**, 1104 (1992)
[92H1] M. Hayashi, T. Myojin, Y. Kasamatsu, Y. Imaeda, T. Ushida, A. Tsujimura, T.H. Hihara, J. Magn. Magn. Mater. **104–107**, 1225 (1992)
[92H2] N.M. Hong, P.N. Minh, N.P. Thuy, Solid State Commun. **82**, 211 (1992)
[92I1] K. Ichinose, F. Maruyama, M. Misawa, A. Tsujimura, H. Nagai, J. Magn. Magn. Mater. **104–107**, 1159 (1992)
[92I2] H. Ido, K. Sugiyama, H. Hachino, M. Date, S.F. Cheng, K. Maki, Physica B **177**, 265 (1992)

[92I3] H. Ido, K. Konno, T. Ito, S.F. Cheng, S.G. Sankar, W.E. Wallace, J. Magn. Magn. Mater. **104–107**, 1361 (1992)

[92J1] E. Jedryka, M. Wojcik, P. Panissod, J. Magn. Magn. Mater. **104–107**, 1405 (1992)

[92K1] Cz. Kapusta, M. Rosenberg, K. H. J. Buschow, J. Alloys Comp. Comp. **187**, 409 (1992)

[92K2] M. Kawakami, S. Satohira, J. Magn. Magn. Mater. **104–107**, 1313 (1992)

[92K3] M. Kawakami, J. Watarizaki, J. Phys. Soc. Jpn. **61**, 3361 (1992)

[92M1] F. Maruyama, K. Ichinose, M. Misawa, A. Tsujimura, H. Nagai, K. Adachi, J. Magn. Magn. Mater. **104–107**, 1165 (1992)

[92M2] T. Myojin, M. Hayashi, T. Ohno, Y. Imaeda, T. Ushida, A. Tsujimura, J. Magn. Magn. Mater. **104–107**, 1195 (1992)

[92P1] M.L.Plumer, A.Caille, Phys. Rev. B **46**, 203 (1992)

[92R1] M. Rosenberg, T. Sinnemann, M, Mittag, K.H.J. Buschow, J. Alloys Comp. **182**, 145 (1992)

[92T1] N.P. Thuy, N.M. Hong, J.P. Liu, X. Li, J.J.M. Franse, F.R. de Boer, Physica B **177**, 270 (1992)

[92Z1] L.Y. Zhang, E.B. Boltich, W.E. Wallace, J. Magn. Magn. Mater. **111**, 75 (1992)

[92Z2] Z.G. Zhao, Y.P. Ge, J.Y. Wang, X.K. Sun, Y.C. Chuang, F.R. de Boer, J. Alloys Comp. **187**, 249 (1992)

[92Z3] Z.G. Zhao, Y.P. Ge, J.Y. Wang, X.K. Sun, Y.C. Chuang, J. Magn. Magn. Mater. **104–107**, 1289 (1992)

[92Z4] T. Zhao, X.K. Sun, Z.D. Zhang, Q. Wang, Y.C. Chuang, F.R. de Boer, J. Magn. Magn. Mater. **104–107**, 2119 (1992)

[92Z5] G.F. Zhou, X. Li, F.R. de Boer, J. Magn. Magn. Mater. **109**, 265 (1992)

[92Z6] G.F. Zhou, X. Li, F.R. de Boer, K.H.J. Buschow, Physica B **177**, 286 (1992)

[93B1] R. Ballou, E. Burzo, V. Pop, A. Pentek, J. Appl. Phys. **73**, 5695 (1993)

[93B2] R. Ballou, E. Burzo, A. Mincic, V. Pop, J. Magn. Magn. Mater. **118**, L285 (1993)

[93C1] J.M. Cadogan, S.J. Campbell, X.L. Zhao, E. Wu, Aust. J. Phys. **46**, 679 (1998)

[93C2] S.J. Campbell, J.M. Cadogan, X.L. Zhao, E. Wu, Hyp. Inter. **78**, 463 (1993)

[93C3] Y. Chen. J. Liang, Q. Lu, X. Chen, F. Huang, J. Alloys. Comp. **288**, 170 (1999)

[93C4] C.N. Christodoulou, T.B. Massalski, W.E. Wallace, J. Phase Equilib. **14**, 31 (1993)

[93C5] C.N. Christodoulou, T.B. Massalski, W.E. Wallace, J. Magn. Magn. Mater. **125**, 177 (1993)

[93D1] N. H. Duc, T. H. Hien, D. Givord, J. J. M. Franse, F. R. de Boer, J. Magn. Magn. Mater. **124**, 305 (1993)

[93D2] T.T. Dung, N.P. Thuy, T.D. Hien, N.M. Hong, D.C. Khan, J.J.M. Franse, Int. J. Mod. Phys. **7**, 798 (1993)

[93G1] I.B. Gubych, P.Yu. Zavalij, Y.B. Kuzma, Neorg. Mater. **29**, 289 (1993)

[93H1] N.M. Hong, N.P. Thuy, T.D. Hien, G. Hilscher, T.S. Zhao, R. Grössinger, J. Appl. Phys. **73**, 5917 (1993)

[93H2] N.M. Hong, N.P. Thuy, T.D. Hien, J. Voiron, E. Gratz, R. Grössinger, G. Hilscher, T. Holubar, A. Lindbaum, P. Hundergger, H. Sassik, Int. J. Mod. Phys. B **7**, 733 (1993)

[93I1] H. Ido, H. Ogata, K. Maki, J. Appl. Phys. **73**, 6269 (1993)

[93I2] T. Ito, H. Ogata, H. Ido, K. Kido, J. Appl. Phys. **73**, 5914 (1993)

[93J1] H.M. Jin, Y. Yan, Phys. Rev. B **48**, 1022 (1993)

[93K1] J. Kong, B.P. Cheng, Z.G. Zhao, X.K. Sun, Y.C. Chuang, Solid State Commun. **85**, 211 (1993)

[93K2] A. Kowalczyk, Phys. Stat. Solidi (b) **177**, K17 (1993)

[93K3] A. Kowalczyk, Phys. Stat. Solidi (b) **180**, K19 (1993)

[93L1] L.X. Liao, Z. Altounian, D.H. Ryan, Phys. Rev. B **47**, 11230 (1993)

[93L2] A. Lindbaum, N.M. Hong, H. Muller, E. Gratz, N.P. Thuy, T.D. Han, J. Appl. Phys. **73**, 6153 (1993)

[93M1] F. Maruyama, H. Nagai, K. Adachi, J. Phys. Soc. Jpn. **62**, 3741 (1993)

[93M2] T. Myojin, M. Hayashi, T. Ohno, Y. Imaeda, T, Ushida, A. Tsujimura, T. Hihara, J. Alloys Comp. **193**, 252 (1993)

[93O1] H. Ogata, H. Ido, H. Yamauchi, J. Appl. Phys. **73**, 5911 (1993)
[93O2] H. Onodera, S.G. Kang, H. Yamauchi, Y. Yamaguchi, J. Magn. Magn. Mater. **127**, 298 (1993)
[93Y1] C.J. Yang, W.Y.Lee, S.D.Choi, J. Koreean Magn. Soc. **3**, 284 (1993)
[93Z1] Z.G. Zhao, J.Y. Wang, Y.P. Ge, F.M. Yang, X.K. Sun, Y.C. Chuang, K.H.J. Buschow, F.R. de Boer, J. Appl. Phys. **73**, 5875 (1993)
[94B1] E. Burzo, C. Gusa, B. Kevorkian, T. Veres, J. Alloys Comp. **209**, L33 (1994)
[94B2] E. Burzo, C. Gusa, B. Kevorkian, T. Veres, Rom. Repts. Phys. **46**, 201 (1994)
[94B3] E. Burzo, V. Pop, C.C. Borodi, R. Ballou, IEEE Trans. Magn. **30**, 628 (1994)
[94I1] H. Ido, M.Nanjo, M. Yamada, J. Appl. Phys.**75**, 7140 (1994)
[94I2] H. Ido, O. Nashima, T. Takahashi, K. Oda, K. Sugiyama, J. Appl. Phys. **76**, 6165 (1994)
[94I3] T. Ito, H. Asano, H. Ido, G. Kido, J. Appl. Phys. **75**, 6283 (1994)
[94K1] A. Kowalczyk, Phys. Stat. Solidi (b) **181**, K73 (1994)
[94K2] A. Kowalczyk, Acta Phys. Polonica, A **85**, 623 (1994); J. Magn. Magn. Mater. **136**, 70 (1994)
[94K3] A. Kowalczyk, P. Stefanski, IEEE Trans. Magn. **30**, 648 (1994)
[94L1] J.P. Liu, F.R. de Boer, P.F. de Chatel, R. Coehoorn, K.H.J. Buschow, J. Magn. Magn. Mater. **132**, 159 (1994)
[94M1] T. Myojin, T. Ohno, A. Tsujimura, T. Hihara, Physica B **194–196**, 227 (1994)
[94W1] G. Wiesinger, R. Pavlovec, N. M. Hong, N.T. Hien, N.P. Thuy, Hyp. Int. **93**, 1531 (1994)
[94W2] E. Wu, J.M. Cadogan, S.J. Campbell, X.L. Zhao, Hyp. Int. **94**, 1903 (1994)
[94Z1] P.Yu.Zavalij, S.I.Mykhalenko, Y.B.Kuzma, J. Alloys Comp. **208**, 55 (1994)
[94Z2] Z.G. Zhao, R de Boer, K.H.J. Buschow, Y.P. Ge, J.Y. Wang, J. Appl. Phys. **76**, 6843 (1994)
[95B1] R. Ballou, E.Burzo, V. Pop, J. Magn. Magn. Mater. **140–144**, 945 (1995)
[95B2] S.L. Budko, M.B. Fontes, D. Aliaga-Guerra, E.M. Baggio-Saitovich, Phys. Rev. B **52**, 305 (1995)
[95B3] S.I.Budko, B.Giordanengo, A.Sulpice, M.B.Fontes, E.M.Baggio-Saitovitch, Solid State Commun. **94**, 119 (1995)
[95B4] E. Burzo, J. Magn. Magn. Mater. **140–144**, 2013 (1995)
[95B5] E. Burzo, R. Tetean, V. Pop, Rom. Repts. Phys. **47**, 749 (1995)
[95B6] E. Burzo, V. Pop, M. Miheţ, I. Lupşa, Balkan Phys. Letters **3**, 60 (1995)
[95C1] S.A. Carter, B. Batlogg, R.J. Cava, J.J. Krajewski, W.F. Peck, Phys. Rev. B **51**, 12829 (1995)
[95D1] N.H. Duc, D. Givord, J. Magn. Magn. Mater. **151**, L13 (1995)
[95F1] J.J.M. Franse, F.R. de Boer, J. Magn. Magn. Mater. **140–144**, 789 (1995)
[95G1] A.N. Gangopadhyay, A. J. Schultz, J. S. Schilling, Physica C **246**, 317 (1995)
[95H1] N.M. Hong, R. Hauser, G. Hilscher, E. Gratz, R. Ballou, H. Ido, J. Magn. Magn. Mater. **140–144**, 933 (1995)
[95I1] H. Ido, Kotai Butsuri **30**, 875 (1995) (in Japanese)
[95I2] H. Ido, H. Asano, G. Kido, N.M. Hong, G. Hilscher, R. Ballou, J. Magn. Magn. Mater. **140–144**, 955 (1995)
[95I3] T. Ito, H. Asano, H. Ido, M. Yamada, G. Kido, J. Magn. Magn. Mater. **140–144**, 943 (1995)
[95K1] A. Kowalczyk, Phys. Stat. Solidi (a) **151**, K49 (1995)
[95K2] A. Kowalczyk, V. Ivanov, J. Baszynski, J. Magn. Magn. Mater. **146**, 182 (1995)
[95K3] A. Kowalczyk, G. Czechowski, V. Ivanov, J. Baszynski, J. Magn. Magn. Mater. **139**, 19 (1995)
[95M1] H.Michor, T.Holubar, C.Dusek, G.Hilscher, Phys. Rev. B **52**, 16165 (1995)
[95M2] F.M. Mulder, J.H.V.J. Brabers, P. Coehoorn, R.C.Thiel, K.H.J. Buschow, F.R.De Boer, J. Alloys Comp. **217**,118 (1995)
[95N1] K. Nahm, C.K. Kim, M. Mittag, Y.H. Jeong, J. Appl. Phys. **78**, 3980 (1995)
[95O1] K. Oda, K. Sugiyama, K. Kindo, N.M. Hong, H. Asano, O. Nashima, H. Ido, J. Magn. Magn. Mater. **140–144**, 947 (1995)

[95T1] C.V. Tang, N.P. Thuy, J.P. Liu, N.T. Hien, T.D. Hien, J. Magn. Magn. Mater. **147**, 55 (1995)
[95T2] N.P. Thuy, N. X. Phuc, N.M. Hong, T. Holubar, G. Hilscher, J. Magn. Magn. Mater. **140–144**, 943 (1995)
[95Y1] H. Yoshie, M. Hoshino, Y. Amako, H. Nagai, H. Wada, M. Shiga, Y. Nakamura, J. Phys. Soc. Jpn. **64**, 2243 (1995)
[95Z1] X. Zhao, J.M. Cadogan, S.J. Campbell, J. Magn. Magn. Mater. **140–144**, 959 (1995)
[95Z2] Z.G. Zhao, F.R. de Boer, K. H. J Buschow, P.F. de Chatel, Physica B **211**, 124 (1995)
[95Z3] Z.G. Zhao, F.R. de Boer, K.H.J. Buschow, P.F. de Châtel, J. Magn. Magn. Mater. **140–144**, 957 (1995)
[96B1] M. Buchgeister, G. Fuchs, J. Zawadzki, Mat. Letters, **26**, 273 (1996)
[96B2] E. Burzo, V. Pop, R. Ballou, J. Magn. Magn. Mater. **157–158**, 631 (1996)
[96D1] G.H.O. Daalderop, P.J. Kelly, M.F.H. Schuurmans, Phys. Rev. B **53**, 14415 (1996)
[96G1] A.M. Gabay, A.G. Popov, V.S. Gaviko, Ye.V. Belozerov, A.S. Yermolenko, N.N. Shchegoleva, J. Alloys Comp. **237**, 101 (1996)
[96G2] W. Gong, G.C. Hadjipanayis, IEEE Trans. Magn. **32**, 4392 (1996)
[96I1] T. Ito, M. Asano, H. Ido, M. Yamada, J. Appl. Phys. **79**, 5507 (1996)
[96K1] Cz. Kapusta, J. Magn. Magn. Mater. **157–158**, 71 (1996)
[96K2] A. Kowalczyk, V. Ivanov, Phys. Stat. Solidi (b) **193**, 155 (1996)
[96L1] G.J. Long, F. Grandjean, J. Magn. Magn. Mater. **162**, 162 (1996)
[96M1] T. Myojin, T. Ohno, K. Mizuno, A. Tsujimura, K. Kojima, Z. Naturforsh. A **51**, 726 (1996)
[96P1] W. Perthold, N.M. Hong, M. Michor, G. Hilscher, H. Ido, H. Asano, J. Magn. Magn. Mater. **157–158**, 649 (1996)
[96P2] N. Plugaru, D.P. Lazar, A. Galateanu, J. Magn. Magn. Mater. **162**, 50 (1996)
[96T1] R. Tetean, E. Burzo, J. Magn. Magn. Mater. **157–158**, 633 (1996)
[96T2] C.V. Thang, J. Stanek, P.E. Brommer, N.M. Hong, J.M. Franse, N.M. Thuy, J. Magn. Magn. Mater. **157–158**, 645 (1996)
[96T3] C.V. Thang, P.E. Brommer, J.H.P. Colpa, N.P. Thuy, J.J.M. Franse, Physica B **228**, 205 (1996)
[96W1] L. Withanawasam, I. Panagiotopoulos, G.C. Hadjipanayis, J. Appl. Phys. Phys. **79**, 4837 (1996)
[96Z1] X.L. Zhao, S.J. Campbell, J.M. Cadogan, H.S. Li, P.W. Thompson, Proc. ICAME Conf. **50**, 211 (1996)
[97B1] E. Burzo, A. Mincic, V. Pop, G. Borodi, Balkan Phys. Lett. **5**, 1179 (1997)
[97B2] E. Burzo, P. Vlaic, V. Pop, G. Borodi, Balkan Phys. Lett. **5**, 490 (1997)
[97D1] F.R. de Boer, K.H.J. Buschow, J. Alloys Comp. **258**, 1 (1997)
[97G1] A. Galatanu, A. Kottar, M. Artigas, N. Plugaru, D.P. Lazar, J. Alloys Comp. **262–263**, 356 (1997)
[97I1] Y. Iwamoto, T. Oda, K. Ueda, T. Kohara, Physica B **230–232**, 886 (1997)
[97K1] J. Kamarad, Z. Arnold, O. Mikulina, V. Sechovsky, H. Ido, N.M. Hong, Physica B **237–238**, 527 (1997)
[97K2] A. Kowalczyk, J. Magn. Magn. Mater. **171**, 113 (1997)
[97K3] A. Kowalczyk, J. Magn. Magn. Mater. **175**, 279 (1997)
[97K4] A. Kowalczyk, V. Ivanov, Phys. Stat. Solidi (b) **201**, 443 (1997)
[97K5] A. Kowalczyk, L. Smardz, Acta Phys. Polonica, A **92**, 285 (1997)
[97K6] A. Kowalczyk, L. Smardz, J. Alloys Comp. **259**, 59 (1997)
[97K7] S. Kramp, M. Febri, J. C. Joubert, J. Solid State Chem. **133**, 145 (1997)
[97L1] S.P. Lee, C.K. Kim, K. Nahm, M. Mittag, Y.H. Jeong, C.M. Ryu, J. Appl. Phys. **81**, 2454 (1997)
[97L2] Q.A. Li, C.H. de Groot, F.R. de Boer, K.H.J. Buschow, J. Alloys Comp. **256**, 82 (1997)
[97M1] W. F. Miao, J. Ding, P. G. McCormick, R. Street, J. Appl. Phys. Phys. **82**, 4439 (1997)
[97M2] T. Myojin, T. Ohno, K. Mizuno, A. Tsujimura, K. Kojima, Physica B **237–238**, 581 (1997)
[97P1] N. Plugaru, A. Galatanu, D.P. Lazar, Solid State Commun. **102**, 23 (1997)
[97P2] V. Pop, G. Suciu, E. Burzo, Balkan Phys. Lett. **5**, 1239 (1997)

[97S1] H. Schmidt, H. F. Braun, Phys. Rev. B **55**, 8497 (1997)
[97S2] Y. Suzuki, T. Ito, T. Uchida, O. Nashima, N.M. Hong, H. Ido, J. Appl. Phys. **81**, 5141 (1997)
[97T1] C.V. Thang, P.E. Brommer, N.P. Thuy, J. J. M. France, J. Magn. Magn. Mater. **171**, 237 (1997)
[97Y1] H. Yoshie, T. Umezawa, H. Ishizone, Y. Amako, H. Nagai, H. Wada, M. Shiga, Physica B **237–238**, 590 (1997)
[98B1] E. Burzo, Rep. Progr. Phys. **61**, 1099 (1998)
[98C1] E.N. Caspi, H. Pinto, M. Kuznietz, H. Ettegui, M. Melamud, H. Shaked, J. Appl. Phys. **83**, 6733 (1998)
[98I1] K. Ichinose, N. Watanabe, K. Shimizu, H. Ido, J. Magn. Magn. Mater. **177–181**, 1113 (1998)
[98I2] H. Ido, Y. Suzuki, T. Suzuki, J. Appl. Phys. **83**, 7133 (1998)
[98J1] A. Jezierski, A. Kowalczyk, S. Mähl, M. Neumann, G. Borstel, J. Phys.: Condens. Matter, **10**, 6277 (1998)
[98J2] Z. Jin, W. Tang, H. Qin, J. Zhang, Y. Du, J. Magn. Magn. Mater. **177–181**, 1115 (1998)
[98K1] J. Kitagawa, N. Suzuki, J. Magn. Magn. Mater. **177–181**, 1357 (1998)
[98K2] C.O. Kim, J.S. Park, E.S. Choi, Y.W. Park, Y.H. Jeong, C.M. Ryu, C.K. Kim, K. Nahm, Phys. Stat. Solidi (b) **208**, 129 (1998)
[98K3] D.H. Kim, G. Hadjipanayis, J. Appl. Phys. **83**, 7124 (1998)
[98K4] D.H. Kim, Y. Zhang, G.C. Hadjipanayis, J. Magn. Magn. Mater. **190**, 302 (1998)
[98K5] A.Kowalczyk, A. Jezierski, J. Magn. Magn. Mater. **188**, 361 (1998)
[98K6] A. Kowalczyk, A. Jezierski, J. Magn. Magn. Mater. **182**, 137 (1998)
[98K7] A. Kowalczyk, L. Smardz, A. Jezierski, J. Magn. Magn. Mater. **190**, 205 (1998)
[98K8] A. Kowalczyk, L. Smardz, A. Jezierski, Acta Phys. Slovaca, **48**, 619 (1998)
[98M1] F. Maruyama, H. Nagai, Y. Amako, H. Yoshie, K. Adachi, Jpn. J. Appl. Phys. **37**, 1825 (1998)
[98M2] F. Maruyama, H. Nagai, Y. Amako, H. Yoshie, K. Adachi, Jpn. J. Appl. Phys. **37**, 2492 (1998)
[98S1] T.Shishido, J.Ye, K.Obara, T.Fukuda, J. Ceramic Soc. Jpn. **106**, 299 (1998)
[98S2] A. Szajek. J. Magn. Magn. Mater. **185**, 322 (1998)
[98Y1] H. Yamada, K. Terao, N. Nakagawa, I. Kitagawa, N. Suzuki, H. Ido, J. Magn. Magn. Mater. **183**, 94 (1998)
[98Y2] J.Ye, T.Shishdo, T.Matsumoto, T.Fukuda, J. Alloys Comp. **275–277**, 76 (1998)
[98Y3] H. Yoshie, W. Takahashi, H. Ishizone, Y. Amako, H. Nagai, M. Shiga, H. Wada, K. Adachi, J. Magn. Magn. Mater. **177–181**, 481 (1998)
[98Z1] Z.G. Zhao, F.R. De Boer, K.H.J. Buschow, J. Alloys Comp. **279**, 69 (1998}
[99B1] A. Belger, G. Zahn, B. Wehner, P. Paufler, G. Graw, G. Behr, J. Alloys Comp. **283**, 26 (1999)
[99C1] Y. Chen, J.K. Liang, X.L. Chen, Q.L. Liu, J. Alloys Comp. **289**, 96 (1999)
[99C2] Y. Chen, J. Liang, Q. Liu, X. Chen, F. Huang, J. Alloys Comp. **288**, 170 (1999)
[99C3] Y. Chen, J.K. Liang, X.L. Chen, Q.L. Liu, B.G. Shen, Y.P. Shen, J. Phys.: Condens. Matter **11**, 8251 (1999)
[99C4] Y. Chen, Q.L. Liu, J.K. Liang, X.L. Chen, B.G. Shen, F. Huang, Appl. Phys. Lett. **74**, 856 (1999)
[99I1] H. Ido, O. Nashima, T. Ito, T. Kaneda, K. Saito, K. Konno, H. Yoshida, M. Motokawa, J. Appl. Phys. **85**, 4865 (1999)
[99M1] F. Maruyama, Jpn. J. Appl. Phys. **38**, 1380 (1999)
[99P1] A.T. Pedziwiatr, M. Artigas, B.F. Bogacz, R. Wielgosz, J. Rubin, J. Bartolomé, C. Rillo, J. Fernandez, J. Magn. Magn. Mater. **196–197**, 639 (1999)
[99R1] R.E.Rapp, M.El Massalami, Phys.Rev. B **60**, 3355 (1999)
[99T1] R. Tetean, V. Pop, E. Burzo, I.G. Deac, Modern Phys. Lett. **13**, 905 (1999)
[99T2] A. A. Tulapurkar, S. N. Mishra, Z. Hossain, Hyp. Int. **120–121**, 225 (1999)

[99Y1] V.A.Yartys, R.V.Denys, O.Gutfleisch, I.I.Bulyk, Y.B.Kuzma, I.R.Harris, Int. J. Hydrogen Energy, **24**, 189 (1999)
[00C1] E. N. Caspi, H. Pinto, M. Melamud, J. Appl. Phys. **87**, 416 (2000)
[00C2] C. Chacon, O. Isnard, Physica B **276–278**, 652 (2000)
[00C3] C. Chacon, O. Isnard, J. Solid State Chem. **154**, 242 (2000)
[00C4] Y. Chen, J. Liang, X. Chen, Q. Liu, J. Alloys. Comp. **296**, L1 (2000)
[00C5] Y. Chen, X. Li, X.L. Chen, J.K. Liang, G.H. Rao, Q.L. Liu, J. Alloys Comp. **305**, 216 (2000)
[00C6] Y. Chen, X. Li, X.L. Chen, J.K. Liang, G.H. Rao, B.G. Shen, Q.L. Liu, L.P. Jin, M.Z. Wang, Chem. Mater. **12**, 1240 (2000)
[00C7] Y. Chen, X. Li, L. Chen, J.K. Liang, B.G. Shen, Q.L. Liu, Phys. Rev. B **61**, 3502 (2000)
[00C8] Y. Chen, X. Li, J.K. Liang, X.L. Chen, B.G. Shen, G.H. Rao, Q.L. Liu, IEEE Trans. Magn. **36**, 2037 (2000)
[00E1] M. ElMassalami, M. S. Da Costa, R.E.Rapp, F.A.B.Chaves, Phys. Rev. B **62**, 8942 (2000)
[00I1] H. Ido, M. Yamada, N. Nashima, T. Ito, H. Yoshida, J. Appl. Phys. **87**, 4912 (2000)
[00K1] A. Kowalczyk, Phys. Stat. Solidi (b) **218**, 495 (2000)
[00K2] A. Kowalczyk, A. Jezierski, L. Smardz, J. Baszynski, Acta Phys. Polonica, **97**, 803 (2000)
[01B1] E. Burzo, P. Vlaic, V. Pop, Mat. Sci. Forum, **373–376**, 637 (2001)
[01B2] E. Burzo, A. Mincic, J. Voiron, Mat. Sci. Forum, **373–376**, 629 (2001)
[01C1] C. Chacon, O. Isnard, J. Phys.: Condens. Matter, **13**, 5841 (2001)
[01C2] C. Chacon, O. Isnard, J. Appl. Phys. **89**, 71 (2001)
[01C3] C. Chacon, O. Isnard, J.F. Berar, Mat. Sci. Forum, **378–381**, 434 (2001)
[01C4] Y. Chen, X. Li, X.L. Chen, J.K. Liang, G.H. Rao, B.G. Shen, Q.L. Liu, L.P. Jin, M.Z. Wang, Chem. Mater. **12**, 1240 (2001)
[01C5] W.G. Chu, G.H. Rao, H.F. Yang, G.Y. Liu, J.K. Liang, J. Appl. Phys. **90**, 1931 (2001)
[01C6] B. A. Cook, J. L. Harringa, F. C. Laabs, R. W. Dennis, A. M. Russell, R. W. McCallum, J. Magn. Magn. Mater. **233**, L136 (2001)
[01E1] M.ElMassalami, E.F.Chagas, R.E.Rapp, J. Magn. Magn. Mater. **226–230**, 1058 (2001)
[01G1] Z.F. Gu, Z.Y. Liu, D.C. Zeng, S.Z. Liang, J.C.P. Klaasse, E. Brück, F.R. de Boer, K.H.J. Buschow, Physica B **304**, 283 (2001)
[01I1] S. Imada, S. Suga, T. Sato, R. Yung, H. Ido, J. Magn. Magn. Mater. **226–230**, 1193 (2001)
[01J1] A. Jezierski, A. Kowalczyk, Acta Phys. Polonica, **100**, 565 (2001)
[01K1] A. Kowalczyk, G. Chelkowska, A. Szajek, Solid State Commun. **120**, 407 (2001)
[01K2] M. Kuznietz, J. Magn. Magn. Mater **233**, 155 (2001)
[01L1] J.K. Liang, Y. Chen, X.L. Chen, Progr. Natural Sci. **11**, 881 (2001)
[01M1] F. Maruyama, J. Alloys Comp. **320**, 7 (2001)
[01P1] S. Pandian, V. Chandrasekaran, K.J.L. Iyer, K.V.S. Rama Rao, J. Mat. Sci. **36**, 5903 (2001)
[01S1] A. Szajek, A. Kowalczyk, L. Smardz, J. Magn. Magn. Mater. **223**, 119 (2001)
[01V1] P. Vlaic, E. Burzo, Rom. Repts. Phys. **53**, 643 (2001)
[01V2] P. Vlaic, E. Burzo, in *Proceedings of the 3rd International Workshop Materials for Electrotechnics*, vol 1, 22 (2001)
[01Z1] W.Y. Zhang, J. Zhang, S.Y. Zhang, B.G. Shen, J. Magn. Mater. **234**, 284 (2001)
[01Z2] W.Y. Zhang, J. Zhang, Z.H. Cheng, S.Y. Zhang, B.G. Shen, J. Phys.: Condens. Matter, **13**, 3859 (2001)
[02B1] E. Burzo, A. Mincic, J. Magn. Magn. Mater. **242–245**, 809 (2002)
[02B2] E. Burzo, P. Vlaic, J. Magn. Magn. Mater. **242–245**, 826 (2002)
[02C1] J.M. Cadogan, S.J. Campbell, X.L. Zhao, H.S. Li, P.W. Thompson, Hyp. Int. C **119** (2002)
[02C2] C. Chacon, O. Isnard, Appl. Phys. A **74**, S831 (2002)
[02C3] C. Chacon, O. Isnard, L. Chioncel, C. Giorgetti, F. Baudelet, E. Dartyge, J. Magn. Magn. Mater. **242–245**, 861 (2002)
[02K1] A. Kowalczyk, T. Toliński, J. Baszynski, Chech. J. Phys. **52**, 239 (2002)
[02M1] Z.X. Mu, H.R. Wang, Z.Z. Wu, J. Magn. Magn. Mater. **241**, 131 (2002)
[02V1] P. Vlaic, E. Burzo, Mold. J. Phys. Sciences, **1**, 40 (2002)

[02Y1] H. Yoshie, E. Hirachi, H. Honda, W. Takahashi, Y. Amako, H. Nagai, M. Shiga, K. Yoshimura, J. Phys. Soc. Jpn. **71**, 2253 (2002)
[02Z1] C. Zlotea, C. Chacon, O. Isnard, J. Appl. Phys. **92**, 7382 (2002)
[03A1] Z. Arnold, N.M. Hong, Y. Skorokhod, Z. Machatova, Physica B **327**, 211 (2003)
[03A2] Z. Arnold, J. Kamarad, Y. Skorokhod, N.M. Hong, N.P. Thuy, C.V. Tang, J. Magn. Magn. Mater. **262**, 382 (2003)
[03B1] K.H.J. Buschow, F.R. de Boer, *Physics of Magnetism and Magnetic Materials* (Plenum Press, 2003)
[03K1] A. Kowalczyk, G. Chelkowska, T. Toliński, V. Ivanov, Phys. Stat. Solidi (a) **196**, 297 (2003)
[03M1] A. Mincic, V. Pop, Phys. Stat. Solidi (b) **237**, 540 (2003)
[03M2] H. Miyagawa, Y. Watanabe, N. Hiraoka, Y. Sakurai, S. Nanao, J. Phys. IV **104**, 503 (2003)
[04A1] M. Akiba, H. Nagai, H. Yonezawa, Y. Amako, J. Magn. Magn. Mater. **272–276**, e1651 (2004)
[04A2] A. Amirabadizadeh, N. Tajabor, M.R. Alinejad, F. Pourarian, J. Magn. Magn. Mater. **272–276**, 2081 (2004)
[04B1] L.P. Barbosa, H. Takiishi, R.N. Faria, J. Magn. Magn. Mater. **268**, 132 (2004)
[04B2] L. Bessais, S. Sab, C. Djega-Mariadassou, N. H. Dan, N. X. Phuc, Phys. Rev. B **70**, 134401 (2004)
[04B3] E. Burzo, Mol. Cryst. Liq. Cryst. **415**, 43 (2004)
[04B4] E. Burzo, L. Chioncel, J. Opt. Adv. Mater. **6**, 917 (2004)
[04B5] E. Burzo, R. Tetean, L. Chioncel, M. Văleanu, Phys. Stat. Solidi (c) **1**, 3357 (2004)
[04C1] C. Chacon, O. Isnard, J. F. Berar, J. Phys. IV **118**, 419 (2004)
[04E1] M. ElMassalami, H.A.Borges, H.Takeya, R.E.Rapp, A.Chaves, J. Magn. Magn. Mater. **279**, 5 (2004)
[04G1] I. Goscianska, T. Tolinski, B. Szymanski, A. Kowalczyk, Chech, J. Phys. **54**, D241 (2004)
[04I1] H. Ido, Y. Yamaguchi, J. Magn. Magn. Mater. **272–276**, e487 (2004)
[04K1] A. Kowalczyk, A. Szajek, G. Chelkowska, T. Tolinski, Solid State Commun. **132**, 225 (2004)
[04L1] Z.W. Liu, H.A. Davies, IEEE Trans. Magn. MAG **40**, 2898 (2004)
[04M1] H. Mori, H. Ido, M. Yamada, H. Yoshida, J. Magn. Magn. Mater. **272–276**, e453 (2004)
[04S1] Y. Saito, H. Ido, H. Yoshida, J. Magn. Magn. Mater. **272–276**, e357 (2004)
[04S2] A. Szajek, A. Kowalczyk, Chech. J. Phys. **54**, D343 (2004)
[04V1] I.V. Veremchuk, N.F. Chaban, V.N. Davydov, Y.B. Kuzma, Inorg. Mater. **40**, 1301 (2004)
[05C1] E.N. Caspi, M. Dubman, H. Ettegui, H. Shaked, S. Short, J.D. Jorgensen, Physica B **359–361**, 944 (2005)
[05D1] M. Dubman, E.N. Caspi, H. Ettegui, L. Keller, M. Melamud, H. Shaked, Phys. Rev. B **72**, 024446 (2005)
[05M1] F. Maruyama, J. Mat. Sci. **40**, 4761 (2005)
[05M2] A. Mincic, J. Alloys Comp. **395**, 57 (2005)
[05P1] N. Plugaru, J. Rubin, J. Bartolomé, V. Pop, Phys. Rev. B **71**, 024433 (2005)
[05P2] N. Plugaru, J. Rubin, J. Bartolomé, C. Piquer, J. Magn. Magn. Mater. **290–291**, 1563 (2005)
[05S1] Y. Saito, H. Ido, H. Yoshida, T. Sakon, M. Motokawa, J. Appl. Phys. **97**, 10M520 (2005)
[05S2] O. Sigalov, A.I. Shames, E.N. Caspi, M. Dubman, H. Ettegui, S.D. Goren, H. Shaked, J. Appl. Phys. **98**, 074105 (2005)
[06B1] E. Burzo, P. Vlaic, V. Pop, I. Creanga, J. Opt. Adv. Mater. **8**, 484 (2006)
[06C1] E.N. Caspi, M. Dubman, H. Ettedgui, H. Shaked, M. Melamud, L. Keller, M. Avdeev, Physica B **385–386**, 339 (2006)
[06D1] A. Duran, S. Bernes, R. Falconi, R. Escudero, O. Laborck, M. Guillot, Phys. Rev. B **74**, 134513 (2006)
[06I1] T. Ito, H. Ido, J. Appl. Phys. **99**, 08J304 (2006)
[06K1] N. V. Kudrevatykh, S. A. Andreev, M. T. Bartashevich, A. N. Bogatin, O. A. Milyaev, P. E. Markin, I. S. Tereshina, T. Palewski, E. A. Tereshina, J. Magn. Magn. Mater. **300**, e488 (2006)

[06L1] G.J. Long, R.P. Hermann, F. Grandjean, C. Chacon, O. Isnard, J. Phys.: Condens. Matter, **18**, 10765 (2006)
[06M1] F. Maruyama, Y. Amako, Solid State Commun. **140**, 130 (2006)
[06M2] H. Mayot, O. Isnard, Z. Arnold, J. Kamarad, High Pressure Res. **26**, 489 (2006)
[06M3] D. Music, J.M. Schneidner, J. Phys.: Condens. Matter, **18**, 4071 (2006)
[06N1] S. Notsu, Y. Shimoji, M. Hedo, Y. Uwatoko, T. Nakama, H. Ido, K. Yagasaki, J. Phys. Soc. Jpn. **75**, 044714 (2006)
[06P1] C. Paduani, J.A. Valcanover, J.D. Ardisson, M.J. Yoshida, Phys. Stat. Solidi (b), **243**, 1347 (2006)
[06Q1] P. Qian, Q.L. Wang, N.X. Chen, J. Shen, J. Phys. D: Appl. Phys. **39**, 1197 (2006)
[06S1] A. Szajek, Phys. Stat. Solidi (c) **3**, 238 (2006)
[06S2] A. Szajek, Mat. Science.-Poland, **24**, 843 (2006)
[06S3] A. Szajek, J.A. Morkowski, Mat. Science–Poland, **24**, 839 (2006)
[06T1] W. Tang, Y.Q. Wu, K.W. Dennis, M.J. Kramer, I.E. Anderson, P.W. McCallum, J. Appl. Phys. **93**, 08B510 (2006)
[06T2] R. Tetean, D. Andreica, I.G. Deac, E. Burzo, L. Chioncel, A. Amato, Physica B **374–375**, 188 (2006)
[06V1] P. Villars, *Landolt-Börnstein Handbook,* vol. III/43A3 (Springer Verlag, 2006)
[06Z1] C. L. Zhang, D. H. Wang, Z. D. Han, S. L. Tang, B. X. Gu, Y. W. Du, Appl. Phys. Lett. **39**, 122503 (2006)
[07B1] J. Bartolomé, N. Plugaru, J. Campo, J. Rubin, E.K. Hlil, C. Rello, A. Arauzo, J. Alloys Comp. **442**, 11 (2007)
[07B2] E. Burzo, AIP Conf. Proc. **899**, 239 (2007)
[07B3] E. Burzo, R. Tetean, L. Chioncel, V. Pop, J. Magn. Magn. Mater. **316**, e379 (2007)
[07C1] G. Cordier, R. Klemens, B. Albert, Zeit. Anorg. Allgem. Chem. **633**, 1603 (2007)
[07C2] W.B. Cui, W. Liu, X.K. Lv, X.G. Zhao, Z.D. Zhang, Physica B **400**, 273 (2007)
[07F1] V. Franco, A. Conde, V.K. Pecharsky, K.A. Gschneinder, Eur. Phys. Lett. **79**, 47009 (2007)
[07G1] F. Grandjean, R.P. Hermann, E. Popiel. G.J. Long, J. Appl. Phys. **101**, 023917 (2007)
[07H1] J. Herrero-Albillos, F. Bartolomé, L. M. Garcia, A. T. Young, T. Funk, J. Campo, G. J. Cuello, Phys. Rev. B **76**, 094409 (2007)
[07H2] N.M. Hong, Commun. Physics **17**, 241 (2007)
[07II] O. Isnard, C. Chacon, J. Alloys Comp. **442**, 22 (2007)
[07K1] E.N. Kablov, A.F. Petrakov, V.P. Piskorskii, V.A. Valeev, N.V. Nazarova, Metal. Sci. Heat Treat. **49**, 159 (2007)
[07M1] F. Maruyama, Y. Amako, H. Nagai, J. Alloys Comp. **429**, 72 (2007)
[07M2] M. Mizumaki, H. Ido, J. Magn. Magn. Mater. **310**, 1871 (2007)
[07P1] N. Plugaru, J. Rubin, J. Bartolomé, J. Campo, G.J. Cuello, M. Tovar, O. Prokhnenko, J. Magn. Magn. Mater. **316**, e438 (2007)
[07R1] C. Rillo, A. Arauzo, N. Plugaru, E.K. Hlil, J. Rubin, J. Bartolomé, J. Magn. Magn. Mater. **316**, 166 (2007)
[07Y1] S. Yehia, AIP Conf. Proc. **888**, 8 (2007)
[08B1] E. Burzo, P. Vlaic, J. Opt. Adv. Mater. **10**, 750 (2008)
[08B2] E. Burzo, I. Creangă, L. Chioncel, J. Opt. Adv. Mater, **10**, 798 (2008)
[08G1] H. Gencer, N. Kervan, A. Gencer, A. Gunes, S. Atalay, J. Alloys Comp. **466**, 1 (2008)
[08I1] O. Isnard, M.D. Kuzmin, M. Richter, M. Loewenhaupt, R. Bewley, J. Appl. Phys. **104**, 013922 (2008)
[08K1] N. Kervan, S. Kervan, A. Gencer, J. Magn. Magn. Mater, **320**, 2839 (2008)
[08K2] N. Kervan, S. Kervan, A. Gencer, J. Phys. Chem. Solids, **69**, 2791 (2008)
[08M1] F. Maruyama, Y. Amako, Solid State Commun. **148**, 206 (2008)
[08M2] H. Mayot, Ph.D. Thesis Univ. J. Fourier, Grenoble (2004)
[08M3] H. Mayot, O. Isnard, Z. Arnold, J. Kamarad, J. Phys.: Condens. Matter, **20**, 135207 (2008)
[08M4] H. Mayot, O. Isnard, F. Grandjean, G.J. Long, J. Appl. Phys. **103**, 093917 (2008)
[08M5] M. Misek, Z. Arnold, O. Isnard, H. Mayot, Y. Skorokhod, J. Kamarad, Acta Phys. Polonica, A **113**, 263 (2008)

[08M6] P. Morales, R. Escudero, A. Duran, J. Low Temp. Phys. **153**, 15 (2008)
[08R1] O. Rivin, R. Osborn, A.I. Kolesnikov, E.N. Caspi, H. Shaked, Phys. Rev. B **78**, 184424 (2008)
[08S1] Y. Skorokhod, Z. Arnold, O. Isnard, H. Mayot, M. Misek, J. Kamarad, Acta Phys. Polonica, A **113**, 259 (2008)
[09A1] E.J. Wolfson, E.N. Caspi, H. Ettegui, H. Shaked, M. Avdeev, J. Magn. Magn. Mater. **320**, L97 (2008)
[09A1] H. Ağıl, N. Kervan, S. Kervan, H. Sözeri, A. Gencer, J. Alloys Comp. **478**, 437 (2009)
[09A2] A. Arauzo, J. Bartolomé, C. Rillo, J. Rubin, Solid State Phen. **152–153**, 466 (2009)
[09B1] E. Burzo, I.G. Pop, V. Pop, P. Vlaic, Opt. Adv. Mater., Rapid Commun. **3**, 481 (2009)
[09E1] M. ElMassalami, R. Moreno, R.M. Saeed, F.A.B. Chaves, C.M. Chaves, R.E. Rapp, H. Takeya, B. Ouladdiaf, M. Amara, J. Phys.: Condens. Matter, **21**, 216006 (2009)
[09E2] M. ElMassalami, R. Moreno, H. Takeya, B. Ouladdiaf, J.W. Lynn, R.S. Freitas, J. Phys.: Condens. Matter, **21**, 436006 (2009)
[09E3] M. ElMassalami, R.E. Rapp, F.A.B. Chaves, R. Moreno, H. Takeya, B. Ouladdiaf, J.W. Linn, Q. Huang, R.S. Freitas, N.F. Oliveira, J. Phys.: Condens. Matter, **21**, 046007 (2009)
[09G1] F. Grandjean, M.T. Sougrati, H. Mayot, O. Isnard, G.J. Long, J. Phys.: Condens. Matter, **21**, 186001 (2009)
[09I1] T. Ito, H. Ido, J. Appl. Phys. **105**, 07E511 (2009)
[09K1] Y.Katsura, H.Ogino, Y.Matsumura, S.Horii, J.I.Shimoyama, K.Kishio, J. Phys.: Conf. Series, **176**, 012016 (2009)
[09K2] N. Kervan, S. Kervan, H. Sözeri, A. Gencer, J. Alloys Comp. **480**, 216 (2009)
[09K3] N. Kervan, S. Kervan, H. Sözeri, A. Gencer, Mat. Chem. Phys. **116**, 586 (2009)
[09K4] D.S. Khoshnoud, N. Tajabor, D. Fruchart, D. Gignoux, S. Miraglia, F.Pourarian, J. Alloys Comp. **480**, 198 (2009)
[09K5] V.H.Ky, L. Bessais, C. Djega-Mariadassou, N.X. Phuc, J. Appl. Phys. **105**, 103905 (2009)
[09L1] L.Li, K.Nishimura, J. Appl. Phys. **106**, 023903 (2009)
[09L2] H.Li, K.Nishimura, J. Phys. D.: Appl. Phys. **42**, 145003 (2009)
[09L3] L.Li, K.Nishimura, H.Yamane, Appl. Phys. Lett. **94**, 102509 (2009)
[09M1] F. Maruyama, Y. Amako, Solid State Commun. **149**, 2036 (2009)
[09M2] H. Mayot, O. Isnard, F. Grandjean, G.J. Long, J. Appl. Phys. **105**, 113908 (2009)
[09O1] S. Okada, K. Kudou, T. Shishido, Trans. Kokushikan Univ., 21 (2009)
[10A1] Z. Arnold, O. Isnard, H. Mayot, M. Misek, J. Kamarad, J. Magn. Magn. Mater. **322**, 1117 (2010)
[10B1] E. Burzo, R. Tetean, I. Balasz, J. Magn. Magn. Mater. **322**, 1109 (2010)
[10M1] H. Mayot, O. Isnard, Z. Arnold, J. Kamarad, J. Appl. Phys. **107**, 103921 (2000)
[10N1] N. Nagaosa, J. Sinova, S. Onoda, A. H. MacDonald, N. P. Ong, Rev. Modern. Phys. 1539 (2010)
[10P1] Q. Ping, J.J. Liu, S. Jiang, L.J. Bai, Q. Ran, L.Y. Wang, Chin. Phys. B **19**, 126001 (2010)
[10P2] V.P. Piskorskii, G.S. Burkhanov, O.G. Ospennikova, R.A. Valeev, I.S. Tereshina, E.A. Davydova, Russ. Metallurgy, **1**, 53 (2010)
[10T1] I. S. Tereshina, E. A. Tereshina, G. S. Burkhanov, S. V. Dobatkin, Rev. Adv. Mater. Sci. **25**, 82 (2010)
[10W1] E.J. Wolfson, E.N. Caspi, H. Ettedgui, H. Shaked, M. Avdeev, J. Phys.: Condens Matter, **22**, 026001 (2010)
[11B1] E. Burzo, L. Chioncel, R. Tetean, O. Isnard, J. Phys.: Condens. Matter, **23**, 026001 (2011)
[11K1] S. Kervan, N. Kervan, A. Özdemir, H. Sözeri, J. Rare Earths, **29**, 466 (2011)
[11L1] L.Li, D.Huo, Z.Qian, K.Nishimura, J. Phys.: Conf. Series, **263**, 012017 (2011)
[11L2] L. Li, D. Huo, H. Igawa, K. Nishimura, J. Alloys Comp. **509**, 1796 (2011)
[11L3] L. Li, H. Igawa, K. Nishimura, D. Huo, J. Appl. Phys **109**, 083901 (2011)
[11L4] L.Li, K.Nishimura, D.Huo, Z.Chan, J. Appl. Phys. **110**, 083915 (2011)
[11L5] L. Li, K. Nishimura, H. Igawa, D. Huo, J. Alloys Comp. **509**, 4198 (2011)
[11M1] G.I. Miletic, Z. Blazina, J. Magn. Magn. Mater. **323**, 2340 (2011)
[11M2] H. Mizoguchi, T. Kuroda, T. Kamiya, H. Hosono, Phys. Rev. Lett. **106**, 237001 (2011)

[11T1] T. Tolinski, M. Falkowski, A. Kowalczyk, K. Synoradzki, Solid State Sci. **13**, 1865 (2011)
[11V1] P. Villares et al., *Landolt Bornstein Handbook*, vol. 43A10 (Springer Verlag, 2011)
[11W1] G. Wang, M. Zhang, Z. Li, Z. Yang, Eur, Phys. Lett. **95**, 17001 (2011)
[11X1] L. Xu, Y. Liu, D. Guo, L. Zhou, F. Wang, H. Zhang, X. Zhang, J. Phys. D: Appl. Phys. **44**, 355002 (2011)
[11Z1] C.I. Zhang, J. Chen, T.Z. Wang, G.X. Xie, C. Zhu, D.H. Wang, Physica B **406**, 2840 (2011)
[12A1] Z. Arnold, O. Isnard, H. Mayot, Y. Skorokhod, J. Kamarad, M. Misek, Solid State Commun. **152**, 1164 (2012)
[12E1] M. ElMassalami, H. Takeya, B. Ouladdiaf, R. Maia Filho, A.M. Gomes, T. Paiva, R.R. dos Santos, Phys. Rev. B **85**, 174412 (2012)
[12I1] O. Isnard, Y. Skourski, L.V.B. Diop, Z. Arnold, A.V. Andreev, J. Wosnitza, A. Iwasa, A. Kondo, A. Matsuo, K. Kindo, J. Appl. Phys. **111**, 093916 (2012)
[12L1] L. Li, G. Hu, I. Umehara D. Huo, T. Namiki, K. Nishimura, J. Phys. Soc. Jpn. **81**, 073701 (2012)
[12L2] L.Li, K.Nishimura, G.Usui, D.Huo, Z.Qian, Intermetallics, **23**, 101 (2012)
[12N1] K. Nakayama, E. Ieki, Y. Tanaka, T. Sato, T. Takahashi, K. Kuroda, H. Mizoguchi, S. W. Kim, H. Hosono, Phys. Rev. B **86**, 014503 (2012)
[12S1] S. Seong, S.Y. Choi, T.A. Albright, Theor. Chem. Acc. **131**, 1091 (2012)
[12S2] J. Ph. Soulie, N. Penin, K. Yvon, J. Alloys Comp. **526**, 63 (2012)
[13A1] O.A. Arinicheva, A.S. Lileev, M. Reissner, A.A. Lukin, A.S. Starikova, Hyp. Int. **219**, 89 (2013)
[13B1] E. Burzo, P. Vlaic, Rom. Repts. Phys. **15**, 778 (2013)
[13D1] L.V.B. Diop, O. Isnard, Y. Skourski, G. Ballou, J. Appl. Phys. **113**, 203911 (2013)
[13D2] L.V.B. Diop, O. Isnard, N.R. Lee-Hone, D.H. Ryan, J.M. Cadogan, J. Phys.: Condens. Matter, **25**, 316001 (2013)
[13E1] M. ElMassalmi, A. M. Gomes, T.Paiva, R. R. dos Santos, H. Takeya, J. Magn. Magn. Mater. **335**, 163 (2013)
[13J1] A. Jain, S. P. Ong, G. Hautier, W. Chen, W. D. Richards, S. Dacek, S. Cholia, D. Gunter, D. Skinner, G. Ceder, S. Cholla,, Appl. Phys. Lett. Materials, **1**, 011002 (2013)
[13L1] N.R. Lee-Hone, D.H. Ryan, O. Isnard, L.V.B. Diop, J.M. Cadogan, J. Appl. Phys. **113**, 12E119 (2013)
[13L2] L. Li, K. Nishimura, D. Huo, Z. Qian, T. Namiki, J. Alloys Comp. **572**, 205 (2013)
[13L3] J.C. Li, P. Qian, Z.F. Zhang, Y. Liu, X.J. Yuan, J. Shen, N.X. Chen, Computer Phys. Commun. **184**, 342 (2013)
[13S1] E.J. Skoug, M.S. Meyer, F.E. Pinkerton M.M. Tessema, D. Haddad, J.F. Herbst, J. Alloys. Comp. **574**, 552 (2013)
[14C1] R.J. Caraballo Vivas, D.L. Rocco, T. Costa Soares, L. Caldeira, A.A. Coelho, M.S. Reis, J. Appl. Phys. **116**, 063907 (2014)
[14D1] L.V.B. Diop, PhD Thesis, Univ. J. Fourier, Grenoble (2014)
[14D2] L.V.B. Diop, Z. Arnold, O. Isnard, J. Kamarad, J. Alloys Comp. **593**, 163 (2014)
[14D3] L.V.B. Diop, O. Isnard, J. Phys.: Condens. Matter, **27**, 026004 (2014)
[14E1] M. ElMassalami, H. Takeya, B. Ouladdiaf, A.M. Gomes, T. Paiva, R.R. dos Santos, J. Magn. Magn. Mater. **372**, 74 (2014)
[14J1] X. Jiang, B. Balamurugan, J.E. Shield, J. Alloys Comp. **617**, 479 (2014)
[14K1] S. Kervan, N. Kervan, H. Sözeri, J. Magn. Magn. Mater. **354**, 17 (2014)
[14P1] N. Plugaru, M. Văleanu, R. Plugaru, J. Campo, J. Appl. Phys. **115**, 023907 (2014)
[14P2] J.Pospisil, H.Kitazawa, A.Gupta, S.Toyoizumi, A.Tamaki, M.Divis, V.Sechovsky, J. Phys. Soc. Jpn. **83**, 054713 (2014)
[14R1] O. Rivin, H. Shaked, A. Gukasov, E.N. Caspi, Phys. Rev. B **89**, 174423 (2014)
[14T1] T. Tolinski, K. Synoradzki, Acta Phys. Polonica, A **126**, 160 (2014)
[15B1] E. Burzo, AIP Conf. Series, **1694**, 030001 (2015)
[15B2] E. Burzo, P. Vlaic, D.P. Kozlenko, Rom. J. Phys. **60**, 200 (2015)

[15L1] R. Li, Y. Zhang, F. Xu, T. Wang, F. Li, in *IEEE International Magnetic Conference*, (2015), p.1
[15M1] F. Masquita, PhD Thesis, Univ. Federal, Rio Grande do Sul, Brasil (2015)
[15M2] F. Mesquita, L.V.B. Diop, G. Fraga, O. Isnard, P. Pureur, IEEE Magn. Letters, **6**, 3800304 (2015)
[15S1] K. Shimizu, K. Kakiuchi, T. Ito, H. Ido, Phys. Procedia, **75**, 853 (2015)
[15W1] Y. Wang, W. J. Ren, T. F. Duan, D. Li, J. Li, Z. D. Zhang, J. Appl. Phys. **117**, 17E116 (2015)
[15Z1] C.N. Zhang, X.X. Wang, in *Proceedings of the 2nd Asian Pacific Conference*, (APEESD, 2015), p. 377
[16H1] G.H. Hu, L.W. Li, I. Umehara, Chinese Phys. B **25**, 067501 (2016)
[16L1] X.B. Liu, Z. Altounian, D.H. Ryan, J. Alloys Comp. **688**, 118 (2016)
[16M1] M. Majumder, A. Ghoshray, P. Khuntia, C. Mazumdar, A. Poddar, M. Baentiz, K. Ghoshray, J. Phys.: Condens. Matter, **28**, 345701 (2016)
[17B1] E. Burzo, J. Phys.: Conf. Series, **848**, 012004 (2017)
[17G1] E. Galego, M.M. Serna, L.V. Ramanathan, R.N. Faria, J. Magn. Magn. Mater. **424**, 298 (2017)
[17H1] J. F. Herbst, L. G. Hector, J. Alloys Comp. **693**, 238 (2017)
[18B1] E. Burzo, L. Chioncel, Rom. J. Phys. **63**, 601 (2018)
[18C1] X. Chi, Y. Li, X.H. Han, J.B. Sung, Y. Zhang, C.X. Cui, Physica B **545**, 176 (2018)
[18C2] X. Chi, Y. Li, X. H. Han, J. B. Sun, Y. Zhang, C. X. Cui, J. Magn. Magn. Mater. **465**, 524 (2018)
[18J1] A. Jimenez-Vazquez, R. Falconi, H. Takeya, B. Ouladdiaf, M. ElMassalami, J. Mat. Sci.: Mat. Electron. **29**, 15411 (2018)
[18M1] F. Mesquita, L. V. Diop, O. Isnard, J. Alloys Comp. **763**, 355 (2018)
[18Z1] X. Q. Zheng, J. W. Xu, H. Zhang, S.C. Wang, Y. Zhang, Z.H. Xu, L. C. Wang, B. B. Shen, AIP Adv. **8**, 056432 (2018)
[19B1] E. Burzo, P. Vlaic, D.P. Kozlenko, N. O. Golosova, A. V. Rustkauskas, S. E. Kichanov, B. N. Savenko, Intermetallics, **110**, 106489 (2019)
[19C1] A. Candan, G.Suruku, a.Gencer, Phys. Scripta, **94**, 5710 (2019)
[19C2] P. Chen, L. Jiang, J. Li, H. Chen, J. He, Y. Wang, J. Appl. Phys. **126**, 085501 (2019)
[19C3] X. Chi, J. B. Sun, S. Wang, H. W. Wang, Y. Zhang, J. Appl. Phys. **125**, 235106 (2019)
[19D1] L.V.B. Diop, Z. Arnold, J. Kamarad, O. Isnard, J. Magn. Magn. Mater **476**, 106 (2019)
[19D2] X.L. Duan, J.B. Sun, L.Z. Wang, K. Guo, L.X. Cui, J. Magn. Magn. Mater. **489**, 165450 (2019)
[19R1] F. Z. Rachid, L. H. Omari, Z. Yamkane, H. Lassri, S. Derkaoui, K. Nouri, W. Bouzidi, Mater. Chem. Phys. **228**, 60 (2019)
[19S1] T. Saito, D. Nishio-Hamane, Intermetallics, **107**, 6 (2019)
[19V1] M. G. Vergniory, L. Elcoro, G. Felser, N. Regnault, B. A. Bernvig, Z. Wang, Nature, **566**, 480 (2019) and references
[20A1] O. Amhoud, N. Zaim, M. Kerouad, A. Zaim, Phys. Lett. **384**, 126443 (2020)
[20J1] J. Jatmika, H. Maruyama, M. S. Rahman, A. Sakai, S. Nakatsuji, A. Iyo, T. Ebihara, Supercond. Sci. Techn. **33**, 125006 (2020)
[20L1] L. Q. Lai, Y. H. Li, F. Bao, W. Zhang, J. Iron Steel Res. Internat. 27, 471 (2019)
[20M1] Materials Research Project, Lawrence Berkeley Nat. Lab. Berkeley, USA (2020)
[20M2] F. Mesquita, J. O. Magalhaes, P. Pureur, L. V.Diop, O. Isnard, Phys. Rev. B **101**, 224414 (2020)
[20S1] T. Saito, D. Nishio-Hamane, J. Magn. Magn. Mater. **513**, 167189 (2020)
[20Z1] Y. Zhang, D. Guo, B. Wu, H. Wang, R. Guan, X. Li, Z. Ren, J. Alloys Comp. **817**, 152780 (2020)
[21E1] G. Gomez-Eslava, M. Ito, C. V. Colin, M. Yano, T. Shoji, A. Kato, E. Suard, N. M. Dempsey, D. Givord, J. Alloys Comp. **851**, 156168 (20021)
[22B1] E. Burzo, P. Vlaic, D. P. Kozlenko, A. V. Rutkauskas, J. Solid State Chem. **160**, 110330 (2022)
[22B2] E.Burzo, L.Chioncel, Rom. Report Phys. **74**, 501 (2022)

Chapter 9
Rare–Earths–Nickel–Boron Compounds

9.1 General Problems

The phase diagrams of R–Ni–B systems, where R is a rare-earth or yttrium, were investigated including those with R = La [72K1, 73K2], R = Ce [71K1], R = Pr [85B1], R = Nd [82B1, 83K3], R = Sm [83B1], R = Eu [81K2, 82C1], R = Gd [83K1, 83K3], R = Tb [85D1], R = Dy [99C3], R = Ho [93G1], R = Er [83C1, 04C1, 04V1], R = Yb [06V1], R = Lu [89G2] and R = Y [75K1, 95C3]. Reviews on the matter were published [79K2, 84P1, 84R1]. Some of the initially reported phase diagrams, were later improved, and new compounds or correcting the earlier proposed compositions, were evidenced. As example, the two partial isothermal sections of the La–Ni–B system, at 800 °C (0–33 at.% La) and 400 °C (33–100 at.%) [72K1] identified initially six ternary compounds, the structure of the hypothetical $LaNi_{12}B_2$ compound being not determined. Latter studies, evidenced that this compound is not formed. Also, the earlier reported $LaNi_5B$ composition was shown to be really $La_3Ni_{13}B_2$ [81K1, 84R1]. The phase diagram of Er–Ni–B system, has been investigated in the composition range 0–35 at.% Er, at 1070 K and 35–100 at.% Er, at 870 K [83C1]. Eleven ternary borides have been initially found. The compositions and crystal structures of seven compounds were determined: $Er_3Ni_{13}B_2$, $Er_2Ni_7B_2$, $ErNi_4B$, $Er_2Ni_{21}B_6$, $ErNiB_4$ (in both high and low temperature modifications) and Er_4NiB_{13} [60K1]. The approximate compositions were also proposed for $ErNi_8B_3$, $ErNi_6B_3$, $Er_2Ni_4B_3$, $ErNi_5B_4$, $ErNi_2B_3$. Later on [88K1, 91G2–91G6], new Er–Ni–B ternary borides were synthesized and their crystal structures reported. Finally, the compounds present in Er–Ni–B system were carefully analyzed [04C1]. There are now 17 structure types in this system. The above data suggest, that additional R–Ni–B compounds, to those included in Table 9.1 can be further synthesized.

There are 28 type structures in R–Ni–B systems, including RNi_2B_2C borocarbides, forming homologous series with rare earths, many of them with less than 40 at.% R—Table 9.1. In some cases, only one compound is formed as in case of $LaNi_3B$, $Lu_5Ni_{19}B_6$, YNi_3B_2, $ErNi_{7.9}B_2$, Ho_4NiB_{14} or $Ce_3Ni_5B_6$ structure types.

E. Burzo, *Rare Earths-Transition Metals-Boron Compounds*,
https://doi.org/10.1007/978-3-030-99245-3_9

Table 9.1 Ternary R–Ni–B compounds[a]

Compound	La	Ce	Pr	Nd	Sm	Eu	Gd	Tb	Dy	Ho	Er	Tm	Yb	Lu	Y	Coordination polyhedral, CP, of boron atoms	Boron atoms bonding type
$R_3Ni_{13}B_2$	○	○	○	○	○	○	○	○	○	○	○	○	○		○	TP	i
$R_{1-x}Ni_{4+x}B$	○	○									○		○		○	TP	i
RNi_4B	○	○	○	○	○	○	○	○	○	○	○	○	○	○	○	TP	i
$R_3Ni_7B_2$					○	○	○	○	○	○	○	○	○	○	○	TP	i
$R_5Ni_{19}B_6$														○		TP	i
RNi_3B	○																
$R_2Ni_{21}B_6$		○								○	○	○	○	○		TA	i
RNi_7B_3											○		○	○		TP, TA	p, i
$RNi_{7.9}B_2$											○						
RNi_3B_2															○		
$R_2Ni_{15}B_6$									○	○					○	TA	i
$RNi_{6.5}B_3$										○	○		○				
$R_2Ni_{10}B_5$							○	○	○	○	○	○	○	○	○	TP, TA	p, i
$R_3Ni_{19}B_{10}$							○	○	○	○	○		○		○	TP, TA	p, i
$RNi_{12}B_{6-x}$						○	○	○	○	○					○	TP	i
$RNi_{12}B_6$	○	○	○			○	○	○	○	○					○	TP	i
$R_2Ni_{15}B_9$								○	○	○	○	○	○	○	○	TP, TA	p, i
$R_2Ni_5B_4$	○	○		○													
RNi_2B_2					○		○	○	○	○	○				○	TP	p
RNi_2B_3		○													○		
$R_3Ni_5B_6$		○															

(continued)

Table 9.1 (continued)

Compound	La	Ce	Pr	Nd	Sm	Eu	Gd	Tb	Dy	Ho	Er	Tm	Yb	Lu	Y	Coordination polyhedral, CP, of boron atoms	Boron atoms bonding type
$R_2Ni_3B_6$								○	○	○	○	○	○	○	○	TP	net
$RNiB_4$ (LT)											○	○	○	○		TP	net
$RNiB_4$ (HT)		○	○	○	○		○	○	○	○	○				○	TP, TA	netw
R_4NiB_{13}							○	○	○	○	○	○	○	○	○	K, TP	netw
R_4NiB_{14}										○							
R_2NiB_{10}		○	○	○	○		○	○	○	○					○	TP	netw
RNi_2B_2C[b)]	○	○	○	○	○	○	○	○	○	○	○	○	○	○	○		

[a]The data reported for some heavy rare-earths compounds [06V1] were extended and also included those with light rare earths; TP-trigonal prism, TA-tetragonal antiprism, K-square; the B atoms are: i-isolated, p-pair, net-B net and netw-network;
[b]Calculated structure for R = Eu; YNi_3B_2 compounds were mentioned with unknown crystal structure

A complex interaction between the components is evidenced in R–Ni–B systems with heavy rare-earths. Thus, the number of ternary borides with heavy rare-earths, ranges between 17 for R = Ho to 10 when R = Tm, while those with light rare earths are generally 5–6 (R = La, Pr, Nd, Sm, Eu), except for R = Ce, when 10-type structures are formed (Table 9.1).

Most of the crystal structures of R–Ni–B borides with less than 40 at.% R have structures with trigonal-prismatic or tetragonal-antiprismatic coordination of the boron atoms [77K1, 06V1]. As the boron content increases, the coordination polyhedra of B, join together by bases or by lateral faces [06V1]. This, in turn, leads to changes in the mode of connection of the B atoms. The presence of isolated boron atoms (0–40 at.% B) is an evidence that the metallic type of bonds is prevalent in the borides, as in $Dy_3Ni_7B_2$ compound—Table 9.1. For (27–47) at.% B, one can observe a transition from isolated B-atoms to B_2-pairs (e.g. RNi_7B_3, $R_2Ni_{15}B_9$). In the crystal structure of compounds having (47–67) at.% B, the boron atoms are situated in front of the lateral faces of the trigonal prisms, which causes the formation of B-nets or B network (e.g. R_4NiB_{13}). In addition, there were predicted and identified borides with compositions $R_{1-x}Ni_{4+x}B$, $RNi_{6.5}B_3$ or $R_3Ni_{19}B_{10}$ [06V1]. Two homologous series $R_{n+2}M_{3n+4}B_{2n}$ and $R_{m+n}M_{5m+3n}B_{2n}$ were evidenced when M = Co—Chap. 8. When M = Ni, there are present only one structure type, with $n = 1$ ($R_3Ni_7B_2$), in the first series and three structure types, with m = 1, n = 1 (RNi_4B), m = 2, n = 1 ($R_3Ni_{13}B_2$) and m = 2, n = 3 ($R_5Ni_{19}B_6$), in the second series. The cell volumes, of the homologous compounds, decrease linearly with the atomic number of rare-earth, in agreement with lanthanide contraction [83K3]. The rare-earths ions, are mainly in R^{3+} valence state (except cerium). The stability of some R–Ni–B compounds were also analyzed [15K1].

The Sects. 9.2, 9.3, 9.4, 9.5, 9.6, 9.7, and 9.8 are devoted to the analysis of physical properties of R–Ni–B compounds. In Sect. 9.4, in addition to those of RNi_2B_2 series, the RNi_2B_2C borocarbides are also presented. The RNi_2B_2C compounds have complex physical properties ranging from superconductivity to magnetic ordering. Since of large volume of scientific results in the field, these are presented according to their common physical properties. In the first part, their crystal structures and elasc properties are analyzed. Then, the non-superconducting borocarbides (R = La, Pr, Nd, Sm, Gd, Tb, Yb) as well as those which show only superconducting behavior (R = Lu, Y) are presented. The physical properties of borocarbides which are both superconducting and magnetic ordered (R = Dy, Ho, Er, Tm) are correlated with those which are only superconducting or show only a magnetic order.

9.2 R–Ni–B Compounds with High Boron Content

The R–Ni–B compounds having ratio B/Ni ≥ 2 and low rare-earth content, including series R_2NiB_{10}, R_4NiB_{14}, R_4NiB_{13}, $RNiB_4$ (low and high temperature forms) and $R_2Ni_3B_6$ or $R_2(NiB_2)_3$, have interesting physical properties. Some of them are situated in ternary region, near Ni_2B, were compounds with lower B content are also located. The crystal structures and lattice sites are listed in Table 9.2, while the lattice parameters are given in Table 9.3.

The R_2NiB_{10} compounds crystallize in an orthorhombic type lattice, *Pbam*-space group. The R atom is bonded in a 12-coordinated geometry to B atoms and Ni in 8-coordinated geometry to B atoms. There are four inequivalent B sites. The structure is closely related to that of RB_6 borides [91V1]. Two B atoms bonding B_6–B_6 clusters in RB_6 structure, are replaced by one Ni atom, in ternary R_2NiB_{10} compounds—Fig. 9.1 and Table 9.2a [00J1]. This replacement would weaken the bonding between clusters and thus it is expected to reduce the lattice thermal conductivity. The large R atoms fill the voids of the three-dimensional network formed by nickel and boron atoms.

The crystal structure of R_4NiB_{13} series consists of infinite planar nets of fused B_7 and B_4 rings. The R atom is located between the 7-membered rings. The Ni and additional B atoms are situated between 4-membered rings [83K1, 12V1]—Table 9.2b. The R_4NiB_{14} is derivative of this structure, where a single B3 atom in *2a* site is replaced by two B atoms in the *4e* site (B_2 dumbbells) [01G1].

The $RNiB_4$ compounds crystallize both in orthorhombic and tetragonal type structures. The orthorhombic crystal structure of $RNiB_4$ compounds ($YCrB_4$-type) contains planar boron nets perpendicular to [001], composed of five- and seven-membered rings (LT form)—Chap. 5. The metal atoms are located between boron nets [05V1]. The R atoms are situated between the seven-membered rings and the Ni atoms are located between the five-membered rings—Table 9.2c. The structure features of the high temperature form of $RNiB_4$ series, space group *I4/mmm* is characterized by the formation of boron 2D-framework with planar B_8 rings. The R and Ni atoms are situated in the larger and smaller voids, respectively [84K1]—Table 9.2d.

The crystal structure of $R_2Ni_3B_6$ compounds is closely related to that of $ThMoB_4$, with 1/3 of Ni atoms replaced by B–B dumb-bells—Chap. 5. The structure consists of two equidistant planar layers which are alternatively stacked along the **c**-axis [00G3]—Table 9.2e. Within the R–Ni1 network, each Ni1 atom is coordinated by four R atoms, leading to a distorted square-planar arrangement. Both symmetry-independent B atoms are located inside trigonal prisms, formed by 4R atoms and 2Ni1 atoms. Within the Ni2–B1–B2 network, each B2 atom is coordinated by two B1 and one B2 atoms, whereas the Ni2 atom is surrounded by four B1 atoms. The two layers are connected by the Ni2 atom, with a square-planar coordination, by Ni1 atoms. There is a strong Ni–Ni bonding. The thermal expansion coefficients of $Ho_2Ni_3B_6$ single crystal, in the temperature range $100\ K \leq T \leq 300\ K$, are $\alpha_a = 8.5 \cdot 10^{-6}\ K^{-1}$, $\alpha_b = 6.7 \cdot 10^{-6}\ K^{-1}$ and $\alpha_c = 1.4 \cdot 10^{-5}\ K^{-1}$ [00G3].

The magnetic susceptibility of Y_2NiB_{10}, at $T > 100$ K, is nearly temperature independent, suggesting the presence of Pauli-type paramagnetism. When decreasing temperature, below 100 K, the susceptibility increases, behavior probably due to the presence of magnetic impurity [00J1].

The temperature dependence of the magnetic susceptibility, χ, in field of $\mu_0H = 0.5$ T, of $Ce_2NiB_{9.7}$, at $1.5\ K \leq T \leq 4\ K$, follows a $\chi = \chi_0 - \alpha T^{-0.5}$ relation [03M3], characteristic for Non-Fermi-Liquid (NFL)-type behavior [01S4]. Application of magnetic fields promotes the recovery of the temperature independent susceptibility.

Table 9.2 Lattice sites and their coordinates of compounds with high boron content

(a) Ce_2NiB_{10} having orthorhombic structure, *Pbam* space group [00J1]

Atom	Sites	x	y	z
Ce	4*h*	0.1911(3)	0.3633(2)	1/2
Ni	2*a*	0	0	0
B1	8*i*	0.212(4)	0.115(2)	0.313(3)
B2	4*g*	0.056(5)	0.184(3)	0
B3	4*g*	0.350(6)	0.190(2)	0
B4	4*g*	0.380(5)	0.036(3)	0

(b) Er_4NiB_{13} having tetragonal structure, *P4/mnc* space group [83K2, 01G1, 12V1][a]

Atom	Sites	x	y	z	Atomic environment
Er	8*h*	0.1703	0.3081	0	Non-coplanar triangle B_3
Ni	2*b*	0	0	1/2	Cuboctahedron B_8Er_4
B1	16*i*	0.177	0.039	0.267	Non-coplanar triangle B_3
B2	8*g*	0.091	0.591	1/4	Coplanar triangle B_3
B3	2*a*	0	0	0	Cuboctahedron B_8Er_4

(c) $YbNiB_4$ having orthorhombic structure, *Pbam* space group [05V1]

Atom	Sites	x	y	z
Yb	4*h*	0.12738(5)	0.15020(2)	0
Ni	4*h*	0.63401(2)	0.08839(8)	0
B1	4*g*	0.785(1)	0.1841(7)	1/2
B2	4*g*	0.386(1)	0.0475(7)	1/2
B3	4*g*	0.861(2)	0.0332(7)	1/2
B4	4*g*	0.479(1)	0.1929(7)	1/2

(d) $ErNiB_4$ having tetragonal structure, *I4/mmm* space group [84K1, 11V1][b, c]

Atom	Sites	x	y	z	Atomic environment
Er1	4*e*	0	0	0.2969	Non-coplanar square B_4
Er2	4*d*	0	1/2	1/4	square prism (cube) B_8
Ni	8*i*	0.2269	0	0	square prism (cube) B_6Ni_2
B1	16*m*	0.231	0.231	0.151	square pyramid B_3Ni_2
B2	16*l*	0.226	0.389	0	trigonal bypiramid B_4Ni

(e) $Ho_2Ni_3B_6$ having orthorhombic structure, *Cmmm* space group [00G3]

Atom	Sites	x	y	z
Ho	4*j*[d)]	0	0.27673(5)	1/2
Ni1	4*h*	0.2251(2)	0	1/2
Ni2	2*a*	0	0	0
B1	8*p*	0.2345(9)	0.1497(9)	0
B2	4*g*	0.386(1)	0	0

[a]Transformed from [83K2] data, with origin shift 1/2 1/2 0
[b]Transformed from published data [84K1], with origin shift 0 0 1/2
[c]Stable phase at T > 1540 K;
[d]Occupation 0.951(5)

Table 9.3 Space groups and lattice parameters of R-Ni-B compounds with high boron content

Compound	T (K)	Space group	Lattice parameters (nm)			References
			a	b	c	
Ce_2NiB_{10}	RT	*Pbam*	0.5654(1)	1.1258(1)	0.4196(5)	[00J1]
$Ce_2NiB_{9.7}$	RT	*Pbam*	0.5649(1)	1.1244(2)	0.4199(1)	[03M3]
Ce_2NiB_{10}	RT	*Pbam*	0.4161	0.5602	1.147	[20M1]
Pr_2NiB_{10}	RT	*Pbam*	0.4202	0.5643	1.1254	[20M1]
Pr_2NiB_{10}	RT	*Pbam*	0.5619(2)	1.1227(3)	0.4192(1)	[00J1]
Nd_2NiB_{10}	RT	*Pbam*	0.5614(1)	1.1192(2)	0.4177(1)	[00J1]
Nd_2NiB_{10}	RT	*Pbam*	0.4196	0.5637	1.1234	[20M1]
Sm_2NiB_{10}	RT	*Pbam*	0.4174	0.5606	1.1164	[20M1]
Sm_2NiB_{10}	RT	*Pbam*	0.5605(2)	1.1133(2)	0.4177(1)	[00J1]
Gd_2NiB_{10}	RT	*Pbam*	0.5576(1)	1.1119(3)	0.4149(1)	[00J1]
Gd_2NiB_{10}	RT	*Pbam*	0.4151	0.5589	1.1098	[20M1]
Tb_2NiB_{10}	RT	*Pbam*	0.5556(2)	1.1109(5)	0.4150(1)	[00J1]
Dy_2NiB_{10}	RT	*Pbam*	0.5549(2)	1.1094(2)	0.4144(1)	[00J1]
Dy_2NiB_{10}	RT	*Pbam*	0.4144	0.5553	1.1087	[20M1]
Ho_2NiB_{10}	RT	*Pbam*	0.4132	0.5632	1.1040	[20M1]
Ho_2NiB_{10}	RT	*Pbam*	0.5545(4)	1.1086(4)	0.4139(2)	[00J1]
Y_2NiB_{10}	RT	*Pbam*	0.5557(1)	1.1109(2)	0.4150(1)	[00J1]
$Y_2NiB_{9.4}$	RT	*Pbam*	0.55513(6)	1.1093(1)	0.41446(5)	[03M3]
Ho_4NiB_{14}	RT	*P4/mnc*	0.72097(8)		0.74587(9)	[01G1]
Gd_4NiB_{13}	RT	*P4/mnc*	0.7257(3)		0.7578(5)	[83K2]
Gd_4NiB_{13}	RT	*P4/mnc*	0.7358		0.6980	[20M1]

(continued)

Table 9.3 (continued)

Compound	T (K)	Space group	Lattice parameters (nm)			References
			a	b	c	
Tb_4NiB_{13}	RT	*P4/mnc*	0.7227(1)		0.7535(3)	[83K2]
Dy_4NiB_{13}	RT	*P4/mnc*	0.7210(4)		0.7516(7)	[83K2]
Dy_4NiB_{13}	RT	*P4/mnc*	0.7317		0.6907	[20M1]
Ho_4NiB_{13}	RT	*P4/mnc*	0.7197(4)		0.7466(3)	[83K2]
Er_4NiB_{13}	RT	*P4/mnc*	0.7186(3)		0.7446(3)	[83C1, 83K2]
Er_4NiB_{13}	RT	*P4/mnc*	0.7282		0.6845	[20M1]
Tm_4NiB_{13}	RT	*P4/mnc*	0.7246		0.6785	[20M1]
Tm_4NiB_{13}	RT	*P4/mnc*	0.7178(3)		0.7411(6)	[83K2]
Yb_4NiB_{13}	RT	*P4/mnc*	0.7163(3)		0.7377(6)	[83K2]
Yb_4NiB_{13}	RT	*P4/mnc*	0.7165(3)		0.7379(5)	[06V1]
Yb_4NiB_{13}	RT	*P4/mnc*	0.7367		0.6753	[20M1]
Lu_4NiB_{13}	RT	*P4/mnc*	0.7220		0.6754	[20M1]
Lu_4NiB_{13}	RT	*P4/mnc*	0.7151(3)		0.7356(6)	[83K2, 90D1]
Y_4NiB_{13}	RT	*P4/mnc*	0.7223(5)		0.7541(7)	[83K2]
Y_4NiB_{13}	RT	*P4/mnc*	0.7348		0.6942	[20M1]
$ErNiB_4$	RT	*Pbam*	0.5792(7)	1.1544(11)	0.3435(6)	[81C1]
$ErNiB_4$	RT	*Pbam*	0.5787(3)	1.1543(2)	0.3436(4)	[04C1]
$TmNiB_4$	RT	*Pbam*	0.5793(24)	1.1456(71)	0.3457(25)	[81C1]
$TmNiB_4$	RT	*Pbam*	0.3391	0.5879	1.1353	[20M1]
$YbNiB_4$	RT	*Pbam*	0.3371	0.5878	1.1403	[20M1]
$YbNiB_4$	RT	*Pbam*	0.5888	1.154	0.3412	[85D2]
$YbNiB_4$	RT	*Pbam*	0.58645(2)	1.13680(5)	0.33850(3)	[06V1]

(continued)

Table 9.3 (continued)

Compound	T (K)	Space group	Lattice parameters (nm)			References
			a	b	c	
$YbNiB_4$	RT	*Pbam*	0.58647	1.1371	0.33877	[09P1]
$LuNiB_4$	RT	*Pbam*	0.5789(24)	1.1340(71)	0.3460(25)	[81C1]
$LuNiB_4$	RT	*Pbam*	0.3362	0.5840	1.1332	[20M1]
$CeNiB_4$	RT	*I4/mmm*	0.7626(1)		0.8600(3)	[84K1]
$CeNiB_4$	RT	*I4/mmm*	0.6882 $\alpha = 112.832°$	0.6882 $\beta = 112.832°$	0.6882 $\gamma = 102.941°$	[20M1]
$PrNiB_4$	RT	*I4/mmm*	0.6973 $\alpha = 112.765°$	0.6895 $\beta = 112.765°$	0.6895 $\gamma = 103.067°$	[20M1]
$PrNiB_4$	RT	*I4/mmm*	0.7683(5)		0.8670(9)	[84K1]
$NdNiB_4$	RT	*I4/mmm*	0.7667(3)		0.8649(6)	[84K1]
$SmNiB_4$	RT	*I4/mmm*	0.7627(3)		0.8601(7)	[84K1]
$SmNiB_4$	RT	*I4/mmm*	0.6895 $\alpha = 112.868°$	0.6895 $\beta = 112.868°$	0.6895 $\gamma = 102.872°$	[20M1]
$GdNiB_4$	RT	*I4/mmm*	0.6850 $\alpha = 112.888°$	0.6850 $\beta = 112.888°$	0.6850 $\gamma = 102.835°$	[20M1]
$GdNiB_4$	RT	*I4/mmm*	0.7580(2)		0.8561(6)	[84K1]
$TbNiB_4$	RT	*I4/mmm*	0.7549(4)		0.8540(7)	[84K1]
$TbNiB_4$	RT	*I4/mmm*	0.6834 $\alpha = 112.927°$	0.6834 $\beta = 112.987°$	0.6834 $\gamma = 102.761°$	[20M1]
$DyNiB_4$	RT	*I4/mmm*	0.6824 $\alpha = 112.953°$	0.6824 $\beta = 112.953°$	0.6824 $\gamma = 102.711°$	[20M1]
$DyNiB_4$	RT	*I4/mmm*	0.7536(2)		0.8522(4)	[84K1]

(continued)

Table 9.3 (continued)

Compound	T (K)	Space group	Lattice parameters (nm)			References
			a	b	c	
$HoNiB_4$	RT	*I4/mmm*	0.6802 α = 112.962°	0.6882 β = 112.962°	0.6802 γ = 102.694°	[20M1]
$ErNiB_4$	RT	*I4/mmm*	0.6792 α = 112.987°	0.6792 β = 112.987°	0.6792 γ = 102.646°	[20M1]
$HoNiB_4$	RT	*I4/mmm*	0.7505(2)		0.8506(6)	[84K1]
$ErNiB_4$	RT	*I4/mmm*	0.7505(2)		0.8496(3)	[84K1]
$YNiB_4$	RT	*I4/mmm*	0.7535(3)		0.8526(7)	[84K1]
$YNiB_4$	RT	*I4/mmm*	0.6830 α = 112.963°	0.6830 β = 112.969°	0.6830 γ = 102.694°	[20M1]
$Tb_2Ni_3B_6$	RT	*Cmmm*	0.7723(4)	0.8636(3)	0.3480(2)	[89G2]
$Dy_2Ni_3B_6$	RT	*Cmmm*	0.7732(4)	0.8632(6)	0.3476(4)	[89G2]
$Ho_2Ni_3B_6$	RT	*Cmmm*	0.76865(9)	0.86679(9)	0.34742(4)	[00G3]
$Ho_2Ni_3B_6$	RT	*Cmmm*	0.7720(5)	0.8639(4)	0.3465(2)	[89G2]
$Ho_2(NiB_2)_3$	RT	*Cmmm*	0.5800 α = 90°	0.5800 β = 90°	0.3467 γ = 96.72°	[20M1]
$Er_2(NiB_2)_3$	RT	*Cmmm*	0.5768 α = 90°	0.5768 β = 90°	0.3458 γ = 94.500°	[20M1]
$Er_2Ni_3B_6$	RT	*Cmmm*	0.7701(4)	0.8632(3)	0.3462(2)	[89G2]
$Tm_2Ni_3B_6$	RT	*Cmmm*	0.7645(4)	0.8625(5)	0.3452(3)	[89G2]
$Tm_2(NiB_2)_3$	RT	*Cmmm*	0.3452 α = 96.638°	0.5773 β = 90°	0.5773 γ = 90°	[20M1]
$Yb_2(NiB_2)_3$	RT	*Cmmm*	0.5756 α = 90°	0.5756 β = 90°	0.3471 γ = 96.302°	[20M1]

(continued)

Table 9.3 (continued)

Compound	T (K)	Space group	Lattice parameters (nm)			References
			a	b	c	
$Yb_2Ni_3B_6$	RT	*Cmmm*	0.7627	0.8604	0.3443	[91G2]
$Yb_2Ni_3B_6$	RT	*Cmmm*	0.7622(2)	0.8607(2)	0.3440(1)	[06V1]
$Lu_2Ni_3B_6$	RT	*Cmmm*	0.7630(3)	0.8625(3)	0.3439(2)	[86D1, 89G2]
$Lu_2(NiB_2)_3$	RT	*Cmmm*	0.5734 $\alpha = 90°$	0.5734 $\beta = 90°$	0.3430 $\gamma = 96.551°$	[20M1]
$Y_2Ni_3B_6$	RT	*Cmmm*	0.7750(4)	0.8644(4)	0.3486(3)	[89G2, 95C3]

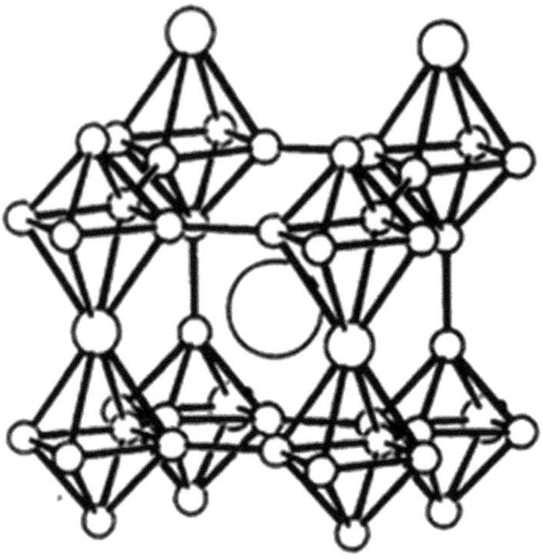

Fig. 9.1 The Ce_2NiB_{10} crystal structure [00J1]. Notation: B (small), Ni(medium) and Ce (large) spheres

At T ≥250 K, the reciprocal susceptibility follows a Curie–Weiss dependence, the effective magnetic moment being nearly the same as that of Ce^{3+} free ion—Table 9.4.

The temperature dependences of the electrical resistivity, ρ(T) and thermopower, S(T), of $Ce_2NiB_{9.7}$ are plotted in Fig. 9.2 [03M3]. In the temperature range 50 K ≤ T ≤ 300 K, there is a small diminution of ρ(T) values, as temperature decreases, the corresponding change being more rapidly at T < 25 K. At low temperatures, ρ(T) shows a linear temperature dependence down to T = 0.4 K, behavior characteristic for NFL system. In the temperature range 200 K < T < 300 K, the magnetic contribution to the resistivity follows lnT dependence, ascribed to incoherent Kondo scattering. A broad maximum at T = 70 K was attributed to the Kondo scattering in the presence of crystalline electric field scattering. The thermopower, S(T), has a maximum at T ≅ 100 K and a shoulder at T ≅ 40 K, behavior evidenced also in other Ce-based heavy fermion (HF) system. The maximum at T = 100 K was attributed to a combined Kondo and CEF effects. The CEF ground state of Ce^{3+} in $Ce_2NiB_{9.7}$ was shown to be a doublet, the first excited level being located at T = 250 K. The specific heat, C/T, at 1.4 K ≤ T ≤ 10 K, follows, in zero field, also a lnT law. The strongly depressed C/T values, at low temperatures, on external fields and level off towards a constant value, were explained qualitatively by the Kondo disorder model, the estimated average Kondo temperature being T_K ≅ 26 K. Atomic disorder induced by boron defects was considered to be the most probable origin of the NFL behavior [03M3]. The thermoelectric performance of CeB_6 is quite low, since thermal conductivity is very large, due to strong B–B bonds. The thermal conductivity of the $Ce_2NiB_{9.7}$, is reduced by a factor 7 compared to that of RB_6 boride. The Ni substitution does not so much reduce the lattice thermal conductivity, the strong scattering of phonons by lattice defects, contributing mainly to this behavior.

Table 9.4 Magnetic properties of R-Ni-B compounds with high boron content

Compound	Magnetic structure and magnetic moments	M_s (μ_B/f.u.)	T_c (T_N) (K)	M_{eff} (μ_B/R)	θ (K)	References
Ce_2NiB_{10}	Mixed valent (?)			2.03(3)	−34(2)	[00J1]
$Ce_2NiB_{9.7}$	Non-Fermi liquid $1.5\ K \leq T \leq 4\ K$, $\chi = \chi_0\text{-}aT^{-0.5}$, at $\mu_0H = 0.5$ T, T > 250 K, C–W behavior	0.76[a]		2.55	−111	[03M3]
Pr_2NiB_{10}	AFM		8(4)	3.51(2)	−27(2)	[00J1]
Nd_2NiB_{10}	AFM		9(1)	3.03(3)	−20(3)	[00J1]
Sm_2NiB_{10}	AFM; Van Vleck paramagnet		11(1)	1.82(2) (T = 400 K)		[00J1]
Gd_2NiB_{10}	AFM		33(2)	7.31(5)	−28(3)	[00J1]
Tb_2NiB_{10}	Metamagnetic	2.65(1)[b]	25(2)	8.92(9)	−27(3)	[00J1]
Dy_2NiB_{10}	Metamagnetic	6.40(2)[b]	21(2)	9.77(7)	−13(1)	[00J1]
Ho_2NiB_{10}	Metamagnetic	10.18(4)[b]	11(1)	9.16(2)	−7(1)	[00J1]
Y_2NiB_{10}	Pauli paramagnet, $\chi = 1\cdot10^{-9} m^3$/f.u					[00J1]
$YbNiB_4$	AFM	0.41‖**c**[c] 0.27⊥**c**[c]	T_{N1} = 5.4 T_{N2} = 4.0	4.64‖**c** 5.06⊥**c**	−96‖**c** −159⊥**c**	[09P1]
$YbNiB_4$ polycrystalline	AFM	2[d]		4.73[e]	−124	[09P1]
Lu_4NiB_{13}	PM, χ little decreases with temperature $\chi = 4\cdot10^{-9}$ m^3/kg, at T = 4 K					[90D1]

[a] T = 1.5 K, μ_0H = 15 T, magnetization not saturated, 5% CeB_4
[b] T = 2 K, μ_0H = 5.5 T
[c] T = 2 K, μ_0H = 5 T
[d] Polycrystalline sample, extrapolation at 50 T;
[e] Effective moment higher than Yb^{3+} free ion value. When it is assumed that Yb^{3+} effective moment is given by free ion value, an effective Ni moment $\mu_{eff} \cong 2.0\ \mu_B$/atom is obtained

The R_2NiB_{10} compounds with R = Pr, Nd and Gd are antiferromagnetically ordered, and those with R = Tb, Dy and Ho show metamagnetic transitions [00J1]. The reciprocal susceptibilities, at T >100 K, follow Curie–Weiss type dependences, the effective moments being close to those of R^{3+} free ion values. The paramagnetic Curie temperatures are negative, as expected for an AFM-type ordering—Table 9.4. The magnetic susceptibility of Sm_2NiB_{10}, at T > T_N, is dominated by the Sm^{3+} Van-Vleck-type paramagnetism. The effective Sm magnetic moment at T = 400 K is 1.82(2) μ_B. This value is close to the theoretical one, for Sm^{3+} ion, calculated with a screening constant σ = 33 [32V1].

Fig. 9.2 Thermal variations of resistivity and thermopower for $Ce_2NiB_{9.7}$. In inset the temperature dependence of resistivity at T < 8 K [03M3]

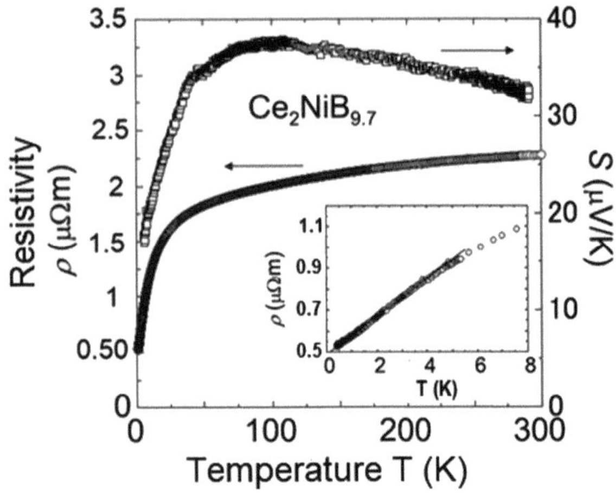

The magnetizations of R_2NiB_{10} compounds with R = Tb, Dy, Ho, at T = 2 K, after metamagnetic transitions, are not saturated in field of $\mu_0H = 5.5$ T. In all cases the magnetic moments are smaller than those expected for parallel alignment of R magnetic moments. The resistivities of R_2NiB_{10} series increase with temperature, as expected for a metallic system [00J1]. Some discontinuities are evidenced in their thermal variations, at temperatures close to T_N values.

The coatings on steel substrate by wear resistant Ni-based clad with addition of 3 wt% Nd_2O_3 improved by 27% hardness, due to presence of Ni_4B_3, Nd_2NiB_{10}, NdC_2 and CrSi phases, respectively [20S1].

The Yb ion, in $YbNiB_4$ compound, is in trivalent state and order antiferromagnetically at $T_{N1} = 5.4$ K, A second transition, to another AFM state, occurs at $T_{N2} = 4.0$ K [09P1]. The magnetic susceptibilities, χ, of $YbNiB_4$ single crystal are weakly anisotropic. A maximum of χ values was shown, at T ≅ 8 K, above the magnetic ordering temperature. A linear increase of the magnetization with external field was evidenced at T = 2 K. Extrapolating the magnetizations obtained for **H**||**c**, from 5 T up to 50 T, a magnetic moment of 2 μ_B/Yb atom was obtained, as expected for a Yb^{3+} CEF doublet [00T1]. The specific heat, C/T, as well as their magnetic part C_m/T, shows lambda type anomalies at T_{N1} and T_{N2} temperatures. Below T_{N2}, the C_m/T values decrease and merge, at T < 1 K, into a constant value. The magnetic entropy $S_m(T)$, reaches a value 0.8*ln*2 around T = 20 K, where it starts to saturate, indicating that the first excited crystal field doublet is well separated from the ground state doublet. The increase of the $YbNiB_4$ resistivity, ρ(T), between 50 and 16 K, as well as a broad maximum at T > T_{N1}, provide evidence for the presence of Kondo interaction. The broad maximum in χ(T), at T ≅ 8 K, suggests the presence, of low dimensional antiferromagnetic correlations. Both possible mentioned features lead to strong fluctuations in an extended temperature range above T_{N1}, up to a characteristic energy, corresponding to Kondo temperature $T_K = 16$ K, as estimated from an analysis of the temperature dependence of the entropy.

The ^{11}B NMR spectra, on polycrystalline $YbNiB_4$, are shifted and broadened as the temperature decreases and show pronounced quadrupole splitting [09S1]. The small negative shift indicates the relevance of the conduction electron polarization by the Yb moments.

9.3 R–Ni–B Multi Phases and Amorphous Alloys

The $Y_xNi_yB_z$ alloys, with various compositions, were prepared and crystal structures, as well their properties, were investigated [05M1, 05N1]. The multiphase system contained the YNi_2B_3, YNi_3B_2 and YNi_4B_3 compounds. No single phases of the above compounds were obtained, although the presence of YNi_3B_2 has been previously reported without indicating their crystal structure [95C3].

The formation of R–Ni–B glasses, by melt-spun technique, their physical properties as well as technical applications in addition to the compounds formed in this system, were also analyzed [94H1, 96O2, 14S2, 15C1]. The compositional range for the formation of amorphous phases, in La–Ni–B alloys, is relatively wide (12–30 at.% B, La < 12 at.%). Five intermetallic compounds are also formed in this composition region [96O2]. Particularly, the amorphous phases, nearest $LaNi_5$-compound have been investigated in connection with hydrogen storage. Although amorphous $LaNi_{6.4}B_{1.2}$ was shown that does not adsorb hydrogen, the ribbons with the same composition, crystallized at $T = 1073$ K, start to adsorb hydrogen.

The atomistic structures of Nd–Ni–B amorphous alloys has been investigated, starting from the $Nd_xNi_{65-x}B_{35}$ and $Nd_{10}Ni_{90-x}B_x$ alloys series [94H1]. The compositions of glass forming, where a single amorphous phase can be found, is 40 at.% $\leq x \leq$ 82 at.%, when the boron content is higher than that of neodymium. The mass densities, of these amorphous alloys, for a fixed nickel or boron concentration, change monotonously with neodymium content. These changes, qualitatively agree with those calculated for mixing of the constituent elements with specific atomic masses and volumes. The crystallization temperatures, T_g, show a definite maximum at $\cong$10 at.% Nd, regardless the nickel or boron content. For a fixed neodymium content, T_g increases gradually with the boron concentration. The amorphous structure of Nd–Ni–B is constructed, by random packing of trigonal prisms of the type Nd_3Ni_6B [94H1].

The yttrium was shown to play a dominant role in glass forming in Y–Ni–B system, due to large atomic radius and highly negative heat of mixing [14S2]. The nickel containing alloys, possess the highest glass forming ability, in the wide range of compositions, comparatively with Y–Fe–B and Y–Co–B alloys. The cerium improves the thermal stability and glass forming ability of $(Y_8Ni_5Al_{87})_{99}Ce_1$ alloy ribbon, produced by melt spinning process [15C1]. The above ribbon showed a glass transition at $T_g = 500$–750 K, in three exothermic reaction stages. The nano-sized Al particles, precipitated within the amorphous matrix, during the first exothermic reaction. When 15 at.% boron was added to ductile $Al_{87}Y_8Ni_5$ amorphous ribbons, their thermal and amorphous forming stabilities increase [13C1]. An increase of activation energy for crystallization was shown in $(Al_{87}Y_8Ni_5)_{85}B_{15}$ ribbons. When annealing temperatures increase, from 506 to 746 K, the Al, $B_{2.5}Ni_2Y$, Al_3Y and Ni_xY_y phases are formed sequentially. Since boron is present also as interstitial element, these alloys achieve higher hardness than those of non-boron counterpart, besides the formation of $B_{2.5}Ni_2Y$ phase.

The amorphous Ni–Co–B–R alloys with R = Ce, Nd and Gd, on nickel foam (NF), can be used as a catalyst for hydrogen evolution reaction [20Z1]. Among them, Ni–Co–B–Gd/NF electrode shows good catalytic activity and kinetics in alkaline solution, due to the synergetic effect of Gd and Ni–Co–B.

The magnetic properties of some R–Ni–B magnetic glasses were investigated [83S1] staring from the Hamiltonian $H = -\mathrm{I}\sum_{i,j} \boldsymbol{J}_\mathrm{i}\boldsymbol{J}_\mathrm{j} - \mathrm{D}\sum_\mathrm{i} (\boldsymbol{n}_\mathrm{i}\boldsymbol{J}_\mathrm{i})^2$ [73H1]. By I is denoted the exchange parameter, assumed to be constant and $\mathrm{D} = \mathrm{Kz}/\mathrm{k_B}$, where K is the anisotropy constant and z the number of magnetic ions. The $\boldsymbol{n}_\mathrm{i}$ gives the direction of the local random magnetic anisotropy. Starting from the above relation, the effect of random anisotropy on the character of the magnetic transitions, the phase diagram and double transitions seen in glasses, were described. As example, the Curie temperature of $Gd_{58}Ni_{32}B_{10}$ glass was found to be $T_c = 124$ K and downturn temperature $T_s = 9$ K. Values D = 0.93 K and I = 8.9 K were determined starting from above relation [83S1].

9.4 RNi_2B_2 and RNi_2B_2C Compounds

The chapter, in addition to physical properties of RNi_2B_2 compounds, analyses also those of RNi_2B_2C borocarbides, where R is a rare earth and yttrium.

The RNi_2B_2 compounds are formed mainly with heavy rare-earths, R = Sm, Gd, Tb, Dy, Ho, Er and Y [91G3, 20M1].

The RNi_2B_2C system is a member of homologous series $(RC)_m(Ni_2B_2)_n$, where sheets of RC layers are located between Ni_2B_2 ones [97S8, 98K2, 05M1]. The series include the RNi_2B_2C (m = 1, n = 1), RNiBC (m = 2, n = 1) which form with heavy R elements [94S4, 95E1], $La_3Ni_2B_2N_3$ (m = 3, n = 1), when C is replaced by N [94C3, 95S3], $R_4Ni_2B_2C_4$ or R_2NiBC_2 (m = 4, n = 1), formed with R = Lu [94Z1] and Y [95L6] borocarbides. The folowing investigations have shown that the four layers RC structure [96Z1] incorporates a B layer between two pairs of RC layers, giving $Lu_4Ni_2B_3C_4$, composition, suggesting that a four layers RC as a unit, is perhaps unstable as such and needs a B layer for stability. Formation of a member of series $Y_3Ni_4B_4C_3$ (m = 3, n = 2) has been reported [97K1], although this phase was not obtained in the pure form [05M1]. The members of the series $Y_5Ni_6B_6C_5$ (m = 5, n = 3) and $Y_5Ni_8B_8C_5$ (m = 5, n = 4), have been reported to exist locally as metastable phases, in as cast/annealed $YNi_{2-x}B_{2+x}C$ alloys [02Y1]. From the observation of stacking faults, possible occurrence of structures $Y_2Ni_3B_3C_2$ (m = 4, n = 3) and $Y_4Ni_7Bi_7C_4$ (m = 8, n = 7), have been also claimed [02Y1].

The RNi_2B_2C borocarbides can be obtained with all rare-earths, except Eu [91R1]. The lattice parameters of $EuNi_2B_2C$ were only calculated [94S5]. In RNi_2B_2 structure, which does not form with all rare-earths, the insertion of C in the R layers modifies the c/a ratio, thereby stabilizing the structure [05M1]. The stability of $ScNi_2R_2C$ structure is affected, due to small scandium radius and the borocarbide forms only as a metastable phase, by using a quenching method [94K2].

The physical properties of RNi_2B_2 compounds were little investigated, unlike those of RNi_2B_2C series, payed a great attention, particularly due to the coexistence of superconductivity and magnetic ordering. The signature of superconductivity was observed in YNi_4B sample prepared by arc melting [93M1]. Latter studies, evidenced the presence of the carbon in the sample composition, as in $YNi_2B_3C_{0.2}$ multiphase material, which showed bulk superconductivity [94N1, 95N2]. The superconductivity was then evidenced, in single phases and often single crystals of RNi_2B_2C borocarbides with R = Dy, Ho, Er, Tm, Lu, Y [94C4, 98C1, 01C1]. Among the above compounds, those with R = Dy Ho, Er, Tm are also magnetically ordered [94E1].

As function of their physical properties, the RNi_2B_2C borocarbides can be classified in five groups: (a) having Pauli type paramagnetism, when R = La, Ce; (b) magnetic ordered with R = Pr, Nd, Gd, Tb; (c) heavy fermion type behavior, as in $YbNi_2B_2C$ compound; (d) superconductors when R = Lu, Y and (e) magnetic ordering which coexist with superconductivity, if R = Dy, Ho, Er, Tm.

9.4.1 Crystal Structures and Elastic Properties

The RNi_2B_2 compounds with R = Sm, Gd, Tb, Dy, Ho and Er crystallize in a monoclinic structure, space group *C2/c* [91G7, 91G8]. The structure contains infinite columns of base linked $B(R_2Ni_4)$ trigonal prisms that shared alternatively faces and edges to form infinite non-planar slabs. The structure is analogous to that of AlB_2-type. If layers of trigonal M_6 prisms, where M is a metal, are distorted and filled each third one by B atoms, the RNi_2B_2-type structure is obtained. The coordination numbers for R, Ni and B atoms, in this structure, are CN = 20, 13 and 8(6 + 2), respectively. Boron pairs are located in distorted trigonal prisms.

The RNi_2B_2C borocarbides, where R is a rare-earth, crystallize, at ambient conditions, as the homologous cobalt series, in a tetragonal structure having *I4/mmm* space group [94S2–94S5]—Fig. 8.3. There are infinite slabs of edge-linked NiB_4 tetrahedra, interconnected via B–C–B linear units, parallel to [001] axis, to form a 3D framework. The structure consists basically of two types of layers, Ni_2B_2 and RC. In Ni_2B_2 layer, the Ni is tetrahedrally coordinated by four boron atoms. The tetrahedra, all share edges, resulting of short Ni–Ni distances. In the RC layer, each R atom is in a square planar coordination by four carbon atom and vice-versa. The two layers are connected via the carbon atoms, defining a linear boron–carbon–boron unit. Thus, the structure can be regarded as a layered system with a "hard" and "soft" layers, accomodating the strain produced by the different sizes of the R atoms. The Ni_2B_2 layer serves as a cushion, allowing the structure with all rare-earths (except Eu) and yttrium. The crystal structure of RNi_2B_2C borocarbides can be considered as a

Table 9.5 Atomic sites in RNi_2B_2 and RNi_2B_2C compounds

(a) $HoNi_2B_2$ having monoclinic structure, space group *C2/c* [91G8, 94P1]

Atom	Site	Coordinates		
		x	y	z
Ho	4*e*	0	0.1197(1)	1/4
Ni	8*f*	0.3718(2)	0.0937(3)	0.3001
B	8*f*	0.171(2)	0.326(2)	0.022(2)

(b) $LuNi_2B_2C$ having tetragonal structure, space group *I4/mmm* [94S5]

Atom	Site	Symmetry	Coordinates			Atomic environment[1]
			x	y	z	
Lu	2*b*	4/*mmm*	0	0	0	Coplanar square CNi_4
Ni	4*d*	-4*m*2	1/2	0	1/4	8-vortex polyhedron B_4Ni_4
B	4*e*	4*mm*	0	0	0.3618(8)	Square pyramid $[Ni]_4$
C	2*a*	4/*mmm*	1/2	1/2	0	Colinear B_2

[1]In [11V1] the coordinates were transformed; origin shift 001/2, Lu(0,0,1/2), Ni(0,1/2,1/4), B(0,0,0.1382), C(0,0,0)

"stuffed" derivative of the $ThCr_2Si_2$-type lattice, with C atoms inserted in R planes, having alternatively RC and Ni_2B_2 layers. The c/a ratio is $\cong 3$. The atomic sites and their coordinates of RNi_2B_2 and RNi_2B_2C compounds are given in Table 9.5 and the lattice parameters are listed in Table 9.6.

The volumes of the RNi_2B_2C unit cells, increase as the lanthanide radius increases. The a lattice parameters exhibit the lanthanide contraction, while the c parameter varies in opposite direction [94S4]. Thus, in-plane basal area, of the RC layer, increases with increasing lanthanide radius. The B–C bond distance along the **c**-axis shows no significant variation, indicating that this bonding is very strong. Thus, the **c**-axis layer defined by the B–C–B atoms, with the rare earth in the middle, does not change height, as the lanthanide radius is varied, while the basal plane area changes according to lanthanide contraction. This, requires that the R–B distance vary in the same manner as the R–C one [94S5, 97L6]. The Ni–B bond distance does not exhibit much variation with lanthanide radius. As a result, there is a change in the B–Ni–B bond angles and Ni-B layer thickness, as function of R partner. The superconducting RNi_2B_2C members with R = Dy, Ho, Er, Tm, Lu and Y have ideal or relaxed NiB_4 tetrahedra, whereas the non-superconducting borides have compressed tetrahedra. A strong sensivity of superconducting transition temperatures, T_s, towards the bonding angle of the NiB_4 tetrahedron was shown [94S5]. The correlation between T_s values, with crystal chemical parameters, suggest the possibility of coexistence of superconductivity and magnetism, due the sandwich of charge carriers layers and dielectric ones.

The crystal structures, of some magnetic ordered RNi_2B_2C compounds, are orthorhombically distorted, at temperatures smaller than the magnetic ordered ones [02M2]. The lattice distortions were directed along [110] axis for R = Dy and Ho

Table 9.6 Space group and lattice constants

(a) RNi_2B_2 compounds

Compound	T (K)	Space group	Lattice parameters (nm)			β°	References
			a	b	c		
$SmNi_2B_2$	RT	*C2/c*	0.8478(1)	0.5354(1)	0.6902(1)	127.09(1)	[91G7]
$GdNi_2B_2$	RT	*C2/c*	0.8420(1)	0.5278(1)	0.6886(1)	126.80(1)	[91G7]
$TbNi_2B_2$	RT	*C2/c*	0.8389(1)	0.5245(1)	0.6890(1)	126.90(1)	[91G7]
$HoNi_2B_2$	RT	*C2/c*	0.8411(4)	0.5199(1)	0.6911(1)	126.93(2)	[91G8]
YNi_2B_2	RT	*C2/c*	0.8388(3)	0.5212(1)	0.6887(1)	127.17(1)	[91G7]
$Sm(NiB)_2$	RT	*C2/c*	0.5009 α = 59.408°	0.5009 β = 59.408°	0.6855 γ = 64.914°		[20M1]
$Tb(NiB)_2$	RT	*C2/c*	0.4972 α = 59.459°	0.4972 β = 59.459°	0.6848 γ = 64.383°		[20M1]
$Dy(NiB)_2$	RT	*C2/c*	0.4948 α = 59.289°	0.4948 β = 59.289°	0.6873 γ = 64.133°		[20M1]
$Er(NiB)_2$	RT	*C2/c*	0.4933 α = 59.322°	0.4933 β = 59.322°	0.6883 γ = 63.797°		[20M1]
$Y(NiB)_2$	RT	*C2/c*	0.4977 α = 59.261°	0.4977 β = 59.261°	0.6885 γ = 64.160°		[20M1]

(b) RNi_2B_2C compounds

Compound	T (K)	Space group	Lattice parameters (nm)			References
			a	b	c	
$LaNi_2B_2C$	RT	*I4/mmm*	0.3801(1)		0.9861(3)	[94S5]
$LaNi_2B_2C$	RT	*I4/mmm*	0.3793(1)		0.9824(2)	[05S2]
$CeNi_2B_2C$	RT	*I4/mmm*	0.3645(1)		1.0251(3)	[94S5]
$CeNi_2B_2C$	RT	*I4/mmm*	0.363783(6)		1.02273(2)	[97L6]
$PrNi_2B_2C$	RT	*I4/mmm*	0.3712(1)		1.0036(3)	[94S5]
$PrNi_2B_2C$	RT	*I4/mmm*	0.37066(1)		0.99993(2)	[97L6]
$PrNi_2B_2C$	RT	*I4/mmm*	0.3696(2)		0.99885(8)	[06D1]
$PrNi_2B_2C$	RT	*I4/mmm*	0.3696(1)		1.0033(1)	[02D3]
$NdNi_2B_2C$	RT	*I4/mmm*	0.3686(1)		1.0097(3)	[94S5]
$NdNi_2B_2C$	RT	*I4/mmm*	0.36780(2)		1.00814(4)	[97L6]
$SmNi_2B_2C$	RT	*I4/mmm*	0.3627(1)		1.0270(3)	[94S5]
$SmNi_2B_2C$	RT	*I4/mmm*	0.36232(5)		1.02437(5)	[00E1]
$SmNi_2B_2C$	RT	*I4/mmm*	0.36193(4)		1.02454(4)	[00E1]
$SmNi_2B_2C$	RT	*I4/mmm*	0.36197(3)		1.0254(1)	[95P2]

(continued)

Table 9.6 (continued)

(b) RNi_2B_2C compounds

Compound	T (K)	Space group	Lattice parameters (nm)			References
			a	b	c	
$EuNi_2B_2C$[a]	RT	*I*4/*mmm*	0.3612(1)		1.0319(3)	[94S5]
$GdNi_2B_2C$	RT	*I*4/*mmm*	0.3588(1)		1.0392(3)	[94S5]
$TbNi_2B_2C$	RT	*I*4/*mmm*	0.3560(1)		1.0463(3)	[94S5]
$TbNi_2B_2C$	RT	*I*4/*mmm*	0.355361(7)		1.04352(7)	[97L6]
$TbNi_2B_2C$	90	*I*4/*mmm*	0.354529(18)		1.04504(4)	[99J1]
$TbNi_2B_2C$	300	*I*4/*mmm*	0.355305(14)		1.04489(3)	[99J1]
$DyNi_2B_2C$	RT	*I*4/*mmm*	0.3542(1)		1.0501(3)	[94S5]
$HoNi_2B_2C$	RT	*I*4/*mmm*	0.3527(1)		1.0560(3)	[97S5]
$HoNi_2B_2C$	RT	*I*4/*mmm*	0.3515(1)		1.0518(2)	[94S2]
$ErNi_2B_2C$	RT	*I*4/*mmm*	0.3500(1)		1.0533(2)	[94S2]
$ErNi_2B_2C$	RT	*I*4/*mmm*	0.3509(1)		1.0582(3)	[94S5]
$TmNi_2B_2C$	RT	*I*4/*mmm*	0.3494		1.0613(3)	[94S5]
$YbNi_2B_2C$	RT	*I*4/*mmm*	0.3483(1)		1.0633(3)	[94S5]
$YbNi_2B_2C$	RT	*I*4/*mmm*	0.34782(3)		1.0607(1)	[97L6]
$YbNi_2B_2C$	RT	*I*4/*mmm*	0.3474		1.0598	[96D3]
$YbNi_2B_2C$[b]	RT	*I*4/*mmm*	0.3479(5)		1.0617(2)	[04A3]
$YbNi_2B_2C$[c]	RT	*I*4/*mmm*	0.3487(5)		1.0643(2)	[04A3]
$LuNi_2B_2C$	RT	*I*4/*mmm*	0.3472(1)		1.0658(3)	[94S5]
$LuNi_2B_2C$	RT	*I*4/*mmm*	0.3462(1)		1.0629(2)	[94S5]
$LuNi_2B_2C$	RT	*I*4/*mmm*	0.3463(1)		1.0626(3)	[05S2]
YNi_2B_2C	RT	*I*4/*mmm*	0.3533(1)		1.0566(3)	[94S5]
YNi_2B_2C	RT	*I*4/*mmm*	0.3525(1)		1.0536(2)	[94S2]
YNi_2B_2C	RT	*I*4/*mmm*	0.35257(1)		1.05336(4)	[94C5]
YNi_2B_2C	300	*I*4/*mmm*	0.35285(4)		1.05384(12)	[96Y3]
YNi_2B_2C	100	*I*4/*mmm*	0.35269(2)		1.05525(7)	[96Y3]
YNi_2B_2C	14	*I*4/*mmm*	0.35249(2)		1.05511(7)	[96Y3]
YNi_2B_2C	RT	*I*4/*mmm*	0.35266(1)		1.05375(4)	[95P2]

[a]Calculated
[b]As grown
[c]Annealed

[98G1, 99K2] or [010] axis when R = Tb or Er [99D1, 99S4]. After the structure transition, at T_N = 14.3 K, the mismatch between a and b lattice parameters of $TbNi_2B_2C$ increases with decreasing temperature, at T = 8.6 K having a value a/b – 1 = 0.55% [99S4, 01K1]. The orthorhombic structure transition, when R = Er, takes place concomitantly with the presence of basal plane transverse spin wave ordering, with Er moments along one of the basal plane axis [97D2, 99D1, 01K1]. As in case of Tb compound, the mismatch between a and b lattice constants coincides with the onset of long-range AFM order and increase with decreasing temperature, reaching a value $a/b - 1 = 0.2\%$ at $T = 3.7$ K [97D2]. The structure distortions for R = Dy [98G1, 01K1] and R = Ho [99K2], as above mentioned, are along [110] axis. The holmium moments are aligned in this direction, below the magnetic ordering temperature. The lattice distortions are proportional to the squared ordered moments [01K1]. Large magnetostrictive effects, at the magnetic transition temperature, T_N, were observed in $GdNi_2B_2C$ borocarbide, but no symmetry breaking. The above behavior has been called *magnetoelastic paradox* [06R1]. Such behavior, shown in zero field, may be lifted in the presence of small magnetic fields. The lattice distortion disappears in large fields ($\mu_0H > 12.5$ T), as expected for a ferromagnetic spin-induced arrangement.

No magnetoelastic effects were observed in compounds with R = Nd and Sm [96H2, 97D1]. In $SmNi_2B_2C$ borocarbide, the moments are aligned parallel to [001] axis, so that the magnetostrictive effect would change the c/a ratio without breaking the crystal symmetry [97D1].

According [00I1], the local structure of YNi_2B_2C is distorted within the Ni layers, at $T \leq 60$ K. The experimental data were modeled assuming a two site Ni–Ni pair distribution, at $T = 5$ K and 9 K and anomalously broadened one-site distribution at $13\ K \leq T \leq 60$ K. There is a correlation motion of Ni–Ni pairs in the vicinity of T_s. This support evidence for spin polaron formation at low temperatures [95K2].

The crystal chemistry of RNi_2B_2C (R = Lu, Y) [94S4] and the first principles study on the structural stability and mechanical properties under pressure [96W1, 98C2, 17L1] were investigated. No structural transitions were found for YNi_2B_2C up to $p = 16.4$ GPa [96M1] or $p = 6$ GPa [98M2]. The elastic constants, of YNi_2B_2C, have been experimentally determined [03M2, 03R4], or computed [12W1]—Table 9.7. The elastic constant c_{44}, in normal state showed a softening at $T < 65$ K, while in the superconducting state a hardening [01I1]. This behavior was related to the softening of the acoustic Δ_4 mode. The calculated elastic constants agree well with experimentally determined values [12W1, 17L2]. The pressure dependences of the bulk modulus and thermal expansions of YNi_2B_2C [18L2] and $LuNi_2B_2C$ [19L1] were theoretically investigated. A correlation between crystal chemical parameters and superconducting properties of RNi_2B_2C borocarbides has been made [02V1].

The thermal variations of lattice parameters of RNi_2B_2C polycrystalline compounds with R = Tb, Dy, Ho, Er, u, Y, at $90\ K \leq T \leq 360$ K [02J1], R = Dy [99S2], R = Ho, at $2\ K \leq T \leq 200$ K [07S2], R = Ho, Ce at $1.6\ K \leq T \leq 300$ K [97L6], as well as single crystals, as $HoNi_2B_2C$ at $2\ K \leq T \leq 300$ K [07S2], R = Er, at $2\ K \leq T \leq 300$ K [06B6], $YbNi_2B_2C$ at $0.5\ K \leq T \leq 300$ K [09S2], or R = Y at $25\ K \leq T$ 300 K [98B1] were investigated. The temperature dependences of lattice parameters of YNi_2B_2C in the temperature range $25\ K \leq T \leq 300$ K were fitted with the polynomial functions: $a = 0.35217(1) - 5.54(175) \times 10^{-7}\ T + 1.45(12) \times 10^{-8}\ T^2 - 1.48(24) \times 10^{-11}\ T^3$ nm and $c = 1.05512(3) - 4.08(515) \times 10^{-7}\ T - 5.32(348) \times 10^{-9}\ T^2 + 1.51(69) \times 10^{-11}\ T^3$ nm, where T is given in K [98B1]. The a lattice parameters and the unit cell volumes increase with temperature, while the c lattice

Table 9.7 Elastic constants and bulk modulus[a]

Compound	Method	T (K)	Elastic constants (GPa)						Bulk modulus (GPa)	References
			c_{11}	c_{33}	c_{44}	c_{66}	c_{12}	c_{13}		
$DyNi_2B_2C$									200	[01F1]
$HoNi_2B_2C$	Pressure, $p < 15$GPa	RT							199	[03O1]
$HoNi_2B_2C$	Sound velocity	77	294.7	282.9	60.2	151.5	165.3		230	[03M2]
$LuNi_2B_2C$	Sound velocity	77	293.9	306.3	59.7	157.1	163.4	231.5	202.7	[03M2]
YNi_2B_2C	Sound velocity	77	279.4	256.1	64.3	140	143.9	178.1		[03M2]
YNi_2B_2C	Ultrasound meas.	300	294(6)	261(5)	64.4(4)	142(1)	157(7)	125(6)		[03R4]
YNi_2B_2C	Ultrasound meas.	4.2	220(20)	211(2)	54(5)	131(10)	100(2)			[01I2]
YNi_2B_2C	Time of flight	300	284(7)		67(1)	143(3)	145(7)			[03R4]
YNi_2B_2C	Time of flight	2	290(20)		67(4)	130(2)	149(8)			[03R4]
YNi_2B_2C	Calc	0	292.7	282.4	68.6	129.8	133.6	138.6	190.78	[12W1]
YNi_2B_2C	Calc	0	282.6	269.91	62.33	140.68	147.22	155.68		[17L2]
YNi_2B_2C	Calc								270	[96M1]
YNi_2B_2C	Calc								207.85	[98C2]
YNi_2B_2C	Pressure, $p < 16.4$GPa	RT							200	[96M1, 98M2]

[a]For compressibilities see [02O1]

constants and of c/a ratio decrease, although a minimum in c values is present at T = 250 K. A similar behavior was shown in other RNi_2B_2C compounds [02J1]. The minimum in the temperature dependence of the c lattice parameters moves towards lower temperatures with increasing the atomic number of the rare earth element. This behavior might be due to the splitting of the d parameter-free Ni position values [00I1], which in turn should give rise to a double-well potential [02J1]. This affects c(T) only. The evolution with temperature of lattice parameters in borocarbides with R = Ho, Ce were correlated with the changes in the distances between atoms [97L6]. In case of $HoNi_2B_2C$ single crystal, the thermal expansion, at T < 20 K, is dominated by magnetic contributions and becomes strongly anisotropic [07S2]. When increasing temperature, a large expansion in the (ab)-plane, is accompanied by a moderate expansion in the **c**-direction. At 20 K $\leq$ T $\leq$ 200 K, the linear expansion remains small and almost isotropic. A strong magnetoelastic coupling was shown in $DyNi_2B_2C$, where large field-dependent length changes, as well as negative thermal expansion coefficient, at 10 K $\leq$ T $\leq$ 30 K, have been observed [99S2]. Large irreversible longitudinal magnetostriction in YNi_2B_2C was seen in both in plane and along **c**-axis directions of applied field, in mixed superconducting state [06B7]. Quantum oscilations of magnetostriction were observed at low temperatures for **H** || **c**, starting at fields H < 0.7 H_{c2}.

The RNi_2B_2C borocarbides, were generally prepared by melting the constituent elements. The YNi_2B_2C compound was also obtained by mechanical alloying [97E1]. The RNi_2B_2C single crystals were also grown [00D3]. The as grown YNi_2B_2C single crystal has different properties as compared with optimal annealed one [02A1].

The time-of flight powder diffraction study on YNi_2B_2C, revealed that the boron and carbon sites are indeed disordered to a level of 8.6% [02H2]. The existence of carbon vacancies and their clusters in YNi_2B_2C were evidenced by positron-life time measurements [96S7]. A thin film superconductor YNi_2B_2C, deposited on an MgO(001) substrate, consists of isolated rectangular grains distributed within a second phase [05C1, 05S3]. This phase has been identified to be $Y_2Ni_{15}B_6$.

The effects of boron and carbon deviations, from 1/2/2/1 composition, were analysed in $HoNi_2B_{2+x}C$ and $HoNi_2B_2C_{1+y}$ series [08S2]. When x, y > 0, caused primary phase HoB_2C_2 precipitation, prior to $HoNi_2B_2C$ growth, whereas depletion of B or C (x, y < 0), resulted in an increased fraction of eutectics containing additional Ni-rich phases. The shift of melt composition also leads to changes in the stoichiometry, within their homogeneity range. A pronounced re-entrant superconducting behavior was shown in the $0 \leq x \leq 0.1$ composition range, whereas small deviations of carbon from stoichiometry (0 < y < 0.03) results in a loss of superconducting properties [08S2].

The dislocation types and stacking faults in RNi_2B_2C borocarbides with R = Tb, Ho, Er, $Y_{0.6}Tb_{0.4}$ and $Er_{0.8}Tb_{0.2}$ were investigated [99Y1, 00Y2]. The lattice defects are correlated with annealing treatment. Two different slip systems were evidenced. Also, the annealing temperature and their duration are important in obtaining single RNi_2B_2C phases [97G2]. The YNi_2B_2C should be annealed at T > 1050 °C [95B1]. However, at such temperature, even in incapsulated samples, some oxidation will take place. The main impurities, as RB_2C_2, were identified. The HoB_2C_2 impurity essentially disappears, when $HoNi_2B_2C$ is annealed two weeks at T = 700 °C, but reappears by annealing at higher temperatures (800 °C), their content growing up at 900 °C [95G2, 97G2]. In case of $GdNi_2B_2C$, the GdB_2C_2 impurity disappears when annealed at T = 900 °C, for two weeks [97G2]. The $TmNi_2B_2C$, as prepared and annealed at T = 800 °C, lowers TmB_2C_2 impurity content, more than when was annealed at T = 900 °C [97G2]. The $CeNi_2B_2C$ needs to be annealed at T = 1000 °C, in order to obtained a single phase, but with a shorter period, to avoid deep oxidation [95A2]. Thus, the equilibrium of RNi_2B_2C and RB_2C_2 phases apparently does not occur in the same way, along the R series, and seems that the annealing temperature should be reduced from the starting toward end members of R series. The impurities introduced, at the Ni site, influence in a larger extent, mainly the superconducting properties [97G2].

The hydrogen in $YNi_2B_2CH_x$ is absorbed to a low molar concentration $x \cong 0.2$ [96G5]. Their presence has no significant effect on the lattice parameters and superconducting transition temperature.

Some of the physical properties of RNi_2B_2C carboborides were reviewed [96G2, 96J1, 98G3, 00R2, 01M3, 02M4, 04T2, 06A1, 06B5, 08M2, 10G2, 12M1].

9.4.2 RNi_2B_2 Compounds

The physical properties of RNi_2B_2 with R = Sm, Gd, Tb, Dy, Er and Y were little investigated. Band structure calculations show that the above compounds are mainly paramagnetic [20M1].

9.4.3 Non Superconducting Borocarbides with R = La, Ce, Pr, Nd, Sm, Gd and Tb

$LaNi_2B_2C$ borocarbide

The $LaNi_2B_2C$ is a Pauli paramagnet [94C3, 94M2, 95F1, 96F2, 96K2]—Table 9.8a. A small temperature dependent term, superposed on the above behavior was sometimes observed and attributed to the presence of magnetic impurities. The compound is not superconductor, at temperatures, T > 0.02 K [96K2]. It was suggested that this is due to the large size of the La^{3+} ion, resulting in flattening of NiB_4 tetrahedra [94M2] and an increased Ni–Ni distance in the layered structure [95L1]. The resistivity follows a power law, close to T^5, at lower temperatures, crossing to a linear dependence, at higher temperatures, consistent with electron–phonon scattering mechanism [95F1]. The electronic thermopower is negative of −0.0134(1) TμV/K, twice that of superconducting YNi_2B_2C compound.

Table 9.8 Magnetic and suprerconducting properties of RNi_2B_2C compounds

(a) Magnetic and superconducting properties

Compound	Magnetic structure and magnetic moments, magnetic susceptibility (emu/f.u.)	M_s (μ_B/Ratom) or (μ_B/f.u.)	T_c (T_N) (K)	T_s (K)	M_{eff} (μ_B/f.u.)	θ (K)	References
$LaNi_2B_2C$	Pauli paramagnet[a], $\chi_0 = 0.995(4)\cdot10^{-4}$						[95F1, 96F2, 98E2]
$CeNi_2B_2C$	PM, Ce in mixed valent state $\chi = \chi_0 + C/(T - \theta)$; $\chi_0 = 6(2)\cdot10^{-4}$ [95C2] $\chi = 6.9\cdot10^{-4}$ [95A2] Pauli paramagnet at T > 0.35(3) K [98E2]			0.1 [98E2]	1% Ce^{3+}		[95A2, 95C2, 98E2]
$PrNi_2B_2C$	AFM **M** ǁ [110]: **q** = [0,0,1] Pr: T = 1.5 K, M = 0.81(9) μ_B		4				[97L6]
$PrNi_2B_2C$	AFM 100 K ≤ T ≤ 300 K, Curie Weiss law		4.3		3.76 (average)	−44 (average)	[99N4, 00N1]
$PrNi_2B_2C$	T = 2 K, AFM 4 K < T < 15 K, weak ferromagnetism	2.7 $\mu_B \perp \mathbf{c}$[c] 1.4 $\mu_B \parallel \mathbf{c}$	4(T_N) 15(?)(T_{WF})		3.465(4)⊥**c** 2.380(13)ǁ**c** 3.142(7) powder		[02D3]
$NdNi_2B_2C$	AFM, **q** = [1/2,0,1/2][d], $\mathbf{M_{Nd}} \parallel \mathbf{a}$ [97L6] T = 1.5 K, M_{Nd} = 2.10(7) μ_B $\chi = C/(T - \theta)$ [95G4]		4.78(2) [97L6] 4.5 [95D1] 5 [95G4]		3.6	−2.3	[95G4, 97L6]
$SmNi_2B_2C$	AFM, **q** = [1/2,0,1/2][d)], $\mathbf{M_{Nd}} \parallel \mathbf{c}$ [97D1] $\chi = \chi_0 + C/(T - \theta)$, $\chi_0 = 11.1\cdot10^{-4}$ [00E1]		9.86 [95P1] 10 [95H4] 10.2 [00E1]		0.59(4)	−23	[00E1]

(continued)

Table 9.8 (continued)

(a) Magnetic and superconducting properties

Compound	Magnetic structure and magnetic moments, magnetic susceptibility (emu/f.u.)	M_s (μ_B/Ratom) or (μ_B/f.u.)	T_c (T_N) (K)	T_s (K)	M_{eff} (μ_B/f.u.)	θ (K)	References
$GdNi_2B_2C$	$T < T_N$, incommensurate **H**‖**a**, T < 13.6 K, elliptical 2**q**, **q** = (0.55,0, 0), (0,0.55,0) **H** ‖ **c** T < 13.6 K elliptical 2**q** 13.6 K < T ≤ 19.5 K non-eliptical 2**q**		19.4 (T_N) 13.6 (T_{sR})		8.10(3) [98C3]	−1.0(6) [98C3]	[08J1]
$GdNi_2B_2C$	incommensurate double **q** structure **q** = (0.551,0,0) or (0,0.551,0) 13.6 K ≤ T ≤ 19.4 K, **q** = (0.550,0,0) or (0,0.550,0), modulation vector decrease when decreasing T T < 13.6 K additional component along **c**-axis, the modulation vector increasing when decreasing T		19.4 (T_N) 13.6 (T_{sR})				[96D2]
$GdNi_2B_2C$	Double **q** model **q** = (q_a,2q_b,0), $q_a = q_b = 0.55$, **q** = (3q_a,0,0) Magnetic order: eliptically polarized at base temperature						[13N1]
$GdNi_2B_2C$	25 K ≤ T ≤ 300, χ = C/(T − θ)	7.2[e]	19.5 (T_N) 13.6 (T_{sR})		8.33(4)‖**a** 8.12(4)‖**c**	1.30(20)‖**a** 0.15(2)‖**c**	[03E1]
$GdNi_2B_2C$	χ = C/(T − θ) 30 K ≤ T ≤ 60 K				8.3‖**a** 8.3⊥**c**	−6‖**c** −2⊥**c**	[95C1]

(continued)

Table 9.8 (continued)

(a) Magnetic and superconducting properties

Compound	Magnetic structure and magnetic moments, magnetic susceptibility (emu/f.u.)	M_s (μ_B/Ratom) or (μ_B/f.u.)	T_c (T_N) (K)	T_s (K)	M_{eff} (μ_B/f.u.)	θ (K)	References
$GdNi_2B_2C$	$\chi = C/(T - \theta)$ $100\ K \le T \le 300\ K$				8.10(5)	−1.0(6)	[95C1]
$TbNi_2B_2C$	Metamagnetic transition **H**‖[110], $H_m = 1.1, 1.65, 2.3$ T T = 2 K, M = 9.5 μ_B, H > 2.3 T				9.7(1) ‖**a** 9.9(1)‖**c** 9.8(1) powder	16(1)‖**a** −60(1)‖**c** 0(1) powder	[96C6]
$TbNi_2B_2C$	T < 15.4 K incommensurate,**q** = (0.551,0,0), **M**‖**a,** T = 2.3 K, **q** = (0.545, 0, 0) T < 8 K, squared SDW, T = 1.7 K, Tb: $M_1 = 9.90(36)$ μ_B, $M_3 = 2.28(13)$ μ_B, ⟨M⟩ = 7.78(28) T = 7 K, $M_3 = 1.65(15)$ μ_B		15.4 (T_N) 8 (T_{WF}) 5 (T_{sR})				[96D1, 97L6]
$TbNi_2B_2C$	T ≤ 15 K, incommensurate, **q** = (0.55,0,0), **q** decreasing from T_N up to 8 K T < 8 K, **q =** (0.5454,0,0) close to commensurate (q = 6/11) T < 4.5 K, integrated intensities of squaring up saturate		15 (T_N) 8 (T_{WF}) 5 (T_{sR})				[03K2]
$TbNi_2B_2C$	T = 1.5 K, **q** = (0.566,0,0), $\mathbf{M_{Tb}}$‖**a**, $M_1 = 9.5(1)$ μ_B, $M_3 = 2.1(1)$ μ_B T = 5 K, **q** minimum value; 5.5 ≤ T ≤ 8 K, **q** increases little; T > 8 K, **q** increases nearly linearly		15.5 (T_N)				[97T1]

(continued)

Table 9.8 (continued)

(a) Magnetic and superconducting properties

Compound	Magnetic structure and magnetic moments, magnetic susceptibility (emu/f.u.)	M_s (μ_B/Ratom) or (μ_B/f.u.)	T_c (T_N) (K)	T_s (K)	M_{eff} (μ_B/f.u.)	θ (K)	References
$TbNi_2B_2C$	LSDW, **q** = (0.55,0,0) T = 1.5 K, M = 7.8 μ_B	7.4(1)[b]	15				[12E1, 13E1]
$Tb_xY_{1-x}Ni_2B_2C$							[01C3]
x = 0.05				14.4	9.60(60) ⊥ **c** 10.20(40) ∥ **c**		
x = 0.10				13.4	9.80(80) ⊥ **c** 9.90(40) ∥ **c**	14.71(1.06)⊥**c** −45.19(45) ∥ **c**	
x = 0.20		7.64[c]		11.4	9.75(35) ⊥ **c** 9.70(35) ∥ **c**	2.46(26) ⊥ **c** −54.66(28) ∥ **c**	
x = 0.40		6.76[c]	6.0 (T_N) 2 (T_{WF})		10.18(55) ⊥ **c** 9.75(60) ∥ **c**	5.71(74) ⊥ **c** −42.50(60) ∥ **c**	
x = 0.60		7.99[c]	9 (T_N) 4 (T_{WF})		9.65(37) ⊥ **c** 10.07(45) ∥ **c**	8.79(32) ⊥ **c** −54.00(55) ∥ **c**	
x = 1.0			15 (T_N) 10.6 (T_{WF})		9.70(10) ⊥ **c** 9.90(10) ∥ **c**	16.00(1.0) ⊥ **c** −60.00(1.0) ∥ **c**	
$Tb(Ni_{0.8}Co_{0.2})_2B_2C$	AFM, **q** = (1/2,0,1/2) T = 1.5 K, M = 7.6(1) μ_B	4.1(1)[b]	10.2(2)				[12E1, 13E1]
$Tb(Ni_{0.6}Co_{0.4})_2B_2C$	AFM, **q** = (1/2,0,1/2) T = 1.5 K, M = 3.7(2) μ_B	7.6(1)[b]	11.0(2) (T_N) 4.8(3) (T_{WF})		9.3	16	[12E1, 13E1]
$Tb(Ni_{0.4}Co_{0.6})_2B_2C$	SDW, **q** = (0,0,1/3) T = 1.5 K, M = 8.5(2) μ_B	7.7(1)[b]	7.6(2)		9.9	18	[12E1, 13E1]
$Tb(Ni_{0.2}Co_{0.8})_2Ni_2C$	FM, **q** = (0,0,0) T = 1.5 K, M = 8.7(2) μ_B	7.6(1)[b]	5.9(2)		9.72	8	[12E1, 13E1]
$TbCo_2B_2C$	FM, **q** = (0,0,0) T = 1.5 K, M = 7.6 μ_B	7.2[b]	6.6				[12E1, 13E1]

(continued)

Table 9.8 (continued)

(a) Magnetic and superconducting properties

Compound	Magnetic structure and magnetic moments, magnetic susceptibility (emu/f.u.)	M_s (μ_B/Ratom) or (μ_B/f.u.)	T_c (T_N) (K)	T_s (K)	M_{eff} (μ_B/f.u.)	θ (K)	References
$TbNi_2B_2C$	$150 \le T \le 300$ K, $\chi = \chi_0 + C/(T - \theta)$	9.18[f]	15 (T_N) 5 (T_{WF})		10.48⊥**c** 8.02‖**c** 9.91 (average)	6.01 −8.3 −0.53	[96T2]
$TbNi_2B_2C$	$20\ K \le T \le 300$ K				9.70(10)⊥**c** 9.90(14)‖**c**	16.0(1.0)⊥**c** −60.0(1.5) ‖**c**	[98C3]
$TbNi_2B_2C$	T < 15.5 K incommensurate magnetic structure T = 1.5 K; q = (0.556,0,0); Tb:M_1 = 9.5(1) μ_B, M_3 = 2.1 μ_B		15.5 (T_N) 8 (T_{WF})				[97T1]
$DyNi_2B_2C$	AFM **M**‖[110] [97G1], **q** = (0,0,1) T = 1.7 K. M_{Dy} = 8.62(6) μ_B		10.6	6			[97L6]
$DyNi_2B_2C$	AFM: FM in (ab) plane, AFM along **c**-axis		10 10.18 [96L5]	6.2(1)			[95D2, 96L5]
$DyNi_2B_2C$	AFM $2\ K \le T \le 300$ K, $\chi = C(T - \theta)$				9.84(11)⊥**c** 10.40(12)‖**c**	25.0(7) −82.0(1.7)	[98C3]
$DyNi_2B_2C$	AFM	9.55[g]	10.5 (T_N)	6.5	12.38⊥**c** 9.00‖**c**	16.4⊥**c** −3‖**c**	[96T3]
$DyNi_2B_2C$	AFM T < 10 K		10 (T_N)	≅5	10.4	−9.8	[94E1]
$HoNi_2B_2C$	T = 2.4 K, commensurate AFM, M = 9.16(3) P_1nnm magnetic space group, M_{Ho}‖[100] or C_Amca magnetic space group, M_{Ho}‖[110]						[19S1]

(continued)

Table 9.8 (continued)

(a) Magnetic and superconducting properties

Compound	Magnetic structure and magnetic moments, magnetic susceptibility (emu/f.u.)	M_s (μ_B/Ratom) or (μ_B/f.u.)	T_c (T_N) (K)	T_s (K)	M_{eff} (μ_B/f.u.)	θ (K)	References
$Pr_{1-x}Dy_xNi_2B_2C$							
x = 0			15.0(3) (T_N) 1.8 (T_{N1})		3.46(1)	−22(1)	[99T2]
x = 0.09			14.2(3) (T_N) 3.5(2) (T_{N1})		4.48(2)	−19(2)	[99T2]
x = 0.25			4.5(3) (T_N)		5.93(6)	−26(5)	[99T2]
x = 0.5			5.0(2) (T_N)		7.67(1)	0.5(5)	[99T2]
x = 1.0			16.3(3) (T_N) 10.4(3) (T_{N1})		10.21(1)	0.6(5)	[99T2]
$HoNi_2B_2C$	T = 5.3 K, three independent magnetic structures $\mathbf{q_1}$ = (0,0,1), M_{Ho} = 4.34(3) μ_B $\mathbf{q_2}$ = (0,0,0.915), M_{Ho} = 4.09(4) μ_B $\mathbf{q_3}$ = (0.585,0,0), M_{Ho} = 3.95(7) μ_B‖**b** T = 2 K, **q** = (0,0,1), M_{Ho} = 9.06(8) μ_B Ho moments distributed in two or three different types of domains	10.5(1)	8				[06S1, 94G4]
$HoNi_2B_2C$	10 K ≤ T ≤ 350 K, χ = C/(T − θ)				9.90(7)⊥**c** 10.46(14)‖**c**	17.0(1.5)⊥**c** −57.0(2.5)‖**c**	[98C3]

(continued)

Table 9.8 (continued)

(a) Magnetic and superconducting properties

Compound	Magnetic structure and magnetic moments, magnetic susceptibility (emu/f.u.)	M_s (μ_B/Ratom) or (μ_B/f.u.)	T_c (T_N) (K)	T_s (K)	M_{eff} (μ_B/f.u.)	θ (K)	References
$HoNi_2B_2C$	$5 K \le T \le 7 K$ incommensurate AFM coexistence **c**-axis basal plane, spiral like structure and **a**-axis modulation: $\mathbf{q}_{h1}$ = (0,0,0.905),$\mathbf{q}_h$ = (0,0,0.915), $\mathbf{q}_m$ = (0.585,0,0) $T < 5 K$ long range commensurate $T > 5.5 K$ short range ordered fluctuations of the commensurate phase, interpreted in terms of local strains		7 (T_c) 5.2 (T_N)	8.5			[96H2]
$HoNi_2B_2C$	$T < 5 K$, commensurate FM in (ab) plane, AFM along **c**-axis, canted 15(6)° from (ab) plane; $M_{Ho} = 8.9(2)\ \mu_B$ $5 K < T < 8.6 K$ modulated along **c**-axis $5 K < T < 7 K$ modulated **a**-axis	8.9(2)	8.5 5 (T_{WF})	9			[95T3]
$HoNi_2B_2C$	$T < 5 K$ FM in (ab) plane, AFM along **c**-axis $T<T_N$ incommensurate,AFM, **q** = (0,0, 0.054), canted 17° from low T, AFM	8.25[h]					[95G3]

(continued)

Table 9.8 (continued)

(a) Magnetic and superconducting properties

Compound	Magnetic structure and magnetic moments, magnetic susceptibility (emu/f.u.)	M_s (μ_B/Ratom) or (μ_B/f.u.)	T_c (T_N) (K)	T_s (K)	M_{eff} (μ_B/f.u.)	θ (K)	References
$HoNi_2B_2C$	100 K ≤ T ≤ 350 K, $\chi = C/(T - \theta)$				10.67(3)⊥**c** 10.34(2)‖**c**	8.5(7)⊥**c** −33.5‖**c**	[96C7]
$HoNi_2B_2C$	T = 1.6 K, AFM $\varphi^i = 75(5)°$, $M_{Ho} = 8.9(2)\ \mu_B$, T < 5 K commensurate 5 K < T < 8.5 K; **q** = (0.414,0,0), **q** = (0,0,0.078)						[96C4]
$Ho_{0.9}Y_{0.1}Ni_2B_2C$	T = 1.6 K, AFM $\varphi^i = 57(1)°$, $M_{Ho} = 7.41(8)\ \mu_B$ T = 3.5 K incommensurate **q** = (0,0,0.09)						[96C4]
$Ho_{0.8}Y_{0.2}Ni_2B_2C$	T = 1.6 K, AFM $\varphi^i = 90°$, **q** = (0,0,0.104)						[96C4]
$HoNi_2B_2C$	AFM, **q** = (0,0,1), T = 1.5 K, $M_{Ho} = 8.6(1)\mu_B$ spiral **q** = (0,0,0.92), $M_{Ho} = 6.7\mu_B$	−8.6[j]					[14E1]
$Ho(Co_{0.2}Ni_{0.8})_2B_2$	AFM, **q** = (0,0,1),T = 1.5 K, $M_{Ho} = 8.7(1)\mu_B$ spiral **q** = (0,0,0.85), $M_{Ho} = 4.8(2)\ \mu_B$	9.0[k]					[14E1]
$Ho(Co_{0.4}Ni_{0.6})_2B_2C$	Spiral, **q** = (0,0,0.49), T = 1.5 K, $M_{Ho} = 5.3\ \mu_B$ **q** = (0,0,0.49), M = 5.2μ_B	8.8[k]					[14E1]

(continued)

Table 9.8 (continued)

(a) Magnetic and superconducting properties

Compound	Magnetic structure and magnetic moments, magnetic susceptibility (emu/f.u.)	M_s (μ_B/Ratom) or (μ_B/f.u.)	T_c (T_N) (K)	T_s (K)	M_{eff} (μ_B/f.u.)	θ (K)	References
$Ho(Co_{0.6}Ni_{0.4})_2B_2C$	Spiral, **q** = (0,0,0.26), M = 7.4(2) μ_B	8.3^k					[14E1]
$Ho(Co_{0.8}Ni_{0.2})_2B_2C$	FM, **q** = (0.0,0), T = 15 K, M = 8.3 μ_B	8.3^k					[14E1]
$HoCo_2B_2C$	FM, **q** = (0,0,0), T = 1.5 K, M_{Ho} = 7.2 μ_B	7.2^k					[14E1]
$HoNi_{1.985}Co_{0.015}B_2C$	T = 2 K, FM in (a,b) planes AMF along **c**-axis, M = 8.26(6)		8	0			[96H4]
$ErNi_2B_2C$	$T \leq T_N$ incommensurate SDW, **M** ‖**b** $\mathbf{q}_1$ = (0.533,0,0) $T \leq 2.2$ K WFM, squared up noncompensated long period AFM T = 1.6 K, $\mathbf{q}_1$ = 0.552(1), $\mathbf{q}_3$ = 0.345(1), $\mathbf{q}_5$ = 0.758(1) M_1 = 7.0(1)μ_B, M_2 = 1.9(2) μ_B, M_3 = 1.0(2) μ_B		6.7(7) (T_N) 2.2(1) (T_{WF})	11.0(5)			[05E1]

(continued)

Table 9.8 (continued)

(a) Magnetic and superconducting properties

Compound	Magnetic structure and magnetic moments, magnetic susceptibility (emu/f.u.)	M_s (μ_B/Ratom) or (μ_B/f.u.)	T_c (T_N) (K)	T_s (K)	M_{eff} (μ_B/f.u.)	θ (K)	References
$ErNi_2B_2C$	$T < T_N$ incommensurate SDW **q** = (0.5526,0,0) or (0,0.5526,0) with spins directed along **a**(**b**), T = 1.5 K: $M_1 = 9.15(12)\mu_B$, $M_3 = 2.77(30)\ \mu_B$, $M_5 = 1.53(40)\ \mu_B$ Low T: squared magnetic structure $M_{Er} = 7.19(10)\ \mu_B$	7.19	6.8 (T_N)	11			[96L5, 97L6]
$ErNi_2B_2C$	$T < T_N$ antiferromagnetic, spin density wave **q** = (0.55,0,0) T < 1.3 K disrupped square wave, with periodicity 20a, $M_{Er} = 0.57(10)\ \mu_B$		6 (T_N) 2.3 (T_{WF})	11			[01C5]
$ErNi_2B_2C$	T = 2 K Squared-up sine wave Er: $M_1 = 7.8\ \mu_B$, $M_3 = 0.30\ \mu_B$, $M_5 = 0.16\ \mu_B$ Ni: $M_{Ni} = -0.35\mu_B$,		6.8 (T_N)	11			[95S4]
$ErNi_2B_2C$	T = 4 K incommensurate SDW $\mathbf{q}_1$,2$\mathbf{q}_1$,3$\mathbf{q}_1$,5$\mathbf{q}_1$ with $\mathbf{q}_1$ = (0.553,0,0) SDW squared for $T < T_{WF}$ T = 2.1 K commensurate **q =** (0.55,0,0), peaks (n/20,0,0) with n even, FM sheets with periodicity 10a (=0.35 nm)	7.8	2.5[1] (T_{WF})				[02K2]

(continued)

Table 9.8 (continued)

(a) Magnetic and superconducting properties							
Compound	Magnetic structure and magnetic moments, magnetic susceptibility (emu/f.u.)	M_s (μ_B/Ratom) or (μ_B/f.u.)	T_c (T_N) (K)	T_s (K)	M_{eff} (μ_B/f.u.)	θ (K)	References
$ErNi_2B_2C$	T < 6 K incommensurate structure, squared **q** = (0.553,0,0), M_{Er} = 8 μ_B, M_{Ni} = 0 μ_B, **M**‖**b**			11			[95Z1]
$ErNi_2B_2C$ (orthorhombic)	T = 4.2 K: incommensurate transverse wave **q** = (0.5537,0,0), **M**‖**b**		6 (T_N)	10.5			[99D1]
$ErNi_{1.96}Fe_{0.04}B_2C_{0.99}$	T < 8 K, transverse modulated **q** = (0.566, 0,0) or (0,0.566,0) T < 5 K, 3**q**; T < 3 K, 5**q** T = 1.4 K, **q** = (0.559,0,0), M_1 = 10.4(1)μ_B, M_3 = 3.4(1) μ_B, M_5 = 1.9(1) μ_B		5.9(2) (T_N) 2.2(2) (T_{WF})				[03A1]
$ErNi_{1.9}Co_{0.1}B_2C$	T < 2.0 K, squared up noncompensated long period AFM, T = 1.6 K: $\mathbf{q}_1$ = (0.570,0,0), $\mathbf{q}_2$ = (0.291,0,0), $\mathbf{q}_5$ = (0.848,0,0) Er: M_1 = 6.8(1)μ_B, M_2 = 1.7(2) μ_B, M_5 = 1.4(2) μ_B		6.2(2) (T_N) 2.0(2) (T_{WF})	3.0(5)			[05E1]
$Er_{0.8}Tb_{0.2}Ni_2B_2C$	Incommensurate structure T = 1.6 K, $\mathbf{q}_1$ = (0.561,0,0), M_1 = 5.3(1) μ_B		6.9(1) (T_N)	3.6(9)			[05E1]

(continued)

Table 9.8 (continued)

(a) Magnetic and superconducting properties

Compound	Magnetic structure and magnetic moments, magnetic susceptibility (emu/f.u.)	M_s (μ_B/Ratom) or (μ_B/f.u.)	T_c (T_N) (K)	T_s (K)	M_{eff} (μ_B/f.u.)	θ (K)	References
$Er_{0.8}Lu_{0.2}Ni_2B_2C$	Incommensurate structure T = 1.6 K, $\mathbf{q}_1$ = (0.566,0,0), M_1 = 4.8(1) μ_B		5.2(2) (T_N)	10.6(9)			[05E1]
$TmNi_2B_2C$	0.35 K ≤ T ≤ 1.5 K transverse incommensurate modulation, considerable squared T = 0.35 K, **q** = (0.094(1),0.094(1),0) **M**‖**c**		1.5 (T_N)	11			[97S7]
$TmNi_2B_2C$	T < 1.5 K incommensurate SDW squared up, **q** = (0.093,0.093,0), **M**‖**c** T ≅ 0.4 K: M_1 = 4.81(18)μ_B, M_2 = 0.93(19) μ_B, M_{Tm} = 3.78 μ_B		1.53 (T_N)	11			[97L6]
$TmNi_2B_2C$	T < 1.5 K incommensurate, **M**‖**c** in the (110) planes, moments modulated sinusoidaly along [110] direction T = 1.2 K: M_{Tm} = 3.74(2)μ_B; T = 0.05 K: M_{Tm} = 4.8(1) μ_B		1.50 (T_N)	11			[96C3]
$TmNi_2B_2C$ (MS)	T = 0.3 K, M_{Tm} = 4.3(1) μ_B, M_{Tm} ≅ 0.1(1) μ_B (due to C vacancies)						[96M4]
$TmNi_2B_2C$	200 K ≤ T ≤ 300 K	5.0[l]	1.52(5) (T_N)	11	7.63(2) ⊥ **c** 7.51(3) ‖ **c** 7.54(2) (powder)	−36.0(6) ⊥ **c** 20.8(3) ‖ **c** −11.6(4)(powder)	[95C5, 98C3]

(continued)

Table 9.8 (continued)

(a) Magnetic and superconducting properties

Compound	Magnetic structure and magnetic moments, magnetic susceptibility (emu/f.u.)	M_s (μ_B/Ratom) or (μ_B/f.u.)	T_c (T_N) (K)	T_s (K)	M_{eff} (μ_B/f.u.)	θ (K)	References
$TmNi_2B_2C$ (μSR)	T = 0.5 K: $M_{Tm} = 4.3\ \mu_B$ $M_{Tm} = 0.1\ \mu_B$		1.5 (T_N)				[98G2, 98M3, 98M4]
$YbNi_2B_2C$	Heavy fermion T > 0.34 K T > 150 K, $\chi = C/(T - \theta)$				4.85⊥**c** 4.67‖**c** 4.72 (powder)	−191.0⊥**c** −63.6‖**c** −129.6(pow.)	[96Y2]
$YbNi_2B_2C$	150 K ≤ T ≤ 300 K $\chi = C/(T - \theta)$				4.40	−101	[96D3]
$LuNi_2B_2C$	Pauli paramagnet, $\chi_0 = 3.9{\cdot}10^{-4}$			16.6 [94T1]			[95L1]
$Lu_{0.976}Ho_{0.024}Ni_2B_2C$	T > 250 K, $\chi = C/(T - \theta)$			16.6 [94C3]	10.63⊥$\mathbf{c}^n$ 10.58‖$\mathbf{c}^n$	20(1) (⊥**c**) −35.0(7) (‖**c**)	[96C7]
$YNi_2B_2C^m$	Pauli paramagnet, χ anisotrop $\chi = 2.056(3)\ 10^{-4}$ [95F1] T = 300 K: $\chi_\perp \cong 2.05{\cdot}10^{-4}$, $\chi_\parallel = 1.55{\cdot}10^{-4}$, $\chi_{iz} = 1.9{\cdot}10^{-4}$			15.5 [94C3]			[96S5]
YNi_2B_2C (BS)	$\chi = 1.82{\cdot}10^{-4}$ [94L1]; $\chi = 1.93{\cdot}10^{-4}$ [96S5]						

(continued)

Table 9.8 (continued)

(b) Pressure dependences of the magnetic transition temperatures

Compound	T_N (K)	dT_N/dp (K/GPa)	T_{WF} (K)	dT_{WF}/dp (K/GPa)	References
$Ho_{0.6}Dy_{0.4}Ni_2B_2C$	7.2	0.48			[01C6]
$HoNi_2B_2C$	5.9	1.3			[10N1]
$ErNi_2B_2C$	6.1	−12	2.3	−14∥**a** −2∥**c** −30(powder)	[06B6]

[a] Small C–W dependence, due to magnetic impurity, corresponding to $M_{eff} = 5.6 \cdot 10^{-2}$ μ_B/f.u.
[b] $T = 2$ K, $\mu_0 H = 9$ T
[c] $T = 2$ K, $\mu_0 H = 5$ T
[d] With an equivalent domain with **q** = [0,1/2,1/2]
[e] $T = 1.5$ K, $\mu_0 H = 16$ T
[f] $T = 2$ K, $\mu_0 H = 12$ T
[g] $T = 1.8$ K, $\mu_0 H = 2.5$ T
[h] $T = 2$ K, $\mu_0 H = 7$ T
[i] Orientation with respect to **c**-axis
[j] [97C3]
[k] $T = 2$ K, $\mu_0 H = 9$ T;
[l] $T = 2$ K, $\mu_0 H = 5$ T;
[m] Little magnetic impurity with C–W contribution
[n] Per Ho ion

Band structure calculations evidenced that the $LaNi_2B_2C$ has a smaller DOS at E_F [94M1, 01F2]. The E_F energy, falls just in the gap between two small peaks. The contribution of Ni3d states at E_F was found to be large. The position of the antibonding metal band, is nearly the same in the RNi_2B_2C series, due to larger Ni–Ni distance in all compounds [01F2]. The core level spectra reflect their highly covalent bonding character [96K4]. Satellites have been observed due to two-hole bound states in the Ni core-level and valence-band spectra. It was suggested that electron correlation and/or electron–phonon interaction may play a significant role in the low-energy excitations in the Ni borocarbides.

$CeNi_2B_2C$ borcarbide

The $CeNi_2B_2C$ is a mixed valent borocarbide [95A2, 98E2, 99T3]. The XAS measurements evidenced that the average valence, v, of the cerium ion is $v = 3.15(2)$, at $T = 300$ K and 3.21(2), at $T = 10$ K [95A2], below the saturation value of $v = 3.3$ [65V1]. A possible mixed valence state of cerium ion, more close to 4+, was suggested from the analysis of RNi_2B_2C lattice parameters [94S5]. The local susceptibility, $\chi_{loc}(T)$ of ^{140}Ce, using TDPAC method, also indicates non-integral valence of Ce ions, dependent on temperature [99T3].

The magnetic susceptibility, χ, of $CeNi_2B_2C$ compound, is little temperature dependent, as described by the relation $\chi = \chi_0 + n_i\mu_{eff}^2/T$, where n_i is the content of trivalent Ce ions [95A2]. The fraction of Ce^{3+} ions was estimated at 1% [95A2] or <4% [95C2]—Table 9.8a. According to [98E2], the magnetic susceptibility is dominantly of Pauli paramagnetic type, at $T \geq 0.2$ K. A diamagnetic shielding signal appears at $T < 0.177(8)$ K, which is complete at $T < 0.025$ K. The possible occurrence of superconductivity, at $T_s \cong 0.1$ K, was suggested [98E2].

The specific heat of $CeNi_2B_2C$, in the temperature range $0.2\ K \leq T \leq 1$ K, follows a $C_p = \gamma T + \beta T^3$ dependence with $\gamma = 28(2)$ mJ/molK^2 and $\beta = 0.23(1)$ mJ/molK^4. A peak at $T = 0.087(4)$ K was associated with a superconducting phase transition [98E2]. The electronic heat capacity of $CeNi_2B_2C$, at high temperatures, has a T^3 electron–phonon term, which rapidly change at $T < 20$ K [95C2]. The large resistivity ratio of $CeNi_2B_2C$, their monotonously increase with temperature, were also found in non-Ce borocarbides [95C2]. The physical properties of $CeNi_2B_2C$ appear to be of d-like in character.

The cobalt substitution in $Ce(Ni_{1-x}Co_x)_2B_2C$ solid solutions, decreases the unit cell volume, the Sommerfeld constant, γ and the Pauli-type susceptibility. The magnetoresistivity data suggest that the replacement of Ni by Co increases the strength of the CEF gradient at the Ce-site [97E2].

$PrNi_2B_2C$ borocarbide

The $PrNi_2B_2C$ borocarbide, is antiferromagnetically ordered at $T < T_N = 4$ K. The praseodymium moment, $M_{Pr} = 0.81(3)\ \mu_B$, is smaller than the free ion value, and oriented along the [110] direction [97L5, 97L6]. The magnetic structure contains ferromagnetic sheets in the (ab) plane, these sheets being antiferromagnetically coupled along **c**-axis—Fig. 9.3a. The T_N temperature, of $PrNi_2B_2C$, is higher than that expected from De Gennes scaling in the series RNi_2B_2C ($T_N \cong 1$ K). This

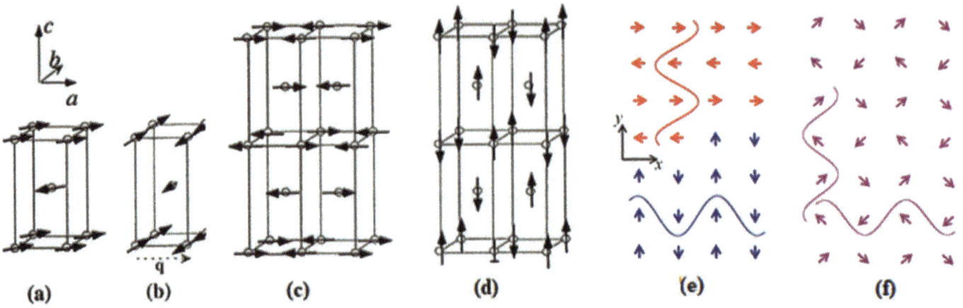

Fig. 9.3 RNi_2B_2C: magnetic structures: **a** simple AFM as evidenced for R = Pr, Dy and Ho, with magnetic moments in (ab) plane; **b** transverse spin density wave, as for R = Er with moments along **b**; **c** R = Nd commensurate magnetic structure; **d** R = Sm commensurate magnetic structure [99S3]; **e** and **f** R = Gd, two dimensional representations of single-**q** (e) and double-**q** (f) AFM order associated with a commensursate propagation wave vector **q** = 0.5. The arrows represent ordered magnetic moments positioned in a square array. The lattice parameter is given by the separation between two adjacent moments [13N1]

behavior was attributed to a large single ion anisotropy and/or Pr4f-conduction band hybridization. A second magnetic ordering temperature, in $PrNi_2B_2C$, was located at T = 15 K [99T2, 02D3]. This magnetic transition was correlated with two interpenetrated magnetic structures formed by Pr^{3+} ions, one of which having a canted spin [02D3].

The temperature dependence of the magnetic susceptibility, of the $PrNi_2B_2C$ single crystal, evidences a high degree of anisotropy [97G1, 99N4, 00N1, 01N1, 01N2, 02D3, 04E2]. At T = 2 K, the magnetic anisotropy is $\chi_{[110]}/\chi_c \cong 11$ and decreases rapidly with increasing temperature. The reciprocal susceptibility follows a linear temperature dependence at T > 100 K, but deviate from a Curie–Weiss law, at $T_N < T < 100$ K. The corresponding paramagnetic Curie temperature, $|\theta_{av}| = 40$ K, one order of magnitude higher than $T_N \cong 4$ K, is an indication of the increased indirect exchange interactions between Pr ions as compared to those between R ions, in the RNi_2B_2C series. The above behavior can be connected with some hybridization between Pr4f level and the conduction electrons, as already suggested [01M3].

The crystal field parameters of $PrNi_2B_2C$ were determined by extrapolation of the values determined in some RNi_2B_2C borocarbides [96G2, 97G1] or by inelastic neutron scattering [08M1]—Table 9.9. The CEF level scheme, below 150 K, comprizes a singlet ground state, a closely spaced doublet at 1 meV and a singlet at 5.2 meV. The calculations suggest also a doublet at 24.3 meV and a singlet at 25.1 meV [08M1]. It was also shown that the magnetic behavior of $PrNi_2B_2C$, can be described by the standard model of rare-earth magnetism. Since the Pr ion has CEF split singlet nonmagnetic ground state, the ordered magnetic moment is induced due to the mixing of higher lying states into the ground state [01R2].

The temperature dependence of the specific heat, C_p (T), has a broad maximum at T $\cong$ 4.3 K, close to the magnetic transition temperature [99N4]. The shape of maximum is quite different from the λ-type anomaly, characteristic of AFM ordering. For temperatures 20 K ≤ T ≤ 30 K, the specific heat and γ value, obtained by extrapolation

Table 9.9 Crystal field parameters in units K (a) and meV (b) of RNi_2B_2C compounds with R = Pr, Nd, Sm, Tb, H_D, Er, Tb and Yb

(a)

Compound	Crystal field parameters (K)					Direction	References
	B_2^0	$B_4^0 \cdot 10^{-3}$	B_4^4	$B_6^0 \cdot 10^{-5}$	$B_6^4 \cdot 10^{-3}$		
$PrNi_2B_2C$	3.48	3.48	−0.858	176	−5.68		[08M1]
$TbNi_2B_2C$[b]	1.50(4)[a]					**H⊥c**	[98C3]
$TbNi_2B_2C$	1.2	4.1	−0.065	−3.6	−0.1		[07E1]
$DyNi_2B_2C$	1.42(15)[a,b]					**H⊥c**	[98C3]
$HoNi_2B_2C$	0.40(4)[a,b]					**H⊥c**	[98C3]
$HoNi_2B_2C$	0.397	−3.70	0.0369	−2.16	−0.0667		[96C7]
$ErNi_2B_2C$	−0.022(17)[a,b]					**H⊥c**	[98C3]
$TmNi_2B_2C$	−2.9(5)[a,b]					**H‖c**	[98C3]
$TmNi_2B_2C$	−1.15(2)[a,b]						[96C4]
$TmNi_2B_2C$	−2.8(5)						[96M4]
$YbNi_2B_2C$	−4.6	−40	1.17	−200	24		[00R1]

(b)

Compound	Crystal field parameters (meV)					References
	A_2^0	A_4^0	A_4^4	A_6^0	A_6^4	
$PrNi_2B_2C$[c]	−22.2	5.82	−170.2	−5.17	61.0	[97G1]
$NdNi_2B_2C$[c]	−20.5	4.92	−143.8	−4.03	47.5	[97G1]
$SmNi_2B_2C$[c]	−18.0	3.86	−113	−2.85	33.6	[97G1]
$TbNi_2B_2C$[c]	−15.4	2.89	−84.4	−1.92	22.6	[97G1]
$DyNi_2B_2C$[c]	−14.7	2.69	−78.6	−1.75	20.7	[97G1]
$DyNi_2B_2C$	−17(2)	5.0(1.5)	−85(3)	−0.1(6)	13.0(2.3)	[98G1]

(continued)

Table 9.9 (continued)

(b)

Compound	Crystal field parameters (meV)					References
	A_2^0	A_4^0	A_4^4	A_6^0	A_6^4	
$HoNi_2B_2C$	−15(6)	2.3(8)	−71.4(3.3)	−0.4(3)	11.7(1.4)	[97M1]
$HoNi_2B_2C$[d]	−6.3(1.3)	3.3(1.3)	−73.9(1.3)	−0.61(12)	10.1(3.1)	[96G2, 97G1, 97M1]
$ErNi_2B_2C$[d]	−14.5(6)	2.3(8)	−71.4(3.3)	−0.42(3)	11.7(1.4)	[96G2, [98G1]
$TmNi_2B_2C$[d]	−13.0(3)	2.16(14)	−63.1(1.2)	−1.32(5)	15.6(1.1)	[96G2, 97G1, 97M1, 98G1]
$TmNi_2B_2C_{0.85}$[d]	−13.7(4)	2.0(2)	−63.8(1.6)	−1.45(5)	12.2(1.5)	[96G2, 98G1]
$YbNi_2B_2C$[c]	−12.5	2.01	−58.7	−1.20	14.2	[97G1]

[a]Experimental value
[b] $\alpha_J = -0.01$(Tb), −0.006(Dy), −0.002(Ho), 0.003(Er), 0.01(Tm) [98C3]
[c]Extrapolation of the data
[d]From neutron CEF spectroscopy

at T = 0 K, are higher than that in heavy fermion $YbNi_2B_2C$ borocarbide. The linear term may be atributed to an increased electronic contribution, as for example, due to Pr4f-conduction electrons hybridization. A possible Schottky contribution, connected with low-lying levels of the Pr^{3+} multiplet, splitted by CEF, can also give a considerable effective linear contribution to the specific heat.

A gradual drop of the resistivity in the polycrystalline $PrNi_2B_2C$ was shown at T < 20 K [99N4]. This behavior was correlated with some peculiarities in scattering of conduction electrons by Pr ions, at low temperatures [01N2]. The smooth change of resistivity at T = 8 K [00N1] is not affected by pressure [01F1] Under applied pressure, the magnetic scattering at low temperatures tend to be suppressed, suggesting a spin fluctuations-type behavior [01F1, 02F1, 11F1].

The $PrNi_2B_2C$ does not exhibit superconductivity down to 0.1 K, at ambient pressure [04E2] or at p = 5.3 GPa [01F1]. The substantial deviations of Pr–Ni–B tetrahedral angle from the ideal value [94S5] and the decreased DOS at E_F [94M1, 00D2], may be the reasons for the absence of superconductivity. The density of states, at E_F, is rather close to the value obtained for $LaNi_2B_2C$ borocarbide [00D2]. The theoretical studies evidenced no Pr4f-conduction band hybridization [94M1, 00D2]. An anomalous heavy fermion type behavior was also proposed [99N4, 00N1]. Two effects were latter shown to be responsible for the absence of superconductivity in $PrNi_2B_2C$, namely strong conduction electron-Pr moment interaction and a lower density of states at E_F, as compared with those determined in RNi_2B_2C (R = Y. Lu) borocarbides [08M1].

$NdNi_2B_2C$ borocarbide

The $NdNi_2B_2C$ borocarbide has a commensurate antiferromagnetic structure, as determined by neutron diffraction studies [97L6, 99S3] or by X-ray scattering measurements [97D1]. The magnetic unit cell is double the chemical unit cell along **a**- and **c**-directions, while it is the same along **b** axis. The magnetic structure of Nd sublattice is antiferromagnetic along **a**-axis, with the moment direction along **a**. The coupling is ferromagnetic along the **b**-axis, while along **c**-axis, one adjacent neighbour is ferromagnetic, while the other is antiferromagnetically aligned, doubling the **c**-axis periodicity—Fig. 9.3c. The modulation vector is **q** = (1/2, 0, 1/2) with an equivalent domain **q** = (0,1/2,1/2) and the moment direction along **b** [97L6, 99S3]. The bilinear exchange interactions between adjacent planes, along **c**-axis, cancel for this spin configuration, due to *bcc* symmetry. The reported magnetic ordering temperatures are T_N = 4.78 K [97L6], 4.8 K [99S3] or 5 K [95G4]. The magnetic susceptibility of $NdNi_2B_2C$, at T > T_N, follows a Curie–Weiss type behavior [99N4, 00N1, 02D3]. The effective neodymium moment is close to the free ion value—Table 9.8a.

The heat capacity measurements confirmed the magnetic ordering [07L2]. The calculated heat capacity curves, with determined crystal field parameters, exhibit sharp peak around T_N = 4.8 K [05D3, 07L2].

$SmNi_2B_2C$ borocarbide

The magnetic structure of $SmNi_2B_2C$, determined by X-ray Emission Spectroscopy (XRES), is similar with that of Nd borocarbide (Fig. 9.3d), but with the magnetic moments oriented along **c**-axis [97D1]. The magnetic unit cell is double of the chemical unit cell along **a** and **c** directions and the same along **b**-axis. The wave vector is **q** = (1/2, 0, 1/2). Above T_N = 9.86 K [95P1], 10 K [95H4] or 10.2 K [00E1], the magnetic susceptibility follows a modified Curie–Weiss law. The determined Sm effective moment, $M_{eff} \cong 0.60\ \mu_B$, is somewhat lower than that of Sm^{3+} free ion value (0.85 μ_B) [00E1]. A Van Vleck contribution, due to the small J multiplet spacing and the coupling of the J = 5/2 ground multiplet, with J = 7/2 state, is present.

The specific heat measurements evidenced a second anomaly, associated with a spin reorientation transition at T_{sR} = 8.4–9.4 K, sensitive to tiny changes of the composition [00E1].

The zero-field muon spin relaxation has been employed to study the physical properties of $SmNi_2B_2C$ borocarbide [95P1, 97C5]. Coherent ordering of Sm moments appeared at T_N = 9.86(3) K, leading to a local field at the muon site of 0.10562(2) T, at T = 3.3 K. The critical behavior of the low temperature ordered phase can be described, by a 2D Ising model, close to the freezing transition. The local field at the muon site is essentially unaffected by 20% Lu substitution, but the local field spread increases drastically [97C5].

$GdNi_2B_2C$ borocarbide

The magnetic structure of $GdNi_2B_2C$ has been first determined by resonant and non-resonant X-ray magnetic scattering techniques [96D2]. The AFM ordering set in at T = 19.4 K. A second AFM–AFM phase transition, occurred at T_{sR} = 13.6 K. In the temperature range 19.4 K ≥ T ≥ 13.6 K, the magnetic structure is of the transverse sine-modulation type with an incommensurate propagation vector **q** = (±0.55,0,0) or **q** = (0, ± 0.55,0). The magnetic "moment" direction was reported to be in the (ab) plane and perpendicular to the propagation vector (transversaly polarized AFM ordering) up to T_{sR}. Below T_{sR} an out of phase (**c**-axis) component, associated with the same propagation wave vector, **q** = (0.55,0,0) or **q** = (0,0.55,0), was found to develop. The magnetic structure depends on the phase shift between these modulations. When the **b** and **c** components are in phase, the magnetic structure is of the sine-modulated type with gadolinium moments in the (bc) plane. Their directions, within this plane, depend on the relative magnitude of both components. The stronger decrease of the **c**-component with increasing temperature leads the moments to lock in along the **b**-axis, at T = 13.6 K. When the phase shift, between the two modulations is different from zero, a spiral-like structure results. The different magnitudes of the **b** and **c** components, lead to an elliptical projection of the gadolinium moments onto the (bc) plane [96D2, 98T1]. The types of magnetic orderings, at T < T_N, in $GdNi_2B_2C$ borocarbide has been studied also by ^{155}Gd Mössbauer spectroscopy [98T1]. It was concluded that at T < T_{sR}, the borocarbide orders in the bunched spiral-like structure with the gadolinium moments rotating within the (bc) plane. The occurrence of a sine-modulated-type structure with the gadolinium moments perpendicular to the tetragonal axis, in the temperature range T_{sR} < T < T_N, was confirmed. The modulation was shown to be almost squared.

The model [96D2], did not determined the phase relationship between the *in* and *out* of (ab) components, nor the relative sizes, hence the polarization of the low temperature AFM order (transverse or elliptical). The designation of "moment" direction rather than of (magnetic) "Fourier component" [96D2] lead to assumption of a single-**q** scenario, in which the ordering wave vector breaks the equivalence of **a**-and **b**-axes [13N1]. The crystal structure is orthorhombic; however no signs of lattice distortions were subsequently observed in $GdNi_2B_2C$ (magnetoelastic paradox) [06R1]. A double-**q** scenario was then predicted, through Landau mean field theory [08J1]. The Fig. 9.3e shows two-dimensional representations of a pair of orthogonal single-**q** domains and in Fig. 9.3.1f, a double-**q** domain. The model explains how the double-**q** scenario leads to a smaller site variation in ordered moment than single-**q** ordering. The double **q**-model carries no expectation of lattice distortion, when no applied field is present and thus an explanation for the magnetoelastic paradox. The resonant elastic X-ray scattering was then employed to test the double-**q** model [13N1]. A satellite reflection associated with a mixed-order component propagation wave vector, viz., (q_a, $2q_b$, 0) with $q_b = q_a \cong 0.55$ was revealed. Their presence is incompatible with single **q** ordering but it is expected from the double-**q** model. A ($3q_a$, 0, 0) wave vector satellite, was also observed, in line with double-**q** model. The temperature dependences of the satellite reflections, along with that of a first order satellite, were compared with the data obtained by calculations based on the double-**q** model and a good agreement was found. The magnetic ordering was shown, as predicted, elliptically polarized at base temperature. The temperature dependence of the "out of (ab) plane" moment component, was also in agreement with calculations. These data provide qualitative support for the double **q**-model and thus in turn corroborate the explanation for the magnetoelastic paradox. The 2**q** model [08J1, 13N1] has been also extended in order to include cystal field effects [19B1].

The phase diagram of $GdNi_2B_2C$, in the presence of external field applied along the **a**- or **c**-axis, has been determined staring from magnetization or magnetostriction measurements and analysed in single-**q** [03E1] and double-**q** model [08J1]—Fig. 9.4 According to [03E1], when the field is applied along the **c**-axis, two phase lines are observed, one at the field at which the polarization changes between an elliptical and a linear one and one between the antiferromagnetic and the paramagnetic phase. When the field is directed along **a**-axis, an additional low-field phase line is observed, interpreted as a domain alignment transition [03E1]. The existance of four equivalent ordering wave vectors, $\mathbf{q} = (\pm 0.55, 0, 0)$ and $\mathbf{q} = (0, \pm 0.55, 0)$ implies the presence of two domains, if the ordering is assumed to be single **q**. The domain in which the moments are perpendicular to the **a**-axis is then favoured by the field. According [08J1], at the ordering wave vector $\mathbf{q} = (0.55, 0, 0)$, the transverse **b** components are more strongly coupled than **a** and **c** components and the magnetic moments order in a linearly polarized sinusoidal wave, just below T_N [08J1]. In this situation, the fourth-order term in the Landau expansion favours the double-**q** ordering. Thus, when the field is oriented along **a**-axis, the domain in which the elliptically polarized moments are perpendicular to the field is favoured by the field and the other double-**q** component of this structure vanishes, at relative low field (1.0–2.0 T). When the field is applied along the **c**-axis, the field does not affect the stability of the double-**q** arrangement, but the elliptical polarization is removed well below the field at which

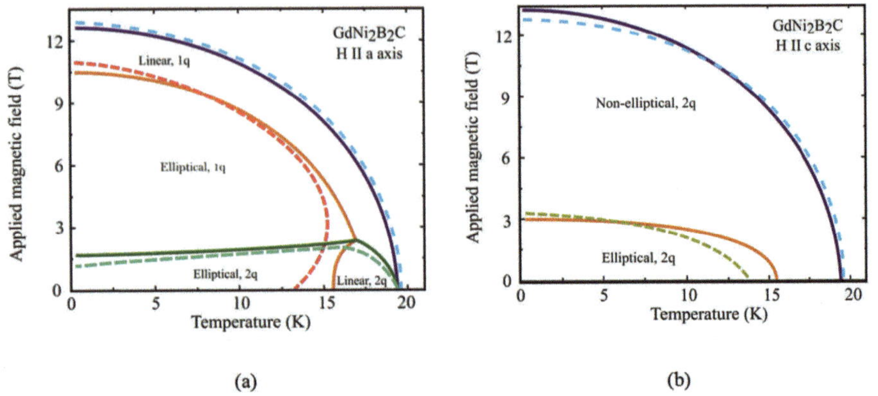

Fig. 9.4 $GdNi_2B_2C$ single crystal: magnetic phase diagram for (a) **H** || **a** and (b) **H** || **c** as function of temperature [08J1]. The data obtained by magnetization and magnetostriction measurements are indicated by solid lines [03E1]. The calculated trend in 2**q** model, by using a mean field analysis, describes well the experimental evidence (dashed lines) [08J1]. The difference between experimental and calculated T_{SR} values is of 2 K

the AFM ordering is destroyed. The phase, at intermediate fields, in the **c**-axis phase diagram, was named non-elliptical 2**q** since the **c**-axis component of the moments is still modulated at the even harmonics [08J1]. In this case the **c**-axis component of the moments is still modulated at the even harmonics of $\mathbf{q}_1$ = (0.55,0,0) and $\mathbf{q}_2$ = (0,0.55,0) and at linear combinations of the two wave vectors. The transition from the "elliptical" to the "nonelliptical" phase is characterized by the disappearance of the two first harmonics (and the higher-order odd ones) in the spatial variation of the **c**-axis moments, which gives rise to a kink in the magnetization curve, as experimentally observed [03E1]. If the elliptical polarizations, were equally shared by the two single-**q** components, at low temperatures, the double 2**q** configuration would not lead to any orthorhombic distortion of the tetragonal (ab) plane, at zero field. In spite of the asymmetry predicted by the model, the correlation function, introduced by [08J1] accounts for the orthorhombic distortion induced by the **a**-axis field.

The gadolinium spin moment in $GdNi_2B_2C$ was also determined by self-consistent density functional theory [96Z3, 00D2].

Magnetic measurements were made on $GdNi_2B_2C$ single crystals and polycrystalline samples [95B2, 95C1, 95E2, 95E3, 95E4, 95G4, 95M2, 96J1, 97S2, 98C3, 98T1, 02H3, 02M3, 03B4, 03E1, 06B5, 09K1, 10E1]. The magnetization isotherms, at low temperatures, are nearly linear on the external field, for μ_0H <11–12 T. The saturation moment of 7.2 μ_B/f.u., at T = 1.7 K, was obtained in external fields μ_0H >12.5 T [03E1]. The reciprocal susceptibilities of $GdNi_2B_2C$ single crystals follow linear temperature dependences, with somewhat different effective moments and paramagnetic Curie temperatures, when field was applied along **c**- or perpendicular to **c**-axis [95C1, 96C6, 03E1]—Table 9.8a. Weak exchange couplings were observed. The presence of anisotropic interactions between localized Gd moment and the conduction electrons, at $T > T_N$, has been evidenced by ESR [09K1]. The above exchange interactions were shown to depend on the conduction electrons momentum transfer, as also evidenced in $Gd_xR_{1-x}Ni_2B_2C$ systems with R = Lu or Y [98P1].

Large values of the electric field gradient were evidenced by ^{155}Gd Mössbauer spectroscopy, due to the presence of carbon surrounding the Gd atoms [95M2]. The investigations by ^{57}Fe Mössbauer spectroscopy on $RNi_{2-x}Fe_xB_2C$ with R = Gd, Tb, Dy, Ho and Er, where 1% [97S2] or 0.5% Ni [00B1] has been substituted by ^{57}Fe, evidenced large quadrupole doublets, ΔQ, dependent on the R partner. The ΔQ values were correlated with c'/a ratio, where c' is the distance of RC layers between which the Ni_2B_2 layer is sandwiched and a lattice parameter, respectively.

Zero field specific heats measurements on RNi_2B_2C single crystals, with R = Gd, Tb, Dy, Ho and Er, were made in the temperature range 0.1 K < T < 25 K [03E2]. The magnon specific heat contribution was analysed, taking into account the effective exchange couplings and anisotropic interactions.

The temperature dependences of the thermal conductivity have been investigated in RNi_2B_2C compounds with R = Gd, Er, Tm, under various magnetic fields [00C4]. In zero magnetic field, small peaks were observed, at T_N, for R = Gd, Er, which disappeared under applied magnetic fields. At $T < T_N$ the thermal conductivity under applied field was larger than that in zero field. The thermal conductivity of RNi_2B_2C (R = Gd to Er) is dominated by electrons and the high temperature thermal conductivity is approximately linear in temperature and anomalous [02H1].

The temperature dependence of the resistivity, ρ(T), of $GdNi_2B_2C$ has only a small change in slope, at T_N [96N1]. The resistivity data show the loss of scattering at T_N, for borocarbides with R = Gd, Tb, Dy and the thermoelectric power, exhibits an enhancement at $T < T_N$ [02H1]. The thermopower, S(T), of $GdNi_2B_2C$ goes to zero linearly, at low temperatures, but there is no indication for magnetic transitions, at T_N and T_{sR} [96N1]. This is consistent with weaker features in ρ(T), at these transitions. There is an abrupt change in slope at T ≅ 90 K. At T > 130 K, the S(T) is almost linear with relative small dS/dT slope.

The substitution of Ni by Co, in $Gd(Ni_{1-x}Co_x)_2B_2C$ series, decreases the T_N and T_{sR} temperatures, due to the increase of the distance between GdC layers and the change of the DOS at E_F, associated with different contributions of Ni and Co3d bands [95B2]. The ^{59}Co and ^{11}B NMR studies evidenced hyperfine fields, at both sites B, around 6 T, assumed as transferred from gadolinium.

$TbNi_2B_2C$ borocarbide

The magnetic structure of $TbNi_2B_2C$ borocarbide has ben studied by neutron diffraction [96D1, 97L6, 97T1, 99S3, 03K2, 03W1], X-ray magnetic circular dichroism [01S3] and high-resolution resonant magnetic X-ray scattering [01S2]. The $TbNi_2B_2C$ borocarbide shows two magnetic transitions at T_N = 15.5 K and T_{WF} = 5–8 K. The magnetic transition, at T_N, is accompanied by a lattice distortion from tetragonal ($T > T_N$), to orthorhombic symmetry ($T < T_N$) [99S3]. Below T_N, the Tb^{3+} moments order antiferromagnetically in a longitudinally modulated structure with propagation vector **q** ≅ (q, 0, 0) [96D1, 97L6, 99S3]. The magnetic modulation vector varies from q = 0.551(1) at T = 15 K to q≅ 0.545(1) at T = 2.3 K

[96D1]. The spin density wave (SDW) is longitudinally polarized, the direction of the moments being parallel to **q**. The unpolarized neutron scattering experiments, at T < 8 K, evidenced an increase in the scattering, consistent with the presence of weak ferromagnetism [96D1]. The magnetic measurements [96C6], ^{57}Fe Mössbauer spectroscopy and μSR experiments [98S1] indicated the onset of weak ferromagnetism at a temperature $T_{WF} \cong 8$ K.

Higher order satellites were evidenced at low temperatures, showing that the SDW was squared up, at T = 1.7 K, with moments $M_1 = 9.90(36)\ \mu_B$ and $M_3 = 2.28(13)\ \mu_B$ [97L6]. At T = 7 K, the essentially the same intensity for the first order satellite was shown, while for the third order satellite, a value, $M_3 = 1.65(15)\ \mu_B$, has been reported. According [97T1], the magnetic moment of Tb^{3+}, estimated from first-order satellites, at T = 1.5 K, was shown to be $M_1 = 9.5(1)\ \mu_B$ and that calculated from the third order satellite $M_3 = 2.1(1)\ \mu_B$. The last one is smaller than one third of the magnetic moment of first-order satellites, suggesting that the longitudinal spin wave was not completely squared. The moment direction is close to the **a**-axis (at an angle of ± 3°) [97T1]. The wave vector of the spin density wave decreases, with lowering temperature, in the SDW phase, until the transition to the weak ferromagnetic phase is reached, at which point the SDW wave vector becomes approximately constant, with **q** = (0.545,0,0). This behavior is characteristic of a continuous lock-in transition to a commensurate phase, occurring at T_{WF}, the commensurate wave vector being **q** = (6/11, 0, 0) [96D1]. This value for the locked in wave vector has been confirmed by high resolution magnetic X-ray scattering studies [01S2]. The X-ray magnetic circular dichroism measurements, on $TbNi_2B_2C$, evidenced an increase in the dichroic signal at TbL_{III} edge, below $T \cong 8$ K, consistent with the onset of weak ferromagnetism involving Tb ions [01S3]. At an external field of 0.05 T, the determined ferromagnetic order parameter agree with the bulk magnetization data. The resonant magnetic X-ray scattering measurements showed that the longitudinal modulation of the magnetic moment is along the longer basal plane axes of the orthorhombic phase [01S2]. The normal-state electronic band-structure calculation, carried out on $TbNi_2B_2C$ borocarbide, having orthorhombic symmetry, showed that stronger nesting is indeed present along the longer basal plane axes [01S2]. According [03W1], the weak ferromagnetism necessarily accompanies the lock-in transition, in the SDW state, of $TbNi_2B_2C$, only if the commensurate wave vector is of the form q = ((m/n)q, 0, 0), with m even integer and n odd integer.

Later on, the neutron diffraction experiments on $TbNi_2B_2C$, at T < 20 K, evidenced magnetic satellite reflections of 2nd order and the observation of satellite reflections close to nuclear ones instead of ferromagnetic signals [03K2]. These satellite reflections could indicate an antiferromagnetically ordered component with a very small propagation vector in the **a**-direction. A reanalysis of magnetic structure was proposed. The creation of structural micro-domains or twins, in a regular pattern, due to magnetoelastic effects, can also result in the presence of such satellites.

Magnetic measurements were performed on $TbNi_2B_2C$ borocarbide [94C3, 94C4, 94E1, 94E2, 95E2, 95H3, 95H4, 96C6, 96N1, 96T1, 96T2, 97C1, 97L6, 97S5, 98C3, 98M5, 98R1, 98S1, 99J1, 99S3, 01M3, 02H1, 03V1, 04K2, 06G1, 07E1, 08R1, 09L1, 09L2, 09R1, 09V2, 11A1, 12E1, 13E1, 16Z1]. The magnetization isotherms, of $TbNi_2B_2C$, show that the borocarbide is highly anisotrop. The magnetization along [001] direction, at T = 2 K, increases nearly linearly with the field. The magnetization in (ab) plane evidenced successive metamagnetic transitions, at fields H_{WF}, H_2 and H_s [96C6, 96T2, 07E1]. The magnetostriction attains its almost full value in a field H_D. This is associated with the domain wall motion. The saturation magnetization is reached in a field H_s and field H_{WF} is related to weak ferromagnetic state. The transformation, at $H_2(T)$, is related to a metamagnetic transition, involving, presumably a change in the modulation wave vector [07E1]. The magnetic phase diagram of $TbNi_2B_2C$ borocarbide has been also investigated starting from field induced modifications of the crystal structure [07E1, 18T1]. The wave vectors of the field-induced lattice deformations suggest a range of commensurate spin-slip-type magnetic structures, at low temperatures,with wave vectors $\mathbf{q}$ = (q,0,0) with q = 6/11 and 5/9. The proposed magnetic structures yield values of magnetization well in line with experiments. The scattering intensity due to magnetoelastic deformation exhibits a jump at the phase boundary, at $\mu_0H = 1.3$ T and low temperatures.

The magnetic ordering temperature T_N, decreases as the applied fields in (ab) plane is increased above $\mu_0H = 1.0$ T [96T2]. At $\mu_0H = 2.0$ T, a value $T_N = 10$ K was reported. As the field further increased, the peak of magnetization, characteristic for antiferromagnetic ordering of Tb moments, disappears and is replaced by a response indicating the presence of ferromagnetic correlation between the Tb moments, at low temperatures. The low temperature magnetic transition is suppressed, for external fields $\mu_0H \geq 0.5$ T. When the field is applied along the **c**-axis, the magnetization is much smaller and shows the peak due to Tb ordering, at $T_N = 15$ K, for external field $\mu_0H \leq 5.5$ T [96T2].

A two sublattices antiferromagnetic mean field model was also used to describe the magnetic properties of $TbNi_2B_2C$ borocarbide [09L2, 09L3]. The magnetic flux structure of $TbNi_2B_2C$ has been investigated [05V2, 09V2]. Bitter decoration, revealed structures associated with a long range magnetic order.

The magnetic susceptibility of $TbNi_2B_2C$ is anisotrop [96C6, 96T2, 98C3, 01C3, 07E1], with a larger value for **H** || [100] than for **H** || [001]. Nearly linear behavior is evidenced at T > 100 K, while powder average sample reveals linear behavior at T > 15 K. The effective moments and paramagnetic Curie temperatures, are little different, according the field direction—Table 9.8a. The anisotropy arises from large splitting between the crystalline electric field levels. The crystal field parameters are given in Table 9.9.

The specific heat measurements were performed on $TbNi_2B_2C$ [95H4, 96T2]. In a single crystal, the magnetic contribution, $C_M(T)$, to the specific heat, shows a jump at $T_N = 15$ K [96T2]. A much sharper λ-like first order transition follows at T = 13.8 K. The third transition is evidenced by an anomaly with a broad peak at T ≅ 5 K, correlated with changes in magnetic structure. The thermal conductivities, κ, of RNi_2B_2C borocarbides with R = Gd, Tb, revealed both electron and phonon scattering mechanisms [02H1]. The high temperature thermal conductivities are approximately linear and anomalous. In $TbNi_2B_2C$, κ values are dominated by the electron portion κ_e, at T_N, while for $GdNi_2B_2C$ the κ_e is about one-half of the total measured κ, at T_N. The electrons appear to be responsible for roughly one half of the thermal conductivity, at 1.5 K $\le$ T $\le$ 300 K. There is no sign of a phonon or electron peak, in the high temperature thermal conductivities of the above borocarbides. The thermal conductivity, for $TbNi_2B_2C$, starts to increase at T_N, due to the diminution in scattering of the charge carriers.

The $TbNi_2B_2C$ is not a superconductor, at T > 0.3 K [96T1] or T > 0.007 K [99P1]. The dispersion of low energy phonon, has been studied by neutron inelastic-scattering [04K2, 11A1]. The absence in superconductivity is a result of pair breaking, due to magnetic order [11A1].

The $TbNi_2B_2C$ resistance, measured along the (ab) plane evidenced a kink, at T_N = 15 K, due to magnetic ordering of Tb moments, which shifts towards lower temperatures as the applied field is increased [96T1, 96T2]. The resistance decreases, as the field is increased, confirming ferromagnetic nature of the correlation between Tb moments within the (ab) planes. When the fields is applied along **c**-direction, there is an increase of resistance, in the same way as the external field, as expected for a simple antiferromagnet. No anomalies were observed in magnetoresistance, which can be associated with the first metamagnetic transition at 0.06 T. The large negative value of the magnetoresistance, at μ_0H > 2.5 T, shows that there is a metamagnetic transition to a ferromagnetic state. According to [96N1], the resistivity and thermopower measurements in RNi_2B_2C series reflect differences in the behavior from non-magnetic to magnetic ordered and heavy fermion borocarbides. The relative cooling power of $TbNi_2B_2C$, in field $\mu_0H \le 7.0$ T, was RCP = 296 J/kg [16Z1].

The electronic structure of $TbNi_2B_2C$ was calculated [00D2]. The stretched exponential muon spin relaxation was observed in the paramagnetic phase [98H1]. This result suggests anomalous broadly rare-earth spin dynamics. There is a slowing of spin fluctuations and an increase in spin correlations, as the magnetic transition temperature is approached.

The neutron diffraction measurements in $Tb(Co_4Ni_{1-x})_2B_2C$ series evidenced successive magnetic modes, when the composition is changed [12E1, 13E1]. The longitudinal modulation, **q** = (0.55,0,0) at x = 0, is transferred into a collinear **q** = (1/2,0,1/2) antiferromagnetic state at x = 0.2 and 0.4, then into a transverse **c**-axis modulation, **q** = (0,0,1/3) mode at x = 0.6 and finally into a simple ferromagnetic structure at x = 0.8 and 1.0—Table 9.8a. Concomitantly, the low temperature orthorhombic distortion of the tetragonal unit cell, at x = 0, is reduced smoothly such that for x $\ge$ 0.4, only a tetragonal unit cell is observed. The above features were attributed to competing influences of the magnetic couplings and crystalline electric field effects, respectively.

9.4.4 Heavy Fermion Borocarbide $YbNi_2B_2C$

The $YbNi_2B_2C$ is a heavy fermion system [96D3, 96Y2]. No magnetic ordering has been observed at T $\ge$ 0.023 K [99B4] on 0.34 K [96Y2]. The temperature dependence of the magnetic susceptibility, of a single crystal, is anisotropic; at T > 150 K follows a Curie–Weiss type behavior with effective moment, determined for

H ⊥ c, little higher than when **H ∥ c** [96Y2]. The paramagnetic Curie temperatures, θ, are negative, indicating that the antiferromagnetic correlations are important. The presence of spin fluctuatioms can be also related to negative θ values. Deviations from Curie–Weiss behavior, at T < 100 K, suggest that the Yb4f levels may be significantly hybridized with conduction band.

The specific heat of $YbNi_2B_2C$, rapid increases at T > 20 K, due in part to phonons [96Y2]. A shoulder was observed below 10 K, belived to be a consequence of spin fluctuations. At T < 1 K, the Sommerfeld coefficient is $\gamma \cong 0.53$ J/molK^2, indicative of a heavy-fermion system.When using the single impurity relation, the Kondo temperature is estimated at $T_K \cong 11$ K [96Y2]. By using inelastic neutron scattering, on $YbNi_2B_2C$, a Kondo temperature, $T_K = 3.8$ K has been also determined [99S1]. Very broad crystal field excitations were found.

The Yb moment is absent or less than (0.05–0.1) μ_B, down to T = 0.023 K, as evidenced by ^{170}Yb Mössbauer spectroscopy and magnetic measurements [99B4]. The Kondo coupling is rather strong, in order to compensate the local moment of Yb ions and to prevent the onset of magnetic order. The ^{11}B NMR study showed that at T > 50 K both the Knight shift and the nuclear spin lattice relaxation rate, T_1^{-1}, can be accounted for by the presence of localized 4f moments at the Yb^{3+} site, which polarize the conduction electrons via RKKY mechanism [97S1]. At T < 5 K, the relaxation rate T_1^{-1} obeys a Korringa-like law with a constant value T_1T, typical of normal metal with high DOS at E_F and no localized moment. These data suggest that $YbNi_2B_2C$ manifests local moment, at T > 50 K and nonmagnetic, itinerant correlated electron behavior for T < 5 K. By ^{172}Yb PAC measurements of $YbNi_2B_2C$, the presence of a sizeable thermal variation of the quadrupole interaction and consequently the existence of crystal-field splittings was shown. Due to strong 4f-conduction electron hybridization, the magnitude of this thermal variation is relatively small [00R1]. The Kondo temperature is strongly renormalized by crystal-field effects and it ranges from high-temperature values $T_K^0 = 100$–200 K up to a low-temperature, $T_K \cong 4$ K [00R1]. The neutron scattering experiments on $YbNi_2B_2C$ [03B3], suggest an Yb^{3+} ground state, with predominantly localized 4f electrons subject to: (1) a crystalline electric field potential and (2) a Kondo interaction, which at low temperatures, is about one order of magnitude smaller than the CEF interaction. The static and dynamic magnetic properties of $YbNi_2B_2C$ were reconciled considering an approximation scheme to the Anderson impurity model and this procedure also indicates that the effective Kondo interaction varies with temperature due to the crystal field splitting. It was concluded that this borocarbide might be close to a quantum critical point on the non-magnetic side [03B3].

The resistivity studies were made on polycrystalline $YbNi_2B_2C$ and also of as grown or thermally treated single crystals [96L1, 96Y2, 97H1, 97Y3, 99N2, 99Y2, 00N2, 02A1, 04A3, 05B4, 06O1]. The resistivity decreases steady below RT and then more sharply below a temperature range 10–40 K, depending on the annealing treatment of the samples [02A1]. At T < 1.5 K, the temperature dependence of the resistivity, exhibits a positive curvature, reminiscent of T^2 variation, characteristic of strongly correlated Fermi liquid ground state. The thermodynamic properties (magnetic susceptibility, $\chi(T)$, heat capacity, C(T)), remained unchanged upon annealing, whereas the zero field transport properties (resistivity, $\rho(T)$ and thermoelectric, $S(T)$) showed changes [02A1]. The evolution of the magnetic properties, upon annealing, was rationalized in terms of redistribution of local Kondo temperatures, associated with ligandal disorder for a small fraction of Yb sites. The lattice dislocations, affecting the surrounding Yb ions, were found to be the dominant defect type [04A3].

The magneto-transport properties of $YbNi_2B_2C$ single crystals, were also investigated [96L1, 97Y3, 99Y2, 01C7]. A pronounced anisotropic magnetoresistance was found when applying magnetic field parallel or perpendicular to the **c**-axis. A well defined maximum, in the magnetoresistance, was observed at $T < 10$ K and $\mu_0H =$ 5 T. At higher temperatures, the magnetoresistance, is always negative and weakers as the temperature is increased. The resistivity, $\rho(T)$, is little influenced by pressure with a maximum in the magnetic contribution, at $p = 2$ GPa and $T = 75$ K [06O1]. The magnetoresistance (MR), at $T = 4.2$ K and $p = 3$ GPa, shows a maximum at a characteristic field, H_{max}, which decreases with increasing pressure. The pressure dependence of MR was discussed in connection with Kondo temperature.

The polycrystalline $YbNi_2B_2C$ samples, show negative Hall coefficient, in the temperature range 1.5 K and 300 K [99N2, 00N2]. The absolute value of Hall coefficient, R_H increases with decreasing temperature, suggesting a moderate heavy fermion-like trend. More complex behavior was shown in $YbNi_2B_2C$ single cystals [05B4]. Annealing the single crystals, caused drastic changes in the Hall coefficient, $R_H(T)$ [05B4]. Whereas for the as grown sample, the Hall coefficient is negative, at $2\ K \le T \le 300$ K, with a pronounced minimum at $T \cong 22$ K, for the sample annealed at $T = 950$ °C, 150 h, $R_H(T)$ changes its sign twice in the same temperature range from negative to positive, on cooling at $T \le 100$ K, and back to negative at $T < 10$ K, with a clear maximum at $T = 45$ K. Intermediate temperature dependences, between the above ones, can be achieved by reducing the annealing time. The changes in $R_H(T)$, upon annealing, were connected with those of the zero-field resistivity and attributed to the redistribution of local Kondo temperatures, for a small number of ytterbium sites [02A1, 04A3, 05B4].

The non-superconducting behavior of $YbNi_2B_2C$ was correlated with the evolution of heavy fermion state [96D3]. Latter on [01M1], by XANES experiments, was suggested that a reduced Ni3d occupancy in the Ni_2B_2 layer must be responsible for the absence of superconductivity.

The evolution from a local moment, magnetically ordered ground state ($T <$ 20 K) in $GdNi_2B_2C$, to a Kondo lattice type, in $YbNi_2B_2C$ has been analyzed on the $Gd_{1-x}Yb_xNi_2B_2C$ single crystals, in correlation with the $Gd_{1-x}Lu_xNi_2B_2C$ system [03B4]. The two series evolve from long-range magnetic order to the spin glass state in a very similar way, with $x = 0.70$–0.75, being the critical concentration at which the spin-glass state forms. Hybridization, in the case of $Gd_{1-x}Yb_xNi_2B_2C$, as evidenced from resistivity and heat capacity data, apparently does not affect this evolution.

9.4.5 Superconducting RNi_2B_2C Borocarbides with R = Lu and Y

The RNi_2B_2C borocarbides with R = Lu, Y are superconductors. On cooling below the superconducting transition temperatures, T_s, Cooper pairs are formed, where two electrons are bound to each other by an attractive interaction. For conventional superconductors, the pairing mechanism is based on electron–phonon coupling. The phonons have a momentum and energy dependent coupling strength to the electronic states, near the Fermi energy, E_F. The interaction is often cast into a single constant λ, which represent the electron phonon coupling strength of individual phonon modes, integrated over all phonon branches and the whole Brillouin zone. The RNi_2B_2C with R = Lu, Y have similar magnetic and superconducting properties, with superconducting transition temperatures $T_s = 16.6$ K and 15.5 K, respectively. Since of their interesting properties these superconductors were intensively investigated. In the following the evolution of the knowledge in the field will be reviewed.

Magnetic measurements were performed on RNi_2B_2C borocarbides with R = Lu [94I1, 94K1, 94P2, 95E2, 96B2, 96F2, 96N1, 99N3, 00S4, 05P1, 15M1] or Y [94C5, 94F1, 94N1, 94P4, 94R1, 94X1, 94X2, 95F1, 95G2, 95H2, 95K2, 95N1, 96C2, 96F1, 96F2, 96M2, 96N1, 96S5, 96S6, 97S8, 97S9, 98K2, 98L1, 99B1, 99N2, 99N3, 99S5, 99S6, 00C6, 00S4, 01T1, 13D1, 15M1]. The RNi_2B_2C (R = Lu, Y) compounds are Pauli-type paramagnets [95L1, 96S5]. Since the presence of some paramagnetic impurities, frequently, a temperature dependent contribution to the susceptibility, χ^{imp}, was shown. The magnetic susceptibilities, χ, of both borocarbides are anisotropic, the values determined for $\mathbf{H} \perp \mathbf{c}$ being higher than for $\mathbf{H} \parallel \mathbf{c}$ direction. The Stoner enhancement factor, $\alpha = 1.4$, is close to that determined from band structures of YNi_2B_2C ($\alpha = 1.82$) [94L1] or $LuNi_2B_2C$ ($\alpha = 1.3$) [94P2]. In the above studies [96S6], the Van Vleck, χ^{vv} and isotropic spin, χ^{spc}, susceptibilities were also calculated. These studies seem to suggest non-existence of a local Ni moment or antiferromagnetic correlations. In contrast with the above suggestion, the presence of AFM correlation was suggested by additional experiments as will be further presented.

The generalised susceptibility for RNi_2B_2C borocarbides has been calculated using the normal-state electronic structure [95R2]. The presence of a peak, in the calculated $\chi(q)$ of $LuNi_2B_2C$, around $\mathbf{q} \cong (0.6, 0, 0)$, was attributed to strong Fermi surface nesting [95R2] and was observed from two-dimensional angular correlation measurements on $LuNi_2B_2C$, using electron positron annihilation radiation technique [99D4]. A magnetic ordering wave vector $\mathbf{q_m} \cong (0.55, 0.0)$, very close to nesting vector, was also evidenced in antiferomagnetic RNi_2B_2C compounds [94G3, 94G4, 95Z1].

The temperature dependences of YNi_2B_2C single crystal magnetizations, in the (ab) plane and along [110] direction, at $T < T_s$, in zero field cooled (ZFC) and field cooled (FC) do not follow the same path, with reversibility region decreasing when increasing field [01T1, 13D1]. The magnetization measurements, in the superconducting state, on both $LuNi_2B_2C$ [97M2] and YNi_2B_2C [99C5], show the presence

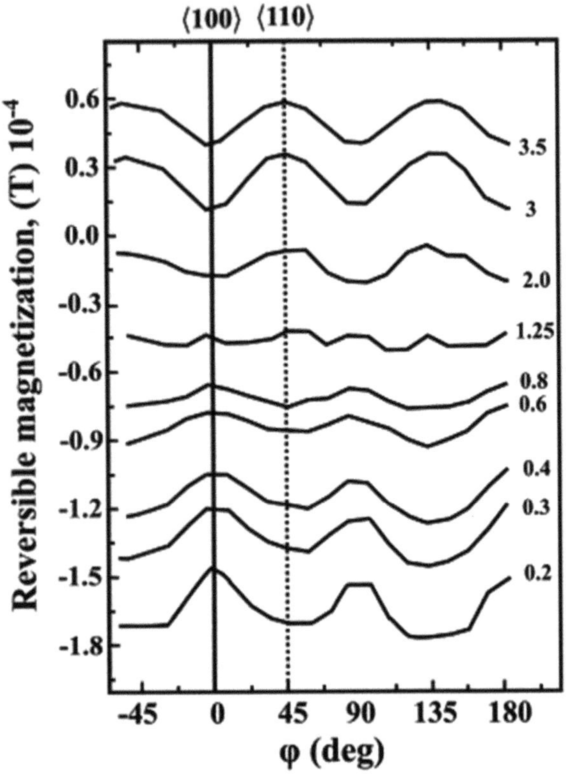

Fig. 9.5 YNi_2B_2C: reversible magnetization, M_{eq}, at T = 7 K, as function of the angle φ between the applied field situated in the (ab) plane with **a**-axis. The field values (T) are indicated in figure [99C5]

of the fourfold anisotropy in the (ab) plane—Fig. 9.5. The upper critical fields are also anisotrop, as shown in $LuNi_2B_2C$ borocarbides [96G4, 97M2]. The anisotropy is 1.16 in out of plane and not temperature dependent, while that in the (ab) plane decrease, from 1.1, at T = 4.5 K, up to 1.0 at T_s. Near T_s, the upper critical fields, H_{c2}, along all directions show a strong upward curvature. This unusual positive curvature, in H_{c2} vs T dependences, is present both in single crystals YNi_2B_2C [94X2] or polycrystalline YNi_2B_2C [96G4, 96K1, 99N6] as well as in $LuNi_2B_2C$ [94T2] and attributed to non local effects arising from the anisotropic Fermi surface and low electron scattering [97M2].

Additional informations on the physical properties were obtained by ^{11}B NMR on RNi_2B_2C borocarbides with R = Lu, [94I1, 96O1] and R = Y [95H1, 95K2, 96O1, 96S6] or at ^{89}Y [95K2, 96S6]. The ^{11}B NMR investigations on RNi_2B_2C with R = Y, Lu, in powdered form, have suggested the occurrence of antiferromagnetic fluctuations of the nickel magnetic moments (correlations) [94B1, 94I1, 95H1, 95K2, 96O1]. The enhancement of nuclear spin–lattice relaxation rate, attributed to these correlations, can been also associated with the increase of the s-band spin susceptibility when decreasing temperature. At $T > T_s$, the ^{11}B nuclear spin–lattice relaxation rate drops rapidly without a coherence [96S6],

The RNi_2B_2C borocarbides with R = Y [94N1, 94S3, 94X2, 95A1, 95F1, 95L1, 96B5, 96E1, 96F2, 96K1, 96M2, 97K2, 97M1, 97N1, 97R1, 98K2, 98W1, 00C6, 00S4, 05M1, 13K1] and R = Lu [94G1, 94S4, 95E2, 96B2, 96B5, 96E1, 96F2, 97F2, 97R1, 98N1, 98W1, 00S4] are type II superconductors. A comparison of superconducting properties of borocarbides with those of cuprates was made [01D1]. The effects of hydrostatic pressure, on the superconducting transition temperature,T_s, of $(Lu_{1-x}Y_x)Ni_2B_2C$ compounds were investigated—Table 9.10. The T_s value of the $LuNi_2B_2C$ borocarbide increases under external hydrostatic pressure, with a rate dT_s/dp = 0.188(12) K/GPa and for YNi_2B_2C decreases with a rate dT_s/dp = −0.058(21) K/GPa [94S2] or increases with 0.03 K/GPa up to p = 0.4 GPa, then decreasing [95A1]. The same trend was reported in following studies—Table 9.10. A polynomial fits of $T_s(p)$ has been also made [95A1]. The dT_s/dp values are also dependent on the used pressure range. The absolute value of dT_s/dp decreases with increasing the volume of the unit cell, reaching a value almost zero for R = Y and changes sign abruptly for borocarbides with R = Er and Tm. The above results can explain the differences in the reported dT_s/dp values for YNi_2B_2C borocarbide, their volume being close to that where a change in sign of the rate is shown. Anisotropic uniaxial pressure dependences of T_s were evaluated using results of zero field thermal expansion measurements [06B7]. A fast decrease in the superconducting transition temperatures, T_s, was shown in polycrystalline YNi_2B_2C sample, during neutron irradiation [13K1].

Representative values characterising the superconducting state will be only mentioned. Thus, the RNi_2B_2C (R = Lu, Y) superconductors have coherence lengths $\xi(0)$ = 8–10 nm [94P4, 99H1] for R = Y and ≅6 nm [94T1, 99H1] for R = Lu. The penetration depths are $\lambda(0)$ = (120–350) nm [94P4, 99H1] for R = Y and (71–130) nm [94T1, 99H1] when R = Lu. The Ginzburg–Landau parameter is $\kappa(0) = \lambda(0)/\xi(0)$ = 15–35 for YNi_2B_2C [94P4, 99H1] and 12–22 for $LuNi_2B_2C$ [94T1, 99H1] borocarbides. Some gap energies $\Delta(0)$ for above compounds are listed in Table 9.11. The gap energies are anisotrop and in the range 2–3 meV, with the ratio $\Delta(0)/k_BT_s$ = 1.7–2.1 [95M1, 96E1]. The mean free paths for R = Y is l ≅33 nm [98D1] and of (29–70) nm when R = Lu [94T1, 98D1]. The ratio $\Delta C/\gamma T_s$ = 1.77 (R = Y) and 2.21 (R = Lu) [95M1], or 3.4 (R = Lu) [99H1]. These data suggest that these borocarbides are conventional intermediate to strong coupling superconductors.

The boron isotope effects are present in polycrystalline YNi_2B_2C [95L2, 99C4] and RNi_2B_2C (R = Y, Lu) single crystals [99C4]. Values $\Delta T_s = T_s(^{10}B) - T_s(^{11}B)$ of 0.29(8) K (R = Y) and 0.16(7) K (R = Lu) were reported [99C4]. It is to be noted that a large anisotropic thermal amplitude in RC plane, has been found for C [95G2], suggesting that elements as B and C may be involved in the superconductivity mechanism of borocarbides.

The type II superconductors, at $H_{c1} < H < H_{c2}$, are in mixed state, in which flux penetrates in quantized units, in nonsuperconducting domains, forming vortex or flux lines (VL). The analysis of the magnetic flux lattice topology in RNi_2B_2C borocarbides, give useful informations on the anisotropy and symmetry of the superconducting gap system. This matter has been investigated by using SANS experiments on borocarbides with R = Lu [97E3, 01E1, 09D2] and R = Y [97E3, 97Y4, 98M1,

Table 9.10 Superconducting transition temperatures and their pressure dependences

Compound	$T_s(0)$ (K)	dT_s/dp (K/GPa)	A (K/GPa)	B (K/GPa)	References
$DyNi_2B_2C^b$	5.8 (onset) 4.8(compl)	−0.7			[01F1, 02F1]
$HoNi_2B_2C^a$	8.665(3)		−0.187(10)	−0.0452(64)	[95A1]
$HoNi_2B_2C^a$	8.674(8)	-0.187^c	−0.255(9)		[95A1]
$HoNi_2B_2C^b$	8.4–8.8	−0.5			[96U1, 10N1]
$Ho_{0.77}Ni_2B_2C^b$	7.5	−0.6(2)			[95S1]
$ErNi_2B_2C^a$	10.961(11)	0.17^d	0.168(86)	−0.089(21)	[95A1]
$ErNi_2B_2C^b$	10.56–10.14	−0.082(17)			[94S2]
$ErNi_2B_2C^b$	10.1	−0.16‖**a** −1.3‖**c** −0.98‖(powder)			[06B6]
$TmNi_2B_2C^a$	10.876(10)		−0.410(30)	−0.0262(17)	[95A1]
$TmNi_2B_2C^b$	10.885(9)	-0.410^e	−0.453(1)		[95A1]
$TmNi_2B_2C^b$	10.25–9.74	−0.178(22)			[94S2]
$TmNi_2B_2C^b$	11	−0.5			[99O1]
$LuNi_2B_2C^{b,c}$	16.35–15.92	0.188(12)			[94S2]
$LuNi_2B_2C^b$	16.75	0.188, $p \leq 0.5$GPa −0.13, $0.5 \leq p \leq 1.67$ GPa			[94M6]
$YNi_2B_2C^{a,f}$	15.764(4)	0.03	0.032(11)	−0.0313(6)	[95A1]
$YNi_2B_2C^{b,c}$	15.40–15.23	−0.058(21)			[94S2]
$YNi_2B_2C^{b,g}$	15.50	−0.09			[94M6]
$YNi_2B_2C^h$	15.9	−0.102 (ab) plane 0.083 ‖ **c**			[06B7]
$YNi_2B_2C^h$	15.1	−0.074 (ab) plane 0.074 ‖ **c**			[06B7]
$YNi_2B_2C^{b,i}$	15.48	−0.09(3)			[95L7]
$YNi_{1.94}Co_{0.06}B_2C^{b,i}$	12.32	−0.19(3)			[95L7]
$YNi_{1.7}Cu_{0.3}B_2C^{b,i}$	10.9	−0.15(3)			[95L7]
$YNi_{1.5}Cr_{0.2}B_2C^{b,i}$	6.7	−0.18(2)			[95L7]

[a] $T_s(p) = T_s(0) + Ap + Bp^2$
[b] Linear fit
[c] $p < 1.5$ GPa, non-linear temperature dependences with peaks at
[d] 0.94 GPa
[e] 0.52 GPa
[f] maximum at 0.52 GPa, $p \leq 1.5$ GPa
[g] $p \leq 1.67$ GPa;
[h] from thermal expansion
[i] $p \leq 0.7$ GPa

98Y1, 05D2], STM method when R = Lu [97D3, 10U1] or R = Y [97A1, 00S2, 09N2, 10U1], decoration procedure for $LuNi_2B_2C$ [97A1, 00V1, 01V1, 01V2] or by other methods, including theoretical studies, as for R = Lu [97K3, 98P2, 01G2,

Table 9.11 Superconducting energy gaps for RNi_2B_2C compounds with R = Lu and Y

Compound	T (K)	Method	Field orientation	Superconducting gap (meV)			References
				Δ(0)	$\Delta_1(0)$	$\Delta_2(0)$	
$LuNi_2B_2C$	1.48	PC[a],s.g.[b]	(ab) plane c-axis	2.6 2.2			[07N4]
$LuNi_2B_2C$	≅2	PCARS[c],s.g.[b]	[001] [110] [100]	2.4 2.6 2.3			[11L4]
$LuNi_2B_2C$	4.2	PC[a],s.g.[b]		3.3			[95J2]
$LuNi_2B_2C$	1.47	PC[a],t.g.[d]	(ab) plane c-axis		2.2 1.8	3 2.8	[07N4]
$LuNi_2B_2C$	1.48	PC[a],s.g.[b]	(ab) plane c-axis	2.16 1.94			[06B4]
$LuNi_2B_2C$[e]	1.47	PC[a],t.g.[d]	(ab) plane c-axis		3.03 3.2	2.0 1.82	[05B3, 06B4]
$LuNi_2B_2C$[e]	1.47	PC[a],s.g.[b]	(ab) plane c-axis	2.74 2.66			[05B3]
$LuNi_2B_2C$[f]	1.47	PC[a],t.g.[d]	(ab) plane c-axis		2.92 2.9	1.82 1.65	[05B3]
$LuNi_2B_2C$[f]	1.47	PC[a],s.g.[b]	(ab) plane c-axis	2.34 1.88			[05B3]
YNi_2B_2C[e]	3.5	PC[a],s.g.[b]		2.80			[06B1]
YNi_2B_2C	1.47	PC[a],s.g.[b]	**a**-axis c-axis [110]	1.5 2.3 2.5			[07N4]
YNi_2B_2C	≅2	PCARS[c],s.g.[b]	[001] [110] [100]	2.0 2.1 2.5			[11L4]
YNi_2B_2C	4.2	PC[a],s.g.[b]		3.2			[95J2]
$Y(Ni_{0.8}Pt_{0.2})_2B_2C$	3.5	PS[g],s.g		1.75			[06B1]

[a]Point contact spectroscopy
[b]Single gap
[c]Point contact Andreev reflection spectroscopy
[d]Two gaps
[e]In a model [03B1]
[f]In a model [94P3]
[g]Photoemission spectroscopy (PS)

09G1, 15M1] and R = Y [98P2, 09D1, 12I1, 15M1]. The vortex motion has been also investigated [09O1, 10O2].

The flux lines in a superconductor are expected to be in an hexagonal array, since those repel each other and a hexagonal lattice of the lines maximizes the distance between them, this being the lowest energy configuration. In RNi_2B_2C borocarbides with R = Lu and Y, the (nearly) hexagonal VL, undergoes a rhombic distorsion having long diagonal along [110] direction (apex angle < 60°). At a field $H_1(T)$, the VL undergoes a first-order 45° reorientation and spontaneous jump of apex angle to a value greater than 60°. When increasing the field, the VL is further distorted, until a square lattice is formed, above a field $H_2(T)$ via continuous transition [97D3, 98M1, 01E1]. In RNi_2B_2C compounds with R = Lu or Y, a square vortex lattice (VL), with one side parallel to[110] axis, in fields **H** ∥ [001], higher than ≅ 0.2 T, was shown [96K5, 97D3, 97E3, 97K3, 97Y4]. As the field is reduced, below a characteristic temperature and material dependent value, $\mu_0 H_2 = (0.03 - 0.1)$ T, the vortex lattice transforms into near hexagonal lattice, made of rhombic cells with diagonals along [100] and [010] directions. The transformation proceeds via a rhombohedral distortion of the square, which preserves the orientations of diagonals along [100] and [010] and the unit cell area, due to flux quantization [97E3, 97K3]. The transition field increases with increasing temperature. The above behavior was correlated with an in-plane anisotropy of the upper critical field H_{c2} and analysed within the framework of Ginzburg–Landau (GL) theory [97D3]. The stability of the square vortex lattice has been further investigated in generalized GL model, by including higher order terms to reduce the rotational symmetry to tetragonal [98P2]. They suggested that $H_2(T)$ should rise with increasing temperature and make a finite intercept with upper critical field $H_{c2}(T)$. By incorporating the effects of thermal fluctuations in the nonlocal London theory, find that $H_2(T)$ and $H_{c2}(T)$ do not cross [01G2]. Strong thermal fluctuations of vortices near $H_{c2}(T)$ are thought to supress the anisotropy induced by the nonlocality resulting in a re-entrance of the rhombic VL phase. When increasing cobalt content in $Lu(Ni_{1-x}Co_x)_2B_2C$ series, there is a decrease of the mean free path, *l*, reducing the nonlocal effects and as a result, the increase of $H_2(T)$ [99G2, 00K1]. The $H_2(T)$ also rises with increasing temperature due to weakened nonlocality as effect of thermal fluctuations [01E2]. The small angle neutron diffraction measurements on YNi_2B_2C with **H** ∥ **c** has been used to determine the structural phase diagram of the vortex lattice [05D2]. The first-order reorientation transition of the rhombic VL, $H_1(T)$, decreases while the second order transition to a square VL, $H_2(T)$ increase with increasing temperature. No evidence was shown that $H_2(T)$ curve upward or become re-entrant for temperatures approaching $H_{c2}(T)$, which may be expected when thermal fluctuations suppress nonlocal effects at high temperature. These data suggested that an underlying electronic asymmetry, other than FS and nonlocal effects and likely due to an anisotropic superconducting gap, controls the VL structure close to $H_{c2}(T)$.

When the current density in YNi_2B_2C increases, the vortex dynamics change from coherent flow of the vortex-lattice to pinning-dominated plastic flow [09O1, 10O2]. Several phases, determined by the balance among pinning force, elastic force and thermal fluctuations can be shown in the magnetic field-temperature diagram of vortices [09O1, 10U1]. In the intermediate magnetic field, vortex lattices are almost perfect at low temperatures, because the vortex-vortex interaction is strong. In higher or lower magnetic field, vortex lattices lose long range order [97E3, 04N1]. The observation of vortex motion in real time and space, has been made by STM measurements on YNi_2B_2C [10U1]. The dislocations, in vortex lattices, determine the characteristics of vortex creep in high magnetic fields. There are vortex bundles separated by glide planes of edge dislocations. The size of vortex bundles, in which the vortices are pinned collectively, correspond to the distances between glide planes of dislocations. The motion of vortices is based on the glides of dislocations, their direction tends to be parallel to the glide planes. The ultrasound velocity measurements, on the YNi_2B_2C, in the vortex state, evidenced the elasticity of the flux line lattice [12I1]. Elastic anomalies, due to a first order 45° reorientation transition of the hexagonal flux line lattice and a second-order hexagonal-to-square lattice transition of the flux line lattice, were shown.

Altought the anisotropic vortex core must to reflect the gap symmetry [96I1], in earlier studies the vortex core determined by STS in RNi_2B_2C with R = Lu and Y, at T = 4.2 K, did not exhibited anisotropy [97D3, 00S2]. Latter on, from the analysis of local quasi-particle density of states, $\eta_s(E, \mathbf{r})$, in the isolated vortex of YNi_2B_2C, at T = 0.46 K, was found that the vortex core is fourfold symmetric star-shaped in the real space [04N1]. Near E = 0 meV, $\eta_s(E,\mathbf{r})$ extents toward [100] direction (**a**-axis) and has a peak at the centre of vortex core. With increasing energy, $\eta_s(E, \mathbf{r})$ comes to extend toward [110] axis and the peak of $\eta_s(E, \mathbf{r})$, splits into four peaks toward [110]. The $\eta_s(E,\mathbf{r})$, in the vortex core, exhibits electron–hole asymmetry.

The quasiparticle DOS in the vortex state in $Y(Ni_{1-x}Pt_x)_2\ B_2C$ has been probed by specific heat measurements under magnetic field [99N6]. The quasiparticle DOS per vortex is H-depentent in the clean limit superconductors, while it is H-independent in the dirty superconductors. These data were discussed in terms of the shrinking of the vortex core radius with increasing magnetic field.

The superconducting mechanism and the pairing symmetry in RNi_2B_2C compounds have been intensively investigated. There are many experimental observations of strong electron–phonon interaction, while considerable doubt has been expressed regarding the adequacy of a solely conventional electron–phonon (EP) mechanism to account for their high T_s. The INS measurements on $LuNi_2B_2C$, along [q00] direction, evidenced that the electron–phonon interaction is quite strong and causes an incipient lattice instability, behavior typical of strongly coupled conventional superconductors [95D1]. Phonon anomalies occured at wave vectors, $\mathbf{q_m}$, close to that determined in analysing the magnetic susceptibility, $\chi(q)$.

Originally, the brocarbides have been considered as a simple mediated s-wave superconductors [94E3, 94M1, 94M5, 94P2]. The thermodynamic [02L1], tunneling [03M1], ultrasound [04W1], as well as point contact measurements [04R1] suggest scenarios such as anisotropic s- or d-wave [02L1, 02M1, 03M1], or (s + g)-wave pairing [02M1, 04R1, 04W1]. The RNi_2B_2C with R = Y, Lu, have large SC gap anisotropy with nodes [97N2, 00Y1, 00Y4, 01B1]. Such an extremely large gap anisotropy cannot be obtained from a simple s-wave pairing wave function that is assumed for phonon-mediated superconductor. Then, the so called s + g wave spin singlet gap, was shown to be consistent with experimental results. Here the second g-wave contribution is given by a fourth degree fully symetric (A_{1g}) basic function. Within BCS theory, by an appropriate ansatz for the pairing interaction and selection of the interaction parameters, have been shown that a stable coexistence of s- and g-wave as well as the nodal state with equal amplitudes of them is possible [03Y1]. Starting from Fermi surfaces, with open and closed sheets [94M1, 94P2, 95K1,

99D4], multiband behavior was suggested [98S2, 05B3, 05M2, 06Y1, 07N4]. The observation of a fourfold symmetry in the anisotropy of the upper critical field of $LuNi_2B_2C$ [97M2], enhanced also the study of Fermi surface (FS) topology.

The phonon properties of RNi_2B_2C compounds with R = Lu [96D1, 97S6, 97Y1] and Y [96K2] were further investigated. Dramatic changes in the phonon spectra near the nesting vector $\mathbf{q}_m$ were shown as they entered in the superconducting state. Just above T_s, there is a broad soft phonon response, which changes dramatically at $T < T_s$. In both compounds a sharp peak near 4 meV was present with a broad sholder on the high-energy side. The neutron studies at higher temperatures showed a strong coupling between the Δ_4 acoustic (A) and Δ_4 optic (O) modes for wave vectors near the nesting vector [98B4]. A strong temperature dependence of frequencies for both modes was also shown, these decreasing in energy as temperature is lowered [99S6, 99Z1]. The mode interaction in Lu- and Y-borocarbides was later investigated up to T = 1000 K [02Z1]. In YNi_2B_2C the optic and acoustic Δ_4 modes are separated at low temperatures and become nearly degenerate at T > 600 K. The modes in $LuNi_2B_2C$ remain separated at all temperatures. In $Y_{0.5}Lu_{0.5}Ni_2B_2C$ the modes are separated at T < 300 K and degenerate at higher temperatures [02Z1]. This behavior is inconsistent with a simple O–A coupling scheme [97A2], because one would expect the temperature dependence to show a crossing at some temperatures and become separated at higher temperatures, rather than remaining degenerate. These data suggested that other interactions must to be involved in the mode coupling scheme. The strongest phonon couplings occur in the vicinity of the Fermi surface nesting vector $q_m \cong (0.5, 0, 0)$. The presence of phonon line around T_s and their dependence on the intensity of magnetic field, similar with critical field H_{c2}, also evidenced that the mechanism of superconductivity is the electron–phonon interaction. The line was interpreted either as a new phonon excitation [96K2] or as a narrowing of phonon width due to increase of phonon life time, by opening up of superconducting gap [97S6, 98B4, 02Z1, 15M1].

The Raman and IR-active phonon frequencies and eigenvectors were theoretical investigated in RNi_2B_2C with R = La, Lu and Y [03R2]. The electron–phonon coupling was analysed and hence superconductivity. One cannot conclude that the coupling is weak along the entire zone. The evolution through T_s of the low temperature lineshape of the phonon at the end point of the transverse acoustic branch in the [110] direction, has been analysed [08W1]. When cooling just above T_s the phonon softens and broadens, indicative of a strong electron–phonon coupling. On cooling through T_s the lineshape deviates from a Lorentzian form and a step-like increase at a certain energy E_s appears, the E_s energy increasing when decreasing temperature. A part of the low energy tail is pushed up in energy to form a narrow spike at 2Δ, where Δ is the energy gap [97A2, 08W1]. This effect is not restricted to phonons with wave vectors close to an extremum vector of the Fermi surface. By using the inelastic neutron scattering investigations, the dispersion surface and linewidths of A_{1g} ($\cong$ 102 meV) and A_u ($\cong$ 159 meV) type phonon modes over the Brillouin zone were investigated in YNi_2B_2C [12W2]. The A_{1g} modes do not strongly contribute to the electron–phonon coupling constnt λ. It has been shown that the high energy vibrations couple indeed very strongly to the electrons, but this coupling is nevertheless irrelevant to superconductivity [12W2]. According [12W2, 14W2], the strong coupling of certain low energy modes, was linked to the presence of large displacements of the light atoms (B, C), which is unusual in view of the rather low phonon energies. Specific modes, exhibiting a strong coupling to the electronic quasiparticles were investigated, as function of temperature. Their energies and linewidths showed marked changes on cooling from RT to just above T_S. It was also shown that YNi_2B_2C has temperature-dependent phonon anomalies without nesting and anharmonicity. If the FS is nested for a particular q_z and the matrix elements for a particular phonon are very strong at that q_z, then the phonon effectively sees a nested two-dimensional FS [14W2]. Thus has been proposed that, for example, the phonon at (0.56, 0, 0) couples strongly to the nested part of FS with $q_z = 0.5$ and this nesting is responsible for the Kohn anomaly at this wave vector.

From the spectral analysis throughout the irreductible Brilouin zone of $LuNi_2B_2C$ was found that the highest contribution to the electron–phonon coupling parameter derives from the acoustic and low-frequency optical modes, characterized by anomalous dispersion, instead of A_{1g} optical-phonon mode [15T1]. By using the average electron–phonon coupling parameter $\lambda = 0.83$, the superconducting critical temperature was found in agreement with experimental value. The magnitude of superconducting gap is a fundamental parameter and its momentum, q, dependence reflects the symmetry of the pairing wave function of Cooper pairs as well as interactions of electrons with elementary excitations. The superconducting order parameter is proportional to the gap function, which in turn, is proportional to the amplitude of the wavefunction for a Cooper pair [06M1].

Band structure calculations were performed on RNi_2B_2C borocarbides with R = Lu [94C6, 94M1, 94P2, 95K1, 95R2, 99D4, 01D1, 08B2, 08S3, 09B2, 09D3, 15T1] and R = Y [94L1, 95R2, 96S3, 04Y1, 09N1,17K1]. The Fermi surfaces for RNi_2B_2C compounds with R = Y and Lu are given in Fig. 9.6.

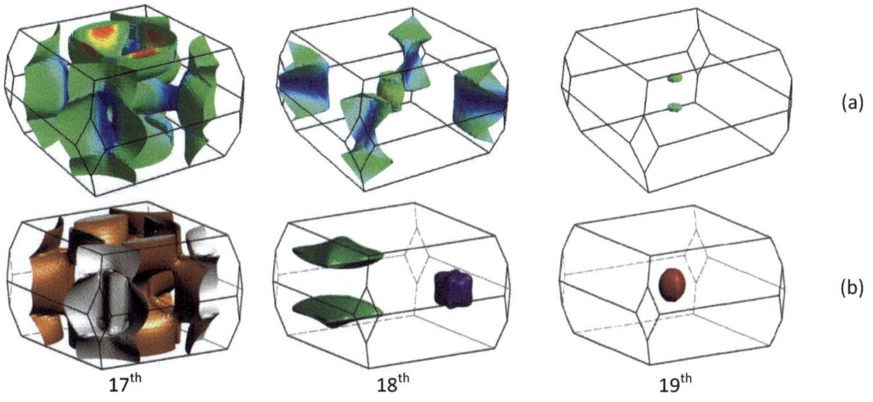

Fig. 9.6 YNi_2B_2C (**a**) [16K1, 17K1] and $LuNi_2B_2C$ (**b**) [09B2]: perspective view of the calculated FS obtained from the band structures. The FS of 17th, 18th and 19th bands are shown

The electronic structures calculations of RNi_2B_2C borocarbides with R = Lu and Y showed that their Fermi surfaces comprised three sheets. Three bands were obtained in YNi_2B_2C (17th,18th, 19th), with mainly Ni3d and Y4d character which cut across the Fermi level, by using the linearized muffin-tin orbital method [94L1] and the general potential linearized augmented plane wave method [96S3], respectively. The YNi_2B_2C band structure, calculated by FLAPW method [04Y1], reproduced correctly the FS obtained by dHvA experiments [04Y1]. The two closed Fermi surfaces along Λ axis, from 19th band, become connected to form one ellipsoidal pocket around the Γ point. The long cigar-like FS, from the 18th band, along the W axis, was split into two parts, while the "calabash"—like FS, centred at the Γ point, is unchanged. The 17th band brings a large electron FS, multiply connected by the necks [04Y1]. The DOS from 17th band, which is related to multiple connected electron FS, has a sharp peak at Fermi energy. The peak arises from a van Hove singularity around (1/5,1/5,0) point in **k** space. It was also suggested that this singularity may lead electron–phonon coupling locally and give rise to anisotropic gap behavior in the superconducting state. The electrons of 17th band predominantly contribute to the superconductivity, the DOS at E_F for the 17th, 18th and 19th bands being 48.64, 7.88 and 0.38 states/Ry, respectively [04Y1].

The field-angle dependences of the heat capacity [03P1] and thermal conductivity [02I1], having oscillating behavior, were suggested that can be used to determine the gap symmetry. For this aim, the field-angle dependent zero-energy DOS for YNi_2B_2C was calculated on the FSs from 17th band [07N1] and taking into account also the 18th band [09N1]. The oscillation amplitude was found to be of the order of 8% [09N1] and does not deviate so much from those experimentally determined [02I1, 03P1]. It was suggested that the gap structure on the 17th FS is appropriated for the in-plane properties [09N1].

The origin of highly anisotropic superconducting gap in YNi_2B_2C was investigated on the framework of the density functional theory for superconductors [16K1, 17K1]. The Fermi surface of YNi_2B_2C with the distribution of the Fermi velocity is shown in Fig. 9.6a. The Fermi velocity varies largely over the Fermi surface, the ratio of its maximum to minimum was about 100. On the Fermi surface there are no regions dominated by B2s2p and C2s2p orbitals. The Fermi velocity is particularly small in the regions where the contributions of Ni3d orbitals are dominant. The anisotropic superconductivity was traced back to the variation of the rate of the Ni3d orbital on the Fermi surface. As the component of the Ni3d orbital increases, the electron–phonon coupling of the electronic states becomes weak and its superconducting gap function becomes small. Because of this effect, the superconducting gap significantly varying over the Fermi surface, emerges. It was established that the orbital character variation on the Fermi surface is the key factor of the anisotropic gap.

The electronic properties of $LuNi_2B_2C$ were investigated by linear augmented-plane-vawe (LAPW) method [94M1, 94M2, 00D2, 17L1], augmented-spherical-wave method [94C6] and plane-wave pseudopotential method [95K1]. The dHvA measurements combined with the full-potential local-orbital (FPLO) calculations were made also on $LuNi_2B_2C$ [08B2, 09B2]. As in case of YNi_2B_2C borocarbide, the Fermi surface comprised three sheets—Fig. 9.6b. Evidence for highly anisotropic band-dependent superconducting coupling strengths was shown. The LMTO calculations [09D3] confirmed the presence of these sheets, in common with previous calculations [94M1, 94P2] and broadly similar to FPLO results [08B2, 09B2]. All the bands at the Fermi level are predominantly of Lu and Ni3d character with some admixture of C2p and B2p. The first sheet is spheroidal in shape and contribute approximately 0.24% of the Fermi level DOS (19th band). The second sheet with two regions, namely a cuboid around the Γ point and flat "cushions" around the P point, contributes 22.64% of the Fermi level DOS. The third sheet, contributing 77.1% to the total Fermi level DOS is multiply connected. This is the sheet which exhibits the strongly nested sections [95R2, 99D4]. The Fermi surface topology of RNi_2B_2C borocarbides was shown to vary little with rare-earth elements, such as for R = Er, Tm and Yb, suggesting that this topology is broadly common [09D3].

The complex Fermi surface structures, with multiple sheets, of RNi_2B_2C with R = Lu and Y, as above presented, involves three bands crossing E_F, all having electron-like character [04Y1, 08B2, 09B2, 11L4, 17K1]. While the small ellipsoidal FS, from the 19th band contributes very little to the DOS, both the 17th and 18th bands, comprised the major Fermi surface sheets, although their topologies are quite different from each other. The "cushion"-like FS from the 18th band, coming exclusively from $Ni3d_{x^2-y^2}$ and $3d_{xy}$ derived states, is not affected by the magnetic moments of R ions. The superconductivity originating from this band can survive under the development of magnetic order [04D1, 08B2, 09B2, 11L4].

The evidence of a multiband nature of superconductivity has been suggested also from the temperature dependence of the upper critical field in RNi_2B_2C samples with R = Lu, Y [98S2]. The photoemission study, on YNi_2B_2C, evidenced a prominent Ni3d band centred at ≅1.5 eV below E_F, whose top crosses E_F and B, C2sp derived states, at higher energies [94G2].

In order to obtain more insight in which bands are mainly involved in the supraconductibility, dHvA experiments were performed, in combination with band structure calculations. Thus, from angular dependent effective masses for different bands, the mass enhancement factors were determined. The involved coupling strenght, mainly due to electron–phonon coupling [99C4], can be extracted from a comparison with the bare masses. By using de Haas-van Alfen effect, each band was associated with a characteristic frequence, f_i. With reference to Fig. 9.6, f_α is associated with nearly spheroidal Fermi surface. Two other independent oscillation frequences appear, f_ε and f_η ascribed to a cube-like and and cushion-like Fermi surfaces, respectively [08B2, 09B2]. The other peaks in the Fourier transformation were associated with the second harmonics of the above dHvA frequencies [09B2]. A dHvA frequency f_x (f_φ) has been assigned to an extremal orbit of the above branched Fermi surface [08I1] or caused by the complicated branched Fermi surface [08B2]. A larger number of dHvA frequencies were already reported. Thus, in YNi_2B_2C, there are five (α, β, δ, ε, γ) [95T1], or eight (α, β, ε, γ_1, γ_2, κ, η_1, η_2) [96N2] branches of dHvA frequencies, in normal state. These were correlated with band structures [04Y1]. The frequencies f_a to f_g, evidenced in $LuNi_2B_2C$, were associated with the most complicated piece of the Fermi surface [06B2].

The three sheets are differently involved in the superconducting pairing, given evidence for the multi-band nature of superconductibility as alredy mentioned. With the additional knowledge of the effective masses of the various bands it is possible to calculate the superconducting properties including anisotropy of the superconducting gap.

For the most investigated α-band, by using either the torque or the field-modulation method, controversial results were obtained in particular on YNi_2B_2C and $LuNi_2B_2C$ borocarbides. The additional damping of the dHvA signal was found in line with the opening of a weak-coupling gap [95T1, 96G6, 03B2], yielding an unexpectedly small gap [08I1, 12B1] or an abrupt vanishing of oscillations below H_{c2} [05I1]. By using ultrasound measurements on YNi_2B_2C, below the upper critical field, a very weak additional damping appeared, due to field inhomogeneity caused by the flux-line lattice. It was concluded on the existence of a gapless or, at least marginally small gaped band in mixed state [17N1].

From angular dependent effective masses for different bands, the mass enhancement factors were determined. The involved coupling strenght, mainly due to electron–phonon coupling [99C4] can be extracted from a comparison with the bare masses. The mass enhancement values, λ, are directly related to the Cooper-pair coupling parameter. For all the observed bands, prominent anisotropies and largely different coupling strengths were shown [95M1, 03M1, 08B2, 09B2]. The weakest coupling was found for the spheroidal Fermi surface [08B2, 09B2]. For the f_η frequency, originating from the cushion-like Fermi surface, λ is 1 in the (100) plane decreasing to $\cong 0.23$ towards [110] and is belived to allow superconductivity. For dHvA frequency f_ε, from the cube-like FS, the coupling strength is highly anisotropic, reaching $\lambda = 2.7$ for field aligned along [100] direction down to $\cong 1$ for other directions [09B2]. These data are further support for the multiband superconductivity. With the additional knowledge of the effective masses of the various bands, as alredy mentioned, it is possible to calculate the superconducting properties including anisotropy of the superconducting gap. The oscilations in the SC state originate from both gapless regions of the Fermi surface caused by a center-of-mass motion of Cooper pairs, in high magnetic fields and regions where a gap exists [97G3].

Unconventional superconductivity is characterized by anisotropic superconducting gap functions, which may have zeros (nodes) along certain directions in the Brillouin zone. The superconductivity in RNi_2B_2C with R = Y, Lu borocarbides has been associated with both the 17th and 18th bands [08B2]. The claimed nesting structure on the 17th band FS has been experimentally observed [08S3, 09D3]. The nested part, only occupies a small portion of the Fermi surface [04D1]. The AFM fluctuations, on the nested parts of the FS, could induce a point node-like-gap minimum, along the nesting wave vector [04K1]. The superconducting gap was observed on multiple bands (17th and 18th) with different momentum dependencies [10B1]. The gap is highly anisotropic on the 17th band having two minima, whereas the 18th band has nearly constant gap. The two minima of 1.5 and 2.3 meV were on the different parts of the 17th band FS. Their momentum direction is the [100] axis [02I1, 03P1, 04P1, 06M1, 11L4]. It was claimed that the points on the FS, where the gap shows a minimum at 1.5 meV, can be connected by the nesting vector **q** = (0.55, 0, 0), implying a connection between the minimal gap and the FS nesting.

Further data showed that the origin of the large SC gap anisotropy is connected with the direction and type of nodal structure and its relation with nesting vector [95R2, 99D4]. The fraction of FS participating to nesting was determined to be 4.4(5)% [99D4]. By Angle Resolved Photoemission Spectroscopy (ARPS), was shown that FS exhibits large parallel regions with a nesting vector [08S3], in agreement with theoretical predictions [95R2] or experimental studies [99D4]. The calculated FS confirmed the existence of large nested parts, with nesting vector in agreement with ARPES results [08S3].

The magnetic field orientation dependence of thermal conductivity [02I1], in-plane anisotropy of ultasonic attenuation [04W1] and STM/spectroscopy in the vortex state [04N1] have reported point nodes located along [100] and [010] directions, while field-angle-dependent heat capacity suggested a line-like nodal structure [03P1]. It was also shown that q point-like minimum of SG gap in YNi_2B_2C is located on a cilindrical FS sheet around X–P and were found to be connected by the known nesting vector [10B1]. The diminishing oscillations amplitude of thermal conductivity and specific heat, in YNi_2B_2C, as the field direction changes from the (ab)-plane, to the **c**-axis, suggested also the possible existence of point-like nodes [02I1, 06M1]. This may be consistent with a large gap anisotropy. Latter on [10B1], by photoemission study, within measured q regions of FS sheets, a point-like minimum of SC gap was shown, whose q positions can be connected by the nesting vector. The strong anisotropy of the gap has also been indicated by broad peak in the tunneling conductance spectrum [03M1] or strong antiferromagnetic spin fluctuations [95K2, 04K1].

The magnitude of a superconducting energy gap is a fundamental parameter and its momentum, **q**, dependence reflects the symmetry of the pairing wave function of a Copper pair, as well as interactions of electrons with elementary excitations. Sharp minima in the gap function of YNi_2B_2C, along certain **q** directions, with possible point nodes along [100] and [001] axes, were experimentally found [00Y4, 01B2, 01I2, 02I1, 03M1, 03P1, 04R1, 04W1]. Based on the shape of the gap function, an order parameter symmetry, with mixed angular momentum, mainly of s + g symmetry, has been analysed [02M1]. In this model, where a roughly spherical Fermi surface was implicitly assumed, the multiband nature of the Fermi surface is ignored. An alternative model has been also proposed, where the gap anisotropy could originate from different bands, on the FS, having distinct gap values, due to the differences in their coupling strength. The contribution from a particular band, in a given **q** direction, depends on a weighted average of the Fermi velocity and the density of states.

The superconducting gap, in RNi_2B_2C borocarbides with R = Lu, Y, has been determined by different methods, mainly the point contact (PC) spectroscopy, when R = Lu [95J2, 05B3, 07N4, 09L4, 10K2, 11L4] and R = Y [95J2, 05M2, 07N4, 11L4]. The field (at T = 2.3 K) and temperature dependences of the superconducting energy gap, along the two principal crystallographic axes ([001] and [100]) of YNi_2B_2C single crystal, by injecting current, I, either along **a** or **c** directions were investigated [05M2]. For both current directions, the magnetic field was applied parallel to the current. In the field range $\mu_0 H < 2.25$ T, the $\Delta_I \parallel \vec{\mathbf{a}}$ decreases linearly with the applied field. The extrapolation of the above data shows that $\Delta \parallel \mathbf{a}$ vanishes at $\mu_0 H = 3.25$ T, well below the upper critical field, $\mu_0 H_{c2} \cong 6.0$ T ($\mathbf{H} \parallel \mathbf{a}$), at T = 2 K. For I ∥ **c**, Δ decreases weakly with magnetic field. The temperature dependence of superconducting gap for I ∥ **c** shows a nearly BCS type behavior, vanishing at $T_s \cong 14.5$ K. For I ∥ **a**, there it is a more rapid decrease. It was assumed that the two gaps originate from two weakly coupled bands with no interband scattering, having $T_s = 14.5$ K and 4.6 K, respectively.

The superconducting energy gap in $LuNi_2B_2C$ has been investigated by point contact spectroscopy in a large temperature range [05B3, 06B4] and data analysed in two gap approximation, by using two models [94P3, 03B1]. The temperature dependences of the large and the small gaps have been estimated for (ab) plane and **c**-direction—Fig. 9.7. In addition to the small Δ_1 and large Δ_2 gaps, an average gap, Δ_M, which allows for the partial contributions to the conductivity of the contact made by the Fermi surface areas with different gaps, is shown [06B4]. It was found that in the BCS extrapolation of the critical temperature for the small gap is $T_s = 10$ K in the (ab) plane and $T_s = 14.5$ K in **c**-direction. For the large gaps, the critical temperature coincides with the bulk T_s and their absolute values are close in both orientations, about 3 meV. In the **c**-direction the contributions to the conductivity, from the small and the large gaps, remain near identical, up to 10–11 K. In the (ab)

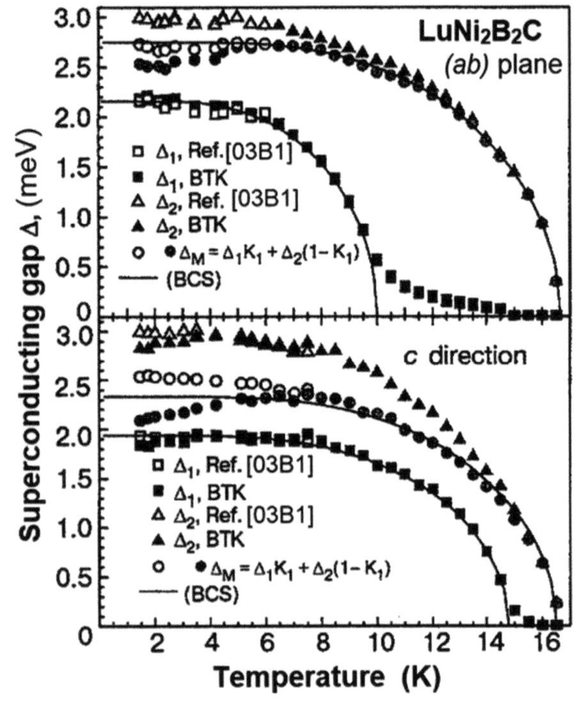

Fig. 9.7 $LuNi_2B_2C$: temperature dependences of the gap energy calculated in the two gap approximation [03B1] and the BTK model [94P3] in the (ab) plane (**a**) and **c**-direction (**b**). Solid lines BCS extrapolation. The large Δ_2 and small Δ_1 gaps as well as their medium Δ_M values are given [06B4]

plane, the contribution from the smaller gap is much smaller and decreases rapidly as temperature rises.

The epitaxial **c**-axis oriented $LuNi_2B_2C$ borocarbide was also studied by point contact spectroscopy [10K2]. The average value of the superconducting gap was $\Delta = 2.6(2)$ meV in one gap approach, whereas in the two gap approach, values $\Delta_1 = 2.14(36)$ meV and $\Delta_2 = 3.00(27)$ meV, were obtained. This study evidenced also the presence of multiband superconductivity in $LuNi_2B_2C$ compound. In YNi_2B_2C films, a distribution of Δ values from $\Delta_{min} = 1.5$ meV up to $\Delta_{max} = 2.4$ meV with a BCS-type temperature dependence was shown [05B2].

The superconducting gaps in RNi_2B_2C borocarbides were also investigated by inelastic light scattering (R = Lu, Y) [00Y1], point contact Andreev reflection spectroscopy (R = Lu, Y) [96R4, 10K2, 11L4], photoemission spectroscopy (R = Y) [06B1, 10B1], de Hass-Van Alphen effect (R = Lu) [12B1] and R = Y [03B2], or computed (R = Y) [11J1]. The ARPES results evidence the importance of momentum dependent superconducting gap [06Y1]. Some gap values determined in one or two gaps models are listed in Table 9.11.

The heat capacities of RNi_2B_2C borocarbides with R = Lu [94C2, 94K1, 94T1, 95M1, 97N2, 99N6, 01S1, 04P1] and R = Y [94M5, 94T1, 95G2, 95H2, 95H3, 95M1, 99N6, 00N3, 01I2, 03P1, 03S2] were investigated. The specific heats, C_p, in zero field, show jumps at the superconducting transition temperatures, T_s [94C2, 94K1, 94M5, 95G2, 95H2, 95M1, 03S2]. The jumps disappear if the measurements are made in fields, excluding H_{c2} values. In the normal state, the heat capacities of the above borocarbides follow a temperature dependence $C_p/T = \gamma + \beta T^2$, where γ is the coefficient of electronic specific heat capacity (Sommerfeld parameter) and β the lattice heat capacity [94C2]. The zero field specific heat of $LuNi_2B_2C$ at $T < T_s$, was described by the relation $C_p(T,0) = \gamma_{imp}T + \beta T^3 + C_s(T,0)$, where $\gamma_{imp}T$ term is due to an impurity contribution and C_s is the specific heat in the superconducting state [01S1]. The $C_s(T,0)$ was shown to follow an exponential temperature dependence $C_s(T,0) = \delta\exp(-\alpha T_s/T)$ with $\delta = 6(3)$ J/molK and $\alpha = 1.65(12)$. The above data were correlated with a behavior characteristic to a strong coupling supraconductivity and a weakly anisotropic s-wave energy gap. The electronic heat capacity, C_{es}, of YNi_2B_2C single crystal, in the superconducting state, was described by a relation $C_{es} \propto (T/T_s)^3$ [03S2], indicating the presence of a gap anisotropy with point nodes, as already showed [00N3, 01I1, 02I1]. A value $\Delta(0)/k_BT_s = 1.82$ and a parabolic temperature dependence of thermodynamic critical field H_c were reported [03S2].

The heat capacity of the YNi_2B_2C, in the field range $H_{c1} < H < H_{c2}$, at T = 2.5 K, follows a $H^{1/2}$ law [00N3, 01I2, 03P1]. This dependence persists even after the introduction of columnar defects which change the electronic structure of the vortex core regime and destroy the regular vortex lattice [01I2]. The Sommerfield coefficient, determined in $LuNi_2B_2C$ single crystal, follows a H^{ε} dependence, on external field, with $\varepsilon = 0.63(12)$ [01S1], close to $\varepsilon = 0.5$, as before mentioned [01I2].

The field directional dependence of the heat capacity in YNi_2B_2C single crystal, at T = 2 K, showed the presence of fourfold oscillations [03P1]. The angular variation, together with $H^{1/2}$ dependence of heat capacity, at T = 2.5 K, suggested that the gap function is anisotropic. The experimental data were in agreement with a model in which the fourfold pattern arises from Doppler-enhanced, fully 3D nodal quasiparticles with momenta in [100] directions [03P1].

The thermal conductivity of RNi_2B_2C borocarbides with R = Lu [96S3, 97R1, 01B2, 02C1, 02M1, 03C1], R = Y [96S3, 97R1, 02C1, 02I1, 02M1, 06M1, 08S1, 09A1], as well as of the thermoelectric power of borocarbides with R = Y [95F1, 97R1, 02I1, 08S1] and R = Lu [97R1] give also valuable information on their physical properties. The thermal conductivity of $LuNi_2B_2C$ has been investigated in the temperature range $0.07\ K \leq T \leq 0.30\ K$, and fields $0 \leq \mu_0 H \leq \mu_0 H_{c2} = 7\ T$, oriented parallel to the **c**-axis and heat current along **a**-axis [01B2]. As soon as vortices enter the sample, the thermal conduction, at $T \to 0$, grows rapidly, showing that the delocalized quasiparticles are present at the lowest energies. The quasiparticle transport grows, as a function of magnetic field, in the same way as in an unconventional superconductors that have a line of nodes in the gap and not at all like s-wave superconductors. It was concluded that the gap function must have nodes in the gap, or at least deep minima [01B2]. The angular variation of **c**-axis thermal conductivity in YNi_2B_2C, at T = 0.43 K, in a field $\mu_0 H = 1.0$ T, rotating within the (ab) plane, shows fourfold oscillations with narrow cusps [02I1]. The amplitude of the fourfold oscillation becomes very small, when **H** is rotated conically around the **c**-axis with a tilt angle of 45°. These data were also correlated with a gap function having point nodes, located along **a**- and **b**-axes. The semiclassical (Doppler shift) approximation was used to calculate magnetic field angle-dependent DOS and thermal conductivity, κ_{zz}, for YNi_2B_2C superconductor, with a quasi-two-dimensional Fermi surface and line nodes along $q_x = 0$ and $q_y = 0$ [09A1]. A quantitative agreement with the experimental data [02I1] was shown. The thermal conductivities of YNi_2B_2C, exhibits a significant anisotropy between the basal plane and **c**-axis by a factor $\cong 2$ [08S1]. For the in plane thermal conductivity, a kink near the superconducting transition was observed, consistent with an anisotropic gap or a multiband description [08S1].

The thermoelectric power, S, in RNi_2B_2C with R = Y, Lu is negative in the normal state indicating an electron character of the charge and just above the superconducting transition temperature, it rapidly drops to zero [95F1, 97R1, 02I1, 08S1]. The thermoelectric power exhibits a relative small anisotropy and can be well described by electron-diffusion and phonon-drag contributions over a wide temperature range. The different S values, reported by various authors, were correlated with samples quality [08S1]. The thermoelectric power of the Pauli-type paramagnet $LaNi_2B_2C$ is also negative, but has a factor of two larger electron-diffusion contributions, comparatively to that of Y-borocarbide [95F1]. Selected Debye temperatures are given in Table 9.12.

A large number of studies were devoted to the analysis of the resistive properties of RNi_2B_2C compounds with R = Y [94N1, 95F1, 95L1, 96B5, 96E1, 96F2, 96K1, 96L3, 96M2, 96R2, 97F2, 97R1, 00C6, 00S4, 05M1, 13K1] or R = Lu [94S1, 96E1, 96F2, 97F2, 97N2, 97R1, 98N1]. In normal state, at RT, these borocarbides are metallic with nearly isotropic resistivity [96R2, 97E1]. The carriers are electrons, as indicated by the Hall efect [99N3] and thermopower [97R1] measurements.

The magnetoresistance (MR) of $LuNi_2B_2C$, at $T_s < T < 40$ K, is small and positive and increases as temperature decreases [98N1]. The in plane magnetoresistance of

Table 9.12 Debye (θ_D) and Enstein (θ_E) temperatures

Compound	Method	θ_D (θ_E) (K)	References
$LaNi_2B_2C$	Specific heat; 1.5 K ≤ T ≤ 100 K	495(8)	[95M1]
$GdNi_2B_2C$	Thermal conductivity	355–360	[02H1]
$TbNi_2B_2C$	Thermal conductivity	354–359	[02H1]
$DyNi_2B_2C$	Thermal conductivity	352–357	[02H1]
$HoNi_2B_2C$	Thermal conductivity	350–356	[02H1]
$ErNi_2B_2C$	Thermal conductivity	349–354	[02H1]
$TmNi_2B_2C$	Thermal conductivity	348–353	[02H1]
$TmNi_2B_2C$	Specific heat; 1.5 K ≤ T ≤ 400 K	320	[94M5]
$LuNi_2B_2C$	Specific heat; 1 K ≤ T < 20 K	350(10)	[94K1]
$LuNi_2B_2C$	Specific heat; 2 K ≤ T < 400 K	345(10)	[94C2]
$LuNi_2B_2C$	Specific heat; 1.5 K ≤ T ≤ 100 K	360(3)	[95M1]
YNi_2B_2C	Specific heat; 2 K ≤ T < 300 K	537	[94H3]
YNi_2B_2C	EXAFS; 80 K ≤ T ≤ 300 K Specific heat 1 K ≤ T ≤ 20 K	350(10)[a]	[94K1, 01M2]
YNi_2B_2C	Specific heat; 1.5 K ≤ T ≤ 100 K	490(5)	[95M1]
YNi_2B_2C	Specific heat; 5 K ≤ T ≤ 22 K	415	[95G2]
YNi_2B_2C	Specific heat; 1.5 K ≤ T ≤ 400 K	489(5)	[94M5]
YNi_2B_2C	Specific heat; 2 K ≤ T ≤ 300 K	495	[03S2]
YNi_2B_2C	RVS; T = 300 K	525(10)	[03R4]
YNi_2B_2C	Pressure study	549.7	[12W1]
YNi_2B_2C	Computed	535.15	[17L1]
YNi_2B_2C	EXAFS; 80 K ≤ T ≤ 300 K	350[a]	[01M2]
$Y_{0.99}Yb_{0.01}Ni_2B_2C$	EXAFS; 80 K ≤ T ≤ 300 K	310[a]	[01M2]
$Y_{0.975}Yb_{0.025}Ni_2B_2C$	EXAFS; 80 K ≤ T ≤ 300 K	300[a]	[01M2]
$Y_{0.95}Yb_{0.05}Ni_2B_2C$	EXAFS; 80 K ≤ T ≤ 300 K	340[a]	[01M2]
$Y_{0.90}Yb_{0.1}Ni_2B_2C$	EXAFS; 80 K ≤ T ≤ 300 K	370[a]	[01M2]

[a] θ_E value

YNi_2B_2C changes from negative to positive, as the temperature decreases [00C6]. The sign reversal occurs at T = 80 K, when $\mu_0H = 4.0$ T. A negative MR supports the presence of magnetically related scattering and could result from suppression, by external field, of spin fluctuations often associated with nearly magnetic nickel.

The RNi_2B_2C borocarbides were also investigated by De Haas-Van Alphen method (R = Lu) [09B2, 09B3, 12B1], Hall effect (R = Lu, Y) [99N3], Raman scattering (R = Lu, Y) [95L6, 00Y1], photoemission spectroscopy (R = Y) [94G2, 00Y4], X-ray absorption spectroscopy (R = Lu) [01M1] or soft X-ray emission (R = Lu, Y) [95S2] spectroscopy, positron lifetime experiments (R = Y) [96S7], ultrasound attenuation (R = Y) [04W1] and microwave (R = Y) [95J1].

The magnetic and superconducting properties of large number of Y- and Lu-based pseudo-quaternary borocarbides were investigated. A short survey concerning the properties of series with paramagnetic (La, Ce), magnetic ordered (Pr, Nd, Sm, Gd, Tb), as well as with the heavy fermion (Yb) borocarbides will be only made. These includes $Lu_xR_{1-x}Ni_2B_2C$ with R = La [05S2], R = Gd [96C5, 97C2, 98P1, 00M2, 03B4, 03R1, 10B3], R = Tb [98G3, 00M2], R = Dy [06I1, 07L1], R = Ho [98F1, 98K3], R = Er [04T1], R = Yb [97B2, 98G1], R = Y [98F2, 01E2, 02Z1, 03R1, 04K1]; $Y_xR_{1-x}Ni_2B_2C$ with R = La [99S3, 00S3], R = Ce [95A2, 96L3, 97L4], R = Pr [99N4, 00F1, 00N1, 01N1, 08M1], R = Nd [99C3, 00F1], R = Sm [94T2, 00F1, 11C1], R = Gd [94E4, 95E3, 98L1, 98P1, 99F1, 00E2, 00F1, 00M2, 03D1, 10B3], R = Tb [98B2, 98L1, 99F1, 99J1, 00E2, 00F1, 00M2, 00Y2, 01C3], R = Dy [98L1, 99F1], R = Ho [96C4, 96E3, 96M5, 97C2, 97L3, 97M3, 98L1, 04E1], R = Er [94M4, 96E2, 98L1, 99F1], R = Yb [97H1, 01M2]. The substitutions at Ni sites have been also made as $Lu(Ni_{1-x}Co_x)_2B_2C$ [97S4, 98C4, 99G2, 00E3]; $Y(Ni,Fe)_2B_2C$ [96Z2, 97Z2], $Y(Ni,Co)_2B_2C$ [94S3, 95G3, 96G1, 96H1, 96Z2], $Y(Ni,Cu)_2B_2C$ [96G1], $Y(Ni,Pt)_2B_2C$ [00N3, 09M2], $Y(Ni,Ru)_2B_2C$ [96Z2], $Y_xLu_{1-x}(Ni,Pt)_2B_2C$ [01D1], $Y_xTb_{1-x}(Ni,Pt)_2B_2C$ [01D1].

The superconducting transition temperatures decrease in $Ce_xY_{1-x}Ni_2B_2C$ system, as the cerium content increases up to x = 0.4. No superconducting behavior is shown for x > 0.4 [97L4]. At $T > T_s$, the borocarbides are Pauli paramagnets. The average cerium valence increases, with x, at T = 300 K, while the reverse trend is observed at T = 10 K [95A2].

No drasting changes in the CEF level scheme, as compared with that of $PrNi_2B_2C$, was shown in $Pr_xY_{1-x}Ni_2B_2C$ series [08M1]. The superconducting transition temperatures decrease linearly when increasing Pr content. The samples with $x \geq 0.35$ do not exhibit superconductivity down to T = 2 K, as also in $Nd_xY_{1-x}Ni_2B_2C$ series for $x \geq 0.4$. The linear diminution of superconducting transition temperature, in $Sm_xY_{1-x}Ni_2B_2C$ series, when Sm content increases up to x = 0.4, has been attributed to structural deformation [11C1].

The magnetic phase diagram of $Gd_xR_{1-x}Ni_2B_2C$, with R = Lu, Y, evidenced that the long range magnetic order, observed at end series compound (x = 1), evolves, for $x \leq 0.3$ into a spin glass state (SG) which coexist with supraconductivity at lower temperatures [10B3]. For both series, T_s is suppressed by 30–35 at.% Gd addition. The evolution with composition, of the superconducting properties has been analysed within Abrikosov-Gorkov theory of magnetic impurities in superconductors [10B3].

The magnetic and superconductivity phase diagram, of $Tb_xY_{1-x}Ni_2B_2C$ series, evidences the presence of a steep slope of T_s, when increasing the Tb content from x = 0 to x = 0.4; around the last composition the superconductivity disappears [98B2]. The AFM transition temperatures decrease from $T_N = 15.4$ K (x = 1) down to $T_N = 5.5$ K (x = 0.45). Beyond that value, the antiferromagnetic behavior was suppressed and T_N could not be determined. The temperature T_{WF}, where the weak ferromagnetism occurs, also decreases up to x = 0.75, when disappears.

The Kondo temperature $T_K = 10$ K, characteristic for $YbNi_2B_2C$ borocarbide, seems to be relatively invariant across the $Yb_xLu_{1-x}Ni_2B_2C$ series [97B2]. A monotonous change from superconducting regime, to single-impurity Kondo regime and finally to a Kondo lattice-type behavior was shown, when the ytterbium content increases. The superconductivity is suppressed for $x > 0.025$ and local moment occurs at $x = 0.05$ [01M2].

The composition dependence of the superconducting transition temperatures in $R_xLu_{1-x}Ni_2B_2C$, with R = La, Y, were correlated with the structural parameter *c'/a* [05S2]. According to these authors, along the series, there is a deviation from the ideal tetrahedral symmetry, of NiB_4 tetrahedra, which determine the T_s reduction.

The superconducting transition temperatures, in $YNi_{2-x}Pt_xB_2C$ single crystals, monotonically decrease from $T_s = 15.85$ K and saturate to a value $T_s \cong 13$ K, for $x \geq 0.14$ [09M2]. The upper critical fields H_{c2}, determined both $\parallel$ **c** and $\perp$ **c** directions decrease when increasing Pt content. For large doping ($x \geq 0.14$), the borocarbides of this series change from a multiband superconductor and behaves as an effective single band superconductor, due to large interband scattering.

A decrease of the superconducting transition temperature, T_s, was evidenced in $Lu(Ni_{1-x}Co_x)_2B_2C$ system, as the cobalt content rises [97S4]. At $x = 0.15$, a value $T_s = 4.8$ K was reported. The observed trend was attributed to a decrease of DOS, at E_F, accompanied by a transition from clean to dirty limit type behavior. The experimental data were analysed in a model which includes nonlocal corrections to London model, due to Fermi surface anisotropy [99G2, 00E3]. As the doping increases, the mean free path, *l*, decreases, the coherence length, $\xi(0)$ grows and the transition from square to hexagonal FLL structural phase transition moves to higher fields.

9.4.6 Magnetic Ordered and Superconducting RNi_2B_2C Borocarbides with R = Dy, Ho, Er and Tm

Four of the RNi_2B_2C borocarbides (R = Dy, Ho, Er, Tm) exhibit coexistence of magnetic order and superconductivity. The exchange interactions between the localized R4f magnetic moments are mediated by conduction electrons, as described by the RKKY model. Their magnetic ordering temperatures, T_N, are well scaled with De Gennes factor, $G = (g_J - 1)^2 J(J + 1)$, with a rate $dT_N/dG \cong 0.9$ K/G. The near linear dependence of the superconducting transition temperatures, T_s, with G, with a rate $dT_s/dG = -1.46$ K/G is consistent with predictions of the Abrikosov-Gorkov (AG) theory [61A1] of pair breaking effect, by magnetic impurities [98C1]. The R magnetic moments perturb the singlet Cooper pairs formed between conduction electrons, influencing their superconducting properties. The ratio T_s/T_N, in the above RNi_2B_2C borocarbides, decreases from 7.2 (R = Tm), to 1.8 (R = Er), 1.6 (R = Ho) and finally to 0.6 (R = Dy). The $DyNi_2B_2C$, is the unique member of the RNi_2B_2C series with $T_N > T_s$. There seems to be a crossover between T_s and T_N, located in the intermediate composition of the pseudoternary $Dy_xHo_{1-x}Ni_2B_2C$ system [96C5, 98C1, 04P2]. The RNi_2B_2C (R = Dy, Ho, Er, Tm) borocarbides have highly anisotropic magnetizations and anisotropy and suppressed superconducting properties, which indicate the existence of significant interaction between the local magnetic moments and superconducting electrons [94C1, 95C4–95C6, 96C6]. Their crystal structures and lattice parameters are listed in Table 9.6. The bulk modulus of RNi_2B_2C with R = Dy and Ho is given in Table 9.7.

$DyNi_2B_2C$ borocarbide

The magnetic structure, of $DyNi_2B_2C$, has been determined by neutron diffraction [95D2, 96L5, 97L6, 97S5, 98G1, 99S3]. Bellow $T_N \cong 10.6$ K, the borocarbide orders in a commensurate antiferromagnetic structure with a wave vector **q** = (0,0,1). The Dy magnetic moments, located in (ab) plane, are ferromagnetically aligned and antiferromagnetically disposed along **c**-axis—Fig. 9.3a. The easy axis of magnetization is along [110] direction [96L4, 96C7]. At T = 1.7 K, a value $M_{Dy} = 8.47(9)\ \mu_B$/atom was determined [97L4]–Table 9.8. The superconducting transition temperature is $T_s \cong 6.2$ K [95C4, 95L5, 95T2, 95T4, 96H3].

The magnetic properties of $DyNi_2B_2C$, both polycrystallines and single crystals, were intensively investigated [94E1, 95C4, 95D2, 95H4, 95L5, 95T2, 95T4, 96K8, 96N1, 96S2, 96T3, 97C1, 97G1, 97L3, 97S5, 98C3, 98N2, 98P3, 99K3, 99S2, 99W1, 00K2, 01B1, 01R1, 01W1, 02H3, 03R3, 05S2, 05T1, 07J1, 09S3, 11L1, 12L1, 12L3, 15M1, 15S1]. The magnetic ordering temperature, of $DyNi_2B_2C$, in the greatest part of reports, has been located at $T_N \cong 10$ K. Some authors suggested the presence of an additional magnetic transition at $T^*_M = 16.3$ K [01K2]. The electron–quasiparticle interaction spectral function, of $DyNi_2B_2C$, in normal state, has been measured by Point Contact Spectroscopy (PCS) [00Y3]. A low frequency peak was found around E $\cong 5$ meV. It becomes measurable at $T^*_M \cong 15$ K and grows in intensity, as temperature is lowered. Similar behavior was shown in $HoNi_2B_2C$. The low frequency peak do not changes, at T^*_M, suggesting that its origin is not due to the spin-density wave excitations (magnons). The peak arises from the strong interaction of conduction electrons with coupled crystal electric field-phonon excitations, whose branches cross at low energy. The peak is visible by PCS, whereas the inelastic neutron magnetic scattering does not feel them. Latter on [03R3], a careful analysis of the data, obtained by magnetic measurements, resistivity and specific heat, confirmed that no magnetic phase transition is present at T$\cong$ 16 K.

The $DyNi_2B_2C$ borocarbide has an extreme magnetic anisotropy, associated with the crystalline electric field (CEF) splitting of the Hund's rule ground state J-multiplet. There is a strong in-plane anisotropy with moments confined in the [110] direction [95D2, 97C1, 97G1, 98G1]. This behavior can be observed even at T $\cong$ 100 K, well above the magnetic ordering temperature. Field induced magnetic transitions were evidenced in polycrystalline samples, the saturation being obtained in magnetic fields $\mu_0H > 6$ T. At T = 2 K, a value $M_{Dy} = 8.9\ \mu_B$ was obtained [95L5, 97C1]. The evolution with external field of the magnetization isotherms [95L5, 96H3, 96L4, 96K8, 96T3, 97C1, 97L2, 98N2, 99S2, 99W2], including their angular dependence in (ab) plane, have been studied in $DyNi_2B_2C$ single crystal [97C1, 99S2]. Depending on the angle between the applied field and the [110] axis, three transitions were observed at T = 2 K [97C1, 99S2, 99W2]. When the field was applied along [110] direction ($\theta = 0°$), a metamagnetic transition with a quarter of the saturation Dy moment, $M_s/4$, takes place in the field range $1.0\ T \le \mu_0H \le 1.25$ T. At higher fields, $\mu_0H \geq 1.20$ T, the borocarbide is ferromagnetically ordered. In the [100] direction ($\theta = 45°$), and $\mu_0H \geq 1.25$ T, the magnetization is $M_s/\sqrt{2}$ [97C1]—Fig. 9.8a. The corresponding magnetic structure is characteristic for an angular range $18° < \theta \leq 45°$. A spin structure for these phases was proposed [99W2]. The phase diagram was extended for **H** || [100], by using resistivity measurements [98P3]. Large hysteresis effects, for $T \leq 1$ K, including a re-entrant effect of superconductivity was shown. A model to describe the magnetic phase diagram in [110] direction with two domains state was also proposed [99S2]. Two domains can exist for the collinear AFM structure of $DyNi_2B_2C$, due to tetragonal symmetry of the (ab) plane, with moments along the [110] and [$\overline{1}\overline{1}$ 0] direction (||-*domains*), or along the [1 $\overline{1}$ 0] and [$\overline{1}$ 10] directions ($\perp$-*domains*), These two domains behave differently with respect to an applied field **H** || [110]—Fig. 9.8b [99S2]. On increasing H, the moments in the ||-*domains* undergoes discrete spin flips, whereas the moments in $\perp$-*domains* are turned continuously in the field direction. For a critical field all perpendicular moments turn completely into the field direction together with the remaining "down" moments of the other domain. Four magnetically ordered states were proposed: AF, AF1, AF2 and FM. The transition SC + AF/AF1 coincides with vanishing of the superconductivity, which implies that ferromagnetic component of the phase AF1 is responsible for pair breaking effect. In the H vs T phase diagram, for [100] direction, only one intermediate phase was shown [97C1], unlike that in [110] direction, where the AF1 and AF2 intermediate phases are present.

The thermal variations of magnetizations in polycrystalline $DyNi_2B_2C$ borocarbide as well as in single crystal were analysed [94E1, 95D2, 96K8, 96L2, 97T1, 98C3]. The low field magnetizations of $DyNi_2B_2C$ single crystal show a transition to a superconducting state, at $T_s = 6.5$ K. For **H** || **c**, the ZFC magnetization in field of 10^{-3} T corresponds to 100% for the shielding fraction [96T3]. When the crystal was FC, the magnetization it is positive. This anomalous paramagnetic Meissner signal was observed for fields $\mu_0H < 0.004$ T, above which becomes negative. For **H** $\perp$ **c**,

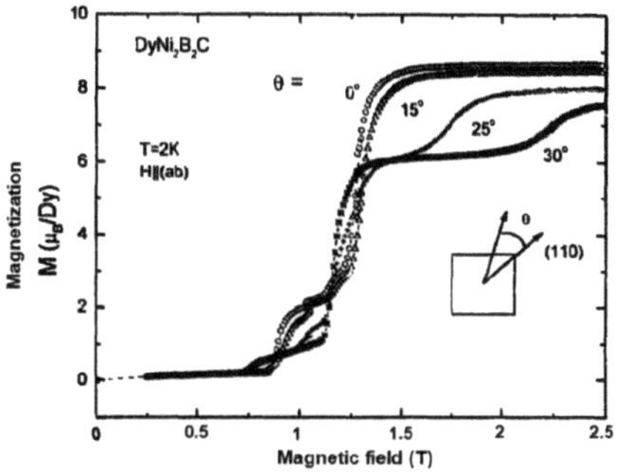

(a)

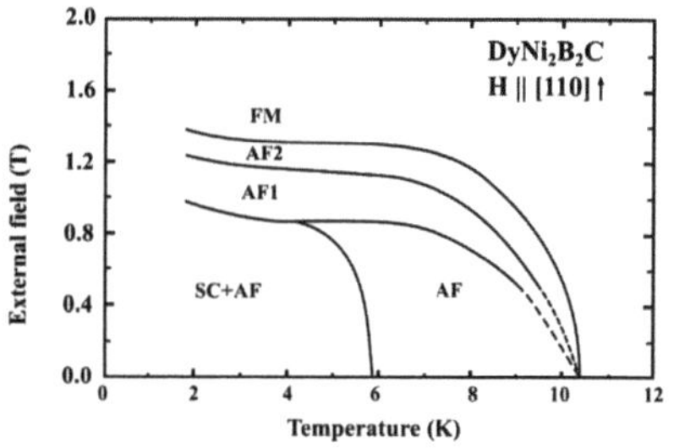

(b)

Fig. 9.8 $DyNi_2B_2C$ single crystal: **a** magnetization isotherms as T = 2 K determined at different angle, θ, as compared to [110] axis [97C1]. **b** Field-temperature phase diagram for **H** || [110], at increasing magnetic field [99S2];

the magnetization gives a smaller fraction for shielding (80%) and Meissner effect (20%), due to a strong paramagnetic contribution from the Dy moment [96T3]. The magnetic susceptibilities, at $T_N = 10.5$ K, for **H** ⊥ **c**, has a peak, whereas for **H** ∥ **c**, it is weak and has a broad maximum at T ≅ 100 K. The anisotropy and broad maximum in susceptibility may be a manifestation of CEF, which persists up to T = 300 K. The crystal field parameters of $DyNi_2B_2C$, are given in Table 9.9.

The ^{161}Dy Mössbauer spectroscopy study on $DyNi_2B_2C$, indicates that the magnetic transition, at T_N, is of first order and the Dy moment is confined in the basal plane [96S2]. The ground state crystal field doublet is composed primarily of the $|\pm1/2>$ component of the J = 15/2 free ion term. Higher-order crystal field terms should be included in the exchange-crystal field Hamiltonian, for a detailed account of the experimental results. The value of the isomer shift is close to those in non-metallic Dy systems, suggesting that the DyC layer basically is non-conducting and thus the superconductivity arises from NiB layer. The ^{11}B NMR study, on $DyNi_2B_2C$ single crystal, showed that the Knight shift is very large and highly anisotropic and proportional to the magnetic susceptibilities [16K2]. This indicates that the hyperfine field at B site originates from the Dy4f magnetic moment. At $T > T_N$, the spin–lattice ($1/T_1$) and spin–spin ($1/T_2$) relaxation rates are very large, while below T_N these were supressed significantly because of the slowing of the 4f spin fluctuations, confirming the hugue change in Dy4f spin dynamics across the AFM transition.

The crystal field (CF) excitations and spin-phonon interactions, in $DyNi_2B_2C$, were investigated by polarized Raman scattering [10R1]. There are a Ni-B_{1g} phonon mode, at 199 cm^{-1} and additional Raman modes, at 124, 151 and 221 cm^{-1}. The 124 cm^{-1} mode exhibits mostly B_{1g} scattering symmetry and was correlated with the excitation to a CF transition. The 151 and 221 cm^{-1} modes correspond to zone-folded phonons. The 119 cm^{-1} mode, observed at higher temperatures ($T > T_N$), corresponds to an excited CF transition from a low lying CF level, at 5 cm^{-1}, to higher CF level at 124 cm^{-1}. The behavior of Ni-B_{1g} mode, at 199 cm^{-1}, in the vicinity of T_N indicates the presence of strong spin-phonon interactions.

A normal giant/large magnetocaloric effect was observed under high magnetic field change, in $DyNi_2B_2C$, related to a field-induced first-order metamagnetic transition [11L1]. For a field change of $\mu_0H = 5$ T, the maximum value of the magnetic entropy change, reached a value of −17.6 J/kgK, corresponding to a maximum adiabatic temperature change of 9.7 K.

The magnetic properties of $DyNi_2B_2C$ nanoparticles were studied also by using a quantum simulation model [12L3].

A large number of experimental data and theoretical analyses were reported on the correlation between the superconducting and magnetic properties of $DyNi_2B_2C$ [94C1, 94C4, 94E1, 94E2, 95C4, 95E2, 95H4, 95L5, 95T2, 95T4, 96H3, 96S1, 96T3, 98P3, 99W1, 99W2, 00F1, 00K2, 01B1, 01R1, 01W1, 02H1, 02H3, 03J1, 03N2, 05S2, 06B5, 06G1, 06S2, 07J1, 07N4, 09N2, 09S3, 15S1]. The $DyNi_2B_2C$ borocarbide has a superconducting transition temperature $T_s \cong 6.5$ K, provided that the stoichiometry is maintained close to nominal composition [95T2, 96H3]. Excess of C or Dy, as compared to 1/2/2/1 stoichiometry, deteriorates the superconducting properties [95T2]. The temperature dependences of the in-plane resistivity, ρ_{ab}, of

RNi_2B_2C single crystals, with R = Dy, Ho, Er, Tm show linear variations at 100 K < T < 300 K [97B1]. The departure from the above behavior, becomes significant at T < 50 K. The low temperatures resistivities follow a T^p dependence with $p = 3.0$ (Dy), 2.6 (Ho), 2.0 (Er) and 1.4 (Tm) [97B1]; a T^3 dependence is consistent with phonon scattering. The decrease of the resistivity, at T_N, was attributed to the diminution of the electron scattering, by the disordered spin structure at $T > T_N$.

The upper critical fields, H_{c2}, measured with magnetic fields parallel and perpendicular to the **c**-axis, in earlier study, show little or no anisotropy [95T4]. Latter on, the temperature dependence of the upper critical field $H_{c2}(T)$ was determined, from **c**-axis resistivity, $\rho_c(H,T)$, with applied field $\mathbf{H} \perp \mathbf{c}$, [14L1] and compared with the values obtained from $\rho_{ab}(H,T)$ with $\mathbf{H} \parallel \mathbf{c}$ [95C4]. It was concluded that the anisotropy and its temperature dependence might be originated from that of anisotropic magnetic Dy^{3+} sublattice [14L1].

The temperature dependence of resistivity in $DyNi_2B_2C$ single crystal shows a drop at the PM to AFM phase transition (T_N = 10.3 K), as well as at the superconducting transition-temperature—Fig. 9.9a. In the temperatures range 0.05 K ≤

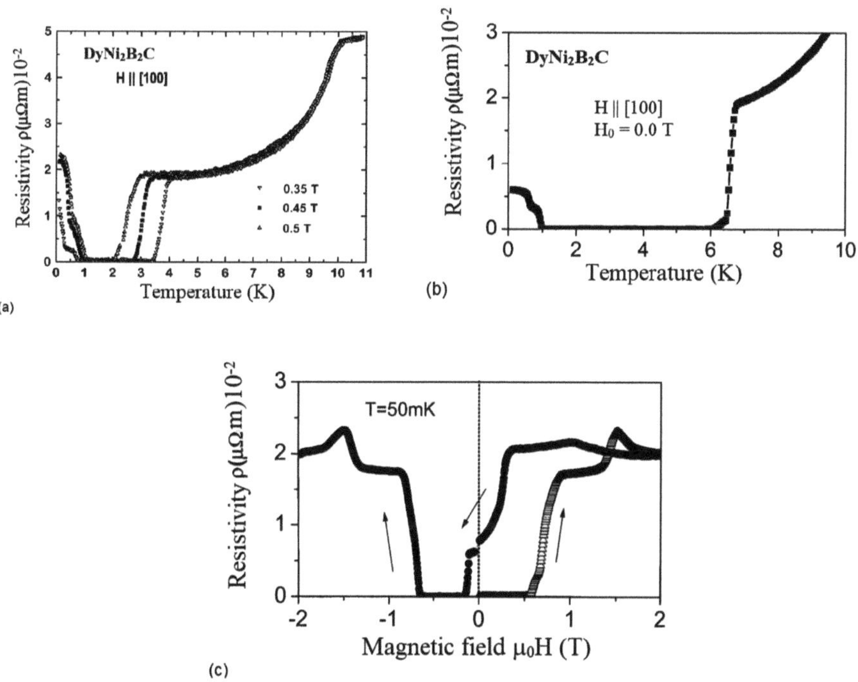

Fig. 9.9 $DyNi_2B_2C$ single crystal: **a** temperature dependence of the resistivity, after the field $\mu_0\mathbf{H} \parallel [100]$, increases from 0 to 2 T, at T = 0.05 K and then lowered to final values for $\mu_0 H_s$ = 0.35, 0.45 and 0.5 T [98P3]; (**b** and **c**) after a field sweep up to 2 T and down to zero at T = 0.05 K (**b**) temperature dependence of the resistivity, when increasing temperature and (**c**) field dependence of resistive state, ρ (H = 0) and superconductivity after field reversal [01W1]

$T \leq 2$ K, large hysteresis effects were observed in the field dependence of the resistivity, when **H** || [100] and also for superconducting transition temperatures. A re-entrant behavior was also evidenced [98P3]—Fig. 9.9a,b. The hyseteresis effect is anisotropic in (ab) plane with a maximum value for **H** || [100] and almost zero for **H** || [110]. At low temperatures the ρ(H) curves, for **H** || [100], show up to three step-like changes which belong to the superconducting or metamagnetic transitions, respectively [99W1, 99W2, 01W1]—Fig. 9.9c [01W1]. In some extreme cases, the metamagnetic state can be frozen after a cycle of field sweeping, up to 2.0 T, down to zero at T = 0.05 K. In this case, the superconductivity is suppressed. The superconducting state can be restored, either with increasing temperature or by applying a small magnetic field in opposite direction. With increasing temperature, the sample undergoes a superconducting transition at $T_{s1} \cong 1$ K and reentres into the normal state at $T_s \cong 6.5$ K—Fig. 9.9b. By applying a small magnetic field, $\mu_0H \cong 0.1$ T, having opposite direction to the magnetization, at T = 0.05 K, the frozen metamagnetic state and the corresponding resistive state are destroyed and the superconductivity occurs, in the field range $0.1\ T < \mu_0H < 0.7\ T$—Fig. 9.9c. The re-entrant behavior and the large hysteresis indicate a strong interplay between magnetism and superconductivity.

The pressure effects on the electrical resistivity of $DyNi_2B_2C$, were analysed [96K7, 01F1, 02F1, 03J1]. The superconducting transition temperature was shown to decrease with a rate $dT_s/dp = -0.7$ K/GPa, up to p = 1.8 GPa, a drastically decrease being evidenced at higher pressures [01F1]—Table 9.10. The feature, related with the antiferromagnetic order, was shown that remain almost constant as effect of pressure. Latter on [03J1], confirmed that T_s values move slowly towards lower temperatures, as effect or pressure, ($p \leq 0.98$ GPa), whereas T_N increases simultaneously.

The temperature dependence of the specific heat of $DyNi_2B_2C$, evidenced peaks at T_N, but any jump at T_s [95L5, 95T4, 01R1]. The determined magnetic entropy S_m = 10.3 J/molK, is of 90% of the R*ln*4 and indicates AFM ordering of dysprosium moments [95L5]. The magnetic contributions to the specific heat were measured in the coexistence regime [01R1]. Assuming an anisotropic AFM ground state structure and using spin wave theory, an explicit expression for the energy gap was obtained. The point contact spectroscopy (PC) studies on $DyNi_2B_2C$ evidenced that the superconducting gap has a BCS like dependence [07N4, 09N2].

The thermal conductivity κ, and thermopower, S, in $DyNi_2B_2C$ borocarbide, is dominated by electrons and the high temperature thermal conductivity is approximately linear in temperature and anomalous [96N1, 97B1, 02C1, 02H1, 03N2]. The low temperature thermal conductivity exhibits a marked loss of scattering at the AFM ordering temperature and thermoelectric power exhibits an enhancement below T_N.

The ^{57}Fe Mössbauer spectroscopy has been used to study the RNi_2B_2C borocarbides, doped by 1 at.% ^{57}Fe [97Z2] or 0.5 at.% ^{57}Fe [00B1]. The main quadrupole doublets, ΔQ, in RNi_2B_2C series, were shown to have linear dependence on the c'/a ratio, where c' is the distance of RC layers between which the Ni_2B_2 layer is sandwiched [00B1]. These data suggest that the structural parameters are important for the occurrence of the superconductivity in borocarbides. Band structure calculations were made on RNi_2B_2C with R = Y, La, Pr-Tm and Lu [00D2]. The total energy is very sensitive to the c/a ratio. A model was proposed to study the coexistence of the superconductivity and the AFM order in RNi_2B_2C borocarbides with R = Dy, Ho, Er, Tm, by taking s-wave pairing into account [15S1].

The pseudoquaternary borocarbides, involving dysprosium were studied as $Dy_xLa_{1-x}Ni_2B_2C$ [05S2], $Dy_xPr_{1-x}Ni_2B_2C$ [99T1, 99T2, 04E2, 05T1], $Dy_xTb_{1-x}Ni_2B_2C$ [05C2], $Dy_xLu_{1-x}Ni_2B_2C$ [96C5, 00M2, 01R2, 05S2, 06I1, 07L1], $Dy_xY_{1-x}Ni_2B_2C$ [97L1, 98L1, 00E2, 00M2, 01D2, 02D2, 05S2, 08W2], $DyNi_{2-x}Cr_xB_2C$ [12L1], $DyNi_{2-x}Fe_xB_2C$ [12L2], $DyNi_{2-x}Co_xB_2C$ [97S4, 12L1], $DyNi_{2-x}Pt_xB_2C$ [09S3].

The substitution of Tb, in $Dy_{1-x}Tb_xNi_2B_2C$ system, leads to a linear decrease of the superconducting transition temperature, T_s, with a rate of −0.12 K/ at.% Dy and for x = 0.4, no superconductivity was observed, at T > 2 K [05C2]. The T_N values were almost constant, for $x \leq 0.6$. In the $Dy_xY_{1-x}Ni_2B_2C$ pseudoquaternary compounds, the T_s decreased with increasing Dy content, in the composition range $0 \leq x \leq 0.55$, with a rate of −0.182 K/ at.% Dy; at x = 0.55 a value $T_s = 4.4$ K has been reported [97L1]. The superconductivity was destroyed for x > 0.6. The magnetic ordering temperature, T_N, increases nearly linearly, from $T_s \cong 5$ K ($x \cong 0.6$), when dysprosium content rises. The XPS study, for samples with $x \geq 0.55$, showed that the valence band is predominantly determined by Dy4f states, whereas Ni3d states govern the valence band of yttrium rich compounds [08W2]. The specific heat jumps, ΔC, at T_s, in $Y_{1-x}Dy_xNi_2B_2C$ compounds follow a relation $\Delta C \propto T_s^2$ [00E2]. This behavior is evidenced also in $Y_{1-x}R_xNi_2B_2C$ system with R = Gd, Ho, Er. The superconducting transition temperatures, in $Dy_{1-x}Lu_xNi_2B_2C$, decrease rapidly with increasing x and the superconductivity disappears at $x \simeq 0.2$ and then increases with further increase of x [07L1]. The magnetic ordering temperatures also gradually decrease. The ^{57}Fe Mössbauer spectroscopy on $Dy_{1-x}R_xNi_{1.99}Fe_{0.01}B$ with R = Lu, Y, La and $x \leq 0.2$, evidenced the presence of the magnetic hyperfine field, H_{hf}, at the ^{57}Fe(Ni) site, for x = 0.2 [05S2]. The T_s reduction, as effect of substitution, was explained assuming that H_{hf} acts as a pair breaking field. The superconducting transition temperature T_s is supressed in $Dy(Ni_{1-x}Pt_x)_2B_2C$ with x = 0.03 down to T_s < 2 K [09S3]. The large magnetic entropy changes in $DyNi_{2-x}A_xB_2C$ with A = Co, Cr [12L1] or A = Fe [12L2], were related to a field induced first order metamagnetic transitions.

$HoNi_2B_2C$ borocarbide

The $HoNi_2B_2C$ borocarbide displays complex magnetic and superconducting properties and is a model system wherein magnetism, superconductivity and their interplay are manifested [95L4]. The borocarbide becomes superconducting at $T_s \cong 8.5$ K [94C1, 94E1], while developing different types of long range magnetic order at T < T_s. The low temperature real and imaginary parts of the magnetic susceptibility, $\chi(T) = \chi'(T) + i\chi''(T)$, evidence phase transitions, at superconducting transition temperature, T_s, accompanied by a diamagnetic real part signal. The two other transitions, were related with changes in the type of magnetic ordering—Fig. 9.10. The real part of *ac* signal indicates a nearly reentrant behavior, starting from the temperature, T_M, reaching a small positive peak at T_N, then dropping sharply back to diamagnetic signal, at T < T_N. From magnetic and resistive measurements, an upper critical

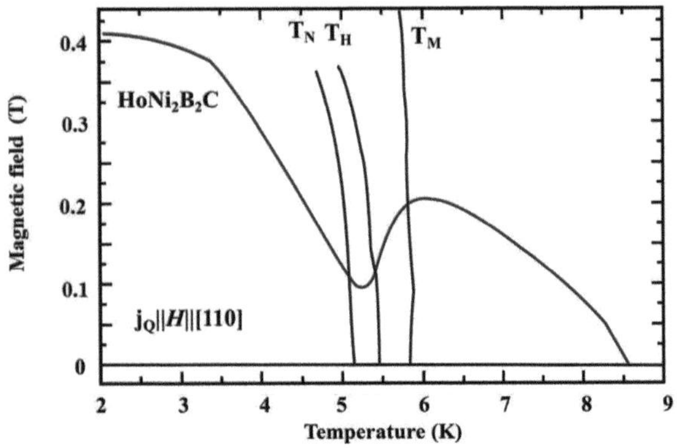

Fig. 9.10 $HoNi_2B_2C$: magnetic phase diagram evidencing field dependences of the characteristic temperatures, determined from magnetic measurements, resistive studies as well as thermal conductivity, with heat current and **H** || [110] [09S2]. The temperature dependence of the critical field H_{c2} is also plotted (green)

field, $\mu_0H_{c2}(0) = 0.35(5)$ T, a lower critical field, $\mu_0H_{c1}(0) = 0.025(10)$T, and a Ginzburg–Landau parameter $\kappa = 3.5$ were determined [95L4].

At low temperatures, magnetoelastic effects were observed and correlated with the existence of a commensurate **c**-axis modulated antiferromagnetic ordering [99K2]. The tetragonal lattice, at low temperatures, is distorted along the [110] direction, in which the long-range ordered Ho moments are aligned. The length of the unit cell, in [110] direction, is shortened by ≅ 0.19%, compared to its length in [$\overline{11}$ 0] direction.

The evolutions with temperature and magnetic field of the $HoNi_2B_2C$ magnetic structure have been investigated mainly by neutron diffraction [94G3, 94G4, 95G3, 95T3, 95V1, 96C4, 96H2, 96L5, 97L6, 97M3, 99K2, 99S3, 00A1, 00C2, 00D1, 06S1, 10A1, 14E1, 19S1]. At $T < T_s = 8.5$ K, three distinct anomalies, at temperatures $T_N = 5.2$ K, $T_H = 5.6$ K and $T_M = 6$ K were observed, connected with changes of the magnetic structure—Fig. 9.10. At $T < T_N$, the tetragonal $HoNi_2B_2C$ has commensurate magnetic structure with wave vector $q_c = (0, 0, 1)$. In this structure, the Ho moments are parallelly aligned in (ab) plane, along the [110] direction and alternate in an antiferromagnetically arrangement along the **c**-axis— Fig. 9.3a [94G3, 94G4, 06S1]. Above $T_N = 5.2$ K the commensurate structure transforms in two incommensurate configurations which coexist, an amplitude modulated with $\mathbf{q}_m = (0.585,0,0)$ that disappears at $T_M \cong 6$ K and a helical structure [94E1] with $\mathbf{q}_h = (0,0,0.915)$ that persists up to $T \cong 8$ K. It is to be noted that the $\mathbf{q}_m$ is the nesting vector of Fermi surface. Magnetic X-ray scattering combined with neutron scattering [96H2] revealed in the temperature range $5.2\text{ K} \le T \le 5.6$ K the existence of an incommensurate structure with $\mathbf{q}_{h1} = (0, 0, 0.905)$ [09S2, 10A1]—Fig. 9.10. The muon-spin rotation and relaxation studies on $HoNi_2B_2C$ borocarbide evidenced also the presence of magnetic transitions at $T_N \cong 5$ K and $T_M \cong 6$ K [95L3, 96L2]. The phase at $T < 5$ K was shown to be commensurate and that at higher temperatures incommensurately modulated. The interaction between magnetism and superconductivity in $HoNi_2B_2C$ manifests as a depression of the upper critical field, $H_{c2}(T)$ between 5.2 and 6 K, leading to a re-entrant or nearly re-entrant behavior in the thermal variation of the resistivity and magnetic susceptibility. The magnetic structure, having $\mathbf{q}_m = (0.585,0,0)$, at $T_N \le T \le T_M$, was assumed to be responsible for the suppression of H_{c2} [94C1, 96C4, 96K7, 97M3]. The neutron diffraction and the specific heat confirmed that the depression of $H_{c2}(T)$, at T = 6.1 K, arises from the $\mathbf{q}_m$ = (0.586,0,0) modulated magnetic structure. The second depression in $H_{c2}(T)$ below T = 5.7 K was ascribed to $\mathbf{q}_{h1}$ = (0, 0, 0.905) magnetic helical structure [10A1].

The magnetic structure of $HoNi_2B_2C$ was also directly determined using spherical neutron polarimetry [19S1]. The magnetic ordering at low temperatures was accompanied by a structural phase transition to orthorhombic lattice type, space group *Fmmm*. Only noncharecteristic orbits were occupied in this space group and the symmetry reduction was due to a lifting of the symmetry induced restrictions on the anisotropic displacemrnt parameters. Based on magnetic symmetry analysis, two types of commensurate AFM structures with different magnetic space groups can be found at $T < T_N$. These where P_Innm where Ho moments are along the [100] direction of the parent structure and C_Amca with Ho moments along [110] axis of the parent structure. Two types of 90° AFM domains are present in the structure having C_Amca space group. The critical exponent related to orderd phase was found to be anomalously small, indicating an effective lowering of dimensionality. The magnetic phase transition was found to be accompanied by the significant change of mosaicity which becomes strongly anisotropic at low temperatures.

A large number of studies were focused on the magnetic properties of $HoNi_2B_2C$ borocarbide [94C1, 94E1, 94G3, 95G3, 95L4, 95R2, 95S1, 95T3, 96C7, 96E3, 96G2, 96L4, 96M3, 96N1, 96R2, 96R3, 96S3, 96Z3, 97C2, 97C3, 98C3, 98K1, 00C2, 00D3, 01O1, 04P2, 07J1, 10A1, 14R1]. Some of basic magnetic parameters are listed in Table 9.8. The $HoNi_2B_2C$ borocarbide has a high magnetic anisotropy, associated with the crystalline electric field (CEF) splitting of the Hund's rule ground state J = 8 multiplet. The local moment is confined in the basal plane of the lattice, for temperatures roughly up to 100 K, well above the magnetic ordering temperature [97C1]. The CEF parameters set of Ho^{3+}, in $HoNi_2B_2C$, have been determined by inelastic neutron scattering [95G3, 96G2, 97L3, 01R2]. At low temperatures, there are two crystal field excitations from ground state to the first CF level, at 11.28(3) meV and from ground state to the second level at 16.00(2) meV [95G3]. A transition of 4.70(9) meV, between these two levels, was observed at elevated temperatures. Latter on [96G2], inelastic features were observed between 8 and 18 meV. At low temperatures, the peaks were found to be rather sharp, singly thermally occupied (low lying state), but at T = 15 K only broad transitions were observed, due to a larger number of low-lying states. The CEF splitting in the antiferromagnetically ordered state was calculated. Singlet–singlet transitions, at 11.2, 12.3, 12.8 and 16.6 meV, with transition probabilities 5.1, 0.8, 0.3, and 1.6, respectively have been obtained. These were in qualitative agreement with experimental data. The CEF parameters were also obtained by fitting single-crystal magnetization data [96C7]. Their set of CEF parameters cannot reproduce the spectroscopic data, but described well the paramagnetic properties—Table 9.9.

In addition to the anisotropy found between magnetizations, collected with the field applied parallel and perpendicular to the **c**-axis, a substantial anisotropy of in plane magnetization, was also shown, when the easy axis is the [110] direction. The in plane anisotropy of $HoNi_2B_2C$ borocarbide leads to a characteristic angular dependence of the metamagnetic transition, as evidenced at T = 2 K. Up to three metamagnetic transitions, were shown, whose angular dependences of the critical field, can be described by μ_0H_{t1} = 0.41 T/cos θ, μ_0H_{t2} = 0.84 T/cos (θ − 45) and μ_0H_{t3} = 0.66 T/sin (θ − 45). By θ is denoted the angle between the applied magnetic field in plane, with [110] easy direction [96C1, 97C1, 97C3]. The two lower field transitions are bounded, while the third transition diverges as θ → 45°. The locally, saturated magnetizations, between the transitions were explained, assuming the alignement of moments along the four crystallographically similar [110] axes [97C1, 97C3].

The ^{57}Fe Mössbauer spectra, of RNi_2B_2C with R = Tb, Dy, Ho, Er, doped with 1 at.% ^{57}Fe show the presence of magnetic hyperfine field at ^{57}Fe nucleus for R = Tb at $T < T_N < 15$ K and R = Ho, for 4.2 K < T < 5.5 K [96S1]. No hyperfine field was observed in borocarbide with R = Dy, Er. These data were correlated with their magnetic structures; the noncolinear AFM spin structure of the rare-earths acts as a pair-breaking field at the Ni site.

The analysis of the dependence of the heat capacity on temperature and in-plane field angle in $HoNi_2B_2C$ confirmed the presence of three magnetic transitions and the superconducting one [04P2]. The transition, at $T_N \cong 5.2$ K, shows hysteresis with temperature, indicating their first-order nature, at H = 0. The transition at $T_M \cong 6$ K, namely the onset of the long-range ordering, display a dramatic in-plane anisotropy. The T_M temperature increases with increasing magnetic field, for H || [$\bar{1}$ 00], while it decreases when increasing the field, for H || [110]. The anomalous anisotropy in T_M indicates that the transition is related to the **a**-axis structure. The transition at $T_H \cong$ 5.5 K, shows similar behavior to that at T_N, i.e. small in plane anisotropy and scaling with Ising model. The last transition is ascribed to the change from $\mathbf{q}_a$ dominant phase, to $\mathbf{q}_c$ dominant phase.

Theoretical models were elaborated in order to describe the metamagnetic transitions in $HoNi_2B_2C$ [98A1, 98K1, 00A1]. The relative simple mean field treatment ignored the **a**-axis modulation, so all the phases were FM coupled in basal plane and also ignoring any small **c**-axis component of the magnetic moment [98A1]. The model [98K1] incorporates couplings between non-nearest neighbours in the direction perpendicular to the ferromagnetic planes, The model, at a proper choice of the coupling constants, yields several metamagnetic phases, in the presence of magnetic field, at T = 0 K. The metamagnetic transitions in $HoNi_2B_2C$, are more complex [00D1] than described by the uniaxial models [98A1, 98K1]. The model [98A1] was latter extended to incorporate interactions with superconducting electrons [00A1]. The onset of the helical magnetic order depresses superconductivity via the reduction of the interactions between phonons and electrons, caused by the formation of magnetic Bloch states. At a mean field level, no additional suppression of superconductivity is introduced by the incommensurability of the helical phase. A generalized susceptibility, for RNi_2B_2C borocarbides, has been calculated using the normal state electronic structure [95R2].

The phase diagrams, of $HoNi_2B_2C$, for magnetic field applied parallel to the [100, 110] and [001] crystal directions were reported [05K1]. The occurrence of magnetic order, heavily suppresses the superconducting parameters, such as $H_{c2}(T)$

and the irreversibility line, IL. This was associated with increased pair breaking of the Copper pairs, due to metamagnetic behavior between $T \cong 5$ and 6 K and the disappearance of this effect below the transition temperature, T_N, into the antiferromagnetically ordered state [94G3]. The field and temperature dependent behavior resulting from SANS experiment, at $T < 5$ K, appeared to be consistent with the presence of metamagnetic transitions [14R1].

The neutron diffraction method was used to analyse the metamagnetic phases, as function of the external field and temperature ($T \leq 5.5$ K) [00C2]. At $T = 2$ K, when the field was applied along [110] direction, metamagnetic transitions were shown at $\mu_0H \cong 0.5$ T, and $\cong 1.2$ T. When the field was oriented along [010] direction, the metamagnetic transitions are located at 0.7 T and 0.96 T, respectively. Both transitions are of first order. For both orientations, the first transition is to a state in which the magnetic structure is modulated along **c**-axis, with wave vector $\mathbf{q_2} = [0, 0, 0.667]$. With the field applied parallel to the $[1\,\overline{1}\,0]$ direction, the second transition is to a saturated paramagnetic state, in which the moments are aligned ferromagnetically by applied magnetic field. With the field parallel to [010] direction, the second transition leads to a state, in which the magnetic structure is modulated along the **b**-axis, with wave vector $\mathbf{q_3} = (0, 0.610, 0)$.

The spin-polarized Density Functional embedded cluster studies, of magnetic RNi_2B_2C compounds, evidenced a magnetic coupling of R4f and 5d electrons, as described by the 4f–5d–3d exchange interaction model [72C1, 10B4]. This provides the pathway for long range interaction, of the 4f moment, with atoms at distance of several coordination shells [96Z1].

A highly anisotropic magnetic behavior of $HoNi_2B_2C$, at $T > T_N$, with a much larger magnetic susceptibilities, for $\mathbf{H} \perp \mathbf{c}$ than for $\mathbf{H} \parallel \mathbf{c}$ was shown [96C7, 98C3]. A broad maximum, in $\chi(T)$ occurs for $\mathbf{H} \parallel \mathbf{c}$, at $T = 8.6$ K. The χ^{-1} vs T, for $\mathbf{H} \parallel \mathbf{c}$, at $T < 100$ K, indicated that the magnetic state of Ho^{3+} ion changes. The main contributions to the anisotropy of magnetic susceptibility, comes from a CEF splitting of the ground $J = 8$ multiplet, of the Ho^{3+} ion, as already mentiond. The analysis of the experimental data, particularly on $Ho_{0.024}Lu_{0.976}Ni_2B_2C$, allowed to determine the crystal field parameters—Table 9.9. The CEF energy level scheme consists of two singlets and one doublet at $T < 10$ K, with next higher levels near $T = 100$ K [96C7]. The magnetization isotherms were well described starting from the determined CEF level scheme.

The resistive properties of $HoNi_2B_2C$ borocarbide and their pressure dependence were also investigated [94E1, 94T2, 95A1, 95E1, 95F1, 95J1, 95L4, 96F2, 96K3, 96N1, 96R2, 96U1, 97B1, 97F2, 98N1, 00D3, 01M3, 01O1, 03J1, 03R3, 07J1, 10A1, 10N1, 10O1, 11L1]. The temperature dependences, of in plane and **c**-axis resistivities, of RNi_2B_2C borocarbides with R = Ho, Er, evidenced that these are fundamental isotropic, but with small anisotropies developing at $T < 150$ K [97F2]. The low temperature deviations from an isotropic resistivity were associated with scattering from rare-earths moments, which might be anisotropic due to crystal field effects. A large anisotropy of the magnetoresistance (MR), depending on the direction of the magnetic field, was also seen [01O1]. When field is oriented perpendicular to the **c**-axis, at $T = 4.2$ K, the superconducting state is broken near $\mu_0H = 0.15$ T.

A discontinuous change occurs at $\mu_0H = 0.5$ T, followed by a small maximum at $\mu_0H \cong 1.0$. These fields correspond to those where metamagnetic transitions are observed. No such behaviour can be seen for **H** || **c**-axis, where only a discontinuous change, at $\mu_0H = 0.15$ T, can be shown.

The effect of pressure on the magnetic, T_N, and superconducting, T_s, transition temperatures, in $HoNi_2B_2C$ were investigated [94S2, 95A1, 96U1, 03J1, 10N1]. According to [96U1], in the pressure range 0 GPa $\leq p \leq$ 2.2 GPa, the superconducting transition temperature decreases, with a rate $dT_s/dp = -0.5$ K/GPa and antiferromagnetic transition temperature increases with $dT_N/dp = 0.7$ K/GPa. A value $dT_N/dp = 1.3$ K/GPa was also reported [01C6]. In the pressure range 0 GPa $\leq p \leq$ 1.65 GPa a value $dT_s/dp = -0.255$ K/GPa has been obtained [95A1]—Table 9.10. At ambient pressure and a current density $j = 5.6.10^4$ Am^{-2}, the resistivity shows sudden drop, at $T_s \cong 8.8$ K and becomes zero below 8.0 K. At $p = 2.0$ GPa, it drops below $T = 8.1$ K, becomes zero, at $T < 7.2$ K and shows a peak at $T \cong$ 5.9 K, due to AFM order [10N1]. Then, it becomes zero again, at $T < 5.7$ K. The superconductivity is affected strongly by changing current density, j. The T_s values, for $j = 5.6.10^4$ A/m^2, decrease with increasing pressure, but increased when a current density is close to the critical value, $j_c = 63.9.10^4$ A/m^2, with increasing pressure up to $p = 3$ GPa. The $H_{c2}(T)$ values decrease with increasing pressure and essentially disappears completely at $p > 0.83$ GPa [03J1].

The matter of the interplay between magnetism and superconductivity, in $HoNi_2B_2C$, has been investigated [94C1, 94E1, 94G3, 95H3, 95L4, 96K7, 96S1, 97M3, 97Y1, 97Y2, 99D2, 01D1, 03J1, 04K2, 05V3, 06S1, 07J1, 07N3, 09S2, 09V2, 10A1, 14C2]. As alredy discussed, the two most important effects are the Anderson-Suhl screening of the indirect exchange interactions and the reduction of the number of Cooper pair states by the superzone gaps created by antiferromagnetic ordering [07J1]. A microscopic theory of interplay between superconductivity and magnetism in $HoNi_2B_2C$, compatible with experimental data has been developed [14C2].

The vortex structure of $HoNi_2B_2C$, at $T < T_N$, studied by decoration technique, evidenced the presence of antiferromagnetic domains, induced by magnetoelastic interactions. According [99D2], the vortex lattice, in $HoNi_2B_2C$ is weakly pinned and dominated by surface barriers, at $T_M < T < T_s$ and $T < T_N$, while in the $T_N \leq T \leq T_M$ temperature range, there is evidence of bulk pinning. The bulk pinning only becomes important in the region where is present the incommensurate magnetic structure with $\mathbf{q}_m = (0.585, 0, 0)$, strong suppression of H_{c2}, respectively. It was also proposed, that at the enhanced pinning, in addition to pair breaking, contributes also a direct interaction between the vortex lattice and **a**-axis ordered magnetic states with $\mathbf{q}_m$ wave vector.

The pinning contribution caused by domain twin boundaries has been analysed [03V1, 05V2, 05V3, 09V2]. In $HoNi_2B_2C$ the tetragonal-orthorhombic distortion caused by magnetoelastic effects occur along [110] direction with the onset of incommensurate–commensurate transition and favours the formation of crystallographic domains with domain walls parallel to the **c**-axis and the [100]([010]) direction. In this direction was observed vortex bands/chains, at low temperatures [05V3]. At higher temperatures ($T < T_N$), but above incommensurate-commensurate transition, a regular triangular vortex lattice was observed.

The superconducting gap and electron–phonon interaction in $HoNi_2B_2C$ borocarbide were mainly investigated by point contact (PC) spectroscopy [96R4, 97S5, 97Y1, 97Y2, 04K2, 05N2, 07N2–07N4, 09S2]. The strong softening, shows that the electron–phonon interaction is quite strong [97S5]. The magnetism further enhances the electron–phonon interaction [97Y1]. The softening occurs at the wave vector close to that of the zone boundary point G_1. This value is close to that of incommensurate wave vector $\mathbf{q} = (0.55, 0, 0)$ and in quite good agreement with the value of the Fermi surface nesting wave vector, obtained from band structure calculations of the generalized electronic susceptibility [95R2]. A strong softening of two phonon branches, similar to that evidenced in others RNi_2B_2C borocarbides was also evidenced [04K2]. The ratio $2\Delta/k_BT_s = 3.7$, corresponds to a moderate electron–phonon coupling strength in $HoNi_2B_2C$ [96R4]. At $T \leq T_s = 8.5$ K, two different SC states, separated at $T_H = 5.6$ K, were reached [07N3]. Below $T_H = 5.6$ K, the superconducting gap, Δ, exhibits a standard single band BCS like behavior, with $2\Delta/k_BT = 3.9$. Above $T_H = 5.6$ K, the gap features, in PC spectra, were strongly suppressed, pointing to a peculiar SC state at $5.6\text{ K} < T < T_s$, when incommensurate order develops [05N2, 07N3].

The thermal conductivity of $HoNi_2B_2C$ borocarbide, mainly is dominated by the electrons and the high temperature thermal conductivity is approximately linear in temperature and anomalous [96S3, 02H1, 03N2, 09S2]. The low temperature thermal conductivity exhibits a marked loss of scattering at T_N, in this way increasing by 25% [02H1, 03N2]. The phase transitions, to superconducting or magnetic states, are evidenced in their temperature dependence. The evidence for gapless superconductivity, at $T > T_N$, was also reported [02A1, 03N2]. Based on thermal conductivity data, different scenarios for the occurrence of small energy gaps below and above the magnetic ordering temperatures were discussed [09S2].

Two transitions, at $T = 5.7$ K and 5.2 K were evidenced by specific heat of $HoNi_2B_2C$ [94C1, 95C2, 95L4, 96S3, 98E1, 00E2, 01C4, 01R1, 03E2, 04P2]. The discontinuity, at T_s, of the order of mJ/molK, too small to be shown, indicates that Ho^{3+} ions act as pair breakers [04P2] The spin-wave contribution to the specific heat follows a $0.29T^3\exp(-5.3/T)$ $Jmol^{-1}$ K^{-1} dependence, indicative of magnon excitation in a 3D AFM state with an energy gap $\Delta \cong k_BT_N$ [01R1]. The magnetic entropy, estimated at $2\text{ K} \leq T \leq 10$ K, is $S = 10.4$ J/molK. The magnetocaloric effect in $HoNi_2B_2C$ was also investigated [05D1, 11L1]. The Debye temperatures are listed in Table 9.12.

The thermoelectric powers of RNi_2B_2C compounds are negative. Their absolute values at $T > 100$ K, show a nearly linear temperature dependence [95F1, 97F2, 02H1]. These have a large intercept when the linear portion is extrapolated down to $T = 0$ K. The presence of disordered magnetic moments adds a term linear in temperature and proportional to the De Gennes factor of rare-earths ions, to the high-temperature thermoelectric power [95F1, 96N1, 97B1, 02H1]. The thermoelectric power exhibit an enhancement below T_N. A possible explanation for this behavior could result from a magnon drag effect due to electron-magnon contribution or the reduction of scattering at T_N due to spin loss scatter.

The positron lifetime measurements evidenced in RNi_2B_2C series, an increase in the bulk lifetime with unit cell volume [96B1].

The magnetic properties of pseudoternary Ho-based borocarbides with R = Nd, Dy, Lu, and Y were also investigated: $Ho_{1-x}Nd_xNi_2B_2C$ [99C2], $Ho_{1-x}Dy_xNi_2B_2C$ [96C5, 97C2, 97K4, 98C1, 99D3, 01C6, 02C2, 03K1, 06S2, 07K1, 08L2, 11L2], $Ho_xLu_{1-x}Ni_2B_2C$ [97K4, 97L3, 98F1, 98K3, 00M2], $Ho_xY_{1-x}Ni_2B_2C$ [96C2, 96C4, 96E3, 96M5, 97C2, 97K5, 97L3, 97M3, 98L1, 99F1, 00C2, 00E2, 00F1, 00M2, 01R2, 04E1, 11O1], as well as $Ho(Co_xNi_{1-x})_2B_2C$ [96H4, 97S4, 14E1]. The presence of neodymium, in $Ho_{1-x}Nd_xNi_2B_2C$ system, supress the superconductivity, due to breaking Copper pairs [99C1], as in case of aplied magnetic field. The temperatures T_N, T_M as well as T_s are rapidly lowered by Nd substitution for Ho.

The magnetic properties, of pseudoternary $Ho_{1-x}Dy_xNi_2B_2C$ borocarbides, are influenced only when the magnetic field acts perpendicular to **c**-axis [08L2]. For $x < 0.2$, the T_s values decrease when Dy content increases, due to antiferromagnetic and modulated spin structure fluctuations. For $x > 0.2$, there is only a slight increase of T_s. The magnetic ordering temperatures are nearly linear dependent on composition, De Gennes factor, respectively. The presence of two magnetic and two superconducting order parameters were assumed. The magnetic phase diagram of $Ho_{1-x}Dy_xNi_2B_2C$, at T = 2 K, in the field-composition plane, shows that the AFM magnetic type ordering, attributed to Dy^{3+}, starts at x = 0.2 and the magnetic phase diagram becomes analogous to that of $DyNi_2B_2C$, as the Dy content increases [08L2]. In the $Ho_{1-x}Dy_xNi_2B_2C$ system, the superconducting transition temperature T_s, for x = 0.1, lies above T_N and for x = 0.4 it is located below T_N [01C6]. The behavior of T_s, at low pressures and of T_N at all pressures for $Ho_{0.9}Dy_{0.1}Ni_2B_2C$ follows the prediction of Abrikosov-Gorkov theory and RKKY model, respectively [03K1]. For the sample with x = 0.4, the T_s is not affected much by pressure, while T_N values increase. The pressure dependence of the upper critical field, H_{c2}, supports the phenomenological theory of multiband superconductivity, in the AFM state.

A large entropy change, ΔS, was observed in the AFM state of $Ho_{1-x}Dy_xNi_2B_2C$ borocarbides, related to field induced first order metamagnetic transitions from AFM to FM state [11L2]. With the magnetic field change of 5 T, the maximum -ΔS values reaches 17.1 (x = 1.0), 20.2 (x = 0.7) 18.5 (x = 0.5), 18.7 (x = 0.3) and 19.2 (x = 0) J/kgK.

Two compositions ranges, in the $Ho_xLu_{1-x}Ni_2B_2C$ system, with different behavior, can be distinguished [98F1]. For $x \leq 0.7$, the superconducting transition temperature T_s, decreases linearly, as the Ho content increases. The relatively strong depression of T_s, with increasing Ho content, was corelated with changes in lattice parameters and consequently with different electronic structures. Only the magnetic structure with incommensurate **c**-axis modulation, which seems to have no influence on the superconducting behavior, was shown in this concentration range. For $x > 0.7$, T_s values remain constant, whereas the temperature characterizing the onset of incommensurate **a**-axis modulated structure, T_M, increases nearly linearly with Ho content (x), up to T_M = 6.5 K. Anomalies, in the temperature dependence of the upper critical field, $H_{c2}(T)$, for $x > 0.8$, were observed. They symbolize the re-entrant behavior appearing in this concentration range, in the same temperature range where the **a**-axis incommensurate AFM structure is present. A linear depression of the superconducting transition temperature was also shown, in $Ho_xY_{1-x}Ni_2B_2C$ system, with increasing Ho content, for $0 \leq x \leq 1.0$. The magnetic ordering temperatures, T_N, increase linearly, for $x \geq 0.5$ [96E3]. The substitution of Y for Ho in $Ho_xY_{1-x}Ni_2B_2C$ borides alters the commensurate nature of the Ho magnetic ordering [95H5, 96C4, 96C5, 96H4, 97M3]. In $Ho_{0.9}Y_{0.1}Ni_2B_2C$, a magnetic structure with wave vectors $\mathbf{q_c}$ = (0, 0, 0.09) is present down to T = 2 K. For x = 0.8 sample, the structure type with $\mathbf{q_c}$ = (0, 0, 0.104) wave vector exists [96C4]—Table 9.8. For the compositions with $x \geq 0.75$, the component having **a**-axis modulated structure was shown in a narrow temperature range, only. The intensity of the spiral magnetic state, along the **c**-axis, for x = 0.85 and 0.75, monotonically increase, as the temperature decreases up to T = 1.6 K. The incommensurate modulated structure along the **a**-axis, was considered responsible for the pair breaking effect. The spiral magnetic structure, with **c**-axis wave vector, was found to coexist with superconductivity [97M3]. A linear suppression of the superconducting transition temperature with increasing Ho content was also reported. The field and temperature dependences of the magnetic specific heat, of $Ho_xY_{1-x}Ni_2B_2C$ system, were analysed in the framework of linearized spin wave theory [04E1]. The magnetocaloric effect has been also reported.

The tetragonal structure of $Ho(Co_xNi_{1-x})_2B_2C$ system with low cobalt content, was maintained, even at low temperatures [14E1]. The thermodynamic studies on these borocarbides showed: (1) the superconductivity is steeply suppresed within $0.005 < x < 0.008$ composition range, reaching re-entrance, at 5 K < T < 6 K and finally quenched at T < 2 K, when x > 0.03 [97S4]; (2) the magnetic mode [95H5, 96H4, 96L4] is little influenced within the $0 \leq x \leq 0.015$ composition range. Such a stability of magnetic modes is not seen, for higher cobalt content, where different magnetic structures are present [14E1]—Table 9.8. These structures can be visualized as a variation in the stacking sequence along the **z**-axis, of the strongly intra-planar FM coupled Ho sheets. Only the **z**-component of the **q** = (0, 0, 0.u*) wave vectors is varied. Thus, an AFM mode with **q** = (0, 0, 1) for x = 0.2, a spiral **q** = (0, 0, 0.49) mode for x = 0.4, a spiral **q** = 0, 0, 0.26) mode for x = 0.6 and a FM **q** = (0, 0, 0) mode for x = 0.8, are present. The critical temperatures, T_N, and the saturated magnetic moments are only weakly influenced by substitutions.

$ErNi_2B_2C$ borocarbide

The $ErNi_2B_2C$ borocarbide, as already mentioned, undergoes at the magnetic transition, the tetrahedral to orthorhombic structural distorsion, due to magnetoelastic effects [97D2]. Generally, the resolution of neutron powder pattern is not sufficient to detect such small distortions of the unit cell. The crystal structure, at low temperatures, was assumed mainly to be tetrahedral.

The magnetic structure of $ErNi_2B_2C$ has been determined by neutron diffraction [95S4, 95Z1, 96L5, 97L6, 97S5, 99D1, 99K1, 99S3, 00C1, 01C5, 02K2, 04J1, 05E1]. The Er magnetic moments start to order at T_N = 6.8 K, with a transversely polarized SDW type structure (with **q** and spin direction orthogonal) having wave vector **q** = (0.5526, 0, 0) and spin direction along **b**-axis—Fig. 9.3b. There are two different waves present. One is a short period, with wave vector **q** = (0.5, 0, 0), which gives a commensurate antiferromagnetic structure. Therefore, most of the nearest-neighbour Er moments are antiparallel aligned along **a**-axis [99S3]. The second one is a longer period, around 10 times larger than the first one, with **q** = (0.0526, 0, 0). This describes the modulation, away from the commensurate structure. At low temperatures, $T_{WF} \cong 2.3$ K, the magnetic structure squared-up and develops higher order harmonies. A lock-in phase transition appears to accompany the weak ferromagnetic (WF) state. The evidence for the lock-in, is that the spin density-wave vector appears to become more or less independent on temperature, at $T < T_{WF}$. The lock-in wave vector is close to **q** = (0.550, 0, 0) or (11/20, 0, 0), respectively [02K2]. The amplitude of the sinusoidal moments are $M_1 = 9.15(12)\ \mu_B$, $M_2 = 2.77(30)\ \mu_B$ and $M_3 = 1.53(40)\ \mu_B$ for $\mathbf{q}_1$, $\mathbf{q}_2$ and $\mathbf{q}_3$, respectively [97L6, 99S3]. As already mentioned, a model in which weak ferromagnetism necessarily accompanies a spin-density-wave look-in transition, in the borocarbide magnetic structure, provided the commensurate phase with wave vector **q** = (m/n, 0, 0), with m even and n odd [03W1]. Values of Er magnetic moments $M_1 = 7.0(1)\ \mu_B$, $M_3 = 1.9(2)\ \mu_B$ and $M_5 = 1.0(2)\ \mu_B$ were also reported at T = 1.8 K [05E1]. The refined wave vectors were $\mathbf{q}_1 = (0.552, 0, 0)$, $\mathbf{q}_3 = (0.345, 0, 0)$ and $\mathbf{q}_5 = (0.758, 0, 0)$, respectively.

A large number of studies were devoted to the analysis of $ErNi_2B_2C$ magnetic properties [94E1, 94S4, 95C6, 95E2, 95G2, 95H3, 95J1, 95N2, 95S4, 95S5, 96C1, 96R1, 96Y1, 97C1, 98C3, 99D1, 99K1, 00B2, 00D2, 00F1, 01C5, 02D1, 02J2, 02K2, 03C2, 03N2, 03T1, 03W1, 04J1, 06B6, 07J1, 07V1, 08B1, 08B3, 08R1, 10B2, 10L2, 11L1, 11L3, 12G1, 13W1, 15S1]. Representative data are included in Table 9.8.

The ^{166}Er Mössbauer spectroscopy on $ErNi_2B_2C$ also evidenced that at T = 1.4 K, there is an incommensurate magnetic structure with a maximum moment, $M_{Er} = 8.2\ \mu_B$ [96B4]. The ground state doublet is close to [J = 15/2; $J_z = \pm 1/2$ >] with a low lying ($\cong$ 10 K) excited doublet. The specific heat measurements provided an improvement of the CEF level scheme which was deduced by neutron scattering [94A1]. The modified CEF level splitting involved 0, 9, 52, 71, 73, 181, 212 and 216 K energy levels. By inelastic neutron scattering, between 0 and 20 meV, two ground state transitions were visible, at 6 and 18 meV [96G2, 97G1]. The corresponding excited transition at 12 meV was observed at higher temperatures. The internal magnetic field at 1.5 K, splits up the ground state doublets.

Metamagnetic transitions were shown in $ErNi_2B_2C$ borocarbide, as function of external field [96C1, 96R1]. The magnetization, at T = 2 K, increases steadily until the field is $\mu_0H = 0.7$ T, where a small jumps occurs. Other metamagnetic transitions occur at $\mu_0H = 1.2$–1.3 T, and at $\mu_0H \cong 2.0$ T [96C1, 97C1]. The angular dependences of lower two critical fields, at T = 2 K are well described by the relations 0.74 T/cosθ and 1.11 T/cosθ, respectively where θ is the angle of field with [100]-axis [97C1]. The field up/field down hysteresis is relatively small. The third transition is only little dependent on field orientation.

The magnetic structures of $ErNi_2B_2C$ were also studied by neutron diffraction at T = 1.7 K and 7 K, in external fields up to $\mu_0H = 2.2$ T, oriented along [010] and [110] directions and up to $\mu_0H = 12.0$ T along [001] direction [04J1]. In zero field, at the lowest temperature, the modulation was confirmed to be close to a square wave. The transitions of wave vector **q** between different commensurable values are observed when varying the field. The commensurable structures were analysed on the basis of mean field model. Experimentally, the minority domain shows no hysteresis and remains stable up to a field close to the upper critical field of superconductivity, when the field was applied along [010]-axis. A peculiar behavior is the rotation of **q** by a small angle, of 0.5° away from the [100] direction, when the field was applied along [110]-axis and it is approximately equal to, or larger than the upper critical field [04J1]. A mean field model was used to further describe the magnetic phase diagram of $ErNi_2B_2C$ borocarbide [02J2, 04J1]. The calculations are based on commensurable structures derived from the basic structure $\mathbf{q} = (0.55, 0, 0)$, which is likely to be the result of the surface nesting effects [95R2, 99D4]. This would have, as a consequence that RKKY interaction should also be dependent, relatively strongly, on the energy gaps in the conduction-electron bands created by the spontaneous field induced by magnetic ordering. The stable structure at low temperatures in zero field was described. A two-ion model, for rare-earth antiferromagnetism was also employed to study the magnetic and thermodynamic properties of $ErNi_2B_2C$ and to derive an analytic formula for the T_N, by using the quantum perturbation theory [11L3]. It was suggested that the magnetic behavior is governed by the low lying CF levels in the temperature region at least up to 100 K. A magnetic hyperfine field at the ^{57}Fe nucleus, in $ErNi_{1.9}Fe_{0.1}B_2C$ sample, provided evidence of pair-breaking field at the Ni sites [97S3]. The magnetoelastic effects in $ErNi_2B_2C$, were also used to construct the H-T diagram of $ErNi_2B_2C$ [02D1].

The magnetic susceptibility, of $ErNi_2B_2C$, is anisotropic, the erbium moments are mainly situated in (ab) plane [98C3]. At T < 100, K for **H**||**c**, a large deviation from a Curie–Weiss behavior, suggests that Er^{3+} changes their magnetic state. The magnetic properties, experimentally determined, are listed in Table 9.8.

There is a large number of studies devoted to the coexistence of superconductivity with different types of magnetic order, in $ErNi_2B_2C$ borocarbide [94E1, 95E2, 95G2, 96B4, 96Y1, 01B3, 02H3, 02K1, 04J1, 06B3, 07J1, 08B3, 12G1, 15S1]. The spontaneous or field induced magnetic ordering influences their superconducting properties through: (a) Anderson-Suhl screening of magnetic susceptibility [59A1] or (b) depletion of the number of Cooper-pair states effectuated by magnetic superzone energy gaps [80M1, 81R1]. Studies of the superconducting gap, Δ, in $ErNi_2B_2C$ have been made [96R4, 00W1, 06B1, 06C1, 07N4, 07N2, 08B1, 08B3, 09B1, 09N2, 10B2, 10S1]. Values of the energy gaps, Δ(0), reported in $ErNi_2B_2C$, were in the range 1.60 meV and 1.82 meV [96R4, 00W1, 06C1, 09B1]. Its temperature dependence agrees on the whole with BCS theory. It has been pointed out [00W1], that at a PM–AFM transition, Δ decreases and the pair-breaking parameter [78D1] reaches maximum in the transition region. These results were interpreted [00W1] using a theory [80M1], which predicted a decrease in Δ, owing to spin-density wave gaps, that open in some parts of the FS. No influence of the AFM transition, on Δ, was observed in subsequent tunnel measurements on $ErNi_2B_2C$ [06C1]. The temperature dependent laser-photoemission spectroscopy showed that above T_N, the Δ values follow the simple BCS model with the SC-gap, $\Delta(0) = 1.40$ meV, corresponding to a reduced gap value $2\Delta(0)/k_BT_s = 3.49$, comparable with that of mean field BCS, of 3.52 [08B1]. At $T < T_N$, the $\Delta(T)$ values deviate from the BCS prediction. This behavior has been discussed in the framework of a model [80M1], which assumes that the energy gaps of SDW opens on the part of FSs where a nesting vector can relate two Fermi momenta. The point contact spectra of $ErNi_2B_2C$ [10B2] were then analysed assuming "one-gap" or "two-gap" approximations, respectively and models which account for the pair breaking effects of magnetic ions (BB) [03B1, 08B3], as well as the GBTK model [95P1], which accounts for the Dynes pair-breaking parameter [78D1]. It was concluded that the "two gap" calculations, in BB model, provided the best agreement with experimental data. The results obtained in the above model, in "two gap" approximation are given in Fig. 9.11 [10B2]. The experimental data as well as the calculations evidenced the anisotropic effect of AFM ordering, regardless the model used. As the temperature is lowered, Δ decreases by $\cong 25\%$ in the **c**-direction and only by $\cong 4\%$ in the (ab)-plane. It was found that the pair-breaking parameter increases in the vicinity of magnetic transitions, the increase being more pronounced in the **c**-direction. The presence of two superconducting gaps was also reported in some RNi_2B_2C compounds, including that with R = Er [09N2]. The temperature dependence of the superconducting gap in $ErNi_2B_2C$, at different external fields, was also theoretically investigated [10S1].

The superconducting upper critical fields, in $ErNi_2B_2C$, were also determined [94E1, 96C1, 99G1, 00B2, 02D1, 04J1, 06B5, 07J1, 10L1, 10L2, 12G1]. The temperature dependences of the critical field, $H_{c2}(T)$ determined by resistivity measurements [04J1] and tunneling spectroscopy [12G1] evidenced features in H_{c2} values, at temperatures T = 6 K, 4 K, 2 K—Fig. 9.12. The features, at T = 6 K and 2 K, are pronounced and $H_{c2}(T)$ becomes nearly flat at these points [12G1]. The low field incommensurate phases with $\mathbf{q}_1 = (0.55–0.56, 0, 0)$ and commensurate phases with $\mathbf{q}_2 = (0.57, 0, 0)$, $\mathbf{q}_3 = (0.58, 0, 0$ and $\mathbf{q}_4 = (0.59, 0, 0)$ were identified at low temperatures (Fig. 9.12). The final transition to the saturated paramagnetic state occurs at $\mu_0H \cong 2$ T along the **a**-axis, field much above the loss of superconducting gap [12G1]. The features at H_{c2} curve, at T = 4 K, measured by tunnelling spectroscopy, coincide with a triple point transition ($\mathbf{q}_1$–$\mathbf{q}_2$–$\mathbf{q}_3$), in the magnetic phase diagram. The remaining $H_{c2}(T)$ curve, obtained by tunnelling spectroscopy follows the magnetic transitions $\mathbf{q}_2$–$\mathbf{q}_3$ and $\mathbf{q}_2$–$\mathbf{q}_4$ from T = 4.5 K to 2 K, respectively. From the data obtained by tunnelling spectroscopy, at T < 2 K, another feature, with a vanishing slope was observed, in $H_{c2}(T)$, probably from a weakly FM state present in low field. Thus, the superconducting phase diagram, closely follows the changes in the local spin arrangements.

In the mixed state, the borocarbide is permeated by an array of quantized flux lines, the superconducting properties being controlled by the flux line lattice (FLL). The studies performed by using small-angle neutron scattering (SANS), separate superconductivity, via the FLL periodicity and the magnetic order, allowing them to be independently investigated. As a result, this method was most used to analyse the vortex structure in $ErNi_2B_2C$ borocarbide [96Y1, 97E4, 99G1, 10K1]. As in RNi_2B_2C with R = Lu or Y, at high fields, a square flux-line lattice was shown, while at low fields, the flux lines order in the hexagonal lattice [96Y1, 97E4]. In

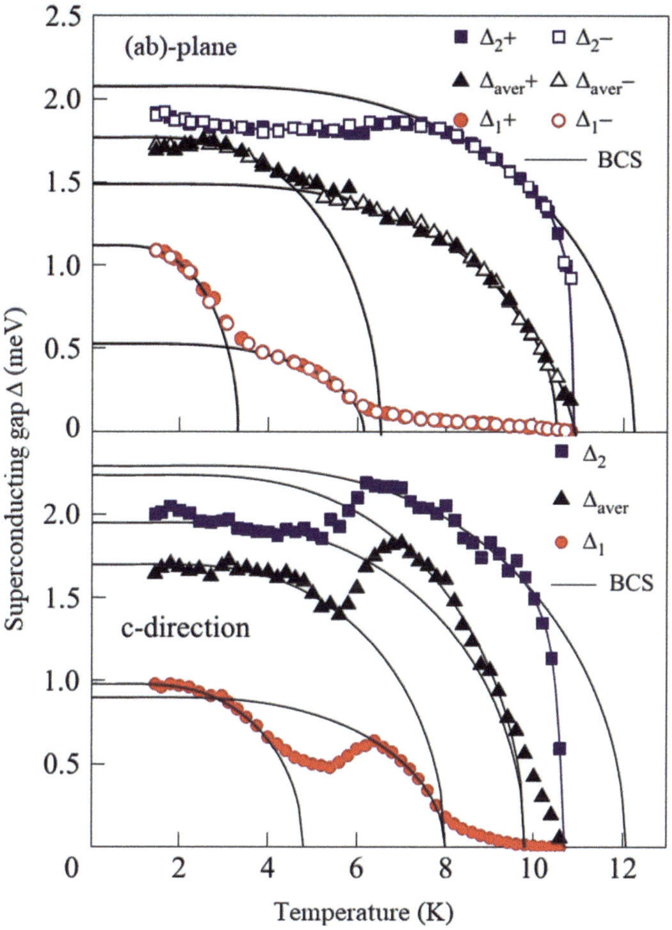

Fig. 9.11 $ErNi_2B_2C$ single crystal: temperature dependences of the superconducting gaps, calculated in two-gap approximation [10B2]. The analysis has been made according to [03B1, 08B3] mode. A polynominal fit is drawn through the points outside the BCS approximation. The scaling factor was S = 0.25

addition, a significant coupling between the magnetic ordering and the flux lines was observed. The development of magnetic order causes the vortex lines to disorder and rotate away from the direction of the applied field, below T_N [96Y1], suggesting that the vortex lattice structure may be coupled to the magnetic order parameter. According to [10K1], as the $ErNi_2B_2C$ enters the weak ferromagnetic phase, at low temperatures, the intensity of the FLL pattern decreases, indicating that the weak magnetic order introduces disorder to the FLL correlations. It was also observed, by Bitter decoration, a residual pinning of vortices on (100) or (010) boundaries, slightly above $T_{WF} = 2.3$ K, which visualizes the weak ferromagnetic domains [07V1].

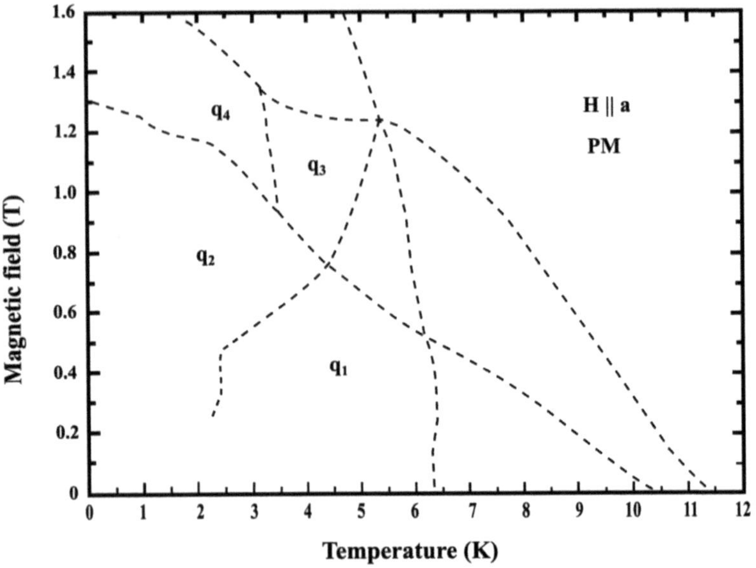

Fig. 9.12 $ErNi_2B_2C$ single crystal: superconducting and magnetic phase diagram at $T \geq 0.1$ K for **H** ǁ **a** [12G1], including the data from [02D1, 04J1, 09G2]. By dashed lines are given the magnetic transitions between the phases with wave vectors $\mathbf{q_1} = (0.55–0.56, 0, 0)$, $\mathbf{q_2} = (0.57, 0, 0)$, $\mathbf{q_3} = (0.58, 0, 0)$, $\mathbf{q_4} = (0.59, 0, 0)$; PM—aramagnetic phase

The vortex lattice structures, at $T < T_s$, evidenced vortex pinning in the {110} twin AFM domain boundaries, in the orthorhombic phase, similar as in $HoNi_2B_2C$ borocarbide [03V1, 05V3, 09V2]. The vortex shows bands/chains, at $T < T_N$, directed along the domain walls. The decoration, at $T > T_N$, still in superconducting state, shows standard triangular vortex lattice. The increase in flux pinning, at $T < T_N$, coincides with the appearance of twin boundaries between antiferromagnetic domains [00S1]. The pinning mechanism is provided by a ferromagnetic component localized at the twin boundaries, parallel to the **c**-axis. The flux lines in $ErNi_2B_2C$ [97E4, 01S3] and vortex structure [10K1] were further analysed. An internal magnetic field, mediated by weak ferromagnetic order, creates new vortices and affects the flux lattice structure [10K1]. The flux line form factor, in small angle neutron scattering and transport data, determines the superconducting lengths scales and critical field in $ErNi_2B_2C$ single crystal [99G1]. For **H** ǁ **c**, the coherence length ξ increases and the penetration depth λ decreases, when crossing T_N. The critical fields show the corresponding anomalies near T_N. For **H** ⊥ **c**, the fourfold modulation of H_{c2} is strongly temperature dependent, changing sign near T_N and can be modelled using the anisotropy of the sublattice magnetization.

The transport and magnetization measurements of the critical current density, in $ErNi_2B_2C$ single crystal, showed strongly enhanced vortex pinning, at T_N, when a low field **H** ǁ **c** was applied, consistent with vortex pinning, at the antiphase boundaries between AFM domains [13W1]. The magnetic penetration depth and their temperature dependence, as the $ErNi_2B_2C$ undergoes magnetic phase transitions in the superconducting regime, was reported [15W1]. The pinning force of stray field induced vortices, shows a temperature dependence related to the existence of various magnetic phases.

The resistive properties of $ErNi_2B_2C$ borocarbide, were investigated [94E1, 94S2, 95A1, 95E2, 95G2, 95J1, 95S5, 96F2, 96N1, 97B1, 97F2, 00B2, 02H1, 02M2, 04T1, 10L1, 10L2, 13W1, 15M1]. The in-plane and **c**-axis resistivities share small but significant deviations from the isotropic behavior below T = 100–150 K [96F2, 97B1, 97F2], associated with scattering from the erbium moment, which might be anisotropic due to crystal-electric-field effects. The magnetoresistivities along **c**-axis, ρ_c(H,T) and in plane, ρ_{ab}(H,T), have been measured with the field **H** perpendicular and parallel to the **c**-axis [10L1, 10L2]. From ρ_c(H,T) curves, the superconducting upper critical fields, H_{c2}(T), were constructed. The anisotropy in H_{c2}(T), was discussed and compared with that determined from ρ_{ab}(H,T), considering magnetic pair breaking and anisotropic field dependence of T_N, originated from Er^{3+} sublattice.

The specific heat, C_p(T) data of $ErNi_2B_2C$ borocarbide, were also reported [94M4, 95C5, 95H2, 00C3, 01R1, 02E1, 03E2, 03N1, 10E1, 15M1]. In the absence of a magnetic field, both antiferromagnetic, T_N and weak ferromagnetic, T_{WF} transition temperatures were observed, as a peak and a kink, respectively [03N1]. Under magnetic fields, parallel to [110] axis, T_N monotonically decreases with increasing field, while T_{WF} increases at low fields and decreases above $\mu_0H = 1.0$ T. The above behavior was explained assuming the competition between magnetocrystalline anisotropy and the weak FM arising from the magnetic moment at the AF domain boundary. Distinctly evolutions, of magnetic specific heat C_m(T), were shown in the temperature range $T_{WF} < T < T_N$ and $T < T_{WF}$, respectively [02E1]. For $T_{WF} < T < T_N$, C_m(T) evolves as $0.45T^2$ J/molK, suggesting a 2D excitations with a gapless linear dispersion relation. For $T < T_{WF}$, C_m(T) is dominated by an exponential factor, indicative of gapped AF excitations. The spin wave (SW) contribution to the specific heat, of $ErNi_2B_2C$, at T < 2 K, was described by the expression: $0.27(\Delta^2 + 4T\Delta + 6T^2)\exp(-\Delta/T)$ J/molK, with $\Delta = 5.9(1)$ K, a gapped dispersion relation of a 2D FM sheet [01R1]. The magnetic contribution to specific heat, C_m(T,H), accros the H-T phase diagram of $ErNi_2B_2C$, was also investigated [10E1]. Within the lower ranges of temperatures and fields, the calculations, based on linearized field-dependent spin-wave theory, reproduced satisfactorily the measured C_m(T,H) curves.

The temperature dependences of the thermal conductivity [00C4, 02C1, 02H1, 03N2] and thermopower [97B1, 02H1], in $ErNi_2B_2C$ borocarbide, are close related to magnetic and superconducting properties. The thermal conductivity, showed a change in slope, at T_s. In zero magnetic field, the thermal conductivity shows small peaks, at T_N, which disappear under applied magnetic fields. At $T < T_s$, the thermal conductivity, in an applied magnetic field, was larger than in zero field [00C4]. A temperature independent contribution to the thermopower, was reported in RNi_2B_2C compounds with R = Tm to Dy, in addition to a linear term in temperature [97B1]. The magnetocaloric (MCE) effect, in $ErNi_2B_2C$, has been studied by investigating the temperature and field dependences of the heat capacity [11L1]. A magnetic entropy change, $\Delta S = -9.8$ J/kgK, was reported, in a field range up to $\mu_0 H = 5.0$ T, correlated with the field induced first order metamagnetic transition. A maximum temperature change, $\Delta T = 4.6$ K, was estimated.

Additional methods were used to characterize the physical properties of $ErNi_2B_2C$ borocarbide. The μSR study, at $T > T_N$, suggested anomalous broadly distributed rare-earth spin dynamics [98H1]. A well-defined internal field, at $T < T_N$, coexisting with SC state, down to 0.1 K, was shown by positive-muon spin relaxation [95L3]. The optical conductivity spectra for RNi_2B_2C compounds with R = Tb, Dy, Er, Y, at T = 300 K, are similar and indicative that R4f states are not actively involved in the optical transitions [01L1]. The pair-breaking, due to the Er local magnetic moment, appear to have no detectable influence on the boron isotope effect of T_s [13H1]. The variation of positron lifetime in RNi_2B_2C (R = Ce, Tb, Ho, Er, Lu, Y) borocarbides was related, as already mentioned, to the trapping behavior of positrons, at carbon vacancies and their clusters [96B1]. The microwave measurements on $ErNi_2B_2C$, do not evidenced the AFM transition in zero field [95J1]. Pair breaking was shown to be significantly present only in the presence of the magnetic fields.

The XANES study indicates a weakly anisotropic Ni3d and anisotropic B2p unoccupied density of states [01M1]. Local scanning tunnelling spectroscopy measurements, were made, at low temperatures, on the surface of $ErNi_2B_2C$, where superconductivity coexists with WFM, AFM and local moment paramagnetism. A finite density of states, at E_F, as high as half the normal state value, over large areas of the surface was shown [06C1]. The DOS shows a striking departure from BCS theory. The DOS peak located at E_F, in $ErNi_2B_2C$, decreases noticeable with pressure [13Y1].

The magnetic and related properties of pseudoquaternary borocarbides, containing erbium, were investigated: $Nd_xEr_{1-x}Ni_2B_2C$ [99C1], $Er_{1-x}Tb_xNi_2B_2C$ [98R1, 99B3, 99P1, 00Y2, 02K3, 03C2, 04T1], $Er_{1-x}Dy_xNi_2B_2C$ [00C3, 05Q1], $Er_{1-x}Ho_xNi_2B_2C$ [20G1], $Er_{0.8}R_{0.2}Ni_2B_2C$ with R = Tb, Lu, [04T1], $Er_xTm_{1-x}Ni_2B_2C$ [00G1], $Er_{1-x}Lu_xNi_2B_2C$ [96B4, 05E1], $Er_xY_{1-x}Ni_2B_2C$ [94M4, 96B4, 96E2, 96M5, 98L1, 99F1, 00E2, 00F1, 01R2], $ErNi_{2-x}Fe_xB_2C$ [03A1, 04A1, 14Z1], $Er(Ni_{1-x}Co_x)_2B_2C$ [96B3, 97S4, 97Z2, 05E1], $ErNi_{2-x}M_xB_2C$ with M = Co, Pd, Pt [96B3, 97F1], $Er(Ni_{1-x}Pt_x)_2B_2C$ [96B3, 97F1, 09M1].

The Nd substitution, in $Er_{1-x}Nd_xNi_2B_2C$ system, leads to a gradual decrease of superconducting transition temperature, the superconductivity being supressed at x = 0.22, behavior correlated with the pair-breaking mechanism [99C1].

The $Er_{1-x}Tb_xNi_2B_2C$ compounds, exhibit a variety of magnetic and superconducting properties, characteristic for end series borocarbides and also to those which interplay between them [96C1, 98R1, 99B3, 02K3, 03C2]. The T_s temperature, is gradually suppressed by increasing Tb content and a crossover for $T_s < T_N$, occurs at x = 0.2. The T_s values follow well the De Gennes scaling, down to $T_s = 2$ K and x = 0.3, where superconductivity breaks down. The T_N and T_{WF} temperatures show distinct behavior in the Er- and Tb-rich sides. The deviation, of T_N, from De Gennes scaling, predicted by the RKKY model, was attributed to crystal field effects. The substitution of R = Tb in $Er_{0.8}R_{0.2}Ni_2B$ borocarbides, enhances the magnetic correlations, leading to an increase of T_N values and the critical fields, while when R =

Lu, the Néel temperatures decrease [04T1]. Two additional effects were also shown: (1) magnetic segmentation (size effects), leads to a modification in the magnon spectrum and (2) there are pinning centres that accommodate the magnetic kinks. As a consequence, the former modifies the thermal evolution of the magnetic specific heat, while the latter hinders the formation of WF state.

In the $Er_{1-x}Dy_xNi_2B_2C$ system, T_N increases monotonically, in the composition range $0 \leq x \leq 1$ [00C3]. A more complicated composition dependence was shown for T_s values. Depending on composition, the coupling between supraconductivity and magnetism in $Er_{1-x}Ho_xNi_2B_2C$ creates several phases, ranging from a re-entrant superconductor with a mixture of commensurate and incommensurate antiferromagnetism to a total incommensurate AFM spin modulation with a weak ferromagnetic state [20G1]. All of these phases coexist with supraconductivity.

The $ErNi_{2-x}Fe_xB_2C$ system, forms solid solutions up to $x = 0.2$ [04A1]. Upon iron doping in $ErNi_{2-x}Fe_xB_2C$ system, the T_N and T_{WF} temperatures moderately decrease, while the superconducting transitions temperature T_s, is strongly depressed, with an average rate $dT_s/dx \cong -150$ K. The $x = 0.04$ sample, with $T_s = T_N = 7.3$ K, shows nearly re-entrant superconductivity in zero applied field [03A1, 04A1]. Parts of the sample reenters the normal state between 5.5 K and 5.7 K. The $x = 0.04$ sample has a modulated magnetic structure with wave vector $\mathbf{q} = (0.559, 0, 0)$, which squares at low temperature and transforms to weak ferromagnetism at $T < 2.2$ K. The magnetocaloric effect was studied in $ErNi_{2-x}Fe_xB_2C$ [14Z1]. The maximum entropy changes, $-\Delta S_m$, are: 14.5 ($x = 0$), 12.7 ($x = 0.1$) and 10.6 ($x = 0.2$) J/kgK, for a range of field changes at $0 \leq \mu_0H \leq 7$ T.

The T_s values, in $Er(Ni_{1-x}Pt_x)_2B_2$ system, strongly decrease in the composition range $0 \leq x \leq 0.1$ and thereafter, a relatively much weaker drop (almost a plateau), with further increase of x [09M1]. The T_N values almost linearly decrease, up to $T_N = 4.7$ K ($x = 0.2$). The partial substitution of nickel, in $ErNi_{2-x}M_xB_2C$, with M = Pd, Pt, up to $x = 0.2$, does not lead to any significant changes in the magnetic ordering and CEF characteristic of Er^{3+} ion [96B3, 97F1]. The superconducting transition temperatures, when Ni is substituted by Pd and Pt, decrease with a rate $dT_s/dx = -12$ K, one order at magnitude smaller than that evidences in cobalt doped samples.

$TmNi_2B_2C$ borocarbide

The $TmNi_2B_2C$ borocarbide is magnetically ordered, at $T \leq T_N = 1.52(5)$ K. The magnetization isotherms evidence the presence of high anisotropy, the easy axis of magnetization, being **c**-direction. The saturated magnetic moment, along this direction, is $M_s = 5.0\ \mu_B$/f.u. In the temperature range $200\ K \leq T \leq 300$ K, the magnetic susceptibility follows a Curie–Weiss type behavior. At $T < 150$ K, the reciprocal susceptibility deviates from linear dependence, in opposite ways for the two field directions [95C5]. The temperature dependence, of the magnetic susceptibility, in polycrystalline samples, was also analysed [94E1].

The anisotropic magnetic behavior, of $TmNi_2B_2C$ single crystal, was further investigated [98C3]—Table 9.8. The inelastic neutron scattering experiments, of the CEF splitting, in $TmNi_2B_2C$, both in paramagnetic and magnetically ordered state, evidenced four ground-state transitions, at 3, 5.4, 10.2 and 12.5 meV, respectively

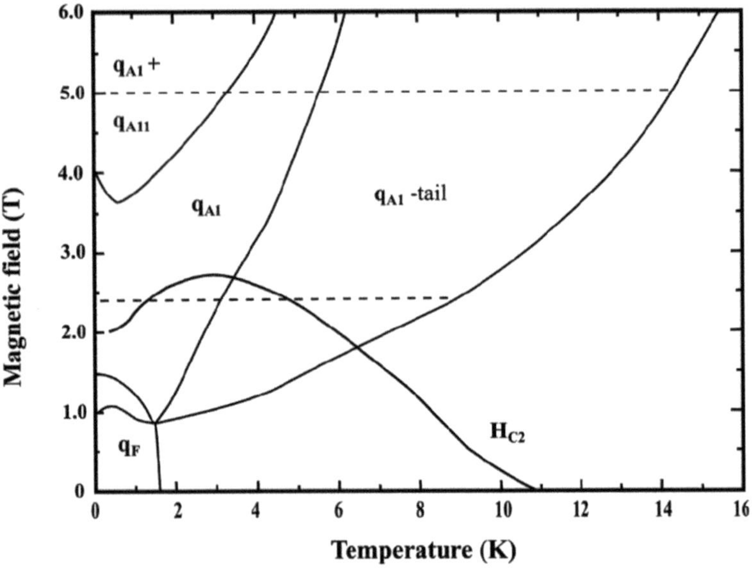

Fig. 9.13 $TmNi_2B_2C$: magnetic phase diagram [04N2]. The temperature and field dependences of superconducting upper critical field H_{c2} is also plotted [99N5]

[96G2, 97G1], at T = 0.1 K. For the peak at 3 meV, a slight asymmetry was observed which could be reproduced by calculating the CEF splitting in an internal field of 0.54 T. The reability of these data, was evidenced by reproducing the experimentally determined magnetic properties.

The magnetic structure of $TmNi_2B_2C$ borocarbide, determined by neutron diffraction, is incommensurate and consists of ferromagnetic (110) planes of Tm moments, aligned along **c**-axis with magnitude of the moments modulated sinusoidally along the [110] direction, with wave vector $\mathbf{q}_F$ = (0.093, 0.093, 0) [96C3, 97L6, 97S5, 97S7, 99S3, 00N4, 04N2]. At low temperatures, a 3**q** peak was observed, indicating that the SDW squared up, with the magnetic structure remaining incommensurate [97L6]. Magnetic moments $M_1 = 4.81(18)\ \mu_B$ and $M_3 = 0.93(19)\ \mu_B$ were determined, at T ≅ 0.34 K for $\mathbf{q}_1$ and $\mathbf{q}_3$, respectively. Assuming a square wave, this gives an ordered moment, $M_{Tm} = 3.78\ \mu_B$· [97L6]. The Tm magnetic moment was also estimated at 3.74(2) μ_B and 4.8(1) μ_B, at T = 1.2 K and 0.05 K, respectively [96C3].

The field-temperature phase diagram, of $TmNi_2B_2C$, in magnetic fields up to at $\mu_0H = 6$ T, applied along **a**-axis, evidences the presence of three types of magnetic ordering with wave vectors $\mathbf{q}_F$ = (0.094, 0.094, 0), $\mathbf{q}_{A1}$ = (0.483, 0, 0) and $\mathbf{q}_{A11}$ = (0.496, 0, 0)—Fig. 9.13 [00N4, 04N2]. In all the above phases, the Tm magnetic moments are oriented along the **c**-axis. In zero and low fields, the Tm moments order in a long wavelength transverse spin density wave, with wave vector $\mathbf{q}_F$. The magnetic structure, with the wave vector $\mathbf{q}_{A1}$, is stabilized by an applied field of $\mu_0H = 1.0$ T and a transition, to $\mathbf{q}_{A11}$ phase, takes place at $\mu_0H = 4.0$ T. For both transitions, there is a broad temperature and field range of overlap between the different magnetic states.

The $\mathbf{q}_A$-type phases persist up to higher temperatures, when the field is increased. The majority of the phase diagram is dominated by the phase having $\mathbf{q}_{A1}$ wave vector. The field and temperature dependence of the supperconducting uper critical field [99N5] is also given in Fig. 9.13. The supression of superconductivity cannot be correlated with the presence of $\mathbf{q}_{A1}$ phase. The origin of the $\mathbf{q}_{A1}$ phase was assumed to be connected with field induced modifications of the Fermi surface. The $\mathbf{q}_{A1}$ and $\mathbf{q}_{A11}$ phases appeared only parallel to the applied field, not perpendicular to it. This symmetry break introduced by the $\mathbf{q}_{A1}$ and $\mathbf{q}_{A11}$ phases, in earlier studies, was not possible to be explained, as well as the fact that low intensity tail of the $\mathbf{q}_{A1}$ and $\mathbf{q}_{A11}$ phases, having a constant correlation length throughout the phase diagram, extend up to $T \cong 15$ K. These features were latter correlated with the occurence of a quadrupolar phase.

The presence of a quadrupolar phase has been evidenced by synchrotron XRD studies [06A2]. The lattice of $TmNi_2B_2C$ was shown to be distorted at $T_q \cong 13.5$ K, in zero field [06A2]. The T_q values increase and the accompanying deformation of the lattice is grossly enhanced when the magnetic field is applied along [100] axis. The distortion occurs, at the same wave vector as that of the antiferromagnetic ordering induced by the **a**-axis field. In $TmNi_2B_2C$, the electron–phonon system is very close to the point of a lattice instability. The transition is occurring because of the additional contributions of magnetoelastic and quadrupolar-quadrupolar interactions. The magnetoelastic interaction, derives from the electrical crystal field, produced by the displacements of the ions, whereas quadrupolar-quadrupolar interaction may be mediated by the same charge-density wave, responsible for the strong electron–phonon interaction and hence may be an additional consequence of the Fermi-surface nesting [06A2]. The quadrupolar ordering has as effect that a field applied along [100], induces AFM ordering, at the same wave vector as the quadrupolar ordering. The Fermi-surface nesting, is presumably important for the strong electron–phonon interaction, at $\mathbf{q}_{Ai}$ phases and possibly also for the quadrupolar-quadrupolar interactions. The RKKY interaction was assumed to be relatively weak at this wave vector. It was concluded that the electrons which are close to the nesting areas, important for creating superconducting phase, are not the same as those involved in RKKY interactions [06A2].

In $TmNi_2B_2C$ there is a clear evidence for a coupling between magnetic and superconducting properties. The magnetic field applied along the **c**-axis induces concurrent changes of the magnetic structure and the symmetry of the flux line lattice [98E3]. The Anderson-Suhl screening is important [00N4]. It explain why this system does order at the small wave vector $\mathbf{q}_F$ instead of zero and why the AFM phase, at $T = 0.1$ K, stay stable up to a **c**-axis field of the order of 0.8 T [07J1]. At the lowest temperature, the $\mathbf{q}_F$ ordering disappears, when applying a field of 1.0–1.5 T along **a** direction and is replaced by a new AFM ordering of **c**-axis moment, of the wave vector $\mathbf{q}_{A1}$ [00N4, 04N2]. The AFM ordering becomes more stable, the larger the field is. This behavior was explained, as already mentioned, by the presence of a quadrupolar phase, below $T_q = 13.5$ K ($H = 0$), showing up as a modulated **c**-axis displacement of the Tm ions, at the wave vector $\mathbf{q}_{A1}$ [06A2]. The AFM ordering is induced by the uniform magnetic field, due to the oscillating anisotropy term, associated with quadrupolar ordering. The order parameter of the quadrupolar phase is sensitive to the **a**-axis fields. When the field is applied along the **c**-axis, the quadrupolar order parameter is small and only induces a minute longitudinally polarized AFM moment. The quadrupolar ordering does not influence the upper critical field in the **c**-axis case, because the superzone energy gaps produced by the quadrupolar moments do not affect the spin degeneracy of the conduction electrons [07J1]. When the field was applied perpendicular to the **c**-axis, the AFM ordering, induced by the field, enhances the superpose energy gaps and gives rise to a linear reduction of the effective density of states, $\eta(0)$. The upper critical field calculated in the above model, describe well the experimental data [07J1].

A model was also proposed to study the coexistence of SC and AFM, in RNi_2B_2C borocarbides, taking s-wave pairing into account [15S1]. The model could explain the interplay of T_s and T_N, with different sets of available realistic parameters.

The temperature dependences of magnetization, for the flux expulsion (FCW) and magnetic shielding (ZFC) effects, for $\mathbf{H} \perp \mathbf{c}$ and $\mathbf{H} \parallel \mathbf{c}$, in $TmNi_2B_2C$ single crystal, at an external magnetic field of $\mu_0H = 0.0010$ T, shows a sharp superconducting transition, with the onset at T = 11 K, transition temperature $T_s = 10.8$ K and transition width (10–90)%, of full diamagnetic signal of about 0.4 K. The FCW values, at T = 2 K, are 80% ($\mathbf{H} \perp \mathbf{c}$) and 40% ($\mathbf{H} \parallel \mathbf{c}$) of perfect superconducting flux expulsion values and the ZFC ones were 150% ($\mathbf{H} \perp \mathbf{c}$) and 420% ($\mathbf{H} \parallel \mathbf{c}$), without correction for diamagnetic effects, indicating bulk superconductivity. In the superconducting state, upper critical fields $\mu_0H_{c2}(0) = 2.12(15)$ T and 2.72(15) T were determined for $\mathbf{H} \parallel \mathbf{c}$ and $\mathbf{H} \perp \mathbf{c}$ directions, respectively. The extrapolated penetration depths, $\lambda(0) = 85.0(6.0)$ nm ($\mathbf{H} \perp \mathbf{c}$) and 78.0(7.0) nm for ($\mathbf{H} \parallel \mathbf{c}$), coherence lengths $\xi(0) = 12.4(5)$ nm ($\mathbf{H} \parallel \mathbf{c}$) and 11.0(3) nm ($\mathbf{H} \perp \mathbf{c}$) were reported [95C5].

The pressure dependences of the superconducting transition temperature can be described by the rates $dT_s/dp = -0.50$ K/GPa [99O1], −0.453(1) K/GPa [95A1] or −0.178(22) K/GPa [94S2]. The experimental data were also fitted by the relation $T_s(p) = 10.876(10)–0.410(30)K/GPa–0.0262(17)K/GPa^2$ [95A1].

A combined ^{169}Tm Mössbauer spectroscopy and μ^+SR studies, evidenced the existence of two types of Tm4f magnetism, in $TmNi_2B_2C$ [95C7, 95L3, 96B2, 96M4, 98G2, 98M3, 98M4]. The μ^+SR measurements, on $TmNi_2B_2C$ [95C7, 95L3, 96B2], revealed that a spontaneous internal field is present, up to $T \cong 30$ K, far above T_N. This spontaneous field saturates, at $T \leq 2.5$ K and shows a T^{-1} dependence above this temperature. The saturation value of this internal field corresponds to a Tm moment of $\cong 0.1\ \mu_B$. The ^{169}Tm Mössbauer spectra also evidenced the coexistence of two types of Tm4f magnetism, at T = 0.3 K [96M4, 98M3, 98M4]. One has a value $M_{Tm} = 4.3(1)\ \mu_B$ and the other of $\cong 0.1\mu_B$. The second type of magnetism was suggested to be caused by carbon vacancies present in $TmNi_2B_2C$ [98M3, 98M4]. These vacancies locally modify the crystalline electric field and therefore the Tm4f magnetism. The occurrence of small magnetic moments, in $TmNi_2B_2C$, from the neutron scattering point of view, was also discussed [98G2]. The absence of a clear second set of CEF transitions, which could belong to the 0.1 μ_B phase, suggested that several CEF surroundings belong to Tm ions, with the small magnetic moment. These additional

CEF transitions are too weak to be distinguished from the background. The carbon-boron disorder cannot be excluded, as suggested by a neutron diffraction study and it might be related to the 0.1 μ_B phase. The nature of magnetism in $TmNi_2B_2C$ was further investigated, on magnetically oriented powdered samples, by μSR [99N1]. The quasistatic magnetic correlations, persists even up to T = 50 K. A peak around 0.0007 T was found in the field dependence of the longitudinal relaxation, attributed to the presence of a level crossing resonance arising from quadrupole levels of the boron nuclei.

The resistive properties of $TmNi_2B_2C$ borocarbide [94E1, 95A1, 95E2, 95J1, 96R2, 97B1, 98N1, 99O1, 02H1, 02M2, 02O1] and magnetoresistivity [98N1], including pressure effects [94S2, 95A1, 99O1, 02O1] were investigated. The resistivity, of $TmNi_2B_2C$, follows the same trend as that for other RNi_2B_2C compounds [97B1]. There is a linear temperature dependence, in the range 100 K $\le$ T $\le$ 300 K. Their thermal variation, at low temperatures, follows a $T^{1/4}$ law. The tunnelling conductance, of $TmNi_2B_2C$, do not changes at T < 1.5 K, when both AFM phase and superconductung state coexist [01S5]. The microwave resistivity $\rho_{\mu w}(T)$, evidenced that the magnetic effects were responsible for the anomalies, in the low temperature surface impedance, at T < 4 K [95J1].

The thermal conductivity [00C4, 02H1, 03N2] and specific heat [94M5] studies on $TmNi_2B_2C$ showed evidence of both electron and phonon scattering mechanisms and the presence of magnetic phase transitions. The low temperature thermal conductivity, exhibits marked loss of scattering, at the T_N temperature. Magnon heat conduction was suggested [02H1]. The temperature dependence of the thermoelectric power, in $TmNi_2B_2C$, as in other RNi_2B_2C compounds, was described as a sum of two contributions, one independent on temperature and other which shows linear thermal dependence [97B1].

The interplay of magnetism and superconductivity, in $TmNi_2B_2C$ was also analysed [94E1, 95E2, 96B2, 97K6, 98E3, 99E1, 99N5, 99O1, 07D1, 07N4, 09N2, 11N1, 12D1]. These studies were mainly connected with the superconducting gap and the flux line lattice peculiarities. The $TmNi_2B_2C$ exhibits the largest difference between T_s and T_N, in the whole RNi_2B_2C series, which points to a relatively weak magnetism, which cannot destroy the superconductivity on the large Fermi surface sheets (FSS's), but only weakens it. The superconductivity is nevertheless affected by the special incommensurate magnetic state and its fluctuations, at T > T_N. In this temperature range, at least for low external magnetic fields, has been expected to detect remnants multi-gap features, as evidenced in non-magnetic borocarbides [98S2, 06B4]. By point-contact (PC) investigations, in normal and superconducting states, the temperature dependence of SC gap was analysed [11N1]. By using the standard single gap approximation (SGA), was shown that this deviates from BCS behavior, displaying a maximum at $\cong T_s/2$. A two gaps analysis has been also made. The presence of a second gap was shown, being about two times larger than the former, which remains close to the small "first" gap. The smaller gap has been assigned to a FSs which has a similar orbital structure and a gap value, as the "cushion" FSs found in 1/2/2/1 borocarbides [11N1]. This not-yet-resolved FSs, for $TmNi_2B_2C$, is almost "protected" from the exchange interactions with magnetic moments of rare-earths. The second larger gap is only a bit smaller than the larger gap in Y/Lu borocarbides. The above behavior interpolates between the "cushion" FSs dominated almost single-band supraconductors (Dy/HoNi_2B_2C) within commensurate AFM state and the nonmagnetic superconducting borocarbides (Lu/YNi_2B_2C). The upper critical field, $H_{c2}(T)$, measured for external fields oriented along **c** direction and within the (ab) plane, behaves similarly, that is, have maxima in their temperature dependences [04N2, 11N1]. This broad maximum is closely related to the re-entrant behavior found in the temperature dependence of the bulk resistivity and magnetic susceptibility in magnetic field [11N1].

The flux line lattice (FLL) behavior, in $TmNi_2B_2C$ borocarbide, in the presence of external field, has been also investigated [98E3, 99E1, 00G1 07D1, 10G1, 12D1]. A hexagonal vortex lattice was found, when the magnetic field ($0.15\ T \leq \mu_0 H \leq 1.4\ T$) was applied along the basal plane, of the tetragonal crystal structure (**H** $\perp$ **c**) and a hexagonal to square transition at $\cong 0.15$ T, when the field was applied along the **c**-axis [10G1]. The intervortex distance seems to decrease, as a function of magnetic field. The magnetic field distribution, around the vortex, in paramagnetic phase ($T > T_N$) of $TmNi_2B_2C$, has been also investigated [07D1]. The vortex form factor, measured by small-angle neutron scattering, was found to be field independent, at $H \leq 0.6\ H_{c2}$, followed by a sharp decrease at higher fields. It has been shown that the conduction electron paramagnetic moments, induced by the exchange interactions around the vortex cores, create nano-tubes of Tm magnetization which act to maintain the field contrast probed by the form factor.

The vortex lattice (VL) in $TmNi_2B_2C$ undergoes the progression of the symmetry transitions as already described in other RNi_2B_2C compounds, at low fields. However, the square phase VL was found to be re-entrant and the VL undergoes the same sequence of transitions but in reverse order as the field is further increased, as already mentiond in others 1/2/2/1 superconductors. This indicates a reduction of the superconducting basal plane anisotropy at high field. The reduction of the basal plane anisotropy was attributed to the strong Pauli paramagnetic effect and the resulting expansion of vortex cores near H_{c2} [12D1]. A model of vortices with a four-fold anisotropy in the vortex-vortex interaction potential has been elaborated [18O1].

The magnetic and superconductive properties of pseudoquaternary borides $Gd_xTm_{1-x}Ni_2B_2C$ [02M3], $Dy_{1-x}Tm_xNi_2B_2C$ [09L1, 10L3], $Tm_{1-x}Yb_xNi_2B_2C$ [04N2], $Tm_{1-x}Y_xNi_2B_2C$ [99F1, 01R2] and $Tm(Ni_{1-x}Co_x)_2B_2C$ [97S4] were investigated. In the $Gd_xTm_{1-x}Ni_2B_2C$ system, the T_s temperature decreases rapidly down to $x = 0.3$ [02M3]. The T_N and T_{sR} temperatures monotonically increase when increasing Gd content. The magnetic and superconducting phase diagrams have been determined.

The superconducting transitions temperatures, in $Dy_{1-x}Tm_xNi_2B_2C$, decrease rapidly when increasing x and shows a minimum at $x \cong 0.15$, then increasing gradually with x [10L3]. The magnetic transition temperatures, T_N, gradually decreases when increasing Tm content. Very small changes in the electronic structures of Dy and Tm ions were evidenced. Doping with Yb in $Yb_xTm_{1-x}Ni_2B_2C$ series, suppresses superconductivity partially, giving $T_s = 4.9$ K, for $x = 0.1$ [04N2]. In the zero field, the magnetic structure of sample with $x = 0.1$ is of $\mathbf{q}_F$-type. The $\mathbf{q}_F \rightarrow \mathbf{q}_{A1}$ phase transition was also observed, but at larger transition field, compared with that of parent Tm compound. The superconducting temperature for sample with x = 0.15 is $T_s = 2$ K; the $\mathbf{q}_F$ phase persists up to at least 1.8 T [04N2]. In $Tm(Ni_{1-x}Co_x)_2B_2C$ system, the coexistence between superconductivity and magnetism, was observed, for Co content up to x = 0.12 [97S4].

The composition dependences of superconducting and magnetic ordering temperatures were analysed in $TmNi_{2-x}M_xB_2C$ system, when Ni is partially substituted by M = Pt and Pd [97F1]. In the composition range $0.05 \le x \le 0.2$, the presence of Pt affects little the magnetic ordering temperatures, while these are reduced by Pd substitutions.

9.5 $R_2Ni_{10}B_5$, $R_3Ni_{19}B_{10}$, $RNi_{12}B_6$ and $R_2Ni_{15}B_9$ Compounds

The rare-earth metal poor part of R–Ni–B phase diagrams (<10 at.% R), revealed the presence of $R_2Ni_{10}B_5$, $R_3Ni_{19}B_{10}$, $RNi_{12}B_6$, and $R_2Ni_{15}B_9$ series. These are formed mainly with heavy rare-earths; those with light rare earths are present only in $RNi_{12}B_6$ series and their derivatives. In addition to the crystal structures of the mentioned series, those of $R_2Ni_5B_4$ and $CeNi_2B_3$ systems are also included in the chapter. The investigations on the mentioned compounds were focused mainly on the crystal structures, their physical properties being little studied.

The $R_2Ni_{10}B_5$ or $R_2(Ni_2B)_5$ compounds with R = Gd, Tb, Dy, Ho, Er, Tm, Yb, Lu and Y crystallize in an orthorhombic type structure, *Pbca* space group [91G3, 91K1]. There are two R sites, the first bonded in 2-coordinate geometry to 14 Ni and 5 B atoms and the second one in a 4-coordinate geometry to 15 Ni and 3 B atoms. In this lattice there are 10 inequivalent Ni positions and 5 boron sites [19M1].

The $R_3Ni_{19}B_{10}$ ($R\,R_2Ni_{19}B_{10}$) series with R = Gd, Tb, Dy, Ho, Er, Yb and Y have a monoclinic type lattice, space group *C2/m* [90K3, 91G5, 19M1]. The structure is tridimensional and consists of two R clusters and one $R_2Ni_{19}B_{10}$ framework. In the $R_2Ni_{19}B_{10}$ framework, R^{3+} is bonded in a 6-coordinate geometry to six B atoms. There are seven inequivalent Ni sites and three B sites.

Two structure types were found for $RNi_{12}B_6$ or $R(Ni_2B)_6$ compounds, the hexagonal $SrNi_{12}B_6$-type having space group $R\overline{3}m$ [71K1, 80J1, 81K2, 85A1, 90K2] and the orthorhombic $CeNi_{12}B_6$-type with non-centrosymmetric space group $Cmc2_1$ [85A1, 87D1, 88G1, 15M2, 19M1]. The $R\overline{3}m$-type was shown for compounds with R = Eu, Gd, Tb, Dy, Ho, Y and $Cmc2_1$-type when R = La, Ce, Pr, Gd, Tb, Dy, Ho and Y. When both modifications occur for the same compound, the hexagonal structure seems to be related with a lower B content than that of stoichiometric composition. No $RNi_{12}B_6$ compounds are formed with rare-earths having small radius (R = Tm, Yb, Lu)—Table 9.13. The hexagonal-type structure contains isolated B atoms, surrounded by distorted trigonal prisms of Ni atoms with one additional Ni upon one of the faces. The prisms are arranged in connected layers consisting of six-

Table 9.13 Crystal structures and lattice parameters of $R_2Ni_{10}B_5$, $R_3Ni_{19}B_{10}$, $RNi_{12}B_6$ and $R_2Ni_{15}B_9$ compounds

Compound	T (K)	Space group	Lattice parameters (nm)			References
			a	b	c	
$Gd_2Ni_{10}B_5$	RT	*Pbca*	1.7638(3)	0.9118(3)	0.9492(2)	[91G3]
$Tb_2Ni_{10}B_5$	RT	*Pbca*	1.7625(6)	0.9064(5)	0.9483(3)	[91G3]
$Tb_2(Ni_2B)_5$	RT	*Pbca*	0.9022	0.9491	1.7592	[20M1]
$Dy_2(Ni_2B)_5$	RT	*Pbca*	0.9011	0.9473	1.755	[20M1]
$Dy_2Ni_{10}B_5$	RT	*Pbca*	1.7602(4)	0.9010(4)	0.9470(3)	[91G3]
$Ho_2Ni_{10}B_5$	RT	*Pbca*	1.7588(4)	0.9001(2)	0.9477(2)	[91G3, 91K1]
$Er_2Ni_{10}B_5$	RT	*Pbca*	1.7545(4)	0.8977(4)	0.9441(3)	[91G3]
$Er_2(Ni_2B)_5$	RT	*Pbca*	0.8968	0.9434	1.7520	[20M1]
$Tm_2(Ni_2B)_5$	RT	*Pbca*	0.8935	0.9429	1.7478	[20M1]
$Tm_2Ni_{10}B_5$	RT	*Pbca*	1.7539(2)	0.8961(3)	0.9435(2)	[91G3]
$Yb_2Ni_{10}B_5$	RT	*Pbca*	1.7518(5)	0.8923(5)	0.9424(3)	[91G3]
$Yb_2Ni_{10}B_5$	RT	*Pbca*	1.7524(7)	0.8943(6)	0.9408(8)	[06V1]
$Lu_2Ni_{10}B_5$	RT	*Pbca*	1.7504(2)	0.8887(2)	0.9418(1)	[91G3]
$Y_2Ni_{10}B_5$	RT	*Pbca*	1.7594(5)	0.9017(3)	0.9478(3)	[91G3]
$Gd_3Ni_{19}B_{10}$	RT	*C2/m*	1.3215(8)	0.8747(5) $\beta =$ 91.50(8)°	0.5846(7)	[91G5]
$Gd_3Ni_{19}B_{10}$	RT	*C2/m*	0.7870 $\alpha = 88.83°$		0.5757 $\gamma =$ 67.179°	[20M1]
$Tb_3Ni_{19}B_{10}$	RT	*C2/m*	1.3208(10)	0.8747(7) $\beta =$ 91.16(8)°	0.5832(6)	[91G5]
$Dy_3Ni_{19}B_{10}$	RT	*C2/m*	1.3172(6)	0.8734(3) $\beta =$ 91.31(3)°	0.5785(1)	[91G5]
$Dy_3Ni_{19}B_{10}$	RT	*C2/m*	0.7864 $\alpha =$ 88.903°		0.5756 $\gamma =$ 67.141°	[20M1]
$Ho_3Ni_{19}B_{10}$	RT	*C2/m*	0.7884 $\alpha = 88.85°$		0.5758 $\gamma = 66.91°$	[20M1]
$Ho_3Ni_{19}B_{10}$	RT	*C2/m*	1.3177(2)	0.8719(2) $\beta =$ 91.33(1)°	0.5780(1)	[91G5]
$Er_3Ni_{19}B_{10}$	RT	*C2/m*	1.3101(9)	0.8674(6) $\beta =$ 91.05(8)°	0.5761(6)	[91G5]
$Er_3Ni_{19}B_{10}$	RT	*C2/m*	0.7858 $\alpha =$ 88.906°		0.5752 $\gamma =$ 66.967°	[20M1]

(continued)

Table 9.13 (continued)

Compound	T (K)	Space group	Lattice parameters (nm)			References
			a	b	c	
$Yb_3Ni_{19}B_{10}$	RT	*C2/m*	1.3139(4)	0.8676(3) β = 91.23(3)°	0.5754(1)	[06V1]
$Y_3Ni_{19}B_{10}$	RT	*C2/m*	1.3253(7)	0.8776(4) β = 91.18(4)°	0.5823(4)	[91G5]
$La_2Ni_5B_4$	RT	*C*12/*m*1	0.9786	0.5158 β = 125.205°	0.67889	[15V1]
$Ce_2Ni_5B_4$	RT	*C*12/*m*1	0.9759	0.5117 β = 125.22°	0.67203	[15V1]
$Nd_2Ni_5B_4$	RT	*C*12/*m*1	0.9783	0.5097 β = 125.687°	0.67648	[15V1]
$LaNi_{12}B_6$	RT	*Cc/mmm*	0.985(2)	0.731(2)	1.111(2)	[72K1]
$La(Ni_2B)_6$	RT	*Cmc*2_1	0.6062		1.1062 γ = 104.737°	[20M1]
$CeNi_{12}B_6$	RT	*Cmc*2_1	0.95897(3)	0.74067(2)	1.10681(3)	[15M2]
$CeNi_{12}B_6$	RT	*Cmc*2_1	0.963(2)	0.739(2)	1.113(3)	[71K1]
$Pr(Ni_2B)_6$	RT	*Cmc*2_1	0.6064		1.038 γ = 104.680°	[20M1]
$EuNi_{12}B_{6-x}$	RT	$R\bar{3}m$	0.9551		0.7404	[81K3]
$GdNi_{12}B_{6-x}$	RT	$R\bar{3}m$	0.9523(6)		0.7419(5)	[81K2]
$Gd(Ni_2B)_6$	RT	*Cmc*2_1	0.6033		1.0973 γ = 104.809°	[20M1]
$Tb(Ni_2B)_6$	RT	*Cmc*2_1	0.6027		1.0961 γ = 104.771°	[20M1]
$TbNi_{12}B_{6-x}$	RT	$R\bar{3}m$	0.9518(5)		0.7411(5)	[81K2]
$TbNi_{12}B_6$	RT	*Cmc*2_1	0.9536	0.7382	1.0991	[90D1]
$DyNi_{12}B_{6-x}$	RT	$R\bar{3}m$	0.9520(6)		0.7410(6)	[81K2]
$DyNi_{12}B_6$	RT	*Cmc*2_1	0.9534	0.7367	1.0983	[90K1, 99C3]
$Dy(Ni_2B)_6$	RT	*Cmc*2_1	0.6038		1.0946 γ = 104.877°	[20M1]
$Ho(Ni_2B)_6$	RT	*Cmc*2_1	0.6025		1.0929 γ = 104.856°	[20M1]
$HoNi_{12}B_6$	RT	*Cmc*2_1	0.9519(4)	0.7364(3)	1.0956(5)	[93G1]

(continued)

Table 9.13 (continued)

Compound	T (K)	Space group	Lattice parameters (nm) a	b	c	References
$HoNi_{12}B_{6-x}$	RT	$R\overline{3}m$	0.9484(3)		0.7356	[93G1]
$YNi_{12}B_{6-x}$	RT	$R\overline{3}m$	0.9526(3)		0.7421(3)	[81K2]
$YNi_{12}B_6$	RT	$Cmc2_1$	0.9597	0.7375	1.1003	[87D1]
$NdNi_{9.7}Si_{1.3}B_4$	RT	*P4/nmm*	1.12260(2)		0.78731(2)	[98W2]
$Nd_3Ni_{29}Si_4B_{10}$	RT	*P4/nmm*	1.12327(7)		0.78754(3)	[98Z1]
$Nd_3(Ni_{1-x}Si_x)_{33}B_{10}$ x = 0.1212	RT	*P4/nmm*	1.12319(2)		0.78730(5)	[99Z2]
$Tb_2Ni_{15}B_9$	RT	*Cmca*	1.5890(5)	1.1609(3)	1.228(4)	[91G6]
$Dy_2Ni_{15}B_9$	RT	*Cmca*	1.5895(5)	1.1573(5)	1.1229(7)	[91G6, 99C3]
$Dy_2(Ni_5B_3)_3$	RT	*Cmce*	0.9816		1.1251 $\gamma =$ 107.781°	[19M1]
$Ho_2(Ni_5B_3)_3$	RT	*Cmce*	0.9818		1.1234 $\gamma =$ 107.670°	[19M1]
$Ho_2Ni_{15}B_9$	RT	*Cmca*	1.6001(4)	1.1656(3)	1.1323(3)	[89O1, 91G6]
$Er_2Ni_{15}B_9$	RT	*Cmca*	1.5874(3)	1.1561(7)	1.1235(8)	[91G6]
$Er_2(Ni_5B_3)_3$	RT	*Cmce*	0.9799		1.1223 $\gamma =$ 107.785°	[19M1]
$Tm_2(Ni_5B_3)_3$	RT	*Cmce*	0.9815		1.1227 $\gamma =$ 107.932°	[19M1]
$Tm_2Ni_{15}B_9$	RT	*Cmca*	1.5856(5)	1.1565(4)	1.1179(5)	[91G6]
$Yb_2Ni_{15}B_9$	RT	*Cmca*	1.5848(6)	1.1560(4)	1.1201(5)	[91G6]
$Yb_2Ni_{15}B_9$	RT	*Cmca*	1.5925(2)	1.1590(1)	1.1232(1)	[05B1]
$Yb_2Ni_{15}B_9$	RT	*Cmca*	1.5865(4)	1.1563(4)	1.1206(3)	[06V1]
$Yb_2(Ni_5B_3)_3$	RT	*Cmce*	0.9767		1.1193 $\gamma =$ 107.564°	[19M1]
$Lu_2(Ni_5B_3)_3$	RT	*Cmce*	0.9765		1.1178 $\gamma =$ 107.871°	[19M1]
$Lu_2Ni_{15}B_9$	RT	*Cmca*	1.5829(7)	1.1544(6)	1.1179(3)	[91G6]
$Y_2Ni_{15}B_9$	RT	*Cmca*	1.5905(5)	1.1615(4)	1.1243(5)	[91G6, 95C3]
$Y_2(Ni_5B_3)_3$	RT	*Cmce*	0.5821		1.1245 $\gamma =$ 107.900°	[19M1]
$La_2Ni_5B_4$	RT	*C2/m*	0.5538 $\alpha =$ 59.364°		0.6812 $\gamma =$ 55.205°	[19M1]

(continued)

Table 9.13 (continued)

Compound	T (K)	Space group	Lattice parameters (nm)			References
			a	b	c	
$Nd_2Ni_5B_4$	RT	*C2/m*	0.952(4)	0.513(2), β = 100.5°	1.082(40)	[82B1, 82B2]
$Nd_2Ni_5B_4$	RT	*C2/m*	0.9783(2)	0.5097(1) β = 136.73(1)°	0.8016(1)	[85K2]
$CeNi_2B_3$	RT	*I4/mmm*	0.762(2)		0.863(2)	[71K1]

and three-membered rings. In orthorhombic-type structure the R is located in the center of a polyhedron formed by 18 Ni and 6 B atoms. The structure contains seven inequivalent Ni sites and four boron sites. A feature of the structure is the presence of slightly puckered 3^34^2 Ni-nets oriented in the (ab)-plane. Such sheets may give rise to highly anisotropic features of the Fermi surface and of the related electronic properties, as revealed in $CeNi_{12}B_6$ compound [15M2]. The remaining nickel atoms are inter-bounded to form pairs of empty [Ni_4] tetrahedra, coupled by common edges.

Since of their possible technical interest the Nd-M-Si-B quaternary systems with M = Co and Ni have been investigated. The existence of $NdNi_8SiB_3$ compound was initially reported [93W1, 97Z1]. The composition $NdNi_{9.7}Si_{1.3}B_4$ has been also mentioned, some impurities being also present in their XRD pattern [98W2]. Latter on [98B3, 98Z1], has been shown the presence of $Nd_3Ni_{29}Si_4B_{10}$ quaternary compound having tetragonal structure, space group *P4/nmm*, with two formula units per unit cell. The unit cell contains 2Nd, 7Ni, 1Si and 4B sites, two sites (Ni2*c*, Si8*i*) having mixed occupancy of Ni and Si. The coordination of R atoms is CN = 20, of Ni and Si, CN = 12 to 14 and of boron atoms CN = 7 and 8. One of Ni sites, Ni2*c*, has a quasi-spherical polyhedral coordination. The above Nd-Ni-Si-B compounds have crystal structure related to that of $NdNi_{12}B_6$-type, linking these ternary borides with $NdNi_9Si_2$ silicides ($BaCd_{11}$-type structure, space group *I41/amd*). The above compounds were shown to be members of quaternary $Nd_3(Ni_{1-x}Si_x)_{33}B_{10}$ series and are formed by replacing Ni with Si in $Nd(Ni_{1-x}Si_x)_{12}B_6$ alloys [99Z2]. For x = 0, the structure is orthorhombic of $CeNi_{12}B_6$-type. Evidence for formation of quaternary phase was shown for x = 0.041, with the new phase dominating at x = 0.083. This phase is not present when x > 0.29. The above compounds can be also formed by adding a suitable amount of B to $NdNi_9Si_2$ phase. With reference to $Nd_3(Ni_{1-x}Si_x)_{33}B_{10}$ series, solid solutions, were shown in the composition range $0.09 \le x \le 0.18$ [99Z2]. As representative of the above series, the ^{57}Fe doped $R_3Ni_{29}Si_4B_{10}$ compounds with R = La to Lu were investigated by Mössbauer spectroscopy [98W2, 00Z1]. The Fe atoms preferentially occupy 2*c* crystallographic site with $4\overline{m}2$ symmetry.

Amorphous alloys with compositions $(LaNi_{12}B_6)_x(LaNi_5)_{1-x}$ and $0 \le x \le 0.3$ were prepared and studied for hydrogen storage [96O2].

The $R_2Ni_{15}B_9$ compounds with R = Tb, Dy, Er, Tm, Lu, Y crystallize in orthorhombic type structure having *Cmca* [89O1, 91G6, 05B1] or *Cmce* [19M1] space-groups. The structure of $Yb_2Ni_{15}B_9$ single crystal was refined [05B1]. There are two R, ten Ni and six B inequivalent sites. The R atoms have high coordination number, CN = 20 or 18 and Ni atoms, CN = 16, 14 and 13, respectively [05B1]. The coordination spheres of the B atoms are distorted Archimedean cubes, which are connected, with one another, by square faces (B2-B3 and B4-B6). The B1 atoms center a trigonal prism composed of nickel atoms. The B5 atoms are also situated in a trigonal prism, in which, one of the apices is occupied by B5 atoms. As a result, the B atoms are either isolated (B1) or form pairs (other boron atoms).

The $R_2Ni_5B_4$ compounds crystallize in a monoclinic structure having *C*12/*m*1 [15V1] or *C*2/*m* [19M1] space groups. The *C*2/*m* structure is three-dimensional. The R is bonded in a 5-coordinate geometry to five B atoms. There are two inequivalent Ni sites and two B sites [19M1].

The $Nd_2Ni_5B_4$ crystallizes in a monoclinic type structure having *C*2/*c* or *C*2/*m* space group [82B1]. The presence of R_2Ni_2B compounds having an unknown crystal structure has been also mentioned [95C3, 99C3, 06V1].

The lattice parameters of the above series are given in Table 9.13.

The band structure calculations on $R_2(Ni_2B)_5$ compounds with R = Tb, Dy, Er and Tm show that these compounds are in paramagnetic state [20M1]. The same behavior was shown in $R(Ni_2B)_6$ series except when R = Gd [20M1]. The $La_2Ni_5B_4$ compound is also non-magnetic—Table 9.14.

The $LaNi_{12}B_6$ is a Pauli-type paramagnet [15M2]. The specific heat $C_p(T)$ has a smooth temperature dependence [18S1]. At T < 4.5 K, the specific heat follows a $C_p(T) = \gamma T + \beta T^3$ behavior with $\gamma = 33.3$ mJmol^{-1} K^{-2}, $\beta = 0.2763$ mJmol^{-1} K^{-4} and $\theta_D(LT) = 304$ K. The electrical resistivity shows a metallic trend, described by the Bloch-Gruneisen model, which accounts for electron–phonon scattering and is complemented by a residual resistivity, arising from static defects and impurities.

The $CeNi_{12}B_6$ has an antiferromagnetic ground state with magnetic ordering temperature $T_N = 1.85$ K [15M2]—Table 9.14. The easy axis of magnetization is the **c** direction, the magnetic moments along **a-** and **b**-axis being essentially quenched at low temperatures. Along **c**-axis the magnetic susceptibility, χ, follows a Curie–Weiss-type dependence, the effective moment, at T > 300 K, being close to that of Ce^{3+} free ion. A rather sharp second order type antiferromagnetic phase transition, at

Table 9.14 Magnetic properties of $RNi_{12}B_6$ compounds[a]

Compound	Magnetic structure and magnetic moments	M_s (μ_B/f.u.)	T_N (K)	M_{eff} (μ_B/f.u.)	θ (K)	References
$CeNi_{12}B_6$	AFM		1.85	$\cong 2.54\ \mu_B$ (T > 300 K)		[15M2]
$GdNi_{12}B_6$ (BS)	FM	7.09				[20M1]
$TbNi_{12}B_6$				10.0	−10	[90D1]

[a]Band structures of some compounds, from the above series, were reported [19M1, 20M1]

T_N, was also evidenced from the heat capacity data. A change in temperature dependence of electrical resistivity, is also shown at T_N. From the magnetic contributions to the specific heat was concluded, that the full entropy of $Ce^{3+}(4f^1)$ crystal field ground state doublet, $S = R\mathit{ln}2$ is reached at $T = 15$ K ($T \cong 8\ T_N$). The entropy, at T_N, has only a value $0.3R\mathit{ln}2$; accordingly, a significant amount of entropy is transferred towards the paramagnetic region. These features relate to a huge single ion anisotropy and possibly highly anisotropic electronic features, arising from nickel sheets oriented parallel to the (ab) plane. At low temperatures, the above features of $CeNi_{12}B_6$ can be described by using a model of gapped antiferromagnetic magnons, yielding a large spin-wave excitation gap, $\Delta \cong 2.2T_N$, which arises from anisotropy with respects to magnetic interactions and/or from single ion anisotropy [01C8, 15M2]. The high pressure study, of electrical resistivity, evidenced an increase of the Néel temperatures, from $T_N = 1.85$ K at ambient pressure to $T_N = 2.3$ K, at $p = 1.4$ GPa [15M2]. These data evidenced a limited relevance of Kondo correlations, which places $CeNi_{12}B_6$ to the magnetically ordered part of the Doniach phase diagram [77D1, 97C4], where RKKY coupling outperforms Kondo correlations.

The paramagnetic Curie temperature of $TbNi_{12}B_6$ is negative, suggesting that this compound is antiferromagnetically ordered [90D1].

9.6 $R_2Ni_{21}B_6$, $R_2Ni_{15}B_6$, RNi_7B_3 and $RNi_{6.5}B_3$ Compounds

The $R_2Ni_{21}B_6$ compounds crystallize in a $Cr_{23}C_6$ type structure [72B1]. In these borides B atoms are occupying the 24*e* position of C atoms, while in the crystallographic positions 4*a*, 32*f* and 48 *h*, occupied by Cr atoms in the prototype, Ni atoms are located—Table 9.15a. The compounds are formed mainly with rare-earths having small atomic radius (R = Ho-Lu) [71K1, 80C1, 89K1, 91G1, 97V1, 04V2]. In the structure, there are columns of Ni polyhedra consisting of tetragonal antiprisms, filled by B atoms, separated by a cubo-octahedron and an empty cube. Boron atoms have tetragonal antiprismatic coordination. The R atoms are coordinated by 4Ni*f* and 12Ni*h*, which form a Friauf polyhedron. The R-Ni*f* distances are shorter by 13% for R = Ho [91G1], 15.5% for R = Er, by 21% for R = Yb [09V1] and by 17% for R = Lu [80C1] than the sum of metallic radii. Thus, the compounds can be formed only with partial occupation of rare-earth site, their composition being described as $R_{2-x}Ni_{21}B_6$. The homogeneity ranges are dependent on the rare earth and situated at $0.1 < x < 0.9$ for R = Er, $0.1 < x < 0.3$ for R = Yb and $x < 0.35$ when R = Lu, due to partial occupancy of the R8*c* position [97V1]. The anisotropic elongation of the Ni*f* atoms displacement ellipsoid, being part of the coordination sphere of the R atoms, in the direction towards the central atom, is related to the above mentioned vacancies [10V1]. The Ni atoms have CN = 13, 14 and 18 and the isolated B atoms are situated in tetragonal antiprism (CN = 8), formed by Ni atoms.

The $R_2Ni_{15}B_6$ compounds with R = Dy, Ho, Y crystallize in a monoclinic $P2_1/c$-type structure [89G1, 91G2].

Table 9.15 Lattice sites and coordinates

(a) $Er_{1.82}Ni_{21}B_6$ structure having $Fm\bar{3}m$ space group [04V2, 09V1]

Atom	Position	x	y	z	Atomic environment
Er	8*c*	1/4	1/4	1/4	16-vertex Frank-Kasper Ni_{16}
Ni1	4*a*	0	0	0	18-vertex polyhedron $Ni_{12}B_6$
Ni2	32*f*	0.3875(1)	0.3875(1)	0.3875(1)	pseudo Frank-Kasper B_3Ni_9Er
Ni3	48*h*	0	0.1690(1)	0.1690(1)	14-vertex polyhedron $B_2Ni_{10}Er_2$
B	24*e*	0.270(2)	0	0	square antiprism Ni_8

(b) $ErNi_7B_3$ structure having $I4_1/amd$ space group [04K4]

Atom	Position	x	y	z	Atomic environment
Er	8*e*	0	0	0.28388	14Ni4B
Ni1	8*c*	0	¼	1/8	2R8Ni4B
Ni2	12*g*	0.1886	x	0	2R8Ni4B
Ni3	16*g*	0	0.2874	0.3932	2R7Ni3B
Ni4	16*h*	0	0.1611	0.5560	2R9Ni2B
B1	8*e*	0	0	0.9422	8Ni1B
B2	16*h*	0.289	1/4	1/4	2R7Ni

(c) $ErNi_{7.9}B_2$ structure having $C2/m$ space group [04V1]

Atom	Position	Occupation	x	y	z	Atomic environment
Er1	4*e*	1	0	0.37727(5)	1/4	18Ni
Er2	8*f*	1	0.34190(3)	0.57946(4)	0.71297(4)	17Ni,2B
Ni1	4*a*	0.83(1)	0	0	1/2	2Er,10Ni
Ni2	8*f*	0.520(8)	0.0209(2)	0.0804(3)	0.3018(4)	2Er,9Ni
Ni3	8*f*	1	0.08283(8)	0.2347(1)	0.5294(1)	2Er,10Ni,2B
Ni4	8*f*	1	0.16447(8)	−0.0026(1)	0.5423(1)	2Er,10Ni,2B
Ni5	8*f*	1	0.17112(9)	0.2051(1)	0.3916(1)	2Er,9Ni,2B
Ni6	8*f*	1	0.26661(7)	0.1607(1)	0.2679(1)	2Er,9Ni,3B
Ni7	8*f*	1	0.32112(8)	0.0483(1)	0.5106(1)	2Er,9Ni,3B
Ni8	8*f*	1	0.32533(8)	0.0851(1)	0.7583(1)	2Er,9Ni,3B
Ni9	8*f*	1	0.54007(7)	0.0987(1)	0.4779(1)	2Er,9Ni,2B
Ni10	8*f*	1	0.58535(7)	0.1397(1)	0.2700(1)	2Er,10Ni,2B
Ni11	8*f*	1	0.68189(8)	0.2309(1)	0.2550(1)	2Er,9Ni,2B
Ni12	8*f*	1	0.91277(8)	0.1889(1)	0.3651(1)	2Er,11Ni
Ni13	8*f*	0.897(7)	0.97086(9)	0.1684(2)	0.6277(1)	2Er,11Ni
B1	8*f*	1	0.2210(8)	0.1855(12)	0.5995(12)	1Er,7Ni
B2	8*f*	1	0.3986(6)	0.0672(10)	0.4005(10)	7Ni
B3	8*f*	1	0.7203(6)	0.1124(10)	0.4087(9)	1Er,7Ni

The RNi_7B_3 compounds crystallize in a tetragonal structure having $I4_1/amd$ space group. The structure is formed, as in $R_2Ni_{21}B_6$ borides, only with rare-earths having small atomic radius, R = Er, Yb and Lu [04K4], as well as with Mg [10L4]. The structural framework is composed of $B1Ni_8$ and $B2Ni_7$ polyhedrons, the R atoms being located in the space between these polyhedrons. The $B1Ni_8$ polyhedrons are connected by face-sharing, forming B_2 pairs while the $B2Ni_7$ polyhedrons are connected with each other by vertex-sharing or edge-sharing; the B2 atoms are isolated one from another. As already mentioned, when increasing boron content in R-M-B borides, the isolated boron atoms are substituted by B_2 pairs [83K1]. The $B1Ni_8$ and $B2Ni_7$ polyhedrons are connected by sharing a triangle face. The R atoms have eighteen neighbors, including fourteen Ni and four B atoms—Table 9.15b. The distance between Er and Ni4 positions are shorter by 9% than the sum of their metallic radii.

The $ErNi_{7.9}B_2$ compound crystallizes in a monoclinic structure, space group $C2/c$ [04V1]—Table 9.15c. The Er atoms have the largest coordination number, CN = 18 and 19. The Ni atoms are in 14- or 13-fold coordinations, except for Ni1 and Ni2 atoms, which reside in partially occupied positions and have smaller CNs (12 and 11, respectively). The B atoms occupy the center position of distorted trigonal prisms of Ni atoms capped on one (B2) or two (B1 and B3) quadrilateral faces. This coordination is typical of B atoms in low boron compounds. Since the prisms share no faces or edges, the B atoms are isolated from each other. The prisms are arranged so that the Ni atom capping one of the quadrilateral faces of a prism containing B2 atom is, at the same time, a corner of a prism containing a B1 atom, while the prims containing B3 atoms are isolated and share no atoms with other prisms. Short distances can be seen between Ni1, Ni2 and Ni3 atoms, located in partially occupied positions [04V1]. There is a significant covalent bonding between Ni and B, as evidenced by the short distances between them.

The $RNi_{6.5}B_3$ (R = Ho, Er, Yb, Lu) compounds crystallize in a cubic type structure [04C1, 06V1]. The lattice parameters of the above mentioned compounds are listed in Table 9.16.

The Ni-based powders containing (wt%): 3–4 B, 4–5 Si, 14–16 Cr with 5 CeO_2, deposited on carbon-steel by flame spray fusing processing, contained $Ce_2Ni_{21}B_6$ phase. The microhardness of the alloy containing this phase, was higher than in the other coatings [16L1].

The $Lu_{1.65}Ni_{21}B_6$ compound was reported to be paramagnetic [12G2, 13K2]. The temperature dependence of the magnetic susceptibility follows a modified Curie law—Table 9.17. The mean effective nickel moment is small, $M_{eff}(Ni) \cong 0.33$ μ_B/atom. Band structure calculations [12G2], suggest the presence of small ordered Ni moment, particularly at 4*a* position. A Pauli-type paramagnetism was shown in $Lu_{1.65}Ni_{20.75}Cu_{0.25}B_6$, where a small nickel fraction was substituted by Cu. When Ni is partially replaced by cobalt, the $Lu_{1.65}Ni_{17.5}Co_{3.5}B_6$ boride becomes magnetic ordered. The low field measurements suggest the presence of a cluster glass type behavior. Assuming an effective Ni moment similar with that in parent compound, the effective cobalt moment, $M_{eff} = 3.12$ μ_B/atom, was estimated from Curie constant. Below T = 6 K, the specific heat, $C_p(T)$, for $Lu_2Ni_{21}B_6$ follows a relation $C_p(T) =$

Table 9.16 Crystal structures and lattice parameters of $R_2Ni_{21}B_6$, $R_2Ni_{15}B_6$, RNi_7B_3 and $RNi_{65}B_3$ compounds

Compound	T (K)	Space group	Lattice parameters (nm)			β°	References
			a	b	c		
$Ce_2Ni_{21}B_6$	RT	*Fm*3*m*	1.0678(6)				[71K1]
$Ce_2Ni_7B_6$	RT	$Fm\bar{3}m$	1.069				[20M1]
$Ho_2Ni_{21}B_6$	RT	*Fm*3*m*	1.0641				[91G1, 93G1]
$Er_2Ni_{21}B_6$	RT	*Fm*3*m*	1.0640(5)				[80C1, 04C1]
$Er_{1.82}Ni_{21}B_6$	RT	$Fm\bar{3}m$	1.06520(2)				[09V1]
$Er_2Ni_{21.8}B_{6.3}$	RT	$Fm\bar{3}m$	1.0637				[13K2]
$Tm_2Ni_{21}B_6$	RT	*Fm*3*m*	1.0633(5)				[80C1]
$Tm_2Ni_{21}B_6$	RT	$Fm\bar{3}m$	1.0643				[13K2]
$Tm_2(Ni_7B_2)_3$	RT	$Fm\bar{3}m$	1.054				[20M1]
$Yb_2Ni_{21}B_6$[a]	RT	$Fm\bar{3}m$	1.0636(5)				[80C1, 13K2]
$Yb_{1.708}Ni_{21}B_6$	RT	*Fm*3*m*	1.06183(2)				[06V1]
$Yb_{1.860}Ni_{21}B_6$	RT	*Fm*3*m*	1.06412(2)				[06V1]
$Yb_{1.70}Ni_{21}B_6$	RT	$Fm\bar{3}m$	1.06243(2)				[09V1]
$Yb_{1.86}Ni_{21}B_6$	RT	$Fm\bar{3}m$	1.06412(2)				[09V1]
$Lu_2Ni_{21}B_6$[a]	RT	*Fm*3*m*	1.0632(5)				[80C1, 90D1]
$Lu_2Ni_{21}B_6$	RT	$Fm\bar{3}m$	1.0624				[13K2]
$Lu_2(Ni_7B_2)_3$	RT	$Fm\bar{3}m$	1.061				[20M1]
$Lu_{1.65}Ni_{21}B_6$	RT	$Fm\bar{3}m$	1.06269(4)				[09V1]
$Lu_{1.60}Ni_{20.7}B_{6.3}$	RT	$Fm\bar{3}m$	1.06256(2)				[12G2]
$Lu_{1.30}Y_{0.3}Ni_{20.6}B_{6.5}$	RT	$Fm\bar{3}m$	1.06328(5)				[12G2]

(continued)

Table 9.16 (continued)

Compound	T (K)	Space group	Lattice parameters (nm)			β°	References
			a	b	c		
$Lu_{1.5}Ni_{17.3}Co_{3.5}B_{6.0.4}$	RT	$Fm\bar{3}m$	1.06306(4)				[12G2]
$Lu_{1.6}Ni_{20.8}Cu_{0.3}B_6$	RT	$Fm\bar{3}m$	1.06329(5)				[12G2]
$Dy_2Ni_{15}B_6$	RT	$P2_1/c$	1.4201(8)	1.0649(5)	0.9578(5)	94.57(5)	[99C3]
$Ho_2Ni_{15}B_6$	RT	$P2_1/c$	1.4223	1.0672	0.9582	94.23	[89G1, 93G1]
$Y_2Ni_{15}B_6$	RT	$P2_1/c$	1.4224(6)	1.0679(4)	0.9577(5)	94.37(4)	[91G1, 95C3]
$ErNi_7B_3$[b]	RT	$I4_1/amd$	0.7665		1.5584		[84K1, 04C1]
$ErNi_7B_3$[c]	RT	$I4_1/amd$	0.76577(2)		1.55798(5)		[04K4]
$ErNi_7B_3$	RT	$I4_1/amd$	0.9468 $\alpha = 132.45°$	0.9468 $\beta = 132.45°$	0.9468 $\gamma = 69.52°$		[20M1]
$YbNi_7B_3$	RT	$I4_1/amd$	0.7674		1.5474		[88K1]
$YbNi_7B_3$	RT	$I4_1/amd$	0.76471(2)		1.55453(8)		[06V1]
$LuNi_7B_3$	RT	$I4_1/amd$	0.7634		1.5546		[90D1]
$ErNi_{7.9}B_2$	RT	$C2/c$	1.6626(2)	0.9525(1) $\beta = 117.76(1)°$	1.0686(1)		[04V1]
$HoNi_{6.5}B_3$	RT	cubic	0.7446(6)				[93G1]
$ErNi_{6.5}B_3$	RT	cubic	0.7735(3)				[04C1]
$YbNi_{6.5}B_3$	RT	cubic	0.7726(3)				[06V1]

[a] $R_2Ni_{21}B_6$ have a = 0.7473 nm (R = Yb) and 0.7505 nm (R = Lu), with $\alpha = \beta = \gamma = 60°$ [20M1]
[b] Lattice parameters a = b = c = 0.9468 nm, $\alpha = \beta = 132.450°$, $\gamma = 69.520°$ or a = 0.763 nm, c = 1.556 nm [20M1]
[c] Composition $Er_4Ni_{29}B_{10}$

Table 9.17 Magnetic properties of $R_2Ni_{21}B_6$ and RNi_7B_3 compounds

Compound	Magnetic structure and magnetic moments	M_s (μ_B/f.u.)	T_c (K)	C (emμ.K/f.u.)	M_{eff} (μ_B/f.u.)	θ (K)	References
$Ce(Ni_7B_2)_3$ (BS)		1.396					[20M1]
$Er_2Ni_{21.3}B_{6.5}$	AFM		0.61		9.60	−5.7	[13K2]
$Er_{1.63}Ni_{21}B_6$	T > 1.8 K, no magnetic ordering 50 K < T < 400 K, $\chi = C(T - \theta)^{-1}$				9.50	−6.02	[09V1]
$Tm_2Ni_{21}B_6$	PM, T > 50 mK				7.54	−1.3	[13K2]
$Yb_2Ni_{21}B_6$	PM, T > 50 mK				4.76	−17	[13K2]
$Yb_{1.7}Ni_{21}B_6$	T > 1.8 K, no magnetic order; χ = C/(T-θ) at 150 K < T < 400 K				6.26	−32.9	[09V1]
$Lu_{1.65}Ni_{21}B_6$	PM, $\chi = \chi_0 + C/T$, $\chi_0 = 1.7 \cdot 10^{-3}$ emu/f.u			0.276[a]	1.5		[12G2]
$Lu_2Ni_{21}B_6$ (BS)	weak ferromagnet Ni:$M_a = 0.2\ \mu_B$, $M_f = 0.03\ \mu_B$, $M_h = 0.05\ \mu_B$	0.96 (theor)					[12G2]
$Lu_{1.65}Ni_{20.75}Cu_{0.25}B_6$	Pauli paramagnet, $\chi_0 = 1.5 \cdot 10^{-3}$ emu/f.u						[12G2]
$Lu_{1.65}Ni_{17.5}Co_{3.5}B_6$	FM, weak feromagnet	2.75[b]	39	4.303[c]	5.86	71	[12G2]
$ErNi_7B_3$					9.8	2	[90D1]

[a]From C value, $M_{eff}(Ni) = 0.33\mu_B$/atom
[b]$T = 1.8$ K, $\mu_o H = 7$ T
[c]Assuming $M_{eff}(Ni) = 0.33\ \mu_B$/atom, the effective cobalt moment is $M_{eff}(Co) = 3.12\ \mu_B$/atom

$\gamma T + \beta T^3$, where γ and β are the electronic and phonon specific heat coefficients, respectively [13K2]. The Debye temperature is $\theta_D = 442$ K.

The $Er_2Ni_{21.8}B_{6.3}$ compound is antiferromagnetically ordered, at $T_N = 0.61$ K [13K2]. At $T > T_N$ the magnetic susceptibility follows a Curie–Weiss temperature dependence, the effective moment per formula unit being close to that of free Er^{3+} ion—Table 9.17. The temperature dependence of the specific heat, $C_p(T)$, has a maximum at T_N. The magnetic contribution to the entropy, $S_{mag}(T)$, reaches a value $R\mathit{ln}2$ at T_N and approaches $R\mathit{ln}4$ at $T > T_N$. The 16-fold degeneracy of the $J = 15/2$ multiplet of Er^{3+} ion, is lifted under the cubic CEF into two doubles and three quartets. The tail in $C_p(T)$, at $T > T_N$, indicates that short range magnetic correlations may contribute to the release of the entropy for the CEF ground state. The CEF ground state was suggested to be two closely separated doublets or a quartet.

The $R_2Ni_{21}B_6$ compounds with R = Tm, Y are not magnetically ordered at $T >$ 0.05 K or $T \geq 1.8$ K [09V1, 13K2]. A broad maximum in $C_p(T)$, was observed in $Tm_2Ni_{21}B_6$, at $T \cong 0.9$ K, explained also by lifting of the degeneracy associated with the CEF ground state [13K2]. The 13-fold degeneracy of the $J = 6$ multiplet, of Tm^{3+} ions, splits into three triplets, a Γ_3 doublet and two singlets, under cubic CEF. The analysis of $S_{mag}(T)$, suggests that the CEF ground state is a non-Kramers Γ_3 doublet, leading at low temperatures, to a constant magnetic susceptibility. The experimentally determined magnetic susceptibility at $T < 3$ K is not temperature dependent, confirming that Γ_3 is the ground state. The non-Kramer Γ_3 doublet, possesses a quadrupolar moment entering into glass state, probably owing the presence of atomic disorder.

No magnetic phase transition was detected in the specific heat $C_p(T)$ of $Yb_2Ni_{21}B_6$ boride, in zero magnetic field. A shoulder like anomaly has been shown at $T \cong 7$ K [13K2]. The eightfold degeneracy of the $J = 7/2$ multiplet of Yb^{3+} ion, split into two doublets (Γ_6 and Γ_7) and a quartet (Γ_8) under cubic CEF. The CEF energy level scheme with Γ_6 doublet ground state accurately reproduces the shoulder like anomaly. The atomic disorder in $R_2Ni_{21}B_6$ compounds with R = Tm and Yb plays an important role in maintaining the long-range nature of multipolar interactions [13K2].

The electrical resistances of $R_2Ni_{21}B_6$ compounds with R = Y [09V1], R = Lu [13K3], and R = Er, Tm, Yb, Lu [13K2] indicated a normal metallic behavior. The high residual resistances, confirm the presence of atomic disordering.

9.7 $R_3Ni_7B_2$ Compounds

The $R_3Ni_7B_2$ phase is the only formed in the $R_{n+2}Ni_{3n+4}B_{2n}$ series (n = 1) and is obtained by the intergrowth of hexagonal $MgZn_2$ Laves phase and $CeCo_3B_2$-type slabs [84P1, 84R1]—Table 9.18. The $R_3Ni_7B_2$ compounds crystallize mainly in a hexagonal type structure, space group *P6₃/mmc* [79K1, 79K2, 80K1, 81K1, 83F1]. Space group *P6/mmm* was also reported for $R_3Ni_7B_2$ borides with R = Sm and Eu [83F1] or $P\bar{6}m2$ when R = Eu [19M1]. In the *P6₃/mmc* type structure the R atoms

Table 9.18 Lattice sites of $Yb_3Ni_7B_2$ having hexagonal structure $P6_3/mmc$ [06V1, 06V2]

Atom	Sites	x	y	z
Yb	*2c*	1/3	2/3	1/4
Yb	*4f*	1/3	2/3	0.03117(9)
Ni	*2a*	0	0	0
Ni	12*k*	0.8344(2)	2x	0.1492(2)
B	2*b*	0	0	1/4
B	2*d*	1/3	2/3	3/4

are located in 2*c* and 4*f* sites, the Ni in 2*a* and 12*k* and boron in 2*b* and 2*d* types of sites. The lattice parameters of $R_3Ni_7B_2$ series are listed in Table 9.19.

The physical properties of $R_3Ni_7B_2$ compounds with R = Sm, Eu, Gd, Tb, Dy, Ho, Er, Tm, Yb, Lu [83F1], R = Y [96F1] and R = Gd Tb, Dy, Ho, Er, [07B2, 08B4] were investigated. According to [83F1], the samples with R = Gd, Tb, Dy, Ho, Er, Tm are antiferromagnetically ordered. Later on, starting from magnetic measurements, XPS and band structure calculations, the compounds with R = Gd, Tb, Dy, Ho and Er were shown to be nearly ferromagnetic [07B2, 08B4]—Fig. 9.14. The Ni moments, determined from band structure calculations are antiparallel oriented to of those rare-earths and contribute only by 1%, to the total magnetizations—Table 9.20. The saturation magnetizations, at T = 4.2 K, for $R_3Ni_7B_2$ compounds with R = Gd or Tb, which have Curie temperatures $T_c > 30$ K are close to the g_JJ values of free R^{3+} ions. The differences between expected magnetizations and measured values at T = 4.2 K, increased as T_c values are lowered, since of thermal effects, these being determined at reduced temperatures $T/T_c = 0.36$ (R = Ho) or $T/T_c = 0.60$ (R = Er). If the experimental magnetizations are extrapolated at T = 0 K, assuming Brillouin type temperature dependences, values close to g_JJ ones of free R^{3+}, were also obtained for these compounds—Table 9.20.

The thermal variations of reciprocal susceptibilities follow Curie–Weiss type dependences with Curie constants little higher than those of rare-earths. Assuming that the Curie constants of rare earths are given by their free ion values, effective nickel moments of (1.2–1.5) μ_B/Ni atom were determined. The Ni magnetic behavior has been described by the spin fluctuations model [91M1, 08B4].

The 4f orbitals of rare-earths, in this series, keep their localized character, as suggested by XPS spectra. The presence of unoccupied Ni3d states has been also shown, in agreement with band structure calculations [08B4]. The computed R5d band polarizations, at both R sites (2*c*, 4*f*), follow linear dependences on the De Gennes factor, with rates determined by their local environments.

The $R_3Ni_7B_2$ compounds with nonmagnetic elements (R = Y, La) are Pauli-type paramagnets [83F1, 96F1]. A paramagnetic behavior was also reported for compounds with R = Sm, Eu and Yb [83F1].

The band structures of $R_3Ni_7B_2$ compounds with R = Sm, Gd, Dy, Er, Tm, Yb, Lu and Y were calculated [19M1].

Table 9.19 Space groups and lattice parameters of $R_3Ni_7B_2$ compounds[a]

Compound	T (K)	Space group	Lattice parameters (m)			References
			a	b	c	
$Sm_3Ni_7B_2$	RT	$P6_3/mmc$	0.5160		1.4418	[19M1]
$Sm_3Ni_7B_2$	RT	$P6/mmm$	0.5015		0.6925	[83F1]
$Sm_3Ni_7B_2$	RT	$P6_3/mmc$	0.5142(2)		1.4348(10)	[79K1, 80K1]
$Eu_3Ni_7B_2$	RT	$P\bar{6}m2$	0.5225		0.7249	[19M1]
$EuNi_7B_2$	RT	$P6/mmm$	0.5040		0.6925	[83F1]
$Gd_3Ni_7B_2$	RT	$P6_3/mmc$	0.5115(2)		1.4342(10)	[79K1, 08B4]
$Gd_3Ni_7B_2$	RT	$P6_3/mmc$	0.5118		1.426	[83F1]
$Gd_3Ni_7B_2$	RT	$P6_3/mmc$	0.5109		1.4338	[19M1]
$Tb_3Ni_7B_2$	RT	$P6_3/mmc$	0.5097(2)		1.4335(10)	[79K1, 08B4]
$Tb_3Ni_7B_2$	RT	$P6_3/mmc$	0.5097		1.431	[83F1]
$Dy_3Ni_7B_2$	RT	$P6_3/mmc$	0.5078(2)		1.4331(10)	[79K1, 81K3, 82C1]
$Dy_3Ni_7B_2$	RT	$P6_3/mmc$	0.5080(5)		1.4315(4)	[08B4]
$Dy_3Ni_7B_2$	RT	$P6_3/mmc$	0.5078		1.4331	[99C3]
$Dy_3Ni_7B_2$	RT	$P6_3/mmc$	0.5071		1.4288	[19M1]
$Ho_3Ni_7B_2$	RT	$P6_3/mmc$	0.5063(2)		1.4285(10)	[79K1, 79K2]
$Ho_3Ni_7B_2$	RT	$P6_3/mmc$	0.5068		1.429	[83F1]
$Ho_3Ni_7B_2$	RT	$P6_3/mmc$	0.5051(9)		1.4307(5)	[08B4]
$Er_3Ni_7B_2$	RT	$P6_3/mmc$	0.5060(2)		1.4276(10)	[79K1, 80K1]
$Er_3Ni_7B_2$	RT	$P6_3/mmc$	0.5045(1)		1.4248(3)	[08B4]
$Er_3Ni_7B_2$	RT	$P6_3/mmc$	0.5041		1.4255	[19M1]
$Tm_3Ni_7B_2$	RT	$P6_3/mmc$	0.5028		1.4235	[19M1]
$Tm_3Ni_7B_2$	RT	$P6_3/mmc$	0.5013(2)		1.4231(10)	[79K1, 80K1]
$Tm_3Ni_7B_2$	RT	$P6_3/mmc$	0.5007		1.431	[83F1]
$Yb_3Ni_7B_2$	RT	$P6_3/mmc$	0.5016(2)		1.4211(2)	[06V1]
$Yb_3Ni_7B_2$	RT	$P6/mmm$	0.4992		0.6975	[83F1]
$Yb_3Ni_7B_2$	RT	$P6_3/mmc$	0.5011		1.4019	[19M1]
$Lu_3Ni_7B_2$	RT	$P6_3/mmc$	0.4983		1.4141	[19M1]
$Lu_3Ni_7B_2$	RT	$P6_3/mmc$	0.5015(2)		1.4180(10)	[79K1]
$Y_3Ni_7B_2$	RT	$P6_3/mmc$	0.5128(2)		1.4343(10)	[79K1, 80K1]
$Y_3Ni_7B_2$	RT	$P6_3/mmc$	0.4997		1.395	[96F1]
$Y_3Ni_7B_2$	RT	$P6_3/mmc$	0.5079		1.4257	[19M1]
$Y_3Ni_7B_2C$	RT	$P6_3/mmc$	0.4985		1.394	[96F1]

[a]see also [84R1]; $Nd_3Ni_7B_2$ was reported to crystallize in $P6_3/mmm$ space group [79K1], but not confirmed [80K1] or observed [82B2]

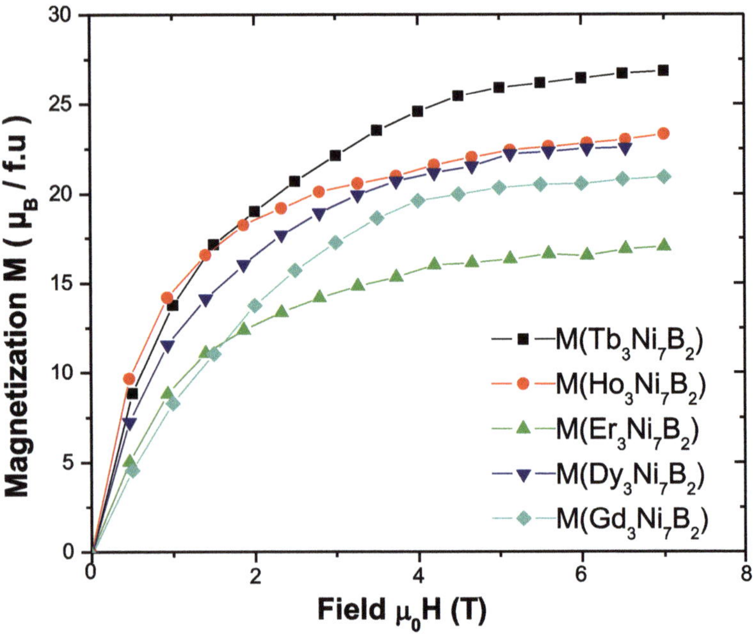

Fig. 9.14 $R_3Ni_7B_2$ compounds with R = Gd, Tb, Dy, Ho, Er: magnetization isotherms at T = 4.2 K [08B4]

The $Ce_2(Co_{1-x}Ni_x)_7B_3$ system forms solid solutions for $x \leq 0.3$ [05S1]. The Ni substitutes preferentially Co, at the 2*a* site. The compounds become Pauli paramagnetic for $x \geq 0.24$ (see also Sect. 8.4).

A cluster glass contribution seems to be superposed over a dominant ferromagnetic behavior of $Y_3Ni_7B_2C$ borocarbide [96F1]. An effective nickel moment $M_{eff}(Ni) \cong 0.90\ \mu_B$/atom has been obtained, at $T > T_c$, close to those determined in $R_3Ni_7B_2$ compounds.

The ^{151}Eu Mössbauer spectra of $Eu_3Ni_7B_2$, evidenced the presence of only one line, as a superposition of those from the Eu 2*c* and 4*f* sites, while two lines were shown in case of ^{155}Gd spectrum in $Gd_3Ni_7B_2$ [83F1]. A high quadrupole coupling constant was shown at Gd2*c* site [83F1]—Table 9.21.

9.8 $R_{m+n}Ni_{5m+3n}B_{2n}$ Series with (m = 1, n = 1), (m = 2, n = 1) and (m = 2, n = 3)

There are three ternary borides, containing ≅ 16.6 at.% rare earths in $R_{m+n}Ni_{5m+3n}B_{2n}$ series: RNi_4B (m = 1, n = 1), $Nd_3Ni_{13}B_2$ (m = 2, n = 1) and $Lu_5Ni_{19}B_6$ (m = 2, n = 3)—Table 9.1. These compounds are formed by stacking m$CaCu_5$ (RM_5) and n$CeCo_3B_2$ layers along **z** direction, keeping a lattice parameter almost unmodified [73K1, 84P1, 84R1].

Table 9.20 Magnetic properties of $R_3Ni_7B_2$ compounds

Compound	Magnetic structure and magnetic moments	M_s (μ_B/f.u.)	T_c (T_N) (K)	C (emuK/f.u.)	M_{eff} (μ_B/R atom)	M_{eff} (μ_B/Ni atom)	θ (K)	References
$Sm_3Ni_7B_2$	FM	0.54[a]	35	0.23	0.79		29	[83F1]
$Eu_3Ni_7B_2$				4.0	3.25		−30	[83F1]
$Gd_3Ni_7B_2$	FIM Ni^b: $M_a = 0.086\ \mu_B$, $M_k = 0.046\ \mu_B$	20.9[c]	38.5	25.8		1.54	39.4	[08B4]
$Gd_3Ni_7B_3$ (BS)	FIM	21.246						[19M1]
$Gd_3Ni_7B_2$		12.96[d]	37	21.7	7.62		46	[83F1]
$Tb_3Ni_7B_2$	FIM Ni^b: $M_a = 0.073\ \mu_B$, $M_k = 0.036\ \mu_B$	26.6[c]	30.6	37.6		1.58	31.2	[08B4]
$Tb_3Ni_7B_2$		4.05[a]	30	37.6	10.0		5	[83F1]
$Dy_3Ni_7B_2$	FIM Ni^b: $M_a = 0.048\ \mu_B$, $M_k = 0.024\ \mu_B$	25.5[c]	22.8	43.67		1.31	22.9	[08B4]
$Dy_3Ni_7B_2$		13.74[a]	12	42.0	10.6		4	[83F1]
$Ho_3Ni_7B_2$	FIM Ni^b: $M_a = 0.038\ \mu_B$, $M_k = 0.02\ \mu_B$	24[c]	14.6	43.25		1.11	14.7	[08B4]
$Ho_3Ni_7B_2$		15.15[a]	8	42.0	10.6		1	[83F1]
$Er_3Ni_7B_2$	FIM Ni^b: $M_a = 0.029\ \mu_B$, $M_k = 0.016\ \mu_B$	18.7[c]	7	35.58		1.21	8.4	[08B4]

(continued)

Table 9.20 (continued)

Compound	Magnetic structure and magnetic moments	M_s (μ_B/f.u.)	T_c (T_N) (K)	C (emuK/f.u.)	M_{eff} (μ_B/R atom)	M_{eff} (μ_B/Ni atom)	θ (K)	References
$Er_3Ni_7B_2$		11.55[a]	7	33.5	9.52		−7	[83F1]
$Tm_3Ni_7B_2$		8.88[a]	5	20.5	7.43		−1	[83F1]
$Yb_3Ni_7B_2$	Pauli paramagnet, $\chi_0 = 2.4 \cdot 10^{-3}$ emu/f.u[e]							[83F1]
$Lu_3Ni_7B_2$	Pauli paramagnet, $\chi_0 = 2.2 \cdot 10^{-3}$ emu/f.u							[83F1]
$Y_3Ni_7B_2$	Pauli paramagnet, $\chi_0 = 1.4 \cdot 10^{-3}$ emu/mol							[96F1]
$Y_3Ni_7B_2C$	FM, 40 K < T < 150 K $\chi = \chi_0 + C(T - \theta)^{-1}$, χ_0 = -3.1·10^{-4} emu/mol	0.28[d]	36			0.90	−38.2	[96F1]

[a] $T = 4.2$ K, $\mu_0 H = 1.8$ T
[b] Band structure calculation
[c] $T = 4.2$ K, $\mu_0 H = 7$ T
[d] $T = 5$ K, $\mu_0 H = 5$ T
[e] 1 emu/g = 1Am^2/kg

Table 9.21 Data obtained by Mössbauer spectroscopy

Compound	T (K)	Nucleus	Site	Hyperfine parameters				References
				δ (mm/s)	B_{eff} (T)	eqQ (MHz)	θ°	
$Gd_2Ni_7B_2$[a]	4.1	^{155}Gd	2*c*	0.22(1)	21.5(5)	722(10)	90(4)	[83F1]
			4*f*	0.02(1)	11.5(4)	154(5)	71(8)	
	77		2*c*	0.30(10)		768(30)		
			4*f*	0.01(1)		120(30)		
$Eu_2Ni_7B_2$[b]	4.1	^{151}Eu		2.70(5)				[83F1]
	300			2.55(3)				

[a] $^{155}SmPd_3$ source
[b] SmF_3 source

The RNi_4B compounds, crystallizes mainly in hexagonal structure, having *P6/mmm* space group [93M1]. In the structure, layers, of $CaCu_5$- and $CeCo_3B_2$-types, are alternatively stacked along **c**-axis. Parallel to **c**-axis, the R atoms form linear chains, which are located at channels having hexagonal cross-section. Each R1*b* atom is coordinated by 6Ni2*c* and 12Ni6*i* atoms, whereas each R1*a* atom is surrounded by 6B and 12Ni6*i* atoms. The B atoms occupy the centers of trigonal prism formed by 6Ni6*i* atoms—Table 9.22a.

The presence of superstructures was shown in some RNi_4B compounds, associated with deviations from ideal composition and described by the $R_{1-x}Ni_{4+x}B$ one [75K1, 93M1, 99B2]. These compounds have lattice parameters $a = na_0$, multiple of the subcell structure, a_0, the same as that of RNi_4B phases. The $R_{1-x}Ni_{4+x}B$ superstructures are also derived from $CaCu_5$ and $CeCo_3B_2$ types. The planar nets that are perpendicular to [00z] direction, form all these structures. The number of planar nets in the structure determine the c-lattice parameters, a-lattice constants being n time larger than those of $CeCo_4B$ type layer.

The YNi_4B compound crystallizes in a superstructure having $CeCo_4B$-type as a subcell, with lattice constant $a = 3a_0$ [73K3, 75K1, 93M1, 99B2]. Superstructures with a_0 sublattice parameters, were also reported for RNi_4B compounds with R = La ($a = 6a_0$) [84S1, 85S1], R = Ce ($a = 8a_0$) [71K1, 75K1] or ($a = 4a_0$) [82C1, 95N2] as well as for R = Er ($a = 3a_0$) [04K4, 13A1] and R = Yb ($a = 3a_0$) [06V1]—Table 9.23. Although the as cast $ErNi_4B$ crystallizes in a superstructure [04K4], the annealed sample has a $CeCo_4B$-type lattice [07S1].

In the $Y_{0.915}Ni_{4.123}B$ superstructure, there are five Y, ten Ni and three B inequivalent sites. From these sites four Y1a, Ni8(2e), Ni9(2e) and Ni10(2e), along the direction [00z] have smaller occupation number than one—Table 9.22b. Contrary to positions corresponding to $CeCo_4B$-type structure, several atoms are displaced in a direction towards the empty site (0, 0, 1/2) [99B2]. The positions 6*l* of the B3 atom and 6*m* of Ni4 one are reversed in comparison with B2*d* and Ni2*c* in YNi_4B substructure. According to [99B2, 06V2] the superstructure features are characterized by the presence of Kagomé-mesh Ni6 layers, which alternate with $(Y, \square)_3Ni_4B_2$ and

Table 9.22 Lattice sites and atomic coordinates

(a) $CeNi_4B$ having *P6/mmm* space group[a] [71K1, 06V2]

Atom	Sites	x	y	z	Atomic environment
Ce1	1*a*	0	0	0	pseudo Frank-Kasper $B_6Ni_{12}Ce_2$
Ce2	1*b*	0	0	1/2	pseudo Frank-Kasper $Ni_{18}Ce_2$
Ni1	2*c*	1/3	2/3	0	14-vertex polyhedron $Ni_9Ce_3B_2$
Ni2	6*i*	1/2	0	0.213	13-vertex polyhedron $B_2Ni_7Ce_4$
B	2*d*	1/3	2/3	0	trigonal prism Ni_6

(b) $Y_{0.915}Ni_{4.123}B$ having *P6/mmm* space group [99B2, 06V2]

Atom	Sites	x	y	z	occ	Atomic environment [06V2]
Y1	6*k*	0.32074	0	1/2	1	Pseudo Frank-Kasper $B_4Ni_{14}Y_2$
Y2	6*j*	0.34448	0	0	1	Pseudo Frank-Kasper $Ni_{16}B_2Y_2$
Y3	2*d*	1/3	2/3	1/2	1	Pseudo Frank-Kasper $Ni_{12}B_6Y_2$
Y4	2*c*	1/3	2/3	0	1	Pseudo Frank-Kasper $Ni_{18}Y_2$
Y5	1*a*	0	0	0	0.465	
Ni1	24*r*	0.16544	0.49779	0.29744	1	Pseudo Frank-Kasper $B_2Ni_7Y_4$
Ni2	12*o*	0.16506	0.33011	0.2673	1	Pseudo Frank-Kasper $B_2Ni_7Y_4$
Ni3	12*n*	0.17083	0	0.1866	1	non-coliniar B_2
Ni4	6*m*	0.09151	0.18301	1/2	1	
Ni5	6*l*	0.22324	0.4465	0	1	Icosahedron Ni_8BY_3
Ni6	6*l*	0.55977	0.11955	0	1	14-vertex Frank-Kasper $Ni_9Y_3B_2$
Ni7	6*i*	1/2	0	0.2904	1	pseudo Frank-Kasper $B_2Ni_7Y_4$
Ni8	2*e*	0	0	0.096	0.325	
Ni9	2*e*	0	0	0.326	0.53	
Ni10	2*e*	0	0	0.400	0.25	
B1	6*m*	0.2152	0.4304	1/2	1	Trigonal prism Ni_6
B2	6*m*	0.5540	0.1081	1/2	1	Trigonal prism Ni_6
B3	6*l*	0.1237	0.2475	0	1	Trigonal prism Ni_6

(c) $Nd_3Ni_{13}B_2$ having *P6/mmm* space group [81K2]

Atom	Sites	x	y	z	Atomic environment [06V2]
Nd1	1*a*	0	0	0	Pseudo Frank-Kasper $B_6Ni_{12}Nd_2$
Nd2	2*e*	0	0	0.328	Pseudo Frank-Kasper $Ni_{18}Nd_2$
Ni1	6*i*	1/2	0	0.134	13-vertex polyhedron $B_2Ni_7Nd_4$
Ni2	4*h*	1/3	2/3	0.323	Anticuboctahedron Ni_9Nd_3
Ni3	3*g*	1/2	0	1/2	Cuboctahedron Ni_8Nd_4
B	2*c*	1/3	2/3	0	Trigonal prism Ni_6

(d) $Lu_5Ni_{19}B_6$[a] having *P6/mmm* space group [85K1]

Atom	Sites	x	y	z	Atomic environment [06V2]
Lu1	1*a*	0	0	0	Pseudo Frank-Kasper $Ni_{12}B_6Lu_2$

(continued)

Table 9.22 (continued)

(d) $Lu_5Ni_{19}B_6$[a] having *P6/mmm* space group [85K1]

Atom	Sites	x	y	z	Atomic environment [06V2]
Lu2	2*e*	0	0	0.204	Pseudo Frank-Kasper $Ni_{18}Lu_2$
Lu3	2*e*	0	0	0.407	Pseudo Frank-Kasper $B_6Ni_{12}Lu_2$
Ni1	6*i*	1/2	0	0.082	13-vertex polyhedron $B_2Ni_7Lu_4$
Ni2	6*i*	1/2	0	0.315	13-vertex polyhedron $B_2Ni_7Lu_4$
Ni3	4*h*	1/3	2/3	0.201	Icosahedron Ni_9Lu_3
Ni4	3*g*	1/2	0	1/2	Rhombic dodecahedron $B_4Ni_6Lu_4$
B	4*h*	1/3	2/3	0.407	Trigonal prism Ni_6
B	2*c*	1/3	2/3	0	Trigonal prism Ni_6

(e) $LaNi_3B$ having orthorhombic type structure *Imma* space group [05F1]

Atom	Sites	x	y	z
La	4*e*	0	1/4	0.56990(6)
Ni1	8*g*	1/4	0.07336(10)	1/4
Ni2	4*a*	0	0	0
B	4*e*	0	1/4	0.1292(12)
B[b]	4*e*	0	1/4	0.8554(15)

[a]Transformed from published data; origin shift (0 0 1/2)
[b]Peak of residual electron density was modeled by this position, 79(4)% occupied by B atoms

$(Y, \square)_3Ni_2B_4$ layers (a Ni_2B or NiB_2 hexagon mesh, part of the hexagons of which are centered by an Y atom), along [001] direction. There are additional disordered Y and Ni along [0 0 z]-axis [06V2].

The crystal structure of $Er_{0.917}Ni_{4.123}B$ has been also refined in space group *P6/mmm* [04K4]. The absence of R atom in the position (0, 0, 1/2) of $CeCo_4B$ structure type and the small occupation, for position (0, 0, 0), determines the channel formed by Ni atoms along the [00z] direction, to be filled only by atoms whose sites are partially occupied; the ≅ 1.4 R atoms are replaced by ≅2.1 Ni atoms. As compared to that of $Y_{0.915}Ni_{4.12}B$, there are two additional partially occupied positions (Ni2*e*, Ni1*b*).

The $NdNi_4B$ boride crystallizes in space group *Imma* [03S1]. The structure is also derived from that of the hexagonal $CeCo_4B$-type, with $a = a_0$, $b = c_0$, and $c = 3a_0\sqrt{3}$. The structure can be regarded as formed from $CeCo_4B$-type fragments, portions of which are shifted with respect to each other by (1/2)c. In $CeCo_4B$-type lattice, the atoms are located on the layers ($z = 0, 0.5, \pm 0.29$), situated normally to the **c**-axis, whereas in the $NdNi_4B$ structure, the atoms are placed in the planes 1/4 and 3/4 with wavelike layers of Ni atoms in between.

The RNi_4B compounds were mainly synthesized by melting the constituent elements. A chemical route was also used for obtaining RNi_4B borides with R = La, Y [96K6]. The mechanical milling of YNi_4B compound leads to a decrease of the

Table 9.23 Lattice parameters of RNi_4B, $R_3Ni_{13}B_2$, $R_6Ni_{19}B_6$ and RNi_3B compounds

Compound	T (K)	Space group	Lattice parameters (nm)			References
			a	b	c	
$LaNi_4B$	RT	*P6/mmm*	0.5122(4)		0.6990(9)	[72K1, 83K3]
$LaNi_4B$[a]	RT		3.071(10)		0.6995(12)	[84S1, 85S1]
$LaNi_4B$[b]	RT		0.5121(1)		0.6995(1)	[94H2]
$LaNi_4BH_{0.9}$[b]	RT		0.5172		0.6997	[85S1]
$CeNi_4B$	RT	*P6/mmm*	0.5005		0.6992	[85S1]
$CeNi_4BH_{0.6}$	RT	*P6/mmm*	0.5127		0.6839	[85S1]
$CeNi_4B$	RT	*P6/mmm*	0.5013(2)		0.6984(4)	[94H2]
$CeNi_4B$[c]	RT		4.00(1)		0.698(1)	[71K1]
$PrNi_4B$[d]	RT		0.5071		0.6972	[85S1]
$PrNi_4BH_{0.2}$[d]	RT		0.5083		0.6960	[85S1]
$PrNi_4B$	RT	*P6/mmm*	0.5065(1)		0.6972(1)	[94H2]
$PrNi_4B$	RT	*P6/mmm*	0.5063(2)		0.6951(6)	[81K1, 83K3]
$NdNi_4B$	RT	*P6/mmm*	0.5054(1)		0.6969(3)	[94H2]
$NdNi_4B$	RT	*P6/mmm*	0.5043(2)		0.6941(6)	[82B2, 83K3]
$NdNi_4B$[e]	RT	*Imma*	0.5057(2)	0.6980(2)	2.6271(3)	[03S1]
$NdNi_4B$	300	*Imma*	0.5052(1)	0.6974(1)	2.6251(3)	[13A1]
$NdNi_4B$	24	*Imma*	0.502(1)	0.695(1)	2.621(1)	[13A1]
$SmNi_4B$	RT	*P6/mmm*	0.5029(2)		0.6956(9)	[80K1, 83K3]
$SmNi_4B$	RT	*P6/mmm*	0.5022		0.6945	[00M1]
$SmNi_4B$	RT	*P6/mmm*	0.5027(3)		0.6954(2)	[94H2]
$SmNi_4B$	RT	*P6/mmm*	0.50305(4)		0.69565(3)	[16Y1]
$SmNi_3MnB$	RT	*P6/mmm*	0.51196(7)		0.69691(5)	[16Y1]
$SmNi_3FeB$	RT	*P6/mmm*	0.50611(3)		0.69484(3)	[16Y1]
$SmNi_3CoB$	RT	*P6/mmm*	0.50375(4)		0.69355(3)	[16Y1]
$SmNi_3CuB$	RT	*P6/mmm*	0.50652(7)		0.69434(5)	[16Y1]
$EuNi_4B$	RT	*P6/mmm*	0.4989(6)		0.6947(16)	[82C1]
$GdNi_4B$[f]	RT	*P6/mmm*	0.4980		0.6933	[82C1, 83K2]
$GdNi_4B$	RT	*P6/mmm*	0.5011		0.6956	[93H1, 02T2]
$GdNi_4B$	RT	*P6/mmm*	0.50116(4)		0.69599(5)	[07S1]
$TbNi_4B$	RT	*P6/mmm*	0.4998(1)		0.6951(3)	[94H2]
$TbNi_4B$	RT	*P6/mmm*	0.4979(1)		0.6941(4)	[81K3, 82C1]
$TbNi_4B$	RT	*P6/mmm*	0.4992(1)		0.6945(1)	[13A1]
$TbNi_4B$	33	*P6/mmm*	0.497(1)		0.692(1)	[13A1]
$TbNi_4B$	RT	*P6/mmm*	0.49943(5)		0.69484(7)	[07S1]
$DyNi_4B$	RT	*P6/mmm*	0.4911(1)		0.6940(1)	[81K3]
$DyNi_4B$	RT	*P6/mmm*	0.4986(1)		0.6950(4)	[94H2]

(continued)

Table 9.23 (continued)

Compound	T (K)	Space group	Lattice parameters (nm)			References
			a	b	c	
$DyNi_4B$	RT	*P6/mmm*	0.5000		0.6966	[02T2]
$HoNi_4B$	RT	*P6/mmm*	0.4962(4)		0.6935(11)	[81K3, 82C1]
$HoNi_4B$	RT	*P6/mmm*	0.49696(4)		0.69419(5)	[00G2]
$HoNi_4B$	RT	*P6/mmm*	0.4975(1)		0.6947(3)	[94H2]
$HoNi_4B$	RT	*P6/mmm*	0.4969(6)		0.69419(5)	[95G1, 00G2]
$HoNi_4B$	RT	*P6/mmm*	0.4967(1)		0.6937(1)	[13A1
$HoNi_4B$	19	*P6/mmm*	0.496(1)		0.692(1)	[13A1]
$ErNi_4B$	RT	*P6/mmm*	0.4949(4)		0.6931(11)	[81K3, 82C1]
$ErNi_4B$	RT	*P6/mmm*	0.4961(1)		0.6937(4)	[94H2]
$ErNi_4B$	RT	*P6/mmm*	0.49614(2)		0.69360(3)	[07S1]
$ErNi_4B$	RT	*P6/mmm*	0.4957(1)		0.6932(1)	[13A1]
$Er_{0.9}Ni_{4.1}B$	RT		1.4870(2)		0.6931(1)	[13A1]
$Er_{0.917}Ni_{4.09}B$	RT	*P6/mmm*	1.48399(3)		0.69194(3)	[04K4]
$TmNi_4B^g$	RT	*P6/mmm*	0.4960(2)		0.6917(5)	[81K3, 82C1]
$YbNi_4B$	RT	*P6/mmm*	0.4938(4)		0.6929(10)	[81K3, 82C1]
$YbNi_4B$	RT	*P6/mmm*	0.4956(1)		0.6928(4)	[94H2]
$YbNi_4B$	RT	*P6/mmm*	0.4935(1)		0.6905(1)	[06V1]
$YbNi_4B$	RT	*P6/mmm*	1.4817(2)		0.6901(1)	[06V1]
$LuNi_4B$	RT	*P6/mmm*	0.4934(3)		0.6918(8)	[81K3, 82C1]
YNi_4B	RT	*P6/mmm*	0.4977(4)		0.6942(5)	[73N1]
YNi_4B	RT	*P6/mmm*	0.4977(1)		0.6940(3)	[81K3]
YNi_4B	RT	*P6/mmm*	0.4987		0.6948	[03P2]
YNi_4B^d	RT		0.4983(1)		0.6949(3)	[94H2]
YNi_4B	RT		1.489(5)		0.691(2)	[73K3, 75K1]
YNi_4B	RT		1.496(5)		0.695(2)	[93M1]
$Y_{0.915}Ni_{4.123}B^h$	RT	*P6/mmm*	1.49085(10)		0.69196(8)	[99B2]
$Y_{0.915}Ni_{4.119}B^h$	RT	*P6/mmm*	1.49124(8)		0.69272(9)	[99B2]
$Y_{0.914}Ni_{4.120}B^h$	RT	*P6/mmm*	1.4897(2)		0.68942(2)	[99B2]
$La_3Ni_{13}B_2$	RT	*P6/mmm*	0.5072(2)		1.0942(8)	[81K1]
$La_3Ni_{13}B_2$	RT	*P6/mmm*	0.50255		1.09590	[05P2]
$La_3Ni_{13}B_2$	RT	*P6/mmm*	0.5094(6)		1.0986(21)	[83S2]
$La_3Ni_{13}B_2$	RT	*Cmcm*	0.5074	0.876	1.1108	[19M1]
$Ce_3Ni_{13}B_2$	RT	*P6/mmm*	0.4954(2)		1.0972(8)	[81K1]
$Pr_3Ni_{13}B_2$	RT	*P6/mmm*	0.5019(1)		1.0929(5)	[81K1]
$Pr_3Ni_{13}B_2$	RT	*P6/mmm*	0.5048		1.0960	[19M1]
$Nd_3Ni_{13}B_2$	RT	*P6/mmm*	0.5023		1.0921	[19M1]
$Nd_3Ni_{13}B_2$	RT	*P6/mmm*	0.5005(3)		1.0904(9)	[81K3, 84P1]

(continued)

Table 9.23 (continued)

Compound	T (K)	Space group	Lattice parameters (nm) a	b	c	References
$Nd_3Ni_{13}B_2$	RT	*P6/mmm*	0.50255(1)		1.0959(0)	[05P2]
$Nd_3Ni_{10}Co_3B_2$	1.5	*P6/mmm*	0.50035(2)		1.0865(1)	[07P1]
$Sm_3Ni_{13}B_2$	RT	*P6/mmm*	0.4990(2)		1.0885(8)	[81K1]
$Sm_3Ni_{13}B_2$	RT	*P6/mmm*	0.5006		1.0968	[19M1]
$Eu_3Ni_{13}B_2$	RT	*P6/mmm*	0.4981(1)		1.0884(2)	[85D1]
$Gd_3Ni_{13}B_2$	RT	*P6/mmm*	0.4976(4)		1.0878(15)	[81K1]
$Tb_3Ni_{13}B_2$	RT	*P6/mmm*	0.4954(2)		1.0893(5)	[83K3]
$Tb_3Ni_{13}B_2$	RT	*P6/mmm*	0.4954(1)		1.0887(4)	[85D1]
$Tb_3Ni_{13}B_2$	RT	*P6/mmm*	0.4939		1.0911	[19M1]
$Dy_3Ni_{13}B_2$	RT	*P6/mmm*	0.4922		1.0906	[19M1]
$Dy_3Ni_{13}B_2$	RT	*P6/mmm*	0.4949(2)		1.0909(7)	[83C2, 83K3]
$Ho_3Ni_{13}B_2$	RT	*P6/mmm*	0.4943(2)		1.090(1)	[83C2, 83K3]
$Er_3Ni_{13}B_2$	RT	*P6/mmm*	0.4938(2)		1.090(1)	[83C2, 04C1]
$Er_3Ni_{13}B_2$	RT	*P6/mmm*	0.4901		1.0905	[19M1]
$Tm_3Ni_{13}B_2$	RT	*P6/mmm*	0.4920		1.0933	[19M1]
$Tm_3Ni_{13}B_2$	RT	*P6/mmm*	0.4925(1)		1.0865(1)	[83C2, 83K3]
$Yb_3Ni_{13}B_2$	RT	*P6/mmm*	0.4976(1)		1.0838(4)	[85D1]
$Yb_3Ni_{13}B_2$	RT	*P6/mmm*	0.4907		1.0725	[19M1]
$Y_3Ni_{13}B_2$	RT	*P6/mmm*	0.4933		1.0852	[19M1]
$Y_3Ni_{13}B_2$	RT	*P6/mmm*	0.4942(3)		1.0886(11)	[81K2, 95C3]
$Y_3Ni_{13}B_2$	RT	*P6/mmm*	0.49579(1)		1.09141(3)	[05P3]
$Y_3Ni_{13}B_2$	RT	*P6/mmm*	0.49524(1)		1.09061(5)	[14P1]
$Lu_5Ni_{19}B_6$	RT	*P6/mmm*	0.4943(1)		1.7161(9)	[85K1, 90D1, 06V2]
$Lu_5Ni_{19}B_6$	RT	*P6/mmm*	0.4916		1.6971	[85K1]
$LaNi_3B$	RT	*Imma*	0.49698(8)	0.71337(8)	0.83001(9)	[05F1]
$LaNi_3BD_{2.7}$	RT	*Cmcm*	1.07709(7)	1.60852(10)	0.76365(5)	[05F1]
$LaNi_3B$	RT		1.235(2)	1.080(2)	0.968(2)	[72K1]
$LaNi_3B$	RT	*Cmcm*	0.5074	0.5062	1.1108	[19M1]

[a] Single crystal, subcell, $a = 6a_0$, $a_0 = 0.5118(17)$ nm
[b] a parameter subcell value
[c] $a = 8a_0$
[d] Anomalous XRD patterns
[e] $a = a_0$, $b = c_0$, $c = 3a_0\sqrt{3}$;
[f] a subcell parameter [94H2]
[g] RNi_4B lattice parameters are also given by [20M1]
[h] $a = 3a_0$

intensities and to broadening of diffraction lines, due to the presence of amorphous state [03T2].

The $R_3Ni_{13}B_2$ compounds crystallize in space group *P6/mmm* [81K1, 83C2]. As already mentioned, the structure is formed by the intergrowth of $CaCu_5$ and $CeCo_3B_2$-type slabs in the ratio 2:1—Sect. 8.7. The R atoms are distributed in two different crystallographic sites (1*a*, 2*e*), the Ni atoms in three different positions (4*h*, 6*i*, 3*g*) and boron in the 2*c* position—Table 9.22c. In the structure, BNi_6 trigonal prisms share edges to form infinite slabs. There are no B-B contacts. The Ni sites 2*c* and 3*g* in $CaCu_5$-type structure, are structurally related to the (2*c*, 4*h*) and (3*g*, 6*i*) sites, respectively, in the 3/13/2 structure, with the 2*c* sites, in the latter filled by boron. The rare-earth 1*a* site in the 1/5 structure, splits into two sites, 1*a* and 2*e* in 3/13/2 structure.

The $Lu_5Ni_{19}B_6$, is the single boride present in $R_5Ni_{19}B_6$ series [85K1]. It is formed by the intergrowth of $CaCu_5$ and $CeCo_3B_2$-type slabs in the ratio 2/3. The BNi_6 trigonal prisms share triangular faces and edges to form infinite triple slabs. There are no B–B contacts The Lu is distributed in three types of sites, Ni in four and B in two positions—Table 9.22d.

The two structural features: local site volume and the presence of disclinations so as to disentangle their role in local magnetism of $CeCo_4B$ and $Nd_3Ni_{13}B_2$ type structure were analyzed [88B1]. It was concluded that factors such as intersite coupling (implicit in the topology of Wigner–Seitz cells) and site volume are both relevant to the local magnetism.

The $Y_{0.915}Ni_{4.12}B$ compound, annealed at T = 1370 K, do not show any structural transition in the temperature range 15 K $\leq$ T $\leq$ 300 K [99B2]. The $Er_{0.917}Ni_{4.09}B$ decomposes, after annealing at T = 1070 K [04C1].

A maximum uptake of 1.5 hydrogen atoms was reported in RNi_4B (R = La, Pr) borides [84S1, 85S1], smaller than those adsorbed in RNi_5 compounds [90B1]. The small hydrogen capacity result from exclusion of hydrogen from sites located in the vicinity of the R-B mid-plane and Ni6*i* layers. The hydrogen occupies sites located between R atoms, in the R-Ni basal plane. The R-B plane should act as a barrier to hydrogen diffusion, along **z** direction. There is a strong La-B interaction in $La_3Ni_{13}B_2$, which will withdraw electrons, which would otherwise be suitable for bonding with hydrogen [83S2]. In the hydrogenated sample, the boron interacts with electrons from adjacent Ni layers, removing electrons from potential bonding with hydrogen. Taking the above into account, the hydrogen uptake was estimated at 13.7 atoms per formula unit.

The hydrogen storage in $(La,Ce)(Ni,Mn,Cu)_5(Fe_{0.43}B_{0.57})_x$ alloys has been investigated [13P1,14S1]. The alloys have $LaNi_5$-matrix phase and as secondary phase the $La_3Ni_{13}B_2$ one. The $La_3Ni_{13}B_2$ content increases when increasing the $Fe_{0.43}B_{0.57}$ fraction. The same behavior was shown when increasing boron content in $La_{15}Ni_{70}Fe_{10-x}Mn_7B_x$ alloys with x $\leq$ 3 [14W1] or in $La_{15}Ni_{72}Fe_2Mn_2B_2Al_2$ with $\leq$5wt% graphene [18L1]. High rate dischargeability has been shown in these composite electrodes.

$LaNi_3B$ compound crystallizes in an orthorhombic-type structure, a defect variant of the hexagonal $CeCo_3B_2$-type [05F1]. The B atoms are coordinated by strongly deformed trigonal Ni prisms with two side faces capped by La atoms, as commonly evidenced in borides structures [92R1]—Table 9.22f. Lanthanum is coordinated by a distorted hexagonal prism of Ni atoms; two of the Ni_4 faces are capped by boron atoms. The La atoms are situated in the distorted hexagonal arrays of nickel prisms of which only half are filled by boron. Compared to the hexagonal $CeCo_3B_2$-type structure, in which all the transition metal prisms are filled by boron, the La atoms in $LaNi_3B$ move away from the center of the nickel prism arrays toward the empty Ni prism and alters those filled by boron atoms along **b**-axis. This fact presumably, causes the orthorhombic distortion of the structure. The Ni atoms have 11-fold (Ni1,[$La_4Ni_5B_2$]) or tenfold (Ni2,[$La_4Ni_4B_2$]) coordinations. Hydrogenation of $LaNi_3BH_y$ induces a symmetry change, from *Imma* to *Bmmb* (*Cmcm* standard setting) space group [04F1, 05F1]. At the transition, the structure expands, in the (ab) plane, by $\cong$ 8% and contracts along **c**-axis (b-axis in standard *Cmcm*) by $\cong$ 3%. The cell contraction of $LaNi_3BH_2$ system can be correlated with the rearrangement of lanthanum atoms in the channels of Ni–B framework. The unit cell in *Bmmb* doubles along **a** and **c** directions. For standard setting *Cmcm*, **b** and **c** need to be interchanged. The $LaNi_3BD_{2.73}$ structure has been described in space group *Cmcm*. Four nearly fully occupied interstitial hydrogen sites were located, having tetrahedral, trigonal-prismatic and trigonal–bipyramidal environments. The lattice parameters reported in an earlier paper [72K1], for $LaNi_3B$, are different from those given by [05F1]—Table 9.23.

The linear expansion coefficients α_a, α_c as well as of the volume, α_v in $Y_{0.915}Ni_{4.12}B$ boride have a maximum located at T $\cong$ 200 K [99B2]. The volume cell, follows a temperature dependence described by the relation v(nm^3) = 1.32737(5) − 1.01(14)10^{-5} T + 1.58(10)10^{-7}T^2 − 2.67(21)10^{-10}T^3, where T is in K [99B2]. The $HoNi_4B$ has a nearly isotropic behavior, the expansion coefficients being α_a = 1.3·10^{-5} K^{-1} and α_c = 1.0·10^{-5} K^{-1} [00G2].

A preferential substitution of Fe atoms at the 2*c* site was shown in $YCo_{4-x}Fe_xB$ compounds [00C5]—Sect. 8.6. The $RCo_{4-x}Ni_xB$ series with R = Y [04C2, 14C1] and R = Pr [09A2] form solid solutions in all the composition range. A more complex superstructure was shown in $YCo_{4-x}Ni_xB$, stabilized by an appropriate annealing treatment. In addition to that reported for YNi_4B with triple a- and b-axis, an additive superstructure was proposed along the **c**-axis, either commensurate or incommensurate, due to presence of B atoms in Co sites [01C2, 04C2]. In $YCoNi_3B$, the Ni prefers the 6*i* sites. The $NdCo_3NiB$ boride crystallizes in $CeCo_4B$-type structure. After hydrogenation and cycling, at T = 373 K, the space group changes to *P6mm* [12S1].Evidence for partial Co/Ni ordering was found in pristine $NdCo_3NiB$ but not in the corresponding deuteride. Solid solutions are formed in $YCo_{4-x}Cu_xB$ for x ≤ 1.5 [04B1]. The presence of solid solutions, in all the composition range was shown in $R_{1-x}R'_xNi_5$ series as well as in $Gd_xY_{1-x}Co_3NiB$ system [04B2].

The structural stability of $R_3Ni_{13-x}Co_xB_2$ series with R = Nd, Sm, Y has been analysed [10P1]. The Co atoms substitute for Ni with a strong preference for 3*g* sites, the order of preference following the sequence 3*g* > 4*h* > 6*i*.

The magnetic properties of RNi_4B compounds with R = La [84S1, 94H2, 95N2, 06I2], R = Ce [94H2, 95N2, 99M1, 02T2, 03T3, 03T6, 04T3, 09T1], R = Pr [94H2,

95N2, 02T2, 02T3], R = Nd [92M1, 94H2, 95N1, 02T2, 08L1, 13A1], R = Sm [92M1, 94H2, 00M1, 02T3, 03T7], R = Eu [95N2], R = Gd [93H1, 94H2, 02T2, 02T4, 04B2, 06V3], R = Tb [94H2, 95N2, 02T3, 06V3, 08L1, 13A1], R = Dy [94H2, 95N2, 02T2, 06V3, 08L1], R = Ho [94H2, 95N2, 02T3, 06V3, 13A1], R = Er [94H2, 95N2, 02T3, 03T4, 06V3, 08L1, 13A1], R = Tm [94H2, 95N2, 03T5], R = Yb [04K3, 05T2, 07T1], R = Y [93H1, 93M1, 94F1, 94H3, 94M3, 94N1, 95H2, 95R1, 99B2, 04A2, 07B4] were investigated.

The $LaNi_4B$ is a Pauli paramagnet, having magnetic susceptibility $\chi = 2.11 \cdot 10^{-4}$ emu/f.u. at T = 300 K and $2.76 \cdot 10^{-4}$ emu/f.u. at T = 90 K [84S1]. The YNi_4B in earlier studies has been reported to be a superconductor having a transition temperature, $T_s \cong 12$ K [93M1, 94M3, 94N1]. The superconducting behavior, in earlier studies, has been attributed to the presence of superstructure. The superconducting fraction was estimated at 2%. Then, it was argued that the observed superconductivity is due to the formation of a second phase with composition YNi_2B_2C [94C4, 94F1, 94M3, 94S7, 04A2], stabilized in the presence of carbon. The YNi_4B has been assumed to be a Pauli-type paramagnet [84S1, 93H1]. The magnetic susceptibility obtained from band structure calculations agrees with that experimentally determined, at low temperatures [95R1]. The following studies evidenced that the temperature dependences of the magnetic susceptibilities of RNi_4B compounds with R = La, Y, Ce at T > (80–100) K, follow a modified Curie–Weiss law $\chi = \chi_0 + C/(T\text{-}\theta)$, with negative paramagnetic Curie temperatures, θ [94H1, 02T2, 03T2, 09T1]—Table 9.24. Supposing that La and Y are non-magnetic, from the Curie constants, effective nickel moments $M_{eff}(Ni) = 0.173\ \mu_B$/atom [94H2] or in the range $(0.18–0.5)\mu_B$/atom were estimated—Table 9.24. The magnetic susceptibilities of $YNi_{4-x}Cu_xB$ ($x \leq 1$) increase up to $T \cong 40$ K, then decreasing and above $T \cong 80$ K, these follow a Curie–Weiss-type dependence [06V3, 08L1], typical for a spin fluctuations-type behavior [91M1].

The XANES studies on $CeNi_4B$ compound revealed that cerium is in a mixed valence state, and the Ni has $3d^9$ configuration in the ground state [99M1]. The mixed valence of cerium has been confirmed by magnetic circular dichroism. The occupation of the f-states is $n_f = 0.83$, as obtained from the analysis of core level spectra. There is a strong hybridization of Ce4f shell with conduction band [09T1]. The Ce and Ni moments are antiparallel oriented—Table 9.24. The magnetic susceptibility of $CeNi_4B$ follows a modified Curie–Weiss dependence, with negative paramagnetic Curie temperature [94H2, 02T2, 03T6, 09T1]—Table 9.24. Both the Ce and Ni atoms contribute to the Curie constant, that of cerium being twice that of nickel, as suggested by magnetic circular dichroism. In the above assumption, effective moments $M_{eff}(Ni) = (0.13–0.15)\mu_B$/atom and $M_{eff}(Ce) = (0.37–0.42)\mu_B$/atom can be estimated from the reported Curie constants [94H2, 02T2, 03T6, 09T1]. The $M_{eff}(Ni)$ values are smaller than those reported in RNi_5 compounds [90B1, 07B4].

The RNi_4B borides with R = Pr, Nd, Sm are ferromagnetically ordered. The saturation moments per formula unit are smaller than those of free ion g_JJ values—Table 9.24. The magnetic structure of orthorhombic $NdNi_4B$ has been determined by neutron diffraction [13A1]—Fig. 9.15. Assuming that Ni is not magnetic, a mean moment, $M_{Nd} = 3.1(1)\ \mu_B$, has been reported, close to saturation moment of free

Table 9.24 Magnetic properties of RNi_4B, $R_3Ni_{13}B_2$ and RNi_3B compounds

Compound	Magnetic structure and magnetic moments magnetic susceptibility (emu/f.u)	M_s (μ_B/f.u.)	T_c (K)	C (emuK/f.u.)	M_{eff}(μ_B/f.u.)	θ (K)	μ_0H_c (T)	References
$LaNi_4B$	PM, spin fluctuations, $\chi = \chi_0 + C/(T - \theta)$, $\chi_0 = 1.42 \cdot 10^{-4}$			0.015	0.35[a]	−14		[94H2]
$CeNi_4B$ (mixed valent)	PM, spin fluctuations, $\chi = \chi_0 + C/(T - \theta)$, $\chi_0 = 7.35 \cdot 10^{-4}$			0.029	0.48[b]	−7.7		[94H2]
$CeNi_4B$ (mixed valent)	PM, spin fluctuations, $\chi = \chi_0 + C/(T - \theta)$, $\chi_0 = 6.39 \cdot 10^{-3}$			0.034	0.52[c]	−10.7		[02T2, 03T6, 04T3]
$CeNi_4B$ (mixed valent) (XMCD)	FIM Ni[d)]:$M^o = 0.0039\ \mu_B$, $M^s = 0.0087\ \mu_B$ Ce[d)]:$M^o = 0.161\ \mu_B$, $M^s = -0.186\ \mu_B$			0.034	0.52	−10.7		[09T1]
$PrNi_4B$	FM	1.8[e]			3.09	68		[02T3, 07T1]
$NdNi_4B$	FM	2.1[f]	11.7					[94H2]
$NdNi_4B$	FM	1.7[e]	12				0.06	[02T2]
$NdNi_4B$	FM		10.5			−45		[95N2]

(continued)

Table 9.24 (continued)

Compound	Magnetic structure and magnetic moments magnetic susceptibility (emu/f.u)	M_s (μ_B/f.u.)	T_c (K)	C (emuK/f.u.)	M_{eff}(μ_B/f.u.)	θ (K)	μ_0H_c (T)	References
$NdNi_4B$ (ND)	FM, T = 1.6 K Nd: M = 3.1(1)μ_B, M_{Ni} = −0.25μ_B, θ^g = 59°		11.0(2)					[13A1]
$NdNi_4B$ (BS)	Nd: M_a = 3.13 μ_B, M_b = 3.15 μ_B Ni: M_c = 0.045μ_B, M_i = 0.014μ_B B: M_h = −0.006 μ_B							[06V3]
$SmNi_4B$	FM	0.15^f	39 31.7 (T_{sR})				2.8 (5 K)	[00M1]
$SmNi_4B$	FM	0.32^h	39				7.0 (5 K)	[02T3, 03T7]
$SmNi_4B$	FM	0.24^e	39					[94H2]
$SmNi_4B^{i1)}$	$T > T_c$, $\chi = \chi_0 + C/(T - \theta)$, $\chi_0 = 5.8 \cdot 10^{-4}$	0.24^{h1}	40		0.76	44	7.4(5 K)	[16Y1]
$SmNi_3MnB^{i2)}$	$T > T_c$, $\chi = \chi_0 + C/(T - \theta)$, $\chi_0 = 3.5 \cdot 10^{-3}$	1.32^{h2}	210		2.56	207	6.9(20 K)	[16Y1]
$SmNi_3FeB^{i3)}$	$T > T_c$, $\chi = \chi_0 + C/(T - \theta)$, $\chi_0 = 8 \cdot 10^{-2}$	1.10^{h3}	325		4.12	325	7.7(60 K)	[16Y1]

(continued)

Table 9.24 (continued)

Compound	Magnetic structure and magnetic moments magnetic susceptibility (emu/f.u)	M_s (μ_B/f.u.)	T_c (K)	C (emuK/f.u.)	M_{eff}(μ_B/f.u.)	θ (K)	μ_0H_c (T)	References
$SmNi_3CoB$[i4)]	$T > T_c$, $\chi = \chi_0 + C/(T - \theta)$, $\chi_0 = 1.1\cdot10^{-3}$	0.42[h4]	90		1.42	94	8.8(20 K)	[16Y1]
$SmNi_4CuB$[i5)]	$T > T_c$, $\chi = \chi_0 + C/(T - \theta)$, $\chi_0 = 6.5\cdot10^{-4}$	0.21[h5]	57		0.47	63	5.2 (5 K)	[16Y1]
$GdNi_4B$	FM	6.4[e]	35				0.028	[02T2]
$GdNi_4B$		7.4[f]	35.1					[94H2]
$GdNi_4B$ (BS)	FIM Gd: $M_a = 7.198\ \mu_B$, $M_b = 7.25\ \mu_B$ Ni: $M_c = -0.067\ \mu_B$, $M_i = -0.037\ \mu_B$ B: $M_h = -0.003\ \mu_B$	7.04 (calc.) 7.0 (exp)	36	9.5	1.79 (μ_B/Ni atom)	45		[06V3]
$TbNi_4B$ (ND)	T = 1.6 K, FIM, $\theta^g = 62(2)°$ Tb: $M = 5.7(1)\ \mu_B$ Ni: $M_c = -0.0(7)\ \mu_B$, $M_i = -0.20(6)\ \mu_B$		18.1(2)					[13A1]
$TbNi_4B$		6.3[e]	21					[02T3, 07T1]
$TbNi_4B$		5.9[e]	19					[94H2]

(continued)

Table 9.24 (continued)

Compound	Magnetic structure and magnetic moments magnetic susceptibility (emu/f.u)	M_s (μ_B/f.u.)	T_c (K)	C (emuK/f.u.)	M_{eff}(μ_B/f.u.)	θ (K)	μ_0H_c (T)	References
$TbNi_4B$			18.5			19.5		[95N2]
$TbNi_4B$ (BS)	FIM Tb: $M_a = 9.16\ \mu_B$, $M_b = 9.20\ \mu_B$ Ni: $M_c = -0.055\ \mu_B$, $M_i = -0.03\ \mu_B$		21					[06V3]
$TbNi_4B$ (BS)	Tb: $M_a = 9.172\ \mu_B$, $M_b = 9.249\ \mu_B$ Ni: $M_c = -0.032\ \mu_B$, $M_i = -0.054\ \mu_B$							[05T2]
$DyNi_4B$		8.2[e]	15				0.012	[02T2, 07T1]
$DyNi_4B$		7.5[e]	12.6					[94H2]
$DyNi_4B$ (BS)	Dy: $M_a = 10.12\ \mu_B$, $M_b = 10.155\ \mu_B$ Ni: $M_c = -0.043\ \mu_B$, $M_i = -0.023\ \mu_B$ B: $M_h = -0.02\ \mu_B$							[06V3]
$HoNi_4B$ (ND)	T = 1.6 K, FIM, $\theta^g = 62°$ Ho: $M = 6.7(1)\ \mu_B$ Ni: $M_i = 0.20(7)\ \mu_B$, $M_c = 0.0(7)\ \mu_B$		6.2(2)					[13A1]

(continued)

Table 9.24 (continued)

Compound	Magnetic structure and magnetic moments magnetic susceptibility (emu/f.u)	M_s (μ_B/f.u.)	T_c (K)	C (emuK/f.u.)	M_{eff}(μ_B/f.u.)	θ (K)	$\mu_0 H_c$ (T)	References
$HoNi_4B$		7.8^h	6					[02T3, 07T1]
$HoNi_4B$		7.5^e	6					[94H2]
$ErNi_4B$ (ND)	T = 1.6 K Er: M = 8.7 μ_B, $M_{Ni} \cong 0$, $\theta^g = 0°$	8.7	10.0(2)					[13A1]
$ErNi_4B$		6.4^h	12					[02T2, 02T3]
$ErNi_4B$	Ni: $M_c = -0.026\ \mu_B$, $M_i = -0.018\ \mu_B$ [06V3] (BS)	7.7^e	9.6					[94H2]
$TmNi_4B$		4.5^j	$T_{c1} = 4.6$ K $T_{c2} = 3.3$ K					[03T5]
$TmNi_4B$			3.5					[95N2]
$YbNi_4B$	PM				3.95			[04K3]
$YbNi_4B$	PM				5.0	−130		[07T1]
$LaNi_4B$ (BS)	PM							[06I2]
YNi_4B	PM, $\chi = \chi_0 + C/(T - \theta)$, $\chi_0 = 1.71 \cdot 10^{-4}$			0.0055	0.21^k	−2.6		[94H2]
YNi_4B	PM, $\chi = \chi_0 + C/(T - \theta)$			0.125	0.98^l			[07B4]

(continued)

Table 9.24 (continued)

Compound	Magnetic structure and magnetic moments magnetic susceptibility (emu/f.u)	M_s (μ_B/f.u.)	T_c (K)	C (emuK/f.u.)	M_{eff}(μ_B/f.u.)	θ (K)	μ_0H_c (T)	References
YNi_4B	PM, $\chi_{exp} = 1.739{\cdot}10^{-4}$, $\chi_{calc} = 1.7745{\cdot}10^{-4}$ [95R1] $\chi_{exp} = 9.1{\cdot}10^{-4}$ [93H1]							[93H1, 95R1]
YNi_3CuB	PM, T = 4.2 K, $\chi_{exp} = 2.1{\cdot}10^{-4}$, $\chi_{calc} = 2.0{\cdot}10^{-4}$			0.093	0.86[l]			[07B4]
YCo_3NiB Ni at 2*c* (BS)	FM Y: $M_a = -0.16\ \mu_B$, $M_b = -0.28\ \mu_B$ Ni: $M_c = 0.32\ \mu_B$, Co: $M_i = 0.80\ \mu_B$ B: $M_h = -0.07\ \mu_B$	1.70 (exp.)[m] 2.40 (calc.)						[04B2, 07B4]
YCo_3NiB Ni at 6*i* (BS)	FIM Y: $M_a = -0.16\ \mu_B$, $M_b = -0.25\ \mu_B$ Co: $M_c = 1.54\ \mu_B$, $M_i = 0.76\ \mu_B$ Ni: $M_i = 0.10\ \mu_B$ B: $M_h = -0.05\ \mu_B$	1.70 (exp.)[m] 2.40 (calc.)						[04B2, 07B4]
$Nd_3Ni_{13}B_2$	FM	6.85[m]	12		3.80μ_B/Nd atom	39		[05P2, 07B1]

(continued)

Table 9.24 (continued)

Compound	Magnetic structure and magnetic moments magnetic susceptibility (emu/f.u)	M_s (μ_B/f.u.)	T_c (K)	C (emuK/f.u.)	M_{eff}(μ_B/f.u.)	θ (K)	μ_0H_c (T)	References
$Nd_3Ni_{10}Co_3B_2$ (ND)	FM T = 1.5 K, $\theta^g = 55(6)°$ Ni: $M_i = M_h = M_g = 0.5\ \mu_B$ Nd: $M_e = 2.2(1)\ \mu_B$, $M_a = 2.0\ (1)\ \mu_B$							[07P1]
$Y_3Ni_{13}B_2$	AFM T = 5 K, $M = M_s + \chi_0 H$, $\chi_0 = 2.73 \cdot 10^{-1}$ $T > T_N$, $\chi = \chi_p + C/(T - \theta)$, $\chi_p = 3.15 \cdot 10^{-1}$	0.29[n]	$T_N = 68(2)$		0.71μ_B/Ni atom	76		[05P2]
$Y_3Ni_{13}B_2$ (BS)	Ni: $M_i = 0.01\ \mu_B$, $M_h = 0.12\ \mu_B$, $M_g = 0.20\ \mu_B$ Y: $M_e = -0.02\ \mu_B$, $M_a = -0.01\ \mu_B$	1.03						[14P1]
$LaNi_3B$ (BS)	paramagnetic							[06O2]
$PrNi_2Fe_2B$	FM	4.04[o]					0.234 (4.2 K)	[88S1]

(continued)

Table 9.24 (continued)

Compound	Magnetic structure and magnetic moments magnetic susceptibility (emu/f.u)	M_s (μ_B/f.u.)	T_c (K)	C (emuK/f.u.)	M_{eff}(μ_B/f.u.)	θ (K)	μ_0H_c (T)	References
$NdNi_2Fe_2B$	FM	5.49[o]					0.15 (4.2 K)	[88S1]
$SmNi_2Fe_2B$	FM	2.45[o]					5.6(4.2 K)	[88S1]
$GdNi_2Fe_2B$	FIM	3.31[o]					0.023(4.2 K)	[88S1]
$DyNi_2Fe_2B$	FIM	3.9[o]					0.28 (4.2 K)	[88S1]
$ErNi_2Fe_2B$	FIM	3.79[o]					3.8 (4.2 K)	[88S1]

[a] $M_{eff} = 0.173\ \mu_B$/Ni atom
[b] $M_{eff} = 0.225\ \mu_B$/Ni atom, if Ce is not magnetic
[c] 0.26 μ_B/Ni atom if Ce is not magnetic
[d] M^o, M^s—orbital and spin moment
[e] $T = 4.2$ K, $\mu_0H = 6$ T
[f] $T = 5$ K, $\mu_0H = 6$ T
[g] θ, angle of moments with **c**-axis
[h] $\mu_0H = 9$ T at $T = 5\ K^1$, $20\ K^2$, $100\ K^3$, $40\ K^4$, $10\ K^5$
[i] metamagnetic transition at $\mu_0H = 1$ T were evidenced at $T = 15\ K^1$, $100\ K^2$, $185\ K^3$, $55\ K^4$, $15\ K^5$
[j] $T = 1.5$ K, $\mu_0H = 9$ T
[k] $M_{eff} = 0.105\ \mu_B$/Ni atom
[l] $M_{eff} = 0.50\ \mu_B$/Ni atom
[m] $T = 4.2$ K, $\mu_0H = 9$ T
[n] $T = 2$ K, $\mu_0H = 5$ T
[o] $T = 4.2$ K, $\mu_0H = 7.5$ T; 1 emu/g = 1 Am^2/kg

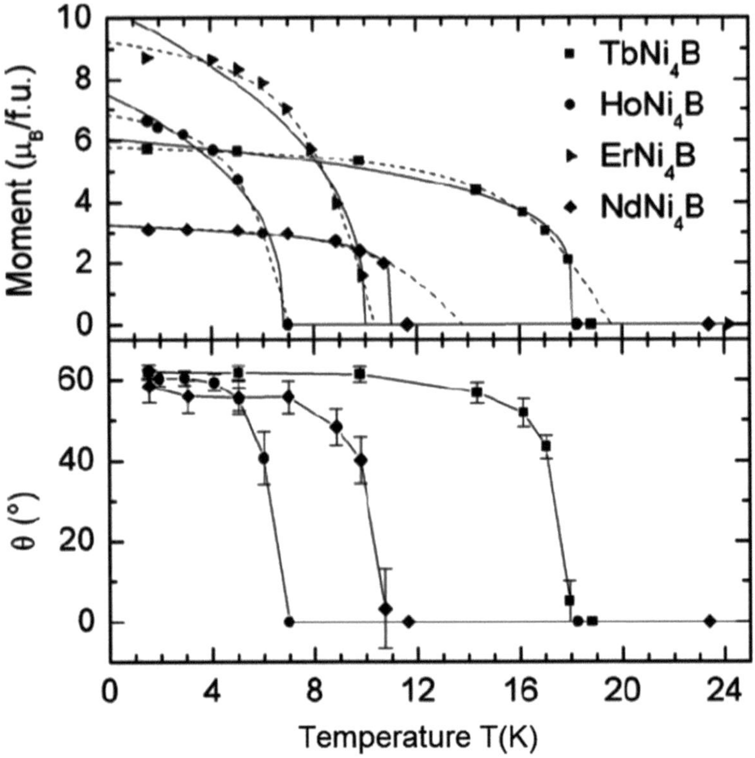

Fig. 9.15 RNi_4B with R = Nd, Tb, Ho and Er: thermal variations of R magnetic moments and of angles θ_i with **c**-axis. The dashed lines indicate the expected evolution described by a Brillouin function [13A1]

ion. The Nd moments are oriented by $\theta_i = 59°$ from the **b**-axis of the orthorhombic structure, which is approximate parallel to **c**-axis of $CeCo_4B$-type lattice. The θ_i angle decreases gradually with temperature, being nil at $T_c = 11$ K. The magnetic transition was suggested to be of first order. In another refinement of ND spectra a small nickel moment was obtained [13A1].

The $SmNi_4B$ is ferromagnetically ordered, the Curie temperature, $T_c = 39$ K [00M1, 03T7], being higher than that of $GdNi_4B$, which according to De Gennes relation must to be the highest. The quadrupole interaction of Sm ions may be responsible for this effect [95N2]. The $SmNi_4B$ boride shows a spin reorientation, at $T_{sR} = 31.7$ K [94H2]. The very low saturation magnetization ($\cong 0.15\mu_B$/f.u.) was attributed to crystal field effects. The boride exhibits a large thermomagnetic irreversibility and also a large magnetic hysteresis, the coercive field, at T = 5 K, being up to $\mu_0H_c \cong 7$ T—Table 9.24. It was suggested that the large coercivity is associated with narrow domain walls [03T7].

The $SmNi_3MB$ borides with M = Mn, Fe, Co and Cu are ferromagnetically ordered and show metamagnetic transition [16Y1]—Table 9.24. Below the magnetic transition temperatures, $SmNi_3MB$ borides, exhibit high coercive field, up to μ_0H_c = 7.7 T. Field induced transitions at μ_0H = 1 T were shown at temperatures T_m, which depend on the M partner. These compounds exhibit negative entropy changes, ΔS_m around T_c and positive around the low temperature metamagnetic-like transitions, T_m. The magnetic susceptibilities at T > T_c follow modified Curie–Weiss law with positive paramagnetic Curie temperatures.

The ^{151}Eu Mössbauer spectroscopy on $EuNi_4B$ compound evidenced an isomer shift δ = 2.3 mm/s (with respect to EuF_3), independent on temperature. The europium ions are in trivalent valence state [95N2].

The RNi_4B compounds with magnetic heavy rare earths (R = Gd, Tb, Dy, Ho, Er, Tm) have been assumed generally to be ferromagnetically ordered, the nickel moment being assumed to be null or nearly null [93H1, 94H2, 02T2, 02T3]. Spin resolved band structure calculations suggest the presence of weak Ni ordered magnetic moment. Their contributions, at the level of the unit cell, decreases from ≅ 0.20 μ_B for R = Gd up to 0.15 μ_B when R = Dy or 0.08 μ_B for R = Er. The Ni magnetic moments are antiparallel oriented to R ones, and have near the same magnitude as the R5d band polarizations [06V3]. Such calculations are supported by the XPS valence bands spectra [03P2, 03T3, 04P3, 04P4, 05T2, 08L1]. The $Ni2p_{1/2}$ and $Ni2p_{3/2}$ core level lines of RNi_4B (R = Tb, Dy, Ho, Er) compounds reproduce well those of pure nickel [08L1]. In all samples, the Ni 6 eV satellite line was shown, indicating the presence of holes in Ni3d band. The R4d core levels are similar to those evidenced in pure R metals. The 4f orbitals keep the localized character and therefore XPS spectra show multiplet structures. The small energy shift of R4d and R4f core level lines, as compared to those of pure R metals, may be correlated with minor changes in shielding, which result from change of electron concentrations, when RNi_4B compounds are formed.

The weak magnetism of nickel in the RNi_4B compounds can be described in models which take into account electron correlation effects in d band [07B3], as spin fluctuations model [91M1] or dynamical mean-field theory [96G3]. Since of small saturation Ni moments, their contributions to the samples magnetizations cannot be evidenced by magnetic measurements, these values being of the order of experimental errors.

Neutron diffraction measurements were performed on RNi_4B borides with R = Tb, Ho, Er [13A1]—Fig. 9.15. The Tb moments, at T = 1.6 K are the same at both sites and make an angle θ = 62° with **c**-axis, decreasing with increasing temperature up to T_c—Table 9.24. The magnetic transition, at T_c, occurs to be of first order [13A1]. A small nickel moment (≅ 0.20 μ_B) was evidenced at 6*i* sites. The thermal variation of $HoNi_4B$ magnetization follows a Brillouin type dependence. The magnetic transition, at T_c, is of second order [13A1]. The Ho magnetic moment, determined at T = 1.6 K (T/T_c = 0.26), is M_{Ho} ≅ 6.7 (1) μ_B. The angle of magnetization with **c**-axis decreases also gradually from 62° up to T_c. The Er moments in $ErNi_4B$ are oriented parallel to **c**-axis of the hexagonal structure [13A1]. A value of 8.7 μ_B was obtained at T = 1.6 K (T/T_c = 0.16). The Curie temperatures of RNi_4B do not follow De Gennes relation, indicating the importance of CEF interactions over the magnetic ones. The RNi_4B compounds with R = Nd, Tb, Ho display a spin reorientation below T_c, which arises from a competition between the second order term and the higher order terms of the CEF Hamiltonian [13A1]. The crystal field interactions in RNi_4B (R = Tb, Ho, Er) compounds controls the magnetic anisotropy and leads to magnetic moments, of the rare-earths, smaller than those of free rare-earth ions. Above the Curie temperatures, the reciprocal susceptibility of RNi_4B (R = Gd, Tb, Dy) show linear dependences [06V3]. The Curie constants are somewhat higher than those of free R^{3+} ions. Effective nickel moments, $M_{eff} \cong 1.80\ \mu_B$/Ni atom were determined assuming that those of R^{3+} are given by the free ion values.

The $TmNi_4B$ boride shows irreversibility of the magnetization, at $T \leq 4.6$ K [03T5]. Two magnetic transitions between two ferromagnetic states were reported—Table 9.24.

The XPS measurements can provide additional information, concerning the physical properties of RNi_4B borides, as for R = Ce [99M1, 03T6, 04T4, 09T1], R = Pr [03T3], R = Nd [02T1, 03T3, 08L1], R = Sm [03T7, 06T1], R = Gd [02T4], R = Tb [05T2, 08L1], R = Dy [02T1, 03T4, 08L1], R = Ho [04P4, 08L1, R = Er [04P3, 08L1], R = Yb [04K3, 07T1] and R = Y [03P2, 03P3]. The cerium was shown to be in mixed valent state and Pr and Nd in + 3 state [03T3]. The hybridization parameter between 4f level and conduction band was shown to decrease in the sequence Ce, Pr, Nd. The Ni3d spin–orbit splitting increases with the number of 4f electrons (R = Ce, Pr, Nd). The valence band of $SmNi_4B$ is well separated from the 4f peak. A small Ni moment is suggested by the presence of a satellite line between $Ni2p_{1/2}$ and $Ni2p_{3/2}$ peaks [03T7]. The Tb4f peaks in $TbNi_4B$ are well localized and the binding energies are similar with those of metallic Tb and Ni [05T2]. The valence band region is dominated by the Ni contribution, at about 4 eV, below E_F. The Dy4f levels in $DyNi_4B$ are also well localized. The spin–orbit coupling energies for Dy and Ni were evaluated [02T1]. In $ErNi_4B$, the valence band at E_F, is determined by Ni3d states below 4 eV [04P3], while in $YbNi_4B$ due to Yb4f states, hybridized with Ni3d ones, the contribution of Yb4f states being dominant [04K3, 07T1]. An intermediate valence state was suggested for ytterbium ion in $YbNi_4B$. In YNi_4B boride, a hybridization of Ni3d states with Y4d and B2p states was shown [03P3].

The band structure calculations were made on RNi_4B compounds with R = La [06I2], Gd [02T4, 06V3], R = Tb [05T2, 06V3], R = Dy [06V3], R = Ho [06V3], R = Yb [04K3] and R = Y [95R1, 03P3]. The above studies revealed that the borides with R = La, Y are paramagnetic. When R is magnetic, small Ni moments, at both 2*c* and 6*i* sites were determined. The computed Ni moments, as well as the R5d band polarizations are linearly dependent on De Gennes factor [06V3]. The R5d band polarizations mediate the exchange interactions as in R-Ni compounds [19B2].

The Curie temperatures of $Nd_{1-x}Gd_xNi_4B$ [11O2], $Sm_{1-x}Gd_xNi_4B$ [11K1, 12O1], $Sm_{1-x}Tb_xNi_4B$ [12N1] and $Gd_xY_{1-x}Co_3NiB$ [04B2] pseudoternary series are linearly dependent on composition. In $Nd_{1-x}Gd_xNi_4B$, at 4.2 K, there is a compensation of magnetizations, at x = 0.3, due mainly antiparallel alignment of Nd and Gd moments [11O2]. The $Gd_xY_{1-x}Co_3NiB$ compounds with x > 0.1, are ferrimagnetically ordered [04B2]. The computed magnetic moments are in agreement with experimental data.

The $PrCo_{4-x}Ni_xB$ with x ≤ 2 are magnetically ordered, while the samples having a higher nickel content are in paramagnetic state [09A2]. The magnetic properties of $YNi_{4-x}Co_xB$ were investigated [04B1, 14C1]. The magnetic anisotropy of $YNi_{4-x}Co_xB$ series can be described starting from a model which takes into account the site occupancy [14C1]—see Sect. 8.6.

The RNi_2Fe_2B compounds, in as cast form, contain 1/4/1 phase as the majority constituent [88S1]. The borides are ferromagnetic when R = Pr, Nd, Sm and ferrimagnetic as R = Gd, Dy or Er. High coercivities were shown in RNi_2Fe_2B compounde with R = Sm and Er—Table 9.24. Stacking faults were observed in $SmNi_2Fe_2B$, which partially may explain their high coercivity.

The specific heat of YNi_4B boride has been described starting from a Debye function with $\theta_D = 458$ K at T < 12 K and T > 170 K [93H1]. In the temperature range 12 K ≤ T ≤ 170 K, the $\theta_D(T)$ decreases, reaching a minimum at T = 32 K ($\theta_D = 355$ K) and then increasing up to T = 170 K. According to [95H2], the combination of one Debye function and two Einstein functions, with different weights, yields together with an electronic specific heat term, a satisfactory description of heat capacity of YNi_4B compound—Table 9.25. The Debye temperature, θ_D, corresponds to the characteristic temperature of the three acoustic branches of phonon spectrum. Theoretically computed θ_D value in YNi_4B [03T4], was close to that experimentally determined.

A λ-type anomaly was shown in the specific heat of $GdNi_4B$ at T_c [93H1]. The magnetic entropy, determined by subtracting that of YNi_4B (phonon contribution) shows a tendency for saturation at ≅ 20–21 J/mol K, higher than the value expected for J = 7/2 multiplet (R*ln* 8 ≅ 17.3 J/mol K). The difference was attributed either to an additional magnetic contribution to specific heat, from magnetically polarized conduction electrons or that $GdNi_4B$ contains a larger phonon contribution than YNi_4B, which in the analysis were considered to be identical. The first assumption is supported by the presence of small Ni moments.

The electrical resistivities of RNi_4B compounds with R = Ce [03T6], R = Ce, Pr, Nd, Sm, Dy, Ho, Er and Y [03T4] and T = Tm [03T5] were reported. The temperature dependence of electrical resistivity, ρ(T), in YNi_4B, was fitted using the Bloch-Grüneisen relation for metal-like compounds, while in $CeNi_4B$, revealed

Table 9.25 Debye, θ_D, Einstein, θ_E, temperatures and bulk modulus, B_o

Compound	Temperature range	B_o (MPa)	θ_D (K)	θ_E (K)	References
YNi_4B	12 K ≤ T ≤ 300 K		240		[03T4]
YNi_4B (calc.)		161.6	273		[03P2]
YNi_4B	T < 12 K T = 32 K T > 170 K		458 355 458		[93H1]
YNi_4B	T < 300 K		458 (LT) 242	$\theta_E^1 = 462(9)$ $\theta_E^2 = 186(6)$	[95H2]
$Y_3Ni_{13}B_2$	0.3 K ≤ T ≤ 300 K		374		[07R1]
$Nd_3Ni_{13}B_2$	0.3 K ≤ T ≤ 300 K		285.9		[09A3]

a shallow Kondo-like minimum, at T = 12 K, with a logarithmic increase of the resistance below this temperature, associated with mixed valent cerium. The magnetic contributions to ρ(T), at $T < T_c$, for compounds with R = Nd, Sm, Dy, Ho, Er were described in the framework of a model including an anisotropy gap in the magnon spectrum [03T4]. Above T_c, a quadratic dependence, typical for spin fluctuations, has been evidenced. The crystal field effects were considered in analysing ρ(T) for $SmNi_4B$. In the paramagnetic region ($7.1 \le T \le 45$ K), the resistivity of $TmNi_4B$ follows a quadratic dependence, connected also with spin fluctuations [03T5]. At T > 45 K, ρ(T) it is linear, as expected for phonon contributions.

The magnetic properties of $R_3Ni_{13-x}Co_xB_2$ with R = Nd, Y and $x \le 12$ were analysed in Sect. 8.7 [05P2, 05P3, 07B1, 07R1, 09A3]. Consequently, only the data on end series $R_3Ni_{13}B_2$ compounds will be mentioned. The temperature dependence of the differential susceptibility of $Y_3Ni_{13}B_2$ shows a cusp at $T \cong 68$ K, characteristic for AFM-PM transition. The presence of a weak ferromagnetic component superposed on the essential antiferromagnetic order, cluster glass behavior, respectively was assumed [05P3]. The Ni in $Y_3Ni_{13}B_2$ has a small ordered magnetic moment. The magnetic susceptibility follows a Curie–Weiss dependence, with a mean effective nickel moment of 0.71 μ_B/Ni atom. The ratio between the mean number of Ni spins determined from Curie constant, S_p, and saturation magnetization, S_0, is $r \cong 7$, suggesting that Ni shows a weak itinerant magnetism [07B1]. The relativistic calculations, within the local spin density approximations, were performed on $Y_3Ni_{13}B_2$ [14P1]. A weak in-plane anisotropy was determined under the assumption of collinear magnetic moments. The Ni magnetic moments, at different sites, decrease in the sequence $g > h > i$—Table 9.24. The Y4d band is negatively polarized.

The $Nd_3Ni_{13}B_2$ compound is antiferromagnetically ordered and show two metamagnetic transitions in fields of $\mu_0H_c = 0.5$ T and 2.0 T [07B1]. These were attributed to AFM to FM state transitions, turning each of the two Nd sublattice magnetization at its own critical field, in the direction of applied field.

The electronic contribution to the low temperature heat capacity of $Y_3Ni_{13}B_2$ yields an electronic constant [07R1] higher than those of YNi_4B [95H2] or YNi_5 [90B1]. The Debye temperature is $\theta_D = 374$ K. The entropy calculation for $Nd_3Ni_{13}B_2$ gives a value $\Delta S = 1.96$ R, close to the theoretical value for Nd ground state ($\Delta S = 2.08$ K). The low temperature heat capacities of $Nd_3Ni_{13-x}Co_xB_2$ evidenced an increase of the DOS at E_F, as the cobalt content is increased [09A3]. For x = 0, the magnetic contribution to the specific heat has rounded shape of anomaly, characteristic for a low dimensionality of the transition [09A3]. The substitution of Ni by Co ($1 \le x \le 3$) induces interaction disorder on the Nd sublattice. The hyperfine contributions to the specific heat were attributed to the interaction between the Co and the Nd nuclei. These are linearly dependent on cobalt content [09A3].

References

[32V1] J.H. van Vleck, The theory of electric and magnetic susceptibilities (Oxford Univ. Press., London, 1932)

[59A1] P.W. Anderson, H. Suhl, Phys. Rev. **116**, 898 (1959)

[60K1] Y.B. Kuzma, N.F. Chaban, Binary and ternary systems which contain boron (Metallurgia, Moscow, 1960) (in Russian)

[61A1] A.A. Abrikosov, I.P. Gorkov, Sov. Phys. JETP **12**, 1243 (1961)

[65V1] E.E. Vainshtein, S.M. Blokhin, Y.B. Paderno, Sov. Phys. Solid State **6**, 2318 (1965)

[71K1] Y.B. Kuzma, N.S. Bilonizhko, Izv. Akad. Nauk SSR, Neorg. Mater. **7**, 620 (1971)

[72B1] A.L. Bowman, G.P. Arnold, E.K. Storms, N.G. Nereson, Acta Crystallogr. **B28**, 3102 (1972)

[72C1] I.A. Campbell, J. Phys. F. : Metal Phys. **2**, L47 (1972)

[72K1] Y.B. Kuzma, N.S. Bilonizhko Sov. Phys. Crystallogr. **16**, 897 (1972)

[73H1] R. Harris, M. Plischke, M.J. Zuckermann, Phys. Rev. Lett. **31**, 160 (1973)

[73K1] Y.B. Kuzma, N.S. Bilonizhko, Kristallogr. 18, **710** (1973)

[73K2] Y.B. Kuzma, N.S. Bilonizhko, E.M. Nimkovich, Dopov. Akad. Nauk, Ukr. RSR A(10) 939 (1973)

[73K3] Y.B. Kuzma, M.P. Khaburskaya, Izv. Akad. Nauk SSSR, Neorg. Mater. (11), 1893 (1973)

[73N1] K. Niihara, Y. Katayama, S. Yajima, Chem. Lett. 613 (1973)

[75K1] Y.B. Kuzma, M.P. Khaburskaya, Inorg. Mater. (10), 1625 (1975)

[77D1] S. Doniach, *Valence instability and related narrow band phenomena*, ed. by R.D. Parks (Plenum Press, New York, 1977)

[77K1] P.I.Kripyakevich, *The structure types of Intermetallic Compounds* (Nauka, Moscow, 1977)

[78D1] R.C. Dynes, V. Narayanamurti, J.P. Garno, Phys. Rev. Lett. **21**, 1509 (1978)

[79K1] Y.B. Kuzma, N.F. Chaban, Dopov. Akad. Nauk. Ukr. RSR A, 88 (1979)

[79K2] Y.B. Kuzma, N.S. Bilonizhko, S.I. Mykhalenko, G.F.Stepanchikova, N.F. Chaban, J. less Common Met. **67**, 51 (1979)

[80C1] N.F. Chaban, Y.B. Kuzma, P.L. Kotovskaya, Dopov. Akad. Nauk. Ukr. RSR A(6), 87 (1980)

[80J1] W. Jung, D. Quentmeier, Zeit. Kristallogr. **151**, 121 (1980)

[80K1] Y.B. Kuzma, N.F. Chaban, Dopov. Akad. Nauk. Ukr. RSR. Ser.A (1), 86 (1980)

[80M1] K. Machida, K. Nokura, T. Matsubara, Phys. Rev. B **22**, 2307 (1980)

[81C1] N.F. Chaban, G.V. Chernyak, Yu.B. Kuzma: Izv. Akad. Nauk SSR, Neorg Mater **17**, 1494 (1981)

[81K1] Y.B. Kuzma, N.S. Bilonizhko, Dopov. Akad. Nauk Ukr. RSR Ser A (10), **88** (1981)

[81K2] Y.B. Kuzma, G.V. Chernyak, N.F. Chaban, Dopov: Akad. Ukr. RSR A (12), 80 (1981)

[81K3] Y.B. Kuzma, N.S. Bilonizhko, N.F. Chaban, G.V. Chernyak, J. less Common Met. **82**, 364 (1981)

[81K4] Y.B. Kuzma, N.S. Bilonizhko, Inorg. Mater. **7**, 542 (1981)

[81R1] T.V. Ramakrishnan, C.M. Varma, Phys. Rev. Lett. **84**, 1800 (1981)

[82B1] N.S. Bilonizhko, B.I. Krik, Yu.B. Kuzma, Dopov. Acad. Nauk. Ukr. RSR, Ser B (4) 21 (1982)

[82B2] N.S. Bilonizhko, B.I. Krik, Yu.B. Kuzma, Dopov. Akad. Nauk, Ukr. RSR, A (4), 22 (1982)

[82C1] G.V. Chernyak, N.F. Chaban, Yu.B Kuzma, Izv. Akad. Nauk. SSSR, Neorg. Mater. **18**, 691 (1982)

[83B1] N.S. Bilonizhko, N.F. Chaban, Yu.B. Kuzma, Dopov. Akad. Nauk. Ukr. Ser B (6), 26 (1983)

[83C1] G.V. Chernyak, Report, cited by [04C1]

[83C2] G.V. Chernyak, N. F. Chaban, Inorg. Chem. **19**, 1565 (1983)

[83F1] I. Felner, J. Phys. Chem. Solids, **44**, 43 (1983)

[83K1] Y.B. Kuzma, *Crystal chemistry of borides* (Vyshcha Shkola, Lvov, 1983) (in Russian)

[83K2] Y.B. Kuzma, G.V. Chernyak, I.G. Akselrud, Kristallogr. **28**, 1210 (1983)

[83K3] Yu.B. Kuzma, N.S. Bilonizhko, N.F. Chaban, G.V. Chernyak, J. less Common Met. 90, 217 (1983)
[83S1] D.J. Sellmyer, M. J. O'Shea, J. less Common Met. **94**, 59 (1983)
[83S2] F. Spada, H. Oesterreicher, J. less Common Met. **90**, L1 (1983)
[84K1] Y.B. Kuzma, N.S. Bilonizhko, V.K. Pecharski, Kristallogr. **29**, 431 (1984)
[84P1] E. Parthé, B. Chabot, *Handbook on the physics and chemistry of rare-earths*, Chap. 48 (Elsevier, 1984), p. 113
[84R1] P. Rogl, *Handbook of physics and chemistry of rare earths*, Chap. 84 (Elsevier, 1984), p. 335
[84S1] F.E. Spada, H. Oesterreicher, R.C. Bowman, M.P. Guse, Phys. Rev. B **30**, 4909 (1984)
[85A1] L.G. Akselrud, Y.B. Kuzma, V.A. Bruskov, Dopov. Akad. Ukr. RSR, B (3), 31 (1985)
[85B1] N.S. Bilonizhko, M.F. Fedyna, Yu.B. Kuzma, Izv. Akad. Nauk SSR, Neorg. Mater. **21**, 148 (1985)
[85D1] O.M. Dub, N.F.Chaban, Yu.B Kuzma, Inorg. Mater. **21**, 994 (1985)
[85D2] O.M. Dub, N.F. Chaban, Yu.B. Kuzma, Izv. AN SSSR Neorg. Mater. **4**, 1967 (1985)
[85K1] Y.B. Kuzma, O.M. Dub, N.F. Chaban, Dopov. Akad. Nauk. Ukr. RSR B (7), 36 (1985)
[85K2] Y.B. Kuzma, L.G.Akselrud, V.A. Bruskov, A.N.Bilonizhko, Krystallogr. **30**, 1005 (1985)
[85S1] F. Spada, H. Oesterreicher, J. less Common Met. **107**, 301 (1985)
[86D1] O.M. Dub, Y.B. Kuzma, M.D. Mazus, N.F.Chaban, T.I. Malinowski, Dopov. AN USSR (2), 34 (1986)
[87D1] O.M. Dub., Y.B. Kuzma, Inorg. Mater. **23**, 42 (1987)
[88B1] L.H. Bennett, R.E. Watson, M. Melamud, J. Physique, **49**, C8–537 (1988)
[88G1] L. B. Gubich, N. F. Chaban, Vestn. Lvov. Ser. Khim. **31**, 14 (1988)
[88K1] Y.B. Kuzma, Yu.M. Dub, V.A. Bruskov, N.F. Chaban, P.Yu. Zavalij, Kristallogr. **33**, 841 (1988)
[88S1] J. Strzeszewski, A. Nazareth, G.C. Hadjipanayis, J. Appl. Phys. **63**, 3978 (1988)
[89G1] T. Glowick, S.V. Orishchin, I.B. Gubich, N.F. Chaban, Y.B. Kuzma, Sov. Phys. Crystallogr. **34**, 602 (1989)
[89G2] I.B. Gubich, Yu.B. Kuzma, N.F. Chaban, Neorg. Mater. **25**, 1317 (1989)
[89K1] Y.B. Kuzma, S.V. Orishchin, T. Glovyak, I.B. Gubich, N.F. Chaban, S.I. Chikhriy, Europ. Cryst. Meet. (Moskow, 1989)
[89O1] S.V. Orishchyn, I.B. Gubich, Yu.B. Kuzma, T. Glowiak, Sov. Phys. Crystallogr. **34**, 1290 (1989)
[90B1] E. Burzo, A. Chelkowski, H.R. Kirchmayr, Landolt Börnstein Handbook, vol. 19d2 (Springer, 1990)
[90D1] O.M. Dub, R.V. Skolozdra, Yu.B. Kuzma, N.S. Dubenko, Neorg. Mater. **26**, 1221 (1990)
[90K1] Y. B. Kuzma, N. F. Chaban, Metallurgya, 320 (1990)
[90K2] Y.B. Kuzma, L.V. Zavalii, Inorg. Mater. **26**, 1906 (1990)
[90K3] Y.B. Kuzma, I.B. Gubich, P.Y. Zavalii, V.A. Bruskov, Kristallogr. **35**, 998 (1990)
[91G1] I.B. Gubich, Visnyk Lviv Univ. **31**, 5 (1991)
[91G2] I.B. Gubich, Yu.B. Kuzma, Inorg. Mater. **27**, 916 (1991)
[91G3] I.B. Gubych, Yu.B. Kuzma, Neorg. Mater. **27**, 2306 (1991)
[91G4] I.B. Gubych, Yu.B. Kuzma, N.F. Chaban, Neorg. Mater. **27**, 2303 (1991
[91G5] I.B. Gubych, Yu.B. Kuzma, Neorg. Mater. **27**, 1621 (1991)
[91G6] I.B. Gubych, Yu.B. Kuzma, N.F. Chaban, Neorg. Mater. **27**, 506 (1991)
[91G7] I.B. Gubich, N. F. Chaban, Yu. B. Kuzma, Inorg. Mater. **37**, 1967 (1991)
[91G8] I.B. Gubich, Y.B. Kuzma, V. A. Bruskova, S.G. Kirsch, Sov. Phys. Crystallogr. **36**, 634 (1991)
[91K1] Y.B. Kuzma, P.V. Zavalii, I.B. Gubich, V.A. Bruskov, Kristallogr. **36**, 490 (1991)
[91M1] T. Moriya, J. Magn. Magn. Mater. **100**, 261 (1991)
[91R1] P. Rogl, *Inorganic reactions and methods*, ed. by J.J. Zuckerman (VCH Publications, Weinheim, vol.13, 1991) p. 85
[91V1] P. Villars, L.D. Calvert, *Pearson's handbook of crystallographic data for intermetallic compounds* (ASM International, Material Park, OH, 1991)

[92M1] C. Mazumdar, R. Nagarajan, C. Godart, L.C. Gupta, B.D. Padalia, R. Vijayaraghavan, *Proc. 37th Conf. Magn. Magn. Mater.,Houston* (1992, BP14)

[92R1] P. Rogl, *Modern perspectives in inorganic crystal chemistry*, NATO ASI Series (Kluwer, 1992), pp. 267–278

[93G1] I.B. Gubich, Yu.B. Kuzma, N.F. Chaban, Neorg. Mater. **29**, 1246 (1993)

[93H1] N.M. Hong, N.P. Thuy, G. Schaudy, T. Holubar, G. Hilscher, J.J. M. Franse, J. Appl. Phys. **73**, 5698 (1993)

[93M1] G. Mazumdar, R, Nagarajan, G. Godart, L.C. Gupta, M. Latroche, S.K. Dhar, C. Levy-Clement, B.D. Padalia, R. Vijayaraghavan, Solid State Commun. **87**, 413 (1993)

[93W1] E. Wu, G.H.J. Wantenaar, S.J. Campbell, H.S. Li, J. Phys.: Condens. Matter, **5**, L457 (1993)

[94A1] L. Afalfiz, C.V. Tomy, J. S. Gardener, E. A. Goremychkin, R. S. Ecclcston, G. Balakrishnan, M. R. Lees, D. M. Paul, Physica C **235–240**, 2555 (1994)

[94B1] F. Borsa, Q. Hu, K.H. Kim, B. J. Suh, V. R. Torgeson, P. Canfield, M. Xu, B. Zhong, Physica C **235–240**, 2547(1994)

[94C1] P. C. Canfield, B. K. Cho, D. C. Johnston, D. K. Finnemore, M. F. Hundley, Physica C **230**, 397 (1994)

[94C2] S. A. Carter, B. Batlogg, R. J. Cava, J. J. Krajewski, W. F. Peck, H. Takagi, Phys. Rev. B **50**, 4216 (1994)

[94C3] R. J. Cava, H. W. Zandbergen, B. Batlogg, H. Eisaki, T. Takagi, J. J. Krajewski, W. F. Peck, E.M. Gyorgy, S. Uchida, Nature, **372**, 245 (1994)

[94C4] R. J. Cava, H. Takagi, B. Batlogg, H. W. Zandbergen, J. J. Krajewski, W.F. Beck, T. Siegrist, K. Mizuhashi, J.O. Lee, H. Eisaki, S. Uchida, Nature, **367**, 252 (1994)

[94C5] B. C. Chakoumakos, M. Paranthaman, Physica C **227**, 143 (1994)

[94C6] R. Coehoorn, Physica C **228**, 331 (1994)

[94E1] H. Eisaki, H. Takagi, R. J. Cava, B. Batlogg, J. J. Krajewski, W. F. Peck, K. Mizuhashi, J. O. Lee, S. Uchida, Phys. Rev. B **50**, 647 (1994)

[94E2] H. Eisaki, H. Takagi, R. J. Cava, K. Mizuhashi, J. O. Lee, S. Uchida, B. Batlogg, J.J. Kajwoki, W. F. Peck, Physica C **235–240**, 2539 (1994)

[94E3] T. Ekino, H. Fuji, M. Kosugi, Y. Zenitani, J. Akimitsu, Physica C **235–240**, 2529 (1994)

[94E4] M. El Massalami, S. L. Budko, B. Giordanengo, M. B. Fontes, J. C. Mondragon, E. M. Baggio-Saitovitch, Physica C **235–240**, 2563 (1994)

[94F1] I. Felner, Phys. Rev. Lett. **72**, 3742 (1994)

[94F2] A. Fujimori, K. Kobayashi, T. Mizokawa, K. Mamiya, A. Sekiyama, H. Eisaki, M. Takagi, S. Uchida, R. J. Cava, J. J. Krajewski, W. F. Peck, Phys. Rev. B **50**, 9660 (1994)

[94G1] L. Gao, X. D. Qiu, Y. Cao, R. L. Meng, Y. Y. Sun, Y. Y. Xue, C. W. Chu, Phys. Rev. B **50**, 9445 (1994)

[94G2] M.S. Golden, M. Knupfer, M. Kielwein, M. Buchgeister, J. Fink, D. Teehan, W. E. Pickett, D. J. Singh, Europhys. Lett. **28**, 369 (1994)

[94G3] A. I. Goldman, C. Stassis, P. C. Canfield, J. Zarestky, P. Dervenagas, B. K. Cho, D. C. Johnston, B. Sternlieb, Phys. Rev. B **50**, R 9668 (1994)

[94G4] T.E. Grigereit, J.W. Lynn, Q. Huang, A. Santoro, R. J. Cava, J. J. Krajewski, W. F. Peck, Phys. Rev. Lett **73**, 2756 (1994)

[94H1] Y. Hayashi, M. Yamada, K. Tanaka, Mater. Trans. **35**, 771 (1994)

[94H2] N.M. Hong, T. Holubar, G. Hilscher, M. Vybornov, P. Rogl, IEEE Trans. Magn. **30**, 4966 (1994)

[94H3] N.M. Hong, H. Michor, M. Vybornov, T. Holubar, P. Hundegger, W. Perthold, G. Hilscher, P. Rogl, Physica C **227**, 85 (1994)

[94I1] K. Ikushima, J. Kikuchi, H. Yasuoka, R. J. Cava, H. Takagi, J. J. Krajewski, W. W. Peck, J. Phys. Soc. Jpn. **63**, 2878 (1994)

[94K1] J. S. Kim, W. W. Kim, G. R. Stewart, Phys. Rev. B **50**, 3485 (1994)

[94K2] H.C. Ku, C.C. Lai, Y. B. You, J. H. Shieh, W. Y. Guan, Phys. Rev. B **50**, 351 (1994)

[94L1] J. I. Lee, T. S. Zhao, I. G. Kim, B. I. Min, S. J. Youn, Phys. Rev. B **50**, 4030 (1994)

[94M1] L. F. Mattheiss, Phys. Rev. B **49**, 13279 (1994)

[94M2] L. F. Mattheiss, T. Siegrist, R. J. Cava, Solid State Commun. **91**, 587 (1994)
[94M3] C. Mazumdar, R. Nagarajan, C. Godart, L.C. Gupta, M. Latroche, S.K. Dhar, C. Levy-Clement, B.D. Padalia, R. Vijayaraghavan, Physica B **194–196**, 1985 (1994)
[94M4] H. Michor, W. Perthold, T. Holubar, N.M. Hong, G. Hilscher, Physica C **235–240**, 2553 (1994)
[94M5] R. Movshovich, M. F. Hundley, J. D. Thompson, P. C. Canfield, B.K. Cho, A. V. Chubukov, Physica C **227**, 381 (1994)
[94M6] C. Murayama, N. Mori, H. Takagi, H. Eisaki, K. Mizuhashi, S. Uchida, R. J. Cava, Physica C **235–240**, 2545 (1994)
[94N1] R. Nagarajan, C. Mazumdar, Z. Hossain, S.K. Dhar, K.V. Gopalakrishnan, L.C. Gupta, C. Godart, B.D. Padalia, R. Vijayaraghavan, Phys. Rev. Lett. **72**, 274 (1994)
[94P1] E. Parthé et al., *Gmelin handbook of inorganic and organometallic chemistry* (Springer, 1994)
[94P2] W. E. Pickett, D. J. Singh, Phys. Rev. Lett. **73**, 3702 (1994)
[94P3] A. Plecenik, M. Grajcar, S. Benacka, P. Seidel, A. Pfuch, Phys. Rev. B **49**, 10016 (1994)
[94P4] R. Prozorov, E.R. Yacoby, I. Felner, Y. Yeshurun, Physica C **233**, 367 (1994)
[94R1] S.B. Roy, Z, Hossain, A.K. Pradhan, C. Mazumdar, P. Chaddah, R. Nagarajan, C. Godart, L.C. Gupta, Physica C **228**, 319 (1994)
[94S1] J.L. Sarrao, M.C. DeAndrade, J. Herrmann, S.H. Han, Z. Fisk, M.B. Maple, R.J. Cava, Physica C **229**, 65 (1994)
[94S2] H. Schmidt, H. F. Braun, Physica C **229**, 315 (1994)
[94S3] H. Schmidt, K. Muller, H.F. Braun, Physica C **235–240**, 779 (1994)
[94S4] T. Siegrist, H.W. Zandbergen, R.J. Cava, J. J. Krajewski , W.F. Peck, Nature, **367**, 254 (1994)
[94S5] T. Siegrist , R.J. Cava, J.J. Krajewski , W.F. Peck, J. Alloys Comp. **216**, 135 (1994)
[94S6] V. Ström, K. S. Kima, A. Mouroux, T. Turkii, A. M. Grishin, K. V. Rao, Physica C **235**, 2537 (1994)
[94S7] S. Süllow, R.W.A. Hendriks, T.J. Gortenmulder, A.A. Menovsky, A. Schenck, J.A. Mydosh, Physica C **233**, 138 (1994)
[94T1] H. Takagi, R. J. Cava, H. Eisaki, J. O. Lee, K. Mizuhashi, B. Batlogg, S. Uchida, J.J. Krajewski, W.F. Peck, Physica C **228**, 389 (1994)
[94T2] C.V. Tomy, L.J. Chang, G. Balakrishnan, D.M.K. Paul, Physica C **235–240**, 2551 (1994)
[94X1] M. Xu, P.C. Canfield, J.E. Ostenson, D.K. Finnemore, B.K. Cho, Z.R. Wang, D.C Johnston, Physica C **227**, 321 (1994)
[94X2] M. Xu, B.K. Cho, P.C. Canfield, D.K. Finnemore, D.C Johnston, D.E. Farrell, Physica C **235–240**, 2533 (1994)
[94Z1] H.W. Zandbergen, R.J. Cava, J.J. Krajewski, W.F. Peck, J. Solid State Chem. **110**, 196 (1994)
[95A1] E. Alleno, J.J. Neumeier, J.D. Thompson, P.C. Canfield, B.K. Cho, Physica C **242**, 169 (1995)
[95A2] E. Alleno, Z. Hossain, C. Godart, R. Nagarajan, L.C. Gupta, Phys. Rev. B **52**, 7428 (1995)
[95B1] M. Buchgeister, A. Handstein, J. Klosowski, N. Mattern, P. Verges, U. Wiesner, Mat. Lett. **22**, 203 (1995)
[95B2] S.L. Budko, M.B. Fontes, D. Aliaga-Guerra, E.M. Baggio-Saitovich, Phys. Rev. B **52**, 305 (1995)
[95C1] P.C. Canfield, B.K. Cho, K.W. Dennis, Physica B **215**, 337 (1995)
[95C2] S.A. Carter, B. Batlogg, R. J. Cava, J. J. Krajewski, W.E. Peck, Phys. Rev. B **51**, 12829 (1995)
[95C3] N.F. Chaban, K.Yu. Valchuk, Yu.B. Kuzma, Izv. Akad. Nauk. Neorg. Mater. **31**, 923 (1995)
[95C4] B.K. Cho, P.C. Canfield, D.C. Johnston, Phys. Rev. B **52**, R3844 (1995)
[95C5] B.K. Cho, M. Xu, P.C. Canfield, L.L. Miller, D.C. Johnston, Phys. Rev. B **52**, 3676 (1995)
[95C6] B.K. Cho, P.C. Canfield, L.L. Miller, D.C. Johnston, W.P. Beyermann, A. Yatskar, Phys. Rev. B **52**, 3684 (1995)

[95C7] D.W. Cooke, J.L. Smith, S. J. Blundell, K.H. Chow, P.A. Pattenden, F.L. Pratt, S.F.J. Cox, S.R. Brown, A. Morrobel-Sosa, R.L. Lichti, L.C. Gupta, R. Nagarajan, Z. Hossain, C. Mazumdar, C. Godart, Phys. Rev. B **52**, R3864 (1995)

[95D1] P. Dervenagas, M. Bullock, J. Zarestky, P. Canfield, B.K. Cho, B. Harmon, A.I. Goldman, C. Stassis, Phys. Rev. B **52**, R9839 (1995)

[95D2] P. Dervenagas, J. Zarestky, C. Stassis, A.I. Goldman, P.C. Canfield, B.K. Cho, Physica B **212**, 1 (1995)

[95E1] M. El Massalami, E. Baggio-Saitovitch, A. Sulpice, J. Alloys Comp. **228**, 49 (1995)

[95E2] M. El Massalami, S.L. Budko, B. Giordanengo, E.M. Baggio-Saitovitch, Physica C **244**, 41 (1995)

[95E3] M. El Massalami, S.L. Budko, B. Giordanengo, M.B. Fontes, J.C. Mondragon, E.M. Baggio-Saitovitch, A. Sulpice, Phys. Stat. Solidi (b) **189**, 489 (1995)

[95E4] M. El Massalami, B. Giordanengo, J. Mondragon, E. M. Baggio-Saitovitch, A. Takeuchi, J. Voiron, A. Sulpice, J. Phys.: Condens. Matter, **7**, 10015 (1995)

[95F1] I. R. Fisher, J. R. Cooper, R.J. Cava, Phys. Rev. B **52**, 15086 (1995)

[95G1] S. Geupel, A. Belger, P. Paufler, G. Graw, Z. Kristallogr. **215**, 4 (1995)

[95G2] C. Godart, L. C. Gupta, R. Nagarajan, S.K. Dhar, H. Noel, M. Potel, C. Mazumdar, Z. Hossain, C. Levy-Clement, G. Schiffmacher, B. D. Padalia, R. Vijayaraghavan, Phys. Rev. B **51**, 489 (1995)

[95G3] T. E. Grigereit, J.W. Lynn, R.J. Cava, J. Krajewski, W. E. Peck, Physica B **248**, 382 (1995)

[95G4] L.C. Gupta, R. Nagarajan, Z. Hossain, Ch. Mazumdar, S.K. Dhar, C. Godart, C. Levy-Clement, B.D. Padalia, R. Vijayaraghavan, J. Magn. Magn. Mater. **140–144**, 2053 (1995)

[95H1] M.E. Hanson, F. Lefloch, W.H. Wong, W.G. Clark, M.D. Lan, C.C. Hoellwarth, P. Klavins, R.N. Shelton, Phys. Rev. B **51**, R674 (1995)

[95H2] G. Hilscher, T. Holubar, N.M. Hong, W. Perthold, M. Vybornov, P. Rogl, J. Magn. Magn. Mater. **140–144**, 2055 (1995)

[95H3] G. Hilscher, H. Michor, N.M. Hong, T. Holubar, W. Perthold, M. Vybornov, P. Rogl, Physica B **206–207**, 542 (1995)

[95H4] Z. Hossain, S.K. Dhar, R. Nagarajan, L.C. Gupta, C. Godart, R. Vijayaraghavan, IEEE Trans. Magn. **31**, 4133 (1995)

[95H5] Q. Huang, A. Santoro, T.E. Grigereit, J.W. Lynn, R.J. Cava, J.J. Krajewski, W. F. Peck, Phys. Rev. B **51**, 3701 (1995)

[95J1] T. Jacobs, B.A. Willemsen, S. Sridhar, R. Nagarajan, L.C. Gupta, Z. Hossain, C. Mazumdar, P.C. Canfield, B.K. Cho, Phys. Rev. B **52**, R7022 (1995)

[95J2] G.T. Jeong, J.I. Kye, S.H. Chun, Z.G. Khim, W.C. Lee, P.C. Canfield, B.K. Cho, D.C. Johnston, Physica C **253**, 48 (1995)

[95K1] H. Kim, C.D. Hwang, J. Ihm, Phys. Rev. B **52**, 4592 (1995)

[95K2] T. Kohara, T. Oda, K. Ueda, Y. Yamada, A. Mahajan, K. Elankumaran, Z. Hossain, L. C. Gupta, R. Nagarajan, R. Vijayaraghavan, C. Mazumdar, Phys. Rev. B **51**, 3985 (1995)

[95L1] C.C. Lai, M.S. Lin, Y.B. You, H. C. Ku, Phys. Rev. B **51**, 420 (1995)

[95L2] D.D. Lawrie, J.P. Franck, Physica C **245**, 159 (1995)

[95L3] L.P. Le, R.H. Heffner, G.J. Nieuwenhuys, P.C. Canfield, B.K. Cho, A. Amato, R. Feyerherm, F. N. Gygax, D.E. MacLaughlin, A. Schenck, Physica B **206–207**, 552 (1995)

[95L4] M.S. Lin, J.H. Shieh, Y.B. You, W.Y. Guan, H.C. Ku, H.D. Yang, J.D. Ho, Phys. Rev. B **52**, 1181 (1995)

[95L5] M.S. Lin, J.H. Shieh, Y.B. You, Y.Y. Hsu, J.W. Chen, S.H. Lin, Y.D. Yao, Y.Y. Chen, J.C. Ho, H.C. Ku, Physica C **249**, 403 (1995)

[95L6] A.P. Litvinchuk, L. Börjesson, N.X. Phuc, N.M. Hong, Phys. Rev. B **52**, 6208 (1995)

[95L7] C. Looney, A.K. Gangopathyay, A.K. Klehe, J.S. Schilling, Physica, C **252**, 199 (1995)

[95M1] H. Michor, T. Holubar, C. Dusek, G. Hilscher, Phys. Rev. B **52** 16165 (1995)

[95M2] F.M. Mulder, J.H.V.J. Brabers, R. Coehoorn, R.C. Thiel, K.H.J. Buschow, F.R. de Boer, J. Alloys Comp. **217**, 118 (1995)

[95N1] R. Nagarajan, L.C. Gupta, S.K. Dhar, C. Mazumdar, Z. Hossain, C. Godart, C. Levy-Clement, B. D. Padalia, R. Vijayaraghavan, Physica B **206–207**, 548 (1995)

[95N2] R. Nagarajan, L.C. Gupta, C. Mazumdar, Z. Hossain, S.K. Dhar, C. Godart, B.D. Padalia, R. Vijayaraghavan, J. Alloys Comp. **225**, 571 (1995)

[95P1] A. Plecenik, M. Grajcar, P. Seidel, A. Pfuch, Phys. Rev. B **51**, 16185 (1995)

[95P2] K. Prassides, A. Lappas, M. Buchgeister, P. Verges, Europhys. Lett. **29**, 641 (1995)

[95R1] P. Ravindran, S. Sankaralingam, R. Asokamani, Phys. Rev. B **52**, 12921 (1995)

[95R2] J.Y. Rhee, X. Wang, B.N. Harmon, Phys. Rev. B **51**, 15585 (1995)

[95S1] H. Schmidt, M. Weber, H. F. Braun, Physica C **246**, 177 (1995)

[95S2] S. Shin, A. Agui, M. Watanabe, M. Fujisawa, Y. Tezuka, T. Ishii, K. Kobayashi, A. Fujimori, M. Hinomaru, H. Takagi, Phys. Rev. B **52**, 15082 (1995)

[95S3] D.J. Singh, W. Pickett, Phys. Rev. B **51**, 8668 (1995)

[95S4] S.K. Sinha, J.W. Lynn, T.E. Grigereit, Z. Hossain, L.C. Gupta, R. Nagarajan, C. Godart, Phys. Rev. B **51**, 681 (1995)

[95S5] S. Sullow, B. Ludoph, S.E. Snel, F.E. Kayzel, E. Bruck, G. J. Nieuwenhuys, A.A. Menovsky, J.A. Mydosh, Z. Phys. B **98**, 17 (1995)

[95T1] T. Terashima, H. Takeya, S. Uji, K. Kadowaki, H. Aoki, Solid State Commun. **96**, 459 (1995)

[95T2] C.V. Tomy, G. Balakrishnan, D.McK. Paul, Physica C **248**, 349 (1995)

[95T3] C.V. Tomy, L.J. Chang, D. McK. Paul, N.H. Andersen, M. Yethiraj, Physica B **213–214**, 139 (1995)

[95T4] C.V. Tomy, M.R. Lees, L. Afalfiz, G. Balakrishnan, D. McK. Paul, Phys. Rev. B **52**, 9186 (1995)

[95V1] T. Vogt, A. Goldman, B. Stemlieb, C. Stassis, Phys. Rev. Lett. **75**, 2628 (1995)

[95Z1] J. Zarestky, C. Stassis, A. I. Goldman, P. C. Canfield, P. Dervenagas, B. K. Cho, D. C. Johnston, Phys. Rev. B **51**, 678 (1995)

[96B1] A. Bharathi, C.S. Sundar, Y. Hariharan, T.S. Radhakrishnan, Z. Hossain, R. Nagarajan, L.C. Gupta, R. Vijayaraghavan, Physica B **223–224**, 123 (1995)

[96B2] S.J. Blundell, S.R. Brown, K.H. Chow, D.W. Cooke, S. F. J. Cox, S.P. Cottrell, C Godart, L.C. Gupta, Z. Hossain, R.L. Lichti, A. Morrobel-Sosa, C. Mazumdar, R. Nagarajan, P.A. Pattenden, F.L. Pratt, J.L. Smith, Physica B **223–224**, 69 (1996)

[96B3] P. Bonville, J.A. Hodges, C. Vaast, E. Alleno, C. Godart, L. C. Gupta, Z. Hossain, R. Nagarajan, Physica B **223–224**, 72 (1996)

[96B4] P. Bonville, J. A. Hodges, C. Vaast, E. Alleno, C. Godart, L. C. Gupta, Z. Hossain, R. Nagarajan, G. Hilscher, H. Michor, Z. Phys. B **101**, 511 (1996)

[96B5] G.S. Burkhanov, S.A. Lachenkov, G.M. Leitus, G.M. Kuzmicheva, E.P. Khlybov, Zh. Neorg. Khim. **41**, 1061 (1996)

[96C1] P.C. Canfield, S.I. Budko, B.K. Cho, Physica C **262**, 249 (1996)

[96C2] T.V. Chandrasekhar Rao, P.K. Mishra, G. Ravikumar, V.C. Sahni, K. Ghosh, G.I. Menon, S. Ramakrishnan, A.K. Grover, G. Chandra, Physica B **223–224**, 86 (1996)

[96C3] L.J. Chang, C.V. Tomy, D. McK. Paul, C. Ritter, Phys. Rev. B **54**, 9031 (1996)

[96C4] L.J. Chang, C.V. Tomy, D. McK. Paul, N.H. Andersen, M. Yethiraj, Physica B **223–224**, 119 (1996)

[96C5] B.K. Cho, P.C. Canfield, D.C. Johnston, Phys. Rev. Lett. **77**, 163 (1996)

[96C6] B.K. Cho, P.C. Canfield, D.C. Johnston, Phys. Rev. B **53**, 8499 (1996)

[96C7] B.K. Cho, B.N. Harmon, D.C. Johnston, P.C. Canfield, Phys. Rev. B **53**, 2217 (1996)

[96D1] P. Dervenagas, J. Zarestky, C. Stassis, A.I. Goldman, P.C. Canfield, B.K. Cho, Phys. Rev. B **53**, 8506 (1996)

[96D2] C. Detlefs, A.I. Goldman, C. Stassis, P.C. Canfield, B.K. Cho, J.P. Hill, D. Gibbs, Phys. Rev. B **53**, 6355 (1996)

[96D3] S.K. Dhar, R. Nagarajan, Z. Hossain, E. Tominez, C. Godart, L.C. Gupta, R. Vijayaraghavan, Solid State Commun. **98**, 985 (1996)

[96E1] T. Ekino, H. Fujii, M. Kosugi, Y. Zenitani, J. Akimitsu, Phys. Rev. B **53**, 5640 (1996)

[96E2] K. Elankumaran, A.V. Mahajan, L C. Gupta, Z. Hossain, R. Nagarajan, R. Vijayaraghavan, Physica **B223–224**, 90 (1996)

[96E3] K. Eversmann, A. Handstein, G. Fuchs, L. Cao, K.H. Müller, Physica C **266**, 27 (1996)

[96F1] I. Felner, J. Magn. Magn. Mater. **157–158**, 622 (1996)
[96F2] I.R. Fisher, J.R. Cooper, K.R. Locherer, R.J. Cava, P.C. Canfield, J. Low Temp. Phys. **105**, 1623 (1996)
[96G1] A.K. Gangopadyay, J.S. Schilling, Physica C **264**, 281 (1996)
[96G2] U. Gasser, P. Allenspach, F. Fauth, W. Henggeler, J. Mesot, A. Furrer, S. Rosenkranz, P. Vorderwisch, M. Buchgeister, Z. Phys. B **101**, 345 (1996)
[96G3] A. Georges, G. Kothar, W. Krauth, M.J. Rosenberg, Rev. Modern Phys. **68**, 13 (1996)
[96G4] K. Ghosh, S Ramakrishnan, A.K. Grover, G. Chandra, T.V. Chandrasekhar Rao, P.K. Mishra, G. Ravikumar, V.C. Sahni, Physica B **223–224**, 109 (1996)
[96G5] C. Godart, Z. Hossain, L.Gaillet, R. Nagarajan, L.C. Gupta, Physica B **223–224**, 76 (1996)
[96G6] G. Goll, M. Heinecke, A. G. M. Jansen, W. Jos, L. Nguyen, E. Steep, K. Winzer, P. Wyder, Phys. Rev. B **53**, R8871 (1996)
[96H1] J. Hartmann, F. Gompf, B. Renker, J. Low Temp. Phys. **105**, 1629 (1996)
[96H2] J.P. Hill, B.J. Sternlieb, D. Gibbs, C. Detlefs, A.I. Goldman, S. Stassis, P.C. Canfield, B.K. Cho, Phys. Rev. B **53**, 3487 (1996)
[96H3] Z. Hossain, L.C. Gupta, R. Nagarajan, S.K. Dhar, C. Godart, R. Vijayaraghavan, Physica B **223–224**, 99 (1996)
[96H4] Q. Huang, J.W. Lynn, A. Santoro, B.C. Chakoumakos, R.J. Cava., J.J. Krajewski, W. F. Peck, Physica C **271**, 311 (1996)
[96I1] M. Ichioka, N. Hayashi, N. Enomoto, K. Machida, Phys. Rev. B **53**, 15316 (1996)
[96J1] D.C. Johnston, F. Borsa, B.K. Cho, P.C. Canfield, P. Dervenagas, A. I. Goldman, B.N. Harmon, L.L. Miller, C. Stassis, B.J. Suh, D.R. Toreson, M. Xu, J. Jarestky, M. F. Hudley, R. Movshovich, J.D. Thompson, A. V. Chubukov, B. Steinlieb, Chin. J. Phys. **34**, P397 (1996)
[96K1] S. Kalavathi, T.G. Kumari, Y. Hariharan, M.C. Valsakumar, M.P. Janawadkar, T.S. Radhakrishnan, Z. Hossain, R. Nagarajan, L. C. Gupta, R. Vijayaraghavan, Physica B **223–224**, 96 (1996)
[96K2] H. Kawano, H. Yoshizawa, H. Takeya, K. Kadowski, Phys. Rev. Lett. **77**, 4628 (1996)
[96K3] K. Krug, M. Heinecke, K. Winzer, Physica C **267**, 321 (1996)
[96K4] K. Kobayashi, T. Mizokawa, K. Mamiya, A. Sekiyama, A. Fujimori, H. Takagi, H. Eisaki, S. Uchida, R.J. Cava, J.J. Krajewski, W. F. Peck, Phys. Rev. B **54**, 507 (1996)
[96K5] V.G. Kogan, A. Gurevich, J.H. Cho, D.C. Johnston, M. Xu, J.R. Thompson, A. Martynovich, Phys. Rev. B **54**, 12386 (1996)
[96K6] S. Kramp, M. Febri, Ph. l'Héritier, P. Ezekwenna, J.C. Joubert, J. Magn. Magn. Mater. **157–158**, 73 (1996)
[96K7] K. Krug, M. Heinecke, K. Winzer, Physica C **267**, 321 (1996)
[96K8] H.C. Ku, M.S. Lin, Y.Y. Hsu, J.H. Shieh, Y.B. You, J. Appl. Phys. **79**, 5863 (1996)
[96L1] A. Lacerda, A. Yatskar, G.M. Schmiedeshoff, W.P. Beyermann, P.C. Canfield, Phil. Mag. B **74**, 641 (1996)
[96L2] L.P. Le, R.H. Heffner, J.D. Thompson, D.E. MacLaughlin, G.J. Nieuwenhuys, A. Amato, R. Feyerhern, F.N. Gygax, A. Schenk, P. C. Canfield, B.K. Cho, Phys. Rev. B **53**, R510 (1996)
[96L3] J. Luo, Y. Wang, Q. Ran, N. Tang, F. Yang, D. Jin, Czech. J. Phys. **46**, S2–819 (1996)
[96L4] J.W. Lynn, Q. Huang, A. Santoro, R.J. Cava, J. J. Krajewski, W.F. Peck, Phys. Rev. B **53**, 802 (1996)
[96L5] J.W. Lynn, Q. Huang, S.K. Sinha, Z. Hossain, L.C. Gupta, R. Nagarajan, C. Godart, Physica B **223–224**, 66 (1996)
[96M1] S. Meenakshi, V. Vijayakumar, R.S. Rao, B.K. Godwal, S. K. Sikka, Z. Hossain, R. Nagarajan, L.C. Gupta, R. Vijayaraghavan, Physica B **223–224**, 93 (1996)
[96M2] S. Miyamoto, H. Takeya, K. Kadowaki, Chech. J. Phys. **46**, 839 (1996)
[96M3] A.I. Morozov, Sov. Phys. JETF **83**, 1048 (1996)
[96M4] A.M. Mulders, P.C.M. Gubbens, K.H.J. Buschow, Phys. Rev. B **54**, 14963 (1996)
[96M5] K.H. Müller, G. Fuchs, A. Handstein, L. Cao, K. Eversmann, Chech. J. Phys. **46**, 829 (1996)

[96N1] D.G. Naugle, K.D.D. Rathnayaka, A.K. Bhatnagar, A.C. Du Mar, A. Parasiris, J.M. Bell, P.C. Canfield, B.K. Cho, Czech. J. Phys. **46**, 3263 (1996)

[96N2] L.H. Nguyen, G. Goll, E. Steep, A.G.M. Jansen, P. Wyder, O. Jepsen, M. Heinecke, K. Winzer, J. Low Temp. Phys. **105**, 1653 (1996)

[96O1] T. Oda, Y. Iwamoto, K. Ueda, T. Kohara, A. Mahajan, K. Elankumaran, Z. Hossain, L.C. Gupta, R. Nagarajan, R. Vijayaraghavan, C. Mazumdar, Physica B **223–224**, 79 (1996)

[96O2] Y. Ohno, N. Sasaki, H. Nakamura, H. Takamura, T. Tanaka, M. Homma, M. Okada, Mater. Trans. **37**, 835 (1996)

[96R1] X.L. Rao, Q.N. Qi, J.M.D. Coey, J. Magn. Magn. Mater. **157–158**, 704 (1996)

[96R2] K.D.D. Rathnayaka, A.K. Bhatnagar, D.G. Naugle, P.C. Canfield, B.K. Cho, Physica B **223–224**, 83 (1996)

[96R3] K.DD. Rathnayaka, D.G. Naugle, B.K. Cho, P.C. Canfield, Phys. Rev. B **53**, 5688 (1996)

[96R4] L.F. Rybaltchenko, I.K. Yanson, A.G.M. Jansen, P. Mandal, P. Wyder, C.V. Tomy, D. McK. Paul, Physica B **218**, 189 (1996)

[96S1] D.R. Sánchez, H. Micklitz, M.B. Fontes, S.L. Bud'ko, E. Baggio-Saitovitch, Phys. Rev. Lett. **76**, 507 (1996)

[96S2] J.P. Sanchez, P. Vulliet, C. Godart, L.C. Gupta, Z. Hossain, R. Nagarajan, Phys. Rev. B **54**, 9421 (1996)

[96S3] M. Sera, S. Kobayash, M. Hiroi, N. Kobayashi, H. Takeya, K. Kadowski, Phys. Rev. B **54**, 3062 (1996)

[96S4] D.J. Singh, Solid State Commun. **98**, 899 (1996)

[96S5] B.J. Suh, F. Borsa, D.R. Torgeson, B.K. Cho, P.C. Canfield, D.C. Johnston, J.Y. Rhee, B.N. Harmon, Phys. Rev. B **53**, R6022, (1996)

[96S6] B.J. Suh, F. Borsa, D.R. Torgeson, B.K. Cho, P.C. Canfield, D.C. Johnston, J.Y. Rhee, B.N. Harmon, Phys. Rev. B **54**, 15341 (1996)

[96S7] C.S. Sundar, A. Bharathi, Y. Hariharan, T.S. Radhakrishnan, Z. Hossain, R. Nagarajan, L.C. Gupta, R. Vijayaraghavan, Phys. Rev. B **53**, R2971 (1996)

[96T1] C.V. Tomy, J.M. Martin, D. McK. Paul, Physica B **223–224**, 116 (1996)

[96T2] C.V. Tomy, L.A. Afalfiz, M.R. Lees, J.M. Martin, D. McK. Paul, Phys. Rev. B **53**, 307 (1996)

[96T3] C.V. Tomy, M.R. Lees, G. Balakrishnan, D.T. Adroja, D. McK. Paul, Physica B **223–224**, 62 (1996)

[96U1] Y. Uwatoko, G. Ooomi, P.C. Canfield, B.K. Ono, Physica B **216**, 329 (1996)

[96W1] R. Weht, O.M. Cappannini, C.O. Rodriguez, N.E. Christensen, Physica C **260**, 125 (1996)

[96Y1] U. Yaron, P.L. Gammel, A.P. Ramirez, D.A. Huse, Nature, **382**, 236 (1996)

[96Y2] A. Yatskar, N.K. Budraa, W.P. Beyermann, P.C. Canfield, S.L. Butko, Phys. Rev. B **54**, R3772 (1996)

[96Y3] W.B. Yelon, Z. Hu, M. Paranthaman, W.J. James, I. Menon, S.K. Malik, Physica B **223–224**, 105 (1996)

[96Z1] H.W. Zandbergen, D. Tang, J. Jensen, R.J. Cava, Ultramicroscopy, **64**, 231 (1996)

[96Z2] Z. Zeng, D.E. Ellis, D. Guensburger, E.M. Baggio-Saitwitch, Phys. Rev. B **53**, 6613 (1996)

[96Z3] Z. Zeng, D.E. Ellis, D. Guenzburger, E. Baggio-Saitovitch, Phys. Rev. B **54**, 13020 (1996)

[97A1] I. Affleck, M. Franz, M. Amin, Phys. Rev. B **55**, R704 (1997)

[97A2]. P. B. Allen, V. N. Kostur, N. Takesue, G. Shirane, Phys. Rev. B **56**, 5552 (1997)

[97B1] A.K. Bhatnagar, K.D.D. Rathnayaka, D.G. Naugle, P.C. Canfield, Phys. Rev. B **56**, 437 (1997)

[97B2] S.L. Budko, P.C. Canfield, A. Yatskar, W.P. Beyermann, Physica B **230–232**, 859 (1997)

[97C1] P.C. Canfield, S.L. Budko, J. Alloys Comp. **262–263**, 169 (1997)

[97C2] P.C. Canfield, S.L. Budko, B.K. Cho, W.P. Beyermann, A. Yatskar, J. Alloys Comp. **250**, 596 (1997)

[97C3] P.C. Canfield, S.L. Bud'ko, B.K. Cho, A. Lacerda, D. Farrell, E. Johnston-Halperin, V.A. Kalatsky, V.L. Pokrovsky, Phys. Rev. B **55**, 970 (1997)

[97C4] A.L. Cornelius, A.R. Gangopadhyay, J.S. Schillimg, W.Assmus, Phys. Rev. B **55**, 14109 (1997)

[97C5] L. Cristofolini, A. Lappas, K. Prassides, K. Vavekis, M. Buchgeister, Hyp. Int. **104**, 61 (1997)
[97D1] C. Detlefs, A. H. M. Z. Islam, A. I. Goldman, C. Stassis, P. C. Canfield, J. P. Hill, D. Gibbs, Phys. Rev. B **55**, R680 (1997)
[97D2] C. Detlefs, A.H.M.Z. Islam, T. Gu, A.I. Goldman, C. Stassis, P.C. Canfield, J.P. Hill, T. Vogt, Phys. Rev. B **56**, 7843 (1997)
[97D3] Y. De Wilde, M. Iavarone, U. Welp, V. Metlushko, A.E. Koshelev, I. Aranson, G.W. Crabtree, P.C. Canfield, Phys. Rev. Lett. **78**, 4273 (1997)
[97E1] J. Eckert, K. Jost, L. Schultz, Mater. Lett. **31**, 329 (1997)
[97E2] M. El Massalami, M.R. Amaral, L. Ghivelder, I. Abrego Castillo, G.J. Nieuwenhuys, C.E. Snel, J. Magn. Magn. Mater. **172**, 139 (1997)
[97E3] M.R. Eskildsen, P.L. Gammel, B.P. Barber, A.P. Ramirez, D.J. Bishop, N.H. Andersen, K. Mortensen, C.A. Bolle, C.M. Lieber, P. C. Canfield, Phys. Rev. Lett. **79**, 487 (1997)
[97E4] M.R. Eskildsen, P.L. Gammel, B.P. Barber, U. Yaron, A.P. Ramirez, D.A. Huse, D.J. Bishop, C. Bolle, C.M. Lieber, S. Oxx, S. Sridhar, N.H. Andersen, K. Mortensen, P.C. Canfield, Phys. Rev. Lett. **78**, 1968 (1997)
[97F1] I. Felner, D. Schmitt, B. Barbara, C. Godart, E. Alleno, J. Solid State Chem. **133**, 5 (1997)
[97F2] I.R. Fisher, J.R. Cooper, P.C. Canfield, Phys. Rev. B **56**, 10820 (1997)
[97G1] U. Gasser, P. Allenspach, J. Mesot, A. Furrer, Physica C **282–287**, 1327 (1997)
[97G2] C. Godart, E. Alleno, E. Tominez, L.C. Gupta, R. Nagarajan, Z. Hossain, J.W. Lynn, P. Bonville, J.A. Hodges, J.P. Sanchez, I. Felner, J. Solid State Chem. **133**, 169 (1997)
[97G3] G. Goll, L. Nguyen, E. Steep, A.G.M. Jansen, P. Wyder, K. Winzer, Physica B **230–232**, 868 (1997)
[97H1] Z. Hossain, R. Nagarajan, S.M. Pattalwar, S.K. Dhar, L. C. Gupta, C. Godard, Physica B **230–232**, 865 (1997)
[97K1] H. Kito, S. Ikeda, S. Takegawa, H. Abe, H. Kitazawa, Physica C **291**, 332 (1997)
[97K2] B. Knigge, A. Hoffman, D. Lederman, C. C. Vier, S. Schultz, I. K. Schuller, J. Appl. Phys. **81**, 2291 (1997)
[97K3] V.G. Kogan, M. Bullock, B. Harmon, P. Miranovic, L. Dobrosavljevic-Grujic, P.L. Gammel, D.J. Bishop, Phys. Rev. B **55**, R8693 (1997)
[97K4] A. Kreyssig, C. Sierks, M. Loewenhaupt, J. Freudenberger, G. Fuchs, K.H. Müller, C. Ritter, Physica B **241–243**, 826 (1997)
[97K5] A. Kreyssig, M. Loewenhaupt, K. H. Müller, G. Fuchs, A. Handstein, C. Ritter, Physica B **234–236**, 737 (1997)
[97K6] M.L. Kulic, A.I. Buzdin, L.N. Bulaevskii, Phys. Lett. A **235**, 285 (1997)
[97L1] M.D. Lan, J.L. Chen, C.W. Hsing, H. Samata, Y. Nagata, J. Phys. Chem. Solids **58**, 1425 (1997)
[97L2] W.C. Lee, E.J. Joo, H.S. Cho, S.E. Lee, H.B. Kim, Physica C **282–287**, 1313 (1997)
[97L3] M. Loewenhaupt, A. Kreyssig, C. Sierks, K.H. Müller, I. Freudenberger, C. Ritter, H. Schober, Annual Report ILL (1997)
[97L4] J. Luo, Y. Wang, Q. Ran, X. Tong, N. Tang, F. Yang, D. Jin, Physica C **282–287**, 1933 (1997)
[97L5] J.W. Lynn, J. Alloys Comp. **250**, 552 (1997)
[97L6] J.W. Lynn, S. Skanthakumar, Q. Huang, S.K. Sinha, Z. Hossain, L. C. Gupta, R. Nagarajan, C. Godart, Phys. Rev. B **55**, 6584 (1997)
[97M1] J. Mesot, P.S. Allenspack, U. Gasser, A. Furrer, J. Alloys Comp. **250**, 559 (1997)
[97M2] V. Metlushko, U. Welp, A. Koshelev, I. Aranson, G. W. Crabtree, P. C. Canfield, Phys. Rev. Lett. 79, 1738 (1997)
[97M3] K.H. Muller, A. Kreyssig, A. Handstein, G. Fuchs, C. Ritter, M. Lowenhaupt, J. Appl. Phys. **81**, 4240 (1997)
[97N1] R. Nagarjan, J. Low Temp. Phys. **107**, 517 (1997)
[97N2] M. Nohara, M. Isshiki, H. Takagii, R. J. Cava, J. Phys. Soc. Jpn. **66**, 1888 (1997)
[97R1] K.D.D. Rathnayaka, A.K. Bhatnagar, A. Parasiris, D.G. Naugle, P.C. Canfield, B.K. Cho, Phys. Rev. B **55**, 8506 (1997)

[97S1] R. Sala, F. Borsa, E. Lee, P.C. Canfield, Phys. Rev. B **56**, 6195 (1997)
[97S2] D.R. Sanchez, H. Mickletz, M.B. Fontes, E. Baggio-Saitovitch, J. Phys.: Condens. Matter, **9**, L299 (1997)
[97S3] D.R. Sánchez, J.C. Tróchez, H. Micklitz, M.B. Fontes, B. Giordanengo, E. Baggio-Saitovitch, Physica C **282–287**, 1941 (1997)
[97S4] H. Schmidt, H. F. Braun, Phys. Rev. B **55**, 8497 (1997)
[97S5] C. Stassis, A.I. Goldman, J. Alloys Comp. **250**, 603 (1997)
[97S6] C. Stassis, M. Bullock, J. Zarestky, P. Canfield, A. I. Goldman, G. Shirane, S. M. Shapiro, Phys. Rev. B **55**, R8678 (1997)
[97S7] B. Sternlieb, C. Stassis, A. I. Goldman, P. Canfield, S. Shapiro, J. Appl. Phys. **81**, 4937 (1997)
[97S8] H. Szillat, P. Majewski, F. Aldinger, J. Alloys Comp. **261**, 242 (1997)
[97S9] H. Szillat, H. Kuhn, Th. Schuster, P. Majewski, M. Seeger, F. Aldinger, H. Kronmüller, Physica C **280**, 43 (1997)
[97T1] C.V. Tomy, L. J. Chang, D. McK. Paul, C. Ritter, Physica B **230–232**, 872 (1997)
[97V1] P. Villars, *Pearson's handbook*, Crystallographic Data, vol.1 (1997)
[97Y1] I.K. Yanson, V.V. Fisun, A.G.M. Jansen, P. Wyder, P.C. Canfield, B.K. Cho, C.V. Tomy, D. McK. Paul, Phys. Rev. Lett. **78**, 935 (1997)
[97Y2] I.K. Yanson, V.V. Fisun, A.G.M. Jansen, P. Wyder, P.C. Canfield, J. Low Temp. Phys. **23**, 712 (1997)
[97Y3] A. Yatskar, S.L. Bud'ko, P.C. Canfield, W. P. Beyermann, G.M. Schmiedeshoff, M.S. Torikachvili, C.H. Mielke, A. Lacerda, Physica B **230–232**, 876 (1997)
[97Y4] M. Yethiraj, D. Mck. Paul, V. V. Tomy, E. M. Forgan, Phys. Rev. Lett. **78**, 4849 (1997)
[97Z1] H. Zhang, S.J. Campbell, S.R. Bulcock, A.V.J. Edge, E. Wu, Physica, B **229**, 333 (1997)
[97Z2] Z. Zeng, D. R. Sánchez, D. Guenzburger, D. E. Ellis, E. M. Baggio-Saitovitch, H. Micklitz, Phys. Rev. B **55**, 3087 (1997)
[98A1] A. Amici, P. Thalmeier, Phys. Rev. B57, 10684 (1998)
[98B1] A. Belger, U. Jaenicke-Rössler, D. Lipp, B. Wehner, P. Paufler, G. Behr, Physica C **306**, 277 (1998)
[98B2] H. Bitterlich, W. Löser, G. Behr, K. Nenkov, G. Fuchs, A. Belger, L. Schultz, Physica C **308**, 243 (1998)
[98B3] S. Bulcock, H. Zhang, E. Wu, S.J. Campbell, J. Mat. Sci. Lett. **17**, 1791 (1998)
[98B4] M. Bullock, J. Zarestky, C. Stassis, A. Goldman, P. Canfield, Z. Honda, G. Shirane, S. M. Shapiro, Phys. Rev. B **57**, 7916 (1998)
[98C1] P.C. Canfield, P.L. Gammel, D.J. Bishop, Phys. Today, **61**, 40 (1998)
[98C2] O.M. Cappannini, C.O. Rodriguez, N. E. Christensen, Physica C **306**, 101 (1998)
[98C3] B.K. Cho, Physica C **298**, 305 (1998)
[98C4] K.O. Cheon, I.R. Fisher, V.G. Kogan, P.C. Canfield, P. Miranovič, P.L. Gammel, Phys. Rev. B **58**, 6463 (1998)
[98D1] A.C. Du Mar, K.D.D. Rathnayaka, D.G. Naugle, R.C. Canfield, Int. J. Modern Phys. B **12**, 3264 (1998)
[98E1] M. El-Hagary, H. Michor, C. Jambrich, R. Hauser, M. Galli, E. Bauer, G. Hilscher, J. Magn. Magn. Mater. **171–181**, 551 (1998)
[98E2] M. El Massalami, R.E. Rapp, G.J. Nieuwenhuys, Physica C **304**, 184 (1998)
[98E3] M.R. Eskildsen, K. Harada, P.L. Gammel, A.B. Abrahamsen, N.H. Andersen, G. Ernst, A.P. Ramirez, D.J. Bishop, K. Mortensen, D.G. Naugle, K.D.D. Rathnayaka, P.C. Canfield, Nature, **393**, 242 (1998)
[98F1] J. Freudenberger, G. Fuchs, K. Nenkov, A. Handstein, M. Wolf, A. Kreyssig, K.H. Müller, M. Loewenhaupt, L. Schultz, J. Magn. Magn. Mater. **187**, 309 (1998)
[98F2] J. Freudenberger, S.L. Drechsler, G. Fuchs, A. Kreyssig, K. Nenkov, S.V. Shulga, K.H. Müller, L. Schultz, Physica C **306**, 1 (1998)
[98G1] U. Gasser, P. Allenspach, A. Furrer, Physica B **241**, 789 (1998)
[98G2] U. Gasser, P. Allenspach, A. Furrer, A. M. Mulders, J. Alloys Comp. **275–277**, 587 (1998)
[98G3] L.C. Gupta, Phil. Mag. **77**, 717 (1998)

[98H1] A.D. Hillier, J.R. Stewart, M.T.F. Telling, R.I. Bewley, Z.P. Han, S.L. Lee, R. Cywinski, J. Magn. Magn. Mater. **177–181**, 1111 (1998)
[98K1] V.A. Kalatsky, V.L. Pokorovsky, Phys. Rev. B **57**, 5485 (1998)
[98K2] K.S. Kim, S.C. Yu, V. Ström, K.V. Rao, J. Magn. Magn. Mater. **177–181**, 553 (1998)
[98K3] A. Kreyssig, C. Sierks, M. Loewenhaupt, J. Freudenberger, G. Fuchs, R.H. Müller, C. Ritter, Physica B **241–243**, 826 (1998)
[98L1] M.D. Lan, T.J. Chang, C.S. Liaw, J. Phys. Chem. Solids, **59**, 1285 (1998)
[98M1] D. McK. Paul, C.V. Tomy, C.M. Aegerter, R. Cubitt, S.H. Lloyd, E.M. Forgan, S.L. Lee, M. Yethiraj, Phys. Rev. Lett. **80**, 1517 (1998)
[98M2] S. Meenakshi, V. Vijayakumar, R.S. Rao, B.K. Godwal, S.K. Sikka, P. Ravindran, Z. Hossain, R. Nagarajan, L.C. Gupta, R. Vijayaraghavan, Phys. Rev. B **58**, 3377 (1998)
[98M3] A.M. Mulders, P.C.M. Gubbens, U. Gasser, C. Baines, K.H.J. Buschow, Phys. Rev. B **57**, 10320 (1998)
[98M4] A.M. Mulders, P.C.M. Gubbens, V. Gasser, C. Baines, K.H.J. Buschow, J. Magn. Magn. Mater. **177–181**, 555 (1998)
[98M5] K.H. Müller, A. Handstein, D. Eckert, G. Fuchs, K. Nenkov, J. Freudenberger, M. Richter, M. Wolf, Physica B **246–247**, 226 (1998)
[98N1] V.N. Narozhnyi, V.N. Kochetkov, A.V. Tsvyashchenko, L.N. Fomicheva, J. Alloys Comp. **275–277**, 484 (1998)
[98N2] D.G. Naugle, K.D.D. Rathnayaka, P.C. Canfield, Int. J. Modern Phys. B **12**, 3174 (1998)
[98P1] P.G. Pagliuso, C. Rettori, S.B. Oseroff, P.C. Canfield, E.M. Baggio-Saitovitch, D. Sanchez, Phys. Rev. B **57**, 3668 (1998)
[98P2] K. Park, D. A. Huse, Phys. Rev. B **58**, 9927 (1998)
[98P3] Z.Q. Peng, K. Krug, K. Winzer, Phys. Rev. B **57**, R8123 (1998)
[98R1] A. Rustom, A.D. Hillier, R. Cywinski, J. Magn. Magn. Mater. **177–181**, 1153 (1998)
[98S1] D.R. Sanchez, S.L. Budko, E.M. Baggio-Saitovitch, M. Hillberg, W. Wagener, H.H. Klauss, G.H. Walf, F.J. Litterst, Phys. Rev. B **57**, 10268 (1998)
[98S2] S.V. Shulga, S.I. Drechsler, G. Fuchs, K.H. Müller, K. Winzer, M. Heinecke, K. Krug, Phys. Rev. Lett. **80**, 1730 (1998)
[98T1] K. Tomala, J.P. Sanchez, P. Vulliet, P.C. Canfield, Z. Drzazga, A. Winiarska, Phys. Rev. B **58**, 8534 (1998)
[98W1] G.Wang, K. Maki, Phys. Rev. B **58**, 6493 (1998)
[98W2] E. Wu, H. Zhang, S.J. Campbell, S.R. Bulcock, J. Alloys Comp. **264**, 232 (1998)
[98Y1] M. Yethiraj, D. McK. Paul, C. V. Tomy, J. R. Thompson, Phys. Rev. B **58**, R 14767 (1998)
[98Z1] H. Zhang, E. Wu, S.J. Campbell, S.J. Kennedy, H.S. Li, A.J. Studer, S.R. Bulcock, A.D. Rae, J. Alloys Comp. **278**, 239 (1998)
[99B1] G. Behr, W. Löser, G. Graw, K. Nenkov, U. Krämer, A. Belger, B. Wehner, J. Mater. Res. **14**, 16 (1999)
[99B2] A. Belger, G. Zahn, B. Wehner, P. Paufler, G. Graw, G. Behr, J. Alloys Comp. **283**, 26 (1999)
[99B3] H. Bitterlich, W. Löser, G. Behr, K. Nenkov, G. Fuchs, A. Gümbel, L. Schultz, Physica C **321**, 93 (1999)
[99B4] P. Bonville, J. A. Hodges, Z. Hossain, R. Nagarajan, S.K. Dhar, L.C. Gupta, E. Alleno, C. Godart, Eur. Phys.J. B **11**, 377 (1999)
[99C1] S. Cao, K. Okamoto, K. Nishimura, K. Mori, J. Low Temp. Phys. **117**, 1611 (1999)
[99C2] S. Cao, T. Ookami, K. Nishimura, K. Mori, Physica C **316**, 273 (1999)
[99C3] N.F. Chaban, Y.B. Kuzma, Neorg. Mater. **35**, 577 (1999)
[99C4] K.O. Cheon, I.R. Fisher, P.C. Canfield, Physica C **312**, 35 (1999)
[99C5] L. Civale, A.V. Silhanek, J. R. Thompson, K. J. Song, C. V. Tomy, D. Mc.K Paul, Phys. Rev. Lett. **79**, 1138 (1999)
[99D1] C. Detlefs, D.L. Abernathy, G. Grübel, P.C. Canfield, Europhys. Lett. **47**, 352 (1999)
[99D2] C. D. Dewhurst, R. A. Doyle, E. Zeldov, D. McK. Paul, Phys. Rev. Lett. **82**, 827 (1999)
[99D3] H. Doh, M. Sigrist, B.K. Cho, S.J. Lee, Phys. Rev. Lett. **23**, 5350 (1999)

[99D4] S.B. Dugdale, M.A. Alam, I. Wilkinson, R.J. Hughes, I.R. Felner, P.C. Canfield, J. Jarlborg, G. Santi, Phys. Rev. Lett. **83**, 4824 (1999)

[99E1] M. R. Eskildsen, K. Harada, P. L. Gammel, N. H. Andersen, G. Ernst, A. P. Ramirez, D. J. Bishop, K. Mortensen, P. C. Canfield, Physica B **259–261**, 582 (1999)

[99F1] J. Freudenberger, G. Fuchs, K.H. Müller, K. Nenkov, S.L. Drechsler, A. Kreyssig, H. Rosner, K. Koepernik, D. Lipp, L. Schultz, J. Low Temp. Phys. **117**, 1623 (1999)

[99G1] P.L. Gammel, B.P. Barber, A.P. Ramirez, C.M. Varma, D.J. Bishop, P.C. Canfield, V.G. Kogan, M.R. Eskildsen, N.H. Andersen, K. Mortensen, K. Harada, Phys. Rev. Lett. **82**, 1756 (1999)

[99G2] P.L. Gammel, D.J. Bishop, M.R. Eskildsen, K. Mortensen, N.H. Andersen, I.R. Fischer, K.O. Cheon, P.C. Canfield, V.G. Kogan, Phys. Rev. Lett. **82**, 4082 (1999)

[99H1] G. Hilscher, H. Michor, Studies on high temperature superconductors, Nova Sci. **28**, 241 (1999)

[99J1] U. Jaenicke-Roessler, A. Belger, G. Zahn, B. Wehner, P. Paufler, H. Bitterlich, Physica C **314**, 43 (1999)

[99K1] H. Kawano, H. Takeya, H. Yoshizawa, K. Kadowaki, J. Phys. Chem. Solids, **60**, 1053 (1999)

[99K2] A. Kreyssig, M. Loewenhaupt, J. Freudenberger, K.H. Müller, C. Ritter, J. Appl. Phys. **85**, 6058 (1999)

[99K3] M. Kuznietz, H. Takeya, J. Phys.: Condens. Matter, **11**, 6561 (1999)

[99M1] C. Mazumdar, Z. Hu, G. Kaindl, Physica B **259–261**, 89 (1999)

[99N1] R. Nagarajan, E. Alleno, S.J. Blundell, C. Mazumdar, D.W. Cooke, S.P. Cottrell, S.F.J. Cox, C. Godart, L.C. Gupta, Z. Hossain, W.L. Hults, Th. Jestadt, E.J. Peterson, F.L. Pratt, J.L. Smith, Physica B **259–261**, 588 (1999)

[99N2] V.N. Narozhnyi, V.N. Kochetkov, A.V. Tsvyashchenko, L.N. Fomicheva, Solid State Commun. **111**, 259 (1999)

[99N3] V.N. Narozhnyi, J. Freudenberger, V.N. Kochetkov, K.A. Nenkov, G. Fuchs, A. Handstein, K.H. Müller, Phys. Rev. B **59**, 14762 (1999)

[99N4] V.N. Narozhnyi, J. Freudenberger, G. Fuchs, K.A. Nenkov, D. Eckert, A. Czopnik, K.H. Müller, J. Low Temp. Phys. **117**, 1599 (1999)

[99N5] D.G. Naugle, K.D.D. Rathnayoka, K. Clark, P.C. Canfield, Int. J. Mod. Phys. B **13**, 3715 (1999)

[99N6] M. Nohara, M. Isshiki, F. Sakai, H. Takagi, J. Phys. Soc. Jpn. **68**, 1078 (1999)

[99O1] G. Oomi, M. Matsuda, F, Honda, T. Kagayama, K. Honda, B.K. Cho, P.C. Canfield, Physica B **259–261**, 601 (1999)

[99P1] Z. Q. Peng, K. Klug, K. Winzer, Physica C **317–318**, 441 (1999)

[99S1] C. Sierks, M. Loewenhaupt. P. Tils, J Freudenberger, K.H. Müller, C.K. Loong, H. Schober, Physica B **259–261**, 592 (1999)

[99S2] C. Sierks, M. Doerr, A. Kreyssig, M. Loewenhaupt, Z.Q. Peng, K. Winzer, J. Magn. Magn. Mater. **192**, 473 (1999)

[99S3] S. Skanthakumar, J. W. Lynn, Physica B **259–261**, 576 (1999)

[99S4] C. Song, Z. Islam, L. Lottermoser, A.J. Goldman, P.C. Canfield, C. Detlefs, Phys. Rev. B **60**, 6223 (1999)

[99S5] K. J. Song, J. R. Thompson, M. Yethiraj, D. K. Christen, C. V. Tomy, D. McK. Paul, Phys. Rev. B **59**, R6620 (1999)

[99S6] C. Stassis, J. Zarestky, A.I. Goldman, P.C. Canfield, G. Shirane, S. Shapiro, Physica C **317–318**, 127 (1999)

[99T1] H. Takeya, M. Kuzneitz, Physica B **259–261**, 596 (1999)

[99T2] H. Takeya, M. Kuzneitz, J. Magn. Magn. Mater. **195**, 366 (1999)

[99T3] A.A. Tulapurkar, S.N. Mishra, Z. Hossain, Hyp. Interact. **120–121**, 225 (1999)

[99W1] K. Winzer, Z.Q. Peng, K. Krug, Physica B **259–261**, 586 (1999)

[99W2] K. Winzer, Z.Q. Peng, K. Krug, Adv. Solid. State Phys. **38**, 497 (1999)

[99Y1] W. Yang, U. Krämer, G. Behr, G. Graw, Cryst. Res. Techn. **34**, 1213 (1999)

[99Y2] A. Yatskar, C.H. Mielke, P.C. Canfield, A.H. Lacerda, W.P. Beyermann, Phys. Rev. B **60**, 8012 (1999)
[99Z1] J. Zarestky, C. Stassis, A.I. Goldman, P.C. Canfield, G. Shirane, S. Shapiro, Phys. Rev. B **60**, 11932 (1999)
[99Z2] H. Zhang, S.J. Campbell, H.S. Li, E. Wu, J. Alloys Comp. **284**, 155 (1999)
[00A1] A. Amici, P. Thalmeier, P. Fulde, Phys. Rev. Lett. **84**, 1800 (2000)
[00B1] E.M. Baggio-Saitovich, D.R. Sanchez, H. Micklitz, Physica C **341–348**, 703 (2000)
[00B2] S.L. Budko, P.C. Canfield, Phys. Rev. B **61**, R14932 (2000)
[00C1] A.J. Campbell, D.McK. Paul, G.J. McIntyre, Solid State Commun. **115**, 213 (2000)
[00C2] A.J. Campbell, D.McK. Paul, G.J. McIntyre, Phys. Rev. B **61**, 5872 (2000)
[00C3] S. Cao, K. Nishimura, K. Mori, Physica C **341–348**, 759 (2000)
[00C4] S. Cao, S. Sakai, K. Nishimura, K. Mori, Physica C **341–348**, 751 (2000)
[00C5] C. Chacon, O. Isnard, Physica B **276–278**, 662 (2000)
[00C6] R.K. Chu, W.K. Chu, Q. Chen, Z.H. Zhang, J.H. Miller, J. Phys.: Condens. Matter **12**, 275 (2000)
[00D1] C. Detlefs, F. Bourdarot, P. Burlet, P. Dervengas, S.L. Budko, P.C. Canfield, Phys. Rev. B **61**, R14916 (2000)
[00D2] M. Divis, K. Schwarz, P. Blaha, G. Hilscher, H. Michor, S. Khmelevskyi, Phys. Rev. B **62**, 6774 (2000)
[00D3] A. Duran, E. Munoz, S. Bernes, R. Escudero, J. Phys.: Condens. Matter, **12**, 7595 (2000)
[00E1] M. El-Hagary, H. Michor, G. Hilscher, Physica B **284–288**, 1489 (2000)
[00E2] M. El-Hagary, H. Michor, G. Hilscher, Phys. Rev. B **61**, 11695 (2000)
[00E3] M. R. Eskildsen, L. R. Fisher, P. L. Gammel, D. J. Bishop, N. H. Anderson, K. Mortensen, P. C. Canfield, Physica C **332**, 320 (2000)
[00F1] J. Freudenberger, G. Fuchs, K. Nenkov, S. L. Drechsler, K. H. Müller, L. Schultz, Physica C **339**, 195 (2000)
[00G1] P.L. Gammel, D. Lopez, D.J. Bishop, M. R. Eskildsen, N. H. Andersen, K. Mortensen, I.R. Fisher, K.O. Cheon, P.C. Canfield, J. Appl. Phys. **87**, 5544 (2000)
[00G2] S. Geupel, A. Belger, P. Paufler, G. Graw, Z. Kristallogr. **215**, 481 (2000)
[00G3] S.Geupel, G. Zahn, P. Paufler, D. Graw, Z. Kristallogr. **215**, 477 (2000)
[00I1] A. Yu. Ignatov, N. Ali, P.V. Konarev, M. Tishcher, A.V. Tsvyashchenko, L.N. Fomicheva, Phys. Rev. B **61**, 3274 (2000)
[00J1] W. Jeitschko, T. Konrad, K. Hartjes, A. Lang, R.D. Hoffman, J. Solid State Chem. **154**, 246 (2000)
[00K1] V.G. Kogan, S.L. Budko, I.R. Fisher, P.C. Canfield, Phys. Rev. B **62**, 9077 (2000)
[00K2] M. Kuznietz, H. Takeya, Physica B **281–282**, 1004 (2000)
[00M1] C. Mazumdar, R. Nagarajan, L.C. Gupta, B.D. Padalia, R. Vijayaraghavan, Appl. Phys. Lett. **77**, 895 (2000)
[00M2] H. Michor, M. El-Hagary, L. Naber, E. Bauer, G. Hilscher, Phys. Rev. B **61**, R6487 (2000)
[00N1] V.N. Narozhnyi, G. Fuchs, J. Freudenberger, K. Nenkov, K.H. Müller, Physica B **284–288**, 535 (2000)
[00N2] V.N. Narozhnyi, V.N. Kochetkov, A.V. Tsvyashchenko, L.N. Fomicheva, Ferroelectrics **247**, 163 (2000)
[00N3] M. Nohara, M. Suzuki, N. Mangkorntong, H. Takai, Physica C **341–348**, 2177 (2000)
[00N4] K. Norgaard, M.R. Eskildsen, N.H. Andersen, J. Jensen, P. Hedegard, S.N. Klausen, P.C. Canfield, Phys. Rev. Lett. **84**, 4982 (2000)
[00R1] M. Rams, K. Krolas, P. Bonville, J.A. Hodges, Z. Hossain, R. Nagarajan, S.K. Dhar, L.C. Gupta, E. Alleno, C. Godart, J. Magn. Magn. Mater. **219**, 15 (2000)
[00R2] H. Rosner, S.L. Drechsler, S.V. Shulga, K. Koepernik, I. Opahle, H. Eschrig, Adv. Solid State Phys. **40**, 713 (2000)
[00S1] N. Saha, R. Surdeanu, M. Marchevsky, G.J. Nieuwenhuys, C.D. Dewhurst, R.J. Wijngaarden, D. Mck. Paul, P.H. Kes, Phys. Rev. B **63**, R020502 (2000)
[00S2] H. Sakata, M. Oosawa, K. Matsuba, N. Nishida, H. Takeya, K. Hirata, Phys. Rev. Lett. **84**, 1583 (2000)

[00S3] D.R. Sanchez, S.L. Budko, E.M. Baggio-Saitovich, J. Phys.: Condens. Matter, **12**, 9941 (2000)

[00S4] G.M. Schmiedeshoff, C. De Boer, M.V. Tompkins, W.P. Beyermann, A.H. Lacerda, J.L. Smith, P.C. Canfield, J. Super. Novel Magn. **13**, 847 (2000)

[00T1] O. Trovarelli, C. Geibel, S.Mederle, C.Langhammer, F.M. Grosche, P. Gegenwart, M. Lang, G. Sparn, F.Steglich, Phys. Rev. Lett. **85**, 626 (2000)

[00V1] L.Y. Vinnikov, T.L. Barkov, K.O. Cheon, P.C. Canfield, V.G. Kogan, Physica B **284–288**, 813 (2000)

[00W1] T. Watanabe, K. Kitazawa, T. Hasegawa, Z. Hossain, R. Nagarajan, L.C. Gupta, J. Phys. Soc. Jpn. **69**, 2708 (2008)

[00Y1] I.S. Yang, M.V. Klein, S.L. Cooper, P.C. Canfield, B.K. Cho, S.I. Lee, Phys. Rev. B **62**, 1291 (2000)

[00Y2] W. Yang-Bitterlich, V. Kramer, Cryst. Res. Techn. **35**, 369 (2000)

[00Y3] I.K. Yanson, N.L. Bobrov, C.V. Tomy, D. McK Paul, Physica C **334**, 152 (2000)

[00Y4] T. Yokoya, T. Kiss, T. Watanabe, S. Shin, M. Nohara, H. Takagi, T. Oguchi, Phys. Rev. Lett. **85**, 4952 (2000)

[00Z1] H. Zhang, S.J. Campbell, H.S. Li, M. Hofmann, A.V.J. Edge, J. Phys.: Condens. Matter, **12**, 5021 (2000)

[01B1] S.N. Behera, B.N. Panda, G.C. Rout, P. Entel, Int. J. Modern Phys. **15**, 2519 (2001)

[01B2] E. Boaknin, R.W. Hill, C. Proust, C. Lupien, L. Taillefer, P.C. Canfield, Phys. Rev. Lett. **87**, 237001 (2001)

[01B3] A. Buzdin, *Rare-earth transition metal borocarbides*, ed. by K.H. Müller, V. Narozhnyi, Kluwer Acad. Publ. (2001), p. 303

[01C1] P.C. Canfield, I.R. Fisher, J. Cryst. Growth, **225**, 155 (2001)

[01C2] C. Chacon, O. Isnard, J. F. Berar, Mater. Sci. Forum, **378–381**, 434 (2001)

[01C3] B.K. Cho, H.B. Kim, S.I. Lee, Phys. Rev. B **63**, 144528 (2001)

[01C4] J.H. Choi, H. Doh, E.M. Choi, H. J. Kim, S.I. Lee, T. Yamamoto, T. Kawae, K. Takeda, J. Phys. Soc. Jpn. **70**, 3077 (2001)

[01C5] S.M. Choi, J.W. Lynn, D. Lopez, P.L. Gammel, P.C. Canfield, S.L. Budko, Phys. Rev. Lett. **87**, 107001 (2001)

[01C6] J.H. Choi, H. Doh, E.M. Choi, S.I. Lee, M. Ohashi, T. Mori, Phys. Rev. B **65**, 024520 (2001)

[01C7] A.D. Christianson, S.L. Budko, G.M. Schmiedeshoff, W.P. Beyermann, P.C. Canfield, G.S. Boebinger, A.H. Lucerda, Physica B **294–295**, 225 (2001)

[01C8] M.A. Continentino, S.N. de Medeiros, M.T.D. Orlando, M.B. Fontes, E.M. Baggio-Saitovich, Phys. Rev. B **64**, 012404 (2001)

[01D1] S.L. Drechsler, H. Rosner, S.V. Shulga, I. Opahle, H. Eschrig, J. Freudenberger, G. Fuchs, K. Nenkov, K.H. Müller, H. Bitterlich, W. Löser, G. Behr, D. Lipp, A. Gladun, Physica C **364–365**, 31 (2001)

[01D2] Z. Drzazga, D. Kaczorowski, A. Winiarska, A. Winiarski, J. Alloys Comp. **323–324**, 562 (2001)

[01E1] M.R. Eskildsen, A.B. Abrahamsen, V.G. Kogan, P.L. Gammel, K. Mortensen, N.H. Andersen, P.C. Canfield, Phys. Rev. Lett. **86**, 5148 (2001)

[01E2] M.R. Eskildsen, A.B. Abrahamsen, D. Lopez, P.L. Gammel, D.J. Bishop, N.H. Andersen, K. Mortensen, P.C. Canfield, Phys. Rev. Lett. **86**, 320 (2001)

[01F1] R. Falconi, A. Duran, R. Escudero, Phys. Rev. B **65**, 024505 (2001)

[01F2] C. Felser, J. Solid State Chem. **160**, 93 (2001)

[01G1] S. Geupel, G. Zahn, P. Paufler, G. Graw, Z. Kristallogr. **216**, 175 (2001)

[01G2] A. Gurevich, V. G. Kogan, Phys. Rev. Lett. **87**, 177009 (2001)

[01I1] S. Ishida, A. Matsushita, H. Takeya, M. Suzuki, Physica C **349**, 150 (2001)

[01I2] K. Izawa, A. Shibata, Y. Matsuda, Y. Kato, H. Takeya, K. Hirata, C. J. van der Beek, M. Konczykowski, Phys. Rev. Lett. **86**, 1327 (2001)

[01K1] A. Kreyssig, A. Schneidewind, M. Loewenhauft, C. Ritter, J. Fraidenberger, C. Fuchs, K. Muller, in *Rare Earth Transition Metal Borocarbides* (Kluwer, 2001), p. 181

[01K2] M. Kuznietz, J. Magn. Magn. Mater. **233**, 155 (2001)
[01L1] S.J. Lee, B.K. Cho, P.C. Canfield, D.W. Lynch, Phys. Rev. B **63**, 233103 (2001)
[01M1] C. Mazumdar, Z. Hu, H. von Lips, M.S. Golden, J. Fink, P.C. Canfield. G. Kaindl, Phys. Rev. B **64**, R020504 (2001)
[01M2] A.P. Menushenkov, P.V. Konarev, A.V. Tsvyashchenko, W. Meyer-Klaucke, P. Cortes, J. Synchrotron Rad. **8**, 910 (2001)
[01M3] K.H. Müller, V.N. Narozhnyi, Rep. Progr. Phys. **64**, 943 (2001)
[01N1] V.N. Narozhnyi, G. Fuchs, J. Freudenberger, K. Nenkov, K.H. Müller, Physica C **364–365**, 571 (2001)
[01N2] V.N. Narozhnyi, G. Fuchs, J. Freudenberger, K.A. Nenkov, D. Eckert, A. Teresiak, A. Czopnik, K.H. Müller, *Rare earth metal borocarbides* (Springer Science, 2001), p. 171
[01O1] G. Oomi, T. Kagayama, H. Mitamura, T. Goto, B.K. Cho, P.C. Canfield, Physica B **294–295**, 229 (2001)
[01R1] R.E. Rapp, F.A. Chaves, M. El Massalami, C.M. Chaves, J. Flores, H. Takeya, Physica C **354**, 178 (2001)
[01R2] M. Rotter, C. Sierks, M. Loewenhaupt, H. Schober, *Rare earth transition metal borocarbides*, ed. by K.H. Müller, V. Narozhnyi (Springer, 2001) p. 137
[01S1] G.M. Schmiedeshoff, J.A. Detwiler, W. Beyermann, A.H. Lacerda, P.C. Canfield, J.L. Smith, Phys. Rev. B **63**, 134519 (2001)
[01S2] C. Song, D. Wermeille, A.I. Goldman, P.C. Canfield, J.Y. Rhee, B.N. Harmon, Phys. Rev. B **63**, 104507 (2001)
[01S3] C. Song, J.C. Lang, C. Detlefs, A. Letoublon, W. Good, J. Kim, D. Wermeille, S.L. Budko, P.C. Canfield, A.I. Goldman, Phys. Rev. B **64**, 020403 (2001)
[01S4] G.R. Stewart, Rev. Mod. Phys. **73**, 797 (2001)
[01S5] H. Suderow, P. Martinez-Samper, S. Vieira, Phys. Rev. B **64**, R020503 (2001)
[01T1] J.R. Thompson, A.V. Silhanek, L. Civale, K.J. Song, C.V. Tomy, D.McK. Paul, Phys. Rev. B **64**, 024510 (2001)
[01V1] L.Y. Vinnikov, T.L. Barkov, P.C. Canfield, S.L. Budko, V.G. Kogan, Phys. Rev. B **64**, 024504 (2001)
[01V2] L.Y. Vinnikov, T.L. Barkov, P.C. Canfield, S.L. Budko, J.E. Ostenson, F.D. Laabs, V.G. Kogan, Phys. Rev. B **64**, R220508 (2001)
[01W1] K. Winzer, K. Krug, Z. Q. Peng, J. Magn. Magn. Mater. **226–230**, 321 (2001)
[02A1] M.A. Avila, S.L. Budko, P.C. Canfield, Phys. Rev. B **66**, 132504 (2002)
[02C1] S. X. Cao, K. Mori, X. L. Qin, K. Nishimura, J. C. Zhang, AIP Conf. Proc. **614**, 1067 (2002)
[02C2] J. H. Choi, H. Doh, E. M. Choi, S. I. Lee, M. Ohashi, N. Mori, Phys. Rev. B **65**, 024520 (2002)
[02D1] M. Doerr, M. Rotter, M. El Massalmi, S. Sinning, H. Takeya, J. Phys.: Condens. Matter, **14**, 5609 (2002)
[02D2] Z. Drzazga, A. Winiarska, D. Kaczorowski, J. Magn. Magn. Mater. **242–245**, 829 (2002)
[02D3] A. Duran, S. Bernes, R. Escudero, Phys. Rev. B **66**, 212510 (2002)
[02E1] M. El Massalmi, R.E. Rapp, E.F. Chagas, H. Takeya, J. Flores, C.M. Chaves, J. Phys. Soc. Jpn. **71**, 582 (2002)
[02F1] R. Falconi, A. Duran, R. Escudero, Phys. Rev. B **65**, 024505 (2002)
[02H1] B.D. Hennings, D.G. Naugle, P.C. Canfield, Phys. Rev. B **66**, 214512 (2002)
[02H2] A. Hillier, R. Smith, R. Cywinski, Appl. Phys. A **74**, S823 (2002)
[02H3] G. Hilscher, M. Michor, M. El-Hagary, M. Divis, Egypt. J. Sol. **23**, 153 (2002)
[02I1] K. Izawa, K. Kamata, Y. Nakajima, Y. Matsuda, T. Watanabe, M. Nohara, H. Takagi, P. Thalmeier, K. Maki, Phys. Rev. Lett. **89**, 137006 (2002)
[02J1] U. Jaenicke-Roessler, P. Paufler, G. Geupel, G. Behr, H. Bitterlich, J. Alloys Comp. **13**, 28 (2002)
[02J2] J. Jensen, Phys. Rev. B **65**, 140514 (2002)
[02K1] H. Kawano-Furukawa, H. Yoshizawa, H. Takeya, K. Kadowski, Phys. Rev. B **66**, 212503 (2002)

[02K2] H. Kawano-Furukawa, H. Takeshita, M. Ochiai, T. Nagata, Y. Oshizawa, N. Furukuwa, H. Takeya, K. Kadowaki, Phys. Rev. B **65**, 180508 (2002)

[02K3] C.A. Kim, B.K. Cho, Phys. Rev. B **66**, 214501 (2002)

[02L1] D. Lipp, M. Schneider, A. Glaudun, S. L. Drechsler, J. Freudenberger, G. Fuchs, K. Nenkov, K. H. Miller, T. Cichorek, P. Gegenwart, Europys. Lett. **58**, 435 (2002)

[02M1] K. Maki, P. Thalmeier, H. Won, Phys. Rev. B **65**, 140502 (2002)

[02M2] X.Y. Miao, S.I. Budko, P.C. Canfield, J. Alloys Comp. **338**, 13 (2002)

[02M3] K. Mori, S. Sakai, S.X. Cao, K. Nishimura, AIP Conf. Proc. **614**, 1074 (2002)

[02M4] K.H. Muller, G. Fuchs, S.E. Drechsler, V.N. Narozhnyi, *Handbook of magnetic materials* **14**, 199 (2002)

[02O1] G. Oomi, I. Minamitake, M. Ohashi, T. Eto, B.K. Cho, P.C. Canfield, High Pressure Res. **22**, 195 (2010)

[02T1] T. Tolinski, G. Chelkowska, A. Kowalczyk, Solid State Commun. **122**, 145 (2002)

[02T2] T. Tolinski, A. Kowalczyk, A. Szlaferek, M. Timko, J. Kovac, Solid State Commun. **122**, 363 (2002)

[02T3] T. Tolinski, A. Kowalczyk, A. Szlaferek, B. Andrzejewski, J. Kovac, M. Timko, J. Alloys Comp. **347**, 31 (2002)

[02T4] T. Tolinski, M. Pugaczowa-Michalska, G. Chelkowska, A. Szlaferek, A. Kowalczyk, Phys. Stat. Solidi (b) **231**, 446 (2002)

[02V1] L.M. Volkova, S.A. Polyshchuk, S.A. Magarill, F.E. Herbeck, J. Supercond. **15**, 663 (2002)

[02Y1] W. Yang-Bitterlich, H. Bitterlich, G. Zahn, U. Krasner, J. Alloys Comp. **347**, 131 (2002)

[02Z1] J.L. Zarestky, C. Stassis, A.I. Goldman, P.C. Canfield, G. Shirane, S.M. Shapiro, J. Phys. Chem. Solids, **63**, 811 (2002)

[03A1] E. Alleno, C. Godart, G. André, S.K. Dhar, S.M. Pattalwar, P. Bonville, R. Nagarajan, L.C. Gupta, Phys. Rev. B **68**, 214518 (2003)

[03B1] S. I. Beloborodko, Low Temp. Phys. **29**, 650 (2003)

[03B2] D. Bintley, P. J. Meeson, Physica C **388–389**, 181 (2003)

[03B3] A.T. Boothroyd, J.P. Barratt, P. Bonville, P.C. Canfield, A. Murani, A.R. Wildes, R.I. Bewley, Phys. Rev. B **67**, 104407 (2003)

[03B4] S.L. Budko, J.D. Strand, N.E. Anderson, R.A. Ribeiro, P.C. Canfield, Phys. Rev. B **68**, 104417 (2003)

[03C1] S. Cao, J. Zhang, X. Qin, K. Nishimura, K. Mori, Physica C **388–389**, 195 (2003)

[03C2] B.K. Cho, C.A. Kim, H.C. Ri, J. Appl. Phys. **93**, 8662 (2003)

[03D1] Z. Drzazga, G. Fuchs, A. Handstein, K. Nenkov, K.H. Müller, Physica C **383**, 421 (2003)

[03E1] M. El Massalami, H. Takeya, K. Hirata, M. Amara, R.M. Galera, D. Schmitt, Phys. Rev. B **67**, 144421 (2003)

[03E2] M. El Massalami, R. E. Rapp, F. A. B. Chaves, H. Takeya, C. M. Chaves, Phys. Rev. B **67**, 224407 (2003)

[03J1] Y. Jo, J.G. Park, C.A. Kim, B.K. Cho, H.C. Kim, H.C. Ri, H. Doh, S.I. Lee, J. Korean Phys. Soc. **43**, 263 (2003)

[03K1] H. J. Kim, J. H. Choi, H. Doh, E. M. Choi, S. I. Lee, M. Ohashi, N. Mori, Physica B **327**, 438 (2003)

[03K2] A. Kreyssig, O. Stockert, A. Dreyhaupt, E. Ressouche, B. Grenier, C. Ritter, H. Bitterlich, G. Behr, P.C. Canfield, M. Loewenhaupt, J. Low Temp. Phys. 131, 1129 (2003)

[03M1] P. Martínez-Samper, H. Suderow, S. Vieira, J. P. Brison, N. Luchier, P. Legay, and P. C. Canfield, Phys. Rev. B **67**, 014526 (2003)

[03M2] E. A. Masalitin, V. D. Fil, K. R. Zhekov, A. N. Zholobenko, T. V. Ignatova, Low Temp. Phys. **29**, 72 (2003)

[03M3] E.Matsuoka, X. Umeo, S.Tsuji, M.Sera, F.Iga, T.Takabatake, J. Phys. Soc. Jpn. **72**, 2344 (2003)

[03N1] S. Nakao, T. Hanaguri, K. Hashimoto, M. Nohara, H. Takagi, K. Kitazawa, Physica C **388–389**, 179 (2003)

[03N2] D.G. Naugle, B.I. Belevtsev, B.D. Hennings, Int. J. Modern Phys. B **17**, 3454 (2003)

[03O1] G.Oomi, N. Matsuda, T.Kagayama, C. K. Cho, P. C. Canfield, J. Modern Phys. B **17**, 3664 (2003)
[03P1] T. Park, M. B. Salamon, E. M. Cho, H. J. Kim, S. I. Lee, Phys. Rev. Lett. **90**, 177001 (2003)
[03P2] M. Pugaczowa-Michalska, G. Chelkowska, A. Kowalczyk, Acta Phys. Polonica, A **104**, 487 (2003)
[03P3] M. Pugaczowa-Michalska, G. Chelkowska, T. Tolinski, A. Kowalczyk, Molec. Phys. Repts. **38**, 107 (2003)
[03R1] K.D.D. Rathnayaka, D.G. Naugle, A.C. Dumar, M.P. Anatska, P.C. Canfield, Int. J. Mod. Phys. B **17**, 3493 (2003)
[03R2] P. Ravindran, K. Kjekshua, H. Fjellvag, P. Puschnig, C. Ambrosch-Draxl, L. Nordstrom, B. Johansson, Phys. Rev. B **67**, 104507 (2003)
[03R3] R.A. Ribeiro, S.L. Budko, P.C. Canfield, J. Magn. Magn. Mater. **267**, 216 (2003)
[03R4] P.M.C. Rourke, J. Paglione, F. Ronning, L. Taillefer, K. Kadowski, Physica C **397**, 1 (2003)
[03S1] P. Salamakha, O. Sologub, Ch. Mazumdar, E. Alleno, H. Noel, M. Potel, C. Godart, J. Alloys. Comp. **351**, 190 (2003)
[03S2] K.J. Song, C. Park, S.S. Oh, Y.K. Kwon, J. R. Thompson, D.G. Mandrus, D. McK. Paul, C. V. Tomy, Physica C **398**, 107 (2003)
[03T1] H. Takeshita, M. Ochiai, E. Habuta, T. Nagata, H. Kawano-Furukawa, H. Yoshizawa, N. Furukawa, H. Takeya, K. Kadowaki, Physica C **388–389**, 193 (2003)
[03T2] M. Timko, A. Kowalczyk, A. Szlaferek, J. Kovac, A. Zentko, T. Tolinski, J. Briancin, Phys. Stat. Solidi (a) **196**, 201 (2003)
[03T3] T. Tolinski, A. Kowalczyk, G. Chelkowska, Phys. Lett. A **308**, 75 (2003)
[03T4] T. Tolinski, A. Kowalczyk, V. Ivanov, Phys. Stat. Solidi (b) **240**, 153 (2003)
[03T5] T. Tolinski, B. Andrejewski, A. Kowalczyk, Z. Trybula, J. Baszynski, J. Magn. Magn. Mater. **267**, 402 (2003)
[03T6] T. Tolinski, A. Kowalczyk, M. Pugaczowa-Michalska, G. Chelkowska, J. Phys.: Condens. Matter, **15**, 1397 (2003)
[03T7] T. Tolinski, G. Chelkowska, B. Andrzejewski, A. Kowalczyk, M. Timko, J. Kovac, Phys. Stat. Solidi (a) **196**, 294 (2003)
[03V1] L.Y. Vinnikov, J. Anderegg, S.L. Budko, P.C. Canfield, V.G. Kogan, JETP Lett. **77**, 502 (2003)
[03W1] M.B. Walker, C. Detlefs, Phys. Rev. B **67**, 132407 (2003)
[03Y1] Q. Yuan, P. Thalmeier, Phys. Rev. B **68**, 174501 (2003)
[04A1] E. Alleno, S. Vivet, C. Godart, E. Leroy, Physica C **411**, 59 (2004)
[04A2] B. Andrzejewski, A. Kowalczyk, J. Stankowski, A. Szlaferek, J. Phys. Chem. Solids, **65**, 623 (2004)
[04A3] M.A. Avila, Y.Q. Wu, C.L. Condron, S.L. Budko, M. Kramer, G.J. Miller, P.C. Canfield, Phys. Rev. B **69**, 205107 (2004)
[04B1] E. Burzo, L. Chioncel, J. Opt. Adv. Mater. **6**, 917 (2004)
[04B2] E. Burzo, R. Tetean, L. Chioncel, M. Văleanu, Phys. Stat. Solidi (c) **1**, 3357 (2004)
[04C1] N.F. Chaban, I.V. Veremchuk, Yu.B. Kuzma, J. Alloys Comp. **370**, 129 (2004)
[04C2] C. Chacon, O. Isnard, J. F. Berar, J. Physique IV **118**, 419 (2004)
[04D1] S.L. Drechsler, H. Rosner, I. Opahle, S.V. Shulga, H. Eschrig, Physica C **408–410**, 104 (2004)
[04E1] M. El Massalami, H. Takeya, C.M. Chaves, Phys. Rev. B **70**, 014429 (2004)
[04E2] M. El Massalami, H.A. Borges, H. Takeya, R.E. Rapp, F.A.B. Chaves, J. Magn. Magn. Mater. **279**, 5 (2004)
[04F1] Y. Filinchuk, K. Yvon, Acta Cryst. A **60**, s219 (2004)
[04J1] A. Jensen, K. Norgaard Toft, A.B. Abrahamsen, D.F. McMorrow, M.R. Eskildsen, N.H. Andersen, J. Jensen, S. Danilkin, K. Prokes, V. Sikolenko, P. Smeibidl, S.I. Budko, P.C. Canfield, Phys. Rev. B **69**, 104527 (2004)
[04K1] H. Kontani, Phys. Rev B **70**, 054507 (2004)

[04K2] A. Kreyssig, O. Stockert, D. Reznik, F.M. Woodward, J.W. Lynn, W. Reichardt, D. Souptel, G. Behr, M. Loewenhaupt, Physica B **350**, 69 (2004)

[04K3] A. Kowalczyk, A. Szayek, G. Chelkowska, T. Tolinski, J. Magn. Magn. Mater. **272–276**, e477 (2004)

[04K4] Y.B. Kuzma, V. Babizhetskyj, I. Veremchuk, N. F. Chaban, J. Solid State Chem. **177**, 425 (2004)

[04N1] H. Nishimori, K. Uchiyama, S. I. Kaneko, A. Tokura, H. Takeya, K. Hirata, N. Nishida, J. Phys. Soc. Jpn. 73, 3245 (2004)

[04N2] K. Norgaard Toft, A.B. Abrahamsen, M.R. Eskilsen, K. Lefmann, N.H. Andersen, P. Vorderwisch, P. Smeibidl, M. Meissner, P.C. Canfield, Phys. Rev. B **69**, 214507 (2004)

[04P1] T. Park, E.E.M. Chia, M.B. Salamon, E.D. Bauer, I. Vekhter, J.D. Thompson, E.M. Choi, H.J. Kim, S.I. Lee, P.C. Canfield, Phys. Rev. Lett. **92**, 237002 (2004)

[04P2] T. Park, M.B. Salamon, E. M. Choi, H. J. Kim, S. I. Lee, Phys. Rev. B **69**, 054505 (2004)

[04P3] M. Pugaczowa-Michalska, A. Kowalczyk, G. Chelkowska, T. Tolinski, J. Alloys Comp. **385**, 44 (2004)

[04P4] M. Pugaczowa-Michalska, T. Tolinski, A. Kowalczyk, G. Chelkowska, Czech. J. Phys. **54**, D347 (2004)

[04R1] P. Raychaudhuri, D. Jaiswal-Nagar, Goutam Sheet, S. Ramakrishnan, H. Takeya, Phys. Rev. Lett. **93**, 156802 (2004)

[04T1] H. Takeya, M. El Massalami, Phys. Rev. B **69**, 024509 (2004)

[04T2] P. Thalmeier, G. Zwicknagi, Handbook on the Physics and Chemistry of Rare Earths, **34**, 135 (2004)

[04T3] T. Tolinski, A. Kowalczyk, V. Ivanov, G. Chelkowska, M. Timko, Czech. J. Phys. **54**, 287 (2004)

[04V1] I. V. Veremchuk, I. V. Babizhetskyy, N. J. Chaban, Yu. B. Kuzma, Inorg. Mater. **40**, 604 (2004)

[04V2] P. Villars et al., *Landold Börnstein handbook*, vol. 43A1 (Springer, 2004)

[04W1] T. Watanabe, M. Nohara, T. Hanaguri, H. Takagi, Phys. Rev. Lett. **92**, 147002 (2004)

[04Y1] K. Yamauchi, H. Katayama-Yoshida, A. Yanase, H. Harima, Physica C **412–414**, 225 (2004)

[05B1] V. Babizhetskyy, I. Veremchuk, N. Chaban, Yu. B. Kuzma, Z. Kristallogr. **220**, 1 (2005)

[05B2] D.L. Bashlakov, Yu. G. Naidyuk, I. K. Yanson, S. C. Wimbush, B. Holzapfel, G. Fuchs, S.L. Drechsler, Supercond. Sci. Technol. **18**, 1094 (2005)

[05B3] N.L. Bobrov, S.I. Beloborodko, J.V. Tyutrina, I.K. Yanson, D.G. Naugle, K.D.D. Rathnayaka, Phys. Rev. B **71**, 014512 (2005)

[05B4] S.L. Budko, P.C. Canfield, Phys. Rev. B **71**, 024409 (2005)

[05C1] G.H. Cao, P. Simon, W. Skrotzki, S.C. Wimbush, B. Holzapfel, Appl. Phys. A **81**, 583 (2005)

[05C2] J.H. Choi, H.J. Kim, H.B. Kim, H.J. Doh, S.I. Lee, B.K. Cho, Phys. Rev. B **72**, 054516 (2005)

[05D1] I.G. De Oliveira, P.J. von Ranke, M. El Massalami, C.M. Chaves, Phys. Rev. B **72**, 174420 (2005)

[05D2] C.D. Dewhurst, S. J. Levett, D. McK Paul, Phys. Rev. B **72**, 014542 (2005)

[05D3] M. Divis, J. Rusz, H. Michor, G. Hilscher, P. Blaha, K. Schwartz, J. Alloys Comp. **403**, 29 (2005)

[05E1] M. El Massalami, R.M. Galera, D. Schmitt, B. Ouladdiaf, H. Takeya, Phys. Rev. B **72**, 144521 (2005)

[05F1] Y.E. Filinchuk, K. Yvon, Inorg. Chem. **44**, 4398 (2005)

[05I1] O. Ignatchik, T. Coffey, J. Hagel, M. Jäckel, E. Jobiliong, D. Souptel, G. Behr, J. Wosnitza, J. Magn. Magn. Mater. **290–291**, 426 (2005)

[05K1] C. Krutzler, R. Fuger, M. Eisterer, G. Fuchs, G. Behr, H.W. Weber, Phys. Rev. B **72**, 144508 (2005)

[05M1] C. Mazumdar, R. Nagarajan, Curr. Sci. **88**, 83 (2005)

[05M2] S. Mukhopadhyay, G. Sheet, P. Raychaudhuri, H. Takeya, Phys. Rev. B **72**, 014545 (2005)

[05N1] R. Nagarajan, C. Muzamdar, Z. Hossain, L. Gupta, in Superconducting Materials, Spinger Berlin, 2005, p. 393

[05N2] Y.G. Naidyuk, O.E. Kvitnitskaya, I.K. Yanson, G. Fuchs, K. Nenkov, S.L. Drechsler, G. Behr, D. Souptel, K. Gloos, Physica B **359–361** (2005)

[05P1] T. Park, A. Malinowski, M. F. Hundley, J. D. Thompson, Y. Sun, M.B. Salamon, E. M. Choi, H.J. Kim, S.I. Lee, P.C. Canfield, V. G. Kogan, Phys. Rev. B **71**, 054511 (2005)

[05P2] N. Plugaru, J. Rubin, J. Bartolome, C. Piquer, J. Magn. Magn. Mater. **290–292**, 1963 (2005)

[05P3] N. Plugaru, J. Rubin, J. Bartolome. V. Pop, Phys. Rev. B **71**, 024433 (2005)

[05Q1] X. Qin, S. Cao, L. Yu, J. Zhang, K. Nishimura, K. Mori, Internat. J. Modern Phys. B **19**, 3945 (2005)

[05S1] Y. Saito, H. Ido, H. Yoshida, T. Sakon, M. Motokawa, J. Appl. Phys. **97**, 10M520 (2005)

[05S2] D.R. Sanchez, E.M. Baggio-Saitovich, H. Micklitz, S. I. Lee, Phys. Rev. B **71**, 024509 (2005)

[05S3] K. Subba Rao, R. Tamm, S.C. Wimbush, G.H. Cao, C.G. Oertel, W. Skrotzki, B. Holzapfel, Mat. Sci. Forum **495–497**, 1425 (2005)

[05T1] H. Takeya, M. Kuznietz, K. Hirata, Physica B **359–361**, 473 (2005)

[05T2] T. Tolinski, M. Pugaczowa-Michalska, G. Chelkowska, Phys. Stat. Solidi (b) **242**, 474 (2005)

[05V1] I.V. Veremchuk, Yu. Prots, A.Leithe-Jasper, Yu.B. Kuzma, Z. Kristallogr. **220**, 135 (2005)

[05V2] I.S. Veschunov, L.Y. Vinnikov, S.L. Budko, P.C. Canfield, *Proc. Third Int. Symp. Magnetism* (2005), p. 105

[05V3] L.Y. Vinnikov, J. Anderegg, S.L. Budko, P.C. Canfield, V.G. Kogan, Phys. Rev. B **71**, 224513 (2005)

[06A1] M. Akhavan, Phys. Stat. Solidi (a) **203**, 2956 (2006)

[06A2] N.H. Andersen, J. Jensen, J.B.S. Jensen, M.Y. Zimmermann, R. Pinholt, A.B. Abrahamsen, K. Norgaard Toft, P. Hedegard, P.C. Canfield, Phys. Rev. B **73**, R020504 (2006)

[06B1] T. Baba, T. Yokoya, S. Tsuda, T. Kiss, T. Shimojima, S. Shin, T. Togashi, C.T. Chen, C.Q. Zhang, S. Watanabe, T. Watanabe, M. Nohara, H. Takagi, Physica C **445–448**, 46 (2006)

[06B2] A.D. Bianchi, B. Bergk, O. Ignatchik, J. Wonitza, J. Perenboom, P.C. Canfield, J. Phys.: Conf. Series, **51**, 263 (2006)

[06B3]. H. Bluhm, S. Sebastian, J. Guikema, I. Fisher, A. Kathryn, K. Moler, Phys. Rev. B73, 014514 (2006)

[06B4] N.L. Bobrov, S.I. Beloborodko, L.V. Tyutrina, V.N. Chernobay, I.K. Yanson, D.G. Naugle, K.D.D. Rathnayaka, Low. Temp. Phys. **32**, 489 (2006)

[06B5] S. Budko, P. Canfield, C. R. Physique, **7**, 56 (2006)

[06B6] S. Budko, G. Schmiedeshoff, P.C. Canfield, Solid State Commun. **140**, 281 (2006)

[06B7] S.L. Budko, G. M. Schmiedeshoff, G. Lapertot, P. C. Canfield, J. Phys.: Cond. Matter, **18**, 8353 (2006)

[06C1] M. Crespo, H. Suderow, S. Vieira, S. Budko, P.C. Canfield, Phys. Rev. Lett. **96**, 027003 (2006)

[06D1] A. Duran, S. Bernes, R. Falconi, R. Escudero, O. Laborde, M. Guillot, Phys. Rev. B **73**, 134513 (2006)

[06G1] L.C. Gupta, Adv. Phys. **55**, 691 (2006)

[06I1] E. Ikeda, K. Mori, K. Nishimura, AIP Conf. Proc. **850**, 689 (2006)

[06I2] T. Ito, H. Ido, J. Appl. Phys. **99**, 08J304 (2006)

[06M1] Y. Matsuda, K. Izawa, I. Vekhter, J. Phys.: Condens. Matter, **18**, R 705 (2006)

[06O1] G. Oomi, H. Akiyama, Y. Sakigawa, S. Kaji, Y. Uwatoko, B.K. Cho, Physica B **378–380**, 473 (2006)

[06O2] E. Orgaz, A. Aburto, J. Chem. Phys. **125**, 144708 (2006)

[06R1] M. Rotter, A. Lindbaum, A. Barcza, M. El Massalami, M. Doerr, M. Loewenhaupt, H. Michor, B. Beuneu, Europhys. Lett. **75**, 160 (2006)

[06S1] M. Schneider, O. Zaharko, U. Gasser, A. Kreyssig, P.J. Brown, P.C. Canfield, Phys. Rev. B **74**, 104426 (2006)

[06S2] A.O. Shorikov, I.A. Nekrasov, V.I. Anisimov, M. Sigrist, Fiz. Metal. Metall. **101**, S21 (2006)

[06T1] T. Tolinski, G. Chelkowska, A. Kowalczyk, Physica B **378–380**, 1114 (2006)

[06V1] I. Veremchuk, N. Chaban, Y.B. Kuzma, J. Alloys Comp. **413**, 127 (2006)

[06V2] P. Villars, *Landolt Börnstein handbook*, vol. 43A3 (Springer, 2006)

[06V3] P. Vlaic, N. Bucur, C. Lazăr, E. Burzo, J. Opt. Adv. Mater. **8**, 490 (2006)

[06Y1] T. Yokoya, T. Baba, S. Tsuda, T. Kiss, A. Chainani, S. Shin, T. Watanabe, M. Nohara, T. Hanaguri, H. Takagi, Y.Takano, H. Kito, J. Itoh, H. Harima, T. Oguchi, J. Phys. Chem. Solids, **67**, 277 (2006)

[07B1] J. Bartolome, N. Plugaru, J. Campo, J. Rubin, E.K. Hlil, C. Rillo, A. Arauzo, J. Alloys Comp. **442**, 11 (2007)

[07B2] N. Bucur, P. Vlaic, E. Burzo, AIP Conf. Proc. **899**, 766 (2007)

[07B3] E. Burzo, Rom. Rep. Phys. **59**, 337 (2007)

[07B4] E. Burzo, R. Tetean, L. Chioncel, V. Pop, J. Magn. Magn. Mater. **316**, e379 (2007)

[07D1] L. DeBeer-Schmitt, M.R. Eskildsen, M. Ichioka, K. Machida, N. Jenkins, C.D. Dewhurst, A.B. Abrahamsen, S.L. Budko, P.C. Canfield, Phys. Rev. Lett. **99**, 167001 (2007)

[07E1] M. El Massalami. M. Amara, R.M. Galera, D. Schmitt, H. Takeya, Phys. Rev. B **76**, 104410 (2007)

[07J1] J. Jensen, P. Hedegaard, Phys. Rev. B **76**, 094504 (2007)

[07K1] J.D. Kim, H.J. Kim, H. Doh, S.I. Lee, J. Phys. Soc. Jpn. **76**, 144 (2007)

[07L1] L. Li, K. Nishimura, E. Ikeda, K. Mori, Jpn J. Appl. Phys. **46**, 6604 (2007)

[07L2] Z. Liu, M. Divis, V. Sechovsky, Phys. Rev. Lett. A **371**, 344 (2007)

[07N1] T. Nagai, Y Kato, N. Hayashi, K. Yamauchi, H. Harima, Phys.Rev. B **76**, 214514 (2007)

[07N2] Y.G. Naidyuk, O.E. Kvitnitskaya, I.K. Yanson, G. Fuchs, K. Nenkov, A. Wälte, G. Behr, D. Souptel, S.L. Drechsler, Phys. Rev. B **76**, 014520 (2007)

[07N3] Y.G. Naidyuk, O.E. Kvitnitskaya, I.K. Yanson, G. Fuchs, K. Nenkov, A. Wälte, G. Behr, D. Souptel, S.L. Drechsler, Physica C **460–462**, 105 (2007)

[07N4] Y.G. Naidyuk, D.L. Bashlakov, N.L. Bobrov, V.N. Chernobay, O.E. Kvitnitskaya, I.K. Yanson, G. Behr, S.L. Drechsler, G. Fuchs, D. Souptel, D.G. Naugle, K.D.D. Rathnayaka, J.H. Ross, Physica C **460–462**, 107 (2007)

[07P1] N. Plugaru. J. Rubin, J. Bartolomé, J. Campo, G. J. Cuello, M. Tovar, O. Prokhnenko, J. Magn. Magn. Mater. **316**, e438 (2007)

[07R1] C. Rillo, A. Arauzo, N. Plugaru, E.K. Hlil, J. Rubin, J. Bartolomé, J. Magn. Magn. Mater. **316**, 166 (2007)

[07S1] P.S. Salamakha, O.L. Sologub, J.R. Hester, C. Rizzoli, A.P. Gonçalves, M. Almeida, J. Alloys Comp. **439**, 162 (2007)

[07S2] M. Schreider, A. Gladun, A. Kreyssig, J. Wosnitza, D. Souptel, G. Behr, J. Magn. Magn. Mater. **311**, 489 (2007)

[07T1] T. Tolinski, Modern Phys. Lett. B **21**, 431 (2007)

[07V1] I.S. Veschunov, L. Ya Vinnikov, S.L. Budko, P.C. Canfield, Phys. Rev. B **76**, 174506 (2007)

[08B1] T. Baba, T. Yokoya, S. Tsuda, T. Kiss, T. Shimojima, K. Ishizaka, H. Takeya, K. Hirata, T. Watanabe, M. Nohara, H. Takagi, N. Nakai, K. Machida, T. Togashi, S. Watanabe, X.-Y. Wang, C. T. Chen, S. Shin, Phys. Rev. Lett. **100**, 017003 (2008)

[08B2] B. Bergk, V. Petzold, H. Rosner, S. L. Drechsler, M. Bartkowiak, O. Ignatchik, A. D. Bianchi, I. Sheikin, P. C. Canfield, J. Wosnitza, Phys. Rev. Lett **100**, 257004 (2008)

[08B3] N.L. Bobrov, V.N. Chernobay, Yu.G. Naidyuk, L.V. Tyutrina, D.G. Naugle, K.D.D. Rathnayaka, S.L. Budko, P.C. Canfield, I.K. Yanson, Eur. Phys. Lett. **83**, 37003 (2008)

[08B4] E. Burzo, N. Bucur, L. Chioncel, V. Rednic, J. Phys.: Condens. Matter, **20**, 275201 (2008)

[08I1] T. Isshiki, N. Kimura, H. Aoki, T. Tereshina, S. Uji, K. Yamauchi, H. Harima, D. Jaswal-Nagur, S. Ramakrishnan, A. K. Grover, Phys. Rev. B **78**, 134528 (2008)

[08J1] J. Jensen, M. Rotter, Phys. Rev. B **77**, 134408 (2008)

[08L1] C. Lazăr, E. Burzo, M. Neumann, J. Opt. Adv. Mater. **10**, 780 (2008)

[08L2] W.C. Lee, Physica B **403**, 2842 (2008)

[08M1] C. Mazumdar, M. Rotter, M. Frontzek, H. Michor, M. Doerr, A. Kreyssig, M. Koza, A. Hiess, J. Voigt, G. Behr, L.C. Gupta, M. Prager, M. Loewenhaupt, Phys. Rev. B **78**, 144422 (2008)

[08M2] K.H. Muller, M. Schneider, G. Fuchs, Handbook Phys. Chem. Rare-Earths **38**, 175 (2008)

[08R1] K.D.D. Rathnayaka, B.I. Belevtsev, D.G. Naugle, P.C. Canfield, S.I. Lee, J.D. Kim, Physica B **403**, 1050 (2008)

[08S1] M. Schneider, A. Gladun, A. Kreyssig, J. Wosnitza, V. Petzold, H. Rosner, G. Behr, D. Souptel, K. H. Müller, S. L. Drechsler, G. Fuchs, F. Phys.: Condens Matter, **20**, 175221 (2008)

[08S2] D. Souptel, G. Behr, W. Löser, A. Teresiak, J. Alloys Comp. **455**, 60 (2008)

[08S3] P. Starowicz, C. Liu, R. Khasanov, T. Kondo, G. Samolyuk, D. Gardenghi, Y. Lee, T. Ohta, B. Harmon, P. Canfield, S. Budko, E. Rotenberg, A. Kaminski, Phys. Rev. B **77**, 134520 (2008)

[08W1] F. Weber, A. Kreyssig, L. Pintschovius, R. Heid, W. Reichardt, D. Reznik, O. Stockert, K. Hradil, Phys. Rev. Lett. **101**, 1287002 (2008)

[08W2] A. Winiarski, A. Winiarska, Z. Drzazga, Acta. Phys. Polonica **119**, 243 (2008)

[09A1] T.R. Abu Alrub, S.H. Curnoe, J. Phys.: Condens Matter, **21**, 415704 (2009)

[09A2] H. Agil, N. Kervan, S. Kervan, H. Sözeri, A. Gencer, J. Alloys Comp. **478**, 437 (2009)

[09A3] A. Arauzo, J. Bartolome, C. Rillo, J. Rubin, Solid State Phenomena, **152–153**, 466 (2009)

[09B1] T. Baba, T. Yokoya, S. Tsuda, T. Kiss, T. Shimojima, K. Ishizaka, H. Takeya, K. Hirata, T. Watanabe, M. Nohara, H. Takagi, N. Nakai, K. Machida, T. Togashi, S. Watanabe, X.-Y. Wang, C. T. Chen, and S. Shin, Phys. Rev. Lett. **100**, 017003 (2008)

[09B2] B. Bergk, J. Wosnitza, Low Temp. Phys. **35**, 687 (2009)

[09B3] B. Bergk, V. Petzold, H. Rosner, S.L. Drechsler, M. Bartkowiak, O. Ignatchik, I. Sheikin, P.C. Canfield, J. Wosnitza, J. Phys.: Conf. Series, **150**, 052021 (2009)

[09D1] P. Das, C. V. Tomy, H. Takeya, S. Ramakrishnan, A. K. Grover, Physica C **469**, 151 (2009)

[09D2] J.M. Densmore, P. Das, K. Rovira, T.D. Blasins, L. De Beer-Schmitt, N. Jenkins, D. Mck. Paul, C.D. Dewhurst, S.L. Budko, P.C. Canfield, M.R. Eskildsen, Phys. Rev. B **79**, 174522 (2009)

[09D3] S.B. Dugdale, C. Utfeld, I. Wilkinson, J. Laverock, Z. Major, M.A. Alam, P.C. Canfield, Supercond. Sci. Technol. **22**, 014002 (2009)

[09G1] A.A. Gapud, S. Moraes, R.P. Khadka, P. Favreau, C. Henderson, P.C. Canfield, V.G. Kogan, A.P. Reyes, L.L. Lumata, D.K. Christen, J.R. Thompson, Phys. Rev. B **80**, 134524 (2009)

[09G2] I. Guillamon, H. Suderow, A. Fernandez-Pacheco, J. Sense, R. Cordoba, J. M. Teresa, M. Ibarra, S. Vieira, Nature Phys. **5**, 651 (2009)

[09K1] V. Kataev, U. Schaufuss, F. Muranyi, A. Alfonsov, M. Doerr, M. Rotter, B. Büchrer, J. Phys.: Conf. Series, **150**, 042086 (2009)

[09L1] L. Li, K. Nishimura, Appl. Phys. Lett. **95**, 132505 (2009)

[09L2] Z.S. Liu, M. Divis, V. Sechovsky, Chin. Phys. Lett. **26**, 067501 (2009)

[09L3] Z.S. Liu, M. Divis, V. Sechovsky, Chin. Phys. Lett. **26**, 107504 (2009)

[09L4] X. Lu, W. K. Park, K. Y. Choi, B. I. Lee, S. Yeo, S. L. Budko, P. C. Canfield, L. H. Greene, J. Phys.: Conf. Series, **150**, 052143 (2009)

[09M1] C. Mazumdar, L. C. Gupta, K. Nenkov, G. Behr, G. Fuchs, J. Alloys. Comp. **480**, 190 (2009)

[09M2] S Mukhopadhyay, G. Sheet, A.K. Nigam, P. Raychaudhuri, H. Takeya, Phys. Rev. B **79**, 132505 (2009)

[09N1] Y. Nagai, N. Hayashi, Y. Kato, K. Yamauchi, H. Harima, J. Phys.: Conf. Series **150**, 052177 (2009)

[09N2] Y.G. Naidyuk, G. Behr, N.L. Bobrov, V.N. Chernobay, S.L. Drechsler, G. Fuchs, O.E. Kvitnitskaya, D.G. Naugle, K.D.D. Rathnayaka, I.K. Yanson, J. Phys.: Conf. Series, **150**, 052178 (2009)

[09O1] S. Okuma, T. Ichimura, H. Takeya, K. Hirata, Physica C **469**, 1093 (2009)

[09P1] A. Prasad, Z. Hossain, H.S.Jeevan, C. Geibel, J. Phys.: Condens. Matter, **21**, 206003 (2009)

[09R1] K.D.D. Rathnayaka, D.G. Naugle, B.I. Belevtsev, P.C. Canfield, S.L. Budko, J. Appl. Phys. **105**, 07E111 (2009)

[09S1] R. Sarkar, M. Baenitz, F. Steglich, C. Geibel, A. Prasad, Z. Hossain, Verh. Deutsch. Phys. Gesell. **44**, 1 (2009)

[09S2] M. Schneider, G. Fuchs, K.H. Müller, K. Nenkov, G. Behr, D.Souptel, S. Drechsler, Phys. Rev. B **80**, 224522 (2009)

[09S3] M. Shishikura, T. Masubuchi, K. Takase, Y. Takano, T. Watanabe, J. Phys.: Conf. Series, **150**, 052239 (2009)

[09S4] G.M. Schmiedeshoff, L.W. Lounsbury, D.J. Luna, E.W. Okraku, S.J. Tracy, S.L. Bud'ko, P.C. Canfield, J. Phys.: Conf. Series, **150**, 042177 (2009)

[09T1] T. Tolinski, J.C. Cezar, H. Wende, A. Kowalczyk, K. Baberschke, Acta Phys. Polonica, A **115**,129 (2009)

[09V1] I. Veremchuk, R., Gumeniuk, Yu. Prots, W. Schnelle, U. Burkhardt, Solid State Sci. **11**, 507 (2009)

[09V2]. L.Y. Vinnikov, I.S. Yeshchunov, S.L. Budko, P.C. Canfield, V.G. Kogan, J. Phys.: Conf. Series, **150**, 052279 (2009)

[10A1] E. Alleno, S. Singh, S. D. Dhar, G. Andre, New J. Phys. **12**, 043018 (2010)

[10B1] T. Baba, T. Yokoya, S. Tsuda, T. Watanabe, M. Nohara, H. Takagi, T. Oguchi, S. Shin, Phys. Rev. B **81**, 180509 (2010)

[10B2] N.L. Bobrov, V.N. Chernobay, Yu.G. Naidyuk, L.V. Tyutrina, I.K. Yanson, J. Low Temp. Phys. **36**, 990 (2010)

[10B3] S.L. Budko, V.G. Kogan, H. Hodovanets, S. Ran, S.A. Moser, M.J. Lampe, P. C. Canfield, Phys. Rev. B **82**, 174513 (2010)

[10B4] E. Burzo, L. Chioncel, R. Tetean, O. Isnard, J. Phys.: Condens. Matter, **23**, 026001 (2010)

[10E1] M. El Massalami, R. M. Saeed, C. M. Chaves, H. Takeya, M. Doerr, H. Michor, M. Rotter, J. Magn. Magn. Mater. **322**, 523 (2010)

[10G1] I. Guillamon, M. Crespo, H. Suderow, S. Vieira, J.P. Brison, S.L. Budko, P.C. Canfield, Physica C **470**, 771 (2010)

[10G2] L.C. Gupta, Adv. Phys. **55**, 691 (2010)

[10K1] H. Kawano-Furukawa, Y. Ishida, F. Yano, R. Nagatomo, A. Noda, T. Nagata, S. Ohira-Kawamura, C. Kobayashi, H. Yoshizawa, K. Littrell, B, L.Winn, N. Furukawa, H. Takeya, Physica C **470**, S716 (2010)

[10K2] O.E. Kvitnitskaya, Y.G. Naidyuk, I. K. Yanson, T. Niemeier, G. Fuchs, B. Holzapfel L. Schultz, Super. Sci. Techn. **23**, 115001 (2010)

[10L1] W.C. Lee, J. Appl. Phys. **107**, 09E132 (2010)

[10L2] W.C. Lee, Int. J. Mod. Phys. B **24**, 5861 (2010)

[10L3] L. Li, K. Nishimura, M. Fujii, K. Matsuda, D. Huo, Phys. Rev. B **81**, 214517 (2010)

[10L4] C.Z. Liao, C. Dong, L.M. Zeng, B. He, W.H. Chao, L.H. Yang, J. Alloys Comp. **493**, 31 (2010)

[10N1] N. Nobori, T. Nakano, G. Oomi, P. C. Canfield, B. K. Cho, J. Phys. : Conf. Series, **200**, 012146 (2010)

[10O1] M. Ohashi, G. Oomi, B. K. Cho, P. C. Canfield, Phys. Stat. Solidi (b) **247**, 599 (2010)

[10O2] S. Okuma, T. Ichimura, H. Takeya, K. Hirata, Physica C **470**, 844 (2010)

[10P1] Q. Ping, J.L. Liu, J. Shen, L.J. Bai, Q. Ran, Y.L. Wang, Chin. Phys. B **19**, 126001 (2010)

[10R1] H. Rho, M.V. Klein, P.C. Canfield, Phys. Rev. B **82**, 064423 (2010)

[10S1] B.K. Sahoo, B.N. Panda, Modern. Phys. Lett. B **29**, 2879 (2010)

[10U1] K. Uchiyama, S. Suzuki, A. Kuwahara, K. Yamasaki, S. Kaneko, H. Takeya, K. Hirata, N. Nishida, Physica C **470**, S796 (2010)

[10V1] P. Villars et al., *Landolt Börnstein handbook* (Springer, Vol. 43 A9, 2010)

[11A1] S. Anissimova, A. Kreyssig, O. Stockert, M. Loewenhaupt, D. Reznik, Phys. Rev. B **84**, 104509, (2011)

[11C1] J.J. Cheng, M. D. Lan, J. Phys. Chem. Solids, **72**, 585 (2011)

[11F1] R. Falconi, A. Duran, M. Nunez-Regueiro, R. Escudero, Phys. Stat. Sol. (a) **208**, 2159 (2011)

[11J1] M. W. Jarosik, R. Szczesniak, Acta Phys. Polonica, **119**, 863 (2011)
[11K1] N. Kantarci, O. Nane, B. Ozcelik, S. Kervan, H. Agil, J. Supercond. Nov. Magn. **24**, 793 (2011)
[11L1] L. Li, K. Nishimura, M. Kadonaga, Z. Qian, D. Huo, J. Appl. Phys. **110**, 043912 (2011)
[11L2] L. Li, K. Nishimura, D. Huo, M. Kadonaga, T. Namiki, Z. Qian, Appl. Physics Express, **4**, 093101 (2011)
[11L3] Z. S. Liu, M. Divis, V. Sechovsky, J. Phys. Chem. Solids, **72**, 983 (2011)
[11L4] X. Lu, W.K. Park, S. Yeo, K.H. Oh, S.I. Lee, S.L. Budko, P.C. Canfield, L.H. Greene, Phys. Rev. B **83**, 104519 (2011)
[11N1] Y.G. Naidyuk, O.E. Kvitnitskaya, L.V. Tiutrina, I.K. Yanson, G. Behr, G. Fuchs, S.L. Drechsler, K. Nenkov. L. Schultz, Phys. Rev. B **84**, 094516 (2011)
[11O1] G. Oomi, Y. Sakikawa, M. Ohmura, M Ohashi, P. C. Canfield, B. K. Cho, J. Phys.: Conf. Series, **273**, 012082 (2011)
[11O2] B. Ozcelik, O. Nane, S. Kervan, J. Supercond. Nov. Magn. **24**, 763 (2011)
[11V1] P. Villars, *Landolt Bornstein Handbook* Vol III/43A10 (2011)
[12B1] B. Bergk, S. L. Drechsler, P. C. Canfield, J. Wosnitza, Eur. Phys. J. B **85**, 57 (2012)
[12D1] P. Das, J.M. Densmore, C. Rastovski, K.J. Schlesinger, M. Laver, C.D. Dewhurst, K. Littrell, S.L. Bud'ko, P.C. Canfield, M.R. Eskildsen, Phys. Rev. B **86**, 144501 (2012)
[12E1] M. El Massalami, H. Takeya, B. Ouladdiaf, R. Maia Filho, A. M. Gomes, T. Paiva, R. R. dos Santos, Phys. Rev. B **85**, 174412 (2012)
[12G1] J.A. Galvis, M. Crespo, I. Guillamon, H. Suderow, S. Vieira, M. Garcia-Hernandez, S. Budko, P.C. Canfield, Solid State Commun. **152**, 1076 (2012)
[12G2] R. Gumeniuk, W. Schnelle, U. Burkhardt, H. Rosner, A. Leithe-Jasper, Intermetallics, **20**, 104 (2012)
[12I1] M.I. Ishikawa, T. Watanabe, J. Phys.: Conf. Series, **391**, 012023 (2012)
[12L1] L. Li, H. Fan, S. Matsui, D. Huo, K. Nishimura, J. Alloys Comp. **529**, 25 (2012)
[12L2] L. Li, K. Nishimura, D. Huo, S. Matsui, Z. Qian, T. Namiki, Phys. Procedia, **27**, 44 (2012)
[12L3] Z.S. Liu, M. Divas, V. Sechovsky, Phys. Stat. Solidi (b) **249**, 202 (2012)
[12M1] K.H. Müller, V. Narozhnyi, *Rare-earth-transition metal borocarbides* (Springer, 2012)
[12N1] O. Nane, B. Ozcelik, S. Kervan, U. Topal, J. Supercond. Nov. Magn. **25**, 1065 (2012)
[12O1] B. Ozcelik, N. Kantarci, O. Nane, M.E. Yakinci, Solid State Phenomena, **190**, 208 (2012)
[12S1] J.Ph. Soulie, N. Penin, K. Yvon, J. Alloys Comp. **526**, 63 (2012)
[12V1] P. Villars et al., *Landolt-Börnstein handbook*, III/43A11 (2012)
[12W1] P. Wang, C.G. Piao, R.Y. Meng, Y. Cheng, F.J. Guang, Physica B **407**, 227 (2012)
[12W2] F. Weber, S. Rosenkranz, L. Pintschovius, J.P. Castellan, R. Osborn, W. Reichardt, R. Heid, K.P. Bohnen, E.A. Goremychkin, A. Kreyssig, K. Hradil, D.L. Abernathy, Phys. Rev. Lett. **109**, 057001 (2012)
[13A1] E. Alleno, C. Mazumdar, J. Solid State Chem. **202**, 15 (2013)
[13C1] S.F. Chen, J.K. Chen, S.L. Liu, Y.L. Lin, J. Alloys Comp. **569**,29 (2013)
[13D1] P. Das, C.V. Tomy, H. Takeya, S. Ramakrishnan, A.K. Grover, Physica C **484**, 81 (2013)
[13E1] M. El Massalami, A. M. Gomes, T. Paiva, R.R. dos Santos, H. Takeya, J. Magn. Magn. Mater. **335**, 163 (2013)
[13H1] H. Hodovanets, S. Ran, P. C. Canfield, S. L. Budko, Phil. Mag. **93**, 1748 (2013)
[13K1] A.E. Karkin, Yu.N. Akshentsev, B.N. Goshchitskii, JETP Lett. **97**, 347 (2013)
[13K2] J. Kitagawa, N.Takeda, M. Ishikawa, J. Alloys Comp. **561**, 101 (2013)
[13K3] J. Kitagawa, M. Ishikawa, Solid State Commun. **153**, 76 (2013)
[13N1] P.S. Normile, M. Rotter, C. Detlefs, J. Jensen, P.C. Canfield, J.A. Blanco, Phys. Rev. B **88**, 05413 (2013)
[13P1] X. Peng, B. Liu, Y. Fan, X. Zhu, Q. Peng, Z. Zhang, J. Power Sources **240**, 178 (2013)
[13W1] M. Weigand, L. Civale, F. J. Baca, J. Kim, S. L. Budko, P. C. Canfield, B. Maiorov, Physical Rev. B **87**, 140506 (2013)
[13Y1] X.Z. Yan, X.Y. Kuang, A.J. Mao, G.G. Kunag, Z.H. Wang, X.W. Sheng, Acta Phys. Sinica, **62**, 107402 (2013)

[14C1] R.J. Caraballo Vivas, D.L. Rocco, T. Costa Suarez, L. Caldeira, A.A. Coelho, M.R. Reis, J. Appl. Phys. **116**, 063907 (2014)
[14C2] P. Chaudhary, A. Nuwal, S.C. Tiwari, R. K. Paliwal, S. L. Kakani, J. Supercond. Nov. Magn. **27**, 1167 (2014)
[14E1] M. El Massalami, H. Takeya, B. Ouladdiaf, A. M. Gomes, T. Paiva, R. R. dos Santos, J. Magn. Magn. Mater. **372**, 74 (2014)
[14L1] W.C. Lee, Physica C **507**, 55 (2014)
[14P1] N. Plugaru, M. Vălеanu, R. Plugaru, J. Campo, J. Appl. Phys. **115**, 023907 (2014)
[14R1] M. Ramazanoglu, M. Laver, A. Yagmurcu, E. M. Choi, S. I. Lee, A. Knigavko, B. D. Gaulin, Low Temp. Phys. **40**, 160 (2014)
[14S1] X. Shang, S. Lu, X. Peng, Y. Fan, B. Zhang, Z. Zhang, B. Liu, Int. J. Electrochem. Sci. **9**, 4913 (2014)
[14S2] Z. Sniadecki, J. Alloys Comp. **615**, S40 (2014)
[14W1] L. Wang, H.Z. Yan, W. Xiong, B.Q. Li, J. Li, F.Q. Kong, J. Power Sources, **259**, 213 (2014)
[14W2] F. Weber, L. Pintschovius, W. Reichardt, R. Heid, K. P. Bohnen, A. Kreyssig, D. Reznik, K. Hradil, Phys. Rev. B **89**, 104503 (2014)
[14Z1] Y.K. Zhang, B.J. Yang, J. Alloys Comp. **610**, 540 (2014)
[15C1] S.F. Chen, C.Y. Chen, C.H. Lin, J. Alloys Comp. **637**, 418 (2015)
[15K1] I. S. Kirklin, J.E. Saal, B. Meredig, A. Thomson, J. W. Deak, M. Aykol, S. Rühl, C. Wolverton, Comput. Mater. **1**, 15010 (2015)
[15M1] C. Mazumdar, R. Nagarajan, Physica C **514**, 173 (2015)
[15M2] H. Michor, T. Markota, I.M. Messner, S. Ozcan, F. Schwarzböck, L. Salamakha, Z.S. Tako, E. Bauer, O. Sologub, J. Phys.: Conf. Series, 592, 012043 (2015)
[15S1] B. K. Sahoo, B. N. Panda, Physica C **513**, 35 (2015)
[15T1] H. M. Tutuncu, H. Y. Uzunok, E. Karaca, G. P. Srivastava, S. Ozer, S. Ugur, Phys. Rev. B **92**, 054510 (2015)
[15V1] P. Villars et al Handbook of Inorganic Substances, De Gruyter (2015)
[15W1] D. Wulferding, I. Yang, J. Yang, M. Lee, H. C. Choi, S. L. Budko, P. C. Canfield, H. W. Yeom, J. Kim, Phys. Rev. B **92**, 014517 (2015)
[16K1] M. Kawamura, R. Akashi, S. Tsuneyuki, arXiv: 1610.07329v2 [cond-mat.supr-con] 27Dec 2016
[16K2] S. G. Kwon, K. Kang, M. Lee, B. Cho, Current Appl. Phys. **16**, 1136 (2016)
[16L1] B. Liang, Z. Zhang, H. Guo, Trans. Indian Inst. Met. **69**, 1943 (2016)
[16Y1] J. Yao, C. Yan, V.O. Yapaskurt, A.V. Morozkin, J. Magn. Magn. Mater. **419**, 176 (2016)
[16Z1] Y. Zang, S. Geng, G. Wilde, J. Supercond. Nov. Magn., **29**, 2681 (2016)
[17K1] M. Kawamura, R. Akashi, S. Tsuneyuki, Phys. Rev. B **95**, 054506 (2017)
[17L1] L. Liu, X. Wu, R. Wang, Z. Hu, Y. Jiang, D. Liu, Crystals **7**, 173 (2017)
[17L2] L. Liu, C. Chen, D. Liu, Z. Hu, G. Xu, R. Wang, Adv. Mater. Sci. Eng. 2017, ID 9038151(2017)
[17N1] J. Nössler, R. Seerig, S. Yasin, M. Uhlarz, S. Zherlitsyn, G. Behr, S. L. Drechsler, G. Fuchs, H. Rosner, J. Wosnitza, Phys. Rev. B **95**, 014523 (2017)
[18L1] S. Li, C. Guo, W. Ying, O. Sha, J. Wang, Chemistry Select **3**, 4536 (2018)
[18L2] L. Liu, X. Wu, R. Wang, Y. He, T. Cai, High Temp-High Press. **47**, 267 (2018)
[18O1] M. W. Olsewski, M. R. Eskilden, C. Reichhardt, C. J. O. Reichhardt, New. J. Phys. **20**, 023005 (2018)
[18S1] S. Steiner, G. Rogl, H. Michor, P. Rogl, G. Giester, A. P. Gonsalves, Dalton Trans. **47**, 12033 (2018)
[18T1] R. Toft-Petersen, T. B. S. Jensen, J. Jensen, M. van Zimmermann, S. Sloth, F. W. Isaksen, N. B. Christiansen, Y. Chen, K. Siemensmeyer, H. Kawano-Furukawa, H. Takeya, A. B. Abrahamsen, N. H. Andersen, Phys. Rev. B **97**, 224417 (2018)
[19B1] D. Betancourth, V. F. Correa, J. J. Facio, J. Fernandez, V. Vildosola, R. Lola-Serrano, J. M. Cadogan, A.A. Aligia, P. S. Cornaglia, D. J. Garcia, Phys. Rev. B **99**, 134406 (2019)
[19B2] E. Burzo, P. Vlaic, Mater. Res. Express, **6**, 125122 (2019)

[19L1] L. Liu, Y. Jiang, L. Chen, Y. Jiang, Y. He, D. Liu, Y. Wen, High Temp.-High Press. **48**, 469 (2019)
[19M1] Materials Project, Lawrence National Lab. (Berkley, USA, 2019)
[19S1] A. Sazonov, V. Hutanu, M. Meven, K. Röwer, G. Roth, E. Lelievre-Berna, G. Fuchs, O. Zaharko, K. Friese, Phys. Rev. B **100**, 014414 (2019)
[20G1] S. Gundogdu, J. P. Chancy, G. Xu, Y. Zhao, P. A. Dube, T. G. Karalar, B. K. Cho, J. W. Lynn, M Ramazanoglu, Mater. Res. Express, **7**, 116002 (2020)
[20M1] Materials Research Project, Lawrence Berkley Nat. Lab. (Berkley, USA , 2020)
[20S1] K. Singh, S. Sharma, Indian J. Eng. Mat. Sci. **27**, 564 (2020)
[20Z1] Y. Zhu, B. Chen, T. Chang, C. Du, S. Zhang, J. Electroanalytical Chem. **878**, 114552 (2020)

Correction to: Rare–Earths–Cobalt–Boron Compounds

Correction to:
Chapter 8 in: E. Burzo, *Rare Earths-Transition Metals-Boron Compounds*, https://doi.org/10.1007/978-3-030-99245-3_8

In the original version of Chapter 8, the following belated corrections have been incorporated: The revised abstract has been inserted.

The correction chapter and the book has been updated with the changes.

The updated version of this chapter can be found at

https://doi.org/10.1007/978-3-030-99245-3_8

E. Burzo, *Rare Earths-Transition Metals-Boron Compounds*,
https://doi.org/10.1007/978-3-030-99245-3_10

MIX
Papier aus verantwortungsvollen Quellen
Paper from responsible sources
FSC® C105338

Printed by Books on Demand, Germany